Lecture Notes in Computer Science 16913

Founding Editors

Gerhard Goos
Juris Hartmanis

The series Lecture Notes in Computer Science (LNCS), including its subseries Lecture Notes in Artificial Intelligence (LNAI) and Lecture Notes in Bioinformatics (LNBI), has established itself as a medium for the publication of new developments in computer science and information technology research, teaching, and education.

LNCS enjoys close cooperation with the computer science R & D community, the series counts many renowned academics among its volume editors and paper authors, and collaborates with prestigious societies. Its mission is to serve this international community by providing an invaluable service, mainly focused on the publication of conference and workshop proceedings and postproceedings. LNCS commenced publication in 1973.

Mingshuai Chen · Max Tschaikowski
Editors

Quantitative Evaluation of Systems and Formal Modeling and Analysis of Timed Systems

Third International Joint Conference, QEST+FORMATS 2026
Liverpool, UK, September 2–4, 2026
Proceedings

Springer

Editors
Mingshuai Chen
Zhejiang University
Zhejiang, China

Max Tschaikowski
Sapienza University Rome
Rome, Italy

ISSN 0302-9743 ISSN 1611-3349 (electronic)
Lecture Notes in Computer Science
ISBN 978-3-032-35297-2 ISBN 978-3-032-35298-9 (eBook)
https://doi.org/10.1007/978-3-032-35298-9

This Springer imprint is published by the registered company Springer Nature Switzerland AG
The registered company address is: Gewerbestrasse 11, 6330 Cham, Switzerland

Preface

This book proudly presents the proceedings of the joint event for the 23rd International Conference on Quantitative Evaluation of SysTems (QEST 2026) and the 24th International Conference on Formal Modeling and Analysis of Timed Systems (FORMATS 2026). The two flagship conferences are pursuing a strategic process of joining forces and sparking a new joint conference for quantitative modeling, analysis, and verification. This year, for the third time, the two conferences formed the QEST+FORMATS joint conference, which was held on 2–4 September 2026 in Liverpool, UK, as part of the CONFEST 2026 umbrella conference.

This edition of QEST and FORMATS continued the strategic merging of the two conferences, initiated in 2024 after each had been held more than 20 times. The major activities of this year for continuing implementing the partnership were (a) forming a joint program committee (PC), (b) having a joint call for papers, (c) having an integrated review process and artifact evaluation adapted to the focus areas of the two conferences, (d) having an integrated PC and review process for artifact evaluation, and (e) having an integrated program.

As a result of maintaining a standard of high-quality publications, reviews, talks, speakers, and discussions despite strong competition, we were delighted to present a strong and diverse QEST+FORMATS 2026 program selected by an expert PC. The 38 PC members, with help from external reviewers, decided to accept 24 papers out of 54 submissions. Each submission received at least 3 single-blind reviews. All accepted papers were strong and most decisions were unanimous. This year, we were able to continue the success story of last year, when the conference received 57 submissions, exceeding the submission numbers from two years ago. The sustained level of interest reflects the community's strong support for the merger of the two conferences. Artifact evaluation is a well-established part of both QEST and FORMATS. QEST+FORMATS 2026 included an artifact evaluation (AE), carried out by 23 AE Committee members (AEC). Three AEC members reviewed each artifact in terms of consistency with and reproducibility of results presented in the paper, completeness, future-proofness, documentation and ease of use, and availability in an online repository with a DOI. Based on these reviews, and following the EAPLS guidelines for artifact badging, every artifact was awarded up to two badges, to be displayed on the front page of the corresponding paper. From 34 artifacts reviewed, 33 received the Available badge, and 26 the Functional badge.

As usual, the purpose of the joint conference was to foster progress in the quantitative evaluation and verification of systems, to promote the study of fundamental and practical aspects of systems with quantitative nature (such as probability, timing, or cost), and to bring together researchers from different disciplines who share interests in the modeling, design, and analysis of computational systems. The particular focus was to stay on the cutting edge of research by welcoming contributions that tap into machine-learning and quantum systems, case studies, and tool papers.

A great highlight was having three excellent invited speakers, Ezio Bartocci, from TU Wien, Austria; Jane Hillston, from the University of Edinburgh, UK; and Kim Guldstrand Larsen, from Aalborg University, Denmark. They are all known worldwide for their strong research on formal and quantitative aspects of systems.

Finally, we have to thank several people sincerely. First, we want to thank all authors who submitted their work to QEST+FORMATS 2026. Likewise, we thank all PC members and all external reviewers for their hard work and timely reviews. The process of evaluating artifacts went smoothly thanks to Emilio Incerto and Matthias Volk, and their excellent work in chairing the artifact evaluation. All the reviewing and evaluation work was supported by EasyChair, which made our lives much easier. Special thanks go to Jie An and Marco Paolieri for serving as the QEST+FORMATS publicity chairs. We also thank the local organising committee of CONFEST, the steering committees of QEST+FORMATS and, last but not least, the co-chair of QEST+FORMATS 2025, Andrea Vandin, who facilitated a smooth transition. Overall, the event was a memorable occasion, and we want to thank everybody who contributed in any way.

July 2026

Mingshuai Chen
Max Tschaikowski

Organization

Program Committee Chairs

Mingshuai Chen	Zhejiang University, China
Max Tschaikowski	Sapienza University of Rome, Italy

Artifact Evaluation Chairs

Emilio Incerto	IMT School for Advanced Studies Lucca, Italy
Matthias Volk	Eindhoven University of Technology, Netherlands

Publicity Chairs

Jie An	Institute of Software, Chinese Academy of Sciences, China
Marco Paolieri	University of Southern California, USA

QEST Steering Committee

Erika Ábrahám	RWTH Aachen University, Germany
Benoît Barbot	Université Paris-Est Créteil, France
Ezio Bartocci	TU Vienna, Austria
Jane Hillston	University of Edinburgh, UK
Nils Jansen	Ruhr University Bochum, Germany
Diwakar Krishnamurthy	University of Calgary, Canada
Andrea Marin	Università Ca' Foscari Venezia, Italy
Guillermo Alberto Perez	University of Antwerp, Belgium
Anne Remke	University of Münster, Germany
Enrico Vicario (Chair)	University of Florence, Italy

FORMATS Steering Committee

Eugene Asarin	Université Paris Cité, France
Martin Fränzle (Chair)	University of Oldenburg, Germany

Joost-Pieter Katoen	RWTH Aachen University, Germany
Kim G. Larsen	Aalborg University, Denmark
Oded Maler (Founding Chair, 1957–2018)	University of Grenoble, France
Pavithra Prabhakar	Kansas State University, USA
Mariëlle Stoelinga	TU Twente, Netherlands
Wang Yi	Uppsala University, Sweden

Program Committee

Erika Ábrahám	RWTH Aachen University, Germany
Étienne André	Nantes Université, France
Giovanni Bacci	University of Aalborg, Denmark
Christel Baier	TU Dresden, Germany
Ezio Bartocci	TU Wien, Austria
Kevin Batz	Cornell University, USA
Lei Bu	Nanjing University, China
Peter Buchholz	TU Dortmund, Germany
Pedro R. D'Argenio	Universidad Nacional de Córdoba, Argentina
Arnd Hartmanns	University of Twente, Netherlands
Jane Hillston	University of Edinburgh, UK
Benny van Houdt	University of Antwerp, Belgium
Antonio Jiménez-Pastor	Universidad Politécnica de Madrid, Spain
Taylor T. Johnson	Vanderbilt University, USA
Sebastian Junges	Radboud University, Netherlands
Christoph Matheja	University of Oldenburg, Germany
Tobias Meggendorfer	Lancaster University Leipzig, Germany
Francesca Randone	TU Wien, Austria
Anne Remke	WWU Münster, Germany
Sabina Rossi	Università Ca' Foscari Venezia, Italy
Ramin Sadre	Université catholique de Louvain, Belgium
Sriram Sankaranarayanan	University of Colorado Boulder, USA
Ocan Sankur	CNRS/Mitsubishi Electric, France
Christian Schilling	Aalborg University, Denmark
Zhikun She	Beihang University, China
Sadegh Soudjani	University of Birmingham, UK/Max Planck Institute for Software Systems, Germany
Miklos Telek	Budapest University of Technology and Economics, Hungary
Mirco Tribastone	IMT School for Advanced Studies Lucca, Italy

Andrea Vandin	Technical University of Denmark, Denmark/IMT School for Advanced Studies Lucca, Italy
Enrico Vicario	University of Florence, Italy
Mahesh Viswanathan	University of Illinois Urbana-Champaign, USA
Masaki Waga	Kyoto University, Japan
Kazuki Watanabe	National Institute of Informatics of Japan, Japan
Verena Wolf	Saarland University, Germany
Naijun Zhan	Peking University, China
Paolo Zuliani	Sapienza University of Rome, Italy

Artifact Evaluation Committee

Adityo Anggraito	Università Ca' Foscari Venezia, Italy
Hrishikesh Balakrishnan	University of Illinois Urbana-Champaign, USA
Ernesto Casablanca	Newcastle University, UK
Nabarun Deka	University of Illinois Urbana-Champaign, USA
Gabriel Dengler	Saarland University, Germany
Marco Esposito	Università di Roma La Sapienza, Italy
Kevin van de Glind	Eindhoven University of Technology, Netherlands
Emanuele Guerrazzi	Sant'Anna School of Advanced Studies, Italy
Chris Johannsen	Iowa State University, USA
Kentaro Kobayashi	University of Tsukuba/AIST, Japan
Martin Kristjansen	Aalborg University, Denmark
Maurice Laveaux	Eindhoven University of Technology, Netherlands
Marco Lewis	Inria, France
Yuntao Li	University of Florida, USA
Luko van der Maas	Radboud University, Netherlands
Dylan Marinho	Sorbonne Université, France
Nausheen Mohammed	University of Illinois Urbana-Champaign, USA
Diletta Olliaro	KTH Royal Institute of Technology, Sweden
Simone Pernice	University of Turin, Italy
Leonardo Picchiami	Sapienza University of Rome, Italy
Lisa Willemsen	University of Twente, Netherlands
Eshita Zaman	Iowa State University, USA
Marco Zamponi	IMT School for Advanced Studies Lucca, Italy

Contents

Stationary and Transient Bounds for Nearly Lumpable or Uncertain Markov Chains

Peter Buchholz(✉)

TU Dortmund University, Informatik 4, 44227 Dortmund, Germany
peter.buchholz@cs.tu-dortmund.de

Abstract. Continuous-time and discrete-time Markov chains often face two major challenges, state space explosion and uncertainty in transition rates or probabilities. State space explosion can be mitigated through state aggregation, which reduces the number of states. However, aggregation typically introduces uncertainty in the transition matrix, resulting in a Markov process with imprecise parameters. This paper defines several variants of parameter uncertainty and presents methods to compute bounds for stationary, transient, and accumulated rewards in Markov chains with uncertain transition matrices and reward structures. For certain types of parameter uncertainty, the problem can be reformulated as an optimization problem for Markov decision processes, allowing the computation of tight bounds. For other variants, however, the resulting optimization problems are non-convex, and tightness of the computed bounds cannot, in general, be guaranteed.

Keywords: Markov Chain · Transient Analysis · Lumpability · Uncertainty · Bound Computation

1 Introduction

Continuous time Markov chains (CTMCs) and *discrete time Markov chains* (DTMCs) are widely used for model-based quantitative analysis. In performance or dependability modeling, CTMCs are more relevant, however, several results apply for both, CTMCs and DTMCs. Consequently, the term *Markov chain* (MC) is used if the distinction between continuous and discrete time is irrelevant. Although the analysis of MCs is well established and a large number of numerical methods for stationary or transient analysis are available [41], practical applications face with two problems, state space explosion and parameter uncertainty.

State space explosion refers to the problem that the number of states of an MC often grows exponentially with the size of the model in terms of components like queues, customers, automata, species etc. Furthermore, the memoryless property of MCs implies that non-exponential distributions have to be

M. Chen and M. Tschaikowski (Eds.): QEST+FORMATS 2026, LNCS 16913, pp. 1–19, 2027.
https://doi.org/10.1007/978-3-032-35298-9_1

approximated by phase type distributions adding additional states to the model. One possibility to mitigate state space explosion is aggregation which means to map the MC on an MC with a smaller state space and compute the solution for the aggregated MC. Results for the aggregated MC are most times only an approximation and it is important to bound the approximation error.

The second problem of modeling are uncertain transition rates or probabilities. Model parameters are often derived from measurements or expert opinions and are not exactly known. The best one can do is to estimate confidence intervals which result in a set of MCs rather than a single one. If analysis considers only one MC out of the set, results are again approximations with an unknown approximation error.

Aggregation and uncertain transition rates or probabilities in an MC are related. For a given aggregation, i.e., a given mapping from a detailed into an aggregated state space, a set of MCs on the aggregated state space can be defined that contains one matrix such that the aggregation is *exact* for stationary analysis and a set of matrices can be defined that describe an inhomogeneous MC for an exact aggregation according to transient analysis.

The abstract problem in both situations, aggregation of MCs and MCs with uncertain parameters, is the analysis of a set of MCs where the transition matrix is from some set of matrices. Performance and dependability measures for an MC are often described by rewards assigned to states. Like transition rates or probabilities, rewards are also often not exactly known and have to be taken from some set. Consequently, the objective is to compute the stationary or transient reward for a DTMC or CTMC which is defined by a set of transition matrices and a set of reward vector. The best we can expect to achieve in this situation are bounds on reward measures.

In this paper, we summarize methods for computing bounds for MCs in the mentioned scenario. Some results are already known from literature, others, in particular for transient analysis, are, to the best of our knowledge, new. The paper is structured as follows. The next section briefly reviews related work. Section 3 introduces the basic model class, aggregation of MCs and different forms of uncertainty. In Sect. 4, methods for bounding the error of stationary analysis are introduced. Then, in Sect. 5, bounds for transient analysis are presented. Subsequently, numerical results for example models are provided. The paper ends with the conclusions and an outlook of possible extensions.

2 Related Work

The literature on MCs is very broad, for a comprehensive introduction we refer to basic textbooks [2,36,41]. Only the most important results that are relevant for this paper are summarized here, many more papers are available in the wider context. A central concept for aggregation is lumpability [4,27] which has been subsequently extended to quasi or near lumpability [3–5,17,22,23]. A new version of near lumpability that distinguishes between sojourn times in a state and transition probabilities has been published recently [39]. Lumpability has been extended to other models beyond MCs [1,13,28] and has also been extended to define a general equivalence between MCs and general stochastic processes [11,12]. Apart from lumpability, which is often denoted as ordinary lumpability, weak lumpability [27] and its specialization exact lumpability [4,6,40] have been defined. However, although for these extensions the idea of near lumpability can be generalized, no method is known to bound the error, as already noticed in [4], whereas the error of an aggregation based on near ordinary lumpability can be bounded as shown in this paper.

As already mentioned, aggregation based on lumpability results in an MC on a smaller state space whereas aggregation based on near lumpability results in an MC where transitions rates or probabilities are defined by intervals as already noticed in the early work on near lumpability [3–5,22,23]. Without aggregation, the model class has been defined as *Markov set chains* [26] if parameters are independent. Moreover, models of more general parameter uncertainty have been published [18,19]. Lumpability has also been defined for MCs with uncertain parameters when parameters are defined as independent intervals [14,15].

If parameter uncertainty is defined by independent intervals, then stationary rewards can be bounded using results from [16]. This approach has been adopted for aggregations resulting from near lumpability and has been improved in [7, 8] based on results for MDPs with uncertain parameters [24]. Similar results have been published for imprecise MCs [20,21]. The problem becomes much harder if the uncertainty is described by complex dependencies and not by simple independent intervals for each transition rate/probability and reward [18].

Bounds for bounded reachability of CTMCs resulting from near lumapbility are developed in [39]. [42] presents an approach to compute bounds for rewards for DTMCs. Computation of bounds for transient rewards of CTMCs is much harder and results that allow one to bound the approximation error are rare. Some recent results allow one to compute an ϵ-approximation of the transient vector [32,33,38] which can then be used to approximate rewards. However, strict bounds cannot be computed with this approach and the error increases with an increasing time bound.

3 Nearly Lumpable or Uncertain Markoc Chains

In this section, the basic definitions, notations and some known results are summarized. Moreover, different variants of parameter uncertainty are defined. Mainly CTMCs are considered which are more important than DTMCs for performance and dependability analysis. However, DTMCs are strongly related to CTMCs such that results for stationary analysis apply to both. For transient analysis bounds for CTMCs are often harder to obtain than for DTMCs.

3.1 Basic Definitions

We consider CTMCs on a finite state space $\mathcal{S}$ with $n = |\mathcal{S}|$, transition rate matrix $\boldsymbol{Q} \in \mathbb{R}^{n \times n}$, initial distribution $\boldsymbol{\pi}(0) \in \mathbb{R}^{1 \times n}$ and reward vector $\boldsymbol{r} \in \mathbb{R}^{n \times 1}$

with the common interpretation [41]. For a CTMC a *uniformized* DTMC with transition probability matrix

$$\boldsymbol{P} = \boldsymbol{Q}/\alpha + \boldsymbol{I} \tag{1}$$

where $\boldsymbol{I}$ is the identity matrix of order n and $\alpha \geq \max_{i \in \mathcal{S}} |\boldsymbol{Q}_{i,i}|$, can be defined and it is well known [41] that the transient distribution of the CTMC at time $t \geq 0$ can be computed as

$$\boldsymbol{\pi}(t) = \boldsymbol{\pi}(0)e^{\boldsymbol{Q}t} = \boldsymbol{\pi}(0)\sum_{k=0}^{\infty}\frac{(\boldsymbol{Q}t)^k}{k!} = \boldsymbol{\pi}(0)\sum_{k=0}^{\infty}\beta(\alpha t, k)\boldsymbol{P}^k. \tag{2}$$

where $\beta(\alpha t, k) = e^{-\alpha t}(\alpha t)^k/k!$ are the Poisson probabilities. It is often preferable to use the latter representation to compute transient distributions. From the transient distribution the reward at time t, $R_t = \boldsymbol{\pi}(t)\boldsymbol{r}$ and the accumulated reward in the interval $[0, \tau]$, $A_t = \int_0^\tau \boldsymbol{\pi}(t)\boldsymbol{r}dt$ can be computed.

If $\boldsymbol{Q}$ is ergodic, then $\boldsymbol{P}$ is irreducible and if $\alpha > \max_{i \in \mathcal{S}} |\boldsymbol{Q}_{i,i}|$, then it is also guaranteed to be aperiodic. In this case, the stationary distribution exists and it is the unique solution of $\boldsymbol{\pi Q} = \boldsymbol{0}$ or $\boldsymbol{\pi P} = \boldsymbol{\pi}$ subject to $\boldsymbol{\pi}\mathbb{I} = 1$ where $\mathbb{I}$ is a column vector of length n consisting of 1s. The expected stationary reward is then given by $R_\infty = \boldsymbol{\pi r}$. These results show that CTMCs and uniformized DTMCs are strongly related.

An important concept to reduce the size of the state space of Markov chains is aggregation. In general, aggregation is described by a mapping from $\mathcal{S}$ into a smaller state space $\tilde{\mathcal{S}}$ with $\tilde{n} = |\tilde{\mathcal{S}}| < n$. We consider only *proper lumpings* which means that one state from $\mathcal{S}$ is mapped onto one state of $\tilde{\mathcal{S}}$. Furthermore, we assume that the mapping is surjective. For $i \in \mathcal{S}$, $\phi(i) \in \tilde{\mathcal{S}}$ is the state on which i is mapped and $[\tilde{i}] \subset \mathcal{S}$ is the set of states that are mapped on $\tilde{i}$.

The mapping can be described by a *collector* matrix $\boldsymbol{V} \in \{0,1\}^{n \times \tilde{n}}$ where $\boldsymbol{V}_{i,\tilde{i}} = 1$ if $i \in [\tilde{i}]$ and 0 otherwise. Furthermore, a corresponding *distributor* matrix $\boldsymbol{W} \in \mathbb{R}_{\geq 0}^{\tilde{i} \times i}$ is a non-negative matrix such that $\boldsymbol{WV} = \boldsymbol{I}$. Matrices $\boldsymbol{V}$ and $\boldsymbol{W}$ can be used to build an *aggregated* CTMC using

$$\tilde{\boldsymbol{\pi}}(0) = \boldsymbol{\pi}(0)\boldsymbol{V},\ \tilde{\boldsymbol{Q}} = \boldsymbol{WQV} \text{ and } \boldsymbol{V}\tilde{\boldsymbol{r}} = \boldsymbol{r}. \tag{3}$$

It can be shown that $(\tilde{\boldsymbol{\pi}}(0), \tilde{\boldsymbol{Q}}, \tilde{\boldsymbol{r}})$ is a CTMC and an ergodic CTMC will result in an ergodic aggregated CTMC if $\boldsymbol{W}_{\tilde{i},i} > 0$ for all $i \in [\tilde{i}]$. Usually, the transient and stationary rewards from the original and aggregated MC differ but the following conditions assure that transient and, in the ergodic case, stationary rewards remain the same [4,12,27].

$$\begin{array}{l} \tilde{\boldsymbol{\pi}}(0) = \boldsymbol{\pi}(0)\boldsymbol{V},\ \boldsymbol{QV} = \boldsymbol{V}\tilde{\boldsymbol{Q}},\ \ \boldsymbol{r} = \boldsymbol{V}\tilde{\boldsymbol{r}} \ \text{ or} \\ \tilde{\boldsymbol{\pi}}(0)\boldsymbol{W} = \boldsymbol{\pi}(0),\ \boldsymbol{WQ} = \tilde{\boldsymbol{Q}}\boldsymbol{W},\ \boldsymbol{Wr} = \tilde{\boldsymbol{r}} \end{array} \tag{4}$$

The resulting concepts are often denoted as ordinary and weak lumpability, respectively. The following relations between the original and the aggregated MC exist, if aggregation is based on lumpability. For ordinary lumpability $\tilde{\pi}_{\tilde{i}}(t) = \sum_{i\in[\tilde{i}]} \pi_i(t)$ and for weak lumpability $\pi_i(t) = \tilde{\pi}_{\tilde{i}}(t) W_{\tilde{i},i}$ for $i \in [\tilde{i}]$. Ordinary lumpability requires that $r_i = \tilde{r}_{\tilde{i}}$ and $\sum_{j\in[\tilde{j}]} Q_{i,j} = \tilde{Q}_{\tilde{i},\tilde{j}}$ for all $i \in [\tilde{i}]$ and all $\tilde{j} \in \tilde{\mathcal{S}}$. No requirements for the initial vector are necessary.

Weak lumpability has no simple requirements based on the sum of transition rates into specific subsets. Therefore, often the easier to prove subform exact lumpability [4] is used. The conditions for exact lumpability equal those of weak lumpability with matrix W where all elements in a row are identical. This implies that $\pi_i(0) = \tilde{\pi}_{\tilde{i}}(0)/|[\tilde{i}]|$ and $\sum_{j\in[\tilde{j}]} Q_{j,i} = \tilde{Q}_{\tilde{j},\tilde{i}}$ for all $i \in [\tilde{i}]$ and all $\tilde{j} \in \tilde{\mathcal{S}}$. No conditions for the reward vector are required.

Uniformization of a CTMC followed by aggregation is identical to aggregation followed by uniformization, if one of the two conditions from (4) holds.

3.2 Markov Chains with Uncertain Parameters

Usually, the parameters of CTMCs result from measurements of physical quantities such that confidence intervals and estimated mean values but no exact values are available. Consequently, bounds and not exact values of transitions rates are known. A second reason for having bounds rather than exact values is the use of aggregation to avoid large state spaces. The original process observed on the aggregated state space is only a Markov chain if one of the conditions in (4) holds. Often conditions for lumpability are only approximately met. This resulted in the definition of near or quasi ordinary lumpability [3,4,22,23] of an aggregation. The following conditions have to hold.

$$\forall \tilde{i}, \tilde{k} \in \mathcal{S},\ \forall i, j \in [\tilde{i}] : \left| \sum_{k\in[\tilde{k}]} Q_{i,k} - \sum_{k\in[\tilde{k}]} Q_{j,k} \right| < \varepsilon \text{ and } |r_i - r_j| < \varepsilon \quad (5)$$

for some $0 \le \varepsilon \ll \max_{i\in\mathcal{S}} |Q_{i,i}|$.

Independently, whether the uncertainty about transition rates resulted from uncertain parameters, aggregation or both, the transition matrix and often also the reward or initial vector are no longer exactly known, if uncertainty is not neglected. Let $\mathbb{Q}$ be a set of generator matrices and $\mathbb{r}$ a set of reward vectors both over state space $\mathcal{S}$. We assume that both sets are convex. Although the sets $\mathbb{Q}$ and $\mathbb{r}$ may contain aggregated matrices and reward vectors, we denote the matrices and vectors from the sets by Q and r, respectively, to avoid an overloading of notation.

Define the following vectors as solutions of the corresponding differential equation.

$$\begin{aligned}
\frac{r^-(t)}{dt} &= \min_{Q\in\mathbb{Q}} \{Q r^-(t)\} && \text{where } r^-(0) = \min_{r\in\mathbb{r}} r \\
\frac{r^+(t)}{dt} &= \max_{Q\in\mathbb{Q}} \{Q r^+(t)\} && \text{where } r^+(0) = \max_{r\in\mathbb{r}} r \\
\frac{g^-(t)}{dt} &= \min_{Q\in\mathbb{Q}, r\in\mathbb{r}} \{Q g^+(t) + r\} && \text{where } g^-(0) = \mathbf{0} \\
\frac{g^+(t)}{dt} &= \max_{Q\in\mathbb{Q}, r\in\mathbb{r}} \{Q g^-(t) + r\} && \text{where } g^+(0) = \mathbf{0}
\end{aligned} \quad (6)$$

The equations are solvable if a matrix or matrix-vector pair exists that minimizes/maximizes all values in the vector on the left side. This usually implies that the rows of matrix $\boldsymbol{Q}$ and the elements of vector $\boldsymbol{r}$ can be selected independently from a set of possible rows and rewards, respectively. If $\boldsymbol{\pi}(0)$ is known, bounds for the reward at time t and the accumulated reward in $[0, t]$ with respect to $\mathbb{Q}$ and $\boldsymbol{r}$ are given by

$$\begin{aligned} R_t^- &= \boldsymbol{\pi}(0)\boldsymbol{r}^-(t) \leq R_t \leq \boldsymbol{\pi}(0)\boldsymbol{r}^+(t) = R_t^+ \text{ and} \\ A_t^- &= \boldsymbol{\pi}(0)\boldsymbol{g}^-(t) \leq A_t \leq \boldsymbol{\pi}(0)\boldsymbol{g}^+(t) = A_t^+. \end{aligned} \tag{7}$$

If also the initial distribution is uncertain, we can define a set $\boldsymbol{p}$ that includes possible initial distributions and the minimum or maximum is built over all three sets. If $\mathbb{Q}$ contains only irreducible matrices, then bounds for the stationary reward are given by

$$\begin{aligned} R_\infty^- &= \min_{\boldsymbol{Q} \in \mathbb{Q}, \boldsymbol{r} \in \boldsymbol{r}} (\boldsymbol{\pi r} | \boldsymbol{\pi Q} = \boldsymbol{0}, \boldsymbol{\pi} \mathbb{1} = 1) \leq R_\infty \leq \\ R_\infty^+ &= \max_{\boldsymbol{Q} \in \mathbb{Q}, \boldsymbol{r} \in \boldsymbol{r}} (\boldsymbol{\pi r} | \boldsymbol{\pi Q} = \boldsymbol{0}, \boldsymbol{\pi} \mathbb{1} = 1). \end{aligned} \tag{8}$$

It should be remarked that $\lim_{t\to\infty} R_t^- \leq R_\infty^-$ and $R_\infty^+ \leq \lim_{t\to\infty} R_t^+$ but not necessarily equality holds. If all matrices in $\mathbb{Q}$ are irreducible, then $R_\infty^- = \lim_{t\to\infty} A_t^-/t$ and $R_\infty^+ = \lim_{t\to\infty} A_t^+/t$ [35].

For a set of generator matrices $\mathbb{Q}$, a corresponding set $\mathcal{P}$ of probability matrices can be defined as follows

$$\mathcal{P} = \left\{ \boldsymbol{P} \mid \boldsymbol{P} = \boldsymbol{Q}/\alpha + \boldsymbol{I}, \boldsymbol{Q} \in \mathbb{Q}, \alpha \geq \max_{\boldsymbol{Q} \in \mathbb{Q}} \max_{i \in \mathcal{S}} |\boldsymbol{Q}_{i,i}| \right\} \tag{9}$$

$\mathcal{P}$ contains for each matrix in $\mathbb{Q}$ the probability matrix resulting from uniformization with parameter α.

Different possibilities exist to define sets $\mathbb{Q}$ and $\boldsymbol{r}$. The simplest form is to assume that all elements are independent and bounds are known for the transition rates and rewards. These bounds may result from confidence intervals for parameters or from a nearly ordinarily lumpable aggregation by defining

$$\boldsymbol{L}_{\tilde{i},\tilde{j}} = \min_{i \in [\tilde{i}]} \sum_{j \in [\tilde{j}]} \boldsymbol{Q}_{i,j} \text{ and } \boldsymbol{U}_{\tilde{i},\tilde{j}} = \max_{i \in [\tilde{i}]} \sum_{j \in [\tilde{j}]} \boldsymbol{Q}_{i,j}. \tag{10}$$

Set $\mathbb{Q}$ contains then all generator matrices $\tilde{\boldsymbol{Q}}$ with $\boldsymbol{L} \leq \tilde{\boldsymbol{Q}} \leq \boldsymbol{U}$. Similarly, set $\boldsymbol{r}$ contains all vectors $\tilde{\boldsymbol{r}}$ with $\min_{i \in [\tilde{i}]} \boldsymbol{r}_i \leq \tilde{\boldsymbol{r}}_{\tilde{i}} \leq \max_{i \in [\tilde{i}]} \boldsymbol{r}_i$.

We denote by $\mathbb{Q}_i$ and $\boldsymbol{r}_i$ the sets of generator matrices and reward vectors, where the entries may vary independently and bounds are used to define the range of the entries. If the sets result from a nearly lumpable aggregation, then $\boldsymbol{r}_i$ contains all available information and the set cannot be reduced based on the aggregation. This does not hold for $\mathbb{Q}_i$. It is known that row $\tilde{i}$ of each matrix $\tilde{\boldsymbol{Q}} \in \mathbb{Q}$ is a convex linear combination of the rows $i \in [\tilde{i}]$ of matrix $\boldsymbol{QV}$, where $\boldsymbol{V}$ is the collector matrix of the aggregation. We denote the corresponding set of matrices as $\mathbb{Q}_r$, which means that each row of a matrix $\boldsymbol{Q} \in \mathbb{Q}_r$ is characterized by a convex linear combination of a set of candidate rows.

If parameters of a CTMC are estimated from data, then often parameters of systems like arrival rates, service rates, failure rates or repair rates are estimated. These parameters appear at several positions in matrix $\boldsymbol{Q}$. Let $\boldsymbol{\mu} \in \mathbb{R}^{1\times p}$ a parameter vector with the rates for the p parameters of the CTMC, then matrix $\boldsymbol{Q}$ is given by

$$\boldsymbol{Q}[\boldsymbol{\mu}] = \sum_{k=0}^{p-1} \boldsymbol{\mu}_k \boldsymbol{Q}^{(k)} \tag{11}$$

where each $\boldsymbol{Q}^{(k)}$ $(0 \le k < p)$ is a generator matrix. We assume that matrix $\boldsymbol{Q}^{(k)}$ is normalized such that $\max_{i\in\mathcal{S}} \left|\boldsymbol{Q}^{(k)}_{i,i}\right| = 1$ which can always be achieved by an appropriate scaling of $\boldsymbol{\mu}_k$. If $\boldsymbol{\mu}_k$ can be chosen independently from some interval $[\mu_k^-, \mu_k^+]$, we denote the resulting set of matrices by $\mathbb{Q}_p$. The independent choice of parameters is correct, if confidence intervals are computed independently for all parameters resulting in lower and upper confidence limits.

If dependent confidence intervals are computed, then parameter vector $\boldsymbol{\mu}$ results from a convex linear combination of a set of base vector $\boldsymbol{\mu}^{(k)}$ $(0 \le k < q)$. We denote the resulting set of matrices as $\mathbb{Q}_c$. Matrices from $\mathbb{Q}_c$ can be represented for some vector $\boldsymbol{\gamma} \in \mathbb{R}^{1\times q}_{\ge 0}$ with $\boldsymbol{\gamma}\mathbb{I} = 1$ and

$$\boldsymbol{Q}[\boldsymbol{\gamma}] = \sum_{k=0}^{q-1} \gamma_k \boldsymbol{Q}[\boldsymbol{\mu}^{(k)}] \tag{12}$$

where $\boldsymbol{\mu}^{(k)} \in \mathbb{R}^{1\times p}_{\ge 0}$ is a vector of basic transition rates and matrix $\boldsymbol{Q}[\boldsymbol{\mu}^{(k)}]$ is built as in (11).

The sets $\mathbb{Q}_p$ and $\mathbb{Q}_c$ may also be used if the vectors $\boldsymbol{\mu}$ or $\boldsymbol{\gamma}$ describe control parameters of the system (e.g., service rates) that can be chosen in some range to improve the performance or availability of the system. In this situation, one is interested the best parameter setting rather than in computing performance bounds from parameter bounds. However, the applied methods are similar.

Sets of reward vectors can be defined similarly. In this paper, rewards defined by lower and upper bounds as defined in set $\boldsymbol{r}_i$ are sufficient.

The differential equations in (6) can be solved for matrices from $\mathbb{Q}_i$ and $\mathbb{Q}_r$, but not for matrices from $\mathbb{Q}_p$ and $\mathbb{Q}_c$. This distinction becomes important when computing bounds for R_t and A_t.

For the different sets $\mathbb{Q}_\cdot$, the corresponding sets $\mathscr{P}_i$, $\mathscr{P}_r$, $\mathscr{P}_p$ and $\mathscr{P}_c$ of the uniformized matrices can be computed. Computation of the matrices in $\mathscr{P}_i$ and $\mathscr{P}_r$ is straightforward using (1). For $\mathscr{P}_p$ and $\mathscr{P}_c$ a more detailed view is required to compute partial derivatives which are later exploited for bound computations. We begin with $\mathscr{P}_p$ and define

$$\nu(\boldsymbol{\mu}) = \sum_{k=0}^{p-1} \boldsymbol{\mu}_k \text{ and } \alpha = \sum_{k=0}^{p-1} \mu_k^+. \tag{13}$$

The value of the parameter α is required for uniformization. Then

$$\boldsymbol{P}[\boldsymbol{\mu}] = \sum_{k=0}^{p-1} \frac{\mu_k}{\mu_k^+}\boldsymbol{P}^{(k)} + \frac{\alpha-\nu(\boldsymbol{\mu})}{\alpha}\boldsymbol{I}, \text{ where } \boldsymbol{P}^{(k)} = \frac{\mu_k^+}{\alpha}\left(\boldsymbol{Q}^{(k)} + \boldsymbol{I}\right). \tag{14}$$

Observe that $\boldsymbol{P}[\boldsymbol{\mu}]$ and $\sum_{k=0}^{p-1}\boldsymbol{P}^{(k)}$ are stochastic matrices. The representation allows one to separate transitions depending on different values of the parameter vector.

For matrices in $\mathbb{Q}_p$, the value of α is defined as

$$\alpha = \max_{k\in\{0,\ldots,q-1\}}\left(\sum_{i=0}^{p-1}\boldsymbol{\mu}_i^{(k)}\right). \tag{15}$$

Matrices from $\mathcal{P}_c$ can then be represented as follows.

$$\boldsymbol{P}(\gamma) = \sum_{k=0}^{q-1}\gamma_k\boldsymbol{P}^{(k)} \text{ where } \boldsymbol{P}^{(k)} = \frac{1}{\alpha}\boldsymbol{Q}^{(k)}[\boldsymbol{\mu}^{(k)}] + \boldsymbol{I} \tag{16}$$

4 Bounds for Stationary Results

Computation of bounds for stationary results is based on the work of Courtois and Semal [16] which has been adopted to compute bounds for aggregated MCs resulting from near ordinary lumpability in [3,4,22]. The basic idea of bound computation is that the stationary vector of any stochastic matrix $\boldsymbol{P}$ for which a lower bound bound $\boldsymbol{L} \le \boldsymbol{P}$ is known, can be represented as a convex linear combination of the rows of matrix $\overline{(\boldsymbol{I}-\boldsymbol{L})^{-1}}$. $\overline{\boldsymbol{A}}$ of some some non-negative matrix $\boldsymbol{A}$ equals the original matrix where each row is normalized to 1. A similar, slightly more restrictive, result holds if an upper bound $\boldsymbol{P} \le \boldsymbol{U}$ is known. Consequently, bounds are computed for a set $\mathcal{P}_i$ which contains all stochastic matrices that are larger than a lower bound or a set that contains all matrices that are smaller than some upper bound. It should be mentioned that the lower and upper bound matrices defined in (10) sometimes can be improved for stationary analysis as shown in [4,16,31].

However, the main problem of the basic approach is that bounds are computed from the lower and upper bound only and not for the set $\mathcal{P}_i$ that results from both, upper and lower bounds together. This has been done in [7,8] where bounds are computed based on *Markov Decision Processes* (MDPs). For the sets $\mathcal{P} \in \{\mathcal{P}_i, \mathcal{P}_r\}$ let $\mathcal{P}(i:)$ be the set of ith rows of the matrices in $\mathcal{P}$. The optimization of the expected reward can then be expressed by the following Bellman equations for the average reward of an MDP [35] as shown in [8].

$$\begin{aligned} R_\infty^- + \boldsymbol{h}_i^- &= \boldsymbol{r}_i^- + \min_{\boldsymbol{p}\in\mathcal{P}(i:)}\sum_{j\in\mathcal{S}}\boldsymbol{p}_j\boldsymbol{h}_j^- \text{ and} \\ R_\infty^+ + \boldsymbol{h}_i^+ &= \boldsymbol{r}_i^+ + \max_{\boldsymbol{p}\in\mathcal{P}(i:)}\sum_{j\in\mathcal{S}}\boldsymbol{p}_j\boldsymbol{h}_j^+ \text{ for all } i\in\mathcal{S} \end{aligned} \tag{17}$$

where $r_i^- = \min_{r \in \mathscr{r}} r_i$ and $r_i^+ = \max_{r \in \mathscr{r}} r_i$. The equations have one degree of freedom which is fixed by selecting $h_n^- = h_n^+ = 0$. They can be solved by policy or value iteration or combinations of both [8]. The computed bounds are tight for a given set $\mathscr{P}$ (or $\mathbb{Q}$) that contains only irreducible matrices. The result is introduced in [8] for the set $\mathscr{P}_i$ and in [21] for $\mathscr{P}_r$ without noticing the relation to MDPs.

Equation (17) does not work for the sets $\mathscr{P}_p$ and $\mathscr{P}_c$ because the rows of matrices from the sets cannot be chosen independently which is necessary to set up Bellman equations. This also implies that the optimization problems resulting from these sets do not have the nice properties of MDPs which allow an efficient computation of the optimum using value iteration, policy iteration or even linear programming.

We begin with the optimization according to $\mathscr{P}_p$ and consider computation of the maximum reward, computation of the minimum can be formulated analogously. For a fixed choice of $\boldsymbol{\mu}$, the expected reward and vector $\boldsymbol{h}$ are given by the solution of

$$R_\infty \mathbb{I} = (\boldsymbol{P}[\boldsymbol{\mu}] - \boldsymbol{I})\, \boldsymbol{h} + \boldsymbol{r}^+. \tag{18}$$

Again the equation has one degree of freedom which is fixed by setting $\boldsymbol{h}_n = 0$. The derivative with respect to $\boldsymbol{\mu}_k$ is then given by

$$\frac{dR_\infty}{d\boldsymbol{\mu}_k}\mathbb{I} = \frac{d\boldsymbol{P}[\boldsymbol{\mu}]}{d\boldsymbol{\mu}_k}\boldsymbol{h} + (\boldsymbol{P}[\boldsymbol{\mu}] - \boldsymbol{I})\,\frac{d\boldsymbol{h}}{d\boldsymbol{\mu}_k} = \left(\frac{1}{\mu_k^+}\boldsymbol{P}^{(k)} - \frac{1}{\alpha}\boldsymbol{I}\right)\boldsymbol{h} + (\boldsymbol{P}[\boldsymbol{\mu}] - \boldsymbol{I})\,\frac{d\boldsymbol{h}}{d\boldsymbol{\mu}_k}. \tag{19}$$

This defines again a set of equations with one degree of freedom which is fixed by setting the derivative of $\boldsymbol{h}_n$ to zero. The value $dR_\infty/d\boldsymbol{\mu}_k$ is the gradient which can be used to maximize (or minimize) the value of R_∞ using some gradient descent method [30, 37] and the constraints $\mu_k^- \le \boldsymbol{\mu}_k \le \mu_k^+$.

For matrices from $\mathbb{Q}_c$ the derivative according to the vector components γ_k given by

$$\frac{dR_\infty}{d\gamma_k}\mathbb{I} = \frac{d\boldsymbol{P}[\gamma]}{d\gamma_k}\boldsymbol{h} + (\boldsymbol{P}[\gamma] - \boldsymbol{I})\,\frac{d\boldsymbol{h}}{d\gamma_k} = \boldsymbol{P}^{(k)}\boldsymbol{h} + (\boldsymbol{P}[\gamma] - \boldsymbol{I})\,\frac{d\boldsymbol{h}}{d\gamma_k}. \tag{20}$$

With the derivatives one can proceed as before to maximize (or minimize) the expected reward, keeping in mind that the constraints $\gamma\mathbb{I} = 1$ and $\gamma \ge \boldsymbol{0}$ have to hold.

Gradient descent methods allow one to compute local minima and maxima but do not guarantee that the results are global minima or maxima. Consequently, the bounds are not strict and smaller or larger reward values may result from some matrix in the set which is not found by the optimization method. However, for most models the methods work well and strict bounds are derived. It should be mentioned that for stationary analysis, the sets $\mathscr{P}_.$ can be used to compute bounds for matrices from $\mathbb{Q}_.$ because the stationary solution of a CTMC with matrix $\boldsymbol{Q}$ is identical to the stationary distribution of the uniformized DTMC with matrix $\boldsymbol{P}$.

5 Bounds for Transient Results

For stationary analysis, reward bounds are reached by one matrix or several matrices from set $\mathbb{Q}$ or $\mathscr{P}$. This follows from the results for MDPs with compact action sets where the optimal reward is achieved by a stationary policy [35] that corresponds to a fixed matrix. For transient rewards this does not necessarily hold. It could be better to change the matrix over time to obtain a larger or smaller reward.

For the problem classes considered here, a varying transition matrix can arise when aggregated MCs are analyzed. If the evaluation of an MC is observed at the level of the aggregated states (i.e., detailed states cannot be distinguished), then the aggregated process is only an MC if one of the conditions in (4) holds. Otherwise, the transition rates evolve over time. For distribution $\boldsymbol{\pi}(t)$ at time t, the rate from $\tilde{i}$ to $\tilde{j}$ is given by $\sum_{i\in[\tilde{i}]}\boldsymbol{\pi}_i(t)/\left(\sum_{k\in[\tilde{i}]}\boldsymbol{\pi}_k(t)\right)\sum_{j\in[\tilde{j}]}\boldsymbol{Q}_{i,j}$. Without additional restrictions on vector $\boldsymbol{\pi}(t)$, any matrix from the sets $\mathbb{Q}_i$ or $\mathbb{Q}_r$ may potentially results from aggregation. Consequently, bounds have to be based on all matrices in these sets.

The situation is different if we consider MCs with uncertain parameters. In this case, it is assumed that a unknown transition rate matrix $\boldsymbol{Q}\in\mathbb{Q}$ exists that describes the MC. Consequently, the matrix remains constant in $[0,t]$ and the matrix minimizing or maximizing R_t or A_t has to be found.

In the following we begin with variable transition matrices where the bounds can be computed using results from MDPs. Afterwards, the more complex case of fixed transition rates is introduced, where the solution has to be computed by some gradient descent method.

5.1 Variable Transition Matrices

For computing transient rewards we begin with a set $\mathscr{P}\in\{\mathscr{P}_i,\mathscr{P}_r\}$. Again, $\mathscr{P}(i:)$ selects all possible ith rows from matrices in $\mathscr{P}$ which can be done for matrices in $\mathscr{P}_i$ or $\mathscr{P}_r$. Define the following sequences of vectors.

$$\begin{aligned}\boldsymbol{r}_i^-(k) &= \min_{\boldsymbol{p}\in\mathscr{P}(i:)}\{\boldsymbol{p}\boldsymbol{r}^-(k-1)\},\ \ \boldsymbol{g}_i^-(k)=\tfrac{1}{k}\boldsymbol{r}_i^- + \tfrac{k-1}{k}\min_{\boldsymbol{p}\in\mathscr{P}(i:)}\{\boldsymbol{p}\boldsymbol{g}^-(k-1)\},\\ \boldsymbol{r}_i^+(k) &= \max_{\boldsymbol{p}\in\mathscr{P}(i:)}\{\boldsymbol{p}\boldsymbol{r}^-(k-1)\},\ \ \boldsymbol{g}_i^+(k)=\tfrac{1}{k}\boldsymbol{r}_i^+ + \tfrac{k-1}{k}\min_{\boldsymbol{p}\in\mathscr{P}(i:)}\{\boldsymbol{p}\boldsymbol{g}^+(k-1)\}\end{aligned} \tag{21}$$

for all $k>0$ and all $i\in\mathcal{S}$ where $\boldsymbol{r}^-(0)=\boldsymbol{r}^-$, $\boldsymbol{r}^+(0)=\boldsymbol{r}^+$ and $\boldsymbol{g}^-(0)=\boldsymbol{g}^+(0)=\boldsymbol{0}$. The sets $\mathscr{P}_i(i:)$ and $\mathscr{P}_r(i:)$ contain an infinite number of rows if the conditions for ordinary lumpability are not observed. However, since both sets are convex, the minimum or maximum in (21) results from a finite subset of rows. Thus, the problem of computing an optimal reward value for the expected or accumulated reward at time $t\in\mathbb{N}$ can be interpreted as an MDP with finite action set. Reward bounds are then given by

$$\begin{aligned}R_t^- &= \boldsymbol{\pi}(0)\boldsymbol{r}^-(t)\le R_t\le R_t^+=\boldsymbol{\pi}(0)\boldsymbol{r}^+(t)\quad\text{and}\\ A_t^- &= t\boldsymbol{\pi}(0)\boldsymbol{g}^-(t)\le A_t\le A_t^+=t\boldsymbol{\pi}(0)\boldsymbol{g}^+(t)\ \text{where}\ t\in\mathbb{N}.\end{aligned} \tag{22}$$

These bounds are tight, if the matrix can be chosen from the set $\mathcal{P}$ in each step.

If we consider a CTMC, set $\mathcal{P}$ contains matrices resulting from uniformization and $\boldsymbol{\pi}(t)$ $(t \in \mathbb{R}_{\geq 0})$ is computed using uniformization (2). If the reward bounds are plugged into the equation, the following bounds are computed.

$$\begin{aligned} R_t^- &= \boldsymbol{\pi}(0) \sum_{k=0}^{\infty} \beta(\alpha t, k)\boldsymbol{r}^-(k) \leq R_t \leq R_t^+ = \boldsymbol{\pi}(0) \sum_{k=0}^{\infty} \beta(\alpha t, k)\boldsymbol{r}^+(k) \quad \text{and} \\ A_t^- &= t\boldsymbol{\pi}(0) \sum_{k=0}^{\infty} \beta(\alpha t, k)\boldsymbol{g}^-(k) \leq A_t \leq A_t^+ = t\boldsymbol{\pi}(0) \sum_{k=0}^{\infty} \beta(\alpha t, k)\boldsymbol{g}^+(k) \end{aligned} \tag{23}$$

where α is the parameter used for uniformization of the matrices in set $\mathbb{Q}$. (23) defines reward bounds for the CTMC because at the time of each transition the minimal or maximal reward for each state is chosen. This can be done for matrices from $\mathcal{P}_i$ and $\mathcal{P}_r$.

Unfortunately, the bounds from (23) are not the best available for the given information. The reason is that the optimal policy for a continuous time MDP, which describes the original problem of reward optimization, is piecewise constant [29,34]. This implies that the matrix remains constant for a time interval, will change for the subsequent interval but not necessarily in each step of the embedded DTMC

An optimal policy corresponds to a matrix from $\mathbb{Q}$ and the corresponding uniformized matrix in $\mathcal{P}$ which can be approximated, either by discretization of the interval $[0, t]$ and computing for each sub-interval the optimal matrix using the methods proposed in Sect. 5.2, or by using the uniformization based approach that has been proposed in for optimizing the accumulated reward in continuous time MDPs [9,10]. The latter approach is now be adopted for the problem handled in this paper. It allows one to compute an ε-approximation of $R_t^\pm$ and $A_t^\pm$ for an arbitrary $\varepsilon > 0$ up to numerical precision.

Let $\boldsymbol{r}^\pm(\tau)$ and $\boldsymbol{g}^\pm(\tau)$ $(\tau \in [0, t])$ be defined as in (6) and define for a set $\mathcal{P}$ that results from uniformization of the matrices $\mathbb{Q} \in \{\mathbb{Q}_i, \mathbb{Q}_r\}$

$$\begin{aligned} \mathcal{P}^k &= \left\{\boldsymbol{P} \text{ maximizes } \boldsymbol{P}\boldsymbol{u}(k-1) | \boldsymbol{P} \in \mathcal{P}^{k-1}\right\} \\ \bar{\mathcal{P}}^k &= \left\{\boldsymbol{P} \text{ maximizes } \boldsymbol{P}\boldsymbol{v}(k-1) | \boldsymbol{P} \in \bar{\mathcal{P}}^{k-1}\right\} \end{aligned} \tag{24}$$

where $\mathcal{P}^0 = \bar{\mathcal{P}}^0 = \mathcal{P}$, $\boldsymbol{u}(k) = \boldsymbol{P}\boldsymbol{u}(k-1)$ with $\boldsymbol{u}(0) = \boldsymbol{r}^+(\tau)$ and $\boldsymbol{v}(k) = (k-1)/K\boldsymbol{P}\boldsymbol{v}(k-1) + \boldsymbol{r}^+(\tau)/k$ with $\boldsymbol{v}(0) = \boldsymbol{0}$. Often $\mathcal{P}^1$ or $\bar{\mathcal{P}}^2$ contain only a single matrix and it is known that all matrices from $\mathcal{P}^n$ belong to matrices $\boldsymbol{Q}$ that are the solution of the differential equation for $\boldsymbol{r}^-$ in (6) and matrices $\boldsymbol{Q}$ belonging to matrices in $\bar{\mathcal{P}}^n$ solve the differential equation for $\boldsymbol{g}^-$ in (6) [34]. Similar sets can be defined for $\boldsymbol{r}^-$ and $\boldsymbol{g}^-$ by substituting the maximum by the minimum. As shown in [10], it is not possible to compute the length of the interval for which matrices are optimal such that only bounds for the interval length can be computed.

The objective is to compute bounds for the vectors $\boldsymbol{r}^-(t)$, $\boldsymbol{r}^+(t)$, $\boldsymbol{g}^-(t)$ and $\boldsymbol{g}^+(t)$. The approach is presented for $\boldsymbol{r}^+(t)$ and $\boldsymbol{g}^+(t)$. It works analogously for the lower bounds $\boldsymbol{r}^-(t)$ and $\boldsymbol{g}^-(t)$.

Vectors $\hat{\boldsymbol{r}}^-(t)$, $\hat{\boldsymbol{r}}^+(t)$, $\hat{\boldsymbol{g}}^-(t)$ and $\hat{\boldsymbol{g}}^+(t)$ are computed such that

$$\begin{array}{l} \hat{\boldsymbol{r}}^-(t) \leq \boldsymbol{r}^+(t) \leq \hat{\boldsymbol{r}}^+(t) \text{ and } \left\|\hat{\boldsymbol{r}}^+(t) - \hat{\boldsymbol{r}}^-(t)\right\|_\infty \leq \varepsilon, \\ \hat{\boldsymbol{g}}^-(t) \leq \boldsymbol{g}^+(t) \leq \hat{\boldsymbol{g}}^+(t) \text{ and } \left\|\hat{\boldsymbol{g}}^+(t) - \hat{\boldsymbol{g}}^-(t)\right\|_\infty \leq \varepsilon \end{array} \tag{25}$$

Vectors are computed by mimicking the piecewise constant policies. Starting from $\tau \in [0, t]$ with vectors that $\hat{\boldsymbol{r}}^-(\tau)$, $\hat{\boldsymbol{r}}^+(\tau)$, $\hat{\boldsymbol{g}}^-(\tau)$ and $\hat{\boldsymbol{g}}^+(\tau)$ that observe (25) with error bound $\tau/t\varepsilon$. From these vectors an interval length δ and new vectors $\hat{\boldsymbol{r}}^-(\tau+\delta)$, $\hat{\boldsymbol{r}}^+(\tau+\delta)$, $\hat{\boldsymbol{g}}^-(\tau+\delta)$ and $\hat{\boldsymbol{g}}^+(\tau+\delta)$ are computed that observe (25) with error bound $(\tau+\delta)/t\varepsilon$. At $\tau = 0$, the vectors are initialized $\hat{\boldsymbol{r}}^-(0) = \boldsymbol{r}^+$, $\hat{\boldsymbol{r}}^+(0) = \boldsymbol{r}^+$ and $\hat{\boldsymbol{g}}^-(0) = \hat{\boldsymbol{g}}^+(0) = \mathbf{0}$.

Assume that vectors at $\tau \in [0, t]$ are known and define for all $i \in \mathcal{S}$ the following sequences of vectors

$$\begin{array}{ll} \boldsymbol{u}_i^-(k) = \bar{\boldsymbol{q}}\boldsymbol{u}^-(k-1) & \boldsymbol{u}_i^+(k) = \max_{\boldsymbol{q}\in\mathcal{P}(i:)}\{\boldsymbol{q}\boldsymbol{u}^+(k-1)\} \\ \boldsymbol{v}_i^-(k) = \bar{\boldsymbol{p}}\boldsymbol{v}^-(k-1), & \boldsymbol{v}_i^+(k) = \max_{\boldsymbol{p}\in\mathcal{P}(i:)}\{\boldsymbol{p}\boldsymbol{v}^+(k-1)\} \\ \boldsymbol{w}_i^-(k) = \bar{\boldsymbol{p}}\boldsymbol{w}_i^-(k-1), & \boldsymbol{w}_i^+(k) = \max_{\boldsymbol{p}\in\mathcal{P}(i:)}\{\boldsymbol{p}\boldsymbol{w}^+(k-1)\} \\ \text{where } \bar{\boldsymbol{q}} = \arg\max\limits_{\boldsymbol{p}\in\mathcal{P}(i:)}\{\boldsymbol{p}\hat{\boldsymbol{r}}^-(\tau)\} & \bar{\boldsymbol{p}} = \arg\max\limits_{\boldsymbol{p}\in\mathcal{P}(i:)}\{\boldsymbol{p}\hat{\boldsymbol{g}}^-(\tau)\} \end{array} \tag{26}$$

with $k > 0$, $\boldsymbol{u}^-(0) = \hat{\boldsymbol{r}}^-(\tau)$, $\boldsymbol{u}^+(0) = \hat{\boldsymbol{r}}^+(\tau)$, $\boldsymbol{v}^-(0) = \hat{\boldsymbol{g}}^-(\tau)$, $\boldsymbol{v}^+(0) = \hat{\boldsymbol{g}}^+(\tau)$ and $\boldsymbol{w}^-(0) = \boldsymbol{w}^+(0) = \boldsymbol{r}^+$. Obviously, $\boldsymbol{v}^-(k) \leq \boldsymbol{v}^+(k)$ and $\boldsymbol{w}^-(k) \leq \boldsymbol{w}^+(k)$.

Before upper bounds are computed, the following abbreviations are introduced.

$$\begin{array}{llll} \bar{\beta}(\alpha\delta, K) = 1 - \sum\limits_{k=0}^{K}\beta(\alpha\delta, k), & & \xi(\alpha,\delta,K) = \delta\bar{\beta}(\alpha\delta,K) - \frac{K+1}{\alpha}\bar{\beta}(\alpha\delta,K+1), & \\ \zeta_{K,\delta}^- & = \bar{\beta}(\alpha\delta,K)\min\limits_{i\in\mathcal{S}}\boldsymbol{v}_i^-(K), & \zeta_{K,\delta}^+ & = \bar{\beta}(\alpha\delta,K)\max\limits_{i\in\mathcal{S}}\boldsymbol{v}_i^+(K) \\ \eta_{K,\delta}^- & = \xi(\alpha,\delta,K)\min\limits_{i\in\mathcal{S}}\boldsymbol{w}_i^-(K), & \eta_{K,\delta}^+ & = \xi(\alpha,\delta,K)\max\limits_{i\in\mathcal{S}}\boldsymbol{w}_i^+(K) \end{array} \tag{27}$$

Then

$$\sum_{k=0}^{K}\beta(\alpha\delta,k)\boldsymbol{u}^-(k) + \zeta_{K,\delta}^-\mathbb{1} \leq \boldsymbol{r}^+(\tau+\delta) \leq R_t \leq \sum_{k=0}^{K}\beta(\alpha\delta,k)\boldsymbol{u}^+(k) + \zeta_{K,\delta}^+\mathbb{1} \tag{28}$$

and

$$\begin{array}{l} \sum\limits_{k=0}^{K}\left(\beta(\alpha\delta,k)\boldsymbol{v}^-(k) + \frac{\bar{\beta}(\alpha\delta,k)}{\alpha}\boldsymbol{w}^-(k)\right) + \left(\zeta_{K,\delta}^- + \eta_{K,\delta}^-\right)\mathbb{1} \leq \boldsymbol{g}^+(\tau+\delta) \leq A_t \leq \\ \sum\limits_{k=0}^{K}\left(\beta(\alpha\delta,k)\boldsymbol{v}^+(k) + \frac{\bar{\beta}(\alpha\delta,k)}{\alpha}\boldsymbol{w}^+(k)\right) + \left(\zeta_{K,\delta}^+ + \eta_{K,\delta}^+\right)\mathbb{1}. \end{array} \tag{29}$$

The proofs for (28) and (29) follow from the proofs in [10] for continuous time MDPs with finite action sets. δ is chosen adaptively such that the resulting difference between lower and upper bound is smaller than $\delta/t\varepsilon$. Then $\varepsilon \geq \boldsymbol{\pi}(0)\hat{\boldsymbol{r}}^+(t) - R_t^+ \geq 0$ and $\varepsilon \geq \boldsymbol{\pi}(0)\hat{\boldsymbol{g}}^+(t) - A_t^+ \geq 0$. Improved bounds can be computed using the upper bound from [9].

5.2 Fixed Transition Matrices

To compute bounds for R_t under the condition that a fixed matrix from $\mathscr{P}$ is chosen for the whole interval, we consider $\mathscr{P} \in \{\mathscr{P}_p, \mathscr{P}_c\}$. For a given matrix $\boldsymbol{P} \in \{\boldsymbol{P}[\boldsymbol{\mu}], \boldsymbol{P}[\gamma]\}$, rewards are computed using uniformization [25,41].

$$R_t = \boldsymbol{\pi}(0) \sum_{k=0}^{\infty} \beta(\alpha t, k) \boldsymbol{P}^k \boldsymbol{r} \text{ and } A_t = \boldsymbol{\pi}(0) \sum_{k=0}^{\infty} \frac{\bar{\beta}(\alpha t, k)}{\alpha} \boldsymbol{P}^k \boldsymbol{r} \tag{30}$$

where $\boldsymbol{r} \in \{\boldsymbol{r}^-, \boldsymbol{r}^+\}$ depending whether the minimum or maximum should be computed. Again the infinite sums can be truncated with respect to a predefined truncation bound [25,41].

To compute bounds for R_t or A_t observe that both can be represented as

$$\begin{aligned} &R_t = \boldsymbol{\pi}(0) \textstyle\sum_{k=0}^{\infty} \beta(\alpha t, k) \boldsymbol{r}(k) \text{ and } A_t = \boldsymbol{\pi}(0) \sum_{k=0}^{\infty} \frac{\bar{\beta}(\alpha t, k)}{\alpha} \boldsymbol{r}(k) \\ &\text{where } \boldsymbol{r}(k) = \boldsymbol{P}\boldsymbol{r}(k-1) \text{ and } \boldsymbol{r}(0) = \begin{cases} \boldsymbol{r}^- & \text{for the minimum and} \\ \boldsymbol{r}^+ & \text{for the maximum} \end{cases} \end{aligned} \tag{31}$$

For matrices from $\mathscr{P}_p$ the derivative of R_t and A_t according to $\boldsymbol{\mu}_h$ $(h = 1, \ldots, p)$ is defined as

$$\frac{dR_t}{d\boldsymbol{\mu}_h} = \boldsymbol{\pi}(0) \sum_{k=0}^{\infty} \beta(\alpha t, k) \frac{d\boldsymbol{r}(k)}{d\boldsymbol{\mu}_h} \text{ and } \frac{dA_t}{d\boldsymbol{\mu}_h} = \boldsymbol{\pi}(0) \sum_{k=0}^{\infty} \frac{\bar{\beta}(\alpha t, k)}{\alpha} \frac{d\boldsymbol{r}(k)}{d\boldsymbol{\mu}_h} \tag{32}$$

where

$$\frac{d\boldsymbol{r}(k)}{d\boldsymbol{\mu}_h} = \frac{d\boldsymbol{P}[\boldsymbol{\mu}]}{d\boldsymbol{\mu}_h} \boldsymbol{r}(k-1) + \boldsymbol{P}[\boldsymbol{\mu}] \frac{d\boldsymbol{r}(k-1)}{d\boldsymbol{\mu}_h} \text{ for } k \geq 1 \text{ and } \frac{d\boldsymbol{r}(0)}{d\boldsymbol{\mu}_h} = \mathbf{0}. \tag{33}$$

Matrix $\frac{d\boldsymbol{P}[\boldsymbol{\mu}]}{d\boldsymbol{\mu}_h}$ is given in (19). For matrices from $\mathscr{P}_c$, derivatives with respect to γ_h are computed in the same way, matrix $\frac{d\boldsymbol{P}[\gamma]}{d\gamma_h}$ is given in (20). In both cases, derivatives are computed by vector matrix products.

A gradient descent method can then be applied to find local minima or maxima. The approach can also be used for matrices from the sets $\mathbb{Q}_i$ or $\mathbb{Q}_r$ which can be described using $\mathbb{Q}_p$. E.g., to specify a matrix where element $\boldsymbol{Q}_{i,j} \in [\boldsymbol{L}_{i,j}, \boldsymbol{U}_{i,j}]$ we define $n(n-1)$ values $\boldsymbol{\mu}_{ij} \in [\boldsymbol{L}_{i,j}, \boldsymbol{U}_{i,j}]$ and a matrix $\boldsymbol{Q}^{(ij)}$ with 1 in position (i,j), -1 in position (i,i) and 0 elsewhere. Matrices from $\mathbb{Q}_r$ can be described similarly by matrices from an appropriately chosen set $\mathbb{Q}_p$.

6 Examples

We consider results for two examples.

6.1 An M/M/2/10 Queue with Failures

The first example is a *M/M/2/20* system with failures and repairs. Jobs arrive according to a Poisson process with rate $\lambda \in [0.891, 9.09]$ and are served by one of the two servers with rate $\mu \in [0.99, 1.01]$. Service times are exponentially distributed and jobs arriving when the system is full get lost. Servers may fail. Failure times are exponentially distributed and failure rates are load dependent. The failure rate of a server equals $0.5 \cdot n \cdot \nu$ ($\nu \in [9.9e-5, 1.01e-4]$) for population n in the system. Failed servers are repaired. Repair times have an Erlang-3 distribution with mean ω^{-1} ($\omega \in [9.9e-4, 1.01e-3]$). The CTMC resulting from this model has 273 states.

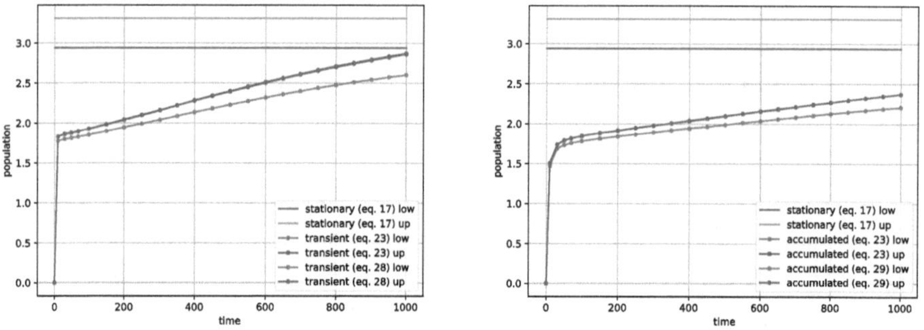

Fig. 1. Population R_t and A_t/t of the M/M/2/20 queue for $\mathscr{P}_i$.

For this model all parameters are uncertain. We first assume that all elements in the transition matrix can vary independently and analyze the transient, accumulated and stationary population. The reward vector to describe the population is exactly known. Figure 1 shows bounds for the stationary and transient population (left side) and the stationary and accumulated population (right side). For the accumulated value the figure shows A_t/t to allow a comparison of the average population in a time interval and the stationary or transient population. For transient analysis the system is initially empty and both servers are up. It can be seen that the transient and accumulated rewards first increase quickly but after some time the rewards grow very slowly. The first step is caused by arrivals and services whereas the latter increase is caused by failures which are rare events such that it takes a long time before a server fails. It can be noticed that the equations (23) and (28/29) give results that are almost identical, differences are not visible in the figure and in the range of $1.0e-3$ which has been used for ε in (28) and (29). Thus, the approximation using the embedded DTMC is for this example sufficient.

For this model the assumption of transition rates that differ in each state and a transition matrix that changes over time is not realistic. The system has four uncertain parameters which take some values in the given intervals. Consequently, we have a vector $\boldsymbol{\mu}$ with four entries that define matrix $\boldsymbol{Q}[\boldsymbol{\mu}]$ and this matrix remains constant over time.

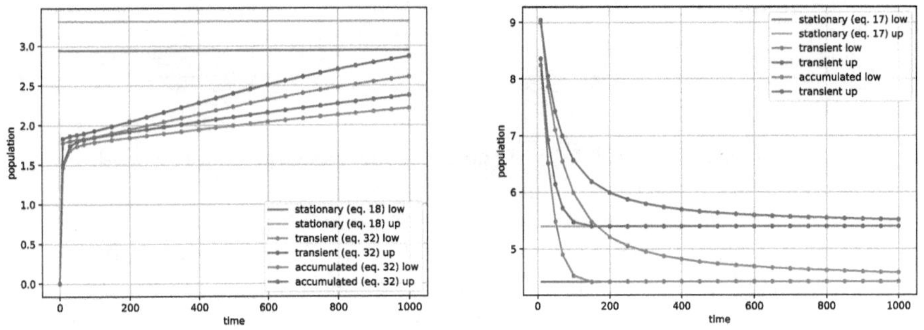

Fig. 2. Population R_t and A_t/t of the M/M/2/20 queue and for the fork-join queue both for $\mathscr{P}_p$.

Figure 2 shows on the left side the transient, accumulated and stationary population for this system. For stationary analysis results are the same as before, because for the lower bound, arrival and failure rates are minimal and service and repair rates are maximal. For the upper bound it is the other way round.

6.2 A Fork-Join Queuing Network

The second example is a fork-join queueing network. The system contains N jobs which cycle between two queues. Jobs are first served at a station with exponential service times and rate $\mu = 0.25$. After leaving the station, each job forks into four threads that are served at the second station. The service time of thread i is exponentially distributed with rate $\lambda_i \in \{0.95, 0.97, 1.03, 1.05\}$ for $i = 1, \ldots, 4$. The second station serves threads of one type in FCFS order but shares its capacity equally among the available thread types. I.e., if threads of all types are at the station, the service rate for thread type i equals $0.25\lambda_i$. After all threads of one job have been served, they join and the resulting job enters the first station again.

For $N = 10$, the resulting MC contains 14641 states. If all threads have the same service rate, the MC is ordinarily lumpable and results in an aggregated MC with only 1001 states. Since the rates are not identical, aggregation is based on near ordinary lumpability and bounds are computed. The population in the first stations is used as result measure. For transient analysis the systems starts with the state where all jobs are in the first station.

Figure 3 shows the results for the example. On the left side the results for matrices $\mathscr{P}_i$ are presented and on the right side results for matrices from $\mathscr{P}_r$. It

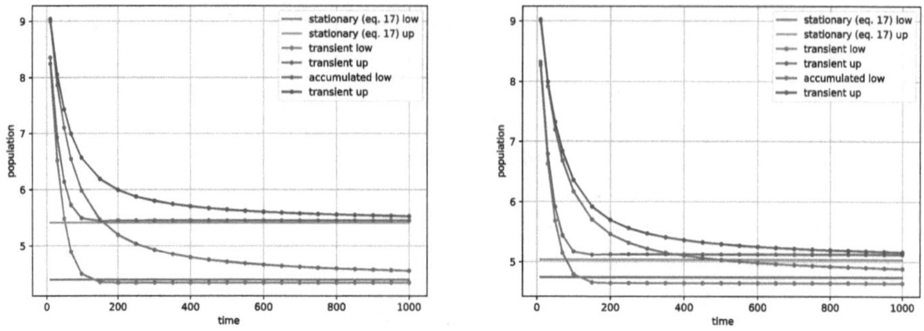

Fig. 3. Population R_t and A_t/t of the fork-join queue queue for $\mathcal{P}_i$ and $\mathcal{P}_r$.

can be noticed that for the matrices $\mathcal{P}_r$ the bounds for transient results do not converge towards the stationary results which is possible as already mentioned.

To show optimization with respect to vector γ, we consider an extended model and assume that threads that have been served and wait for a join, expire with thread type specific rates $\omega_i \in \{0.03, 0.05, 0.07, 0.09\}$. If a thread expires it requires a second service at the corresponding station. To control the system it is now possible to share the capacity of the second station in an unequal way among the thread types by defining vector γ. A configuration with $N = 5$ jobs is considered which results in a CTMC with 1296 states. To reach a maximal throughput in the system distribution $(0.24753, 0.25028, 0.25112, 0.25106)$ is optimal and results in throughput 0.15. If the population of active or waiting threads in the parallel stations is minimized the distribution $(0.24812, 0.25050, 0.25102, 0.25036]$ results in minimal thread population of 15.58.

7 Conclusion

The paper defines different forms of parameter uncertainties in Markov chains that may result from unknown parameters or aggregation of a large Markov chain based on near lumpability. Methods are introduced to compute bounds on stationary, transient and accumulated rewards. Some of the presented methods can already be found in the literature, others are adoptions to the specific problem and some are new. Apart from bound computation, the proposed methods can also be applied to find optimal control parameters.

There are different ways to extend the presented techniques. They can be applied to absorbing Markov chains to compute bounds on the moments of absorption time or the pdf of absorption times. From a theoretical perspective it would be interesting to prove for which models the gradient descent methods result in strict bounds and how good bounds for aggregated models resulting from nearly exactly lumpable partitions can be derived. From a practical perspective efficient implementations of the bounding methods are required.

References

1. Alves, R.B., Saporito, Y.F., Carvalho, L.M.: On the lumpability of tree-valued Markov chains (2024). arXiv:2410.17919 arXiv preprint
2. Bolch, G., Greiner, S., De Meer, H., Trivedi, K.S.: Queueing networks and Markov chains: modeling and performance evaluation with computer science applications. John Wiley & Sons (2006)
3. Buchholz, P.: Die strukturierte Analyse Markovscher Modelle, Informatik-Fachberichte, vol. 282. Springer (1991). https://doi.org/10.1007/978-3-642-76923-8
4. Buchholz, P.: Exact and ordinary lumpability in finite Markov chains. J. Appl. Probab. **31**(1), 59–75 (1994)
5. Buchholz, P.: Lumpability and nearly-lumpability in hierarchical queueing networks. In: Proceedings of 1995 IEEE International Computer Performance and Dependability Symposium, pp. 82–91. IEEE (1995)
6. Buchholz, P.: Exact performance equivalence: an equivalence relation for stochastic automata. Theor. Comput. Sci. **215**(1-2), 263–287 (1999). https://doi.org/10.1016/S0304-3975(98)00169-8
7. Buchholz, P.: An improved method for bounding stationary measures of finite Markov processes. Perform. Evaluation **62**(1-4), 349–365 (2005). https://doi.org/10.1016/J.PEVA.2005.07.002
8. Buchholz, P.: Bounding reward measures of Markov models using the Markov decision processes. Numer. Linear Algebra Appl. **18**(6), 919–930 (2011). https://doi.org/10.1002/NLA.792
9. Buchholz, P., Dohndorf, I., Scheftelowitsch, D.: Optimal decisions for continuous time Markov decision processes over finite planning horizons. Comput. Oper. Res. **77**, 267–278 (2017). https://doi.org/10.1016/J.COR.2016.08.003
10. Buchholz, P., Schulz, I.: Numerical analysis of continuous time Markov decision processes over finite horizons. Comput. Oper. Res. **38**(3), 651–659 (2011). https://doi.org/10.1016/J.COR.2010.08.011
11. Buchholz, P., Telek, M.: Rational processes related to communicating Markov processes. J. Appl. Probab. **49**(1), 40–59 (2012)
12. Buchholz, P., Telek, M.: On minimal representations of rational arrival processes. Ann. Oper. Res. **202**(1), 35–58 (2013)
13. Burgholzer, L., Jiménez-Pastor, A., Larsen, K.G., Tribastone, M., Tschaikowski, M., Wille, R.: Forward and backward constrained bisimulations for quantum circuits using decision diagrams. ACM Trans. Quantum Comput. **6**(2), 13:1–13:21 (2025). https://doi.org/10.1145/3712711
14. Cardelli, L., Grosu, R., Larsen, K.G., Tribastone, M., Tschaikowski, M., Vandin, A.: Lumpability for uncertain continuous-time Markov chains. In: Abate, A., Marin, A. (eds.) Quantitative Evaluation of Systems - 18th International Conference, QEST 2021, Paris, France, August 23-27, 2021, Proceedings. Lecture Notes in Computer Science, vol. 12846, pp. 391–409. Springer (2021). https://doi.org/10.1007/978-3-030-85172-9_21
15. Cardelli, L., Grosu, R., Larsen, K.G., Tribastone, M., Tschaikowski, M., Vandin, A.: Algorithmic minimization of uncertain continuous-time Markov chains. IEEE Trans. Autom. Control. **68**(11), 6557–6572 (2023). https://doi.org/10.1109/TAC.2023.3244093
16. Courtois, P., Semal, P.: Bounds for the positive eigenvectors of nonnegative matrices and for their approximations by decomposition. J. ACM **31**(4), 804–825 (1984). https://doi.org/10.1145/1634.1637

17. Dayar, T., Stewart, W.J.: Quasi lumpability, lower-bounding coupling matrices, and nearly completely decomposable Markov chains. SIAM J. Matrix Anal. Appl. **18**(2), 482–498 (1997). https://doi.org/10.1137/S0895479895294277
18. De Cooman, G., Hermans, F., Quaeghebeur, E.: Imprecise Markov chains and their limit behavior. Probab. Eng. Inf. Sci. **23**(4), 597–635 (2009)
19. Dhople, S.V., Dominguez-Garcia, A.D.: A parametric uncertainty analysis method for Markov reliability and reward models. IEEE Trans. Reliab. **61**(3), 634–648 (2012)
20. Erreygers, A., Bock, J.D.: Imprecise continuous-time Markov chains: Efficient computational methods with guaranteed error bounds. In: Antonucci, A., Corani, G., Couso, I., Destercke, S. (eds.) Proceedings of the Tenth International Symposium on Imprecise Probability: Theories and Applications, Lugano, Switzerland, 10-14 July 2017. Proceedings of Machine Learning Research, vol. 62, pp. 145–156. PMLR (2017). http://proceedings.mlr.press/v62/erreygers17a.html
21. Erreygers, A., Bock, J.D.: Bounding inferences for large-scale continuous-time markov chains: a new approach based on lumping and imprecise Markov chains. Int. J. Approx. Reason. **115**, 96–133 (2019). https://doi.org/10.1016/J.IJAR.2019.09.003
22. Franceschinis, G., Muntz, R.R.: Bounds for quasi-lumpable Markow chains. Perform. Evaluation **20**(1-3), 223–243 (1994). https://doi.org/10.1016/0166-5316(94)90015-9
23. Franceschinis, G., Muntz, R.R.: Computing bounds for the performance indices of quasi-lumpable stochastic well-formed nets. IEEE Trans. Softw.Eng. **20**(7), 516–525 (1994). https://doi.org/10.1109/32.297940
24. Givan, R., Leach, S., Dean, T.: Bounded-parameter Markov decision processes. Artif. Intell. **122**(1–2), 71–109 (2000)
25. Gross, D., Miller, D.R.: The randomization technique as a modeling tool and solution procedure for transient Markov processes. Oper. Res. **32**(2), 343–361 (1984). https://doi.org/10.1287/OPRE.32.2.343
26. Hartfiel, D.J.: Markov set-chains. Lect. Notes Math. (1998). Springer
27. Kemeny, J.G., Snell, J.L.: Finite Markov Chains. Undergraduate Texts in Mathematics, van Nostrand,, Princeton, NJ (1976)
28. Leguizamon-Robayo, A., Jiménez-Pastor, A., Tribastone, M., Tschaikowski, M., Vandin, A.: Approximate constrained lumping of chemical reaction networks. CoRR **abs/2411.14242** (2024). https://doi.org/10.48550/ARXIV.2411.14242
29. Lembersky, M.R.: On maximal rewards and ε-optimal policies in continuous time Markov decision chains. In: The Annals of Statistics, pp. 159–169 (1974)
30. Luenberger, D.G., Ye, Y.: Linear and Nonlinear Programming, Springer (2016). 4 edn
31. Mahévas, S., Rubino, G.: Bound computation of dependability and performance measures. IEEE Trans. Comput. **50**(5), 399–413 (2001). https://doi.org/10.1109/12.926156
32. Michel, F., Siegle, M.: Markov chain aggregation with error bounds on transient distributions. In: Devos, A., Horváth, A., Rossi, S. (eds.) Analytical and Stochastic Modelling Techniques and Applications - 28th International Conference, ASMTA 2024, Venice, Italy, June 14, 2024, Proceedings. Lecture Notes in Computer Science, vol. 14826, pp. 1–17. Springer (2024). https://doi.org/10.1007/978-3-031-70753-7_1
33. Michel, F., Siegle, M.: Formal error bounds for the state space reduction of Markov chains. Perform. Evaluation **167**, 102464 (2025). https://doi.org/10.1016/J.PEVA.2024.102464

34. Miller, B.L.: Finite state continuous time Markov decision processes with a finite planning horizon. SIAM J. Control **6**(2), 266–280 (1968)
35. Puterman, M.L.: Markov Decision Processes: Discrete Stochastic Dynamic Programming. John Wiley & Sons (2014)
36. Rubino, G., Sericola, B.: Markov Chains and Dependability Theory, Cambridge University Press (2014)
37. Ruder, S.: An overview of gradient descent optimization algorithms (2016). arXiv:1609.04747 arXiv preprint
38. Sonnentag, P., Michel, F., Siegle, M.: The Arnoldi aggregation for approximate transient distributions of Markov chains (2025). arXiv:2508.02078 arXiv preprint
39. Spork, T., Baier, C., Katoen, J., Klüppelholz, S., Piribauer, J.: Approximate probabilistic bisimulation for continuous-time Markov chains. In: Piskac, R., Rakamaric, Z. (eds.) Computer Aided Verification - 37th International Conference, CAV 2025, Zagreb, Croatia, July 23-25, 2025, Proceedings, Part II. Lecture Notes in Computer Science, vol. 15932, pp. 56–81. Springer (2025). https://doi.org/10.1007/978-3-031-98679-6_3
40. Sproston, J., Donatelli, S.: Backward bisimulation in Markov chain model checking. IEEE Trans. Softw. Eng. **32**(8), 531–546 (2006). https://doi.org/10.1109/TSE.2006.74
41. Stewart, W.J.: Introduction to the Numerical Solution of Markov Chains, Princeton University Press (1995)
42. Termine, A., Antonucci, A., Facchini, A., Primiero, G.: Robust model checking with imprecise Markov reward models. In: International Symposium on Imprecise Probability: Theories and Applications, pp. 299–309 (2021). PMLR

State-Space Abstractions for Parametric Timed Games

Mikael Bisgaard Dahlsen-Jensen[1](✉), Laure Petrucci[2], and Jaco van de Pol[1]

QEST + FORMATS Artifacts Evaluation Available

[1] Aarhus University, Aarhus, Denmark
{mikael,jaco}@cs.au.dk
[2] Université Sorbonne Paris Nord, LIPN, CNRS UMR 7030, Villetaneuse, France
Laure.Petrucci@lipn.univ-paris13.fr

QEST + FORMATS Artifacts Evaluation Functional

Abstract. Synthesizing controllers for real-time systems under both timing uncertainty and adversarial environments requires exploring prohibitively large symbolic state spaces. While zone inclusion checking has been applied to Parametric Timed Games, more aggressive abstractions from the Parametric Timed Automata and Timed Games literature—double inclusion, zone merging, hull abstractions, and location-based abstraction—have not yet been lifted to the parametric game setting. We define a general abstraction framework for Parametric Timed Games, instantiate it with each of the aforementioned abstractions, and prove that the framework preserves correctness of parameter synthesis and winning strategies. Experimental results on an established production cell benchmark and a novel adversarial IoT case study show that the abstractions significantly improve scalability, solving instances previously intractable for existing techniques.

1 Introduction

State-space abstractions have proven effective for scaling the verification of parametric real-time systems, but a general abstraction framework is lacking for the game-theoretic setting where a controller must win against an adversary. Timed automata (TA) [1] provide a formal model for real-time systems, extended by Parametric Timed Automata (PTA) [2], which replace concrete timing constants with unspecified parameters. PTA support *parameter synthesis*: computing the set of parameter valuations under which a given property holds. Timed Games (TG) [12,28] complement this by introducing a two-player structure, distinguishing *controllable* actions (chosen by a controller) from *uncontrollable* ones (chosen by an adversary), and asking whether a winning strategy exists for the controller. Parametric Timed Games (PTG) [24,25] combine both settings: they synthesize the set of parameters under which the controller can guarantee winning. An on-the-fly PTG reachability algorithm was implemented in [16] and extended to full controller synthesis in [17].

This work was supported by the CNRS International Research Network CLoVe and Innovationsfonden Danmark's DIREC project SIoT (Secure Internet of Things).

M. Chen and M. Tschaikowski (Eds.): QEST+FORMATS 2026, LNCS 16913, pp. 20–38, 2027.
https://doi.org/10.1007/978-3-032-35298-9_2

PTG algorithms suffer from *state-space explosion*: the state space is explored symbolically using *parametric zones*, convex constraints over clocks and parameters. Reducing the number of zones while preserving correctness is the central challenge addressed in this paper. While zone inclusion checking has been applied to PTG [16], more aggressive abstractions remain unexplored: double inclusion and zone merging [4,6] from the PTA setting and location-based abstraction [23] from the TG setting. Lifting these to PTG is non-trivial: the PTA techniques must be shown to preserve winning strategies, not just reachability, while the TG technique must be extended to handle parameters.

Contributions. We make the following contributions:

1. A *general abstraction framework* for PTG that preserves correctness of parameter synthesis and winning strategies for any valid instantiation (Sect. 3).
2. Seven concrete instantiations, recovering known PTA and TG techniques and introducing new abstractions inspired by numerical abstract domains (convex hull, octagonal hull and box hull).
3. An implementation in the IMITATOR model checker [3], evaluated on a production-cell benchmark [27] and a novel adversarial IoT case study, showing reduced state spaces and improved scalability, with several instances going from timeout to terminating in seconds.

Related Work. Difference Bound Matrices (DBMs) [20] are the standard zone representation for timed automata, restricting zones to conjunctions of difference-bound constraints; Parametric DBMs [22] extend this to the parametric setting. Alternatively, symbolic zones can be represented by convex polyhedra [3]. DBMs can be viewed as abstractions, providing over-approximations of arbitrary convex polyhedra. In the non-parametric setting, extrapolation techniques such as LU-bounds [10] are widely used zone abstractions for timed automata, e.g. in UPPAAL [26]. Zone extrapolation was first adapted to the parametric setting in [5,11] and developed further in [7].

Zone inclusion checking was shown to be effective for timed games in [12] and has been used in PTG [16]; our framework provides a general correctness proof subsuming this technique. For PTA, zone merging was introduced in [4] and further developed in [6]. Our hull abstractions draw on classical numerical abstract domains: intervals [13], convex polyhedra [14], and octagons [29]. We now lift these abstractions to PTGs. A form of convex hull approximation for zones was introduced in [9] and is used in UPPAAL: it computes the smallest DBM containing two given zones, which includes the true convex hull but may be strictly larger when the hull is not representable as a DBM. A location-based abstraction was used in [23] to speed up on-the-fly symbolic algorithms with abstractions for timed ATL. We lift this abstraction to parametric timed reachability games. Foundational frameworks for abstracting games [19,21] abstract the moves of the players, requiring an asymmetric treatment of the two players; the SYNTHIA tool [30] synthesizes controllers for (non-parametric) timed games by abstraction refinement. In contrast, our abstractions coarsen only the explored symbolic state space, keeping the winning computation exact (cf. Sect. 3.2).

Outline. Section 2 introduces PTG and the necessary background. Section 3 defines the abstraction framework, adapts the on-the-fly PTG algorithm, proves its correctness, and instantiates it with concrete abstractions. Section 4 presents the experimental evaluation, including the adversarial IoT case study. Section 5 concludes.

2 Preliminaries

A Parametric Timed Game (PTG) is a structure based on timed automata (TA). Like classical automata, it has locations connected by discrete transitions. It also has clocks. Locations are associated a condition on clock valuations (invariant) that must be satisfied while staying in the location. An action in a timed automaton is either to take a discrete transition or to let some time pass. Discrete transitions have a guard that must be satisfied in order to take the transition. In a parametric setting, these conditions use linear terms over clocks and parameters. Parameters are unspecified but constant during a run. A discrete transition has a subset of clocks which are reset when the transition is taken.

A *clock valuation* is a function $v_X \in \mathbb{R}^X_{\geq 0}$ assigning a non-negative real value to each clock. A *parameter valuation* $v_P \in \mathbb{Q}^P_{\geq 0}$ assigns a non-negative rational value to each parameter. A *valuation of a game* is a pair $v = (v_X, v_P)$. The set of all valuations of the game is denoted $V = \mathbb{R}^X_{\geq 0} \times \mathbb{Q}^P_{\geq 0}$. A *linear term* over P is a term defined by the following grammar: $plt \ := \ k \mid kp \mid plt + plt$ where $k \in \mathbb{Q}$ and $p \in P$. *Zones* allow for capturing a set of valuations in a game. The set of *parametric zones* $\mathcal{Z}(X,P)$ is the set of formulas defined inductively by the following grammar: $\phi \ := \ \top \mid \phi \wedge \phi \mid x \sim plt \mid x - y \sim plt \mid plt \sim 0$ where $x, y \in X$, $\sim \in \{<; \leq; =; \geq; >\}$ and plt is a linear term over P.

In a two-player timed game, discrete transitions are partitioned between controllable transitions and uncontrollable (environment) transitions.

Definition 1 (PTG). *A* Parametric Timed Game *is a tuple of the form* $G = (\mathbb{L}, X, P, Act, T_c, T_u, \ell_0, Inv)$ *such that*

- $\mathbb{L}$, X, P, Act *are sets of* locations, clocks, parameters, transition labels.
- *We write* $T = T_c \sqcup T_u$ *for the set of all* transitions, *partitioned into* controllable *(T_c) and* uncontrollable *(T_u) ones, where* $T \subseteq \mathbb{L} \times \mathcal{Z}(X,P) \times Act \times \mathcal{P}(X) \times \mathbb{L}$ *is the set of* transitions *of the form* (ℓ, g, a, Y, ℓ') *where:* ℓ, ℓ' *are source and target locations,* g *is the guard,* a *the label,* Y *the set of clocks to reset.*
- ℓ_0 *is the initial location.*
- $Inv : \mathbb{L} \to \mathcal{Z}(X,P)$ *associates an* invariant *with each location.*

Function v_P is naturally extended to linear terms on parameters, by replacing each parameter in the term with its valuation. With $v \models \phi$, we denote that valuation $v = (v_X, v_P)$ *satisfies* a zone ϕ. Zones can also be seen as a convex set in the space of valuations by considering those satisfying the condition.

2.1 Semantics of Parametric Timed Games

A *state* of a PTG consists of a location and a valuation of clocks and parameters. Transitions modify clock valuations by letting time pass or resetting clocks.

Let $v = (v_X, v_P)$ be a valuation of the game and $\delta \geq 0$ a delay. $\forall x \in X$: $(v_X + \delta)(x) = v_X(x) + \delta$ and $v + \delta = (v_X + \delta, v_P)$. Let $Y \subseteq X$. $v_X[Y := 0]$ is the valuation obtained by *resetting the clocks* in Y, i.e.: $\forall x \in Y : v_X[Y := 0](x) = 0$ and $\forall x \in X \setminus Y : v_X[Y := 0](x) = v_X(x)$, and $v[Y := 0] = (v_X[Y := 0], v_P)$.

The semantics of a Parametric Timed Game is defined as a timed transition system on states, with timed and discrete transitions, and a winning condition. A *state* of a PTG is a pair (ℓ, v) where ℓ is a location and v a valuation of the game satisfying its invariant: $v \models Inv(\ell)$. The *state space* is then $\mathbb{S} = \{(\ell, v) \in \mathbb{L} \times V \mid v \models Inv(\ell)\} = \bigcup_{\ell \in \mathbb{L}} \{\ell\} \times Inv(\ell)$. Let $\delta \in \mathbb{R}_{\geq 0}$ be a time delay. A *timed transition* is a relation $\rightarrow^{\delta} \in \mathbb{S} \times \mathbb{S}$ s.t. $\forall (\ell, v), (\ell', v') \in \mathbb{S} : (\ell, v) \rightarrow^{\delta} (\ell', v')$ iff $\ell = \ell'$ and $v' = v + \delta$. Let $t = (\ell, g, a, Y, \ell') \in T$ be a transition. A *discrete transition* is a relation $\rightarrow^{t} \in \mathbb{S} \times \mathbb{S}$ s.t. $\forall (\ell, v), (\ell', v') \in \mathbb{S} : (\ell, v) \rightarrow^{t} (\ell', v')$ iff $v \models g$ and $v' = v[Y := 0]$.

States are grouped in symbolic states, similar to valuations grouped in zones.

Definition 2 (symbolic state). *A* symbolic state *of a PTG is a pair $\xi = (\ell, Z)$ where ℓ is a location and Z a zone of the game satisfying its invariant: $Z \models Inv(\ell)$. The symbolic state (ℓ, Z) represents the subset of states $\{\ell\} \times Z \subseteq \mathbb{S}$.*

Definition 3 (temporal zone operators). *For a zone Z, its* time-forward closure *and* time-backward closure *are:*

$$Z^{\nearrow} := \{\, v + \delta \mid v \in Z,\ \delta \geq 0 \,\},$$
$$Z^{\swarrow} := \{\, v \mid \exists \delta \geq 0 : v + \delta \in Z \,\}.$$

Definition 4 (discrete zone operators). *Let $t = (\ell, g, a, Y, \ell') \in T$. For a zone Z:*

$$Succ(t, Z) := \{\, v[Y := 0] \mid v \in Z,\ v \models g \,\},$$
$$Pred(t, Z) := \{\, v \mid v \models g,\ v[Y := 0] \in Z \,\}.$$

Definition 5 (safe temporal predecessor). *For zones W and U:*

$$SafePred(W, U) := \{\, v \;\big|\; \exists \delta \geq 0 :\ v + \delta \in W \ \wedge\ \forall \delta' \in [0, \delta] : v + \delta' \notin U \,\}.$$

That is, $v \in SafePred(W, U)$ if time can elapse from v to reach W without passing through U.

Remark 1 (zone operators on symbolic states) The operators above are defined on zones (sets of valuations). They lift to symbolic states by intersecting with the location invariant (so that the zone of the result satisfies the invariant, as required by Definition 2): for $\xi = (\ell, Z)$ and $t = (\ell, g, a, Y, \ell')$:

$$\xi^{\nearrow} := (\ell,\ Z^{\nearrow} \cap Inv(\ell)),$$
$$\xi^{\swarrow} := (\ell,\ Z^{\swarrow} \cap Inv(\ell)),$$
$$Succ(t, \xi) := (\ell',\ Succ(t, Z) \cap Inv(\ell')).$$

Pred and *SafePred* are used directly on zones throughout. We will not distinguish a zone from its corresponding set of valuations in our notation.

Let $\mathbf{0}$ be the clock valuation where all clocks have value 0. The set of possible initial states of the PTG is $\xi_0 = \{(\ell_0, (\mathbf{0}, v_P)) \mid v_P \in \mathbb{Q}^P_{\geq 0} : (\mathbf{0}, v_P) \models Inv(\ell_0)\}$.

Algorithms for PTA (and thus PTG) explore and operate on a Parametric Zone Graph (PZG).

Definition 6 (parametric zone graph). *A* Parametric Zone Graph *(PZG) for a PTG $G = (\mathbb{L}, X, P, Act, T_c, T_u, \ell_0, Inv)$ is a tuple $(\Xi, \xi_I, \Rightarrow^t_c, \Rightarrow^t_u)$ where $\Xi \subseteq 2^{\mathbb{S}}$ is a set of symbolic states, $\xi_I \in \Xi$ is the initial symbolic state, and $\Rightarrow^t_c, \Rightarrow^t_u \subseteq \Xi \times \Xi$ are transition relations labeled by transitions of G.*

Definition 7 (canonical PZG). *The* canonical PZG *of a PTG G is the PZG $(\Xi, \xi_I, \Rightarrow^t_c, \Rightarrow^t_u)$ where $\xi_I = \xi_0^{\nearrow}$ and the transition relations are defined by: $\xi \Rightarrow^t_c \xi'$ iff $\xi' = Succ(t, \xi)^{\nearrow}$ and $t \in T_c$; and $\xi \Rightarrow^t_u \xi'$ iff $\xi' = Succ(t, \xi)^{\nearrow}$ and $t \in T_u$; and Ξ is the smallest set containing ξ_I and closed under these transitions.*

The canonical PZG exists and is unique: Ξ is well-defined as the least fixed point of forward exploration from ξ_I, and the transitions are deterministic (each pair (ξ, t) yields exactly one successor), so the structure is uniquely determined by G.

Definition 8 (zones at a location). *For a location ℓ (with X, P, and Ξ implicit from the PZG under discussion), the associated zone set is*

$$\mathcal{Z}_\ell := \{ Z \in \mathcal{Z}(X, P) \mid (\ell, Z) \in \Xi \}.$$

We study PTGs with a *reachability objective* given by a target set $R \subseteq \mathbb{L}$. The controller is considered winning from a state if there exists a memoryless strategy for the controller such that, regardless of how the environment plays, the game eventually reaches a location in R. Solving a PTG then means characterizing the parameter valuations for which the controller is winning from the initial state ξ_0. A detailed definition of winning strategies can be found in [16].

3 State Space Reduction Techniques

3.1 General Abstraction Framework

We now formally define a general framework for state space abstractions in PTG and prove its correctness in Sect. 3.3. The remaining subsections instantiate the framework with concrete abstraction functions, recovering known techniques and introducing new ones.

Definition 9 (abstraction function). *Given a PZG with symbolic state set Ξ, an* abstraction function *is a function $\alpha : \mathbb{L} \to 2^{\mathcal{Z}(X,P)}$. For every location ℓ, it must over-approximate the original zones:*

$$\forall Z \in \mathcal{Z}_\ell \; \exists Z' \in \alpha(\ell) \text{ such that } Z \subseteq Z'.$$

Equivalently, every zone in $\mathcal{Z}_\ell$ must be covered by some zone in $\alpha(\ell)$.

The simplest instance, which serves as a baseline, is the trivial abstraction function.

Example 1 α_{triv} is the abstraction function that leaves the zone set unchanged:

$$\alpha_{\text{triv}}(\ell) = \mathcal{Z}_\ell$$

To define the reduced PZG, we need to know which abstract zone each original zone maps to. This is captured by the representative map induced by α.

Definition 10 (representative map). *Given an abstraction function α, let $rep_\alpha(\ell, Z)$ be an arbitrary but fixed zone $Z' \in \alpha(\ell)$ such that $Z \subseteq Z'$ (non-empty by coverage). Define the* representative map *as the function taking symbolic states and returning their representative in the reduced symbolic state space:*

$$\rho_\alpha(\ell, Z) := (\ell, rep_\alpha(\ell, Z))$$

Definition 11 (reduced parametric zone graph). *Let $(\Xi, \xi_I, \Rightarrow^t_c, \Rightarrow^t_u)$ be a PZG and $\alpha : \mathbb{L} \to 2^{\mathcal{Z}(X,P)}$ an abstraction function with representative map ρ_α. The reduced transition relation for controllable transitions is*

$$\Rightarrow^t_{c,red} := \{(\rho_\alpha(\xi), \rho_\alpha(\xi')) \mid \xi \Rightarrow^t_c \xi'\}$$

and analogously for uncontrollable transitions:

$$\Rightarrow^t_{u,red} := \{(\rho_\alpha(\xi), \rho_\alpha(\xi')) \mid \xi \Rightarrow^t_u \xi'\}$$

Together with the abstract symbolic states $\{\rho_\alpha(\xi) \mid \xi \in \Xi\}$, these define the reduced parametric zone graph, *itself a PZG for G:*

$$PZG^\alpha = (\{\rho_\alpha(\xi) \mid \xi \in \Xi\},\ \rho_\alpha(\xi_I),\ \Rightarrow^t_{c,red},\ \Rightarrow^t_{u,red})$$

3.2 PTG Reachability Algorithm

In [24,25] a semi-algorithm was introduced and later implemented in [16] that synthesizes the set of parameters for which a winning strategy exists. We recall this algorithm here (Alg. 1), as it forms the basis for the correctness proofs that follow. For the purposes of this paper, the algorithm is presented with the PZG as an explicit input, rather than deriving the canonical PZG from the PTG internally; this makes the abstraction substitution in Sect. 3.3 precise.

We briefly explain the algorithm. The idea is to compute reachable states while propagating winning conditions backward over controllable and uncontrollable transitions. The main loop (ll. 6 and 7) calls EXPLORE or UPDATE until no states remain. Procedure EXPLORE picks a state ξ from *WaitingExplore* (l. 10) and iterates over its PZG transitions (ll. 11-14): each successor ξ' records ξ in *Depends* and, if new, is queued for exploration. A target state is marked winning, queuing its predecessors for update (ll. 15-17); in any case ξ is queued for update (l. 18) and marked explored (l. 19).

Procedure UPDATE back-propagates winning zones for ξ from *WaitingUpdate*. It collects in *UnCtrl* the predecessors (*Pred*) of the non-winning part of the uncontrollable successors (l. 22), and in *WinningMove* the predecessors of the winning part of the controllable successors (l. 23). The safe temporal predecessors (*SafePred*) of $Win[\xi] \cup WinningMove$ that can let time elapse into ξ while avoiding *UnCtrl* (l. 24) are winning. On a change to $Win[\xi]$, the predecessors of ξ are re-queued (ll. 25-27), and the parameter projection at ξ_I is reported (l. 28).

Algorithm 1. For PTG $G = (\mathbb{L}, X, P, Act, T_c, T_u, \ell_0, Inv)$ and PZG $(\Xi, \xi_I, \Rightarrow_c^t, \Rightarrow_u^t)$, and reachability objective R, returns the set of all parameter valuations that win the game.

1: $Explored, WaitingUpdate, WaitingExplore \leftarrow \emptyset, \emptyset, \{\xi_I\}$ ▷ *Symbolic state sets*
2: $Win := \{\}$ ▷ *Map from symbolic states to unions of zones*
3: $Depends := \{\}$ ▷ *Map from symbolic states to sets of symbolic states*
4: $WinningParam := False$

5: **function** SOLVEPTG
6:     **while** $WaitingExplore \neq \emptyset \vee WaitingUpdate \neq \emptyset$ **do**
7:         Choose either EXPLORE() or UPDATE()
8:     **return** $WinningParam$

9: **procedure** EXPLORE
10:     $\xi \leftarrow extract(WaitingExplore)$
11:     **for** $\xi \Rightarrow^t \xi'$ in PZG : **do**
12:         $Depends[\xi'] \leftarrow Depends[\xi'] \cup \{\xi\}$
13:         **if** ξ' not in $Explored$ **then**
14:             $WaitingExplore \leftarrow WaitingExplore \cup \{\xi'\}$
15:     **if** $\xi.\ell \in R$ **then**
16:         $Win[\xi] \leftarrow \xi.Z$
17:         $WaitingUpdate \leftarrow WaitingUpdate \cup Depends[\xi]$
18:     $WaitingUpdate \leftarrow WaitingUpdate \cup \{\xi\}$
19:     $Explored \leftarrow Explored \cup \{\xi\}$

20: **procedure** UPDATE
21:     $\xi \leftarrow extract(WaitingUpdate)$
22:     $UnCtrl \leftarrow \bigcup_{\{(\xi',t) \mid \xi \Rightarrow_u^t \xi'\}} Pred(t, \xi' \setminus Win[\xi'])$
23:     $WinningMove \leftarrow \bigcup_{\{(\xi',t) \mid \xi \Rightarrow_c^t \xi'\}} Pred(t, Win[\xi'])$
24:     $NewWin := SafePred(Win[\xi] \cup WinningMove,\ UnCtrl) \cap \xi$
25:     **if** $NewWin \not\subseteq Win[\xi]$ **then**
26:         $WaitingUpdate \leftarrow WaitingUpdate \cup Depends[\xi]$
27:         $Win[\xi] \leftarrow Win[\xi] \cup NewWin$
28:     $WinningParam \leftarrow (Win[\xi] \cap \xi_0){\downarrow_P}$

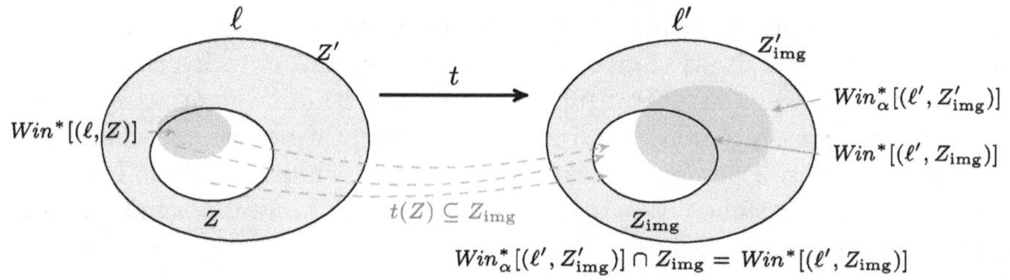

Fig. 1. Consequence of Lemma 1. $Win^*_\alpha[(\ell', Z'_{\text{img}})]$ (light and dark green) extends into the region $Z'_{\text{img}} \setminus Z_{\text{img}}$, but states from Z map only into Z_{img} (image-preserving, dashed arrows). Restricted to Z_{img}, the abstract and full winning sets coincide (dark green), so the predecessor computation restricted to Z recovers exactly $Win^*[(\ell, Z)]$.

Role of the Abstraction. The abstraction changes only which symbolic states are explored and how they are grouped, not the game's moves: UPDATE applies *Pred* and *SafePred* to the original guards and resets, so the reduced PZG is only a skeleton carrying the *exact* backward fixed point. Thus abstract zones never enlarge the winning region (recovered exactly on each original zone, Lem. 1), and, unlike game abstractions [19,21], no asymmetric handling of the players is needed: both predecessors are concrete, so coverage alone gives soundness and completeness.

3.3 Correctness of State Space Abstractions

The goal of this section is to show that the algorithm (Alg. 1) produces the same winning parameter set whether it receives the canonical PZG or the reduced PZG PZG^α as input. The argument rests on two properties. First, every original zone is contained in an abstract zone (coverage requirement). Second, PZG transitions are *image-preserving*: for any symbolic state (ℓ, Z) and transition t from it with successor zone Z_{img}, every $v \in Z$ where t is enabled satisfies $v[Y := 0] \in Z_{\text{img}}$, since Z_{img} is by definition the set of images of t applied to Z. In particular, states in Z always land in Z_{img}, not in the possibly larger abstract successor $Z'_{\text{img}} \supseteq Z_{\text{img}}$.

Throughout this section, whenever a transition $t = (\ell, g, a, Y, \ell')$ appears, we write g, Y, and ℓ' for its guard, reset clocks, and target location; subscripted variants (g_u, Y_u, ℓ'_u for t_u, etc.) follow the same convention.

We write $Win^*[(\ell, Z)]$ for the set of valuations eventually marked winning at symbolic state (ℓ, Z) in the canonical PZG (under fair exploration), and $Win^*_\alpha[(\ell, Z')]$ for the corresponding set computed on PZG^α.

Lemma 1 (Zone correspondence) *Let α be an abstraction function. For every abstract symbolic state $(\ell, Z') \in PZG^\alpha$ and every original symbolic state (ℓ, Z) in the canonical PZG with $\rho_\alpha(\ell, Z) = (\ell, Z')$:*

$$Win^*_\alpha[(\ell, Z')] \cap Z = Win^*[(\ell, Z)]. \tag{1}$$

Proof. We show both inclusions separately, using the correctness theory of [16]. For each location ℓ, write $W(\ell)$ for the set of valuations from which the controller has a winning strategy at ℓ. By [16], the PTG algorithm is sound at every step (Invariant 4: every state marked winning is genuinely winning) and complete in the limit (Theorem 3: under fair exploration, every winning state is eventually discovered). In particular, $Win^*[(\ell, Z)] = W(\ell) \cap Z$ for every symbolic state (ℓ, Z) in the canonical PZG.

Soundness. ($Win^*_\alpha[(\ell, Z')] \cap Z \subseteq Win^*[(\ell, Z)]$). We show that every valuation added to $Win_\alpha[(\ell, Z')]$ is genuinely winning at ℓ, by induction on the order in which valuations are added. At target locations ($\ell \in R$), the algorithm initializes $Win_\alpha[(\ell, Z')] \leftarrow Z'$; every $v \in Z'$ at a target location is winning by definition. At non-target locations, a valuation v is added only through the *SafePred* computation, which requires: (i) a controllable transition t and a delay $\delta \geq 0$ such that $v + \delta \models g$, with $(v + \delta)[Y := 0]$ in a set already marked winning; and (ii) every uncontrollable successor during $[0, \delta]$ is also in a set already marked winning. Since abstract transitions inherit their guards and resets from genuine PZG transitions, conditions (i) and (ii) correspond to real game moves. By the inductive hypothesis, the successor sets contain only genuinely winning valuations, so v is genuinely winning. Therefore, $Win^*_\alpha[(\ell, Z')] \subseteq W(\ell)$. Intersecting with $Z \subseteq Z'$: $Win^*_\alpha[(\ell, Z')] \cap Z \subseteq W(\ell) \cap Z = Win^*[(\ell, Z)]$.

Completeness. ($Win^*[(\ell, Z)] \subseteq Win^*_\alpha[(\ell, Z')] \cap Z$). Let $v \in Win^*[(\ell, Z)] = W(\ell) \cap Z$. Since $Z \subseteq Z'$ (coverage), $v \in Z'$, so v is present in the abstract symbolic state (ℓ, Z'). We verify the completeness argument of [16] (Theorem 3) for PZG^α, by induction on the discrete distance $n(\ell, v)$ to the target: the minimum, over winning strategies from (ℓ, v), of the maximum number of discrete transitions along any play consistent with that strategy.

Base case ($n = 0$): v is at a target location $\ell \in R$. The algorithm sets $Win_\alpha[(\ell, Z')] \leftarrow Z'$ upon exploration, so $v \in Z' \subseteq Win_\alpha[(\ell, Z')]$.

Inductive step ($n > 0$): The winning strategy from v waits time $\delta \geq 0$ and then takes a controllable transition t to location ℓ', reaching state $v' = (v + \delta)[Y := 0]$ with $n(\ell', v') < n$. We show in two parts that v is eventually added to $Win_\alpha[(\ell, Z')]$.

(a) Lifting successors to the abstract graph. Since $v \in Z \subseteq Z'$, every original PZG transition from (ℓ, Z) lifts to an abstract transition from (ℓ, Z') whose successor zone contains the original. For the controllable transition t: $(\ell, Z) \Rightarrow^t_c (\ell', Z_{\text{img}})$ gives $(\ell, Z') \Rightarrow^t_{c,red} (\ell', Z'_{\text{img}})$ with $Z'_{\text{img}} = \rho_\alpha(\ell', Z_{\text{img}}) \supseteq Z_{\text{img}}$. Since $v + \delta \in Z$ (PZG zones include time successors) and $v + \delta \models g$, we have $v' = (v + \delta)[Y := 0] \in Z_{\text{img}} \subseteq Z'_{\text{img}}$ (image-preserving). As $n(\ell', v') < n$, the inductive hypothesis gives that v' is eventually in $Win_\alpha[(\ell', Z'_{\text{img}})]$. For each uncontrollable transition t_u enabled from $v + \delta'$ ($\delta' \in [0, \delta]$), the same lifting applies: the concrete successor $v'' = (v + \delta')[Y_u := 0]$ reaches location ℓ'_u and lies in the original successor zone Z^u_{img} (image-preserving), which lifts to an abstract successor $Z'^u_{\text{img}} \supseteq Z^u_{\text{img}}$. Since v is genuinely winning, v'' has distance $n(\ell'_u, v'') \leq n - 1$ (the uncontrollable move consumes one discrete transition),

so by the inductive hypothesis v'' is eventually in Win_α at the corresponding abstract successor.

(b) Backward propagation. Once all successors are resolved, the algorithm's update step (step l. 24) at (ℓ, Z') computes $NewWin = SafePred(Win_\alpha[(\ell, Z')] \cup WinningMove,\ UnCtrl) \cap Z'$, where:

- *Controllable:*

$$WinningMove = \bigcup_{(\ell,Z') \Rightarrow^{t_c}_{c,red} (\ell'_c, Z'_c)} Pred(t_c,\ Win_\alpha[(\ell'_c, Z'_c)]).$$

  Since $v' \in Win_\alpha[(\ell', Z'_{\text{img}})]$ and $v + \delta \models g$ with $(v + \delta)[Y := 0] = v'$, we get $v + \delta \in Pred(t,\ Win_\alpha[(\ell', Z'_{\text{img}})]) \subseteq WinningMove$.
- *Uncontrollable:*

$$UnCtrl = \bigcup_{(\ell,Z') \Rightarrow^{t_u}_{u,red} (\ell'_u, Z'_u)} Pred(t_u,\ Z'_u \setminus Win_\alpha[(\ell'_u, Z'_u)]).$$

  Since every uncontrollable successor from the waiting interval $[0, \delta]$ is now in Win_α at its abstract successor, $v + \delta' \notin UnCtrl$ for all $\delta' \in [0, \delta]$.

With both conditions met: $v + \delta \in WinningMove$ and $v + \delta' \notin UnCtrl$ for $\delta' \in [0, \delta]$. By definition of *SafePred*, this gives $v \in NewWin$, so v is added to $Win_\alpha[(\ell, Z')]$. Under fair exploration, this update happens in finite time.

Combining both directions gives $Win^*_\alpha[(\ell, Z')] \cap Z = Win^*[(\ell, Z)]$.

Remark 2 (Consequence). Lem. 1 says that merging zones loses no precision on the states that were actually there: $Win^*_\alpha[(\ell, Z')]$ restricted to the original zone $Z \subseteq Z'$ recovers exactly $Win^*[(\ell, Z)]$ (see Fig. 1).

Theorem 1. ***Correctness of reduced PZG.*** *Let α be an abstraction function satisfying the coverage requirement. Under fair exploration, running the PTG algorithm on PZG^α converges to the same winning parameter set as running it on the canonical PZG. In particular, if either algorithm terminates, both yield the same result; if the PZG is infinite, every winning parameter is eventually discovered by the reduced algorithm.*

Proof. Recall $\xi_0 = (\ell_0, Z_{\text{init}})$ (all clocks zero, parameters unconstrained), and let $(\ell_0, Z_0) = \xi_I = \xi_0^{\nearrow}$ be the initial symbolic state of the canonical PZG. Let $(\ell_0, Z'_0) = \rho_\alpha(\ell_0, Z_0)$ be its abstract counterpart. By the coverage requirement, $Z_0 \subseteq Z'_0$. By Lemma 1 applied at (ℓ_0, Z_0):

$$Win^*_\alpha[(\ell_0, Z'_0)] \cap Z_0 = Win^*[(\ell_0, Z_0)].$$

Since $Z_{\text{init}} \subseteq Z_0$ (the zero-clock zone is contained in its time-successor closure):

$$Win^*_\alpha[(\ell_0, Z'_0)] \cap Z_{\text{init}} = Win^*[(\ell_0, Z_0)] \cap Z_{\text{init}}.$$

Projecting onto parameters:

$$\big(Win^*_\alpha[(\ell_0, Z'_0)] \cap Z_{\text{init}}\big){\downarrow_P} = \big(Win^*[(\ell_0, Z_0)] \cap Z_{\text{init}}\big){\downarrow_P}.$$

By the limit-correctness of the PTG algorithm on the canonical PZG [16], the right-hand side equals the true winning parameter set. Since the left-hand side equals the right-hand side, the reduced algorithm converges to the same set. □

3.4 Instantiating the Framework

The remaining subsections instantiate the framework with concrete abstraction methods. Each method is specified as a type of *abstraction step*: a local operation that produces a valid abstraction function from an existing one. This formulation avoids requiring the full zone set $\mathcal{Z}_\ell$ (which may be infinite) and matches the incremental nature of on-the-fly exploration.

Definition 12. (abstraction step). *Let α be an abstraction function. An* abstraction step *at location ℓ produces a new function $\alpha' : \mathbb{L} \to 2^{\mathcal{Z}(X,P)}$ satisfying:*

1. $\alpha'(\ell') = \alpha(\ell')$ *for all* $\ell' \neq \ell$ (locality),
2. *for every* $Z \in \alpha(\ell)$*, there exists* $Z' \in \alpha'(\ell)$ *with* $Z \subseteq Z'$ (coarsening).

An abstraction sequence *is a sequence* $(\alpha_i)_{i \geq 0}$ *where* $\alpha_0 = \alpha_{\text{triv}}$ *and each* α_{i+1} *is obtained from* α_i *by an abstraction step.*

Lemma 2. *Every element of an abstraction sequence is a valid abstraction function.*

Proof. α_{triv} satisfies coverage. Each step preserves coverage: if $Z \in \mathcal{Z}_\ell$ was covered by $Z' \in \alpha(\ell)$, then coarsening gives $Z'' \in \alpha'(\ell)$ with $Z \subseteq Z' \subseteq Z''$. □

Remark 3. For finite PZGs, each abstraction sequence stabilizes after finitely many steps; the closed-form characterizations given alongside each step type below describe these fixed points. In general, the result of applying steps may depend on the order in which they are applied, but every abstraction function produced by a step sequence is sound by Thm. 1.

3.5 Zone Inclusion Check

The idea is to prune redundant symbolic states: if two zones $Z \subseteq Z'$ appear at the same location, Z is already covered by Z' and can be discarded. We consider two variants depending on whether this check is applied in one or both directions.

Definition 13. (inclusion step). *Let α be an abstraction function and $Z \in \alpha(\ell)$. If there exists $Z' \in \alpha(\ell)$ with $Z \neq Z'$ and $Z \subseteq Z'$, then removing Z is an* inclusion step*:*

$$\alpha'(\ell) = \alpha(\ell) \setminus \{Z\}.$$

Coarsening is immediate: Z itself is covered by Z', and all other zones are unchanged.

Remark 4. An inclusion step can be applied in two directions during on-the-fly exploration. *Forward*: when a newly discovered zone Z is already subsumed by an existing $Z' \in \alpha(\ell)$, discard Z immediately. *Backward*: when a newly discovered zone Z' subsumes an existing $Z \in \alpha(\ell)$, retroactively remove Z. Both are the same formal step; they differ only in which zone is new and which is old.

When $\mathcal{Z}_\ell$ is finite, a maximal sequence of inclusion steps applied in *both* directions (*double inclusion*) stabilizes to the *antichain* of $\subseteq$-maximal zones:

$$\alpha_{\subseteq}^{\leftrightarrow}(\ell) = \{Z \in \mathcal{Z}_\ell \mid \nexists Z' \in \mathcal{Z}_\ell : Z \subsetneq Z'\}.$$

Applying steps in only one direction may stabilize earlier: forward-only never removes an existing zone, and backward-only never discards a new zone subsumed by an already-present one.

3.6 Zone Merging

Zone merging, introduced for parametric timed automata in [4,6], reduces the state space by replacing two zones at the same location with their union, when that union is already convex (so no new states are introduced).

Definition 14. (merge step). *Let α be an abstraction function and $Z_1, Z_2 \in \alpha(\ell)$ where $Z_m = Z_1 \cup Z_2$ is convex [4]. A* merge step *replaces them with Z_m:*

$$\alpha'(\ell) = (\alpha(\ell) \setminus \{Z_1, Z_2\}) \cup \{Z_m\}.$$

Coarsening holds: $Z_1, Z_2 \subseteq Z_m = Z_1 \cup Z_2$, and all other zones are unchanged.

3.7 Location-Based

The location-based abstraction is the coarsest instantiation of the framework: every zone at location ℓ is replaced by the invariant $\mathit{Inv}(\ell)$, yielding a single abstract state per location.

Definition 15. (α_{loc}).

$$\alpha_{\text{loc}}(\ell) = \{\mathit{Inv}(\ell)\}$$

Coverage holds trivially: every symbolic state (ℓ, Z) satisfies $Z \subseteq \mathit{Inv}(\ell)$ by definition, so the single zone $\mathit{Inv}(\ell)$ covers all zones at ℓ. Unlike the other instantiations, α_{loc} does not require the abstraction step machinery: the characterization above is well-defined for any PZG, including infinite ones, since it depends only on the invariant of each location, not on enumerating the reachable zones. A similar location-based abstraction is used in [23] for timed ATL.

3.8 Hull Abstractions

Unlike zone merging, which only combines zones when their union is already convex, hull abstractions always collapse zones into a single over-approximating zone, potentially introducing new valuations. The variants differ in the class of zones they produce, trading precision for computational cost.

Definition 16. (hull operator). *A* hull operator *is a function $\mathcal{H} : 2^{\mathcal{Z}(X,P)} \to \mathcal{Z}(X, P)$ satisfying $\bigcup_{Z \in S} Z \subseteq \mathcal{H}(S)$ for every finite set of zones S.*

Definition 17. (hull step). *Let α be an abstraction function, $Z \in \alpha(\ell)$ a newly discovered zone, and $H \in \alpha(\ell)$ an existing zone at ℓ. A* hull step *replaces both with the hull of the pair:*

$$\alpha'(\ell) = (\alpha(\ell) \setminus \{H, Z\}) \cup \{\mathcal{H}(\{H, Z\})\}.$$

Coarsening holds: $H, Z \subseteq \mathcal{H}(\{H, Z\})$, and all other zones are unchanged. Successive hull steps at a location accumulate into a running hull: after processing zones $Z_1, \ldots, Z_k$, the abstract zone at ℓ is $\mathcal{H}(\{\ldots \mathcal{H}(\{Z_1, Z_2\}), \ldots, Z_k\})$. The result is order-independent for associative hull operators (convex hull, octagonal hull, box hull).

Fixed-Point Characterization (finite PZGs). When $\mathcal{Z}_\ell$ is finite, any maximal sequence of hull steps at ℓ stabilizes to a single zone. For a hull operator $\mathcal{H}$, the fixed point is:

$$\alpha_h(\ell) := \{\, \mathcal{H}(\mathcal{Z}_\ell) \,\}.$$

Instances. The three principal instances of $\mathcal{H}$, ordered from finest to coarsest, are:

α_{ch} (convex hull) $\mathcal{H}_{\text{conv}}(S)$ is the smallest convex zone containing $\bigcup S$.
α_{oct} (octagonal hull) $\mathcal{H}_{\text{oct}}(S)$ is the smallest octagonal zone containing $\bigcup S$. Octagonal zones extend difference-bound constraints with terms of the form $x_i \pm x_j \leq c$, capturing more diagonal structures than a bounding box.
α_{box} (box hull) $\mathcal{H}_{\text{box}}(S)$ is the smallest axis-aligned bounding zone containing $\bigcup S$, using only individual-variable bounds $x_i \leq c$ and $x_i \geq c$. It is the cheapest to compute but the coarsest.

These satisfy $\mathcal{H}_{\text{conv}}(S) \subseteq \mathcal{H}_{\text{oct}}(S) \subseteq \mathcal{H}_{\text{box}}(S)$ for any S, so α_{ch} is the finest and α_{box} the coarsest hull abstraction.

4 Experimental Evaluation

We evaluate the practical impact of the abstractions from Sect. 3.4, focusing on state-space reduction and scalability on previously intractable instances.

4.1 Implementation

The PTG synthesis algorithm of [16] (Alg. 1) is implemented in IMITATOR [3], a model checker supporting a wide range of PTA synthesis algorithms, with zones represented using the Parma Polyhedra Library [8]. We extend IMITATOR; the source code is available on GitHub.[1] Our extensions are threefold:

Abstraction Functions. The baseline without subsumption (Baseline) and zone inclusion (Inc) were already supported by [16]. We add double inclusion

[1] https://github.com/imitator-model-checker/imitator, branch `develop`.

(DInc), on-the-fly merge (Mrg), location-based abstraction (Loc), and the convex (CH), octagonal (Oct) and box (Box) hull variants, as described in Sect. 3.4. We also implement two hybrid configurations (OctH, BoxH) that apply convex hull while zones remain below a constraint-count threshold, falling back to octagonal or box hull when they grow more complex.

Layer-Based Exploration with Merge. We adopt the layer-based exploration strategy of [6], in which zones are explored layer by layer with abstraction applied between layers.

Data-Structure Propagation. The main PTG-specific adaptation is propagating the game-theoretic data structures across abstractions. When a zone replaces another, the maps *Win* and *Depends* must be updated by redirecting entries for the absorbed zone to its target, combining values with zone union and set union respectively.

4.2 Case Study: Adversarial IoT Scheduler

We introduce a case study inspired by an adversarial IoT scenario in which a central scheduler coordinates multiple IoT devices under adversarial delays. We refer to this benchmark as *Malicious Synchronization* in the experiments, reflecting the adversary's goal of forcing simultaneous transmissions. The setting is easily scalable in the number of devices and naturally induces the synchronized timing patterns that cause symbolic state explosion.

System Overview. The system consists of a network of battery-powered IoT devices that periodically transmit data to a cloud server, a central scheduler that grants transmission requests to prevent network congestion, and an adversary that can delay transmissions by dropping packets (modelled as a time delay). The adversary's goal is to force enough devices to transmit simultaneously, causing congestion. The adversary observes scheduler state but cannot interfere with scheduler-device communication, and has a limited energy budget bounding the number of delays.

Parametric Timed Game Model. The system is modelled as a network of PTG: one for each IoT device and one for the scheduler. Data transmission is simulated by enforcing a time delay in a `transmission` location of each device. Each device has an uncontrollable action, available when the adversary has remaining energy, that models adversarial delay by injecting additional time before transmission begins.

Three values are parametric: the minimum time between consecutive grants, the maximum adversarial delay per intervention, and the transmission duration. The adversary's energy budget is a discrete constant that bounds the total number of delay interventions. The goal is to reach a state where all devices have completed their data transmission, while avoiding states in which too many devices transmit simultaneously, causing congestion. An overview and the IMITATOR models for the two-device instance are given in the extended version [18].

4.3 Experimental Setup

All experiments were conducted on a single core of a computer with an Intel Core i5-1135G7 CPU @ 2.40GHz with 16 GB of RAM running Ubuntu 22.04.5 LTS. Each configuration was run 5 times; we report the average time. A timeout of 1,200 s was applied per run.

We evaluate the abstractions described in Sect. 3.4 using two case studies: the adversarial IoT scheduler introduced in Sect. 4.2, and a production-cell benchmark [27]. The production cell models a two-armed robot transferring plates between an input conveyor, a press, and an output conveyor. The n-plate goal is for all plates to reach the output; the game is lost if two plates share the buffer. A parameter sets the minimum wait between plates; robot timing is uncontrollable.

Table 1. Experimental results. Each cell shows time in seconds followed by number of explored states. T/O denotes timeout at 1,200 s.

	Prior work				*This paper*															
Instance	Baseline		Inc		DInc		Mrg		Loc		CH		Oct		Box		OctH		BoxH	
Malicious Synchronization																				
(2/2)	2.71	1319	1.33	561	1.01	318	0.93	274	1.29	212	0.85	212	0.74	212	0.66	212	0.81	212	0.83	212
(2/3)	32.9	11079	16.5	4428	9.76	2406	9.11	2158	10.5	1398	7.49	1398	5.96	1398	5.26	1398	7.46	1398	7.54	1398
(3/3)	T/O		679	25694	574	14699	50.2	4416	475	1438	T/O		24.3	1438	T/O		47.5	1438	35.7	1438
(2/4)	485	94504	238	34896	125	17815	128	16484	64.6	8952	T/O		49.8	8952	39.3	8952	61.4	8952	62.6	8952
(3/4)	T/O		T/O		T/O		T/O		T/O		T/O		334	9592	T/O		T/O		T/O	
(4/4)	T/O		T/O		T/O		T/O		T/O		T/O		T/O		T/O		T/O		T/O	
Production Cell																				
1	0.34	771	0.07	77	0.07	73	0.06	59	0.05	91	0.05	55	0.06	55	0.06	55	0.05	55	0.05	55
2	T/O		0.51	507	0.43	424	0.31	306	1.10	544	0.29	222	0.43	222	T/O		0.33	222	0.32	222
3	T/O		2.25	1576	1.19	913	0.60	476	T/O		1.17	293	0.94	293	T/O		0.70	293	0.96	293
4	T/O		23.2	7822	11.5	4054	2.31	973	T/O		T/O		3.91	684	T/O		2.92	684	T/O	
5	T/O		111	26749	63.8	13704	6.14	2073	T/O		453	1360	13.8	1360	T/O		13.5	1360	T/O	

Each Malicious Synchronization instance is denoted (k/n), where n is the number of devices and k the congestion limit; Production Cell instances are numbered by plate count. We compare the ten configurations introduced above, referring to each by its shorthand name.

4.4 Results

The results are shown in Table 1. All abstractions consistently reduce the number of explored states compared to Baseline, often by an order of magnitude. All hull methods and Loc converge to identical state counts on several instances (e.g., 212 on (2/2) and 1,438 on (3/3)), suggesting they reach the same minimal reduced graph at those sizes. DInc consistently improves over Inc, roughly halving the state count on most instances, but does not match the reductions achieved by Mrg or the hull methods. Notably, our new abstractions solve instances intractable for the prior technique (Inc): Oct is the only configuration to resolve Malicious Synchronization (3/4), doing so in 334 s.

Convex and Box Hull. CH performs well on small instances but can struggle on larger ones—though not monotonically: it times out on Production Cell 4 yet solves instance 5 in 453 s. Empirically, convex hull abstractions can accumulate increasingly complex polyhedra, at which point the underlying arithmetic operations stall; whether this occurs appears to depend on the geometry of the specific instance rather than its size alone. Box avoids this by restricting to simpler shapes, but its rectangular enclosure discards diagonal constraints that are structurally important in the Production Cell's geometry, which may lead to larger zones with more successors, potentially explaining the timeouts on all but the smallest instance. BoxH can avoid both pitfalls: on Malicious Synchronization (3/3), it succeeds where CH and Box individually time out.

Octagonal Hull. Oct is the most robust configuration: it solves every instance solved by any other method and is the only one to solve Malicious Synchronization (3/4). It appears to strike the right balance between aggressiveness and geometric fidelity, retaining enough diagonal structure to keep zone operations cheap without over-approximating to the degree that causes the blowup seen with Box. OctH behaves similarly to Oct across most instances, with no consistent advantage in either direction.

Location-Based Abstraction. Loc can reach a state count comparable to the hull methods yet be disproportionately slow: on Malicious Synchronization (3/3) it explores 1,438 states in 475 s, versus 24.3 s for Oct. The larger zones likely introduce spurious transitions in the parametric zone graph, causing repeated propagation through states that do not contribute to convergence.

On-the-Fly Merge. Mrg is the strongest configuration that introduces no geometric over-approximation of zones. It handles all Production Cell instances comfortably and is competitive on Malicious Synchronization, making it a reliable fallback when a tighter zone representation is preferred over raw speed.

Our evaluation is limited to two benchmark families; further case studies are needed to confirm that the relative performance of the abstractions generalises.

5 Conclusion

We presented a general abstraction framework for parametric timed games that reduces the symbolic state space while preserving correctness of both parameter synthesis and winning strategies. The framework subsumes the existing zone inclusion technique for PTG and lifts double inclusion and zone merging from PTA and location-based abstraction from timed games to the PTG setting, while also introducing new hull abstractions inspired by classical numerical abstract domains. Experiments on a production cell benchmark and a novel adversarial IoT case study show order-of-magnitude reductions in explored states, with octagonal hull emerging as the most robust configuration, solving every tractable instance, including instances previously intractable for existing techniques.

Several directions remain open. First, combining abstractions, for instance merge steps followed by hull steps, could exploit the strengths of both. Second, the hybrid configurations showed promise but relied on a fixed constraint-count threshold; a principled heuristic for switching from convex to a coarser hull would make these strategies more broadly applicable. Finally, evaluating the framework on benchmarks with richer clock and parameter structure, would clarify which abstractions suit which problem characteristics.

Acknowledgements. We thank Baptiste Fievet for contributions to the algorithmic ideas, and Étienne André for advice on adapting zone merging to the parametric game setting. Parts of this paper used generative AI tools (Claude by Anthropic) to assist with drafting and editing for style and clarity.

Data Availability. The models, scripts, and tools to reproduce our experimental evaluation are archived and publicly available at [15].

References

1. Alur, R., Dill, D.L.: A theory of timed automata. Theoret. Comput. Sci. **126**(2), 183–235 (1994)
2. Alur, R., Henzinger, T.A., Vardi, M.Y.: Parametric real-time reasoning. In: STOC, pp. 592–601. ACM (1993)
3. André, É.: IMITATOR 3: Synthesis of Timing Parameters Beyond Decidability. In: Silva, A., Leino, K.R.M. (eds.) CAV 2021. LNCS, vol. 12759, pp. 552–565. Springer, Cham (2021). https://doi.org/10.1007/978-3-030-81685-8_26
4. André, É., Fribourg, L., Soulat, R.: Merge and Conquer: State Merging in Parametric Timed Automata. In: Van Hung, D., Ogawa, M. (eds.) ATVA 2013. LNCS, vol. 8172, pp. 381–396. Springer, Cham (2013). https://doi.org/10.1007/978-3-319-02444-8_27
5. André, É., Lime, D., Roux, O.H.: Integer-Complete Synthesis for Bounded Parametric Timed Automata. In: Bojańczyk, M., Lasota, S., Potapov, I. (eds.) RP 2015. LNCS, vol. 9328, pp. 7–19. Springer, Cham (2015). https://doi.org/10.1007/978-3-319-24537-9_2
6. André, É., Marinho, D., Petrucci, L., van de Pol, J.: Efficient convex zone merging in parametric timed automata. In: Bogomolov, S., Parker, D. (eds.) FORMATS, vol. 13465 of Lecture Notes in Computer Science, pp. 200–218. Springer, Cham (2022). https://doi.org/10.1007/978-3-031-15839-1_12
7. Arcile, J., André, É.: Zone extrapolations in parametric timed automata. In: NFM, Lecture Notes in Computer Science, pp. 451–469. Springer, Cham (2022)
8. Bagnara, R., Hill, P.M., Zaffanella, E.: The parma polyhedra library: Toward a complete set of numerical abstractions for the analysis and verification of hardware and software systems. Sci. Comput. Program. **72**(1–2), 3–21 (2008)
9. Balarin, F.: Approximate reachability analysis of timed automata. In: RTSS, pp. 52–61. IEEE Computer Society (1996)
10. Behrmann, G., Bouyer, P., Guldstrand Larsen, K., Pelánek, R.: Lower and upper bounds in zone based abstractions of timed automata. In: TACAS, Lecture Notes in Computer Science, pp. 312–326. Springer, Cham (2004). https://doi.org/10.1007/s10009-005-0190-0

11. Bezděk, P., Beneš, N., Barnat, J., Černá, I.: LTL Parameter Synthesis of Parametric Timed Automata. In: De Nicola, R., Kühn, E. (eds.) SEFM 2016. LNCS, vol. 9763, pp. 172–187. Springer, Cham (2016). https://doi.org/10.1007/978-3-319-41591-8_12
12. Cassez, F., David, A., Fleury, E., Larsen, K.G., Lime, D.: Efficient on-the-fly algorithms for the analysis of timed games. In: CONCUR, Lecture Notes in Computer Science, pp. 66–80. Springer, Cham (2005)
13. Cousot, P., Cousot, R.: Abstract interpretation: A unified lattice model for static analysis of programs by construction or approximation of fixpoints. In: POPL, pp. 238–252. ACM (1977)
14. Cousot, P., Halbwachs, N.: Automatic discovery of linear restraints among variables of a program. In: POPL, pp. 84–96. ACM Press (1978)
15. Dahlsen-Jensen, M.B., Petrucci, L., van de Pol, J.C. : Artifact for "state-space abstractions for parametric timed games. Zenodo (2026). https://doi.org/10.5281/zenodo.19680181
16. Dahlsen-Jensen, M.B., Fievet, B., Petrucci, L., van de Pol, J.: On-the-fly algorithm for reachability in parametric timed games. In: TACAS (3), Lecture Notes in Computer Science, pp. 194–212. Springer, Cham (2024)
17. Dahlsen-Jensen, M.B., Fievet, B., Petrucci, L., van de Pol, J.: Controller synthesis for parametric timed games. In: QEST+FORMATS, Lecture Notes in Computer Science, pp. 314–332. Springer, Cham (2025). https://doi.org/10.1007/978-3-032-05792-1_17
18. Dahlsen-Jensen, M.B., Petrucci, L., van de Pol, J.: State-space abstractions for parametric timed games (2026)
19. de Alfaro, L., Godefroid, P., Jagadeesan, R.: Three-valued abstractions of games: Uncertainty, but with precision. In: LICS, pp. 170–179. IEEE Computer Society (2004)
20. Dill, D.L.: Timing assumptions and verification of finite-state concurrent systems. In: Sifakis, J. (ed.) CAV 1989. LNCS, vol. 407, pp. 197–212. Springer, Heidelberg (1990). https://doi.org/10.1007/3-540-52148-8_17
21. Henzinger, T.A., Majumdar, R., Mang, F., Raskin, J.-F.: Abstract Interpretation of Game Properties. In: Palsberg, J. (ed.) SAS 2000. LNCS, vol. 1824, pp. 220–239. Springer, Heidelberg (2000). https://doi.org/10.1007/978-3-540-45099-3_12
22. Hune, T., Romijn, J., Stoelinga, M., Vaandrager, F.: Linear Parametric Model Checking of Timed Automata. In: Margaria, T., Yi, W. (eds.) TACAS 2001. LNCS, vol. 2031, pp. 189–203. Springer, Heidelberg (2001). https://doi.org/10.1007/3-540-45319-9_14
23. Jensen, N., Larsen, K.G., Lime, D., Srba, J.: On-the-fly symbolic algorithm for timed ATL with abstractions. In: CONCUR, volume 348 of LIPIcs, pp. 25:1–25:19. Schloss Dagstuhl - Leibniz-Zentrum für Informatik (2025)
24. Jovanovic, A., Faucou, S., Lime, D., Roux, O.H. : Real-time control with parametric timed reachability games. In: IFAC WODES, pp. 323–330. Elseviers (2012)
25. Jovanovic, A., Lime, D., Roux, O.H.: A game approach to the parametric control of real-time systems. Int. J. Control **92**(9), 2025–2036 (2019)
26. Larsen, K.G., Pettersson, P., Yi,W.: UPPAAL in a nutshell. Int. J. Softw. Tools Technol. Transf. **1**(1-2), 134–152 (1997)
27. Lewerentz, C., Lindner, T.: Case study "production cell": A comparative study in formal specification and verification. In: Broy, M., Jähnichen, S. (eds.) KORSO: Methods, Languages, and Tools for the Construction of Correct Software. LNCS, vol. 1009, pp. 388–416. Springer, Heidelberg (1995). https://doi.org/10.1007/BFb0015473

28. Maler, O., Pnueli, A., Sifakis, J.: On the synthesis of discrete controllers for timed systems. In: Mayr, E.W., Puech, C. (eds.) STACS 1995. LNCS, vol. 900, pp. 229–242. Springer, Heidelberg (1995). https://doi.org/10.1007/3-540-59042-0_76
29. Miné, A.: The octagon abstract domain. In: WCRE, p. 310. IEEE Computer Society (2001)
30. Peter, H.-J., Ehlers, R., Mattmüller, R.: Synthia: Verification and Synthesis for Timed Automata. In: Gopalakrishnan, G., Qadeer, S. (eds.) CAV 2011. LNCS, vol. 6806, pp. 649–655. Springer, Heidelberg (2011). https://doi.org/10.1007/978-3-642-22110-1_52

Facing the Core: State-Space Compression of Dynamic Fault Trees via PH-Distributions

Nazareno Garagiola(✉) and Holger Hermanns

Saarland University, Saarland Informatics Campus, Saarbrücken, Germany
{garagiola,hermanns}@depend.uni-saarland.de

Abstract. Modern systems exhibit complex, time-dependent failures that demand accurate dependability modelling. Dynamic Fault Trees can capture these behaviours effectively. Often they are paired with failure distribution of basic events that can be represented as acyclic phase-type distributions (APH). However, analysing them often suffers from severe state-space explosion, limiting scalability and applicability. This paper introduces two complementary compression techniques rooted in a common foundational concept. We exploit that the matrix representations of APH are not unique, but can be efficiently compressed by studying their longest core series and dropping redundant phases that otherwise are a notorious side effect of model construction. This provides (i) a novel way to compress the state spaces induced by dynamic fault trees, and it also enables (ii) a situation-specific bounding mechanism that further reduces model size while providing upper and lower probability bounds. Both techniques are orthogonal to existing state-space reduction methods. Empirical results demonstrate substantial compression and improved scalability, highlighting the potential of these approaches for DFT analysis and beyond.

Keywords: Phase Type Distributions · Continuous Time Markov Chains · Dynamic Fault Trees

1 Introduction

Phase-type (PH) distributions constitute one of the most versatile families of probability distributions, widely used as the stochastic support in areas like Queuing Theory [5,35], Network Design [13], Reliability analysis [3,50], Survival Analysis [1,37], and many other applications. The family of PH-distributions encompasses known distributions like the *Exponential*, *Erlang*, *Hypoexponential* or *Coxian* [15] distributions. The family of continuous PH distributions is topologically dense on the positive reals, which implies that PH distributions can

M. Chen and M. Tschaikowski (Eds.): QEST+FORMATS 2026, LNCS 16913, pp. 39–58, 2027.
https://doi.org/10.1007/978-3-032-35298-9_3

be used to approximate any other continuous probability distribution [6,28] – including those based on empirical observation, supported by tools like HyperStar [41], BuTools [27], G-FIT [46] or PhFit [26].

Each PH distribution describes the time until absorption in an absorbing continuous-time Markov chain (CTMC) [35], or, more precise, of an infinitely large set of absorbing CTMCs, all having the same distribution of the time to absorption. These chains differ in size, and it is theoretically interesting and practically relevant to work with a chain having the smallest achievable number of states.

Fault Trees (FT) are a powerful modelling framework in the field of probabilistic risk analysis, offering a systematic method to model and analyse failures in a complex system. A traditional FT embodies a Boolean function, manifesting the interdependence of overall system failures on the malfunction of system components. Dynamic behaviour has been introduced into FTs [18] with dynamic gates, which enable the modelling of dependencies, spare management, failure ordering, and more. This provides usability across diverse domains, including aerospace [44], nuclear power plants [7,47], chemical processes [12], and others. A historical account of FTs can be found in [14], and [43] holds a survey of fault tree variations.

Typical evaluations of Dynamic FTs make use of stochastic frameworks [9, 11,48] across which CTMC-based analysis techniques dominate, provided by model checkers like STORM [25] or PRISM [33]. The tool SAFEST [49] provides a full comprehensive assessment of FTs, together with a graphical interface. Several state-space analysis and reduction techniques have been proposed in this context [4,16,48], among them the inspiring work of Junges *et al.* [29].

The approach developed in this paper addresses the same problem domain, but is orthogonal to all the work mentioned above, meaning that it can enhance any of those approaches. Our contribution is threefold; We (i) develop a method to construct compressed representations of acyclic phase-type distributions, rooted in [39]. We (ii) demonstrate how the proposed method, improves the existing work on PH distributions [38] and, in combination with that work, can efficiently obtain minimal-size representations of failure distributions of complex structures, such as Dynamic Fault Trees, thereby yielding smaller state spaces than current state-of-the-art techniques. Finally, we (iii) capitalize on the setup by presenting a context-specific bounding technique providing sound upper and lower bounds while compressing the state spaces even further.

Organization of the paper: Sect. 2 presents *acyclic Phase-types* and *Dynamic FTs*. Section 3 introduces *Core Series*, the building blocks of our approach. Section 4 illustrates the translation from FTs into compressed representations of acyclic PH, while Sect. 5 how we exploit our developments to construct compact bounds. Finally, Sect. 6 gives an empirical evaluation together with a case study, and Sect. 7 concludes the paper.

2 Preliminaries

We assume the reader to be familiar with basic concepts of (continuous-time) Markov chains, for which we otherwise refer to [10].

Acyclic Phase-type Distributions: Phase-type distributions (PH) are the probability distribution of the time to absorption in a CTMC. If the chain does not contain any cycles, an *Acyclic Phase-type* (APH) distribution describes the absorption time. The family of all APH distributions is denoted by $\mathcal{APH}$. In this situation, the states can be ordered so that state s_{n+1} does not have any outgoing transition, and all other states $s_1, \ldots, s_n$ are transient. Therefore, the generator matrix $\mathbf{Q}$ of the corresponding CTMC $\mathcal{M} = (\mathcal{S}, \mathbf{Q}, (\vec{\alpha}, \alpha_{n+1}))$ can be written as

$$\mathbf{Q} = \begin{bmatrix} \mathbf{A} & \vec{A} \\ \vec{0} & 0 \end{bmatrix}$$

where $\vec{A}$ and α_{n+1} are fixed if $\mathbf{A}$ and $\vec{\alpha}$ are given, and $\mathbf{A}$ is upper-triangular. Matrix $\mathbf{A}$ is called a *PH-generator*, the pair $(\vec{\alpha}, \mathbf{A})$ is a representation of the PH distribution. PH$(\vec{\alpha}, \mathbf{A})$ denotes the phase-type distribution with representation $(\vec{\alpha}, \mathbf{A})$. Its cumulative distribution function (CDF) is $F(t) = 1 - \vec{\alpha} e^{\mathbf{A}t} \vec{\mathbf{1}}, t \in \mathbb{R}_{>0}$, with $\vec{\mathbf{1}}$ being the vector where all entries have value 1. Besides its CDF, a PH distribution is also characterized by its *Laplace-Stieltjes Transform* (LST). The LST of PH$(\vec{\alpha}, \mathbf{A})$ is a rational function:

$$\hat{f}(s) = \int_{-\infty}^{\infty} e^{-st} dF(t) = \vec{\alpha}(s\,\mathbf{I} - \mathbf{A})^{-1}\,\vec{A} + \alpha_{n+1} = \frac{p(s)}{q(s)}, s \in \mathbb{R}_{\geq 0}$$

where $\mathbf{I}$ is the identity matrix of size n. We call the *exit rate* of a state s, to the sum of all rates exiting the state $(\sum_{s' \neq s} \mathbf{Q}(s, s') = -\,\mathbf{Q}\,(s, s))$, and *Rates*(PH$(\vec{\alpha}, \mathbf{A})$) denotes the set of all exit rates in $\mathbf{A}$.

An APH distribution admits multiple representations, all of which share the same LST. A representation is *minimal* if the size of its generator matrix matches the *algebraic degree* of the distribution [39], that is, the degree of the polynomial q in the denominator of the LST when expressed as an *irreducible rational function.*

Since both p and q are polynomial functions, the size differences are linked to common roots of the polynomials p and q. O'Cinneide [36] has pioneered the exploration of this phenomenon, which later has been turned into an algorithm for state-space reduction by Pulungan [39]. The algorithm iteratively checks for and (if possible) removes "redundant" states, while preserving that the APH of the representation remains unchanged.

Bidiagonal Representations: An infinitesimal generator matrix $\mathbf{A}$ with eigenvalues $\lambda_1, \ldots, \lambda_n$, in which only the entries of the diagonal and super-diagonal are non-zero is denoted by $Bi(\lambda_1, \ldots, \lambda_n)$; and $(\vec{\alpha}, Bi(\lambda_1, \ldots, \lambda_n))$ is called a *Bidiagonal representation* [17]. Figure 1 shows the graphical depiction of a bidiagonal representation. Likewise, we denote by $Bi(\lambda_1^{k_1}, \ldots, \lambda_n^{k_n})$ the bidiagonal PH-generator, where k_i is the number of states with exit rate $\lambda_i, 1 \leq i \leq n$. We speak of an *Ordered Bidiagonal Representation* (OBR) if $\lambda_1 \leq \lambda_2 \leq \lambda_3 \leq \cdots \leq \lambda_n$.

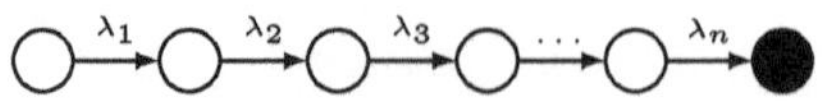

Fig. 1. A bidiagonal APH representation, ordered if the rates are increasing.

Any APH representation can be turned into an OBR of the same size. The process is called "diagonalization", and there are multiple approaches for doing so, some of which depending on certain preconditions to be met by the representation [17,36]. We choose to use the *Spectral Polynomial Algorithm* [24] (SPA), which does not make any assumption on the original representation. The SPA "diagonalizes" representation $(\vec{\alpha}, \mathbf{A})$ with $Rates((\vec{\alpha}, \mathbf{A})) = \{\lambda_1, \ldots, \lambda_n\}$, by constructing a matrix $\mathbf{P}$, so that $\mathbf{AP} = \mathbf{PBi}$, where $\mathbf{Bi} = Bi(\lambda_{(1)}, \ldots, \lambda_{(n)})$ is an ordered bidiagonal generator, and $\mathbf{P}$ has unit row-sums. The SPA then constructs the representation $(\vec{\alpha}\mathbf{P}, \mathbf{Bi})$ such that $PH(\vec{\alpha}, \mathbf{A}) = PH(\vec{\alpha}\mathbf{P}, \mathbf{Bi})$.

Dynamic Fault Trees: A Fault Tree (FT) is a *directed acyclic graph* that models how the failure of single components can propagate in a system, eventually resulting in the failure of the overall system. The leaves of the graph are called *basic events* and intermediate nodes are called *gates*, with the top level event (TLE) representing the failure of the system. Dynamic Fault Trees (DFT) [18] extend "static" FTs by introducing dynamic gates – such as sequence enforcers, spare components, and more. Those can capture complex time-dependent failure mechanisms, enabling accurate modelling of the system in cases where this depends not only on the failing of atomic components, but also on their order of failing, resource sharing, or the interaction between them. In all this, there is an influence of time to be considered, something that can be abstracted away if only working with static FTs. [43] presents an overview on diverse types of FTs. In this time-sensitive setting, it is common to use exponential or, more generally, APH distributions to model the time to failure of individual basic events.

Formally, a DFT is a tuple $\mathcal{F} = (V, \sigma, T, top, \Gamma)$ where

- V is a set of nodes,
- $top \in V$ is the TLE of $\mathcal{F}$,
- $\sigma\colon V \to \mathcal{P}(V)$ maps each node to its set of children,
- $\Gamma\colon V_{\mathsf{BE}} \to \mathcal{APH}$ assigns an APH distribution to each basic event, and
- $T\colon V \to \mathit{Types}$ maps each node to its type.

Additionally, it is required that the graph induced by σ is acyclic, rooted in top (the transitive closure of σ applied to top being V), and that V_{BE} comprises all leaves, and no other nodes. Function T maps a type $\in \mathit{Types}$ to each node, which – apart from basic events at the leaves of the tree – can be one of nine variants of gates, each one with different propagation semantics. We base our semantics of DFTs on the ones from [30]. In particular, our semantics diverge from the ones in that work, but we exclude the *Priority* AND gate, the effect of which cannot be represented with APH (unless supporting multiple absorptions [51]).

We will often work in a subset of DFTs, namely those that are strictly *tree-structured*, meaning that all nodes have at most one parent. We denote them

Fig. 2. A simple tFT (left) and a DFT with shared element B, not a tFT (right).

as *tree-shaped Fault Trees* (tFT)[1]. Figure 2 shows distinguishing FT examples, where AND and OR gates adhere to the usual visualization, and also a SEQ gate appears, denoting a sequential dependency. In the analysis of non-repairable FTs, the most prominent metrics are the system's *unreliability*, which states the probability that the top event occurs (fails) within a fixed time limit t, and the *mean time to failure* (MTTF), which describes the expected time at which the system fails. Several other results enrich the analysis of FTs [43]. We here will only focus on the two metrics mentioned.

3 Construction of APH Representations

This section presents the concept of *Core Series* [40] and shows how to exploit it to construct compressed OBR of APH distributions. Subsequently, we provide an efficient algorithm to turn any absorbing acyclic CTMC into an OBR representation of identical size.

Core Series. Consider an APH representation of size n, with $m \leq n$ distinct exit rates $\lambda_{(m)} > \cdots > \lambda_{(1)} > 0$, strictly ordered by their magnitudes. Let $\mathcal{M}$ be the underlying CTMC, and $Paths(\mathcal{M})$ the set with all sequences of states in $\mathcal{M}$ that start in an initial state, and end in the absorbing one. We denote with $c_\sigma(\lambda)$ the number of occurrences of states with exit rate λ in a path σ of $\mathcal{M}$, and extend it to $Paths(\mathcal{M})$:

$$c_{\mathcal{M}}(\lambda) = max_{\sigma \in Paths(M)}\ c_\sigma(\lambda)$$

Thus, $c_{\mathcal{M}}(\lambda)$ denotes the maximum number of occurrences of states with exit rate λ on any path in $\mathcal{M}$. We call $c_{\mathcal{M}}(\lambda)$ the *multiplicity* of λ in $\mathcal{M}$. When the CTMC is clear from context, we drop the subscript and simply write $c(\lambda)$.

The key observation will be that $c_{\mathcal{M}}(\lambda)$ can be considerably smaller than the number of states with exit rate λ, and that this can be exploited for compression purposes. Let $k_i = \sum_{h=1}^{i} c_{\mathcal{M}}(\lambda_{(h)})$ with $k_0 = 0$ for $1 \leq i \leq m$, and $\ell = k_m$. Thus, k_i denotes the maximum number of occurrences of states with exit rates $\lambda_{(h)}$, for all $1 \leq h \leq i$, in any path of $\mathcal{M}$. Notably, $m \leq \ell \leq n$. If all rates are distinct,

[1] This restriction in particular prevents voting gates (VOT) in tFTs, given that their translation to AND and OR gates does not form a tree structure.

then $m = n$; if the underlying CTMC proceeds from a bidiagonal representation, then $\ell = n$. Usually, m and ℓ are considerably smaller than n.

Definition 1. *(Core Series [40]) Consider the family of ℓ hypoexponential representations $\langle \mathcal{C}_1, \ldots, \mathcal{C}_\ell \rangle$ described by: $\mathcal{C}_\ell = (\vec{e}_1, Bi(\lambda_{(m)}))$, and for $1 \leq i \leq \ell - 1$ $\mathcal{C}_i = (\vec{e}_1, Bi(\lambda_i, \ldots, \lambda_\ell))$, where $\lambda_j = \lambda_{(h)}$ such that $k_{h-1} < j \leq k_h$ for $i \leq j \leq \ell$. This family of hypoexponential representations is called the* core series *of the corresponding CTMC $\mathcal{M}$, and the hypoexponential representation of size ℓ is referred to as the* Longest Core Series *(LCS).*

The LCS has the largest size in the family. Intuitively, we extend with rate $\lambda_{(h)}$ in the series of exit rates $c_{\mathcal{M}}(\lambda_{(h)})$-times, which is, by definition, the maximum number of instances of exit rate $\lambda_{(h)}$ in all paths of the representation.

Example 1. Consider Fig. 3. Note that there are 2 states with exit rate 5 (namely, s_3 and s_4), but both cannot occur in the same path, hence $c_{\mathcal{M}}(5) = 1$. On the contrary, both states s_0 and s_1 have an exit rate 2, and because there is a path that goes through both states, we obtain $c_{\mathcal{M}}(2) = 2$.

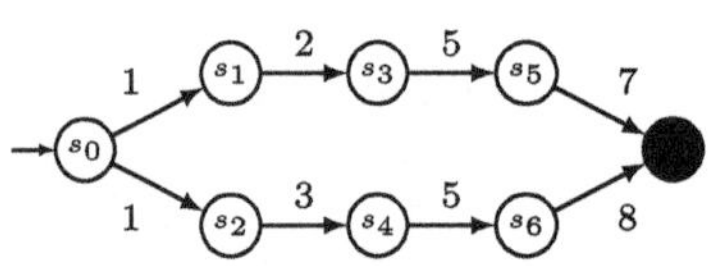

$$\begin{aligned}
\mathcal{C}_6 &= (\vec{e}_1, Bi(8)) \\
\mathcal{C}_5 &= (\vec{e}_1, Bi(7, 8)) \\
\mathcal{C}_4 &= (\vec{e}_1, Bi(5, 7, 8)) \\
\mathcal{C}_3 &= (\vec{e}_1, Bi(3, 5, 7, 8)) \\
\mathcal{C}_2 &= (\vec{e}_1, Bi(2, 3, 5, 7, 8)) \\
\mathcal{C}_1 &= (\vec{e}_1, Bi(2, 2, 3, 5, 7, 8))
\end{aligned}$$

Fig. 3. An APH representation and its core series.

Lemma 2. *(Lemma 4.5 of [38]) For each $\sigma \in Paths(\mathcal{M})$ there exists some convex combination of the core series $\langle \mathcal{C}_1, \ldots, \mathcal{C}_\ell \rangle$ such that the sum of their distribution functions agrees with that of σ.*

As far as the distribution function concerns, a CTMC is a convex combination of its paths. Lemma 2 implicitly states that a CTMC is a convex combination of its core series.

3.1 PH-Generator via the Longest Core Series

We now turn to the central result of the paper, that enables us to compress representations of APH distributions using its LCS. The theorem is a generalization of [38, Lemma 5.8] by Pulungan, and its proof follows from repeated applications of that Lemma.

Theorem 3. *Let $(\vec{\beta}, Bi(\lambda_1^{k_1}, \ldots, \lambda_n^{k_n}))$ be an OBR, where k_i is the number of states with exit rate λ_i, $1 \leq i \leq n$. Let $(\vec{e}_1, Bi(\mathbf{Bi}))$ be the LCS of the representation. Then, there exists a unique sub-stochastic vector $\vec{\beta'}$ such that:*

$$PH(\vec{\beta}, Bi(\lambda_1^{k_1}, \ldots, \lambda_n^{k_n})) = PH(\vec{\beta'}, \mathbf{Bi})$$

Consider representation $(\vec{\alpha}, \mathbf{A})$ with LCS $\mathbf{Bi}$. Recall that the SPA algorithm is used to diagonalize a given APH representation. Due to Theorem 3 we can now use the SPA algorithm to not only diagonalize it, but in addition also compress it. In a nutshell, the system of equations $\mathbf{AP} = \mathbf{PBi}$ now makes matrix $\mathbf{P}$ to be a rectangular matrix, where the number of columns matches the size of the LCS. Thus, representation $(\vec{\alpha}\mathbf{P}, \mathbf{Bi})$ will have the same size as the LCS. Theorem 3 guarantees that the redistribution of initial probability from $\vec{\beta}$ to $\vec{\beta'}$ will succeed. The process of diagonalizing an APH representation with its LCS we will refer to as *compression*. Note that if the size of the LCS matches the algebraic degree of the distribution, the compression yields a minimal representation. Otherwise, it can almost surely be further minimized with the algorithm of Pulungan [39].

3.2 Harvesting Compression in Standard Constructions

To understand how effective the compression introduced above is, we now cover standard operations on bidiagonal representations, focussing on how to obtain the LCS from the inputs of the operations, while avoiding the construction of the monolithic PH-generator. The following facts we take for granted [4,35].

- The convolution $\texttt{Con}(F(t), G(t))$ of two APH-distributed random variables with CDF $F(t)$ and $G(t)$, is APH-distributed. Furthermore, it is the semantics of the binary SEQ operator for tFT.
- The maximum $\texttt{Max}(F(t), G(t))$ of two APH-distributed random variables with CDF $F(t)$ and $G(t)$, is APH-distributed. Furthermore, it is the semantics of the binary AND operator for tFT.
- The minimum $\texttt{Min}(F(t), G(t))$ of two APH-distributed random variables with CDF $F(t)$ and $G(t)$, is APH-distributed. Furthermore, it is the semantics of the binary OR operator for tFT.

Convolution Operation: We can assume that both operands are already in OBR form and compressed. The convolution operation of two OBR generates a bidiagonal representation. Hence, the LCS of the representation is the same bidiagonal representation (modulo reordering needs), and thus, there is no compression.

Maximum Operation: Consider APH distributions A and B, with ordered bidiagonal representations $(\vec{\alpha}, Bi(\lambda_1, \ldots, \lambda_n))$ and $(\vec{\beta}, Bi(\mu_1, \ldots, \mu_m))$, respectively.

Let $\mathcal{M}$ be the underlying CTMC of the maximum of A and B. Figure 4 shows the graphical representation of $\mathcal{M}$, omitting the initial distributions for clarity. Note how "new" exit rates are generated in the cross-product sum of the rates from each representation, together with the original rates from A and B. Let k_i, $k'_j \in \mathbb{N}$ denote the number of rates of λ_i and μ_j respectively, with $1 \leq i \leq m, 1 \leq j \leq n$. Then, for each exit rate $\eta \in \mathit{Rates}(\mathtt{Max}(\mathsf{A}, \mathsf{B}))$, we denote $\Psi_\eta = \{(i,j) \mid \eta = \lambda_i + \mu_j\}$.

Observation 4. *For all paths in $\mathcal{M}$, the maximum number of times a state with exit rate η is visited is determined by $max(\{(k_i + k'_j - 1) \mid (i,j) \in \Psi_\eta\} \cup \{k_i \mid \lambda_i = \eta\} \cup \{k'_j \mid \mu_j = \eta\})$. Hence, $c_{\mathcal{M}}(\eta) = max(\{(k_i + k'_j - 1) \mid (i,j) \in \Psi_\eta\} \cup \{k_i \mid \lambda_i = \eta\} \cup \{k'_j \mid \mu_j = \eta\})$.*

Lemma 5. *Let $\mathcal{C}$ denote the LCS of the standard representation of the maximum of A and B, and* **Bi** *be the ordered bidiagonal PH-generator associated with $\mathcal{C}$. Then, for some sub-stochastic vector $\vec{\delta}$, $(\vec{\delta}, \mathbf{Bi})$ is a representation of* $\mathtt{Max}(\mathsf{A}, \mathsf{B})$.

Proof. From the construction of the core series, and Theorem 3, it is straightforward that $(\vec{\delta}, \mathbf{Bi})$ is a representation of $\mathtt{Max}(\mathsf{A}, \mathsf{B})$. $\vec{\delta}$ provides the parameters of the convex combination, and it can be computed using the SPA algorithm. □

Minimum Operation: Given APH distributions A and B characterized by the representations $(\vec{\alpha}, Bi(\lambda_1, \ldots, \lambda_n))$ and $(\vec{\beta}, Bi(\mu_1, \ldots, \mu_m))$ respectively. Let $\mathcal{M}$ be the underlying CTMC of the minimum of A and B. The representation of $\mathcal{M}$ is close to the one in Fig. 4, except that states (m,i) and (j,n) only have transitions to the absorbing state, for $1 \leq i \leq n, 1 \leq j \leq m$. Let k_i, $k'_j \in \mathbb{N}$ be the amount of rates λ_i and μ_j, respectively, with $1 \leq i \leq m, 1 \leq j \leq n$. For each exit rate $\eta \in \mathit{Rates}(\mathtt{Min}(\mathsf{A}, \mathsf{B}))$, we denote $\Psi_\eta = \{(i,j) \mid \eta = \lambda_i + \mu_j\}$,

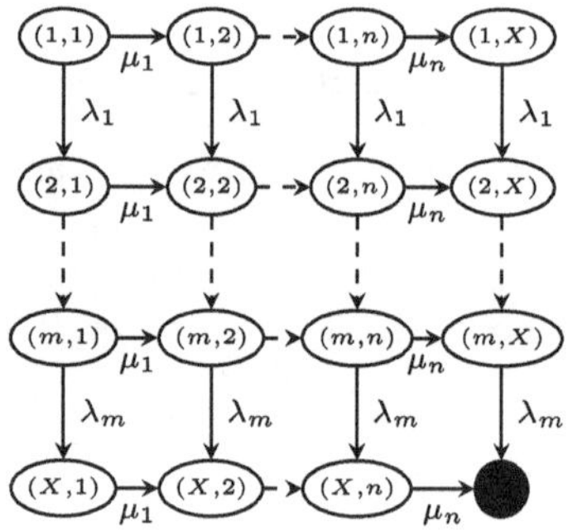

Fig. 4. Graphical representation of the maximum of A and B.

Observation 6. *For all paths in $\mathcal{M}$, the maximum number of times a state with exit rate η is visited is determined by $max(\{k_i + k'_j - 1 \mid (i,j) \in \Psi_\eta\})$. Hence, $c_{\mathcal{M}}(\eta) = max(\{k_i + k'_j - 1 \mid (i,j) \in \Psi_\eta\})$.*

The proof of the following lemma is analogous to the one of Lemma 5.

Lemma 7. *Let $\mathcal{C}$ denote the LCS of the standard representation of the minimum of A and B, and* **Bi** *be the ordered bidiagonal PH-generator associated with $\mathcal{C}$. Then, for some sub-stochastic vector $\vec{\delta}$, $(\vec{\delta}, \mathbf{Bi})$ is a representation of* $\mathtt{Min}(\mathsf{A}, \mathsf{B})$.

Example 2. Consider Fig. 5, that graphically represents the minimum of two hypoexponentials distributions: $(\vec{e}_1, Bi(3,8))$ and $(\vec{e}_1, Bi(2,2,7))$. If we diagonalize the representation shown in the Figure, we obtain the PH-generator $\mathbf{Bi}(5,5,10,10,10,15)$. By Observation 6, we get that $c(10) = max(\{k_3 + k'_7, k_8 +$

$k_2'\}) - 1 = 2$. In a same fashion we compute $c(5)$ and $c(15)$. Thus, by only considering the multiplicities of the rates from the input representations, we can construct the LCS of the final representation, without having to construct the monolithic representation from Fig. 5. Lemma 7 guarantees that the obtained representation, is a representation of the minimum.

3.3 Longest Core Series Beyond tFTs

Theorem 3 applies to arbitrary acyclic CTMCs, but the computation of the LCS as per its definition may get impractical and expensive. In the following, we introduce an algorithm to find the LCS of such CTMCs, while bypassing the exponential search over all paths of the underlying graph otherwise needed. In the proposed algorithm, each state keeps a table of exit rates that can be passed until reaching such state. The procedure is rooted in a topological sorting of the states, which is enabled by the CTMC being acyclic.

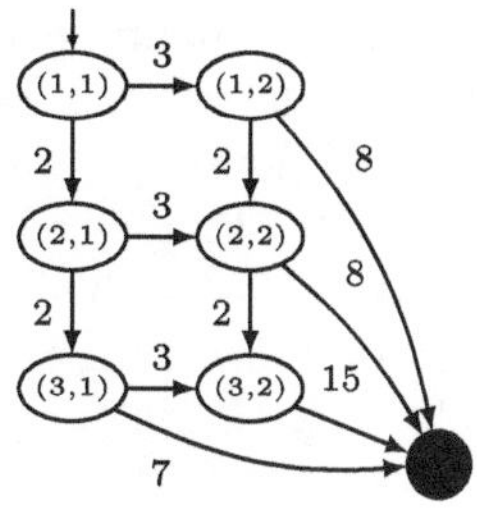

Fig. 5. Representation of $\texttt{Min}((\vec{e}_1, Bi(3,8)),$ *underlying CTMC of the maximum* $(\vec{e}_1, Bi(2,2,7)))$.

The algorithm first finds a topological order of the states, based on Khan's Algorithm [31] for DAGs. Then, it iterates once over all states, where each state adds its own exit rate to its own table with the present exit rates. Then, it attempts to update the tables of its successors, by taking the maximum multiplicity of each of the overlapping rates, and adding the rates that are not in the successors table. This procedure ends once it reaches the absorbing state, for which the table with the exit rates and multiplicities provide the parameters to construct the PH-generator of the LCS. Example 3 illustrates the procedure.

Example 3. Consider the CTMC from Fig. 6. The state order is given by the subindex of each state. The first state to update the table of state s_6 is s_4, which informs the exit rates: 8, 2 and 2, from states s_1, s_2 and s_4 respectively. When s_5 aims at updating the table from s_6, the multiplicity of exit rate 2 is no greater than that previously known by s_6, but it adds the new exit rate 4 from state s_5. Three updates occur to the absorbing state table. From s_4, s_5 and s_6. From those, the tables from s_5 and s_6 have less information than the one from s_4.

The memory footprint of the algorithm grows proportionally to the product of the number of *unique* exit rates, and the number of states in the CTMC.

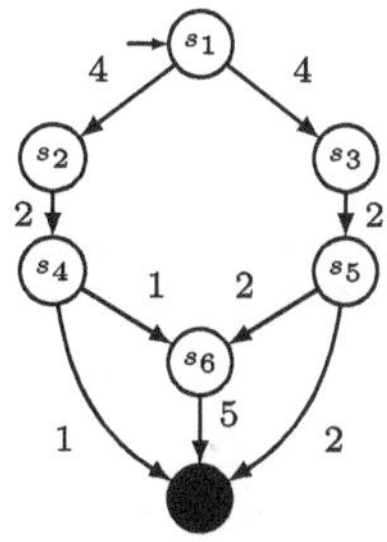

Fig. 6. Graphical representation of a CTMC of irregular structure.

4 Transforming DFT Into APH

This section details the method to translate FTs into APH representations. Specifically, CTMCs are derived from tFTs and DFTs and then compressed to OBRs using the SPA algorithm [24], with its LCS acting as the PH generator.

4.1 Tree-Shape Fault Trees to APH

As per their definition, tFTs are directly translated to APH just by employing the standard APH operations. Because if all basic events, or *components* from now on, have a failure distribution dictated by an APH, we can just replace the gates with the standard operations, and its subsequent LCS compression: AND gates become maximum operations (`Max`) and are compressed with Lemma 5, OR gates become minimum operations (`Min`), subject to compression with Lemma 7, and SEQ gates become convolution operations (`Con`).

Other DFT gates either can be rewritten as other gates (e. g. cold spares as SEQ, hot spares as AND), cannot occur in the tree structure (e. g. FDEP gates), or are not allowed because of their failure semantics are not compatible with the standard operations (e. g. warm spares).

4.2 Dynamic Fault Trees to APH

The translation to APH representations on non-tree-shaped DFT is not straightforward, and it requires to explore the induced state space to set up the representation. The reason behind this is that some dynamic gates, as well as nodes with more than one parent, cannot be expressed by standard operations. The procedure we have developed has some resemblance to the ones from [48], or [11].

However, our ambitions are tempered by the fact that a full mapping from DFTs to APH distributions (and thus absorbing CTMCs) is impossible [9]. This is because in full generality, the semantics of DFTs is inherently nondeterministic, breaking our target model class CTMCs. The phenomenon is by now well-understood, it is related to interference of certain dynamic gates. One solution is to pre-suppose a certain order of events, for instance one that echoes the order in which nodes are introduced (in the input file of the analysis tool). This choice is supported (sometimes as options) by software tools for FT analysis, among them `StormDFT` and SAFEST. It is also the choice we take to make sure an APH can be unambiguously associated with any DFT. Other choices are equally well possible, the approach is not bound to this choice.

We proceed as follows. In each state, transitions are given **only** by the phases of the component nodes. For every component, we represent its *status*, as *Active(phase)* (A_{ph}), *Passive(phase)* (P_{ph}) or *Failed* (F). The *phase* encodes the current phase for the component. Then, a state describes a specific configuration of all component nodes in the tree, together with the *CuC(id)*, which denotes the *Currently in Use Child* node which is exclusively in use for every spare gate in the tree. The phase of each component lets us compute the initial distribution of the state. Other gates' status are not encoded in the *state*.

After identifying a new successor state, such state is *sanitized*, and inserted to the set of states. The sanitization consists in activating triggered children by FDEP gates, and recursively verifying if the DFT is in a failed state. If it is, we conclude that the state is the *Absorbing* state, otherwise, we search for new *usable* children for spare gates, and activating components from SEQ gates. At the initial state all components start active, then SEQ and CSP gates restrict non-primary children by marking them as passive. Invalid failure configurations cannot occur since only active nodes generate new transitions. Finally, states are ordered so the CTMC generator matrix is upper triangular, with component phases or failures increasing from left to right. Lastly, the initial probabilities of each state are computed based on the initial probabilities of each state's active components.

Example 4. Consider the DFT from Fig. 2 (right). Additionally, assume the components A and C to be distributed according to $\mathsf{exp}(\lambda)$, and $\mathsf{exp}(\mu)$ respectively, while B is distributed according to $(\vec{e}_1, Bi(\eta_1, \eta_2))$. The constructed CTMC is shown in Fig. 7. Note how some state configurations (like A_0,F,A_0) are not represented, because they already encode the failure of the TLE, and are merged into the absorbing state. The ordering of the states is induced by ascending along the numbers marked in blue next to each state.

5 Bounding APH Representations

We now turn to a more aggressive compression technique, that is no longer exact, but instead is of approximate nature. Approximate methods are widespread in the analysis of very large systems whenever exact computations are impractical. For instance, for the stochastic models underlying DFTs, *Statistical Model Checking* [2] is a commonly used evaluation approach, sometimes combined with specialized techniques like *Rare Event Simulation* [11].

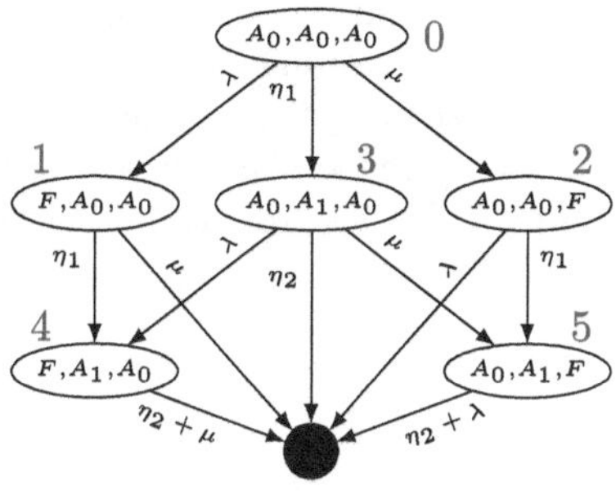

Fig. 7. Constructed CTMC for the DFT from Fig. 2 (right).

We here propose a novel approximate method to obtain sound bounds for APH distributions having a potentially large algebraic degree. It is based on the following four simple observations:

1. Increasing (or decreasing) the exit rate of any state in an OBR increases (resp. decreases) the absorption probability for any time point, and so the resulting distribution will lie above (resp. below) the original.
2. If two distinct exit rates are (very) close to each other in value, then picking the lower (or higher) for both will provide a (tight) lower (resp. upper) bound on the absorption time distribution.
3. If the two distinct exit rates are not on the same path, their unification will lead to a smaller LCS, and hence to a smaller OBR representing the bounding distribution.
4. Two states that are at the same depth (i.e., distance from a given initial state) and have distinct exit rates, cannot be on the same path.

Based on these observations, our method increases (or decreases) certain rates to thereby arrive at upper (resp. lower) bounding distributions. The changes are driven by the intention to induce a more drastic compression through the LCS construction. In the sequel, our bounding notation is inspired by the floating-point rounding modes: Up (upper), Down (lower), and Exact.

5.1 Bounds for APH from Standard Constructions

For APH representations arising from standard operations, we exploit items 1 and 2 from above to tweak one of the input distributions, so that, in combination with the other, it produces redundant phases that will then be compressed by the LCS construction. Because convolutions do not introduce any new paths to absorption in the underlying CTMC, we restrict to developing the technique for the maximum and minimum operations.

Consider distributions $PH(\vec{\alpha}, Bi(\lambda_1, \ldots, \lambda_m))$ and $PH(\vec{\beta}, Bi(\mu_1, \ldots, \mu_n))$. The bounding proceeds by partially constructing the LCS, i.e. using all rates of one input, w.l.o.g $\lambda_1, \ldots, \lambda_n$, but only the first-rate μ_1, from the other representation, and constructing a set of target values $\mathcal{T} = \{\lambda_i + \mu_1 \mid 1 \leq i \leq n\}$. Then, all remaining $\mu_j, 2 \leq j \leq m$ are replaced according to Equation (1), which aims at keeping the original value if, when combined with any other rate λ_i, it generates a rate that already exists in $\mathcal{T}$, or replaces it by another value that does satisfy the property, if existing.

$$R(\mu_j) = \begin{cases} \mu_j, & \text{if } \exists \lambda_i \colon \lambda_i + \mu_j \in \mathcal{T}, 1 \leq i \leq n \\ \mu'_j, & \text{if } \mu'_j \lessgtr \mu_j \wedge R(\mu_{j-1}) < \mu'_j \wedge \exists \lambda_i \colon \lambda_i + \mu'_j \in \mathcal{T}, 1 \leq i \leq n \\ \mu_j, & \text{otherwise} \end{cases} \quad (1)$$

where $\lessgtr$ is either a $<$ or $>$, depending on the rounding method.

5.2 Bounds for Arbitrary CTMCs

Similarly, we can conceive a method to bound the underlying APH of an arbitrary acyclic CTMC, by modifying the exit rates of certain states in a controlled way w.r.t. the LCS of a representation. The procedure is rooted in a topological search, based on the topological order of the states induced by the CTMC's acyclicity. It first computes the *depth* of each state, which is the maximum distance of a state from the initial one, then, each state attempts to update the depth of its direct successors. Termination is guaranteed because the CTMC is acyclic. Later, we use items 3 and 4 introduced above, to normalize the exit rate for each of the depths present in the CTMC. We take the maximum (resp. the minimum) exit rate of each depth level, and use it to replace the other rates present at the same depth, to find an under-approximation (resp. an over-approximation).

Example 5. We return to Fig. 6. Let's consider nodes s_4 and s_5. Both are at the same distance from the initial state s_1, with exit rates 2 and 4 respectively. Using the rounding Up, we increase the exit rate of s_4 to 4, and each of its transitions (to s_6 and to the absorbing state) now have rate 2. Analogously, using Down, the exit rate of s_5 decreases to 2, and its two transitions now have rate 1 each. In both cases, the obtained LCS is one state smaller than the exact LCS, obtained by the method from Sect. 3.3.

The method leverages the construction of the LCS to compress the representation to just it's maximum depth, but larger rate discrepancies widen the bounds. It can be easily extended with constraints like relative tolerance, making this an interesting area for future research.

6 Implementation and Results

We present an empirical evaluation to assess the effectiveness of APH compression on a publicly available DFT benchmark set, comparing our results against the uncompressed *ground truth*. As ground truth, we could have taken what our approach produces when turning off any compression. Instead, we here compare with `StormDFT` [25], a state-of-the-art tool for the analysis of systems involving probabilistic phenomena. Our focus is on the size of the resulting representation, a dimension in which `StormDFT` is more optimized than our approach when compression is turned off. Our intention is not to demonstrate superiority in computing results, but to study how far the compression, as an orthogonal concept, can enhance the ground truth/state-of-the-art, as well as to analyse the versatility of the proposed bounding techniques. For this, we afterwards present a case study that provides deeper insights into the size savings achievable through bounding.

Technical Details. We implemented APH support from scratch in the *Rust* [34] programming language. The prototype, `APH`, is available in GitHub under MIT license. We implemented the reduction algorithm for APH [39], the SPA algorithm, the methods to compute the LCS introduced in Sect. 3, the translation of DFT to APH from Sect. 4 and the bounding methods depicted in Sect. 5. Our implementation takes as input either a DFT in *GALILEO* format [19,45], or a CTMC in *explicit transition file* (`.tra`) [32]. `APH` can also output `.tra` files, or, using the *Prism Language* [33], export models as OBR (`.ma`), or as a *Coxian* [15] representation (`.ctmc`). Moreover, `APH` computes the first moment of the distributions (which is the MTTF metric for FTs). Further analysis, e.g. sampling the CDF at specific time points, can be done with model checkers, like Storm or Prism, on the exported model. Notably, the more samples are needed, the more worthwhile it is to have the model compressed upfront.

Besides, we have adapted the linear time algorithm for modularization [22], a common processing step in FT analysis [9,23,48]. It allows us to solve independent modules and replace them by new *component* nodes, the failure time of which is dictated by the compressed APH obtained from translating the module. Often enough, FTs have a top gate where its children are all independent, then the modularization approach permits us to combine both translations for tFTs and DFTs, first translating the dynamic submodules with state space exploration, and then transforming the top tree using the standard operations.

Results on DFTs. First, we focus on the central compression method (no bounding takes place, the model transformations are exact) and compare against the model checker STORM [25] version 1.11.1. `StormDFT` is a module of STORM that supports DFTs in GALILEO format [45]. It can process both static FTs by employing Binary Decision Diagrams and Minimal Cut Sets [8], as well as DFTs by means of Model Checking [48] or Petri Nets [30]. For the DFT calculations, we executed `StormDFT` with the flag `--firstdep`, that resolves the non-determinism of the models in the same way as `APH` does.

Most benchmarking models are hosted at the University of Twente, as part of FFORT Collection [42]. We also included models used to evaluate APH-based reliability, like *Attack Trees* [4], and the static FT models *processors and memories* (ft3p2m) [21,38]. Overall, we evaluated on 53 diverse FTs, where each of the basic events was assumed failing according to an APH distribution. Experiments ran on an AMD Ryzen 9 5950X (16-Cores) with 125 GB RAM and a time limit set to 150 min. Both tools agreed on the MTTF metric, except in the case of DFTs with Erlang-based events under SEQ gates (e.g. FTPP), due to divergent semantics – our approach strictly sequentializes the two events, while `StormDFT` prevents only the last phase of the second event, until the first one has occurred. All experiments were conducted with modularisation enabled for both tools.

Figure 8 presents, for each model, the final number of transitions (left axis) obtained from each tool, and the additional time needed by `APH` compression relative to the state-of-the-art (right axis), where a positive value means is an increase in runtime, and a negative value, a decrease.

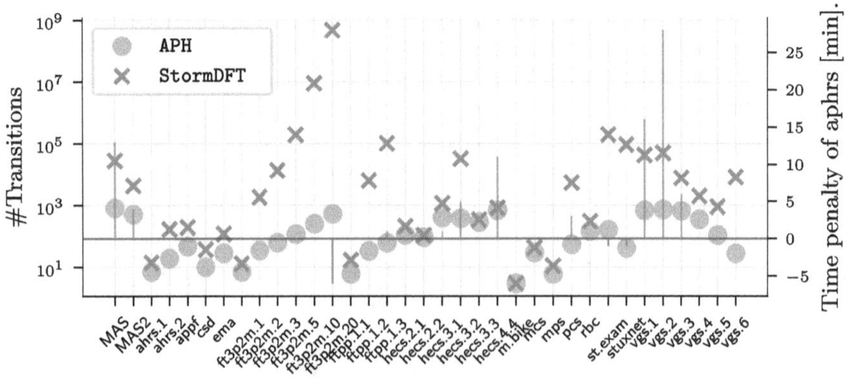

Fig. 8. Reported number of transitions in state-of-the-art compared to `APH`, and additional time taken by `APH` (in minutes).

If a tool was unable to solve a model due to resource limits, incompatibility, or missing outputs, the entry was skipped. Overall, APH compression yields considerably smaller representations than StormDFT, sometimes by several orders of magnitude, at the cost of generally increased runtimes, by a factor of 6 on average. Across all benchmarks, APH is able to construct and compress 38 models within the time limit, while StormDFT is able to arrive at 47 DFTs. We attribute the result to the extensive use, excellent support, and significant contributions that STORM has developed to become a state-of-the-art model checker. The inclusion of our compression approach will evidently further enhance this.

For models successfully compressed by APH, the majority of the runtime is spent on the SPA procedure. This is largely due to the high dimensionality of the CTMCs explored and, consequently, of the generator matrices that appear on the left-hand side of the SPA equations. The remaining procedures – modularization step, LCS computation, and the final reduction – are not negligible in terms of runtime either; however, they are not the limiting factors that would render a model incompressible via the LCS approach.

Several techniques have been developed to improve the efficiency of DFT exploration, such as *Don't Care Propagation* [48] or *Graph-Rewriting* [30]. Using those as a preprocessing step can lead to more compact CTMCs, that in turn enable faster compression via the LCS, as discussed in Sect. 3.1. On the other hand, models that expand into excessively large intermediate CTMCs before any compression can be applied, are not successfully compressed within the time limit. The largest module we are able to handle has 1266432 transitions prior to compression, and 99 after, but due to the modularization we are able to handle state spaces that normally would yield 10^8 transitions, like the ft3p2m.10 instance, that we make fit into less than a hundred. If the parameter 10 (the number of Erlang phases) were further increased, the models would exceed the supported size limits unless compressed, so for those models APH is the enabler for any kind of subsequent analysis.

Case Study: Mission Avionics System. We now analyse the DFT of the *Mission Avionics System* (MAS) [20], a highly redundant, safety-critical system, by employing exact and bounded compressions, detailed in Sects. 4 and 5.

Results are summarized in Table 1, and corresponding CDFs of a selection of the respective failure distributions are shown in Fig. 9. The table reports, for each bounding method, the number of states, the MTTF, and the time spent in the SPA algorithm and overall. The Up and Down models employ the bounding techniques introduced in Sect. 5, while Depth Up and Depth Down rely solely on depth-based bounds for both DFTs and tFTs. An asterisk indicates that bounding is applied only to the TLE. Using only depth-based bounds yields uniform model sizes at the expense of looser bounds, visible in the MTTFs. In contrast, combining both techniques requires a comparable amount of time while producing tighter bounds. Applying bounds exclusively to the TLE consistently moves the MTTF metrics and CDFs closer to the true values, and it reduces computation time to approximately one fifth of that required to compress the exact distribution. Yet, it relies on an exact representation of the distribution failure of each submodule in the tree.

Table 1. MAS compression results.

Experiment	Metrics		Times (s)	
	States	MTTF	SPA	Total
Exact	792	1754.101	232.237	323.847
Up	76	1513.641	17.130	18.941
Down	200	1810.699	17.970	20.830
Depth Up	21	1448.904	15.261	16.721
Depth Down	21	2264.468	15.613	17.244
Up*	144	1738.483	71.197	72.638
Down*	143	1798.326	69.340	70.801
Depth Up*	45	1652.529	70.128	71.603
Depth Down*	45	2065.912	70.683	72.158

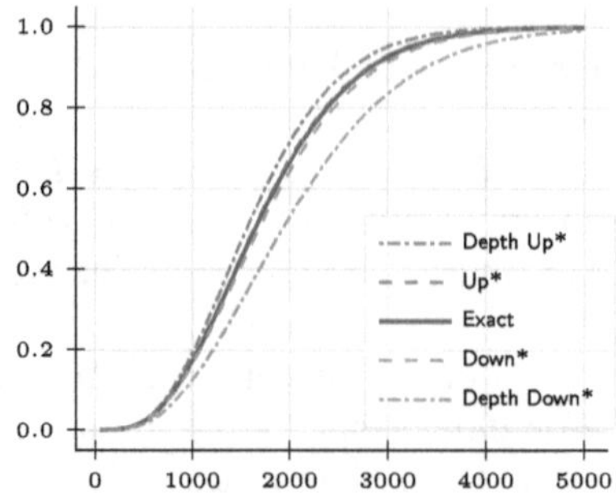

Fig. 9. CDFs of a selection of compressed MAS models. Each plot was obtained by sampling the exported model with STORM.

Looking specifically at the CDFs, the bounds computed by the specific method for the standard operations produce tight bounds around the exact representation along the entire curve. In contrast, the depth-based bounds gets wider, reaching a maximum distance between the bounds of 0.15% at 1850 time units, while the specific bounding has a maximum margin of 3%. Hence, the bounding method can rapidly yield meaningful reliability bounds of a complex system, provided the choice of the bounding method is informed by the underlying DFT structure.

7 Conclusion

The paper has discussed the use of *longest core series* for the compression of acyclic phase-type distributions. In cases where the obtained representation is non-minimal, the approach still enhances existing reduction techniques – such as the one from Pulungan *et al.* [39] – through the early elimination of redundant states. We have demonstrated that considerable savings in state-space sizes can be achieved for Dynamic Fault Trees. The compression concept is orthogonal to other techniques to mitigate excessive growth of Markov chain models underlying DFT modelling, making it an interesting add-on for existing tools.

In our implementation, most of the computation time is spent on the SPA algorithm, which needs to solve an equation system in the size of the state space, and this puts some natural limits in non-modular contexts. Usually however, DFT models exhibit modularity. We also showed how further drastic compression is possible if accepting to work with bounding distributions of the distribution at hand. Nonetheless, orthogonal improvements in state-space exploration – like *partial order reduction*, or *Don't care propagation* – could significantly lighten the required computation time.

As future work, we aim to move beyond our model limitations and make APH support all behaviours of DFTs with APHs. This likely implies the integration of multi-exit phase-type distributions [51] to represent *FailSafe* states.

Acknowledgments. This work has been partially funded by DFG grant 389792660 as part of TRR 248 – CPEC, by the EU under grant agreement No 101008233 – MISSION, and as part of the INTERREG North Sea project STORM_SAFE.

References

1. Aalen, O.O.: Phase type distributions in survival analysis. Scand. J. Stat. **22**(4), 447–463 (1995)
2. Agha, G., Palmskog, K.: A Survey of Statistical Model Checking. ACM Trans. Model. Comput. Simul. **28**(1), 1–39 (2018). https://doi.org/10.1145/3158668
3. Alkaff, A., Qomarudin, M.N.: Modeling and analysis of system reliability using phase-type distribution closure properties. Appl. Stoch. Model. Bus. Ind. **36**(4), 548–569 (2020)
4. Arnold, F., Hermanns, H., Pulungan, R., Stoelinga, M.: Time-Dependent Analysis of Attacks. In: Abadi, M., Kremer, S. (eds.) POST 2014. LNCS, vol. 8414, pp. 285–305. Springer, Heidelberg (2014). https://doi.org/10.1007/978-3-642-54792-8_16
5. Asmussen, S.: Phase-type representations in random walk and queueing problems. Ann. Probab. **20**(2), 772–789 (1992)
6. Asmussen, S., Koole, G.: Marked point processes as limits of Markovian arrival streams. J. Appl. Probab. **30**(2), 365–372 (1993)
7. Baek, S., Heo, G.: Application of dynamic fault tree analysis to prioritize electric power systems in nuclear power plants. Energies **14**(14), 4119 (2021). https://doi.org/10.3390/en14144119
8. Basgöze, D., Volk, M., Katoen, J., Khan, S., Stoelinga, M.: BDDs strike back - efficient analysis of static and dynamic fault trees. In: NFM 2022 (2022). https://doi.org/10.1007/978-3-031-06773-0_38
9. Boudali, H., Crouzen, P., Stoelinga, M.: A rigorous, compositional, and extensible framework for dynamic fault tree analysis. IEEE Trans. Dependable Secur. Comput. **7**(2), 128–143 (2010). https://doi.org/10.1109/TDSC.2009.45
10. Buchholz, P., Kriege, J., Felko, I.: Phase-Type Distributions. In: Input Modeling with Phase-Type Distributions and Markov Models. SM, pp. 5–28. Springer, Cham (2014). https://doi.org/10.1007/978-3-319-06674-5_2
11. Budde, C.E., D'Argenio, P.R., Monti, R.E., Stoelinga, M.: Analysis of non-Markovian repairable fault trees through rare event simulation. Int. J. Softw. Tools Technol. Transfer **24**(5), 821–841 (2022). https://doi.org/10.1007/s10009-022-00675-x
12. CCPS: Guidelines for Hazard Evaluation Procedures, 3rd Edition. Wiley-AIChE (2008), Center for Chemical Process Safety
13. Chakravarthy, S., Ravi, K.: A stochastic model for a computer communication network node with phase type timeout periods. Numer. Sol. Markov Chains 261–285. CRC Press (2021)
14. Clifton, E.: Fault tree analysis-a history. In: Proceedings of the 17th International Systems Safety Conference, pp. 1–9 (1999)

15. Cox, D.R.: A use of complex probabilities in the theory of stochastic processes. Math. Proc. Cambridge Philosop. Soc. **51**(2), 313–319 (1955). https://doi.org/10.1017/S0305004100030231
16. Crouzen, P., Lang, F.: Smart Reduction. In: Giannakopoulou, D., Orejas, F. (eds.) FASE 2011. LNCS, vol. 6603, pp. 111–126. Springer, Heidelberg (2011). https://doi.org/10.1007/978-3-642-19811-3_9
17. Cumani, A.: On the canonical representation of homogeneous markov processes modelling failure-time distributions. Microelectron. Reliab. **22**(3), 583–602 (1982). https://doi.org/10.1016/0026-2714(82)90033-6
18. Dugan, J., Bavuso, S., Boyd, M.: Fault trees and sequence dependencies. In: RAMS 1990, pp. 286–293 (1990). https://doi.org/10.1109/ARMS.1990.67971
19. Dugan, J.B.: Galileo: A Tool for Dynamic Fault Tree Analysis. In: Haverkort, B.R., Bohnenkamp, H.C., Smith, C.U. (eds.) TOOLS 2000. LNCS, vol. 1786, pp. 328–331. Springer, Heidelberg (2000). https://doi.org/10.1007/3-540-46429-8_24
20. Dugan, J.B., Assaf, T.: Dynamic fault tree analysis of a reconfigurable software system (2001). https://api.semanticscholar.org/CorpusID:15626675
21. Dugan, J.B., Doyle, S.A.: New results in fault-tree analysis. In: Reliability and maintainability symposium, pp. 568–573 (1996)
22. Dutuit, Y., Rauzy, A.: A linear-time algorithm to find modules of fault trees. IEEE Trans. Reliab. **45**(3), 422–425 (1996). https://doi.org/10.1109/24.537011
23. Gulati, R., Dugan, J.: A modular approach for analyzing static and dynamic fault trees. In: Annual Reliability and Maintainability Symposium, pp. 57–63. IEEE, Philadelphia, PA, USA (1997). https://doi.org/10.1109/RAMS.1997.571665
24. He, Q.M., Zhang, H.: Spectral polynomial algorithms for computing bi-diagonal representations for phase type distributions and matrix-exponential distributions. Stoch. Model. **22**(2), 289–317 (2006). https://doi.org/10.1080/15326340600649045
25. Hensel, C., Junges, S., Katoen, J.P., Quatmann, T., Volk, M.: The probabilistic model checker Storm. Int. J. Softw. Tools Technol. Transfer **24**(4), 589–610 (2022). https://doi.org/10.1007/s10009-021-00633-z
26. Horváth, A., Telek, M.: PhFit: A General Phase-Type Fitting Tool. In: Field, T., Harrison, P.G., Bradley, J., Harder, U. (eds.) TOOLS 2002. LNCS, vol. 2324, pp. 82–91. Springer, Heidelberg (2002). https://doi.org/10.1007/3-540-46029-2_5
27. Horváth, G., Telek, M.: BuTools 2: a rich toolbox for markovian performance evaluation. In: Puliafito, A., Trivedi, K.S., Tuffin, B., Scarpa, M., Machida, F., Alonso, J. (eds.) 10th EAI International Conference on Performance Evaluation Methodologies and Tools, VALUETOOLS 2016, Taormina, Italy, 25th–28th Oct 2016. ACM (2016). https://doi.org/10.4108/EAI.25-10-2016.2266400
28. Johnson, M., Taaffe, M.: The denseness of phase distributions. Purdue Sch. Ind. Eng. Res. Memoranda **88**, 20 (1998)
29. Junges, S., Guck, D., Katoen, J.-P., Rensink, A., Stoelinga, M.: Fault trees on a diet: automated reduction by graph rewriting. Formal Aspects Comput. **29**(4), 651–703 (2017). https://doi.org/10.1007/s00165-016-0412-0
30. Junges, S., Katoen, J.-P., Stoelinga, M., Volk, M.: One Net Fits All. In: Khomenko, V., Roux, O.H. (eds.) PETRI NETS 2018. LNCS, vol. 10877, pp. 272–293. Springer, Cham (2018). https://doi.org/10.1007/978-3-319-91268-4_14
31. Kahn, A.B.: Topological sorting of large networks. Commun. ACM **5**(11), 558–562 (1962). https://doi.org/10.1145/368996.369025
32. Katoen, J.P., Khattri, M., Zapreevt, I.: A markov reward model checker. In: Second International Conference on the Quantitative Evaluation of Systems (QEST'05), pp. 243–244 (2005). https://doi.org/10.1109/QEST.2005.2

33. Kwiatkowska, M., Norman, G., Parker, D.: PRISM 4.0: Verification of Probabilistic Real-Time Systems. In: Gopalakrishnan, G., Qadeer, S. (eds.) CAV 2011. LNCS, vol. 6806, pp. 585–591. Springer, Heidelberg (2011). https://doi.org/10.1007/978-3-642-22110-1_47
34. Matsakis, N.D., Klock, I.I., F.S.: The Rust language. ACM SIGAda Ada Lett. **34**, 103–104. ACM (2014)
35. Neuts, M.F.: Matrix-geometric solutions in stochastic models - an algorithmic approach, Dover Publications (1994)
36. O'Cinneide, C.A.: On non-uniqueness of representations of phase-type distributions. Commun. Stat. Stochastic Models **5**(2), 247–259 (2007). https://doi.org/10.1080/15326348908807108
37. Payne, K., Marshall, A.H., Cairns, K.J.: Investigating the efficiency of fitting coxian phase-type distributions to health care data. IMA J. Manag. Math. **23**(2), 133–145 (2012). https://doi.org/10.1093/imaman/dpr008
38. Pulungan, R.: Reduction of acyclic phase-type representations, Ph.D. thesis. Germany (2009)
39. Pulungan, R., Hermanns, H.: Effective Minimization of Acyclic Phase-Type Representations. In: Al-Begain, K., Heindl, A., Telek, M. (eds.) ASMTA 2008. LNCS, vol. 5055, pp. 128–143. Springer, Heidelberg (2008). https://doi.org/10.1007/978-3-540-68982-9_10
40. Pulungan, R., Hermanns, H.: The minimal representation of the maximum of erlang distributions. In: Bause, F., Buchholz, P. (eds.) Proceedings 14th GI/ITG Conference on Measurement, Modelling and Evaluation of Computer and Communication Systems (MMB 2008), March 31 - April 2, 2008, Dortmund, Germany, pp. 207–222. VDE Verlag (2008)
41. Reinecke, P., Krauss, T., Wolter, K.: Hyperstar: Phase-type fitting made easy. In: Ninth International Conference on Quantitative Evaluation of Systems, QEST 2012, London, United Kingdom, September 17-20, 2012, pp. 201–202. IEEE Computer Society (2012). https://doi.org/10.1109/QEST.2012.29
42. Ruijters, E., et al.: Ffort: A benchmark suite for fault tree analysis. In: ESREL 2019. pp. 878–885 (2019). https://doi.org/10.3850/978-981-11-2724-3_0641-cd
43. Ruijters, E., Stoelinga, M.: Fault tree analysis: A survey of the state-of-the-art in modeling, analysis and tools. Comput. Sci. Rev. **15**, 29–62 (2015). https://doi.org/10.1016/J.COSREV.2015.03.001
44. Stamatelatos, M., Vesely, W., Dugan, J., Fragola, J., III, J.M., Railsback, J.: Fault Tree Handbook with Aerospace Applications. NASA (2002)
45. Sullivan, K., Dugan, J.B.: Galileo user's manual & design overview. https://www.cse.msu.edu/~cse870/Materials/FaultTolerant/manual-galileo.htm, v2.11 - University of Virginia
46. Thümmler, A., Buchholz, P., Telek, M.: A novel approach for phase-type fitting with the EM algorithm. IEEE Trans. Dependable Secur. Comput. **3**(3), 245–258 (2006). https://doi.org/10.1109/TDSC.2006.27
47. Vesely, W.E., Goldberg, F.F., Roberts, N.H., Haasl, D.F.: Fault Tree Handbook. U.S, Nuclear Regulatory Commission (1981)
48. Volk, M., Junges, S., Katoen, J.P.: Fast Dynamic Fault Tree Analysis by Model Checking Techniques. IEEE Trans. Ind. Inform. **14**(1), 370–379 (2018). https://doi.org/10.1109/TII.2017.2710316
49. Volk, M., Sher, F., Katoen, J.P., Stoelinga, M.: SAFEST: Fault Tree Analysis Via Probabilistic Model Checking. In: 2024 Annual Reliability and Maintainability Symposium (RAMS), pp. 1–7. IEEE, Albuquerque, NM, USA (2024). https://doi.org/10.1109/RAMS51492.2024.10457719

50. Wang, G., Hu, L., Zhang, T., Wang, Y.: Reliability modeling for a repairable (k1, k2)-out-of-n: G system with phase-type vacation time. Appl. Math. Model. **91**, 311–321 (2021)
51. Zimmer, D., Hermanns, H.: There are several ways out of the hospital. and into. In: 18th EAI International Conference, VALUETOOLS 2025, Glasgow, Scotland, December 10–11, 2025, Proceedings. Springer, Glasgow, Scotland (2025), (Forthcoming)

Weighing Timed Regular Languages: The Final Step

Eugene Asarin[1(✉)], Aldric Degorre[1], Cătălin Dima[2], and Bernardo Jacobo Inclán[1,3]

[1] Université Paris Cité, CNRS, IRIF, Paris, France
{asarin,adegorre}@irif.fr
[2] LACL, Université Paris-Est Créteil, Créteil, France
dima@u-pec.fr
[3] LABRI, Université de Bordeaux, Talence, France
bernardo.jacobo-inclan@labri.fr

Abstract. The bandwidth of a timed language characterizes the quantity of information per time unit (with a finite observation precision ε). The asymptotic behavior of the bandwidth as $\varepsilon \to 0$ classifies timed regular languages in three classes: meager, normal, and obese. Normal timed automata have a bounded frequency of events and some non-punctual transitions, and, up to now, were the only class of timed automata for which no algorithm was available for computing their bandwidth. In this article, we compute the bandwidth of any such automaton in the form $\approx \alpha \log 1/\varepsilon$. Our approach reduces this problem to computing the best reward-to-cost ratio in a weighted finite graph constructed from the given timed automaton.

Keywords: timed automata · information theory · bandwidth · entropy

1 Introduction

Studies of entropy (growth rate, information content) of regular and other formal languages raising to [13,20], constitute nowadays a textbook material [10,17], fundamental to symbolic dynamics, theory of codes, and practice of data coding and compression. In the long run, we are extending those studies to timed regular languages, with theoretical and practical ambitions.

In earlier works (see [3], and also [2] for applications to information coding), we have defined and explored the notion of entropy of a timed language. It characterizes the quantity of information *per event* contained in its words observed with some precision ε. In [6–8,15] we take a different, practically more relevant perspective. Given a timed regular language (represented by a timed automaton), observed with a precision ε, we want to know its *bandwidth*, as introduced

M. Chen and M. Tschaikowski (Eds.): QEST+FORMATS 2026, LNCS 16913, pp. 59–76, 2027.
https://doi.org/10.1007/978-3-032-35298-9_4

in [15]—quantity of information in the words of the language measured in bits *per time unit*.

In [6], we have identified three classes of timed regular languages and proved that they constitute an exhaustive classification of timed regular languages. They are illustrated by automata on Fig. 1:

meager with bandwidth $O(1)$, which allows only encoding of information in discrete choices, and not in timing;
normal[1] with bandwidth $\Theta(\log(1/\varepsilon))$, where time delays (in quantity proportional to the duration) can be used to encode additional information;
obese with bandwidth $\Theta(1/\varepsilon)$, events can happen with a very high frequency, and thus a huge amount of information can be encoded in timing.

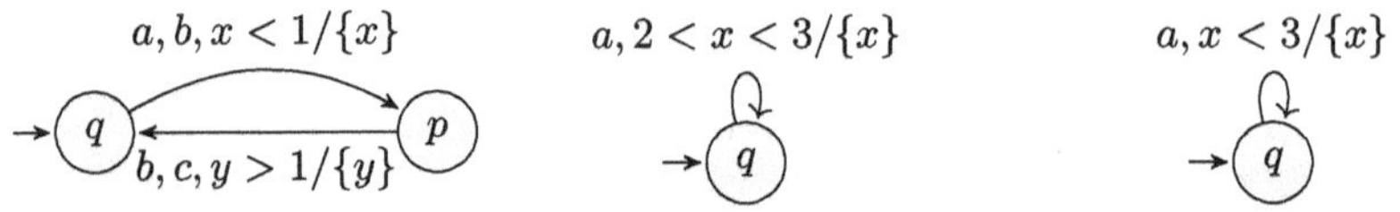

(a) meager: 2 (b) normal: $\frac{1}{2}\log\frac{1}{\varepsilon}$ (c) obese: $\frac{1}{\varepsilon}$

Fig. 1. Timed automata and their approximate bandwidths

As shown in [6], the three classes can be characterized by structural properties (presence or absence of certain patterns), but deciding to which class a TA belongs is PSPACE-complete.

The next research aim is a more quantitative analysis: for each class, find the main asymptotic term of the bandwidth: in the form[2] α, $\alpha\log(1/\varepsilon)$, and α/ε for meager, normal, and obese automata, respectively.

In [7], we solved the problem for (deterministic) meager automata and computed α using a new barycentric abstraction to transform a timed automaton into a (finite-state) simply-timed one, for which α can be characterized in terms of matrix algebra. In [8], we solved it for obese automata by reduction to the maximum reward-per-time in the abstracted timed graph.

In this paper, we solve the problem for the last remaining case of a normal automaton. At first glance, the bandwidth of such a TA is $\alpha\log(1/\varepsilon)$ with α the maximal number of free (non-punctual) transitions per time unit. However, as we show in this work, the freedom of a transition is a subtle notion, which depends on the context (other transitions in the automaton) and the strategic choices (in which directions to move and in which not to move). We formalize these ideas in terms of a refined bi-weighted graph associated with the automaton. Paths in this graph represent only certain runs in the timed automata, but are sufficient to generate its full bandwidth. Finally, α is characterized as the maximal reward-to-cost ratio in the refined graph.

[1] as opposed to meager and obese.
[2] all the logarithms in this article are base 2.

Finding (and proving) matching lower and upper bounds for the bandwidth was a challenging task. To establish the former, we considered only "regular" runs of the automaton, which iterate many times one optimal cycle, and take every now and then another, resetting cycle to avoid a sort of saturation. The proof of the upper bound is much more involved, since it has to deal with all the possible behaviours of the automaton. It makes use of the algebraic version of Ramsey theory [21], advanced reachability theory for timed automata [19], and our previous findings on the bandwidth [6].

The paper is structured as follows. In Sect. 2 we recall some useful notions. In Sect. 3 we define the bandwidth and state the problem of its computation. In Sect. 4 we solve the problem for normal timed languages. We put together the results from [6–8] and this article and discuss the perspectives in Sect. 5. Due to the lack of space, here we present the algorithm, some intuitive explanations, and key lemmas; proof details are given in the arXiv version of this article [9].

2 Background

2.1 Timed Words, Languages and Automata

Timed automata have been introduced in [1] for modeling and verification of real-time systems. We will use the main definitions in the following form:

Definition 1. *Given Σ, a finite alphabet of discrete events, a* timed word *over Σ is an element from $(\Sigma \times \mathbb{R}_+)^*$ of the form $w = (a_1, t_1) \dots (a_n, t_n)$, with $0 \leq t_1 \leq t_2 \cdots \leq t_n$. We denote $\mathrm{dur}(w) = t_n$. A* timed language *over Σ is a set of timed words over the same alphabet.*

For a set of variables X, let G_X be the set of finite conjunctions of constraints of the form $x \sim b$ and $x \sim y + b$ with $x, y \in X$, $\sim \in \{<, \leq, >, \geq\}$ and $b \in \mathbb{N}$.

Definition 2. *A* timed automaton *is a tuple $(Q, X, \Sigma, \Delta, S, I, F)$, where*

- *Q is the finite set of discrete locations;*
- *X is the finite set of clocks;*
- *Σ is a finite alphabet;*
- *$S, I, F : Q \to G_X$ define respectively the starting[3], initial, and final clock values for each location;*
- *$\Delta \subseteq Q \times Q \times \Sigma \times G_X \times 2^X$ is the transition relation, whose elements are called edges. Given an edge $(q, q', a, \mathfrak{g}, \mathfrak{r})$, we call q its origin, q' target, a label, $\mathfrak{g}$ guard and $\mathfrak{r}$ reset.*

A TA is *bounded* if there is a constant M such that all the starting conditions $S(q)$ and the guards of all the edges imply that clock values are smaller than M.

For $\mathbf{x} \in \mathbb{R}^n$ and $d \in \mathbb{R}$ we write $\mathbf{x} + d$ as a shortcut for $(x_1 + d, \dots, x_n + d)$.

The semantics of a timed automaton is given by a *timed transition system* whose states are tuples $(q, \mathbf{x})$ composed of a location $q \in Q$ and a clock valuation

[3] this feature was not present in the original definition [1].

(vector) $\mathbf{x} \in [0, \infty)^X$. Each edge $\delta = (q, q', a, \mathfrak{g}, \mathfrak{r}) \in \Delta$ generates many timed transitions, which are tuples $(q, \mathbf{x}) \xrightarrow[d]{\delta} (q', \mathbf{x}')$ where $\mathbf{x} \models S(q)$, $\mathbf{x} + d \models \mathfrak{g}$ and $x'_c = 0$ whenever $c \in \mathfrak{r}$ and $x'_c = x_c + d$ otherwise, provided that $\mathbf{x}' \models S(q')$. We also denote $\mathbf{x}[\mathfrak{r}]$ the operation of resetting the clocks in $\mathfrak{r}$, hence $\mathbf{x}' = (\mathbf{x} + d)[\mathfrak{r}]$.

Paths are sequences of edges that agree on intermediary locations. At the semantic level, they generate *runs* which are sequences of timed transitions that agree on intermediary configurations. An *accepting run* is a run $\rho = (q_0, \mathbf{x}_0) \xrightarrow[d_1]{\delta_1} (q_1, \mathbf{x}_1) \cdots \xrightarrow[d_n]{\delta_n} (q_n, \mathbf{x}_n)$, in which the first state satisifes $\mathbf{x}_0 \models I(q_0)$ and the last state satisfies $\mathbf{x}_n \models F(q_n)$.

For a given path π, its set of runs is $Runs_\pi(\mathbf{x}, \mathbf{y})$. The language of π: $L_\pi(\mathbf{x}, \mathbf{y})$ is the set of timed words associated with runs in $Runs_\pi(\mathbf{x}, \mathbf{y})$, and the reachability relation $Reach_\pi$ is the set of pairs of $(\mathbf{x}, \mathbf{y}) \in \mathbb{R}_+^X \times \mathbb{R}_+^X$ such that $Runs_\pi(\mathbf{x}, \mathbf{y}) \neq \emptyset$. Also, for $T \in \mathbb{R}_+$, $L_{\pi,T}(\mathbf{x}, \mathbf{y})$ denotes the time bounded language $\{w \mid w \in L_\pi(\mathbf{x}, \mathbf{y}) \wedge \mathrm{dur}(w) \leq T\}$. The sets $Runs_\pi$, L_π and $L_{\pi,T}$ denote, respectively, the unions of $Runs_\pi(\mathbf{x}, \mathbf{y})$, $L_\pi(\mathbf{x}, \mathbf{y})$ and $L_{\pi,T}(\mathbf{x}, \mathbf{y})$ for all possible values of $\mathbf{x}$ and $\mathbf{y}$. The language of a timed automaton is the set of timed words associated with some accepting run and is denoted $L(\mathcal{A})$. We will also need *time-restricted languages* generated by some timed automaton: for any nonnegative real $T \in [0, \infty)$, $L_T(\mathcal{A}) = \{w \in L(\mathcal{A}) \mid \mathrm{dur}(w) \leq T\}$.

Furthermore, if $\delta_i = (q_i, q_{i+1}, a_i, \mathfrak{g}_i, \mathfrak{r}_i)$, then the timed word associated with ρ and initial time t_0 is defined as $word(\rho, t_0) \triangleq (a_1, t_0 + d_1)(a_2, t_0 + d_1 + d_2) \dots (a_n, t_0 + \sum_{i \leq n} d_i)$. We do not mention t_0 when it equals 0. The *length* of ρ is $\mathrm{dur}(\rho) = \mathrm{dur}(word(\rho))$.

Regions and orbit graphs. Many algorithms related to timed automata utilize the finite abstraction of the timed transition system called the *region construction* [1] that we briefly recall here. First, for $\xi \in \mathbb{R}$, we denote $\lfloor \xi \rfloor$ its integral part, and $\{\xi\}$ its fractional part. Then two clock vectors $\mathbf{x}$ and $\mathbf{y}$ are *region-equivalent* iff (1) $\lfloor x_c \rfloor = \lfloor y_c \rfloor$ for any clock c; (2) for any clock c we have that $\{x_c\} = 0$ iff $\{y_c\} = 0$; (3) for any two clocks c_1, c_2 we have that $\{x_{c_1}\} \leq \{x_{c_2}\}$ iff $\{y_{c_1}\} \leq \{y_{c_2}\}$. *Regions* are equivalence classes of the region equivalence.

We will use a special form of TA, similar to the classical region automaton [1], but typed again as a TA. We recall the definition [3,6] in the form of [8].

Definition 3. *A* region-split TA *(or* RsTA*) is a TA* $(Q, X, \Sigma, \Delta, S, I, F)$, *which is bounded and such that, for any location* $q \in Q$*: (1)* $S(q)$ *defines a non-empty region, called the* starting region *of* q*; (2) all states in* $\{q\} \times S(q)$ *are reachable from an initial*[4] *and co-reachable to a final one; (3) for any edge* $(q, q', a, \mathfrak{g}, \mathfrak{r}) \in \Delta$, $(\{S(q) + d \mid d \in \mathbb{R}_+\} \cap \mathfrak{g})[\mathfrak{r}] = S(q')$, *where we utilize the* $(\cdot)[\mathfrak{r}]$ *operator lifted to sets of clock valuations.*

Lemma 1. *Any TA with an upper bound on some clock at every transition can be turned into a bounded region-split TA accepting the same language. This transformation uses the same clocks and an exponentially larger set of locations.*

[4] a state (q, x) is initial whenever $x \models I(q)$ and final whenever $x \models F(q)$.

The construction is sketched in [8, full, B3].

Closed semantics. We define the *closed semantics* of an RsTA, in which an edge $\delta = (q, q', a, \mathfrak{g}, \mathfrak{r})$ generates a transition from $(q, \mathbf{x})$ to $(q', \mathbf{x}')$ with duration $d > 0$, when $\mathbf{x} \models \bar{S}(q)$, $\mathbf{x}' \models \bar{S}(q')$, $\mathbf{x} + d \models \bar{\mathfrak{g}}$ and $\mathbf{x}' = (\mathbf{x} + d)[\mathfrak{r}]$, where $\bar{S}$ and $\bar{\mathfrak{g}}$ are obtained from S and $\mathfrak{g}$ by replacing all "<" by "≤" and all ">" by "≥".

Definitions of $\overline{Runs_\pi}(\mathbf{x}, \mathbf{y})$, and $\overline{L_\pi}(\mathbf{x}, \mathbf{y})$ and variants thereof are obtained similarly to the previous semantics, but based on closed semantics transitions, rather than usual transitions. Additionally, we define the closed reachability relation $\overline{Reach_\pi} = \{(\mathbf{x}, \mathbf{y}) \in \mathbb{R}_+^X \times \mathbb{R}_+^X \mid \overline{Runs_\pi}(\mathbf{x}, \mathbf{y}) \neq \emptyset\}$.

2.2 Orbit Graphs and Simon's Theorem

An important tool for characterizing reachability in timed automata is the orbit graph from [19], which we define using monoid terminology.

Definition 4 (Orbit graph [19] and morphism γ [3]). *Given a non-empty path π from location p to location q in a* RsTA, *its* orbit graph $\gamma(\pi)$ *is the triple $\langle p, G, q \rangle$, where G is the graph whose vertices are the vertices of $\overline{S(p)}$ and of $S(q)$ and such that there is an edge (v, v') whenever $(v, v') \in \overline{Reach(\pi)}$.*

For an empty path, we put $\gamma(\epsilon) = 1$, for a disconnected sequence of edges π by definition $\gamma(\pi) = 0$.

The image of γ (containing orbit graphs that can be composed with each other by interpreting them as relations over vertices, as well as special elements 0 and 1), denoted O, has a natural monoid structure; and $\gamma : \Delta^* \to O$ is a monoid morphism.

We will use a powerful Simon's theorem to deal with such a morphism (see the original article [21] and a modern presentation in [11]).

Theorem 1 (Simon). *Let Γ be a finite alphabet, M a finite monoid, and $\beta : \Gamma^* \to M$ a monoid morphism. Then the sequence of sets*

$$X_0 = \Gamma \cup \{\epsilon\}; \qquad X_{h+1} = X_h \cdot X_h \cup \bigcup_{\substack{e \in M \\ ee=e}} (X_h \cap \beta^{-1}(e))^* \tag{1}$$

stabilizes at some finite level H (in fact $H \leq 3|M|$) with $X_H = \Gamma^*$.

Thus Γ^* is stratified into H levels, on level 0 there are only letters; and every word of level $i + 1$ can be split into two arbitrary words of level i, or into many words of level i all associated with a same idempotent element of M.

2.3 ε-Capacity and ε-Entropy

These two closely-related notions from [16] characterize the quantity of information needed to describe (with precision ε) any element of a given space.

We first recall a couple of definitions:

Definition 5. *Let (X, d) be a metric space, subspace of some Y. A subset $M \subseteq X$ is called* ε-separated *whenever $\forall x \neq y \in M : d(x, y) > \varepsilon$. A set $N \subseteq Y$ is an* ε-net *for X whenever $\forall x \in X \exists y \in N : d(x, y) \leq \varepsilon$.*

For a compact space X and any $\varepsilon > 0$, a finite ε-net must exist within X itself, and cardinalities of ε-separated sets are bounded, which justifies the following definition.

Definition 6 (Kolmogorov & Tikhomirov). *Given a compact metric space $X \subseteq Y$, let $\mathcal{M}_\varepsilon(X)$ be the maximal size of an ε-separated subset of X. Then the* ε-capacity *of X is defined as $\mathcal{C}_\varepsilon(X) = \log \mathcal{M}_\varepsilon(X)$. Let now $\mathcal{N}_\varepsilon(X)$ be the minimal size of an ε-net for X (in Y). Then the* ε-entropy *of X (in Y) is $\mathcal{H}_\varepsilon^Y(X) = \log \mathcal{N}_\varepsilon(X)$.*

As shown in [16], ε-entropy and ε-capacity are related by inequalities

$$\mathcal{H}_\varepsilon^Y(X) \leq \mathcal{C}_\varepsilon(X) \leq \mathcal{H}_{\varepsilon/2}^Y(X). \quad (2)$$

2.4 Optimal Reward-to-Cost Ratio in Finite Graphs

Consider now a finite directed graph where to each edge e is associated its *reward* $r(e)$ and *cost* $c(e)$. For a path π its reward $r(\pi)$ is the sum of rewards of its edges, similarly for costs. A classical optimization problem is maximizing the average reward-to-cost ratio $r(\pi)/c(\pi)$ over long or infinite paths. This ratio can be attained by iterating some optimal simple cycle. We summarize useful facts in a lemma, and refer the reader to [14] for a survey of exact and approximated algorithms for finding the optimal cycle and/or computing the best ratio.

Lemma 2. *Given a finite directed graph with non-negative rewards and costs associated with edges, let α be the maximal reward-to-cost ratio over all simple cycles. Then every cycle π satisfies $r(\pi) \leq \alpha c(\pi)$. Given a graph with $0/1$ rewards and costs, α (as a rational number) is* FNL*-computable*[5].

3 Measuring Timed Regular Languages

3.1 Pseudo-Distance on Timed Words

We are using the proximity measure on timed words introduced in [4].

Definition 7. *The* pseudo-distance *d between timed words is defined as follows: given $w = (a_1, t_1) \ldots (a_n, t_n)$ and $v = (b_1, s_1) \ldots (b_m, s_m)$,*

$$\overrightarrow{d}(w, v) \triangleq \max_{i \in 1..n} \min_{j \in 1..m} \{|t_i - s_j| : a_i = b_j\}; d(w, v) \triangleq \max(\overrightarrow{d}(w, v), \overrightarrow{d}(v, w))$$

(with the standard convention $\min \emptyset = \infty$).

[5] computable by a nondeterministic Turing machine in logarithmic space.

This pseudo-distance allows a meaningful comparison of timed words with a different number of events. It is illustrated on Fig. 2. Intuitively, two words are close to each other when they cannot be distinguished by an observer that reads the discrete letters of the word exactly (they can determine whether or not a letter has occurred), but with some imprecision w.r.t. time. So, when two letters are very close to one another, the observer cannot determine which one came before the other, or not even how many times a letter was repeated within a short interval.

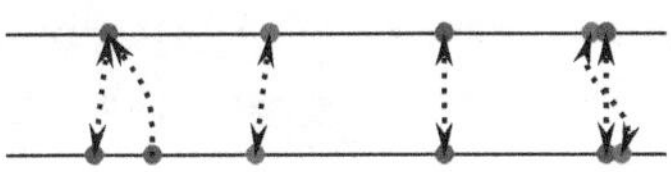

$u = (a, .7), (b, 1.8), (a, 3), (b, 4), (a, 4.1)$

$v = (a, .6), (a, 1), (b, 1.7), (a, 3), (a, 4.1), (b, 4.2)$

Fig. 2. Pseudo-distance between two timed words (dotted arrows represent matching letters). $\overrightarrow{d}(u, v) = 0.2$; $\overrightarrow{d}(v, u) = 0.3$, thus $d(u, v) = 0.3$.

It is possible that d fails to distinguish timed words, that is $d(w_1, w_2) = 0$ but $w_1 \neq w_2$; this is the case for $w_1 = (a, 1)(b, 1)$ and $w_2 = (b, 1)(b, 1)(a, 1)$, that is why d is only a pseudo-distance.

3.2 Bandwidth Definition and Problem Statement

Definition 8. *The* ε-entropic bandwidth *and* ε-capacitive bandwidth *of a timed language L are defined respectively as*

$$\mathcal{BH}_\varepsilon(L) = \limsup_{T\to\infty} \mathcal{H}_\varepsilon(L_T)/T \text{ and } \mathcal{BC}_\varepsilon(L) = \limsup_{T\to\infty} \mathcal{C}_\varepsilon(L_T)/T.$$

We remark that Definitions 5 and 6 were given for metric spaces, but they can be extended verbatim to timed words with our pseudo-distance d. Additionally, ε-nets (and thus ε-entropy) are considered within the universal timed language U. We start by mentioning a few properties. First, it follows from (2) that

$$\mathcal{BH}_\varepsilon(L) \le \mathcal{BC}_\varepsilon(L) \le \mathcal{BH}_{\varepsilon/2}(L). \quad (3)$$

Second, the bandwidth is almost the same for L_π and $\overline{L_\pi}$:

Lemma 3 ([6]). *In an* RsTA, *for any path π, clock vectors x, y, time bound $T > 0$ and precision $\varepsilon > 0$:*

- $\mathcal{H}_\varepsilon(L_{\pi,T}(\mathbf{x}, \mathbf{y})) = \mathcal{H}_\varepsilon(\overline{L_{\pi,T}}(\mathbf{x}, \mathbf{y}))$;
- $\mathcal{C}_\varepsilon(L_{\pi,T}(\mathbf{x}, \mathbf{y})) \le \mathcal{C}_\varepsilon(\overline{L_{\pi,T}}(\mathbf{x}, \mathbf{y}))$;
- *for any* $0 < \varepsilon' < \varepsilon$, $\mathcal{C}_\varepsilon(\overline{L_{\pi,T}}(\mathbf{x}, \mathbf{y})) < \mathcal{C}_{\varepsilon'}(L_{\pi,T}(\mathbf{x}, \mathbf{y}))$.

Third, an upper bound on the bandwidth of timed languages holds, proved in different terms in [4]:

Proposition 1 (Maximal possible bandwidth). *The bandwidth of the universal timed language U_Σ satisfies $\mathcal{BC}_\varepsilon(U_\Sigma) = \Theta(\frac{1}{\varepsilon})$. Hence, for any timed language L, its bandwidth cannot be larger: $\mathcal{BC}_\varepsilon(L) = O(\frac{1}{\varepsilon})$.*

Main Problem. *Given a timed automaton $\mathcal{A}$, compute the ε-bandwidths of its language asymptotically, as $\varepsilon \to 0$.*

As stated in [6] and discussed above, there are three classes of timed regular languages: meager with bandwidth $O(1)$, normal with bandwidth $\Theta(\log \frac{1}{\varepsilon})$, and obese with the maximal possible bandwidth $\Theta(\frac{1}{\varepsilon})$. In the rest of the paper, we compute the bandwidths of **normal** TA. Precisely, given such a TA, we compute α such that both entropic and capacitive bandwidths are $\alpha \log \frac{1}{\varepsilon} + o\left(\log \frac{1}{\varepsilon}\right)$.

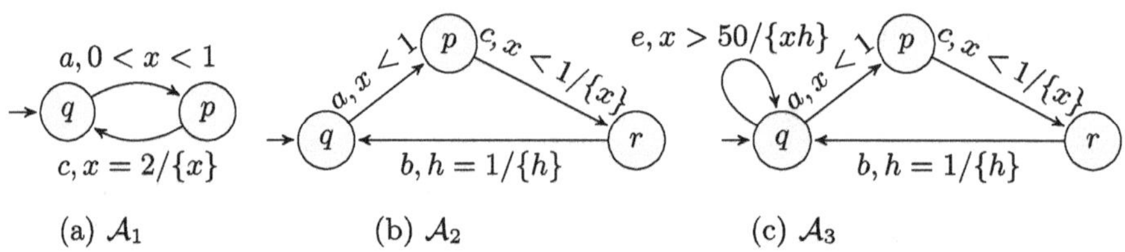

Fig. 3. Three normal automata

4 Bandwidth Computation for Normal Automata

4.1 Informal Discussion and Three Examples

Intuitively, the coefficient α in the bandwidth of a normal automaton corresponds to the number of "degrees of freedom" per time unit. A degree of freedom corresponds to a transition for which the duration can be freely chosen in some interval. A natural idea would be to assign the reward 1 to each "free" transition and compute the maximal reward-per-time ratio as described in [12]. For example, the automaton $\mathcal{A}_1$ on Fig. 3 has one free transition every 2 time units, which yields the bandwidth of $0.5 \log(1/\varepsilon)$.

Unfortunately, it is not always easy to classify transitions into free and non-free ones, as illustrated by two other examples on Fig. 3.

$\mathcal{A}_2$ seems to have 2 free transitions along the cycle of duration 1, but in fact, c is very constrained (it occurs closer and closer to the preceding b), and only a is free, we will show later that the bandwidth is only $\log(1/\varepsilon)$. However, in $\mathcal{A}_3$, thanks to the slow loop labeled by e, taken quite seldom, the two events a, c become free and yield the bandwidth $2\log(1/\varepsilon)$.

To distinguish between free and constrained transitions, we refine the states of the automaton. In a refined state, we commit to moving only in some fixed directions (parallel to some region faces, referred to as *active* ones). A transition is considered free when it can take a variable time while respecting the commitment. Additionally, for each set of active faces in a location, we require the existence of a *resetting cycle* which repositions arbitrarily the clocks in any direction parallel to active faces.

Technically speaking, we will characterize the bandwidth coefficient α of a normal[6] timed automaton $\mathcal{A}$ as the maximal reward-to-cost ratio of a finite graph G obtained by refinement of locations of $\mathcal{A}$. Refined states keep information on "active faces" of timed regions used in the run. Reward corresponds to freedom degrees, and cost roughly corresponds to time. Paths in G represent conservative runs of $\mathcal{A}$ staying at a fixed distance from the active faces, but we will prove that this is sufficient to get the full bandwidth, as non-conservative runs tend to quickly drop some "potential energy" in an unrecoverable way.

4.2 Pre-processing the Timed Automaton

We start with a simple bandwidth-preserving transformation of the TA considered. We refer the reader to [8] for correctness proof and intuitive justification

Construction 1 (Adding heartbeat, [8]**).** *Given a TA, we add a new clock h and a new letter b; next, we add to each transition the constraint $h \leq 1$, and to each location p a self-loop $p \xrightarrow{b,h=1,\{h\}} p$.*

As stated in [8, Lemma 9], this transformation preserves the bandwidth (intuitively, beat events every time unit do not bring any information). At the next step, we transform our TA into the bounded RsTA form (the heartbeat clock serves as a bounded clock required in Lemma 1). Following [8], we call this normal form SFA (*standard-form automata*)[7].

4.3 Refining the Automaton

We recall that geometrically clock regions are simplices in $\mathbb{R}^X$. In a simplex, any non-empty set of vertices C represents a simplex face $\operatorname{conv} C$.

Definition 9. *A* recurrent face list *for location q is a non-empty list $(\ell_1, \ldots, \ell_k)$ with ℓ_i disjoint non-empty sets of vertices of region $S(q)$, satisfying the following property. There exists a cycle σ from q to q, such that its orbit graph $\gamma(\sigma)$ is complete (a clique) on the vertices in each ℓ_i.*

By Puri's theory [19], each clock valuation on the face $\operatorname{conv}(\ell_i)$ is reachable from each other on the same face via a run along the path σ, which should be the same for all faces.

Construction 2. *The* refinement *of an* SFA $\mathcal{A}$ *is a bi-weighted finite graph $G = (Q', \Delta', r, c)$ with*

[6] Our analysis applies to any timed automaton. For a meager one it returns 0, for an obese one ∞.

[7] in [8] we made yet another bandwidth-preserving transformation, not used here.

- Q' *the set of couples* (q, ℓ) *with* $q \in Q$, ℓ *a recurrent face list for* q;
- Δ' *containing an edge* $\delta' = (p, \ell) \to (q, \ell')$ *whenever* $|\ell| = |\ell'|$, *and* Δ *contains* $\delta = p \xrightarrow{a,\mathfrak{g}/\mathfrak{r}} q$, *with orbit graph* $\gamma(\delta)$ *containing for each* $i \leq |\ell|$
  - *an edge from each vertex of* ℓ_i *to some vertex of* ℓ'_i;
  - *an edge to each vertex of* ℓ'_i *from some vertex of* ℓ_i.
- *The reward* $r(\delta')$ *is* 1 *if for some* $i \leq |\ell|$ *exists* $v \in \ell_i$ *and some* $l < u$ *such that for any delay* $d \in (l, u)$ *it holds that* $v + d \models \mathfrak{g}$ *and* $(v + d)[\mathfrak{r}] \subseteq \mathrm{conv}(\ell'_i)$, *otherwise* $r(\delta') = 0$.
- *The cost* $c(\delta')$ *is* 1 *whenever* $\mathfrak{r} \ni h$ *and* 0 *otherwise.*

Informally, the reward for a transition is 1 whenever the transition is non-punctual at least on one recurrent face. The cost is 1 whenever it is a heartbeat transition.

By Lemma 2, one can compute the optimal cycle in G and its reward-to-cost ratio, a rational number. Now we can formulate the main result of this article.

Theorem 2. *Given an* SFA $\mathcal{A}$, *let* G *be its refined graph, with* $\alpha = \alpha(G)$ *its maximal reward-to-cost ratio.*

- *If* G *does not contain cycles or if* $\alpha = 0$ *then* $\mathcal{A}$ *is meager;*
- *if* $\alpha = \infty$, *then* $\mathcal{A}$ *is obese;*
- *otherwise it is normal with*

$$\mathcal{BH}_\varepsilon(L(\mathcal{A})) = \alpha \log \frac{1}{\varepsilon} + o\left(\log \frac{1}{\varepsilon}\right) \text{ and } \mathcal{BC}_\varepsilon(L(\mathcal{A})) = \alpha \log \frac{1}{\varepsilon} + o\left(\log \frac{1}{\varepsilon}\right).$$

A proof sketch is given in Subsects. 4.5 and 4.6, and full proofs can be found in the arXiv version of this article [9]. As for complexity, it can be estimated using Lemma 2 and knowing that the SFA-form and graph G are exponential in the size of the original automaton.

Corollary 1. *Given a timed automaton, the coefficient* α *as in the theorem is a rational number computable in* FPSPACE.

4.4 Examples of Bandwidth Computation

We start by sketching the application of the algorithm above to three normal automata from Fig. 3.

For $\mathcal{A}_1$, we first add a heartbeat (for simplicity, we group together b and c), the result is given on Fig. 4,a. We have also explicitly drawn the starting regions of each location.

In this automaton, trivially, all starting regions are recurrent, which yields the cycle in the refined graph presented on Fig. 4,b. The reward is 1 on the non-punctual a-transition, the cost is 1 on both b-transitions, which yields the ratio 1/2. Other cycles in the graph (not presented on the figure) correspond to

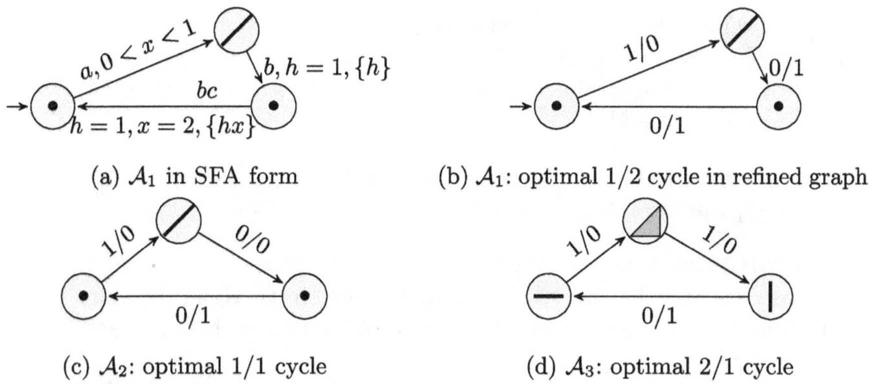

(a) $\mathcal{A}_1$ in SFA form (b) $\mathcal{A}_1$: optimal 1/2 cycle in refined graph

(c) $\mathcal{A}_2$: optimal 1/1 cycle (d) $\mathcal{A}_3$: optimal 2/1 cycle

Fig. 4. Computing the bandwidth of $\mathcal{A}_1, \mathcal{A}_2, \mathcal{A}_3$ from Fig. 3. Starting regions on (a) and active faces on (b-d) are sketched within each location.

the choice of one-point active face in p, which leads to 0 reward everywhere. We conclude that $\alpha = 1/2$, i.e. $\mathcal{BC}_\varepsilon(\mathcal{A}_1) \approx 1/2 \cdot \log(1/\varepsilon)$.

In $\mathcal{A}_2$, the heartbeat is already present, and we can build the refined graph directly. Unfortunately, the starting regions are not recurrent, and the graph G contains only active faces of smaller dimensions. An optimal cycle with $\alpha = 1/1$ is shown in Fig. 4,c.

In $\mathcal{A}_3$, the "slow" loop e makes the starting region of q recurrent, hence the cycle acb (which does not involve e!) provides the reward-to-cost ratio $\alpha = 2/1$.

4.5 Lower Bound: Coding with Two Cycles

The proof idea for the lower bound is as follows. We consider the optimal cycle π from (q, ℓ) to itself in the weighted graph G with reward-to-cost ratio α. We call region faces spawned by ℓ_i active faces. By construction, they are recurrent: there exists a "resetting" cycle σ as in Definition 9 which allows arbitrary movement of the clock vector within each face. We show that runs following $(\pi^\theta\sigma)^*$ for some big value of θ produce the bandwidth arbitrarily close to α. To that aim, we iterate the cycle π keeping the clock vector $\mathbf{x}$ on a cross-section of the region parallel to all active faces. We use all the free transitions (i.e., those with the reward $= 1$) to encode $\approx \log(1/\varepsilon)$ bits of information by choosing an ε-discrete duration. At some moment, this could lead to a saturation, making further encoding impossible, like in $\mathcal{A}_3$. At this moment, we use the cycle σ to reset all clocks and restart. In the long run, we encode $\approx \alpha \log(1/\varepsilon)$ bits/time unit while evolving in π, and less when resetting clocks on σ. However, if ε is small, θ can be chosen big, and the resulting bandwidth will be arbitrarily close to $\alpha \log(1/\varepsilon)$. Technically speaking, the strategy described above is used to build a large ε-separated set for words that are generated following $(\pi^\theta\sigma)^*$.

Hereinafter, we consider an SFA $\mathcal{A}$ with a refinement graph G, and fix notations for their size parameters. So, let N be the number of locations, M a bound on its clock values, $K-1$ the number of clocks; O the monoid of orbit graphs, and finally $H \leq 3\#O$ the stabilization level in Theorerm 1 for this monoid. Later on we will introduce several other parameters $K_0, K_1, \ldots$ depending only on $\mathcal{A}$ and G (and not on ε and T).

Definition 10. *A* route π, π' *is a couple of paths* $\pi = \delta_1 \ldots \delta_n$ *in* $\mathcal{A}$ *and* $\pi' = \delta'_1 \ldots \delta'_n$ *in* G*, where each* δ'_i *refines* δ_i *as described in Construction 2.*

To prove the lower bound, we first relate the ε-capacity of a cycle π in $\mathcal{A}$ to the reward $r(\pi)$ and the cost $c(\pi)$ of its refinement in G. Roughly, along such a cycle in time $\approx c(\pi)$, one can do $r(\pi)$ choices of timing with $\approx 1/\varepsilon$ possibilities for each choice.

For a given location q and face list ℓ containing k elements: The precise setting is as follows.

- π, π' is a cyclic route from (q, ℓ) to itself;
- σ is a cyclic path from q to q in $\mathcal{A}$ such that all ℓ_i are cliques in $\gamma(\sigma)$. (4)

The following lemma formalizes the idea of iterating along the cyclic route π, π' by producing an ε-separated set with the appropriate size:

Lemma 4. *In setting* (4)*, there exist* $\beta > 0$ *(depending on* π, π'*), a polytope* $R \subseteq S(q)$*, and a point* $\mathbf{x}^* \in R$*, such that for all* $\varepsilon > 0$ *small enough there exists an* ε*-separated set* $\mathsf{Sep} \subset L^{\mathcal{A}}_{\pi,t'}(\mathbf{x}^*, R)$ *of cardinality* $\geq \beta/\varepsilon^{r(\pi)}$*. Here* $t' = c(\pi)+1$.

The next lemma then formalizes the idea of "resetting all clocks" on the cyclic path σ, which will allow the previous ε-separated set to be repeated again after resetting:

Lemma 5. *In setting* (4)*, for any* $\mathbf{y} \in R$ *there exists a run* ρ^y *along* σ *of duration* $\leq c(\sigma)+1$ *from* $(q, \mathbf{y})$ *to* $(q, \mathbf{x}^*)$.

Combining the two previous lemmas, we obtain

Corollary 2. *In setting* (4)*, there exist* $\beta > 0$*, and a point* $\mathbf{x}^* \in R$*, such that for all* $\varepsilon > 0$ *small enough there exists an* ε*-separated set* Sep' *in* $L^{\mathcal{A}}_{\pi\sigma,t''}(\mathbf{x}^*, \mathbf{x}^*)$ *of cardinality* $\geq \beta/\varepsilon^{r(\pi)}$*, where* $t'' = c(\pi) + c(\sigma) + 2$.

We apply this to the cycle π^θ (with an appropriately selected large θ), and iterating $\approx T/t''$ times the set Sep' we obtain a large ε-separated set in $L_T(\mathcal{A})$, which allows us to prove the following:

Proposition 2 (Lower bound). *If* $\alpha < \infty$*, then for every* $\varkappa > 0$*, and for* $\varepsilon > 0$ *small enough,* $\mathcal{BC}_\varepsilon(L(\mathcal{A})) \geq (\alpha - \varkappa)\log\frac{1}{\varepsilon}$.

The following proposition complements the results from [6]:

Proposition 3. *If* $\alpha = \infty$*, then* $\mathcal{A}$ *is obese.*

4.6 Upper Bound: Factorizing and Approximating

For the rest of the paper, given a path σ, we denote $\overline{\sigma}$ its *closed semantics*, obtained by transforming each strict inequality into a non-strict one. Results from [6] show that this transformation of a path does not modify its bandwidth.

The proof of the upper bound proceeds in several steps. The first step uses Theorem 1 (of Simon) to factorize every run into conservative cycles (evolving in parallel to some active faces), and arbitrary intermediate lossy chunks (the number and total duration of the latter depend only on ε). The idea is that there is a Lyapunov function, a potential, which is preserved by conservative runs, but irremediably and quickly decreased by the lossy chunks.

By Theorem 1, each path in $\mathcal{A}$ belongs to the level X_H of the sequence (1). Similarly, runs of $\mathcal{A}$ belong to level R_H of the sequence $R_0 = R_\Delta \cup \{\epsilon\}$, $R_{h+1} = R_h \cdot R_h \cup \bigcup_{\substack{e \in O \\ ee=e}} (R_h \cap Runs(\gamma^{-1}(e)))^*$. Note that H is polynomial in the size of the SFA $\mathcal{A}$.

For a given idempotent orbit graph $e \in O$, let C_i be the strongly-connected components (cliques or transient singletons) within its vertices; we enumerate them in a topological order w.r.t. the graph e. Next, we introduce a variant of barycentric coordinates [18]: any point $\mathbf{x} \in S(q)$ can be represented as $\mathbf{x} = \sum_i \lambda_i \mathbf{x}_i$ with $\mathbf{x}_i \in \operatorname{conv}(C_i)$, and $\lambda_i \geq 0$ such that $\sum_i \lambda_i = 1$; the coordinates $\boldsymbol{\lambda}(\mathbf{x}) = (\lambda_i)$ characterize the position of $\mathbf{x}$ with respect to active faces $\operatorname{conv} C_i$. According to [3,19], the coordinates $\boldsymbol{\lambda}(\mathbf{x})$ can only decrease (w.r.t. the lexicographic order $\succeq$) when traversing a run in $Runs(\gamma^{-1}(e))$.

Then, the next ingredient of our reasoning are runs where the inequality $\boldsymbol{\lambda}(\mathbf{x}^1) \succeq \boldsymbol{\lambda}(\mathbf{x}^2)$ turns into equality (or is ε-close to equality). The next definition formalizes these:

Definition 11. *Let $\sigma \in \gamma^{-1}(e)$ be a cycle from q to itself. A run from $(q, \mathbf{x}^1)$ to $(q, \mathbf{x}^2)$ along $\overline{\sigma}$ is a* conservative e-run *whenever $\boldsymbol{\lambda}(\mathbf{x}^1) = \boldsymbol{\lambda}(\mathbf{x}^2)$. It is* ε-conservative *whenever $||\boldsymbol{\lambda}(\mathbf{x}^1) - \boldsymbol{\lambda}(\mathbf{x}^2)|| < \varepsilon$.*

Denote also $f_{\mathcal{A}}$ the real-valued function with $f_{\mathcal{A}}(\varepsilon) = (1 + 1/\varepsilon)^{KH}$. The following result plays a central role in the construction of the desired ε-net:

Proposition 4. *For each $\varepsilon > 0$, any timed word $w \in L(\mathcal{A})$ can be factorized as $u_0 v_1 u_1 \dots v_n u_n$ with $n \leq f_{\mathcal{A}}(\varepsilon)$ and $\sum_{i=0}^{n} \operatorname{dur}(u_i) < f_{\mathcal{A}}(\varepsilon)$ and all v_i are traces of ε-conservative runs.*

Proof (idea). Denote $b = (1 + 1/\varepsilon)^K$. We may prove, by induction on h, that each timed word which is a trace of R_h can be factorized as $u_0 v_1 u_1 \dots v_n u_n$ with $n \leq b^h$, $\sum \operatorname{dur}(u_i) \leq b^h$, and all v_i are traces of ε-conservative runs.

We give here the proof idea for the inductive case, the subcase of w trace of a run $\rho = \rho_0 \dots \rho_m$ with all paths $\rho_j \in (R_h \cap Runs(\gamma^{-1}(e)))$, for some integer m. In this situation, the hypothesis of e being an idempotent, combined with results on Lyapunov functions on the orbit graph [3]. imply that, if $(q, \mathbf{x}_i) \xrightarrow{\rho_i} (q, \mathbf{x}_{i+1})$, then $\lfloor \boldsymbol{\lambda}(\mathbf{x}_i)/\varepsilon \rfloor \succeq \lfloor \boldsymbol{\lambda}(\mathbf{x}_{i+1})/\varepsilon \rfloor$.

On the other hand, for each fixed ε, the quantities $\lfloor \boldsymbol{\lambda}(\mathbf{x}_i)/\varepsilon \rfloor$ may have at most $b = (1 + 1/\varepsilon)^K$ different values. But then ρ can be re-decomposed as

$$(q, \mathbf{y}_0) \xrightarrow{\rho_0'} (q, \mathbf{y}_0') \xrightarrow{\rho_0''} (q, \mathbf{y}_1) \xrightarrow{\rho_1'} (q, \mathbf{y}_1') \cdots (q, \mathbf{y}_j) \xrightarrow{\rho_j'} (q, \mathbf{y}_j'),$$

such that $\mathbf{y}_0 = \mathbf{x}_0$ and $\mathbf{y}_j' = \mathbf{x}_j$, $j \leq b$, $\lfloor \boldsymbol{\lambda}(\mathbf{y}_i)/\varepsilon \rfloor = \lfloor \boldsymbol{\lambda}(\mathbf{y}_i')/\varepsilon \rfloor$ i.e., ρ_i' is ε-conservative, and each of ρ_i'' belongs to R_h. The induction hypothesis applied to ρ_i'' then allows us to conclude. □

Thanks to the following proposition, instead of ε-conservative runs, we can consider only conservative runs, with better properties:

Proposition 5. *There exists $K_0 > 0$ depending only on $\mathcal{A}$, such that any ε-*conservative *e-run is $K_0\varepsilon$-close to a* conservative *e-run.*

Next, we show that conservative runs necessarily follow cycles in the graph G ("follow" is formalized in the next definition). Such runs satisfy the inequality $r(\pi) \leq \alpha c(\pi)$, i.e. the number of free transitions in time Θ is bounded by $\alpha\Theta$. This fact is used to construct a small ε-net for traces of all conservative runs.

Definition 12. *Fix a route π, π' with $\pi = \delta_1 \cdots \delta_n$ and $\pi' = (q_0, \ell^0) \to (q_1, \ell^1) \to \cdots \to (q_n, \ell^n)$. A run $\rho = (q_0, \mathbf{x}^0) \xrightarrow[d_1]{\delta_1} (q_1, \mathbf{x}^1) \cdots \xrightarrow[d_1]{\delta_1} (q_n, \mathbf{x}^n)$ in $\overline{Runs_\pi}$* follows *this route whenever:*

- *all the lists ℓ^j have the same length k;*
- *there exists a decomposition $\boldsymbol{\rho} = \sum_{i=1}^k \lambda_i \boldsymbol{\rho}_i$ with all $\lambda_i > 0$, $\sum \lambda_i = 1$, all ρ_i runs of $\mathcal{A}$ in $\overline{Runs_\pi}$;*
- *each ρ_i hits the i-th active face on all the intermediate states, that is, it has a form $\rho_i = (q_0, \mathbf{x}_i^0) \to (q_1, \mathbf{x}_i^1) \cdots \to (q_n, \mathbf{x}_i^n)$ with $\mathbf{x}_i^j \in \operatorname{conv} \ell_i^j$ for $j = 0..n$.*

Proposition 6. *Every conservative run follows some route.*

Routes of cost and reward 0 need to be handled properly when creating an ε-net of the appropriate size. They satisfy the following pumping lemma (with an appropriate definition of 0-loop):

Lemma 6. *For any* SFA $\mathcal{A}$ *there exists a constant B, such that any route π, π' with $|\pi| > B$ and $c(\pi) = r(\pi) = 0$ contains a 0-loop.*

Definition 13. *A route is* proper *if it does not contain 0-loops.*

Lemma 7. *Given a normal* SFA, *there exist constants K_i such that the number of proper cyclic routes of a cost $\leq T$ is at most $K_1 2^{K_2 T}$. The number of transitions in any such route is at most $K_{10}T + K_{11}$.*

It is possible to transform any route into a proper one by removing all 0-loops. In the following lemma, ϕ is an auxiliary multi-valued function on timed words (it inserts some instantaneous events in w).

Lemma 8. *For every route π, π' exists a proper route $\hat{\pi}, \hat{\pi}'$ such that for any trace w of a run following the former exists $\hat{w}$ trace of a run following the latter such that $d(w, \phi(\hat{w})) = 0$.*

We are now in a position to build the desired ε-net. We do this by defining ε-nets for the conservative factors of runs, which are then interleaved with ε-nets for intermediate chunks.

Construction 3 (ε-net for words following one route) *Given a proper route π, π' from (q, ℓ) to itself, and an initial time interval $[\tau, \tau + 1)$, we build a set $\mathsf{Net}_\varepsilon^{\mathbf{fol}}(\pi, \pi', \tau)$ containing all words $v = (a_1, s_1) \dots (a_n, s_n)$, with ε-discrete timestamps $s_i \in \varepsilon\mathbb{N}$, that satisfy the following properties:*

1. *if $s_i - \tau > 2$ and $a_i = b$, then $t_i \in s_{\mathrm{pb}_i} + [1 - \varepsilon, 1 + \varepsilon]$;*
2. *$s_i \in [\tau + \mathrm{cb}_i - 1 - \varepsilon, \tau + \mathrm{cb}_i + 2 + \varepsilon]$;*
3. *if $s_{i-1} - \tau > M$ and $r(\delta') = 0$, then $s_i \in s_{i-1} + g_i(\mathbf{x}_{i-1}) + [-\nu\varepsilon - \varepsilon, \nu\varepsilon + \varepsilon]$, where $\mathbf{x}_i^k = s_i - s_{\mathrm{r}_i^k}$.*

Here cb_i, pb_i and r_i^k depend on the path π only, and g_i is a uniformly ν-Lipshitz continuous function.

Lemma 9. *The set $\mathsf{Net}_\varepsilon^{\mathbf{fol}}(\pi, \pi', \tau)$ defined in Construction 3 is an ε-net for all words traces of runs following the route π, π' (starting within $[\tau, \tau + 1)$).*

Construction 4 (ε-net for words following proper routes) *We define $\mathsf{Net}_{\varepsilon,\Theta}^{\mathbf{fol}}(\tau) = \bigcup_{\pi,\pi'} \mathsf{Net}_\varepsilon^{\mathbf{fol}}(\pi, \pi', \tau)$, where π, π' range over all proper routes with cost $\leq \Theta$.*

Lemma 10. *The set $\mathsf{Net}_{\varepsilon,\Theta}^{\mathbf{fol}}(\tau)$ defined in Construction 4 is an ε-net for all words starting at $t_0 \in [\tau, \tau + 1)$ and following some proper route of cost $\leq \Theta$.*

Construction 5 (ε-net for conservative words) *Let now*

$$\mathsf{Net}_{\varepsilon,\Theta}^{\mathbf{cons}}(\tau) = \phi\left(\mathsf{Net}_{\varepsilon,\Theta+1}^{\mathbf{fol}}(\tau)\right).$$

Lemma 11. *The set $\mathsf{Net}_{\varepsilon,\Theta}^{\mathbf{cons}}(\tau)$ is an ε-net for all conservative words of $L_\Theta(\mathcal{A})$ starting at $t_0 \in [\tau, \tau + 1)$. It is also a $(K_0 + 1)\varepsilon$-net for ε-conservative words.*

Construction 6 (ε-net for $L_T(\mathcal{A})$) *Given T we proceed as follows:*

- *for each $n \leq f_\mathcal{A}(\varepsilon)$, take all possible sequences ζ of breakpoints in $\varepsilon\mathbb{N}$:*

$$0 = a_0 \leq b_0 \leq a_1 \leq b_1 \cdots \leq a_n \leq b_n \leq T$$

  with cumulative duration of odd intervals $\sum_i (b_i - a_i) \leq f_\mathcal{A}(\varepsilon)$;
- *for each such sequence ζ we construct an ε-net $\mathsf{Net}_\varepsilon^\zeta$ like that:*
  - *on each odd interval $[a_i, b_i]$ put the universal ε-net for duration $b_i - a_i$;*
  - *on each even interval $[b_i, a_{i+1}]$ put $\mathsf{Net}_{\varepsilon/(K_0+1),a_{i+1}-b_i}^{\mathbf{cons}}(b_i - 1/2)$;*
- *let $\mathsf{Net}_{\varepsilon,T} = \bigcup_\zeta \mathsf{Net}_\varepsilon^\zeta$ with the union taken over all breakpoint sequences.*

Lemma 12. *The set* $\mathsf{Net}_{\varepsilon,T}$ *described in Construction 6 is an* ε*-net for* $L_T(\mathcal{A})$.

Finally, we estimate the cardinality of those nets.

Proposition 7. *The cardinalities of the sets defined in the previous constructions can be bounded as follows (with* K_i *depending only on the automaton* $\mathcal{A}$*):*

$$\log \#\mathsf{Net}_{\varepsilon}^{\mathbf{fol}}(\pi,\pi',\tau) \leq (\alpha c(\pi') + K_3)\log\frac{1}{\varepsilon} + K_5 c(\pi') + K_{12}; \tag{5}$$

$$\log \#\mathsf{Net}_{\varepsilon,\Theta}^{\mathbf{cons}}(\tau) \leq \alpha\Theta\log\frac{1}{\varepsilon} + K_3\log\frac{1}{\varepsilon} + K_6\Theta + K_7; \tag{6}$$

$$\limsup_{T\to\infty}\frac{\log \#\mathsf{Net}_{\varepsilon,T}}{T} \leq \alpha\log\frac{1}{\varepsilon} + K_9. \tag{7}$$

We then obtain the upper bound for $\mathcal{BH}_\varepsilon$:

Proposition 8 (Upper bound). *If* $\alpha > 0$*, then exists a constant* K_9 *such that for all* $\varepsilon > 0$ *it holds that*

$$\mathcal{BH}_\varepsilon(L(\mathcal{A})) \leq \alpha\log\frac{1}{\varepsilon} + K_9.$$

The final step is given by the following proposition, which also complements the results from [6]:

Proposition 9. *If the graph* G *is acyclic or* $\alpha = 0$*, then* $\mathcal{A}$ *is meager.*

Now we may combine Proposition 8 with Proposition 2 and the inequalities (3) to get the identities in Theorem 2.

5 Conclusions

5.1 Summarizing Four Papers

Putting together the results from [6–8] and this article, we obtain the following comprehensive result:

Theorem 3 *Given a nondeterministic timed automaton* $\mathcal{A}$ *without* ϵ*-transitions, one of the following holds for its language* L *(for* $\varepsilon \to 0$*):*

- *L is* ***meager****, i.e.* $\mathcal{BC}_\varepsilon(L) = O(1)$ *and* $\mathcal{BH}_\varepsilon(L) = O(1)$*. If moreover* $\mathcal{A}$ *is deterministic, a number* $\alpha \geq 0$ *can be computed (as a logarithm of an algebraic number), such that both* $\mathcal{BC}_\varepsilon(L)$ *and* $\mathcal{BH}_\varepsilon(L)$ *equal* $\alpha + o(1)$*.*
- *L is* ***normal****: a rational* $\alpha > 0$ *can be computed (in* FPSPACE*), such that both* $\mathcal{BC}_\varepsilon(L)$ *and* $\mathcal{BH}_\varepsilon(L)$ *equal* $\alpha\log\frac{1}{\varepsilon} + o\left(\log\frac{1}{\varepsilon}\right)$*.*
- *L is* ***obese****:* $\alpha > 0$ *can be computed (as a logarithm of an algebraic number), such that both* $\mathcal{BC}_\varepsilon(L)$ *and* $\mathcal{BH}_{\varepsilon/2}(L)$ *equal* $\frac{\alpha}{\varepsilon} + o\left(\frac{1}{\varepsilon}\right)$*.*

Given $\mathcal{A}$*, the three decision problems "is it meager?", "is it normal?", "is it obese?" are* PSPACE*-complete.*

5.2 Perspectives

While Theorem 3 is almost exhaustive, a couple of questions should still be explored:

- settling the exact complexity of computing α (exactly and approximately) in each of the three cases;
- computing α for non-deterministic meager automaton;
- and analyzing the bandwidth for timed automata with ϵ-transitions.

More generally, we are also interested in:

- more precise computation of $\mathcal{C}_\varepsilon(L_T), \mathcal{H}_\varepsilon(L_T)$, in the spirit of [5], where not only the asymptotic exponential growth rate, but also the exact generating function was computed (for a different "metrics");
- interpretation of bandwidth in terms of Kolmogorov's complexity and topological entropy;
- timed coding/decoding by transducers;
- applications to information transmission and compression of timed data.

Our results and techniques strongly depend on specific properties of timed automata. More general study of extended automata, especially hybrid ones, from the standpoint of information theory, remains an attractive research subject.

This work was funded by ANR project MAVeriQ ANR-CE25-0012. The authors thank Thomas Colcombet for a valuable advice.

References

1. Alur, R., Dill, D.L.: A theory of timed automata. Theor. Comput. Sci. **126**, 183–235 (1994). https://doi.org/10.1016/0304-3975(94)90010-8
2. Asarin, E., Basset, N., Béal, M.-P., Degorre, A., Perrin, D.: Toward a Timed Theory of Channel Coding. In: Jurdziński, M., Ničković, D. (eds.) FORMATS 2012. LNCS, vol. 7595, pp. 27–42. Springer, Heidelberg (2012). https://doi.org/10.1007/978-3-642-33365-1_4
3. Asarin, E., Basset, N., Degorre, A.: Entropy of regular timed languages. Inf. Comput. **241**, 142–176 (2015). https://doi.org/10.1016/j.ic.2015.03.003
4. Asarin, E., Basset, N., Degorre, A.: Distance on Timed Words and Applications. In: Jansen, D.N., Prabhakar, P. (eds.) FORMATS 2018. LNCS, vol. 11022, pp. 199–214. Springer, Cham (2018). https://doi.org/10.1007/978-3-030-00151-3_12
5. Asarin, E., Basset, N., Degorre, A., Perrin, D.: Generating Functions of Timed Languages. In: Rovan, B., Sassone, V., Widmayer, P. (eds.) MFCS 2012. LNCS, vol. 7464, pp. 124–135. Springer, Heidelberg (2012). https://doi.org/10.1007/978-3-642-32589-2_14
6. Asarin, E., Degorre, A., Dima, C., Jacobo Inclán, B.: Bandwidth of timed automata: 3 classes. In: Proceedings of FSTTCS. LIPIcs, vol. 284, pp. 10:1–10:17 (2023). https://doi.org/10.4230/LIPICS.FSTTCS.2023.10, full version https://arxiv.org/abs/2310.01941

7. Asarin, E., Degorre, A., Dima, C., Jacobo Inclán, B.: Computing the bandwidth of meager timed automata. In: Proceedings of CIAA. LNCS, vol. 15015, pp. 19–34. Springer, Cham (2024). https://doi.org/10.1007/978-3-031-71112-1_2, full version https://arxiv.org/abs/2406.12694
8. Asarin, E., Degorre, A., Dima, C., Jacobo Inclán, B.: Weighing obese timed languages. In: Reachability Problems. LNCS, vol. 16230, pp. 112–125. Springer, Cham (2026). https://doi.org/10.1007/978-3-032-09524-4_8, full version https://arxiv.org/abs/2508.18133
9. Asarin, E., Degorre, A., Dima, C., Jacobo Inclán, B.: Weighing timed regular languages: The final step (long version) (2026). https://doi.org/10.48550/arXiv.2606.11003
10. Berstel, J., Perrin, D., Reutenauer, C.: Codes and Automata. Cambridge University Press (2009). https://doi.org/10.1017/CBO9781139195768
11. Bojańczyk, M.: Factorization Forests. In: Diekert, V., Nowotka, D. (eds.) DLT 2009. LNCS, vol. 5583, pp. 1–17. Springer, Heidelberg (2009). https://doi.org/10.1007/978-3-642-02737-6_1
12. Bouyer, P., Brinksma, E., Larsen, K.G.: Optimal infinite scheduling for multi-priced timed automata. Formal Methods Syst. Des. **32**(1), 3–23 (2008). https://doi.org/10.1007/S10703-007-0043-4
13. Chomsky, N., Miller, G.A.: Finite state languages. Inf. Control **1**(2), 91–112 (1958). https://doi.org/10.1016/S0019-9958(58)90082-2
14. Dasdan, A., Irani, S.S., Gupta, R.K.: Efficient algorithms for optimum cycle mean and optimum cost to time ratio problems. In: Proceedings of DAC, pp. 37–42. ACM (1999). https://doi.org/10.1145/309847.309862
15. Jacobo Inclán, B., Degorre, A., Asarin, E.: Bounded delay timed channel coding. In: Proceedings of FORMATS. LNCS, vol. 13465, pp. 65–79. Springer, Cham (2022). https://doi.org/10.1007/978-3-031-15839-1_4
16. Kolmogorov, A.N., Tikhomirov, V.M.: ε-entropy and ε-capacity of sets in function spaces. Uspekhi Matematicheskikh Nauk **14**(2), 3–86 (1959). https://doi.org/10.1090/trans2/017, http://mi.mathnet.ru/eng/umn7289
17. Lind, D., Marcus, B.: An Introduction to Symbolic Dynamics and Coding. Cambridge University Press (1995). https://doi.org/10.1017/CBO9780511626302
18. Möbius, A.F.: Der barycentrische Calcul. J.A. Barth., Leipzig (1827). http://sites.mathdoc.fr/cgi-bin/oeitem?id=OE_MOBIUS__1_1_0
19. Puri, A.: Dynamical properties of timed automata. Discrete Event Dyn. Syst. **10**(1–2), 87–113 (2000). https://doi.org/10.1023/A:1008387132377
20. Shannon, C.E.: A mathematical theory of communication. Bell Syst. Tech. J. **27**(3), 379–423 (1948). https://doi.org/10.1002/j.1538-7305.1948.tb01338.x
21. Simon, I.: Factorization forests of finite height. Theor. Comput. Sci. **72**(1), 65–94 (1990). https://doi.org/10.1016/0304-3975(90)90047-L

Exact Evaluation of Probabilistic Programs with Cylindrical Algebraic Decomposition

Mohamed Hamza Bandukara, Fredrik Dahlqvist, and Niki Omidvari(✉)

Queen Mary University of London, London, UK
{m.h.bandukara,f.dahlqvist,n.omidvari}@qmul.ac.uk

QEST + FORMATS Artifacts Evaluation Available

Abstract. We present a method for computing the exact output distribution of small programs with random inputs. Specifically, we are interested in inline programs manipulating sensor data such as *e.g.* GPS or inertial measurement sensors whose inputs have a known or well-modelled distribution. These programs typically only include relatively few variables, arithmetic operations, square roots and if-else statements. This small syntax allows us to recast the problem of computing the exact output distribution as a cylindrical algebraic decomposition problem followed by symbolic and/or numerical integration. We present this method in detail and show with two prototypes that it can successfully be applied to benchmarks from the literature on floating-point arithmetic and small programs from open-source sensor libraries.

Keywords: Probabilistic program · Range analysis · Cylindrical Algebraic Decomposition · Symbolic computations

1 Introduction

There are two ways in which randomness can find its way into a computer program. First, it can be inserted there by design, through sampling from a Pseudo-Random Number Generator (PRNG). This kind of randomness is used in a huge range of applications, most notably in Monte-Carlo simulations for tasks as varied as pricing financial instruments, weather forecasting or computer graphics. Second, it can enter the program via interactions with the outside world, for example through sensors forming part of a cyber-physical system like an autonomous vehicle, an industrial robot or a mobile phone. All sensors are necessarily noisy, and any program processing sensor data will therefore exhibit some randomness. Importantly, this noise can be – and often is – modelled to a good level of precision, *e.g.* [14,23,37] for GPS sensors or [40,41] for inertial measurement sensors (gyroscopes, accelerometers, magnetometers).

Irrespective of the source of randomness, we will refer to a program exhibiting randomness as a *probabilistic program*. Our objective in this work is to compute the exact output distribution of probabilistic programs with known continuous

M. Chen and M. Tschaikowski (Eds.): QEST+FORMATS 2026, LNCS 16913, pp. 77–95, 2027.
https://doi.org/10.1007/978-3-032-35298-9_5

input distributions[1]. In principle, our method can be applied to any probabilistic program. In practice, it is aimed at programs manipulating sensor data. Computing the output distribution of a probabilistic program is a very hard problem in general and we therefore restrict our attention to small programs containing (*i*) arithmetic operations, (*ii*) radicals, and (*iii*) if-else statements. This syntax is sufficient to capture many functions in the code bases of various sensor libraries, for example associated with Inertial Measurement Units (IMUs) or GPS sensors. Below is a fairly typical example from an IMU library[2]; based on an input acceleration from an accelerometer it returns the first component of a quaternion representing the orientation of the global w.r.t. the local frame.

```
double getMeasurement(double ax, double ay,
   double az){
  normalizeVector(ax, ay, az);
  if (az >= 0){return sqrt((az + 1) * 0.5);}
  else {return -ay / (2.0 * sqrt((1 - az) * 0.5)
     ;}}
```

We focus on relatively small probabilistic programs like this one (included as `get_meas` in our results, see Table 1) rather than the typically much more complex code bases of Monte-Carlo simulations.

There is of course a very simple way of characterising the output distribution of a probabilistic program: sample repeatedly from the input distributions, run the program on these inputs, collect the outputs, and plot a histogram. In other words, analyse the program by Monte-Carlo simulation. This is a frequent and conceptually simple approach to testing probabilistic programs. This simplicity however comes at the expense of *precision* and *formal guarantees.*

Precision. By definition, Monte-Carlo simulation can only *approximate* the probability of a range of output values. By the central limit theorem, the error incurred by this approximation only decreases as $1/\sqrt{n}$, where n is the number of samples. Thus what may at first seem like a computationally cheap method can turn out to be quite expensive if a high level of precision is required, particularly if the probability of a rare event must be approximated. Beyond the computational cost, the problem of generating a lot of quality random inputs isn't trivial. There are two options: either sample analog data directly from the sensor under consideration, the throughput of samples is then typically limited, or use a much faster PRNG, in which case it is typically impossible to exclude the presence of statistical artifacts beyond a small number of samples (*e.g.* the original Mersenne twister [29] only guarantees the i.i.d. behaviour of successive samples up to $n = 623$). This phenomenon is illustrated perfectly by the poor performance of 'good' PRNGs on physics problem requiring many samples [19].
Formal guarantees. Monte-Carlo simulations cannot offer formal guarantees of the type '*the probability of the output lying in a set X of bad outcomes is less that*

[1] We leave the addition of discrete and mixed distributions to future work.
[2] https://github.com/CCNYRoboticsLab/imu_tools.

α%'. This is partly due to the impossibility of ruling out statistical artifacts when using PRNGs, but even when samples are *bona fide* i.i.d. samples, it is extremely difficult to provide useful upper-bounds to the distance between the empirical output distribution (based on samples) and the actual output distribution.

Computing exact output distributions is much harder than Monte-Carlo simulation but it solves the precision problem and can also be used to provide formal guarantees in the development and certification of cyber-physical systems. As we shall show in Sect. 5, computing exact/analytical output distributions is a practical proposal for benchmarks from the floating-point analysis community (FPBench [17]) and small programs from open-source IMU sensors libraries.

The structure of this paper is as follows: in Sect. 2 we will define arithmetic on random variables, Cylindrical Algebraic Decomposition (CAD), and show how the latter can be used to compute the former. In Sect. 3 we will describe the syntax and semantics of the programs that we consider in this work. This probabilistic semantics (the output distribution) is concretely evaluated using algorithms described in Sect. 4. Finally, we present an empirical evaluation in Sect. 5.

Related work.

Statistical models. Arithmetic operations and if-else branching on random variables is used by the statistical community to develop generative models, *e.g.*

$$X \sim \mathcal{N}(0,1) \qquad Y = \begin{cases} \alpha_1\sqrt{X} + \epsilon & X \geq 0 \\ \alpha_2 X + \epsilon & X < 0 \end{cases} \qquad \epsilon \sim \mathcal{N}(0,\sigma).$$

These models will be implemented as probabilistic programs. Methods for evaluating the Probability Density Function (PDF) of such models are developed in [25,26]. However, their method is purely numerical and cannot deal with arithmetic expressions containing repeated variables. Although we do not focus on this application, our method can easily be applied to statistical models.

Probabilistic Floating-Point Arithmetic. The literature on floating-point arithmetic for random inputs [4,5,12,13,28,36] is closely related to our work, it targets the same kind of programs under the same condition of random inputs. The key differences with our work are that (a) this line of work does not evaluate output distributions exactly, relying on under- and over-approximations like *p*-boxes [4,12], and (b) we do not (yet!) consider rounding errors when computing exact output distributions and work instead in infinite precision (symbolically).

Exact Inference in Probabilistic Programs. Though our focus is on small in-line probabilistic programs without inference, there is broader work on exact inference for probabilistic programs that includes in particular computing output distributions (the *evidence* in the Bayesian perspective). For discrete distributions, exact inference is more tractable and can be done using weighted model counting, as in Dice [24]. Continuous distributions can be discretised to exploit these methods; the recent 'bit-blasting' approach [20] being a particularly efficient example of this approach. Being approximate, these discretisation methods cannot offer the formal guarantees we aim to compute.

Computer algebra can be used to simplify the distributional representations of probabilistic programs, though they may leave results partially unevaluated or finally use numeric methods. Hakaru [8,31] uses the Maple computer algebra system for simplification but may still rely on sampling for full evaluation. Similarly, [3] uses a type-based system to represent probability distributions but uses numerical integration to evaluate them, and so does not produce closed-form solutions. The top solver for exact symbolic inference is PSI [21,22], which supports both discrete and continuous distributions and a rich syntax including conditioning, bounded loops, and higher-order functions. PSI works by incrementally building, simplifying, and integrating the joint probability distribution of a program. PSI is a powerful general-purpose tool targeting a very expressive language, and may therefore return results containing unevaluated integrals or overly complex PDFs. We have designed two tools, one to compute closed-form Cumulative Distribution Functions (CDFs) and PDFs, based on Mathematica's symbolic integration engine, the other, based on C++ libraries, to evaluate output CDFs and PDFs without necessarily constructing a full closed-form solution.

Probabilistic Circuits [9], such as sum-product networks [32,35] and arithmetic circuits [18], represent joint distributions as trees of products and weighted sums of density functions. Their structural constraints enable exact polynomial-time inference for queries such as marginalisation and conditioning but can lead to large circuits with many repetitions. They are designed for queries on joint distributions, rather than for computing output distributions. The SPPL system [34] bridges this gap by translating probabilistic programs into sum-product expressions, enabling exact symbolic inference comparable to PSI. SPPL is faster than PSI on shared benchmarks but targets a much less expressive language (*e.g.* it cannot compute $Z = X \div Y$ or accept infinite-support distributions).

2 Preliminaries

In this section we describe the problem of evaluating arithmetic expressions over random variables, and show how cylindrical algebraic decomposition (CAD) can solve it. We apply this technique in the next sections to evaluate the output distributions, *i.e.* the *probabilistic semantics*, of probabilistic programs.

Arithmetic on Random Variables. We assume familiarity with the notions of random variable, probability measure, and Probability Density (resp. Cumulative Distribution) Functions, *i.e.* PDF (resp. CDF). By *arithmetic expression* we mean an expression given by the grammar:

$$e ::= X \in \mathrm{Var} \mid c \in \mathbb{R} \mid e + e \mid e - e \mid e \times e \mid e \div e$$

where $\mathrm{Var} = \{X_1, \ldots, X_n\}$ is a finite set of variable symbols. A *polynomial expression*, or simply a *polynomial*, is an arithmetic expression not containing the division symbol $\div$.

Independent Variables. Consider first the case of two *independent* real random variables X_1, X_2 described by PDFs f_1 and f_2 respectively. It is well known [38] that the density functions of the distributions of $X_1+X_2, X_1-X_2, X_1\times X_2$ and $X_1 \div X_2$ can be expressed analytically as follows:

$$f_{X_1+X_2}(t) = \int_{-\infty}^{\infty} f_1(x)f_2(t-x)\,dx \qquad f_{X_1-X_2}(t) = \int_{-\infty}^{\infty} f_1(x)f_2(x-t)\,dx$$
$$f_{X_1\times X_2}(t) = \int_{-\infty}^{\infty} \frac{1}{|x|}f_1(x)f_2\Big(\frac{t}{x}\Big)\,dx \qquad f_{X_1\div X_2}(t) = \int_{-\infty}^{\infty} |y|\,f_1(ty)f_2(y)\,dy \quad (1)$$

The independence assumption is crucial as it means that the joint density of X_1, X_2 factors as the product of the densities, *viz.* $f_{X_1,X_2}(x,y) = f_1(x)f_2(y)$, which makes the derivations of the expressions (1) possible. Evaluating independent arithmetic expressions over random expression numerically is discussed in [25] and implemented in the Python library PaCal [26].

The density function of any arithmetic expression over independent random variables $X_1,\dots,X_n$ can thus be computed inductively using (1) *provided that no variable is repeated.* Indeed, variable repetition immediately breaks independence: consider the expression $X\div(X+Y)$, the random variables X and $X+Y$ are not independent since what we usually mean when writing $X\div(X+Y)$ is that the two occurrences of X represent the *same* random element. Put differently, to sample from $X\div(X+Y)$ we sample x from X, y from Y, *copy* x and evaluate $x\div(x+y)$ (using our two copies of x).

Non-independent Variables. If we cannot assume that variables are independent, we must assume a general joint density $f_{X_1,X_2}(x_1,x_2)$ that cannot be factored as above. To evaluate the distribution of $X_1 * X_2$ for $* \in \{+,-,\times,\div\}$ one can compute the CDF as $F_{X_1*X_2}(t) = \int_{\{(x,y)\mid x*y\leq t\}} f_{X_1,X_2}(x_1,x_2)\,dx_1dx_2$ and, if it is analytically tractable, take its derivative to get the PDF. More generally, for an arithmetic expression $\varphi(X_1,\dots,X_n)$ over non-independent random variables with joint density $f(x_1,\dots,x_n)$, the CDF of $\varphi(X_1,\dots,X_n)$ is given by

$$F(t) = \int_{\{(x_1,\dots,x_n)\mid\varphi(x_1,\dots,x_n)\leq t\}} f(x_1,\dots,x_n)\,dx_1\dots dx_n. \quad (2)$$

To concretely evaluate integrals of this shape we will exploit the fact that the set over which the integral above is taken has a very specific shape known as a *semi-algebraic set.* Although it is not essential for our approach to work, we will also assume throughout that the n input variables are independent, meaning that $f(x_1,\dots,x_n) = f_1(x_1)\dots f_n(x_n)$. However, this does not mean that we can fall back on (1), as $\varphi(X_1,\dots,X_n)$ will typically contain variable repetitions.

Cylindrical Algebraic Decomposition (CAD). The class of *semi-algebraic sets* [2, §2] of $\mathbb{R}^n$ is the smallest class of subsets $S\subseteq\mathbb{R}^n$ that contains (*i*) *algebraic sets*, *i.e.* sets $\{x\in\mathbb{R}^n \mid p(x)=0\}$ of zeroes of a polynomial p, (*ii*) sets $\{x\in\mathbb{R}^n \mid p(x)>0\}$ defined by polynomial inequalities, and (*iii*) is closed under intersection, union and complementation. A function $f: S\to T$ between two semi-algebraic sets $S\subseteq\mathbb{R}^m$ and $T\subseteq\mathbb{R}^n$ is *a semi-algebraic function* if its graph $\{(x,f(x))\mid x\in\mathbb{R}^m\}$ is semi-algebraic in $\mathbb{R}^{m+n}$. It is easy to check that the sets over which integrals of the shape (2) are taken are semi-algebraic.

Proposition 1. *For any arithmetic expression* $e(X_1, \ldots, X_k)$*, the set* $\{x \in \mathbb{R}^k \mid e(x) < 0\}$ *is semi-algebraic.*

We can now define the central concept of *Cylindrical Algebraic Decomposition* or CAD, see *e.g.* [1],[2, §5]. A CAD of $\mathbb{R}^n$ is a sequence $\mathcal{S}_1, \ldots, \mathcal{S}_n$ of finite partitions $\mathcal{S}_i$ of $\mathbb{R}^i, 1 \leq i \leq n$ into semi-algebraic subsets called *cells of level* i or i*-cells*. These partitions must satisfy the following properties:

1. Each cell of level 1 in $\mathcal{S}_1$ is either a point or an open interval
2. For each i-cell $S \in \mathcal{S}_i, 1 \leq i < n$, there are finitely many continuous semi-algebraic functions $\xi_{S,1} < \ldots < \xi_{S,k_S} : S \to \mathbb{R}$ such that the *cylinder* $S \times \mathbb{R} \subseteq \mathbb{R}^{i+1}$ is the disjoint union of the $2k_S + 1$ cells of level $i+1$ given by:
   - $\{(x_1, \ldots, x_i, x_{i+1}) \mid (x_1, \ldots, x_i) \in S, x_{i+1} = \xi_{S,j}(x_1, \ldots, x_i)\}$, that is to say the graph of the functions $\xi_{S,j}, 1 \leq j \leq k_S$, or
   - $\{(x_1, \ldots, x_i, x_{i+1}) \mid (x_1, \ldots, x_i) \in S, \xi_{S,j}(x_1, \ldots, x_i) < x_{i+1} < \xi_{S,j+1}(x_1, \ldots, x_i)\}$, that is to say the band bounded from above and below by the graphs of $\xi_{S,j}$ and $\xi_{S,j+1}$ for $0 \leq j \leq k_S$ where $\xi_{S,0} = -\infty$ and $\xi_{S,k_S+1} = \infty$.

Let $\pi_i : \mathbb{R}^{i+1} \simeq \mathbb{R}^i \times \mathbb{R} \to \mathbb{R}^i$ be the canonical projection forgetting the last element in an $i+1$-tuple. Given a cell S of level i, the cells of level $i+1$ decomposing the cylinder $S \times \mathbb{R} = \pi_i^{-1}(S)$ are called the cells *above* S since they project down into S. Note in particular that if $k_S = 0$ in the definition above, then the only cell above S is the entire cylinder $S \times \mathbb{R}$.

A CAD $(\mathcal{S}_i)_{1 \leq i \leq n}$ is said to be *adapted to a finite family* $T_1, \ldots, T_k$ *of semi-algebraic sets* if each $T_j, 1 \leq j \leq k$ is a union of cells of $(\mathcal{S}_i)_{1 \leq i \leq n}$. A CAD $(\mathcal{S}_i)_{1 \leq i \leq n}$ is said to be *adapted to a finite family* $P_1, \ldots, P_k$ *of polynomials in the variables* $X_1, \ldots, X_n$ if for each n-cell S and each polynomial P_i, the sign of P_i is constant on S (*i.e.* >0, <0 or =0).

Example 1. Here is a CAD of $\mathbb{R}^2$ adapted to the closed unit disk $\{(x_1, x_2) \mid x_1^2 + x_2^2 = 1\}$ and the polynomial $x_1^2 + x_2^2 - 1$. The 1-cells are given by $\mathcal{S}_1 \leq \{(-\infty, -1), \{-1\}, (-1, 1), \{1\}, (1, \infty)\}$. There is only one 2-cell above $(-\infty, -1)$ (resp. $(1, \infty)$), namely the cylinder $(-\infty, -1) \times \mathbb{R}$ (resp. $(1, \infty) \times \mathbb{R}$). Above $\{-1\}$ there are three 2-cells: $\{-1\} \times (-\infty, 0)$, $\{-1\} \times \{0\}$ and $\{-1\} \times (0, \infty)$ (and similarly above $\{1\}$)), *i.e.* $\xi_{(-1,1),1} = 0$. Finally, above $(-1, 1)$ there are five 2-cells defined by the two semi-algebraic functions $\xi_{(-1,1),1}(x_1) = -\sqrt{1 - x_1^2}$ and $\xi_{(-1,1),1}(x_1) = \sqrt{1 - x_1^2}$ representing the elements of $(-1, 1) \times \mathbb{R}$ which are strictly above the unit disk, on the positive boundary of the unit disk, in the interior of the unit disk, on the negative boundary of the unit disk and strictly below the unit disk respectively. The closed unit disk is the union of 3 of the 2-cells, and the CAD is therefore adapted as claimed.

The problem of computing a CAD adapted to a finite collection of semi-algebraic sets or polynomials was initially solved by Collins in [10] and later refined in *e.g.* [1,2,11,39]. Algorithmically, CAD does not produce an explicit description of the semi-algebraic functions $\xi_{S,i}, 1 \leq i \leq k_S$, for every cell S in the decomposition (this would in any case be impossible since the roots of polynomials of degree ≥ 5 do not in general have algebraic expressions). Instead, it relies on the fact that it is enough to find a single *sample point* in each cell $S \in \mathcal{S}_i$ to be able to express these functions *implicitly* as the roots of so-called 'projection polynomials' $p : \mathbb{R}^{i+1} \to \mathbb{R}^i$ denoted $\mathrm{Root}_j(p(x_1, \ldots, x_i, \cdot))$ and whose existence is guaranteed by the implicit function theorem and the definition of these polynomials, see [1,2]. Formally, given a finite family polynomials $P_1, \ldots, P_k$ in n variables and of degree bounded by d, the CAD algorithm returns the tree structure of a CAD adapted to $P_1, \ldots, P_k$ and a (rational or algebraic) sample point in each cell [1]. The complexity of this algorithm is $(kd)^{O(1)^{n-1}}$, *i.e.* polynomial in the degree but doubly exponential in the number of variables [2, §11]. In our experimental evaluation, we use Mathematica's implementation of CAD, `CylindricalDecomposition[]` [39] and the C++ implementation QEPCAD [6,11] for which we have developed a Python library.

Using CAD for arithmetic on RVs. Our aim is to compute the integral (2) when $f(x_1, \ldots, x_n) = f_1(x_1) \ldots f_n(x_n)$, *i.e.* when the n input variables are independent, for an arbitrary arithmetic expression $\varphi(x_1, \ldots, x_n)$ (in particular variables can be repeated). To achieve this, we will construct a CAD $(\mathcal{S}_i)_{1 \leq i \leq n}$ adapted to the semi-algebraic set

$$T = \{(t, x_1, \ldots, x_n) \in \mathbb{R}^{n+1} \mid \varphi(x_1, \ldots, x_n) - t < 0\}.$$

Moreover, we will choose this CAD in such a way that the 1-cells describe the possible ranges of values taken by the variable t. Beyond this, we can assume w.l.o.g. that the projections follow the variable ordering $x_1, \ldots, x_n$, *i.e.* the 2-cells describe the ranges of possible values for the variables (t, x_1), etc. We will also assume that we can restrict our attention to the part of the CAD partitioning T, that is to say the cells below the set $\mathcal{S}_{n+1}$ of $(n+1)$-cells partitioning T.

From the definition of a CAD, there are $2k_S + 1$ cells of level $i+1$ above any given i-cell S, $1 \leq i \leq n$. However, k_S of these have dimensionallity i, being the graph a semi-algebraic function $\xi_{S,j} : S \to \mathbb{R}$. They will therefore not contribute to our integration problem (2) since all cells above them will necessarily have a dimension that is smaller than their ambient space and thus be of (Lebesgue) measure zero. We will therefore ignore these cells and only consider the $k_S + 1$ bands above S defined by the semi-algebraic maps $\xi_{S,1}, \ldots \xi_{S,k_S}$; this simplification is considered for the same reason in [39]. Finally, to write down our expression for (2) we introduce the following convention. We will write $S_{i_1}, 0 \leq i_1 \leq k_1$ for the cells of level 1 that are open sets (and defined by $k_1 - 1$ points), we will write $S_{i_1,i_2}, 1 \leq i_2 \leq k_{S_{i_1}}$ for the (two-dimensional) 2-cells above S_{i_1}, and so on until $S_{i_1,\ldots,i_{n+1}}, 1 \leq i_{n+1} \leq k_{S_{i_1,\ldots,i_n}}$ for the cells of level $n+1$ above the cell $S_{i_1,\ldots,i_n}$. With this notation and a CAD adapted to the set T we can write the CDF $F(t)$ in (2) as the piecewise function defined at $t \in S_{i_1}$ by equation (3) below. This expression illustrates the bad complexity of CAD w.r.t. the number of variables. However, we find in Sect. 5 that evaluating the CAD in (3) is manageable in practice, the computational bottleneck being the symbolic evaluation of the integrals.

$$
\begin{aligned}
F(t) &= \int_{\{(x_1,\ldots,x_n)|\varphi(x_1,\ldots,x_n)\leq t\}} f_1(x_1)\ldots f_n(x_n)\ dx_1\ldots dx_n \\
&= \sum_{\substack{0\leq i_2\leq k_{S_{i_1}} \\ 0\leq i_3\leq k_{S_{i_1,i_2}} \\ \ldots \\ 0\leq i_{n+1}\leq k_{S_{i_1,\ldots,i_n}}}} \int_{\xi_{S_{i_1},i_2}(t)}^{\xi_{S_{i_1},i_2+1}(t)} f_1(x_1) \int_{\xi_{S_{i_1,i_2},i_3}(t,x_1)}^{\xi_{S_{i_1,i_2},i_3+1}(t,x_1)} f_2(x_2)\ldots \\
&\qquad \int_{\xi_{S_{i_1,\ldots,x_n},i_{n+1}}(t,x_1,\ldots,x_{n-1})}^{\xi_{S_{i_1,\ldots,i_n},i_{n+1}+1}(t,x_1,\ldots,x_{n-1})} f_n(x_n)\ dx_1\ldots dx_n
\end{aligned} \tag{3}
$$

3 Stochastic Programs

Syntax. We develop our exact/non-Monte-Carlo based evaluation for simple in-line probabilistic programs, specifically if-else statements, square roots and arithmetic expressions over independent continuous random variables. The grammar of programs that we support is given by:

$$
\begin{aligned}
\text{Expressions:}\quad & e ::= X \in \mathrm{Var} \mid c \in \mathbb{R} \mid e + e \mid e - e \mid e \times e \mid e \div e \mid \sqrt{e} \\
\text{Tests:}\quad & B ::= e > 0 \mid \neg B \mid B \wedge B \mid B \vee B \\
\text{Programs:}\quad & p ::= e \mid \mathrm{If}[B, p, p]
\end{aligned} \tag{4}
$$

where $\mathrm{Var} = \{X_1, X_2, ..., X_n\}$ is a finite and enumerated set of variables.

Semantics. We write $\mathrm{AST}(e)$ for the abstract syntax tree of e, $\mathrm{Var}(e)$ for the list of variables in e ordered by the enumeration of Var, $\mathrm{Var}(\mathrm{AST}(e))$ for the list of variables at the leaves of $\mathrm{AST}(e)$ ordered from left to right, and we define $M_e = |\mathrm{Var}(e)|$ and $N_e = |\mathrm{Var}(\mathrm{AST}(e))|$. Thus for $e = X_1 \div (X_1 + X_2)$, $\mathrm{Var}(e) = \{X_1, X_2\}$, $M_e = 2$, $\mathrm{Var}(\mathrm{AST}(e)) = (X_1, X_1, X_2)$ and $N_e = 3$. We overload the syntactic operators $\{+, -, \times, \div, \surd\}$ to denote their usual interpretations on $\mathbb{R}$.

Semantics of Expressions. We define both a deterministic semantics (used for tests) and a probabilistic semantics (used for programs) of an expression e and denote them by $(\!|e|\!)$ and $[\![e]\!]$ respectively. We opt not to follow the familiar approach consisting in updating a global valuation function over all the variable names in the language. Instead, we define $(\!|e|\!)$ as a map $\mathbb{R}^{M_e} \to \mathbb{R}$ (*i.e.* we only consider the variable in e as inputs). For this, we will need a bookkeeping function defined as follows: clearly $M_e \leq N_e$ and by combining the maps $\mathsf{copy} : \mathrm{Var} \to \mathrm{Var} \times \mathrm{Var}, X_i \mapsto (X_i, X_i)$ and $\mathsf{swap} : \mathrm{Var} \times \mathrm{Var} \to \mathrm{Var} \times \mathrm{Var}, (X_i, X_j) \mapsto (X_j, X_i)$ we can construct a map $\varphi_e : \mathrm{Var}^{M_e} \to \mathrm{Var}^{N_e}$ that turns the list $\mathrm{Var}(e)$ into the list $\mathrm{Var}(\mathrm{AST}(e))$. The same formal combination of copy and swap defines a map $\varphi_e : \mathbb{R}^{M_e} \to \mathbb{R}^{N_e}$ that populates the leaves of $\mathrm{AST}(e)$ with M_e concrete values, one for each variable in $\mathrm{Var}(e)$. We now define the semantics in two steps. First we inductively define $(\!|\mathrm{AST}(e)|\!) : \mathbb{R}^{N_e} \to \mathbb{R}$ as:

$$
\begin{aligned}
\text{Base cases:}\quad & (\!|\mathrm{AST}(X_i)|\!) : \mathbb{R} \to \mathbb{R}, x \mapsto x \qquad (\!|\mathrm{AST}(c)|\!) : \mathbb{R}^0 \to \mathbb{R}, 0 \mapsto c \\
\text{Inductive case:}\quad & (\!|\mathrm{AST}(e_1 * e_2)|\!) = * \circ ((\!|\mathrm{AST}(e_1)|\!) \times (\!|\mathrm{AST}(e_2)|\!)) \quad * \in \{+, -, \times, \div\} \\
& (\!|\mathrm{AST}(\sqrt{e})|\!) = \surd \circ (\!|\mathrm{AST}(e)|\!)
\end{aligned}
$$

where $\mathbb{R}^0 = \{0\}$, the 0-dimensional real vector space, and $(\!|\mathrm{AST}(e_1)|\!) \times (\!|\mathrm{AST}(e_2)|\!) : \mathbb{R}^{N_{e_1}} \times \mathbb{R}^{N_{e_2}} \simeq \mathbb{R}^{N_e} \to \mathbb{R} \times \mathbb{R}$ is the pairing of the maps $(\!|\mathrm{AST}(e_1)|\!), (\!|\mathrm{AST}(e_2)|\!)$. We can now define

$$(\!|e|\!) = (\!|\mathrm{AST}(e)|\!) \circ \varphi_e$$

where φ_e is the bookkeeping map defined earlier which 'prepares' the inputs of the function $(\!|e|\!)$ so that they can be fed to $\mathrm{AST}(e)$. This function will introduce perfect correlations/dependencies (when variable names are copied) in the probabilistic semantics that we define next. Thus, unless it is a permutation, φ_e is the mathematical representation of why we cannot simply inductively evaluate the probabilistic semantics using the formulas in (1).

The probabilistic semantics is defined in terms of the deterministic one via:

$$[\![e]\!] : \mathcal{P}(\mathbb{R}^{M_e}) \to \mathcal{P}(\mathbb{R}), \mathbb{P} \mapsto (\!|e|\!)_*\mathbb{P} \tag{5}$$

where $\mathcal{P}(\mathbb{R}^k)$ is the space of probability measures on the Borel space $\mathbb{R}^k$, $\mathbb{P}$ is the input (joint) distribution over the variables in $\mathrm{Var}(e)$ and $(\!|e|\!)_*\mathbb{P}$ is the output distribution defined as the pushforward of $\mathbb{P}$ through $(\!|e|\!)$, i.e. $(\!|e|\!)_*\mathbb{P}(A) = \mathbb{P}((\!|e|\!)^{-1}(A))$ for any Borel measurable $A \subseteq \mathbb{R}$. In practice, we will associate with each $X_i \in \mathrm{Var}(e)$ a probability distribution $\mathbb{P}_i \in \mathcal{P}(\mathbb{R})$, and take the input distribution to be the corresponding product of these distributions (i.e. we will assume that the input random variables are independent). For example, we might be interested in computing a distribution like

$$[\![X_1 \div (X_1 + X_2)]\!](\mathsf{Uniform}(0,1) \otimes \mathsf{Uniform}(2,3)) \tag{6}$$

In the language of random variables, this is the distribution of $X_1 \div (X_1 + X_2)$ when $X_1 \sim \mathsf{Uniform}(0,1)$ and $X_2 \sim \mathsf{Uniform}(2,3)$ are independent. By definition of the pushforward, the CDF of (6) is

$$\begin{aligned} &[\![X_1 \div (X_1 + X_2)]\!](\mathsf{Uniform}(0,1) \otimes \mathsf{Uniform}(2,3))((-\infty, t)) \\ = \ &\mathsf{Uniform}(0,1) \otimes \mathsf{Uniform}(2,3)(\{(x_1,x_2) \mid (\!|X_1 \div (X_1+X_2)|\!)(x_1,x_2) < t\}) \\ = \ &\int_{\left\{(x_1,x_2) \mid \frac{x_1}{x_1+x_2} < t\right\}} \mathbb{1}_{[0,1]}(x_1)\mathbb{1}_{[2,3]}(x_2)\, dx_1 dx_2 \end{aligned}$$

which is of the shape (2). CAD is the tool that will allow us to compute these integrals, and therefore the probabilistic semantics of expressions and programs.

Semantics of Tests. The semantics of tests is defined inductively as expected

Base case: $(\!|e > 0|\!) = \{x \in \mathbb{R}^{M_e} \mid (\!|e|\!)(x) > 0\}$

Inductive case: $(\!|\neg B|\!) = (\!|B|\!)^c, (\!|B_1 \wedge B_2|\!) = (\!|B_1|\!) \cap (\!|B_2|\!), (\!|B_1 \vee B_2|\!) = (\!|B_1|\!) \cup (\!|B_2|\!)$,

Semantics of Programs. We need only define the semantics of conditional branching. We follow [16,27]: every test B defines a measurable subset $(\!|B|\!) \subseteq \mathbb{R}^{M_B}$ and this subset defines a linear operator $z_B : \mathcal{P}_{\leq 1}(\mathbb{R}^{M_B}) \to \mathcal{P}_{\leq 1}(\mathbb{R}^{M_B})$ – where $\mathcal{P}_{\leq 1}(\mathbb{R}^{M_B})$ is the set of *sub*distributions on $\mathbb{R}^{M_B}$ – defined by $z_B(\mathbb{P})(A) = \mathbb{P}(A \cap (\!|B|\!))$. Clearly, $z_B(\mathbb{P})$ is in general a subdistribution since the total mass of $\mathbb{R}^{M_B}$ under $z_B(\mathbb{P})$ is $z_B(\mathbb{P})(\mathbb{R}^{M_B}) = \mathbb{P}(\mathbb{R}^{M_B} \cap (\!|B|\!)) = \mathbb{P}((\!|B|\!))$. In terms of density functions, z_B zeroes out the PDF of $\mathbb{P}$ on $(\!|B|\!)^c$. Thus the density of $z_{X_1>0}(\mathsf{Normal}(0,1))$ for example is the positive half of the bell curve. We define

$$[\![\mathrm{If}[B, p_1, p_2]]\!] = [\![p_1]\!] \circ z_B + [\![p_2]\!] \circ z_{\neg B} \tag{7}$$

Intuitively, z_B and $z_{\neg B}$ split the input into a subdistribution over M_B-tuples that satisfes the guard and another that satisfies its complement; these are passed to their respective branches, then re-combined to produce the output distribution.

4 Algorithms

In this section we describe the function CDF() from Algorithm 2 that takes as input a program $p(X_1, \ldots, X_k)$ from the syntax (4) and for each $X_i, 1 \leq i \leq k$, a PDF $f_i : (a_i, b_i) \to \mathbb{R}^+, a_i, b_i \in \mathbb{R} \cup \{-\infty, \infty\}$, and returns the CDF

$$F_p(t) = [\![p]\!](\mathbb{P}_1 \otimes \ldots \otimes \mathbb{P}_k)\big((-\infty, t)\big).$$

Semi-algebraic Input Set Generation. The first step consists in specifying the semi-algebraic set over which we will integrate the input PDF following (2). This semi-algebraic set will be defined from three types of symbolic inequalities:

1. constraints on the support of the input variables, *viz.* $(a_i < X_i) \wedge (X_i < b_i)$,
2. the guards of the if-else statements and their complements,
3. the expressions $e(X_1, \ldots, X_n) < t$ at the leaves of the program's AST that specify the problem of computing the CDF.

The symbolic semi-algebraic set associated with the input data described above is constructed recursively by the function SAS() in Algorithm 1 applied to the program $p(X_1, \ldots, X_n)$ and the symbolic variables range constraints

$$S = \bigwedge_{i=1}^{k} \big((a_i < X_i) \wedge (X_i < b_i)\big). \tag{8}$$

The function SAS() of Algorithm 1 calls a function to get rid of square root expressions which, strictly speaking, are not part of the syntax defining semi-algebraic sets. It is not hard to check that this can always be done as is proved below and implemented as the function RemoveRoots() in [15, Appendix]. Our syntax only includes square roots, but the result below can easily be generalised to all roots and we could therefore have included them in our syntax.

Proposition 2. *For every inequality of the form* $e(X_1, \ldots, X_k) < 0$, *where* $e(X_1, \ldots, X_k)$ *is constructed from* (4), $(\!|e(X_1, \ldots, X_k) < 0|\!)$ *is semi-algebraic.*

Algorithm 1. Computing the Semi-Algebraic Set associated with the problem

Input: Program $p(X_1, \ldots, X_k)$, semi-algebraic set S initialised as (8)
Output: The symbolic semi-algebraic set associated with computing p's output CDF

```
function SAS(p, S)
    if p = If[B, p1, p2] then
        B ← RemoveRoots(B)
        ifSAS ← SAS(p1, B ∧ S)
        elseSAS ← SAS(p2, ¬B ∧ S)
        return (ifSAS ∨ elseSAS)
    else                                ▷ Base case: p(X1, ..., Xn) is an expression
        return S∧ RemoveRoots(p − t < 0)
    end if
end function
```

Given the semantics of if-else branching described in (7), it may not immediately be clear that aggregating the preconditions, the guards, their complements, and the inequalities $p - t < 0$ at each leaf in the way described by Algorithm 1 will produce the desired semantics. The following result shows that it does.

Proposition 3. *Algorithm 1 is semantically correct in the sense that*

$$[\![p]\!](\mathbb{P}_1 \otimes \ldots \otimes \mathbb{P}_k)\big((-\infty, t)\big) = \mathbb{P}_1 \otimes \ldots \otimes \mathbb{P}_k\big((\!|\mathrm{SAS}(p, S)|\!)\big)$$

CAD. Once the semi-algebraic input set S has been constructed by SAS(), it is fed to CylindricalAlgebraicDecomposition($S, \{t, X_1, \ldots, X_n\}$)) that will construct the associated CAD, using the specified variable ordering $\{t, X_1, \ldots, X_n\}$. CylindricalAlgebraicDecomposition() returns a boolean combination of inequalities that we assign to the variable CADTree in Algorithm 2. Its exact shape is in practice somewhat unpredictable and putting it in DNF allows us to traverse it conveniently. Each disjunct of DNF(CADTree) corresponding to a given 1-cell (c_i, d_i) will be a conjunction of the shape

$$c_i < t < d_i \wedge f_1(t) < x_1 < g_1(t) \wedge \ldots \wedge f_k(t, x_1, \ldots, x_{k-1}) < x_k < f_k(t, x_1, \ldots, x_{k-1}).$$

As discussed in Sect. 2, the semi-algebraic functions $f_i(t, x_1, \ldots, x_{i-1}), 1 \leq i \leq k-1$ in the expression above are defined as the parametrised roots of a (projection) polynomial in x_i. Unless this root is expressible explicitly by radicals, it will be represented implicitly in this way.

Full and Semi-evaluations of CDF. Each of the disjuncts above allow us to compute one of the summands in the CAD-based expression of the CDF (3). This computation, *viz.* piece ← piece + Integration(conjunction, $f_1, \ldots, f_k$), defines the CDF as a piecewise function (the guard of the **if** just above checks whether the conjunction contains the current one-cell $c_i < t < d_i$.). In implementing Integration() in Algorithm 2 there are three options: we can perform it symbolically, numerically at specific values of t, or as a mixture of the two, with some summands of (3) evaluated symbolically and others left unevaluated

– and thus requiring numerical evaluation – using a best effort strategy. In the first case we will talk of full evaluation, in the other cases of semi-evaluation[3].

Algorithm 2. Algorithm to compute the CDF of $p(X_1, ..., X_k)$

Input: Program $p(X_1, \ldots, X_k)$, PDFs $f_1, \ldots, f_k$ with ranges $(a_1, b_1), \ldots (a_k, b_k)$.
Output: Piecewise CDF function F

```
function CDF(p, S, f)
    F ← []                                                    ▷ Empty piecewise function
    InputSAS ← SAS(p, ⋀_{i=1}^{k} (a_i < X_i) ∧ (X_i < b_i))
    CADTree ← CAD(InputSAS, {t, X_1, ..., X_n})
      ▷ Optional: try different variable orderings {t, X_{i_1}, ..., X_{i_n}} to get fewer cells
    CADTree ← RemoveMeasureZeroCells(CADTree)
    OneCells ← GetOneCells(CADTree)             ▷ Defines the piecewise structure F
    CADTree ← DNF(CADTree)
    for cell in OneCells do
        piece ← 0
        for conjunction in CADTree do
            if conjunction ⊂ cell then
                piece ← piece + Integration(conjunction, f_1 ··· f_k)
                                                ▷ Optional: run integrations in parallel
            end if
        end for
        F.append([cell, piece])
    end for
    return F             ▷ The CDF is then plotted and differentiated to get the PDF
end function
```

Performance Optimisation. We have seen that CAD is doubly exponential in the number of variables in the program. In practice however, we observe that the computational bottleneck is not the generation of the CAD, but the nested integrals of (3) that are exponential in the number of variables. To speed up this part of Algorithm 2 we use two strategies. The first is to find a good variable elimination ordering as this can lead to significantly fewer cells, thus reducing the cost of integration. Brown's heuristic [7,33] aims to achieve this by ordering variables from the simplest (low maximal degree and few occurrences) to the most complex (high maximal degree and many occurrences). Second, since every Integration(conjuction, $f_1, \ldots, f_k$) computation is independent of the others, we can parallelise the computation of the integrals.

Implementations. We have developed two implementations of Algorithm 2. The first, written in Mathematica, is designed to compute full evaluations and relies on Mathematica's powerful symbolic integration capabilities. It provides our gold-standard solutions. The second, using only open-source code, aims to

---

[3] Thus $\frac{t^3}{3} + \frac{t^2}{2}$ is a full evaluation of $\int_0^t (x^2 + x)\,dx$ and $\int_0^t x^2\,dx + \frac{t^2}{2}$ is semi-evaluated.

efficiently obtain semi-evaluated solutions that can be evaluated or plotted fast. Both tools implement the performance optimisations described above.

Mathematica Implementation. Mathematica's `CylindricalDecomposition[]` function [39] computes our CADs. Mathematica encodes the root expressions defining a CAD by using `Root`$[p(x_1, \dots x_i, \#)\&, j]$ expressions, standing for the jth real root of the (projection) polynomial $p(x_1, \dots x_i, x_{i+1})$ in the variable x_{i+1}. Its powerful integration engine allows us to compute closed-form solution for nested integrals containing root expressions in their bounds. This allows us to evaluate (3) in closed-form. Mathematica can also differentiate these expressions and produce closed-form expressions for the corresponding PDFs.

Open-Source Implementation. We have developed Py-CAD, a tool written in Python and C++ that performs the (semi-)evaluation of (3). After denominator and radical elimination (Props. 1 and 2) Py-CAD starts by simplifying the input semi-algebraic set S as much as possible using Z3 [30]. The simplified input is fed to QEPCAD [6,11]. QEPCAD produces projection polynomials and constructs cells consisting of an index and a sample point. Py-CAD parses the projection polynomials into SymPy expressions and, for each cell, reconstructs the symbolic boundaries as SymPy `RootOf` objects by solving the projection polynomials at the corresponding cell sample points. These cells are then assembled into a disjunctive normal form representation that is logically-equivalent to that produced by Mathematica. Py-CAD can then assemble the CDF (3) from integrals with exact algebraic boundaries regardless of the existence of closed form radical equivalents. During evaluation, a pass is conducted to identify fully determined `RootOf` nodes and substitute them for their numerical value via root-finding, allowing the integration engine to proceed. Py-CAD (like PSI) uses symbolic integration when possible (*e.g.* via C++ pattern matching) and then performs numerical evaluation on semi-evaluated expressions. In order to improve performance on larger benchmarks, we introduce compositional sub-CADs: when an independent subtree of variables is identified (e.g. y^2+z^2 within $x^2+y^2+z^2 < t$), its distribution is computed via an isolated sub-CAD and injected as a PDF into the outer problem, reducing a three-variable CAD to two two-variable CADs, a significant saving given that the complexity is doubly exponential in the number of variables [2]. For now, Py-CAD is focusing on numerical integration of semi-evaluated expressions, with future plans to include full symbolic integration.

5 Experimental Evaluation

Experimental Setup. We test both our implementations on programs and input distributions of various levels of complexity, from Taylor expansions (of order 5) of well-known functions, to programs from the FPBench benchmark suite [17] used in floating-point analysis, to three functions from IMU libraries (including `getMeasurement` from Sect. 1). Run times and various metrics are presented in Table 1. The column t_{math} (resp. $\hat{t}_{math}$) shows the run time for our Mathematica full evaluation (resp. numerical evaluation over 400 equidistant points), $\hat{t}_{open}$ show the run time of Py-CAD's semi-evaluation over 400 equidistant points and t_{psi} that of PSI's full evaluation (completed by Mathematica if marked by *).

The tools (artifact available at https://doi.org/10.5281/zenodo.20080511) take as input a program as well as a description of its input distributions. When a range of values is already defined in the literature, *e.g.* for the FPBench benchmarks, we stick to this range and place a distribution over it. We choose as input distributions on ranges $[a, b]$ either the uniform $\mathrm{U}(a, b)$, the triangular $\mathrm{T}(a, b)$ or the $\mathrm{Beta}(2, 2, a, b)$ distributions. The latter have PDFs

$$f_{\mathrm{T}}(x) = \begin{cases} \frac{4(x-a)}{(b-a)^2} & \text{if } a < x \leq a + {(a+b)}/{2} \\ \frac{4(b-x)}{(b-a)^2} & \text{if } a + {(a+b)}/{2} < x \leq b. \end{cases} \qquad f_{\mathrm{Beta}}(x) = \mathbb{1}_{[a,b]}(x)\frac{6(x-a)(b-x)}{(b-a)^3}$$

Distributions with infinite or compact supports are supported by our tool, the only practical limitation is the tractability of their PDFs in symbolic integrations.

Evaluation of Correctness. To sanity-check our method, we compare its result to Monte-Carlo sampling, a fool-proof approximation of the ground truth. For this we draw 10^6 samples from each distribution to create an empirical CDF and a histogram of the output distribution. The running time for this procedure is reported in column t_{mc}. We measure the distance between our full evaluation and Monte-Carlo sampling using the KS-distance

$$KS(\mathbb{P}_{full}, \mathbb{P}_{mc}) = \sup_t |F_{full}(t) - F_{mc}(t)|$$

where $\mathbb{P}_{full}$ (resp. F_{full}) is our output distribution (resp. closed-form CDF) and similarly for Monte-Carlo[4]. Examples of comparison between exact and empirical PDFs and CDFs can be found in Table 2 with a complete list of graphs in [15, Appendix] showing that our tools return the correct output distributions.

Discussion of Results. Our Mathematica implementation produces symbolic CDFs and PDFs (see Table 2 and [15, Appendix]) that quickly become extremely complex. For some of the benchmarks in Table 1, *e.g.* the `doppler` programs, the symbolic expressions returned by Mathematica are so large that they cannot be displayed on an A4 page. It is remarkable that CAD and symbolic integration are capable of evaluating these closed-form outputs in a reasonable time, providing a complete probabilistic range analysis for these programs, and in particular the ability to compute the probability of outcomes to arbitrary precision.

It is interesting to note that the full evaluation is often competitive with the semi-evaluated numerical answer, *i.e.* it is often more efficient to solve symbolically once than perform 400 numerical integrations. The running time in either case grows quickly as we move from one- or two-variable programs to three-variable ones. However, it remains doable to semi-evaluate the output distribution of a relatively complex program like get_meas from Sect. 1 at hundreds

[4] We use Mathematica's built-in `KolmogorovSmirnovTest` function. When it takes too long (over 30 seconds), we estimate the KS distance by evaluating the CDFs over 1000 equidistant percentiles: $\max(_{1 \leq i \leq 1000}|(F_a(x_i) - F_{mc}(x_i)|)$.

Table 1. Running time and KS-distance of a collection of benchmarks. The tools are run on an 8-core laptop with Apple M2 chip and 16GB RAM, or 10-core AMD Ryzen AI 9 with 32GB RAM (Py-CAD). The code of the benchmarks can be found in the arXiv version of this paper [15, Appendix] together with plots of the output CDFs and PDFs. Depth is the Mathematica tree-depth of the expression (this is smaller than the tree depth given by our syntax (4)). # vars describes the number of variables in the program, # cells is the total number of cells in the corresponding CAD – and therefore a good proxy for the complexity of the symbolic integration. KS denotes the KS distance between the symbolic distribution and the empirical distribution based on 10^6 input samples. Timings are given as follows. t_{mc} for the Monte-Carlo approach (10^6 samples). t_{math} for fully-evaluated Mathematica, while $\hat{t}_{math}$ is numerically-evaluated Mathematica. $\hat{t}_{open}$ is from semi-evaluated Py-CAD. Finally, t_{psi} is the timing for PSI. An asterix (*) is used where the PSI output contained integrals, which were then symbolically integrated via Mathematica (in parallel, and included in the timing). T-O stands for 'timed out'; timeouts were chosen according to benchmark complexity. X means error (no result given).

benchmark	depth	#vars	#cells	KS $\times 10^{-3}$	t_{mc}	t_{math}	$\hat{t}_{math}$	$\hat{t}_{open}$	t_{psi}
1a. Taylor expansions - uniform inputs									
exp	3	1	2	0.97	1.69	0.86	1.19	2.85	5.49*
sin	3	1	2	1.11	1.74	0.65	0.90	2.82	0.76*
cos	3	1	2	1.81	1.71	0.89	1.04	2.19	0.55*
log	4	1	2	0.71	1.70	0.61	0.87	2.75	0.81*
1b. Taylor expansions - triangular inputs (timeout=5mins)									
exp	3	1	2	0.72	1.74	1.52	1.14	3.84	175.74*
sin	3	1	2	0.54	1.73	1.03	0.87	4.08	0.76*
cos	3	1	2	0.76	1.73	1.29	0.83	4.09	T-O
log	4	1	2	0.62	1.74	1.03	1.10	3.83	163.79*
2a. FPBench [17] benchmarks - uniform inputs (timeout=20mins)									
sum	1	3	9	0.47	2.00	8.36	12.90	6.25	4.89
xbyxy	3	2	7	0.69	1.76	1.10	1.37	2.88	X
nonlin1	3	1	2	0.84	5.16	2.73	2.83	1.95	1.58*
bspline3	2	1	2	0.76	1.72	0.72	0.99	2.08	0.25
cav10	4	1	5	0.67	2.92	0.82	1.16	2.89	0.72
doppler1	4	3	14	0.91	1.81	27.22	53.04	9.33	362.18*
doppler2	4	3	14	0.79	1.77	29.11	35.00	9.85	T-O
doppler3	4	3	14	0.65	1.77	29.20	39.76	9.61	T-O
rigidBody1	2	3	40	0.70	1.90	63.88	X	112.4	T-O
2b. FPBench [17] benchmarks - triangular inputs (timeout=20mins)									
sum	1	3	9	0.73	2.07	79.01	54.72	294.52	9.81
xbyxy	3	2	7	0.47	1.84	13.50	2.87	19.33	T-O
nonlin1	3	1	2	0.80	6.96	1.75	1.67	3.62	11.82*
bspline3	2	1	2	0.49	1.74	1.19	1.00	3.21	0.28
cav10	4	1	5	1.09	2.99	0.99	1.24	5.51	0.87
doppler1	4	3	14	0.75	1.76	T-O	168.03	196.77	T-O
doppler2	4	3	14	0.96	1.83	T-O	208.13	220.26	T-O
doppler3	4	3	14	0.51	1.82	T-O	94.25	390.19	T-O
2b. FPBench [17] benchmarks - Beta(2,2) inputs (timeout=20mins)									
sum	1	3	9	0.76	2.13	45.89	105.34	88.14	8.83
xbyxy	3	2	7	1.04	1.90	3.09	1.85	7.8	X
nonlin1	3	1	2	1.26	6.59	1.90	1.68	2.45	4.87*
bspline3	2	1	2	0.73	1.77	1.02	0.86	2.12	0.26
cav10	4	1	5	0.62	3.08	1.16	1.20	3.18	0.90
doppler1	4	3	14	0.92	1.89	T-O	360.76	41.86	X
doppler2	4	3	14	1.12	1.92	T-O	489.75	49.2	X
doppler3	4	3	14	0.63	1.95	T-O	260.12	60.86	T-O
3a. IMU benchmarks - uniform inputs (timeout=60mins&100mins)									
v_norm	4	3	10	0.76	1.86	134.63	9.72	14.26	T-O
change_gyro	3	3	9	0.78	7.71	T-O	230.93	X	T-O
get_meas	4	3	40	1.57	5.42	T-O	T-O	5653.12	T-O

Table 2. Some examples of PDF and CDF.

CDF	PDF

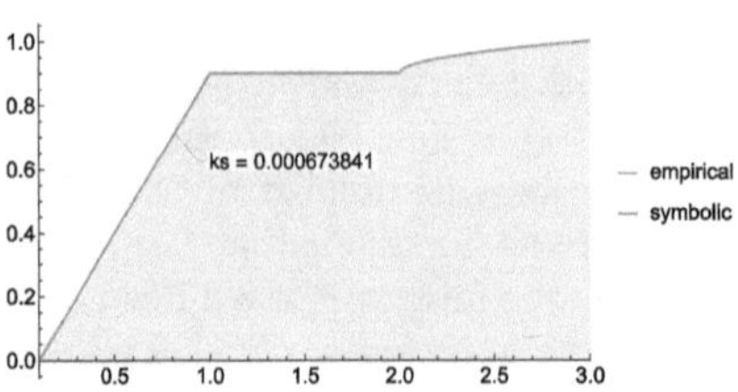

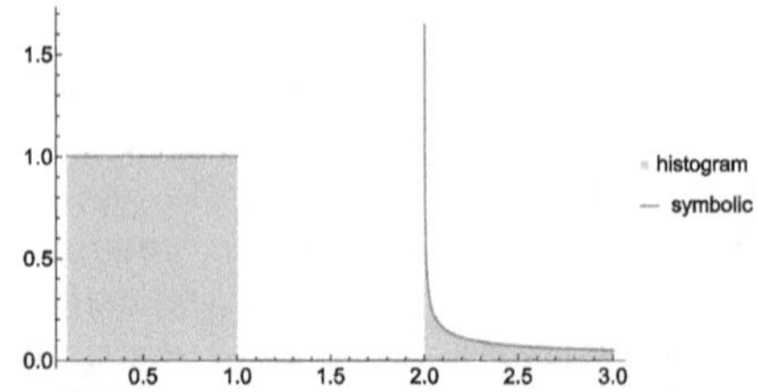

cav10: If$\left[x^2 - x \geq 0, \frac{x}{10}, x^2 + 2\right]$, $x \sim \mathrm{U}(0, 10)$

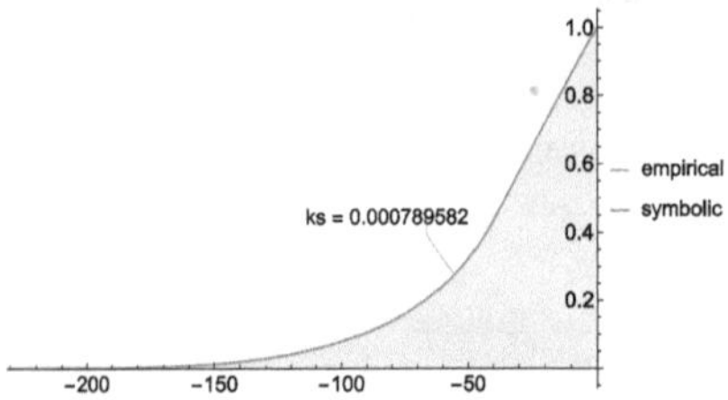

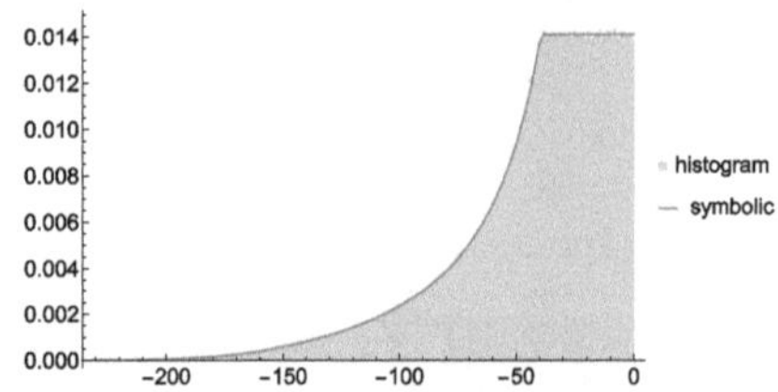

doppler2: $\frac{\left(-\frac{3T}{5} - \frac{1657}{5}\right)v}{\left(\frac{3T}{5} + u + \frac{1657}{5}\right)^2}$, $u \sim \mathrm{U}(-125, 125), v \sim \mathrm{U}(15, 25000), T \sim \mathrm{U}(-40, 60)$

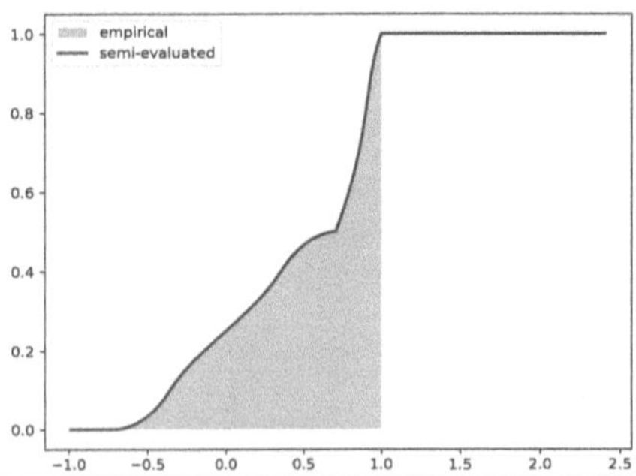

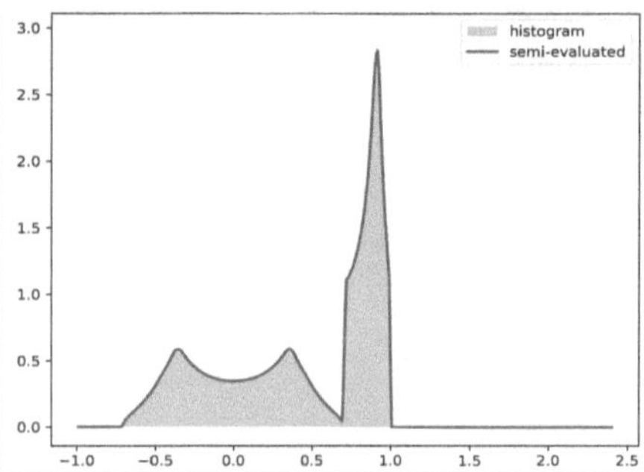

get_meas: c.f. § 1, $x \sim \mathrm{U}(-1, 1), y \sim \mathrm{U}(-1, 1), z \sim \mathrm{U}(-1, 1)$ [*evaluated by Py-CAD*]

of points. In fact, in order to provide formal guarantees like $\mathbb{P}[\mathsf{BadOutcome}] < \varepsilon$, it is typically sufficient to evaluate the CDF at a handful of points (describing the region BadOutcome), which will be orders of magnitude faster than the 400 evaluations of the CDFs and PDFs in our experiments. We can therefore expect that computing these kinds of exact probabilistic bounds should scale beyond the examples in our experiment.

Finally, we note that our tools perform well compared with PSI, which although much more general and able to perform exact inference, sometimes crashes and often times-out due to the complexity of the expressions it returns.

6 Conclusions and Future Work

We have presented a generic method for exactly evaluating the output distribution of programs with random inputs described by density functions. This method relies on Cylindrical Algebraic Decomposition and symbolic integration and handles if-else programs performing arithmetic operations and radicals. We have implemented our method in Mathematica and as Py-CAD, an open-source Python/C++ tool, and applied it successfully to small benchmarks.

Despite its poor worse-case complexity, the main computational bottleneck of this approach is not the computation of the CAD, but the symbolic evaluation of nested integrals. Computing exact output distributions will always scale badly for this reason. However, we have shown that it is a realistic proposition for small programs from industrial application (FPBench) and sensor libraries. Using further refinements and optimisations (such as evaluating independent sub-programs using sub-CADs), we believe that we can further increase the range of programs for which exact output distributions can be either fully or semi-evaluated. Bearing in mind that for the purpose of program verification it is enough to be able to compute the probability of certain outcomes, and that this typically only requires a few evaluations of the output CDF, we are hoping to apply our technique to larger programs, *e.g.* four- or five-variable programs manipulating quaternions in IMU libraries.

Apart from these scalability questions, our next task will be to incorporate floating-point errors to our framework. The model of [12,13] seems particularly well-suited to this task as the error is represented as an independent noise variable applied multiplicatively at each arithmetic operation.

Acknowledgments. This work was supported by the Research Institute on Verified Trustworthy Software (VeTSS) grant "*Probabilistic Precision Tuning*".

References

1. Arnon, D.S., Collins, G.E., McCallum, S.: Cylindrical algebraic decomposition i: The basic algorithm. SIAM J. Comput. **13**(4), 865–877 (1984)
2. Basu, S., Pollack, R., Roy, M.F.: Algorithms in real algebraic geometry. Springer, Cham (2006)
3. Bhat, S., Borgström, J., Gordon, A.D., Russo, C.: Deriving probability density functions from probabilistic functional programs. In: Proceedings of the 19th international conference on Tools and Algorithms for the Construction and Analysis of Systems, pp. 508–522 (2013)
4. Bouissou, O., Goubault, E., Goubault-Larrecq, J., Putot, S.: A generalization of p-boxes to affine arithmetic. Computing **94**(2), 189–201 (2012)
5. Bouissou, O., Goubault, E., Putot, S., Chakarov, A., Sankaranarayanan, S.: Uncertainty Propagation Using Probabilistic Affine Forms and Concentration of Measure Inequalities. In: Chechik, M., Raskin, J.-F. (eds.) TACAS 2016. LNCS, vol. 9636, pp. 225–243. Springer, Heidelberg (2016). https://doi.org/10.1007/978-3-662-49674-9_13

6. Brown, C.W.: Qepcad b: a program for computing with semi-algebraic sets using cads. SIGSAM Bull. **37**(4), 97–108 (2003). https://doi.org/10.1145/968708.968710
7. Brown, C.W.: Companion to the tutorial: Cylindrical Algebraic Decomposition. Presented at ISSAC 2004 (2004). available from https://www.usna.edu/Users/cs/wcbrown/research/ISSAC04/handout.pdf
8. Carette, J., Shan, C.-C.: Simplifying Probabilistic Programs Using Computer Algebra. In: Gavanelli, M., Reppy, J. (eds.) PADL 2016. LNCS, vol. 9585, pp. 135–152. Springer, Cham (2016). https://doi.org/10.1007/978-3-319-28228-2_9
9. Choi, Y., Vergari, A., Van den Broeck, G.: Probabilistic circuits: A unifying framework for tractable probabilistic models. UCLA **6** (2020). http://staraics.ucla.edu/papers/ProbCirc20.pdf
10. Collins, G.E.: Quantifier elimination for real closed fields by cylindrical algebraic decomposition: a synopsis. ACM SIGSAM Bull. **10**(1), 10–12 (1976)
11. Collins, G.E., Hong, H.: Partial cylindrical algebraic decomposition for quantifier elimination. J. Symb. Comput. **12**(3), 299–328 (1991)
12. Constantinides, G., Dahlqvist, F., Rakamarić, Z., Salvia, R.: Rigorous Roundoff Error Analysis of Probabilistic Floating-Point Computations. In: Silva, A., Leino, K.R.M. (eds.) CAV 2021. LNCS, vol. 12760, pp. 626–650. Springer, Cham (2021). https://doi.org/10.1007/978-3-030-81688-9_29
13. Constantinides, G., Dahlqvist, F., Rakamarić, Z., Salvia, R.: Automated roundoff error analysis of probabilistic floating-point computations. ACM Trans. Probab. Mach. Learn. **1**(3), 1–35 (2025)
14. Cutler, T.J.: Dutton's Nautical Navigation, Naval Institute Press Annapolis (2004)
15. Dahlqvist, F., Bandukara, M.H., Omidvari, N.: Exact evaluation of probabilistic programs with cylindrical algebraic decomposition (2026). https://arxiv.org/abs/2606.24514
16. Dahlqvist, F., Silva, A., Kozen, D.: Semantics of probabilistic programming: A gentle introduction (2020)
17. Damouche, N., et al.: Toward a standard benchmark format and suite for floating-point analysis (2016)
18. Darwiche, A.: A differential approach to inference in bayesian networks. J. ACM (JACM) **50**(3), 280–305 (2003)
19. Ferrenberg, A.M., Landau, D.P., Wong, Y.J.: Monte carlo simulations: Hidden errors from "good" random number generators. Phys. Rev. Lett. **69**, 3382–3384 (1992). https://doi.org/10.1103/PhysRevLett.69.3382
20. Garg, P., Holtzen, S., Van den Broeck, G., Millstein, T.: Bit blasting probabilistic programs. Proc. ACM Program. Lang. **8**(PLDI), 865–888 (2024)
21. Gehr, T., Misailovic, S., Vechev, M.: PSI: Exact Symbolic Inference for Probabilistic Programs. In: Chaudhuri, S., Farzan, A. (eds.) CAV 2016. LNCS, vol. 9779, pp. 62–83. Springer, Cham (2016). https://doi.org/10.1007/978-3-319-41528-4_4
22. Gehr, T., Steffen, S., Vechev, M.: λpsi: exact inference for higher-order probabilistic programs. In: Proceedings of the 41st ACM sigplan Conference on Programming Language Design and Implementation, pp. 883–897 (2020)
23. Hofmann-Wellenhof, B., Legat, K., Wieser, M.: Navigation. Springer Science & Business Media (2003)
24. Holtzen, S., Van den Broeck, G., Millstein, T.: Scaling exact inference for discrete probabilistic programs. Proc. ACM Program. Lang. **4**(OOPSLA), 1–31 (2020)
25. Jaroszewicz, S., Korzeń, M.: Arithmetic operations on independent random variables: a numerical approach. SIAM J. Sci. Comput. **34**(3), A1241–A1265 (2012). https://doi.org/10.1137/110839680

26. Korzeń, M., Jaroszewicz, S.: Pacal: A python package for arithmetic computations with random variables. J. Stat. Softw. **57**, 1–34 (2014)
27. Kozen, D.: Semantics of probabilistic programs. In: 20th Annual Symposium on Foundations of Computer Science (sfcs 1979), pp. 101–114. IEEE (1979)
28. Lohar, D., Prokop, M., Darulova, E.: Sound Probabilistic Numerical Error Analysis. In: Ahrendt, W., Tapia Tarifa, S.L. (eds.) IFM 2019. LNCS, vol. 11918, pp. 322–340. Springer, Cham (2019). https://doi.org/10.1007/978-3-030-34968-4_18
29. Matsumoto, M., Nishimura, T.: Mersenne twister: a 623-dimensionally equidistributed uniform pseudo-random number generator. ACM Trans. Model. Comput. Simul. (TOMACS) **8**(1), 3–30 (1998)
30. de Moura, L., Bjørner, N.: Z3: An Efficient SMT Solver. In: Ramakrishnan, C.R., Rehof, J. (eds.) TACAS 2008. LNCS, vol. 4963, pp. 337–340. Springer, Heidelberg (2008). https://doi.org/10.1007/978-3-540-78800-3_24
31. Narayanan, P., Carette, J., Romano, W., Shan, C., Zinkov, R.: Probabilistic Inference by Program Transformation in Hakaru (System Description). In: Kiselyov, O., King, A. (eds.) FLOPS 2016. LNCS, vol. 9613, pp. 62–79. Springer, Cham (2016). https://doi.org/10.1007/978-3-319-29604-3_5
32. Poon, H., Domingos, P.: Sum-product networks: A new deep architecture. In: 2011 IEEE International Conference on Computer Vision Workshops (ICCV Workshops), pp. 689–690. IEEE (2011)
33. del Río, T., England, M.: New heuristic to choose a cylindrical algebraic decomposition variable ordering motivated by complexity analysis. In: Computer Algebra in Scientific Computing (CASC 2022). LNCS, vol. 13366, pp. 300–321. Springer, Cham (2022). https://doi.org/10.1007/978-3-031-14788-3_17
34. Saad, F.A., Rinard, M.C., Mansinghka, V.K.: Sppl: probabilistic programming with fast exact symbolic inference. In: Proceedings of the 42nd ACM sigplan International Conference on Programming Language Design and Implementation, pp. 804–819 (2021)
35. Sánchez-Cauce, R., Paris, I., Díez, F.J.: Sum-product networks: A survey. IEEE Trans. Pattern Anal. Mach. Intell. **44**(7), 3821–3839 (2021)
36. Sankaranarayanan, S., Chou, Y., Goubault, E., Putot, S.: Reasoning about uncertainties in discrete-time dynamical systems using polynomial forms. Adv. Neural. Inf. Process. Syst. **33**, 17502–17513 (2020)
37. Specht, M.: Consistency analysis of global positioning system position errors with typical statistical distributions. J. Navig. **74**(6), 1201–1218 (2021). https://doi.org/10.1017/S0373463321000485
38. Springer, M.D.: The algebra of random variables. Wiley, New York (1979)
39. Strzeboński, A.: Solving systems of strict polynomial inequalities. J. Symb. Comput. **29**(3), 471–480 (2000)
40. Wu, Y., Zou, D., Liu, P., Yu, W.: Dynamic magnetometer calibration and alignment to inertial sensors by kalman filtering. IEEE Trans. Control Syst. Technol. **26**(2), 716–723 (2017)
41. Zheng, T., Xu, A., Xu, X., Liu, M.: Modeling and compensation of inertial sensor errors in measurement systems. Electronics **12**(11) (2023)

Scalable Reachability Analysis of Linear Continuous Systems with Property-Driven Time-Step Adaptation

Mikkel Bjørn, Daniel H. Hansen, Grace Melchiors, Kim G. Larsen, and Christian Schilling(✉)

Aalborg University, Aalborg, Denmark
christianms@cs.aau.dk

QEST + FORMATS Artifacts Evaluation Available

QEST + FORMATS Artifacts Evaluation Functional

Abstract. We study safety verification for linear time-invariant systems with bounded inputs in continuous time. The standard approach reduces to a reachability analysis in two steps: first discretize time and then apply a forward analysis in the discretized system. Existing algorithms use either a fixed time step or an adaptive time step that changes based on the approximation error compared to the underlying continuous system. In this paper, we present an efficient reachability algorithm that adapts the time step based on a given safety property. Essentially, our algorithm makes the largest possible time step such that it can still prove safety. For this approach to be scalable in practice, we discuss several optimizations such as avoiding the repeated expensive calculation of the matrix exponential during discretization and a careful balance how we tame the approximation error stemming from the states and the inputs. This allows our algorithm to yield a moderate approximation error even when using a large time step, thus requiring much fewer steps than prior algorithms. We demonstrate the effectiveness and scalability on the large-scale SLICOT benchmark suite, where our algorithm consistently outperforms other state-of-the-art approaches.

Keywords: Reachability Analysis · LTI System · Safety Verification · Adaptive Time Step · Discretization

1 Introduction

Verification of safety properties for linear time-invariant (LTI) systems is a well-studied problem [1,2,5,8,14,15,21,24]. In this paper, we focus on the common setting with continuous-time dynamics described by ordinary differential equations (ODEs), where the initial states and inputs are uncertain within a bounded domain, and where time is bounded. Formally, our systems have the form

$$\dot{x}(t) = Ax(t) + u(t), \quad t \in [0, T], \tag{1}$$

M. Chen and M. Tschaikowski (Eds.): QEST+FORMATS 2026, LNCS 16913, pp. 96–113, 2027.
https://doi.org/10.1007/978-3-032-35298-9_6

where the initial state $x(0) \in \mathcal{X}_0$ is constrained to a set of initial states, $u(t) \in \mathcal{U}$ is the input signal constrained to an input domain, and $T > 0$ is the time horizon.

Given a safety property in the form of a set of safe states $\mathcal{X}_s$, the safety verification problem is to determine whether all trajectories emerging from $\mathcal{X}_0$ with input signal $u(t) \in \mathcal{U}$ stay safe within time T. The problem thus reduces to computing the set of reachable states $\mathcal{R}_{[0,T]}(\mathcal{X}_0)$ up to time horizon T and checking that $\mathcal{R}_{[0,T]}(\mathcal{X}_0)$ is included in $\mathcal{X}_s$. The set $\mathcal{R}_{[0,T]}(\mathcal{X}_0)$ is generally not computable except for simple subclasses [7,8,19]. Hence, one typically computes an overapproximation $\widehat{\mathcal{R}}_{[0,T]}(\mathcal{X}_0) \supseteq \mathcal{R}_{[0,T]}(\mathcal{X}_0)$; this is sufficient, as $\widehat{\mathcal{R}}_{[0,T]}(\mathcal{X}_0) \subseteq \mathcal{X}_s$ implies $\mathcal{R}_{[0,T]}(\mathcal{X}_0) \subseteq \mathcal{X}_s$. The standard approach to obtaining $\widehat{\mathcal{R}}_{[0,T]}(\mathcal{X}_0)$ is to (1) discretize the continuous system and (2) compute a series of discrete steps in the resulting discrete system [10]. The approximation quality can be increased by reducing the time step of the discretization, which, however, then necessitates a larger number of steps to be taken.

While most approaches use a fixed time step and thus require a single discretization operation, some more advanced approaches instead use an adaptive time step [11,12,17,22,25]. The latter approaches have in common that they analyze the approximation error (based on error estimates) dynamically and choose a time step such that the error is below a user-defined threshold at all times.

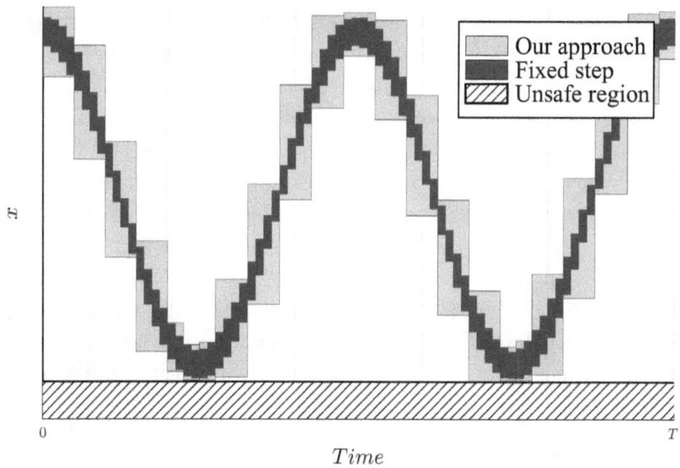

Fig. 1. Two overapproximations $\widehat{\mathcal{R}}_{[0,T]}(\mathcal{X}_0)$ of the reachable states. Each block corresponds to one step in time. The green sequence was obtained by a fixed-step approach with a step size chosen such that safety (staying above the bottom unsafe region) is verified. The brown sequence was obtained by our approach, which chooses the time step depending on the distance to the unsafe region.

In this paper, we propose a new approach that uses neither a fixed time step nor a fixed approximation error. As a motivation, we now discuss the two main drawbacks of these two existing paradigms. First, both paradigms scale poorly when high precision is required *at some point* during the analysis. This is clear for approaches using a fixed time step, as the time step must be chosen small enough for such critical points, even if a larger step would be sufficient most of the time. Approaches using a fixed error can sometimes make larger steps, but they are still held back by the prescribed error bound, and, just like above, this error bound must be small enough; hence, the variance in step sizes is often rather modest in practice. This drawback is illustrated in Fig. 1, where the adaptive algorithm would have to use a similar error as the fixed-step algorithm, since the approximation $\widehat{\mathcal{R}}_{[0,T]}(\mathcal{X}_0)$ gets very close to the error states.

The second drawback is that the user is expected to anticipate the required precision up-front. Either a small enough step size or an allowed approximation error must be chosen, but it is initially unknown how closely the reachable states (and hence the approximation $\widehat{\mathcal{R}}_{[0,T]}(\mathcal{X}_0)$) will approach the boundaries of $\mathcal{X}_s$.

Our approach does not have these two drawbacks, i.e., it can make large time steps for better scalability and does not require an error bound. Specifically, we assume that the user only wants to verify safety as fast as possible and does not care about the approximation error. At a high level, our approach is optimistic and tries to make large time steps. If at some point the precision is insufficient to prove $\widehat{\mathcal{R}}_{[0,T]}(\mathcal{X}_0) \subseteq \mathcal{X}_s$, we backtrack and reduce the step size, which improves the precision. Later, we attempt to increase the step size again to improve scalability.

While this *property-guided* idea is simple, a naive implementation would not scale, especially for high-dimensional systems. In particular, it is costly to repeatedly compute a discretization whenever the time step changes, among other reasons due to the expensive computation of the matrix exponential. Our main contribution is an efficient implementation of this idea. A key insight is to precompute a library of conservative time-discretized systems [10] for different step sizes, which allows us to change the step size fast. Moreover, for efficient calculation of this library, we use specific step sizes (powers of two; in particular, this choice avoids the repeated computation of the matrix exponential). Furthermore, to deal with the uncertain inputs $u(t)$, we propose a hybrid approach that combines small-step and large-step analysis. Finally, we utilize a mixed set representation when calculating $\widehat{\mathcal{R}}_{[0,T]}(\mathcal{X}_0)$; specifically, we combine zonotopes and the support function [1], where we use zonotopes for the discretization and the support function for taking the discrete steps.

As the main limitation of our approach, the precision is selected based on the property; hence, the results are not reusable if the property changes.

Contributions. First, we present a new algorithm for property-guided reachability analysis of LTI systems (1), which significantly reduces the computational verification effort. Second, we show how to implement conservative time discretization for variable time steps in a scalable way by reusing intermediate results to obtain a more precise sets for larger step sizes while only calculating the matrix exponential once. Third, we implement our approach efficiently and demonstrate its scalability on a set of high-dimensional benchmark models in comparison to state-of-the-art reachability algorithms. Our source code and benchmark scripts are publicly available [16].

1.1 Related Work

Fixed Time Step. The standard way to reachability analysis for LTI dynamics discretizes system (1) for a small time step $\delta > 0$ [10]. We shortly sketch the general procedure for the simplified case without inputs $u(t)$ here (and we recall one concrete way in the next section). First, one computes an overapproximation $\widehat{\mathcal{R}}_{[0,\delta]}(\mathcal{X}_0)$, for which a coarse approximation suffices because δ, and hence the approximation error, is small. One also computes the matrix $e^{A\delta}$ (the matrix exponential of $A\delta$); the motivation is that the linear map $e^{A\delta}\mathcal{X}_0$ describes the set of reachable states $\mathcal{R}_{[\delta,\delta]}(\mathcal{X}_0)$ at time point δ. Finally, one propagates the set $\Omega_0 = \widehat{\mathcal{R}}_{[0,\delta]}(\mathcal{X}_0)$ forward with the recurrence $\Omega_{k+1} = e^{A\delta}\Omega_k$ until $k\delta \geq T$. The sets Ω_k are the blocks depicted in Fig. 1.

There are many algorithms implementing the above procedure in different ways, for instance varying in the choice of the representation or the way to obtain $\widehat{\mathcal{R}}_{[0,\delta]}(\mathcal{X}_0)$ [1,5,11,13,14,20,21]. All these approaches have in common that they use a fixed time step δ. As a consequence, the user needs to choose δ small enough for achieving sufficient precision to prove the safety property; hence, k becomes large in practice, even if most of the sets Ω_k are far away from the boundaries of the safe region and would consequently not require high precision.

Adaptive Time Step. Existing reachability algorithms with variable time steps have the goal to achieve a prescribed approximation error [11,12,17,22,25]. For that, these algorithms estimate the error and adapt the time step accordingly. Hence, they typically shine if a good approximation is desired throughout the computation. However, this error-bounded paradigm is decoupled from the verification problem: one first approximates the full reachable states and only checks the safety property afterward. Thus, if one is interested in solving the verification problem, one has to guess which precision will be needed. In contrast, we propose a property-driven algorithm, for which the approximation error does not matter as long as it is small enough to prove the property. We automatically adapt to the error such that the safety property can be proven. While our algorithm sometimes has to backtrack if the precision is insufficient to prove safety, this backtracking is always by a single step only, and we only have to approximate the reachable states once in this way, while error-bounded adaptive algorithms would have to redo the whole work if the error parameter was chosen too large.

Time Discretization. Since the seminal work by Girard [13], efficient reachability algorithms use the paradigm of time discretization followed by set propagation, albeit different approximation models have been proposed for the discretization [10]. Notable approximation models are based on interval matrices [3] or an optimization problem over a forward and a backward approximation [11]. In this paper, we utilize a simplified (forward-only) version of the latter, which avoids the optimization step [5] (see Sect. 2.2). One of our central contributions is a new approach to compute a time discretization for large time steps with good precision and scalability, which has not been demonstrated before.

2 Preliminaries

In this section, we recall the essential steps to derive classical reachability algorithms for LTI systems. We first explain, assuming some discretization method, how to compute a sequence of sets that contain the reachable states. Then, we recall one particular such discretization method. Finally, we discuss two efficient ways to represent and operate with sets: zonotopes and the support function.

2.1 Approximating the Reachable States

Recall that we consider the continuous-time LTI dynamics in (1), which are characterized by the triple $(A, \mathcal{X}_0, \mathcal{U})$, with $A \in \mathbb{R}^{n \times n}$ and $\mathcal{X}_0, \mathcal{U} \subseteq \mathbb{R}^n$:

$$\dot{x}(t) = Ax(t) + u(t), \quad t \in [0, T] \tag{1}$$

Note that this includes systems of the form $\dot{x}(t) = Ax(t) + Bu(t) + c$ by a standard transformation of the input domain $\mathcal{U}' \leftarrow B\mathcal{U} \oplus \{c\}$, where $\oplus$ denotes the Minkowski sum. The unique solution of ODE system (1) at time point t is

$$x(t) = e^{At}x(0) + \int_0^t e^{A(t-s)}u(s)ds, \tag{2}$$

for initial state $x(0)$ and input signal $u(t) \in \mathcal{U}$. We assume that both $\mathcal{X}_0$ and $\mathcal{U}$ are compact and convex sets. Let $\mathcal{R}_{[t_1,t_2]}(\mathcal{X}_0)$ denote the reachable states in a time interval $[t_1, t_2]$ starting from the initial set $\mathcal{X}_0$; formally:

$$\mathcal{R}_{[t_1,t_2]}(\mathcal{X}_0) = \{x(\tau) \mid t_1 \leq \tau \leq t_2, x(0) \in \mathcal{X}_0, u(t) \in \mathcal{U}\}$$

(Our notation omits the implicit dependence of $\mathcal{R}$ on $\mathcal{U}$ for simplicity.) The exact set of reachable states for a time interval is often not computable, and one typically overapproximates it with a sequence of sets $\Omega_0, \ldots, \Omega_{N-1}$ called a *reach tube* such that $\mathcal{R}_{[t_k,t_{k+1}]}(\mathcal{X}_0) \subseteq \Omega_k$ up to time T, where $t_0 = 0$ and $t_N = T$. Thus, from now on we consider the overapproximation $\widehat{\mathcal{R}}_{[0,T]}(\mathcal{X}_0) = \Omega_0 \cup \cdots \cup \Omega_{N-1}$.

Given a set of safe states $\mathcal{X}_s \subseteq \mathbb{R}^n$, the *safety verification problem* is to determine whether $\mathcal{R}_{[0,T]} \subseteq \mathcal{X}_s$ holds. A sufficient condition is $\widehat{\mathcal{R}}_{[0,T]}(\mathcal{X}_0) \subseteq \mathcal{X}_s$, for any overapproximation; however, a failed inclusion check is inconclusive.

As the system is linear, we can utilize the superposition principle of the solution in (2) to compute the *input solution* (derived by letting $x(0) = 0$) separately. For a time step $\delta > 0$, we have that [11, Lemma 1]

$$\mathcal{R}_{[t,t+\delta]}(\mathcal{X}_0) = e^{At}\mathcal{R}_{[0,\delta]}(\mathcal{X}_0) \oplus \mathcal{R}_{[t,t]}(\{0\}). \tag{3}$$

Thus, for any time point $t \geq 0$, to overapproximate the reachable states in the time interval $[t, t+\delta]$, it suffices to overapproximate the two summands, which cover the reachable states in the interval $[0, \delta]$ resp. the input solution at time t. Following notation from [11], we introduce shorthands for these approximations:

$$\begin{aligned} \Omega_{[0,\delta]}(\mathcal{X}_0, \mathcal{U}) &\supseteq \mathcal{R}_{[0,\delta]}(\mathcal{X}_0), \\ \Psi_t(\mathcal{U}) &\supseteq \mathcal{R}_{[t,t]}(\{0\}). \end{aligned} \tag{4}$$

Given a time step δ, we call a triple $(e^{A\delta}, \Omega_{[0,\delta]}(\mathcal{X}_0, \mathcal{U}), \Psi_\delta(\mathcal{U}))$ satisfying (4) a *discretized system*. Letting $\Psi_0 = \{0\}$ and $\delta_k = t_{k+1} - t_k$, Eq. (3) yields a recurrence to compute the reach tube $\Omega_0, \ldots, \Omega_{N-1}$ [11, Eq. (6)]:

$$\begin{aligned} \Psi_{k+1} &= \Psi_k \oplus e^{At_k} \Psi_{\delta_k}(\mathcal{U}), \\ \Omega_k &= e^{At_k} \Omega_{[0,\delta_k]}(\mathcal{X}_0, \mathcal{U}) \oplus \Psi_k \end{aligned} \tag{5}$$

Note that (5) does not require to use a constant time step δ. But when mixing different time steps, one also needs to compute multiple discretized systems. It can be shown that the reach tube obtained from (5) for time steps δ_k with N chosen such that $\sum_{k=0}^{N-1} \delta_k \geq T$ satisfies [11, Proposition 1]

$$\mathcal{R}_{[0,T]}(\mathcal{X}_0) \subseteq \bigcup_{k=0}^{N-1} \Omega_k.$$

2.2 Forward Discretization Method

We use the *forward* approximation model [5] (a simplification of the model in [11] which also goes backward in time) for instantiating $\Omega_{[0,\delta]}$ and Ψ_t in (4).

$$\Psi_\delta(\mathcal{U}) = \delta\mathcal{U} \oplus E_\Psi(\mathcal{U}, \delta), \tag{6}$$

$$\Omega_{[0,\delta]}(\mathcal{X}_0, \mathcal{U}) = CH(\mathcal{X}_0, e^{A\delta}\mathcal{X}_0 \oplus E^+(\mathcal{X}_0, \delta) \oplus \Psi_\delta(\mathcal{U})) \tag{7}$$

where $CH(\mathcal{X}, \mathcal{Y})$ denotes the convex hull of the union of two sets $\mathcal{X}$ and $\mathcal{Y}$ and

$$\begin{aligned} E_\Psi(\mathcal{U}, \delta) &= \boxdot(\Phi_2(|A|, \delta) \boxdot(A\mathcal{U})), \\ E^+(\mathcal{X}_0, \delta) &= \boxdot(\Phi_2(|A|, \delta) \boxdot(A^2\mathcal{X}_0)), \end{aligned}$$

where $\Phi_2(A, \delta) = \sum_{i=0}^{\infty} \frac{\delta^{i+2}}{(i+2)!} A^i$ [11], $|A|$ denotes the component-wise absolute value of A, and $\boxdot(\mathcal{X})$ yields the symmetric interval hull of $\mathcal{X}$, which is the smallest axis-aligned box centered in the origin and containing $\mathcal{X}$.

2.3 Zonotopes

A *zonotope* is the image of a unit hypercube under an affine map. Equivalently, it is described by a center $c \in \mathbb{R}^n$ and a finite number of generators $g_1, \ldots, g_p \in \mathbb{R}^n$:

$$\mathcal{Z} = \left\{ c + \sum_{i=1}^{p} \alpha_i g_i \mid \alpha_i \in [-1, 1] \right\}$$

We use the shorthand notations $\mathcal{Z} = (c, \langle g_1, \ldots, g_p \rangle)$ and $\mathcal{Z} = (c, G)$, where $G \in \mathbb{R}^{n \times p}$ is a matrix whose columns are the generators. The order of a zonotope is $o = \frac{p}{n}$, describing the relative number of generators with respect to the dimensionality of the state space.

Let $\mathcal{Z} = (c, \langle g_1, \ldots, g_p \rangle) = (c, G)$ and $\mathcal{Z}' = (c', \langle g'_1, \ldots, g'_\ell \rangle)$ be two zonotopes. Their Minkowski sum is the zonotope $\mathcal{Z} \oplus \mathcal{Z}' = (c + c', \langle g_1, \ldots, g_p, g'_1, \ldots, g'_\ell \rangle)$. Applying a linear map M to $\mathcal{Z}$ yields the zonotope $M\mathcal{Z} = (Mc, MG)$. The convex hull of (the union of) two zonotopes is generally not a zonotope; we utilize the operator $\overline{CH}(\mathcal{Z}, \mathcal{Z}')$ presented in prior work [13] that yields a zonotope overapproximating the exact convex hull $CH(\mathcal{Z}, \mathcal{Z}')$, e.g., in (7).

The Minkowski sum increases the number of generators, which makes it computationally critical when applied repeatedly. We use two countermeasures. First, we apply a classical order-reduction algorithm [13], which overapproximates a zonotope by another zonotope with fewer generators [18]. Second, we sometimes use the support function (described below) to represent the effect of the Minkowski sum instead of explicitly computing the zonotope.

2.4 Support Function

The *support function* ρ yields the support of a compact convex set $\mathcal{X} \subseteq \mathbb{R}^n$ in a direction $\ell \in \mathbb{R}^n$, which is the scalar $\rho_\mathcal{X}(\ell) = \max\{\ell^\top x \mid x \in \mathcal{X}\}$. Specifically, the support of a zonotope $\mathcal{Z} = (c, \langle g_1, \ldots, g_p \rangle)$ is $\rho_\mathcal{Z}(\ell) = \ell^\top c + \sum_{i=1}^{p} |\ell^\top g_i|$ [1]. Given compatible compact convex sets $\mathcal{X}, \mathcal{Y}$ and a matrix M, it is easy to evaluate the support function for Minkowski sums and linear maps [1, Table II]:

$$\begin{aligned} \rho_{\mathcal{X} \oplus \mathcal{Y}}(\ell) &= \rho_\mathcal{X}(\ell) + \rho_\mathcal{Y}(\ell), \\ \rho_{M\mathcal{X}}(\ell) &= \rho_\mathcal{X}(M^\top \ell). \end{aligned}$$

It has previously been shown that reachability analysis can be performed with the use of the support function [20], also in combination with zonotopes [1]. Using the support function has two main benefits: first, evaluating the support function for a small number of directions ℓ is typically cheaper than applying the full set computations. Second, we do not increase the zonotope order when evaluating the support of the Minkowski sum of two zonotopes; instead, we only add the (scalar) support of the sets in the given direction. We will use zonotopes for the discretization, as they avoid the recursive definition of the support function.

Note that only knowing the support of the sets Ω_k in selected directions is not a restriction for us. We only need to check whether they are contained in the safe region $\mathcal{X}_s$. Typically, $\mathcal{X}_s$ is given as a polyhedron in half-space representation, so it suffices to evaluate Ω_k in one direction per half-space. If $\mathcal{X}_s$ is not given in half-space representation, one can approximate Ω_k with template polyhedra [11].

3 Property-Driven Time-Step Adaptation

In this section, we describe our adaptive reachability approach with the following roadmap. We first select a finite set of time steps (Sect. 3.1). Next, given an LTI system (1), we discretize the continuous dynamics using (6)–(7) for each selected time step (Sect. 3.2). Then, we present our adaptive algorithm to compute the reach tube using the recurrence in (5), where we try to choose the time step as large as possible (Sect. 3.3). During this computation, we check whether the set Ω_k violates the safety property. If so, we backtrack and decrease the time step. This way, we dynamically adjust the time step to verify safety. Finally, we propose an adaptation policy for when to increase the time step again (Sect. 3.4) and discuss how to mitigate high zonotope orders (Sect. 3.5).

3.1 Selection of Candidate Time Steps

We initially make a selection of potential time steps $\Delta = [\delta_0, \delta_1, \ldots, \delta_m]$ (sorted in ascending order). For each time step, we will need to compute the approximation in (7). To avoid excessive computations of the matrix exponential $\phi_{\delta_i} = e^{A\delta_i}$, given a choice for $\delta_0 > 0$, we choose the other step sizes by doubling:

$$\delta_{i+1} = 2\delta_i, \tag{8}$$

where $(0 \leq i < m)$. Thus, Δ is defined by the user parameters δ_0 and m (technically, m is upper-bounded via the time horizon T). Let $\bar{\phi}_i = e^{A\delta_i}$ be the matrix exponential for a given δ_i. Following (8), $\bar{\phi}_{i+1} = e^{A\delta_i \cdot 2} = (\bar{\phi}_i)^2$. Thus, we can compute $\bar{\phi}_i$ from $\bar{\phi}_0$ using matrix multiplication instead of computing the usual approximation of the power series, which is particularly expensive in high dimensions. For convenience, we introduce the aliases $\delta^- = \delta_0$ and $\delta^+ = \delta_m$.

Remark 1. The recurrence (8) makes the time steps grow fast (doubling each time). Using scaling and squaring, one could easily obtain a more fine-grained sequence that still exploits the same idea to avoid explicitly computing the matrix exponential. Still, in our empirical evaluation, we observed that the simpler version in (8) always sufficed. Note that having more candidate step sizes is not always beneficial, as it also increases the choices during backtracking later.

3.2 Discretization for Multiple Time Steps

Given an LTI system (1) of the form $(A, \mathcal{X}_0, \mathcal{U})$ and a selection of time steps $\Delta = [\delta_0, \ldots, \delta_m]$, we compute the discretized systems $(\phi_\delta, \Omega_{[0,\delta]}(\mathcal{X}_0, \mathcal{U}), \Psi_\delta(\mathcal{U}))$ for all $\delta \in \Delta$ using (6)–(7) (recall that $\phi_\delta = e^{A\delta}$).

Remark 2. While we chose to present our discretization algorithm as a variant of the *forward* method (cf. Sect. 2.2), the approach is generic and also applies to other discretization methods [10].

Algorithm 1 shows a naive baseline to compute the discretized systems. The discrete systems are represented as three maps ϕ, X, and V, which for each value $\delta \in \Delta$ store the corresponding matrix exponential ϕ_δ respectively the sets $\Omega_{[0,\delta]}$ and Ψ_δ. Observe that the algorithm is agnostic to the fact that the time step always doubles in line 8. In particular, the matrix exponential (and,

Algorithm 1. Naive Discretization

Input: A, $\mathcal{X}_0$, $\mathcal{U}$, δ^-, δ^+
1: $\delta \leftarrow \delta^-$
2: **while** $\delta \leq \delta^+$ **do**
3: $\quad \phi[\delta] \leftarrow e^{A\delta}$ ▷ Compute matrix exponential for each $\delta \in \Delta$
4: $\quad \Psi \leftarrow$ Compute using (6)
5: $\quad \Omega \leftarrow$ Compute using (7)
6: $\quad X[\delta] \leftarrow \Omega$
7: $\quad V[\delta] \leftarrow \Psi$
8: $\quad \delta \leftarrow 2\delta$
9: **end while**
10: **return** (ϕ, X, V)

Algorithm 2. Optimized Discretization

Input: A, $\mathcal{X}_0$, $\mathcal{U}$, δ^-, δ^+
1: $\delta \leftarrow \delta^-$
2: $\phi[\delta] \leftarrow e^{A\delta}$
3: $\Psi \leftarrow$ Compute using (6)
4: $\Omega \leftarrow$ Compute using (7)
5: **while** $\delta \leq \delta^+$ **do**
6: $\quad X[\delta] \leftarrow \Omega$
7: $\quad V[\delta] \leftarrow \Psi$
8: $\quad \Omega \leftarrow \overline{CH}(\Omega, \phi[\delta]\Omega \oplus \Psi)$ ▷ Improved two-step recurrence (5)
9: $\quad \Psi \leftarrow \Psi \oplus \phi[\delta]\Psi$
10: $\quad \delta \leftarrow 2\delta$
11: $\quad \phi[\delta] \leftarrow (\phi[\frac{\delta}{2}])^2$
12: **end while**
13: **return** (ϕ, X, V)

hidden in the references to (6)–(7), the matrix Φ_2) is computed for each value δ from scratch. Moreover, since the approximation error of the discrete systems grows as the time steps grow larger, the results for larger δ are useless in practice.

Algorithm 2 shows our optimized discretization method, which exploits the knowledge that the time steps always double to improve both efficiency and precision. The algorithm utilizes two crucial insights. First, as mentioned, despite using multiple time steps δ_i, we only need to compute the matrix exponential ϕ_{δ_0} *once* from scratch (line 2) and then can utilize matrix multiplication (line 11). The second insight is that it suffices to use the approximation model (6)–(7) for the smallest time step δ^- (lines 3–4), for which the approximation error is typically small. For larger time steps, we instead propagate the sets for the previous time step one step forward using the recurrence (5) (lines 8–9).

We illustrate the difference between Algorithm 1 and Algorithm 2 in Fig. 2 for the case $\mathcal{U} = \{0\}$. In theory, the outputs are incomparable because the better approximation of the iterative Algorithm 2 may be partially lost due to the additional approximation when taking the zonotope approximation of the

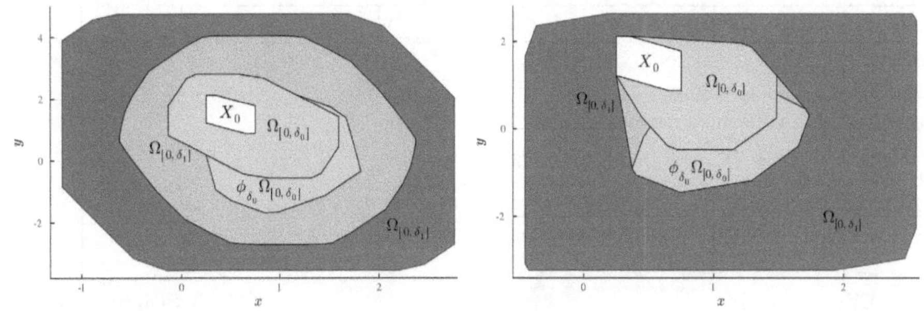

(a) Without zonotope approximation. (b) With zonotope approximation.

Fig. 2. Comparison of the discretization algorithms 1 and 2 for a system with similar dynamics as in Fig. 1 with initial states $\mathcal{X}_0$ (white) and $\mathcal{U} = \{0\}$. On the right, we use the exact convex-hull operator (CH), while on the left, we use the zonotope approximation ($\overline{CH}$). Both algorithms agree on $\Omega_{[0,\delta_0]}$. Algorithm 2 computes $\Omega_{[0,\delta_1]}$ as the convex hull of $\Omega_{[0,\delta_0]} \cup \phi_{\delta_0}\Omega_{[0,\delta_0]}$ (brown). Algorithm 1 computes $\Omega_{[0,\delta_1]}$ from scratch (green).

convex hull in line 8 (cf. Sect. 2.3). In practice, Algorithm 2 yields a much smaller approximation error, in particular for the larger time steps. This is because the common approximation models (like the one described in Sect. 2.2 that we use here) are typically only precise for small time steps [10].

3.3 Reach Tube Approximation with Adaptive Time Step

After obtaining the discrete systems, we use Algorithm 3 to compute the approximation of the reach tube. Additional inputs are the safe region $\mathcal{X}_s$, which we assume to be polyhedral, along with the list of directions L that describe its facets, the time horizon T and the available time steps Δ, and a time-step policy π which we explain later.

Algorithm 3 is inspired by the well-known reachability algorithm based on the support function [20] (however, that algorithm uses a fixed time step). The motivation to use the support function is to avoid (1) the explicit computation of the reach tube $\Omega_0, \dots, \Omega_{N-1}$ and (2) the emergence of high-order zonotopes, which would have to be tamed by order reduction (and thus additional approximation errors). Instead, we only evaluate the support function for the reach tube in selected directions, which is sufficient to check inclusion in the safe region $\mathcal{X}_s$.

The loop in line 6 computes the next set Ω_k of the reach tube by starting with the current time step δ. For each attempted δ, we compute bounds for the candidate set Ω_k (line 9) and perform the safety check (line 11). If Ω_k is safe, we accept the time step δ and update Φ, which stores the cumulative matrix exponential, for the next iteration (line 14). Otherwise, if Ω_k is not safe, we try the next-smaller time step δ (line 17). This process repeats until either $t \geq T$ (successful verification) or $\delta < \delta^-$ (failed verification) holds.

Algorithm 3. Computation of the Reach Tube

Input: (ϕ, X, V): discretized systems,
Input: $\mathcal{X}_s, L$: safe region and collection of corresponding direction vectors,
Input: T, Δ: time horizon and available time steps,
Input: π: policy for increasing time step
1: **for** $\ell \in L$ **do**
2: $\quad \Psi_0[\ell] \leftarrow 0$
3: **end for**
4: $t \leftarrow 0$, $k \leftarrow 0$, $\delta \leftarrow \delta^+$, $\Phi \leftarrow I$
5: **while** $t < T$ **do** ▷ Invariant: $\Phi = e^{At}$
6: $\quad$ **while** $\delta \geq \delta^-$ **do**
7: $\quad\quad$ **for** $\ell \in L$ **do** ▷ Evaluate $\rho_{\Psi_{t+\delta}(\mathcal{U})}(\ell)$ and $\rho_{\Omega_{[t,t+\delta]}(\mathcal{X}_0,\mathcal{U})}(\ell)$
8: $\quad\quad\quad \Psi_{k+1}[\ell] \leftarrow \Psi_k[\ell] + \rho_{V[\delta]}(\Phi^\top \ell)$
9: $\quad\quad\quad \Omega_k[\ell] \leftarrow \rho_{X[\delta]}(\Phi^\top \ell) + \Psi_k[\ell]$
10: $\quad\quad$ **end for**
11: $\quad\quad$ **if** $\Omega_k[\ell] \leq \rho_{\mathcal{X}_s}(\ell)$ for all $\ell \in L$ **then** ▷ Check $\Omega_k \subseteq \mathcal{X}_s$
12: $\quad\quad\quad t \leftarrow t + \delta$
13: $\quad\quad\quad k \leftarrow k + 1$
14: $\quad\quad\quad \Phi \leftarrow \Phi\phi[\delta]$ ▷ $e^{A(t+\delta)} = e^{At}e^{A\delta}$
15: $\quad\quad\quad$ **break**
16: $\quad\quad$ **else** ▷ Decrease time step and try again
17: $\quad\quad\quad \delta \leftarrow \frac{1}{2}\delta$
18: $\quad\quad$ **end if**
19: $\quad$ **end while**
20: $\quad$ **if** $\delta < \delta^-$ **then** ▷ $\Omega_k \not\subseteq \mathcal{X}_s$ at maximal precision
21: $\quad\quad$ **return** "Could not verify safety"
22: $\quad$ **else** ▷ May try a different step size next time
23: $\quad\quad \delta \leftarrow$ Adjust according to π
24: $\quad$ **end if**
25: **end while**
26: **return** "System is safe"

Clearly, Algorithm 3 terminates because there are finitely many choices for δ. The following theorem asserts that the computed result is correct; moreover, the algorithm can prove safety for the same systems as the induced fixed-step approach that uses the same discretization for the step size $\delta = \delta^-$ all the time. Note that this holds even if Algorithm 3 has not used δ^- all the time.

Theorem 1 (Soundness and relative completeness). *If Algorithm 3 returns in line 26, the system is safe. If Algorithm 3 returns in line 21, the standard fixed-step algorithm using δ^- at all times would also fail.*

Proof (Sketch). The first property holds trivially since we check safety for each step. The second property holds because (1) we always backtrack when a safety check fails and (2) right before Algorithm 3 fails, we use the smallest step δ^- and t is a multiple of δ^-. The latter holds because our time steps are multiples of δ^-. Hence, the last set we compute equals precisely the set the fixed-step algorithm would compute at this point, and thus that algorithm would also fail.

3.4 Adaptation Policy

While Algorithm 3 yields sound results under any policy π to adjust the next time step δ_i to a value in Δ, the policy impacts the efficiency. In our implementation, we use a policy that dynamically adjusts δ_i based on previous adjustments. First, our policy never decreases δ_i. Moreover, if the time step did not have to be reduced in the previous four steps, we double the time step (provided this would not exceed δ^+). We found that this policy works well in practice.

3.5 Mitigating the Growth of the Zonotope Order

As mentioned, when using zonotopes, the increase in generators caused by the repeated Minkowski additions can become a bottleneck. We use the support function to represent Ψ_k in Algorithm 3 (line 8). Furthermore, by accumulating Φ in line 14, we avoid additional order increases in line 9, since Ω_k is not recursively defined. However, order reductions can still be necessary in Algorithm 2 for larger time steps, as that algorithm uses the recurrence formulation.

Remark 3. The repeated convex hull computations can also significantly increase the zonotope order when $\delta^+ \gg \delta^-$. As evaluating the support function of a convex hull can be done exactly, this error could also be mitigated. However, empirically we found that this negatively impacts the efficiency of the algorithm.

4 Evaluation

In this section, we evaluate our implementation of the proposed approach, which we refer to as *Algorithm* 3 in the following (but which also includes Algorithm 2), in the Julia programming language. For all results, we use a machine with an Intel i5 11400F CPU and 16 GB RAM running Windows and using Julia v1.12.

4.1 Benchmark Problems and Baseline Algorithms

We consider a subset of the SLICOT benchmark models [9], which represent high-dimensional real-world applications. All models are LTI systems of the form (1) and are summarized in Table 1. The time horizon is always $T = 20$.

We compare to two state-of-the-art reachability algorithms that use a fixed time step. The first algorithm is the one based on the support function [15,20], which we denote as *LGG*. The second algorithm is based on decomposition [5], which we denote as *BFFPSV*. For both LGG and BFFPSV, we apply the same algorithm options as in a prior evaluation [5] and use the *forward* discretization. We employ the implementations of these algorithms in the tool JuliaReach [6], which is also written in Julia, for a fair comparison; for instance, we use the same method for calculating the matrix exponential for all three approaches.

Table 1. SLICOT benchmark statistics. The specifications are taken from [4,23]. Some specifications use a linear combination of the state variables, denoted by y.

Model	Dimension (n)	Safety property ($\mathcal{X}_s$)
Motor	8	$x_1 \in [0.35, 0.4] \vee x_5 \notin [0.45, 0.6]$
Building	48	$x_{25} < 6 \cdot 10^{-3}$
PDE	84	$y < 12$
Heat	200	$x_{133} < 0.1$
ISS	270	$y \in [-7, 7] \cdot 10^{-4}$
Beam	384	$x_{89} < 2100$
MNA1	578	$x_1 < 0.5$

4.2 Reach Tube Computation and Verification of Safety Properties

We report about the results of our evaluation in Table 2. Since LGG and BFF-PSV first compute the full reach tube and then verify safety (whereas in our approach, these steps are interleaved), we report the total time for these tools.

We used the same fixed time steps δ for LGG and BFFPSV as in the prior evaluation [5, Table 4]. We selected the step ranges (δ^- and δ^+) such that $\delta^- = \delta$; by Theorem 1, we thus can verify the same safety properties as the baselines.

We highlight that we often choose large values for δ^+ that would normally seem unrealistic (e.g., $\delta^+ = 1.024$ for the Heat model, which is close to $T = 20$). This highlights the precision we can achieve thanks to Algorithm 2.

Table 2. Evaluation on the SLICOT benchmarks. Bold entries mark the best result and † marks a failure to verify the property. Algorithm 3 uses $\delta^- = \delta$, where δ is the fixed time step, and δ^+ is given as a power of two larger than δ.

Model	$\delta^- = \delta$	Algorithm 3 (ours)			LGG	BFFPSV	
		δ^+/δ^-	Steps	Time (s)	Time (s)	Time (s)	Steps
Motor	$1 \cdot 10^{-3}$	2^3	2503	$\mathbf{1.82 \cdot 10^{-2}}$	$6.45 \cdot 10^{-2}$	$1.94 \cdot 10^{-1}$	20000
Building	$2 \cdot 10^{-3}$	2^9	239	$\mathbf{9.35 \cdot 10^{-3}}$	$2.49 \cdot 10^{-1}$	$1.46 \cdot 10^{-1}$	10000
PDE	$3 \cdot 10^{-4}$	2^{10}	81	$\mathbf{4.01 \cdot 10^{-2}}$	5.67	322	66667
Heat	$1 \cdot 10^{-3}$	2^{10}	28	$\mathbf{1.83 \cdot 10^{-1}}$	9.00	4.71	20000
ISS	$6 \cdot 10^{-4}$	2^5	1042	$\mathbf{3.25 \cdot 10^{-1}}$	2.87	† (88.5)	33334
Beam	$5 \cdot 10^{-5}$	2^5	12501	**4.44**	778	343	400000
MNA1	$4 \cdot 10^{-4}$	2^{11}	971	**2.57**	688	123	50000

Our approach consistently outperforms the baselines across all benchmarks, often achieving dramatic speedups of one or two orders of magnitude. The main reason is that we can effectively employ large time steps most of the time, whereas the fixed-step approaches must use tiny steps all the time. For instance, we are able to verify the 578-dimensional MNA1 model in less than 3 seconds, which is a $\times 47$ speedup over BFFPSV and a $\times 267$ speedup over LGG.

For the ISS model, BFFPSV is not precise enough to prove the property. This is consistent with previous results [5]. LGG is able to verify the property, and does so efficiently due to the sparsity of the system matrix. However, our approach still outperforms LGG with a $\times 8$ speedup.

4.3 Approximation Quality

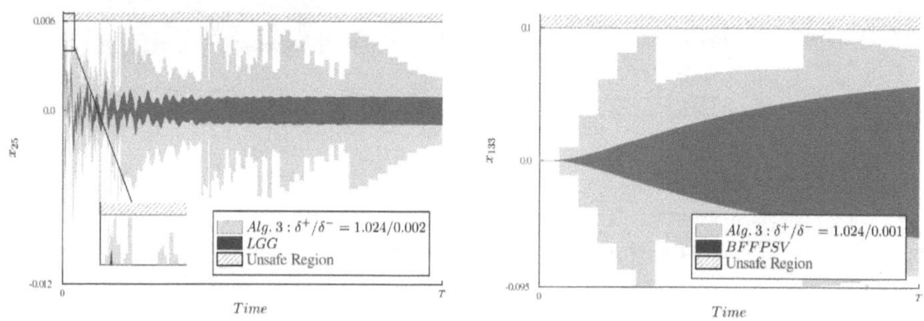

(a) Building model. (b) Heat model.

Fig. 3. Comparison of Algorithm 3 to LGG and BFFPSV. The step sizes for all algorithms are identical with those reported in Table 2.

In contrast to the baselines, our approach is willing to produce large approximation errors as long as the specification can be proven. Hence, a direct comparison of the approximation quality is not meaningful. Nevertheless, we show how the approximations differ empirically for three selected benchmark models. (Since the approximation quality of LGG and BFFPSV is typically similar to each other, we only show one of these approaches per plot.)

For the Building model (Fig. 3(a)), the reach tube computed by LGG is closest to the unsafe region in the first few steps, after which the high precision would not be needed anymore. Algorithm 3 gets close to the unsafe region (and hence switches the time step) many times. The reach tube computed by Algorithm 3 only consists of 239 steps, whereas LGG computes for 10,000 steps.

The Heat model (Fig. 3(b)) highlights multiple benefits of Algorithm 3. We can take large time steps because of the high precision of our discretization. Algorithm 3 verifies the Heat model in 28 steps, while BFFPSV computes for 20,000 steps. Additionally, the adaptive approach allows Algorithm 3 to continuously use the best time step from the selection Δ.

For the ISS model (Fig. 4), the green sets computed by LGG get close to the unsafe region. This explains why BFFPSV, which is generally less precise than LGG, cannot prove this specification. (Note that using smaller time steps

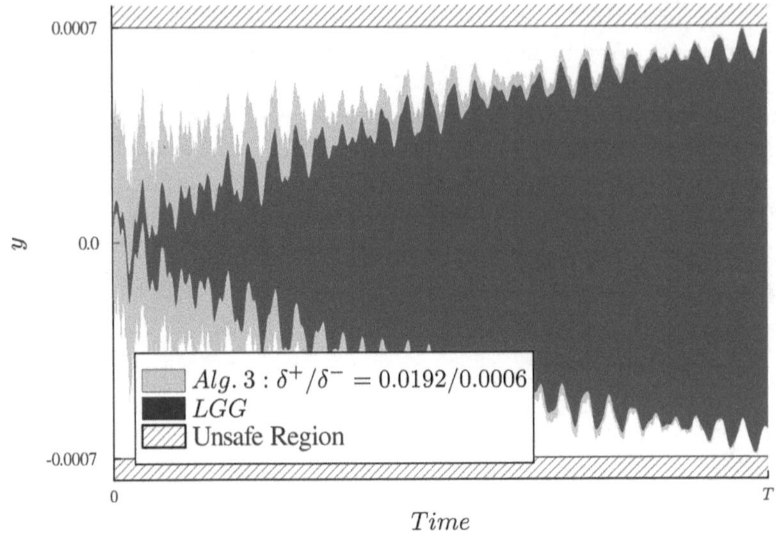

Fig. 4. Comparison of Algorithm 3 to LGG on the ISS model. The step sizes for the algorithms are identical with those reported in Table 2.

would not necessarily help, as that requires more time steps and hence more error accumulation.) Algorithm 3 can generally take larger time steps in the beginning. The reach tube computed by Algorithm 3 only consists of 1,042 steps, whereas LGG computes for 33,334 steps.

4.4 Property-Driven Adaptation

Figure 3(a) demonstrates how our approach benefits from property-driven time-step adaptation on the Building model. We see that the time step evolves over time for a model where precision is crucial early in the time interval $[0, T]$. As time advances, the distance between the green reach tube and the unsafe region becomes larger. This allows our approach to gradually increase δ, which drastically decreases computation time. Notably, the largest time step we use is $\times 512$ larger than the smallest one that is necessary to prove safety.

4.5 Comparison of Algorithm 1 and Algorithm 2

As mentioned, our optimized discretization algorithm empirically achieves lower approximation errors compared to the naive algorithm, even at large time steps. This allows us to sometimes even use the largest time step δ^+ for several steps during the reach tube computation in Algorithm 3. In Fig. 5, we demonstrate the difference of using the optimized discretization (Algorithm 2) over the naive discretization (Algorithm 1). Both algorithms use the smallest time step $\delta^- = 0.001$. With Algorithm 2 and $\delta^+ = 1.024$, our method successfully verifies the

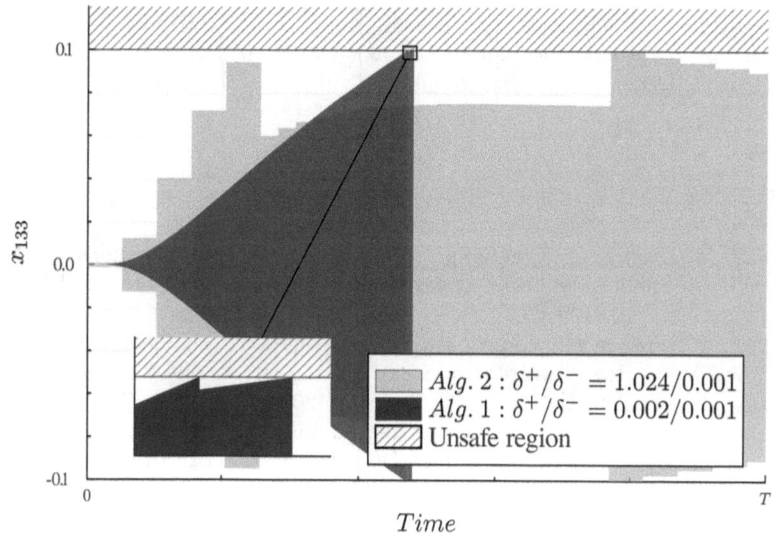

Fig. 5. Comparison of the discretization algorithms 1 and 2 on the Heat model. Algorithm 2 uses $\delta^- = 0.001$ and $\delta^+ = 1.024$, while Algorithm 1 uses $\delta^- = 0.001$ and $\delta^+ = 0.002$.

model, for which it only uses the largest time steps 0.512 and 1.024. Even when limiting Algorithm 1 to $\delta^+ = 0.002$, our method fails within 10 time units because the precision is not high enough. This is due to the accumulation of the approximation errors in Ψ when using Algorithm 1.

5 Conclusion

We have proposed an efficient reachability algorithm for continuous-time linear time-invariant systems with uncertain inputs. Our algorithm is based on time discretization and uses an adaptive step size. In contrast to other adaptive algorithms, ours is property-driven instead of error-driven: the precision is increased only when necessary to prove the specification. To achieve efficient performance, we also developed a novel discretization method that allows us to make large time steps in practice. We have implemented our algorithms efficiently and demonstrated performance improvements over state-of-the-art methods of one or two orders of magnitude on high-dimensional benchmark systems.

In the future, it will be interesting to investigate how our approach can be used for reach-avoid specifications. We also plan to generalize it to hybrid systems. There we can partially reuse our algorithm and treat transition guards as unsafe regions. In contrast, however, once an intersection with the guard is detected, we are in a new setting. Thus, a key challenge will be to avoid that our discretization algorithm becomes a bottleneck after each discrete transition, both in terms of performance and overapproximation.

Acknowledgments. This research was partly supported by the Independent Research Fund Denmark under reference number 10.46540/3120-00041B and the Villum Investigator Grant S4OS under reference number 37819.

References

1. Althoff, M., Frehse, G.: Combining zonotopes and support functions for efficient reachability analysis of linear systems. In: CDC, pp. 7439–7446. IEEE (2016). https://doi.org/10.1109/CDC.2016.7799418
2. Althoff, M., Frehse, G., Girard, A.: Set propagation techniques for reachability analysis. Annu. Rev. Control. Robotics Auton. Syst. **4**, 369–395 (2021). https://doi.org/10.1146/ANNUREV-CONTROL-071420-081941
3. Althoff, M., Stursberg, O., Buss, M.: Reachability analysis of linear systems with uncertain parameters and inputs. In: CDC, pp. 726–732. IEEE (2007). https://doi.org/10.1109/CDC.2007.4434084
4. Bak, S., Duggirala, P.S.: Simulation-equivalent reachability of large linear systems with inputs. In: CAV, pp. 401–420. LNCS, Springer (2017). https://doi.org/10.1007/978-3-319-63387-9_20
5. Bogomolov, S., Forets, M., Frehse, G., Podelski, A., Schilling, C.: Decomposing reach set computations with low-dimensional sets and high-dimensional matrices (extended version). Inf. Comput. **289** (2022). https://doi.org/10.1016/j.ic.2022.104937
6. Bogomolov, S., Forets, M., Frehse, G., Potomkin, K., Schilling, C.: JuliaReach: a toolbox for set-based reachability. In: HSCC, pp. 39–44. ACM (2019). https://doi.org/10.1145/3302504.3311804
7. Chen, S., Ge, X.: Reachability analysis of linear systems. Acta Informatica **61**(3), 231–260 (2024). https://doi.org/10.1007/S00236-024-00458-8
8. Dantam, M., Pouly, A.: On the decidability of reachability in continuous time linear time-invariant systems. In: HSCC, pp. 15:1–15:12. ACM (2021). https://doi.org/10.1145/3447928.3456705
9. Dooren, Y.C.P.V.: A collection of benchmark examples for model reduction of linear time invariant dynamical systems (2002)
10. Forets, M., Schilling, C.: Conservative time discretization: a comparative study. In: iFM. LNCS, vol. 13274, pp. 149–167. Springer (2022). https://doi.org/10.1007/978-3-031-07727-2_9
11. Frehse, G., et al.: SpaceEx: Scalable verification of hybrid systems. In: CAV. pp. 379–395. LNCS. Springer (2011). https://doi.org/10.1007/978-3-642-22110-1_30
12. Frehse, G., Kateja, R., Guernic, C.L.: Flowpipe approximation and clustering in space-time. In: HSCC, pp. 203–212. ACM (2013). https://doi.org/10.1145/2461328.2461361
13. Girard, A.: Reachability of uncertain linear systems using zonotopes. In: HSCC, pp. 291–305. LNCS. Springer (2005). https://doi.org/10.1007/978-3-540-31954-2_19
14. Girard, A., Guernic, C.L., Maler, O.: Efficient computation of reachable sets of linear time-invariant systems with inputs. In: HSCC, pp. 257–271. LNCS. Springer (2006). https://doi.org/10.1007/11730637_21
15. Guernic, C.L., Girard, A.: Reachability analysis of hybrid systems using support functions. In: CAV. pp. 540–554. LNCS. Springer (2009). https://doi.org/10.1007/978-3-642-02658-4_40

16. Hansen, D.H., Melchiors, G., Bjørn, M.: Kyniker/adaptive-algorithm-for-linear-system: React (2026). https://doi.org/10.5281/zenodo.20793291
17. Kochdumper, N., Bak, S.: Fully automated verification of linear time-invariant systems against signal temporal logic specifications via reachability analysis. Nonlinear Anal. Hybrid Syst **53**, 101491 (2024). https://doi.org/10.1016/j.nahs.2024.101491
18. Kopetzki, A., Schürmann, B., Althoff, M.: Methods for order reduction of zonotopes. In: CDC, pp. 5626–5633. IEEE (2017). https://doi.org/10.1109/CDC.2017.8264508
19. Lafferriere, G., Pappas, G.J., Yovine, S.: Symbolic reachability computation for families of linear vector fields. J. Symb. Comput. **32**(3), 231–253 (2001). https://doi.org/10.1006/JSCO.2001.0472
20. Le Guernic, C., Girard, A.: Reachability analysis of linear systems using support functions. Nonlinear Anal. Hybrid Syst. **4**(2), 250–262 (2010). https://doi.org/10.1016/j.nahs.2009.03.002, iFAC World Congress 2008
21. Luo, E., Kochdumper, N., Bak, S.: Reachability analysis for linear systems with uncertain parameters using polynomial zonotopes. In: HSCC, pp. 17:1–17:12. ACM (2023). https://doi.org/10.1145/3575870.3587130
22. Prabhakar, P., Viswanathan, M.: A dynamic algorithm for approximate flow computations. In: HSCC, pp. 133–142. ACM (2011). https://doi.org/10.1145/1967701.1967722
23. Tran, H., Nguyen, L.V., Johnson, T.T.: Large-scale linear systems from order-reduction. In: ARCH, pp. 60–67. EPiC Series in Computing, EasyChair (2016). https://doi.org/10.29007/XK7X
24. Wetzlinger, M., Althoff, M.: Backward reachability analysis of perturbed continuous-time linear systems using set propagation. IEEE Trans. Autom. Control **71**(1), 352–367 (2026). https://doi.org/10.1109/TAC.2025.3592719
25. Wetzlinger, M., Kochdumper, N., Bak, S., Althoff, M.: Fully automated verification of linear systems using inner and outer approximations of reachable sets. IEEE Trans. Autom. Control **68**(12), 7771–7786 (2023). https://doi.org/10.1109/TAC.2023.3292008

Robust Parameter Learning for Uncertain MDPs

Yannik Schnitzer(✉), Alessandro Abate, and David Parker

University of Oxford, Oxford, UK
yannik.schnitzer@cs.ox.ac.uk

Abstract. Learning-based approaches to verifying unknown Markov decision processes (MDPs) often employ uncertain MDPs. These models use, for example, confidence intervals to capture transition uncertainty and allow synthesis of policies that are robust to this uncertainty. However, this approach typically quantifies uncertainty independently for individual transition probabilities, ignoring dependencies due to shared latent quantities. We propose to learn such models using parametric MDPs (pMDPs), where transition probabilities are expressions over a set of parameters. We project statistical uncertainty from empirical transition frequencies onto the pMDP's parameter space, yielding a probably approximately correct (PAC) uncertainty model for the underlying MDP that respects the algebraic dependencies between transitions. The resulting models are algorithmically challenging to solve, so we propose a hierarchy of sound polytopic outer approximations of the induced confidence set. We implement and evaluate our approach, demonstrating substantially tighter uncertainty estimates than classical interval-based uncertain MDP learning techniques.

1 Introduction

Markov decision processes (MDPs) are the standard formalism for sequential decision-making under uncertainty. Their classical formulation assumes that all transition probabilities are known exactly, which is rarely realistic in practice. Two closely related model classes relax this assumption: *uncertain MDPs* (UMDPs) [19,28,43] and *parametric MDPs* (pMDPs) [17,23]. UMDPs represent *epistemic* uncertainty by associating transition probabilities with *uncertainty sets* of admissible values; interval MDPs (IMDPs) are a common example. pMDPs specify transition probabilities symbolically as expressions over parameters that range over an admissible parameter space. Both formalisms thus capture imprecise model knowledge by describing not a single precise model, but a family of models, each representing a possible true underlying system.

Parametric MDPs have found applications in model repair [6,25], sensitivity analysis [4], and quality-of-service control [9], predominantly via the problem of *parameter synthesis* [10,12,13,20,31], which partitions the parameter space into two regions, corresponding to parameter valuations for which the model respectively satisfies or violates a given property. Uncertain MDPs have emerged

M. Chen and M. Tschaikowski (Eds.): QEST+FORMATS 2026, LNCS 16913, pp. 114–133, 2027.
https://doi.org/10.1007/978-3-032-35298-9_7

as a standard model for reasoning about *robustness*: they support the synthesis of *robust* policies, i.e., policies that are optimal under the assumption that epistemic uncertainty is resolved adversarially by the environment, selecting the worst-case admissible distributions with respect to the agent's desired objective [19,28,43].

UMDPs are therefore well suited to *learning-based* methods for the verification of MDPs whose transition probabilities are unknown. Typically, uncertainty sets for transition probabilities are learned from interactions with the true but unknown system, for instance as statistically sound confidence sets such as confidence intervals [1,27,33,36,39]. The resulting learned UMDP then contains the true system with high confidence. Consequently, any robust policy synthesised on this UMDP is guaranteed, with the same confidence, to achieve on the true system at least the robust value certified by the learned model.

Existing approaches to UMDP learning typically treat transition probabilities across the model as independent, learning separate confidence sets for each of them [5,38,39]. In practice, however, these distributions are often coupled through shared latent quantities, such as common failure probabilities, shared reliability parameters or global environmental conditions that remain fixed and affect multiple transitions. Such dependencies are often structurally known and are conveniently represented by pMDPs, even though the precise values of the underlying quantities are unknown and cannot be directly observed.

This work addresses the problem of UMDP learning under known structural parametric dependencies that are captured by a pMDP. We assume a known pMDP for which an unknown parameter instantiation induces the true underlying system. From sampled interactions with this system, we first derive statistical confidence sets for the individual transition probabilities and then exploit the parametric structure of the pMDP to project these sets into the parameter space. This yields a confidence region in parameter space that contains the true underlying parameter instantiation with high confidence.

Example. We illustrate this idea using an example: a Mars-rover navigation domain, inspired by the semi-autonomous vehicle environment in [21,37]. A controller communicates with the rover via two lossy satellite channels while the rover attempts to collect samples and return home. The probability of successful communication depends on the rover's position through known coefficients, but on unknown channel reliabilities. At each position, it is given by $v^\top\theta = v_1\theta_1 + v_2\theta_2$, where $v = (v_1, v_2)$ is position-dependent and $\theta = (\theta_1, \theta_2)$ are the unknown reliabilities. The domain is therefore naturally modelled as a pMDP as shown in Fig. 1a.

From sampled transitions of the rover, the standard approach learns an IMDP by assigning a confidence interval to each transition probability; see Fig. 1b. This yields a robust model with formal confidence guarantees on inclusion of the true unknown system, but treats transition probabilities as independent. In our example, however, they are coupled through the shared parameter instantiation θ.

Our approach exploits this structure by projecting the learned transition-wise intervals through the parametric structure of the pMDP into

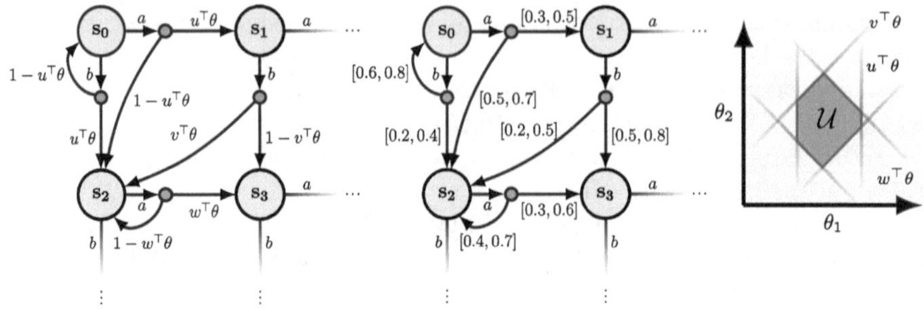

Fig. 1. Construction of the parameter confidence region $\mathcal{U}$ from a learned IMDP.

the parameter space. Each learned transition confidence interval $[l, u]$ induces constraints of the form

$$l \le v_1\theta_1 + v_2\theta_2 \le u.$$

Collecting these constraints yields a parameter confidence region $\mathcal{U}$, as shown in Fig. 1c. With high confidence, the true parameter instantiation lies in $\mathcal{U}$.

The confidence region $\mathcal{U}$, together with the pMDP, induces a UMDP U that contains the true MDP with high confidence. This is tighter than a UMDP constructed from the confidence sets for individual transitions (e.g., the IMDP in Fig. 1b) since it rules out transition functions that are inconsistent with any of the constraints induced by the parametric expressions and the learned intervals, leading to less conservative robust policies and tighter guarantees from the same data.

However, because of its (non-rectangular) structure, the UMDP U is computationally very challenging to solve. We tackle this problem by proposing a hierarchy of relaxations of the induced uncertainty set, yielding UMDPs that can be solved more efficiently to generate robust policies. All relaxations are sound and thus produce *probably approximately correct* (PAC) [40] performance guarantees. This hierarchy illustrates a clear trade-off between the tightness and efficiency of the relaxations. We implement our methods and evaluate them on a range of established benchmarks. The results demonstrate substantially tighter uncertainty estimates than classical interval-based uncertain MDP learning techniques.

2 Preliminaries

For a finite set X, let $\Delta(X) = \{\mu\colon X \to [0,1] \mid \sum_{x\in X} \mu(x) = 1\}$ denote the set of probability distributions over X.

Definition 1 (Markov decision process). *A (finite) MDP is a tuple $M = (S, A, s_0, P)$, where S is a finite set of states, A is a finite set of actions, $s_0 \in S$ is an initial state, and $P : S \times A \to \Delta(S)$ is a transition kernel. We write*

$P(s,a)(s')$ (or $P(s,a,s')$) for the probability to move from state s to state s' under action a.

A *policy* (or *strategy*) is a function $\pi \colon S \to A$ that selects an action $\pi(s)$ in each state s. An (infinite) *path* is a sequence $\rho = s^0 a^0 s^1 a^1 \cdots \in (S \times A)^\omega$ such that $P(s^t, a^t)(s^{t+1}) > 0$ for all $t \geq 0$. We write Paths(s) for the set of paths starting in s and Π for the set of policies. An *objective* is a measurable mapping $R\colon \text{Paths}(s) \to \mathbb{R}$ assigning a return to each path. Typical objectives are based on satisfaction of temporal-logic properties [8] or accumulation of rewards [30]. A policy $\pi \in \Pi$ executed in an MDP M induces a probability measure over Paths(s) [30]. We define the *value* of π in s as $V_M^\pi(s) := \mathbb{E}_{M,s}^\pi[R]$.

2.1 Parametric and Uncertain MDPs

Let $\Theta = \{\theta_1, \dots, \theta_\ell\}$ be an ordered set of parameters. A *parameter instantiation* is a vector $\mathbf{u} \in \mathbb{R}^\ell$. We assume that all parameters are bounded, i.e., $l_i \leq \mathbf{u}_i \leq u_i$ for some $l_i, u_i \in \mathbb{R}$, and refer to $\mathcal{D} := \prod_{i=1}^{\ell} [l_i, u_i]$ as the *parameter space*. Let $\mathbb{Q}[\Theta]$ denote the ring of polynomials over Θ with rational coefficients. Each $f \in \mathbb{Q}[\Theta]$ induces a function $f : \mathbb{R}^\ell \to \mathbb{R}$, where $f[\mathbf{u}]$ denotes the polynomial f evaluated by substituting every occurrence of θ_i with its instantiation $\mathbf{u}_i$. We denote by

$$\Delta_\Theta(X) = \{f \colon X \to \mathbb{Q}[\Theta] \mid \forall \mathbf{u} \in \mathcal{D}.\ f(x)[\mathbf{u}] \in [0,1], \sum_{x \in X} f(x)[\mathbf{u}] = 1\}$$

the set of all *parametric probability distributions* over a finite set X, i.e., the set of all parametric functions that induce a valid probability distribution over X for any parameter instantiation $\mathbf{u} \in \mathcal{D}$.

Parametric Markov decision processes extend standard MDPs by allowing transition probabilities to be specified as parametric probability distributions, where fixing a parameter instantiation yields an ordinary MDP.

Definition 2 (Parametric MDP). *A* parametric MDP *(pMDP) is a tuple $M_\Theta = (S, A, s_0, \Theta, P_\Theta)$, where S, A, s_0 are as for MDPs, Θ is a parameter set, and $P_\Theta : S \times A \to \Delta_\Theta(S)$ is a* parametric *transition probability function. For a parameter instantiation $\mathbf{u} \in \mathbb{R}^l$, we obtain the instantiated MDP $M[\mathbf{u}] = (S, A, s_0, P[\mathbf{u}])$ with $P[\mathbf{u}](s,a)(s') := P_\Theta(s,a)(s')[\mathbf{u}]$, for all $s, s' \in S$, $a \in A$.*

In this work, we assume there is a true but unknown MDP M, obtained by instantiating a *known* pMDP M_Θ with an *unknown* parameter instantiation $\mathbf{u}$. The objective is to learn, from sampled interactions, an approximation of the true dynamics with quantified uncertainty, and to synthesise policies that are robust to the residual uncertainty. To represent partial knowledge, we adopt the framework of UMDPs. A UMDP is an MDP in which transition probabilities are not fixed, but are only required to lie in specified *uncertainty sets.*

Definition 3 (Uncertain MDP). *A UMDP is a tuple $U = (S, A, s_0, \mathcal{P})$, where S, A, s_0 are as for MDPs and $\mathcal{P} \subseteq (\Delta(S))^{S \times A}$ is an* uncertainty set *of admissible transition probability functions $P \in \mathcal{P}$. The UMDP is* (s,a)-rectangular *if $\mathcal{P}$ factorises as the product of local uncertainty sets, i.e., $\mathcal{P} = \prod_{(s,a)\in S\times A} \mathcal{P}(s,a)$, where $\mathcal{P}(s,a) \subseteq \Delta(S)$. Otherwise, the uncertainty is* coupled *(non-rectangular).*

There is a close relationship between pMDPs and UMDPs. Given a pMDP and a set $\mathcal{U}$ of admissible parameter instantiations, we obtain a UMDP by letting the uncertainty set contain exactly the successor distributions induced by instantiations $\mathbf{u} \in \mathcal{U}$. Conversely, a UMDP can be represented as a pMDP together with a set of admissible parameter instantiations inducing precisely the uncertainty set. We formalise this correspondence in the technical report [34].

In this work, the known pMDP serves as a structural model, capturing dependencies across transitions. From sampled interactions with the unknown system, we infer a set of plausible parameter instantiations $\mathcal{U}$ with quantified confidence and, equivalently, a UMDP whose uncertainty sets contain the transition distributions consistent with $\mathcal{U}$. This UMDP is then used for robust policy synthesis under the remaining uncertainty about the true dynamics of M.

Interval MDPs. We recall a standard class of UMDPs that we use throughout. An *interval MDP* (IMDP) specifies, for each $(s, a, s') \in S\times A\times S$, componentwise lower and upper bounds $\ell_{(s,a,s')}, u_{(s,a,s')} \in [0, 1]$. The admissible successor distributions from (s, a) then are those whose components respect these bounds:

$$\mathcal{P}(s,a) = \Big\{\mu \in \Delta(S) \;\Big|\; \forall s' \in S: \; \ell_{(s,a,s')} \le \mu(s') \le u_{(s,a,s')}\Big\}.$$

2.2 Solving Uncertain MDPs

For a UMDP, a *robust* policy is one that optimises worst-case performance over all admissible transition kernels. Following the standard robust MDP interpretation [19,28,43], we view uncertainty as being resolved adversarially by an *environment* (or *nature*) that selects admissible transition probabilities. Formally, an agent policy π and an admissible transition kernel $P \in \mathcal{P}$ induce a probability measure over paths [44]. We denote the corresponding expectation by $\mathbb{E}^{\pi}_{P,s}[R]$ and define the value of state s under (π, P) as $V^{\pi}_{P}(s) := \mathbb{E}^{\pi}_{P,s}[R]$. The *robust value* is the maximal expected return under the worst-case nature:

$$V^*_U(s) := \sup_{\pi\in\Pi} \inf_{P\in\mathcal{P}} V^{\pi}_{P}(s), \tag{1}$$

and a *robust-optimal policy* is any $\pi^* \in \operatorname{argsup}_{\pi} \inf_{P} V^{\pi}_{P}(s)$. In particular, π^* guarantees to achieve at least $V^*_U(s)$ for any admissible transition kernel $P \in \mathcal{P}$.

Computing (1) can be challenging in general, in particular for coupled uncertainty sets [43]. For *rectangular* uncertainty (i.e. $\mathcal{P} = \prod_{(s,a)} \mathcal{P}(s,a)$), the inner minimisation decomposes into local optimisations at each state–action pair, and the robust value satisfies a Bellman-type optimality equation. For example, for reachability with target set $T \subseteq S$, the robust Bellman equations take the form

$$V^*_U(s) = \max_{a\in A} \underbrace{\min_{\mu\in\mathcal{P}(s,a)} \sum_{s'\in S} \mu(s')\, V^*_U(s')}_{\text{Inner Optimisation}}, \quad s \in S \setminus T, \tag{2}$$

together with boundary conditions $V_U^*(s) = 1$ for all $s \in T$. Analogous robust Bellman equations can be given, e.g., for expected or discounted reward objectives [19,28]. These equations can be solved with *robust value iteration*, i.e., by repeatedly applying the robust Bellman operator until convergence [19,28].

2.3 Learning Uncertain MDPs

We review the basics of learning an UMDP from data generated by an unknown MDP M. We assume access to a dataset of transition samples $C = \{(s_i, a_i, s_i')\}_{i=1}^n$, obtained either (i) by collecting trajectories in the MDP, i.e., episodic sampling with restarts from the initial state, or continuing interaction, or (ii) via a *generative model* that allows direct sampling of next states for state–action pairs [36,38].

From C we extract the empirical counts $\#(s,a)$ and $\#(s,a,s')$ for all $s, s' \in S$, $a \in A$, where $\#(s,a)$ is the number of occurrences of (s,a) and $\#(s,a,s')$ the number of observed transitions from s to s' under action a. Whenever $\#(s,a) > 0$, the empirical estimate of the transition probability is $\tilde{P}(s,a,s') := \frac{\#(s,a,s')}{\#(s,a)}$.

A standard UMDP learning technique is to construct an IMDP that over-approximates the unknown transition probabilities by confidence intervals. For each (s,a,s'), we view $\#(s,a,s')$ as a binomial observation with $\#(s,a)$ trials and success probability $P(s,a,s')$, and lift the empirical estimate $\tilde{P}(s,a,s')$ to a confidence interval $I_\gamma(s,a,s') := [\,\ell(s,a,s'),\, u(s,a,s')\,]$, such that $\Pr\big[P(s,a,s') \in I_\gamma(s,a,s')\big] \geq 1-\gamma$, for $\gamma \in (0,1)$. Suitable choices for I_γ include standard binomial confidence intervals such as the Clopper–Pearson interval [11,27]. If $\#(s,a) = 0$, no information is available for that state–action pair, and we therefore use the trivial interval $I_\gamma(s,a,s') = [0,1]$ for all $s' \in S$.

For the objectives considered in this paper, such as reachability probabilities, intervals containing both zero and non-zero values can be handled directly by the robust value iteration in Eq. (2). For more general temporal-logic objectives, however, satisfaction may depend on the graph structure of the UMDP. In that case, intervals containing both zero and non-zero values allow the adversarial nature to alter the graph structure itself, which complicates robust synthesis. This issue is not specific to our approach, but inherent to temporal-logic reasoning over UMDPs in general. Existing techniques therefore often assume *constant support*, i.e., that the set of successors is known a priori [38,39,44]. Under this assumption, one may replace any lower bound 0 in the learned intervals by a sufficiently small $\varepsilon > 0$ in order to preserve the known graph structure.

To obtain an overall confidence level $1-\delta$, with $\delta \in (0,1)$, we distribute the confidence budget across all transition probabilities and set $\gamma := \delta / |S \times A \times S|$. Hence, with probability at least $1-\delta$, the unknown MDP is contained in the learned IMDP (e.g., [33,39]). Consequently, the value of a policy on the learned IMDP yields a $(1-\delta)$-valid lower bound on the performance of the policy in M. Other common constructions of uncertainty sets $\mathcal{P}(s,a)$ with formal confidence guarantees include, e.g., L_1-norm balls based on the Weissman inequality [36,42].

2.4 Parameter Tying

An established method for improving IMDP learning in the presence of known parametric dependencies is *parameter tying* [14,29,32]. Consider two transitions (s, a, s') and (t, b, t') in M_Θ with $s \neq t$ and $P_\Theta(s, a, s') = P_\Theta(t, b, t')$, i.e., both transition probabilities are described by the same expression. Then the samples in C obtained for these transitions correspond to the same Bernoulli experiment and can be *pooled* to derive a single interval, which is then assigned to both transitions. Moreover, the number of components to be learned over which we distribute the overall confidence budget δ reduces to the number of distinct parametric expressions in M_Θ. Details are provided in the technical report [34].

Parameter tying applies only when parametric expressions are *exactly* equivalent. Our approach instead aggregates information from the parametric structure and the collected samples by projecting the learned constraints through the algebraic structure of M_Θ into the parameter space. This jointly accounts for observations across all transitions and can tighten uncertainty even for transitions whose expressions have not been observed in any sample in C. We use parameter tying as a pre-processing step: we pool the observed counts for equivalent parametric expressions and then apply our technique to the pooled data. Let

$$\Lambda = \{ P_\Theta(s, a, s') \mid s, s' \in S,\ a \in A \} \tag{3}$$

be the set of distinct parametric expressions occurring in M_Θ. We distribute the confidence budget only across those expressions that are not constant polynomials: $\Lambda_{\mathsf{unk}} := \{ f \in \Lambda \mid f \notin \mathbb{Q} \}$. For each $f \in \Lambda_{\mathsf{unk}}$, we compute from the pooled counts a confidence interval $[l_f, u_f]$ as per Sect. 2.3 such that, under the true but unknown parameter instantiation $\mathbf{u}$, $\Pr\big[\, l_f \leq f[\mathbf{u}] \leq u_f \,\big] \geq 1 - \frac{\delta}{|\Lambda_{\mathsf{unk}}|}$. For constant expressions $f \in \Lambda \setminus \Lambda_{\mathsf{unk}}$, we set $l_f = u_f = f$. Substituting each occurrence of f in M_Θ by the corresponding interval $[l_f, u_f]$ yields an IMDP that contains the true (unknown) MDP M with probability at least $1 - \delta$.

3 Robust Parametric UMDP Learning

IMDP learning inherently assumes that all transition probabilities are *independent* and learned as individual confidence intervals. We now introduce our approach which leverages dependencies specified by an underlying pMDP model. From the observed transitions, we infer a set of plausible instantiations $\mathcal{U}$ with quantified confidence, exploiting the dependencies captured by shared parameters.

Throughout, we consider a fixed pMDP M_Θ and an unknown MDP M induced by M_Θ under an unknown parameter instantiation $\mathbf{u}$. We further fix a data set of i.i.d. transition samples C obtained from M, as described in Sect. 2.3. We focus initially on the case where the per-transition uncertainty sets are intervals, i.e., we learn an IMDP approximation of M and then project it into the parameter space. The same ideas extend to other classes of uncertainty sets, such as L_1-balls that preserve per-state coupling [36], as we shall discuss in Sect. 3.4.

Given the learned intervals $[l_f, u_f]$ for $f \in \Lambda$, we define the *uncertainty region*

$$\mathcal{U} = \{ \mathbf{v} \in \mathcal{D} \mid \forall f \in \Lambda : \ l_f \leq f[\mathbf{v}] \leq u_f \}. \tag{4}$$

The region $\mathcal{U}$ contains exactly those parameter instantiations that satisfy all learned interval constraints simultaneously. Together with the pMDP M_Θ, it induces a UMDP $U = (S, A, s_0, \mathcal{P}_\mathcal{U})$, where the uncertainty set consists of all transition kernels obtained by instantiating M_Θ with a parameter vector in $\mathcal{U}$:

$$\mathcal{P}_\mathcal{U} := \{ P[\mathbf{v}] \in (\Delta(S))^{S \times A} \mid \mathbf{v} \in \mathcal{U} \}.$$

Theorem 1 (Soundness of $\mathcal{U}$). *With probability at least $1 - \delta$, the true parameter instantiation* $\mathbf{u}$ *satisfies* $\mathbf{u} \in \mathcal{U}$. *Equivalently,*

$$\Pr[\ P[\mathbf{u}] \in \mathcal{P}_\mathcal{U}\ ] \geq 1 - \delta.$$

Consequently, for any policy $\pi \in \Pi$ and state $s \in S$, with probability at least $1 - \delta$ the robust value under the induced UMDP lower-bounds the true value in M, i.e.,

$$\Pr[\ V_U^\pi(s) \leq V_M^\pi(s)\ ] \geq 1 - \delta.$$

□

The induced UMDP U also yields a robust policy π^*. By Theorem 1, π^* achieves, with probability at least $1 - \delta$, at least its robust value $V_U^\pi(s)$ on the unknown MDP $M[\mathbf{u}]$. However, computing the robust value and policy from U remains very challenging. We summarise the key issues below and then describe in more detail how we address them in the remainder of this section.

1. **Non-rectangularity and tractability (Sect. 3.1).** The uncertainty set $\mathcal{P}_\mathcal{U}$ for U is in general *non-rectangular*: shared parameters couple uncertainty across states and actions. As a consequence, the robust optimisation problem in (1) becomes very difficult and robust policy synthesis is NP-hard in general [43]. We therefore introduce a family of *rectangular* relaxations that over-approximate $\mathcal{P}_\mathcal{U}$. These relaxations preserve the soundness of Theorem 1, while enabling efficient synthesis. For each relaxation, we analyse (i) the cost of constructing it, (ii) the resulting complexity of robust policy synthesis, and (iii) its tightness. Further, we establish an inclusion hierarchy among the uncertainty sets, making the tightness–efficiency trade-off explicit.
2. **Polynomial constraints and linearisation (Sect. 3.2).** The uncertainty set $\mathcal{P}_\mathcal{U}$ is in general defined by polynomial constraints, since transition probabilities may be polynomial functions of the parameters. As a result, even after rectangularisation, the inner optimisation in (2) can remain challenging, as it requires optimisation under polynomial constraints. We therefore leverage *linear relaxations* that transform these polynomial constraints into linear ones. Combined with rectangular relaxations, we show that this yields tractable inner problems, while retaining tight uncertainty sets.

3. **Possible emptiness of $\mathcal{P}_{\mathcal{U}}$ (Sect. 3.3).** Finally, we show that, unlike in standard IMDP learning, the induced uncertainty set $\mathcal{P}_{\mathcal{U}}$ may be empty. We give a statistical interpretation of this edge case, derive an upper bound on the probability that it occurs, and provide a principled fallback mechanism.

3.1 Rectangular Relaxations

Robust policy synthesis over $\mathcal{P}_{\mathcal{U}}$ is intractable in general, since $\mathcal{P}_{\mathcal{U}}$ may be non-rectangular. We therefore introduce a range of rectangular relaxations, i.e., uncertainty sets that over-approximate $\mathcal{P}_{\mathcal{U}}$. These relaxations preserve the inclusion guarantee for the unknown MDP $M[\mathbf{u}]$ while enabling tractable synthesis. For each relaxation, we describe the additional effort required to construct it, as well as the resulting complexity of policy synthesis, measured by the cost of the inner minimisation in (2) in each iteration of robust value iteration. We organise the relaxations into an inclusion hierarchy relative to $\mathcal{P}_{\mathcal{U}}$ and to the baseline uncertainty set $\mathcal{P}_{\mathrm{I}}$ induced solely by the intervals, as depicted in Fig. 2.

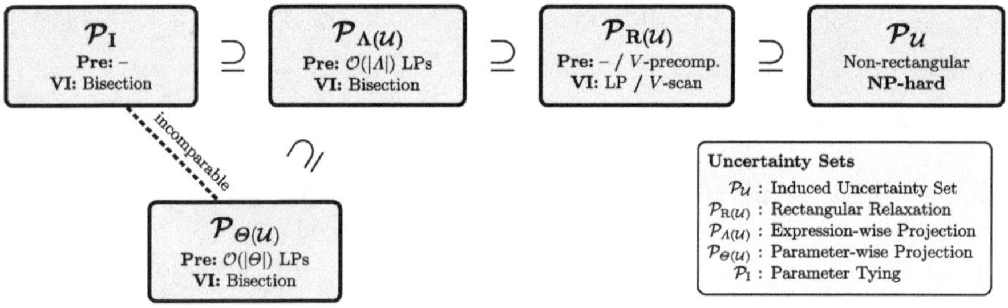

Fig. 2. Inclusion hierarchy of the considered uncertainty sets. For each set, we report the computational effort required to construct the relaxation (**Pre**), in addition to computing the confidence intervals and constructing the model, and the effort per inner minimisation during robust value iteration (**VI**).

From here on, for ease of presentation, we assume that each parametric expression $f \in \Lambda$ is linear in the parameters. This comes without loss of generality, as we will discuss in Sect. 3.2 how to relax polynomial constraints into linear ones while preserving soundness of the resulting uncertainty sets. Recall that we consider the case where the *original* uncertainty sets derived from data are confidence intervals. As we shall argue in Sect. 3.4, our results on rectangular relaxations extend to other uncertainty classes, such as L_1-balls and ellipsoids.

Rectangularisation via Local Projections. The most natural way to relax a non-rectangular uncertainty set into a rectangular one is to allow the adversarial environment to choose its worst-case transition probabilities independently in each state–action pair. Formally, given the coupled uncertainty set $\mathcal{P}_{\mathcal{U}}$, its *rectangular relaxation* $\mathcal{P}_{R(\mathcal{U})}$ is the product of its local projections:

$$\mathcal{P}_{R(\mathcal{U})} := \prod_{(s,a)\in S\times A} \mathcal{P}_{R(\mathcal{U})}(s,a), \quad \text{where} \quad \mathcal{P}_{R(\mathcal{U})}(s,a) := \{ P[\mathbf{v}](s,a) \mid \mathbf{v} \in \mathcal{U} \}.$$

Rather than requiring a single instantiation $\mathbf{v} \in \mathcal{U}$ to be used consistently throughout the pMDP, the environment can select an instantiation from $\mathcal{U}$ separately for each state–action pair. Essentially, this corresponds to renaming each shared parameter, thus eliminating dependencies [18,31]. This is a relaxation as it grants the environment more power: it may choose a different worst-case instantiation from $\mathcal{U}$ for each state–action pair, which can only decrease the robust value.

Since each parametric expression $f \in \Lambda$ is linear and constrained by interval bounds $l_f \leq f[\mathbf{v}] \leq u_f$, the uncertainty region $\mathcal{U}$ is a polytope. Hence, for a state–action pair (s,a), the inner minimisation in (2) reduces to a linear program:

$$\min_{\mathbf{v}\in\mathcal{U}} \sum_{s'\in S} P[\mathbf{v}](s,a,s')\, V_U^*(s').$$

or, equivalently:

$$\min_{\mathbf{v}\in\mathcal{D}} \sum_{s'\in S} P[\mathbf{v}](s,a,s')\, V_U^*(s') \quad \text{s.t.} \quad l_f \leq f[\mathbf{v}] \leq u_f,\ \forall f \in \Lambda.$$

Hence, policy synthesis via robust value iteration requires solving a linear program for each state–action pair in every iteration. In practice, this can often still be efficient for two reasons. First, the feasible region is fixed: the constraints defining $\mathcal{U}$ depend only on the parametric expressions and the learned bounds $[l_f, u_f]$, and thus can be constructed once and reused throughout value iteration. Second, successive iterations induce only small changes in the objective, since the coefficients $P[\mathbf{v}](s,a,s')$ are fixed and the value estimates converge. Using a simplex-based LP solver [7,41], we can *warm-start* from the vertex that was optimal in the previous iteration. Since the changes in the objective become smaller as value iteration converges, the previously optimal vertex is often still optimal, or close to optimal, so only few steps are needed to update the solution.

Alternatively, since each inner optimisation is over the same feasible region $\mathcal{U}$, which remains fixed, we can precompute a vertex representation of $\mathcal{U}$, i.e., its set of extreme points. This is closely related to the enumeration approach used in parameter synthesis when the parameter space is a hyperrectangle [18,31]. The inner minimisation then reduces from solving an LP to evaluating the objective at the vertices. However, converting a polytope to vertex representation can be expensive in general, and the number of vertices may grow exponentially with the dimension of the parameter space. This approach is therefore only suitable for low-dimensional parameter spaces, as we examine in the evaluation in Sect. 4.

Expression-Wise Projections. We next consider a relaxation that projects $\mathcal{U}$ back to interval bounds over the parametric expressions. This aggregates information from all transitions jointly through $\mathcal{U}$ and can yield substantially tighter intervals than those obtained directly through parameter tying. The resulting model is an IMDP, whose inner optimisation in robust value iteration can be solved efficiently via bisection [19,28].

For each expression $f \in \Lambda$, we derive refined bounds l_f^Λ and u_f^Λ as the minimum and maximum values that $f[\mathbf{v}]$ can attain over $\mathbf{v} \in \mathcal{U}$. Concretely, we solve

$$l_f^\Lambda := \min_{\mathbf{v} \in \mathcal{U}} f[\mathbf{v}] \qquad u_f^\Lambda := \max_{\mathbf{v} \in \mathcal{U}} f[\mathbf{v}].$$

We define the corresponding rectangular interval uncertainty set $\mathcal{P}_{\Lambda(\mathcal{U})}$ as

$$\mathcal{P}_{\Lambda(\mathcal{U})}(s,a) := \left\{ \mu \in \Delta(S) \;\middle|\; \forall s' \in S : \; l^\Lambda_{P_\Theta(s,a,s')} \leq \mu(s') \leq u^\Lambda_{P_\Theta(s,a,s')} \right\}.$$

Computing $\mathcal{P}_{\Lambda(\mathcal{U})}$ requires solving $2|\Lambda|$ linear programs. Robust value iteration on the resulting IMDP then has the same per-iteration cost as the standard interval model, but can yield significantly tighter uncertainty sets since the bounds reflect information aggregated across all transitions through $\mathcal{U}$. Moreover, the individual LPs are independent and can be solved in parallel when constructing the model.

Parameter-wise projections. Expression-wise projections can be expensive when the number of distinct expressions $|\Lambda|$ is large. As an alternative, we project $\mathcal{U}$ to interval constraints on the parameters Θ themselves. Concretely, this yields an axis-aligned over-approximation of the polytope $\mathcal{U}$ by projecting onto each parameter dimension and forming the corresponding enclosing hyperrectangle.

For each parameter $\theta_i \in \Theta$, $1 \leq i \leq \ell$, we compute its lower and upper bounds over the uncertainty region $\mathcal{U}$ by $\underline{\theta}_i := \min_{\mathbf{v} \in \mathcal{U}} \mathbf{v}_i$ and $\overline{\theta}_i := \max_{\mathbf{v} \in \mathcal{U}} \mathbf{v}_i$. These define the smallest hyperrectangle that contains $\mathcal{U}$, $B(\mathcal{U}) := \prod_{i=1}^{\ell} [\underline{\theta}_i, \overline{\theta}_i] \supseteq \mathcal{U}$. We then derive bounds for each expression $f \in \Lambda$ by evaluating it over $B(\mathcal{U})$,

$$l_f^\Theta := \min_{\mathbf{v} \in B(\mathcal{U})} f[\mathbf{v}], \qquad u_f^\Theta := \max_{\mathbf{v} \in B(\mathcal{U})} f[\mathbf{v}].$$

Since each f is linear, these extrema are attained at a vertex of $B(\mathcal{U})$. Thus, computing l_f^Θ and u_f^Θ is simpler than the expression-wise projection bounds, as it does not require solving an LP over $\mathcal{U}$, but only selecting, for each parameter θ_i, either $\underline{\theta}_i$ or $\overline{\theta}_i$ according to the sign of its coefficient in f. The resulting interval uncertainty set $\mathcal{P}_{\Theta(\mathcal{U})}$ is then defined analogously to $\mathcal{P}_{\Lambda(\mathcal{U})}$.

Computing $\mathcal{P}_{\Theta(\mathcal{U})}$ requires solving $2|\Theta|$ linear programs over $\mathcal{U}$, which can be significantly cheaper than solving two LPs for every expression in Λ. However, the hyperrectangle $B(\mathcal{U})$ discards dependencies between parameters and may therefore yield considerably looser bounds than expression-wise projection. Moreover, the induced intervals need not be tighter than the original intervals obtained directly from the samples: the two constructions are, in general, incomparable, and either can produce tighter bounds. We provide counterexamples for both directions in the technical report [34]. To recover inclusion of the original intervals, one can simply intersect the resulting bounds.

We now state the inclusion hierarchy of the discussed uncertainty sets. We denote by $\mathcal{P}_I$ the original interval uncertainty set obtained from the samples using parameter tying as described in Sect. 2.4.

Theorem 2 (Inclusion hierarchy). *The following inclusions hold:*

$$\mathcal{P}_I \supseteq \mathcal{P}_{\Lambda(\mathcal{U})} \supseteq \mathcal{P}_{R(\mathcal{U})} \supseteq \mathcal{P}_{\mathcal{U}}, \quad \textit{and} \quad \mathcal{P}_{\Theta(\mathcal{U})} \supseteq \mathcal{P}_{\Lambda(\mathcal{U})}.$$

Moreover, $\mathcal{P}_{\Theta(\mathcal{U})}$ is in general incomparable to $\mathcal{P}_I$. □

The inclusion hierarchy of Theorem 2 implies that every over-approximating relaxation inherits the same high-confidence inclusion guarantee for the true, unknown MDP as $\mathcal{P}_{\mathcal{U}}$ in Theorem 1. Consequently, all such relaxations can be used for robust policy synthesis with high-confidence performance guarantees.

3.2 Linear Relaxations for Polynomial Constraints

Thus far, we assumed that the parametric expressions in Λ are linear, and hence that $\mathcal{U}$ is a polytope in the parameter space. In general, transition probability expressions in a pMDP may be *polynomials* in the parameters, which makes the projection constraints $l_f \leq f[\mathbf{v}] \leq u_f$ non-linear. As a consequence, the rectangular relaxations introduced in Sect. 3.1 no longer reduce to linear programs. To retain tractability while preserving soundness, we construct a linear outer approximation of the feasible region using *McCormick envelopes* [26,35].

McCormick envelopes are linear inequalities that describe the convex hull of a bilinear term $z = xy$ with upper and lower bounds $\underline{x} \leq x \leq \overline{x}$ and $\underline{y} \leq y \leq \overline{y}$ [26]. Concretely, they are given by:

$$\begin{aligned} z &\geq \underline{y}\,x + \underline{x}\,y - \underline{x}\,\underline{y}, & z &\leq \underline{y}\,x + \overline{x}\,y - \overline{x}\,\underline{y}, \\ z &\geq \overline{y}\,x + \overline{x}\,y - \overline{x}\,\overline{y}, & z &\leq \overline{y}\,x + \underline{x}\,y - \underline{x}\,\overline{y}. \end{aligned} \tag{5}$$

While McCormick envelopes provide the exact convex-hull relaxation of a single bilinear term, they can be applied compositionally to obtain a linear outer approximation of an entire polynomial constraint. For instance, for the cubic monomial $\theta_1\theta_2\theta_3$ we introduce auxiliary variables $z_{12} = \theta_1\theta_2$ and $z_{123} = z_{12}\theta_3$ for the intermediate products, and add the McCormick envelopes in (5) for both bilinear terms. Applying this construction to all monomials in the constraint system defining $\mathcal{U}$ in (4) yields a system of linear constraints that over-approximates the original polynomial relations. As an outer approximation, it preserves soundness and can be used within our relaxations. Formal details on the construction and the resulting inequality system, are provided in the technical report [34].

Optimisation-Based Bound Tightening. The McCormick inequalities in (5) require finite lower and upper bounds for every variable. We obtain initial bounds by solving two LPs per variable over the current linear relaxation, and then iteratively refine them: tighter bounds yield tighter McCormick envelopes, which in turn can imply further bound improvements. This iterative procedure is known as optimisation-based bound tightening (OBBT) and is a standard domain-reduction technique in global optimisation [16]. In our implementation, we apply OBBT as a preprocessing step to construct a tight linear relaxation of the uncertainty region $\mathcal{U}$. In the experimental evaluation in Sect. 4, we show that this approach yields tight relaxations and resulting robust guarantees.

3.3 Joint Feasibility and Possible Emptiness of $\mathcal{U}$

Unlike the interval uncertainty sets obtained in standard IMDP learning, the induced uncertainty region $\mathcal{U}$ may be empty. This is a consequence of enforcing *global* consistency with the underlying parametric structure: while each learned interval constraint may be individually feasible, there need not exist a single parameter instantiation that satisfies all of them simultaneously. In that case, the induced uncertainty set $\mathcal{P}_{\mathcal{U}}$ is empty as well, and so are its rectangular relaxations.

If the pMDP M_Θ correctly specifies the underlying system from which the transition samples are obtained, then with probability at least $1-\delta$ the true parameter instantiation $\mathbf{u}$ satisfies all constraints, and hence lies in $\mathcal{U}$. In particular,

$$\Pr[\mathcal{U} \neq \emptyset] \geq \Pr[\mathbf{u} \in \mathcal{U}] \geq 1-\delta, \quad \text{or equivalently,} \quad \Pr[\mathcal{U} = \emptyset] \leq \delta.$$

Thus, under correct specification, emptiness can occur with probability at most δ.

Interpretation. If $\mathcal{U}$ is empty, then no admissible parameter instantiation simultaneously satisfies all learned interval constraints. Statistically, this means that the uncertainty statements obtained from the data are jointly incompatible with the assumed parametric structure at confidence level $1-\delta$. Hence, emptiness can be interpreted as evidence against the chosen pMDP model at significance level δ.

This observation is not just an edge case but also an advantage of our approach. By enforcing joint consistency across all learned constraints, it can reveal model misspecification that would be hidden by purely local interval uncertainty sets.

Fallback to Interval Uncertainty. If $\mathcal{U}$ is empty, one can safely fall back to the original interval uncertainty set learned directly from the samples. In contrast to $\mathcal{U}$, this interval uncertainty set is always non-empty, since for each state-action pair it contains at least the empirical estimate of the corresponding successor distribution, and thus admits at least one feasible distribution per local choice.

3.4 Extension to Other Classes of Uncertainty Sets

We have so far assumed that the base uncertainty sets mapped through the structure of the underlying pMDP into parameter space are intervals. However, our approach and the inclusion hierarchy of Theorem 2 extend beyond intervals. In the following, we discuss two uncertainty classes that illustrate this extension.

L_1 uncertainty Sets. One important class is given by L_1-balls [27,36] based on the Weissman inequality [42]. For each state–action pair, this yields a high-confidence uncertainty set around the empirical estimate $\tilde{P}(s,a)$ in the form of an L_1 ball, with details provided in the technical report [34]. Since L_1-balls are polytopes, they fit seamlessly with our approach. Concretely, given empirical estimates $\tilde{P}(s,a)$ and radii $\varepsilon(s,a)$ for all $(s,a) \in S \times A$, we define the uncertainty region

$$\mathcal{U}_{L_1} = \{\mathbf{v} \in \mathcal{D} \mid \|P_\Theta[\mathbf{v}](s,a) - \tilde{P}(s,a)\|_1 \leq \varepsilon(s,a)\}. \tag{6}$$

Since each $P_\Theta(s,a,s')$ is linear in the parameters, the L_1-constraints can be expressed as linear inequalities. Hence, $\mathcal{U}_{L_1}$ is again a polytope in parameter space, and the rectangular relaxations introduced in Sect. 3.1 carry over directly.

Ellipsoidal Uncertainty Sets. Another relevant class is given by *ellipsoids.* These may arise as local uncertainty sets, for instance from second-order approximations of log-likelihood functions [28], or directly in parameter space [3, 45]. In both cases, the resulting uncertainty region is generally no longer a polytope. Nevertheless, the rectangular relaxations of Sect. 3.1, as well as the inclusion hierarchy, carry over.

In particular, for an ellipsoidal parameter region, the local rectangular relaxation can still be obtained by projecting to state–action-wise uncertainty sets, but the corresponding inner optimisation problems are no longer linear programs. Instead, optimising a linear Bellman objective over ellipsoidal constraints yields second-order cone programs (SOCPs) [24]. Likewise, expression-wise projections can be obtained by solving two SOCPs per expression to derive lower and upper bounds, after which robust value iteration proceeds as in the interval case. Details on the construction are provided in the technical report [34].

4 Experimental Evaluation

We have integrated our new methods for UMDP learning into the PRISM model checker [22]. We evaluate uncertainty sets from the three relaxations introduced in Sect. 3: rectangular ($\mathcal{P}_{\mathrm{R}(\mathcal{U})}$), expression-wise ($\mathcal{P}_{A(\mathcal{U})}$), and parameter-wise ($\mathcal{P}_{\Theta(\mathcal{U})}$) comparing to a baseline of standard interval learning with parameter tying ($\mathcal{P}_{\mathrm{I}}$). A comparison to ellipsoidal uncertainty sets is provided in the technical report [34]. We compare both the tightness of the learned uncertainty sets and the computational efficiency of robust solving across a range of established benchmarks with linear and polynomial parameter structures. Table 1 summarises the salient features, detailed description are in the technical report [34].

Experimental Setup. We evaluate our methods in two settings. First, we compare the tightness of the uncertainty sets and the computational efficiency of constructing and solving the corresponding models from the *same* data, i.e., the same set of collected transition samples C. To this end, we sample a total of 10^5 trajectories from the true model under a uniform sampling policy.

In the second setting, we evaluate the different approaches in an online learning scenario. Here, trajectories are sampled sequentially, the uncertain model is periodically updated from the data collected so far, and the resulting policy is used to guide further exploration. Concretely, we follow the *optimism in the face of uncertainty* principle [2,15,39]: whenever the count $\#(s, a)$ doubles for some state–action pair (s, a), we recompute the uncertain model from all trajectories collected so far and update the sampling policy to the *optimistic* policy, i.e., the policy that is optimal for the current uncertain model under the assumption that the environment resolves uncertainty in the agent's favour. This is a standard mechanism for balancing exploration and exploitation in policy learning. We emphasise, however, that our approach is agnostic to the concrete sampling procedure. Once an overall sampling budget is exhausted, we use all collected trajectories to construct the final uncertain model of the respective class and synthesise the corresponding *robust* policy. So, unlike the first setting, the different approaches are not evaluated on a fixed common data set. Instead, the data collection itself is guided by the evolving uncertainty structure, so tighter uncertainty sets can lead to more effective exploration and, ultimately, to improved final policies and guarantees. All uncertain models are built at PAC confidence level $1 - \delta = 0.999$.

Table 1. Salient characteristics of the benchmark instances. $\mathbb{P}$ denotes a temporal-logic satisfaction property and $\mathbb{E}$ an expected-reward property.

Benchmark	Instance	$\|S\|$	$\|T\|$	$\|\Theta\|$	$\|\Lambda\|$	Prop.
Aircraft	(50, 10)	5463	73033	6	721	$\mathbb{P}$
	(100, 20)	39028	551598	6	1441	$\mathbb{P}$
	(200, 40)	294458	4285228	6	2881	$\mathbb{P}$
Betting Game	(50)	13500	111678	2	2613	$\mathbb{E}$
	(100)	52000	448378	2	5283	$\mathbb{E}$
	(150)	93825	817943	2	7147	$\mathbb{E}$
Engagement	(100)	1994	5962	5	61	$\mathbb{E}$
	(300)	5994	17962	5	61	$\mathbb{E}$
	(1000)	19994	59962	5	61	$\mathbb{E}$
Mars Rover	(50, 50)	191868	1003861	6	643	$\mathbb{P}$
	(75, 75)	550447	2973287	6	943	$\mathbb{P}$
	(125, 125)	1749083	9591771	6	1543	$\mathbb{P}$
Glider	(21, 17)	357	5209	4	777	$\mathbb{E}$
	(51, 47)	2397	37066	4	4977	$\mathbb{E}$
	(106, 99)	4757	74299	4	9768	$\mathbb{E}$
Parallel Betting	(5)	3051	15543	3	477	$\mathbb{E}$
	(10)	21006	187730	3	1077	$\mathbb{E}$
	(20)	151556	1846840	3	185209	$\mathbb{E}$

Results. Table 2 reports the results for the first setting under uniform sampling. For each uncertainty structure, we evaluate how well the resulting UMDP can certify the performance of the policy that is optimal in the true MDP. We report the robust value $\underline{V}$ and the optimistic value $\overline{V}$ of this policy, corresponding to the worst- and best-case value that can be certified under the respective uncertainty structure. This allows a fair comparison of the different approximations. In addition, we report the *relative gap*, i.e., the width of the interval $[\underline{V}, \overline{V}]$ divided by the true value V^*, as well as the total runtime required to construct and solve the respective uncertain models. The technical report [34] extendeds the results, including a breakdown of the runtime spent on model construction and solving.

Figure 3 illustrates the online policy learning process for two representative benchmarks. For each uncertainty structure, we plot the number of processed trajectories against (1) the performance of the robust policy synthesised after

Table 2. Comparison of the final bounds and runtimes obtained after sampling 10^5 trajectories uniformly. For each method, $[\underline{V}, \overline{V}]$ denotes the final lower and upper bound on the value of the optimal policy in the true model, and Rel. gap the interval width relative to the true value V^*. TO indicates timeout at 2 [h].

Benchmark	Instance	$\mathcal{P}_I$			$\mathcal{P}_{\Theta(\mathcal{U})}$			$\mathcal{P}_{\Lambda(\mathcal{U})}$			$\mathcal{P}_{R(\mathcal{U})}$		
		$[\underline{\mathbf{V}}, \overline{\mathbf{V}}]$	Rel. gap	Time [s]	$[\underline{\mathbf{V}}, \overline{\mathbf{V}}]$	Rel. gap	Time [s]	$[\underline{\mathbf{V}}, \overline{\mathbf{V}}]$	Rel. gap	Time [s]	$[\underline{\mathbf{V}}, \overline{\mathbf{V}}]$	Rel. gap	Time [s]
Aircraft	(50, 10)	[0.67, 0.814]	0.19	4.04	[0.618, 0.828]	0.28	4.62	[0.719, 0.767]	0.06	5.72	[0.725, 0.761]	**0.05**	47.12
	(100, 20)	[0.745, 0.906]	0.19	30.45	[0.679, 0.918]	0.29	34.71	[0.805, 0.865]	0.07	39.54	[0.814, 0.858]	**0.05**	353.36
	(200, 40)	[0.628, 0.92]	0.36	688.27	[0.55, 0.928]	0.47	878.04	[0.754, 0.85]	0.12	882.78	[0.769, 0.838]	**0.09**	4464.25
Betting Game	(50)	[14.4, 65.3]	1.88	2.70	[25.8, 28.7]	**0.10**	6.49	[25.8, 28.7]	**0.10**	7.53	[25.8, 28.7]	**0.10**	2.29
	(100)	[16.5, 162]	3.24	13.07	[41.8, 47.7]	**0.13**	12.86	[41.8, 47.7]	**0.13**	26.31	[41.8, 47.7]	**0.13**	17.05
	(150)	[17.5, 263]	3.98	25.18	[56.5, 65.5]	**0.15**	37.98	[56.5, 65.5]	**0.15**	53.16	[56.5, 65.5]	**0.15**	38.34
Engagement	(100)	[38.4, 49.3]	0.26	5.46	[16.6, 5422]	127.35	4.44	[39.8, 45]	**0.12**	1.00	[39.8, 45]	**0.12**	1.82
	(300)	[38.6, 46.8]	0.19	16.55	[18.7, 3615]	84.74	21.30	[40.8, 45.3]	**0.11**	7.95	[40.8, 45.3]	**0.11**	7.27
	(1000)	[39.1, 45.4]	0.15	70.02	[24, 687]	15.63	57.74	[41.3, 43.8]	**0.06**	65.80	[41.3, 43.8]	**0.06**	95.44
Mars Rover	(50, 50)	[0, 0.797]	1.59	92.02	[0.177, 0.833]	1.31	136.87	[0.488, 0.515]	**0.05**	115.15	–	–	TO
	(75, 75)	[0, 0.895]	1.42	521.67	[0.0904, 0.9]	1.28	576.92	[0.619, 0.647]	**0.04**	580.21	–	–	TO
	(125, 125)	[0, 0.877]	1.42	2169.21	[0.0441, 0.922]	1.42	2553.84	[0.599, 0.635]	**0.06**	2306.29	–	–	TO
Glider	(21, 17)	[41.1, 44.6]	0.08	0.25	[42.4, 42.9]	0.01	0.19	[42.5, 42.8]	**0.01**	0.71	[42.5, 42.8]	**0.01**	37.26
	(51, 47)	[62.3, 20001]	293.07	28.92	[58.8, 59.2]	0.01	1.67	[58.8, 59.1]	**0.01**	17.16	[58.8, 59.1]	**0.01**	6268.23
	(106, 99)	[66.4, 35254]	464.04	52.95	[75.6, 76.1]	0.01	3.88	[75.6, 76]	**0.01**	48.50	–	–	TO
Parallel Betting	(5)	[24, 29.4]	0.20	0.31	[25.1, 28.2]	0.12	0.40	[26.1, 27.2]	**0.04**	0.80	[26.1, 27.2]	**0.04**	10.16
	(10)	[26.4, 40.6]	0.44	2.86	[29.7, 34.5]	0.15	3.23	[31.1, 32.9]	0.06	6.42	[31.1, 32.8]	**0.05**	116.52
	(20)	[13, 110]	3.39	21.23	[25.4, 31.7]	**0.22**	666.30	–	–	TO	–	–	TO

that number of trajectories, evaluated in the true (hidden) MDP, shown as solid lines, and (2) the corresponding PAC performance guarantee, shown as dashed lines.

Discussion. In the first setting, where all uncertainty sets are constructed from the same data, the results confirm the behaviour predicted by the inclusion hierarchy of Theorem 2. As expected, the rectangular relaxation $\mathcal{P}_{R(\mathcal{U})}$ of the uncertainty region $\mathcal{U}$ yields the tightest approximation and, consequently, the sharpest certification of policy performance, but at the highest computational cost. In low-dimensional domains such as the Betting Game, $\mathcal{P}_{R(\mathcal{U})}$ remains practically efficient via vertex enumeration, but this becomes ineffective in higher dimensions.

Further, the results show that the expression-wise projection $\mathcal{P}_{\Lambda(\mathcal{U})}$ remains tight while retaining computational efficiency. In many cases, it achieves exactly the same values as $\mathcal{P}_{R(\mathcal{U})}$, but at a cost comparable to the substantially less tight baseline interval uncertainty set $\mathcal{P}_I$ obtained from parameter tying alone. The main limitation arises when the number of distinct expressions $|\Lambda|$ is very large, since constructing $\mathcal{P}_{\Lambda(\mathcal{U})}$ then requires solving a correspondingly large number of LPs. The experiments also illustrate the incomparability of $\mathcal{P}_I$ and $\mathcal{P}_{\Theta(\mathcal{U})}$: the latter can be tighter in some instances at low computational cost, but is overall dominated by the more consistently effective expression-wise projection. Overall, $\mathcal{P}_{\Lambda(\mathcal{U})}$ appears to offer the best balance between computational efficiency and approximation tightness. By leveraging the parametric structure, it achieves substantially tighter approximations than the baseline $\mathcal{P}_I$ based solely on parameter tying, in several cases tightening the bounds by orders of magnitude.

We observe a similar pattern in the second setting, where robust policies are learned online from iteratively collected data and progressively refined

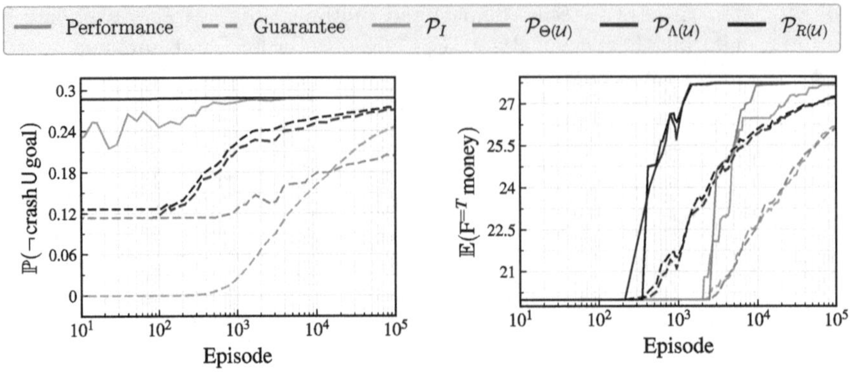

(a) Aircraft

(b) Parallel Betting

Fig. 3. Robust policy learning progress in the online setting. Solid lines show the performance of the current robust policy on the true model. Dashed lines show the corresponding PAC performance guarantee certified for the policy.

uncertainty models. Learning with $\mathcal{P}_{R(\mathcal{U})}$ consistently yields the tightest and strongest performance guarantees throughout the learning process. In particular, it improves sample efficiency, measured by the certified performance guarantee achieved after a given number of processed trajectories, by up to an order of magnitude compared to the baseline approach based on $\mathcal{P}_I$. The expression-wise projection $\mathcal{P}_{\Lambda(\mathcal{U})}$ again closely tracks the guarantees obtained via $\mathcal{P}_{R(\mathcal{U})}$, while the results also empirically illustrate the incomparability of $\mathcal{P}_{\Theta(\mathcal{U})}$ and $\mathcal{P}_I$.

The evaluation demonstrates that exploiting the parametric structure can lead to substantial improvements in both the tightness of the learned uncertainty sets and the quality of the resulting robust policies and their certified performance guarantees, in several cases by orders of magnitude. Overall, this highlights the central benefit of our approach: by lifting statistical uncertainty from the transitions to the parameter space and reasoning about it jointly through the pMDP structure, we obtain substantially less conservative robust learning and synthesis than interval-based uncertain MDP learning alone, while maintaining formal PAC guarantees on performance and robustness.

5 Conclusions

We have presented a framework for robust parameter learning in pMDPs that lifts statistical uncertainty from transitions into parameter space. This yields tighter uncertainty sets than traditional UMDP learning while retaining PAC guarantees, and admits tractable relaxations for robust policy synthesis. Our experiments show substantially improved certified guarantees and more effective robust policies.

Acknowledgements. This work was supported in part by the UKRI Erlangen AI Hub on Mathematical and Computational Foundations of AI (No. EP/Y028872/1).

References

1. Ashok, P., Křetínský, J., Weininger, M.: PAC Statistical Model Checking for Markov Decision Processes and Stochastic Games. In: Dillig, I., Tasiran, S. (eds.) CAV 2019. LNCS, vol. 11561, pp. 497–519. Springer, Cham (2019). https://doi.org/10.1007/978-3-030-25540-4_29
2. Auer, P., Jaksch, T., Ortner, R.: Near-optimal regret bounds for reinforcement learning. In: NIPS, pp. 89–96. Curran Associates, Inc. (2008)
3. Ayoub, A., Jia, Z., Szepesvari, C., Wang, M., Yang, L.: Model-based reinforcement learning with value-targeted regression. In: Proceedings of the 37th International Conference on Machine Learning, vol. 119, pp. 463–474. PMLR (2020)
4. Badings, T., Junges, S., Marandi, A., Topcu, U., Jansen, N.: Efficient sensitivity analysis for parametric robust Markov chains. In: Enea, C., Lal, A. (eds.) CAV (3), Lecture Notes in Computer Science, pp. 62–85. Springer, Cham (2023). https://doi.org/10.1007/978-3-031-37709-9_4
5. Badings, T.S., Abate, A., Jansen, N., Parker, D., Poonawala, H.A. , Stoelinga, M.: Sampling-based robust control of autonomous systems with non-Gaussian noise. In: AAAI, pp. 9669–9678. AAAI Press (2022)
6. Bartocci, E., Grosu, R., Katsaros, P., Ramakrishnan, C.R., Smolka, S.A.: Model Repair for Probabilistic Systems. In: Abdulla, P.A., Leino, K.R.M. (eds.) TACAS 2011. LNCS, vol. 6605, pp. 326–340. Springer, Heidelberg (2011). https://doi.org/10.1007/978-3-642-19835-9_30
7. Bertsimas, D., Tsitsiklis, J.N.: Introduction to linear optimization, volume 6 of Athena scientific optimization and computation series. Athena Scientific (1997)
8. Bianco, A., de Alfaro, L.: Model checking of probabilistic and nondeterministic systems. In: Thiagarajan, P.S. (ed.) FSTTCS 1995. LNCS, vol. 1026, pp. 499–513. Springer, Heidelberg (1995). https://doi.org/10.1007/3-540-60692-0_70
9. Calinescu, R., Ghezzi, C., Johnson, K., Pezzè, M., Rafiq, Y., Tamburrelli, G.: Formal verification with confidence intervals to establish quality of service properties of software systems. IEEE Trans. Reliab. **65**(1), 107–125 (2016)
10. Ceska, M., Dannenberg, F., Paoletti, N., Kwiatkowska, M., Brim, L.: Precise parameter synthesis for stochastic biochemical systems. Acta Informatica **54**(6), 589–623 (2017)
11. Clopper, C.J., Pearson, E.S.: The use of confidence or fiducial limits illustrated in the case of the binomial. Biometrika **26**(4), 404–413 (1934)
12. Cubuktepe, M., et al.: Sequential convex programming for the efficient verification of parametric mdps. In: TACAS (2), Lecture Notes in Computer Science, pp. 133–150 (2017)
13. Cubuktepe, M., Jansen, N., Junges, S., Katoen, J.-P., Topcu, U.: Convex optimization for parameter synthesis in MDPs. IEEE Trans. Autom. Control **67**(12), 6333–6348 (2022)
14. Engelen, K., Pérez, G.A., Suilen, M.: Data-efficient safe policy improvement using parametric structure. CoRR, abs/2507.15532 (2025)
15. Fruit, R., Pirotta, M., Lazaric, A., Brunskill, E.: Regret minimization in MDPs with options without prior knowledge. In: NIPS, pp. 3166–3176 (2017)
16. Gleixner, A.M., Berthold, T., Müller, B., Weltge, S.: Three enhancements for optimization-based bound tightening. J. Glob. Optim. **67**(4), 731–757 (2017)
17. Hahn, E.M., Han, T., Zhang, L.: Synthesis for PCTL in parametric Markov decision processes. In: Bobaru, M., Havelund, K., Holzmann, G.J., Joshi, R. (eds.) NASA Formal Methods, Lecture Notes in Computer Science, pp. 146–161. Springer, Cham (2011). https://doi.org/10.1007/978-3-642-20398-5_12

18. Heck, L., Quatmann, T., Spel, J., Katoen, JP., Junges, S.: Generalized parameter lifting: Finer abstractions for parametric markov chains. In: DÂťSouza, M., Komondoor, R., Srivathsan, B. (eds) ATVA, Lecture Notes in Computer Science, pp. 207–230. Springer, Cham (2025). https://doi.org/10.1007/978-3-032-08707-2_10
19. Iyengar, G.N.: Robust dynamic programming. Math. Oper. Res. **30**(2), 257–280 (2005)
20. Junges, S., Ábrahám, E., Hensel, C., Jansen, N., Katoen, J.-P., Quatmann, T., Volk, M.: Parameter synthesis for Markov models: covering the parameter space. Formal Methods Syst. Des. **62**(1), 181–259 (2024)
21. Junges, S., Jansen, N., Dehnert, C., Topcu, U., Katoen, J.-P.: Safety-Constrained Reinforcement Learning for MDPs. In: Chechik, M., Raskin, J.-F. (eds.) TACAS 2016. LNCS, vol. 9636, pp. 130–146. Springer, Heidelberg (2016). https://doi.org/10.1007/978-3-662-49674-9_8
22. Kwiatkowska, M., Norman, G., Parker, D.: PRISM 4.0: Verification of Probabilistic Real-Time Systems. In: Gopalakrishnan, G., Qadeer, S. (eds.) CAV 2011. LNCS, vol. 6806, pp. 585–591. Springer, Heidelberg (2011). https://doi.org/10.1007/978-3-642-22110-1_47
23. Lanotte, R., Maggiolo-Schettini, A., Troina, A.: Parametric probabilistic transition systems for system design and analysis. Formal Aspects Comput. **19**(1), 93–109 (2007)
24. Lobo, M.S., Vandenberghe, L., Boyd, S., Lebret, H.: Applications of second-order cone programming. Linear Algebra Appl. **284**(1–3), 193–228 (1998)
25. Long, F., Rinard, M.C.: Automatic patch generation by learning correct code. In: POPL, pp. 298–312. ACM (2016)
26. McCormick, G.P.: Computability of global solutions to factorable nonconvex programs: Part I - convex underestimating problems. Math. Program. **10**(1), 147–175 (1976)
27. Meggendorfer, T., Weininger, M., Wienhöft, P.: What are the odds? improving statistical model checking of Markov decision processes. In: Prabhakar, P., Vandin, A. (eds.) Proceedings 2nd International Joint Conference on Quantitative Evaluation of Systems and Formal Modeling and Analysis of Timed Systems (QEST+FORMATS'25), LNCS, pp. 195–218. Springer, Cham (2025). https://doi.org/10.1007/978-3-032-05792-1_11
28. Nilim, A., El Ghaoui, L.: Robust control of Markov decision processes with uncertain transition matrices. Oper. Res. **53**(5), 780–798 (2005)
29. Polgreen, E., Wijesuriya, V.B., Haesaert, S., Abate, A.: Data-Efficient Bayesian Verification of Parametric Markov Chains. In: Agha, G., Van Houdt, B. (eds.) QEST 2016. LNCS, vol. 9826, pp. 35–51. Springer, Cham (2016). https://doi.org/10.1007/978-3-319-43425-4_3
30. Puterman, M.L.: Markov Decision Processes: Discrete Stochastic Dynamic Programming. Wiley Series in Probability and Statistics. Wiley (1994)
31. Quatmann, T., Dehnert, C., Jansen, N., Junges, S., Katoen, J.-P.: Parameter Synthesis for Markov Models: Faster Than Ever. In: Artho, C., Legay, A., Peled, D. (eds.) ATVA 2016. LNCS, vol. 9938, pp. 50–67. Springer, Cham (2016). https://doi.org/10.1007/978-3-319-46520-3_4
32. Schnitzer, Y., Abate, A., Parker, D.: Certifiably robust policies for uncertain parametric environments. In: Gurfinkel, A., Heule, M. (eds.) TACAS (3), vol. 15698 of Lecture Notes in Computer Science, pp. 63–83. Springer, Cham (2025). https://doi.org/10.1007/978-3-031-90660-2_4

33. Schnitzer, Y., Abate, A., Parker, D.: Efficient solution and learning of robust factored mdps. AAAI (2026)
34. Schnitzer, Y., Abate, A., Parker, D.: Robust parameter learning for uncertain MDPs. CoRR, abs/2605.01339 (2026)
35. Scott, J.K., Stuber, M.D., Barton, P.I.: Generalized mccormick relaxations. J. Glob. Optim. **51**(4), 569–606 (2011)
36. Strehl, A.L., Littman, M.L.; A theoretical analysis of model-based interval estimation. In *ICML*, vol. 119 of ACM International Conference Proceeding Series, pp. 856–863. ACM (2005)
37. Stückler, J., Schwarz, M., Schadler, M., Topalidou-Kyniazopoulou, A., Behnke, S.: Nimbro explorer: Semiautonomous exploration and mobile manipulation in rough terrain. J. Field Robot. **33**(4), 411–430 (2016)
38. Suilen, M., Badings, T.S., Bovy, E.M., Parker, D., Jansen, N.: Robust markov decision processes: A place where AI and formal methods meet. In: Principles of Verification (3), vol. 15262 of Lecture Notes in Computer Science, pp. 126–154. Springer, Cham (2024). https://doi.org/10.1007/978-3-031-75778-5_7
39. Suilen, M., Simão, T.D., Parker, D., Jansen, N.: Robust anytime learning of Markov decision processes. In: NeurIPS (2022)
40. Valiant, L.G.: A theory of the learnable. In: STOC, pp. 436–445. ACM (1984)
41. Vanderbei, R.J.: Linear programming - foundations and extensions, volume 4 of Kluwer international series in operations research and management service, Kluwer (1998)
42. Weissman, T., Ordentlich, E., Seroussi, G., Verdú, S., Weinberger, M.J.: Inequalities for the l_1 deviation of the empirical distribution. Technical Report HPL-2003-97(R.1), Hewlett-Packard Laboratories (2003)
43. Wiesemann, W., Kuhn, D., Rustem, B.: Robust Markov decision processes. Math. Oper. Res. **38**(1), 153–183 (2013)
44. Wolff, E.M., Topcu, U., Murray, R.M.: Robust control of uncertain markov decision processes with temporal logic specifications. In: CDC, IEEE (2012)
45. Zhou, D., Gu, Q., Szepesvári, C.: Nearly minimax optimal reinforcement learning for linear mixture markov decision processes. In: Proceedings of the 34th Annual Conference on Learning Theory, vol. 134, pp. 4532–4576. PMLR (2021)

Stationary Analysis of Finite-Capacity M/M/1 Fork-Join Queues

Pierre M. Fiorini(✉)

CF Search Marketing, North Hampton, NH, USA
pierre@cfsearchmarketing.com

Abstract. Fork–join queues are a core abstraction for synchronized parallel services, yet exact stationary analysis remains limited: closed forms for the infinite-buffer $M/M/1$ fork–join queue are known only in the two-server case, and for finite-capacity $M/M/1/K_i$ systems with $N \geq 3$ no broadly usable closed-form characterization exists–direct CTMC solution is exact but quickly infeasible, and the stationary distribution and delay metrics are otherwise approached through approximations, bounds, asymptotics, or simulation. For finite-capacity $M/M/1/K_i$ systems with all-or-nothing admission, this paper develops a regenerative busy-period framework coupled with a deterministic staged-particle solver. The empty state is a regeneration point, busy periods decompose into admitted-arrival stages, and exact stage-occupation identities recover the stationary distribution, blocking probability, effective throughput, queue-length moments, and admitted-arrival Erlang-mixture latency CDFs. The solver propagates the stage via systematic deterministic quantization with particle resolution P and, when fully homogeneous, exploits exact histogram compression from $(K+1)^N$ to $\binom{N+K}{K}$. The exact stage kernel is uniformly absorbing, one-stage errors are $O(P^{-1})$, and a particle-limit theorem shows that as $P \to \infty$ the solver converges in ℓ_1 to the true stationary distribution. Experiments benchmark against exact CTMC baselines and independent discrete-event simulation up to $N = 1000$ synchronized servers, where the exact reduced chain already has $\binom{1010}{10} \approx 2.91 \times 10^{23}$ states.

Keywords: Fork-join queues · Stationary distribution · Finite buffers · All-or-nothing admission · Regenerative analysis · Deterministic particle methods · Performance evaluation · Large-scale systems

1 Introduction

Fork–join queues are the canonical stochastic model for synchronized parallel service: each job fans out into multiple branches and departs only after every required branch finishes. They arise in data-parallel analytics, fan-out microservice calls, erasure-coded storage, serverless orchestration, and assembly-style service pipelines [4,13]; a replicated or erasure-coded store with a bounded queue

M. Chen and M. Tschaikowski (Eds.): QEST+FORMATS 2026, LNCS 16913, pp. 134–151, 2027.
https://doi.org/10.1007/978-3-032-35298-9_8

per replica, admitting a request only when every replica can accept its sub-request and completing it only when all finish, is exactly the finite-capacity all-or-nothing model studied here. The key job-level penalty is synchronization: even when individual branches are lightly loaded, end-to-end latency is governed by the maximum branch delay rather than an average.

Despite their importance, fork–join queues remain among the most analytically resistant models in queueing theory. Exact classical analyses concentrate on the two-server setting [7,8,25]; beyond that regime, the literature is dominated by approximations, bounds, diffusion limits, asymptotics, and simulation [1,10,13,17,18,21,23]. For heterogeneous finite-buffer $M/M/1/K_i$ fork–join queues with $N \geq 3$, no general closed-form stationary characterization is known to us; even the underlying stationary queue-length distribution is widely regarded as intractable [13,17,24]. The difficulty is structural: the state space $\mathcal{S} = \prod_{i=0}^{N-1}\{0, 1, \ldots, K_i\}$ grows combinatorially, synchronization couples all branches, finite capacities couple the blocking boundaries, and the job metric of interest is a maximum rather than a separable sum. Direct solution of $\pi Q = 0$ remains correct for small models and is used that way in Sect. 9, but the fully homogeneous benchmark with $N = 1000$, $K_i \equiv 10$ already yields $\binom{1010}{10} \approx 2.91 \times 10^{23}$ states.

The purpose of this paper is therefore not to claim a new closed form for general fork–join queues, but to show that the finite-capacity model admits exact regenerative identities and a deterministic approximation scheme whose target is the true stationary distribution and whose error can be controlled. The key idea is to work on regenerative busy periods rather than on a global generator. The empty state is a regeneration point and busy periods decompose naturally into admitted-arrival stages; once the stagewise occupation measures are known, the full stationary distribution follows by renewal reward and admitted-arrival latency metrics follow from exact Erlang mixtures. This converts the problem from explicit global matrix construction to stagewise deterministic transport on the embedded jump chain. The resulting staged-particle solver is matrix-free, reproducible, and supported by a particle-limit convergence theorem: as P grows and the stage tail is driven to zero, the solver approaches the true stationary distribution.

2 Contributions

The paper makes five contributions.

(i) A **busy-period renewal–reward representation** (Sect. 5) that expresses the stationary distribution through regenerative busy-period occupation measures and yields exact admitted-arrival Erlang-mixture formulas for waiting and response distributions.

(ii) A **deterministic staged-particle solver** (Sect. 6) that propagates stage-boundary laws on the embedded jump chain by systematic quantization, requires no random number generator, and is naturally matrix-free.

(iii) A **particle-limit consistency theorem** (Sect. 7): the exact stage kernel is uniformly absorbing, one-stage transport and occupation errors are $O(P^{-1})$, and $\widehat{\pi}_{P,S}$ converges in ℓ_1 to the true stationary distribution as $P \to \infty$ under a suitable stage cutoff schedule.
(iv) An **exact homogeneous histogram reduction** (Sect. 8) compressing the fully symmetric labeled space from $(K+1)^N$ to $\binom{N+K}{K}$; the compression is classical, the contribution is its integration with regenerative occupation accounting and staged transport.
(v) **Numerical benchmarks** (Sect. 9) combining exact CTMC baselines on small homogeneous and heterogeneous models, independent DES checks on homogeneous benchmarks up to $N = 1000$ synchronized queues, and a heterogeneous robustness study across load, rate spread, and per-branch capacities.

3 Related Work

Exact fork–join analyses are classical but narrow. Flatto and Hahn derived the fundamental two-server stationary analysis for synchronized arrivals, and Wright studied a related two-processor coupled-input model [7,8,25]; Konstantopoulos and Walrand established stability for broader fork–join networks [19], and Pinotsi and Zazanis gave a product-form characterization for synchronized queues with deterministic arrivals [20]. Beyond two servers the literature turns to approximation: Nelson–Tantawi's homogeneous approximation [10], Baccelli–Makowski–Shwartz synchronization bounds [1], Varki–Merchant–Chen variable-subtask bounds [24], Ko–Serfozo response-time formulas for $M/M/s$ and $G/M/1$ networks [17,18], heterogeneous, multiclass, and tail extensions [15,21], computable martingale bounds [22], diffusion and heavy-traffic limits [16,23], and Thomasian's survey [13].

This paper sits between explicit CTMC solution and classical fork–join approximation: exchangeability-based occupancy compression is itself classical, but its combination with exact regenerative occupation identities, matrix-free stagewise transport, deterministic quantization, and an explicit particle-limit theorem is new. When the (possibly symmetry-reduced) CTMC fits in memory, direct linear algebra remains the right approach; when it does not, the staged solver gives a deterministic, reproducible, controllable-error route to the same objects π, P_{blk}, λ_{eff}, queue-length moments, and latency distributions.

4 Model, Stationary Quantities, and Exact Latency Mixtures

Definition 1 (Fork–Join $M/M/1/K_i$ system). *Consider N parallel queues indexed by $i \in \{0, 1, \ldots, N-1\}$. Arrivals form a Poisson process of rate λ. Every arrival forks one sub-job to each queue. Queue i has exponential service rate μ_i and capacity $K_i \in \{1, 2, \ldots\}$, where K_i counts both waiting jobs and* *any job in service. Each queue uses first-come first-served service. Service times are mutually independent across jobs and branches and are independent of the arrival process.*

An arrival is admitted if and only if $q_i < K_i$ for all i. If any queue is full, the entire fork is blocked. A job completes only when all of its sub-jobs complete.

Because each queue has finite capacity, the joint queue-length process is a finite-state continuous-time Markov chain. We make this explicit next.

Definition 2 (CTMC state and generator). *The state is $\mathbf{q} = (q_0, \ldots, q_{N-1}) \in \mathcal{S}$, where $q_i \in \{0, \ldots, K_i\}$. The nonzero off-diagonal transition rates are*

$$\mathbf{q} \to \mathbf{q} + \mathbf{1}, \qquad \text{rate } \lambda\,\mathbf{1}[q_i < K_i \ \forall i], \tag{1}$$

$$\mathbf{q} \to \mathbf{q} - \mathbf{e}_i, \qquad \text{rate } \mu_i\,\mathbf{1}[q_i > 0], \qquad i = 0, \ldots, N-1. \tag{2}$$

The total exit rate from state $\mathbf{q}$ is

$$R(\mathbf{q}) = \lambda\,\mathbf{1}[q_i < K_i \ \forall i] + \sum_{i:q_i>0} \mu_i. \tag{3}$$

Throughout we assume $\lambda > 0$, $\mu_i > 0$ for all i, and $K_i \geq 1$. Then $\mathcal{S}$ is finite and the CTMC is irreducible: any state empties by service completions alone, and from $\mathbf{0}$ any target state is reached by finitely many admitted arrivals interleaved with selected service completions, every step having positive probability. Hence the chain is positive recurrent with a unique stationary distribution π.

4.1 Basic Stationary Performance Measures

The all-or-nothing admission rule partitions the state space into an admitted set

$$\mathcal{A} = \{\mathbf{q} \in \mathcal{S} : q_i < K_i \ \forall i\}$$

and a blocked set $\mathcal{B} = \mathcal{S} \setminus \mathcal{A}$. Under PASTA, a Poisson arrival sees the pre-arrival state distribution π, so

$$P_{\text{blk}} = \sum_{\mathbf{q} \in \mathcal{B}} \pi(\mathbf{q}), \tag{4}$$

$$P_{\text{adm}} = 1 - P_{\text{blk}}, \tag{5}$$

$$\lambda_{\text{eff}} = \lambda P_{\text{adm}}, \tag{6}$$

$$\mathbb{E}[q_i] = \sum_{\mathbf{q} \in \mathcal{S}} \pi(\mathbf{q}) q_i. \tag{7}$$

4.2 Exact Admitted-Arrival Waiting and Response Mixtures

Under the FCFS discipline and independent exponential services of Definition 1, sub-jobs at different queues evolve independently once the admitted pre-arrival state is fixed, so the maximum waiting and response times of a job are mixtures of products of Erlang CDFs weighted by the stationary admitted-arrival distribution.

For an admitted arrival that sees pre-arrival state $\mathbf{q} \in \mathcal{A}$, the post-arrival state is $\mathbf{q} + \mathbf{1}$. Thus branch i contributes

$$W_i \mid \mathbf{q} \sim \text{Erlang}(q_i, \mu_i), \qquad R_i \mid \mathbf{q} \sim \text{Erlang}(q_i + 1, \mu_i),$$

where $\text{Erlang}(0, \mu)$ is interpreted as a point mass at 0. Both use the memoryless property: at the arrival epoch the residual service at branch i is again $\text{Exp}(\mu_i)$, so the tagged wait is a sum of q_i i.i.d. exponentials (these formulas do not extend verbatim to general service). Conditional independence across branches then yields exact mixtures for the job-level maxima.

Theorem 1 (Exact admitted-arrival latency CDFs). *Let $F_{k,\mu}(t)$ denote the CDF of an Erlang random variable with integer shape $k \geq 0$ and rate μ, with $F_{0,\mu}(t) = 1$ for $t \geq 0$. Then the exact stationary CDFs of the maximum waiting and response times of admitted jobs are*

$$\mathbb{P}\{W_{\max} \leq t\} = \frac{1}{P_{\text{adm}}} \sum_{\mathbf{q} \in \mathcal{A}} \pi(\mathbf{q}) \prod_{i=0}^{N-1} F_{q_i, \mu_i}(t), \tag{8}$$

$$\mathbb{P}\{R_{\max} \leq t\} = \frac{1}{P_{\text{adm}}} \sum_{\mathbf{q} \in \mathcal{A}} \pi(\mathbf{q}) \prod_{i=0}^{N-1} F_{q_i+1, \mu_i}(t). \tag{9}$$

Consequently,

$$\mathbb{E}[W_{\max}] = \int_0^\infty \left(1 - \mathbb{P}\{W_{\max} \leq t\}\right) dt, \tag{10}$$

$$\mathbb{E}[R_{\max}] = \int_0^\infty \left(1 - \mathbb{P}\{R_{\max} \leq t\}\right) dt. \tag{11}$$

Remark 1 (Pre-arrival and post-arrival distributions). The formulas are written for the admitted pre-arrival distribution; the implementation often stores the admitted *post-arrival* distribution instead, equivalently, since $\mathbf{q} \in \mathcal{A}$ maps deterministically to $\mathbf{q} + \mathbf{1}$.

5 Renewal–Reward and Stage Decomposition

Rather than assembling and solving the full global-balance system, our approach exploits the regenerative structure of the CTMC to decompose the busy period into stages that can be analyzed independently.

The empty state $\mathbf{0}$ is a regenerative atom. A renewal cycle consists of an idle period until the first admitted arrival, followed by the busy period started by that arrival.

Definition 3 (Busy period and stages). *A busy period starts immediately after the first admitted arrival into the empty system, i.e., from state* $\mathbf{1} = (1, \ldots, 1)$*, and ends at the first return to* $\mathbf{0}$*. Stage* n *is the sub-epoch between the* n*th and* $(n+1)$*st admitted arrivals. Stage* 1 *therefore begins at* $\mathbf{1}$.

Inside a stage, an admitted arrival exits the stage, while service completions are the only nonabsorbing state-changing transitions. This defines an *embedded discrete-time chain* (EDTC). Exogenous arrivals that occur while some queue is full are blocked and are CTMC self-events; they do not appear as EDTC transitions.

Definition 4 (Within-stage EDTC). *From a nonempty state* $\mathbf{q}$*, the EDTC next-step probabilities are*

$$p_{\text{arr}}(\mathbf{q}) = \frac{\lambda \, \mathbf{1}[q_i < K_i \; \forall i]}{R(\mathbf{q})}, \tag{12}$$

$$p_{\text{dep},i}(\mathbf{q}) = \frac{\mu_i \, \mathbf{1}[q_i > 0]}{R(\mathbf{q})}. \tag{13}$$

The expected holding time of one EDTC visit to $\mathbf{q}$ *is* $1/R(\mathbf{q})$.

Proposition 2 (Exact stage recursion)**.** *Let* D_n *denote the exact post-arrival distribution at the start of stage* n*. Let* $\mathcal{T}$ *be the conditional kernel that maps a stage-start distribution to the post-arrival distribution at the* next admitted arrival, *given that the stage does not empty first, and let* α_n *be the probability that stage* n *ends with another admitted arrival rather than absorption into* $\mathbf{0}$*. If* p_n *is the probability that stage* n *is reached, then*

$$D_{n+1} = \mathcal{T}(D_n), \qquad p_{n+1} = p_n \alpha_n, \qquad p_1 = 1.$$

Thus every exact busy-period statistic can be decomposed stage by stage.

The next result reassembles the per-stage contributions into π.

Proposition 3 (Stationary distribution from stage occupation)**.** *Let* $\mathcal{X} = \{\mathbf{q} \in \mathcal{S} : q_i \geq 1 \; \forall i\}$ *be the stage-start state space. For* $x \in \mathcal{X}$ *and* $q \in \mathcal{S} \setminus \{\mathbf{0}\}$*, let*

$$H(x, q) = \mathbb{E}_x \left[\int_0^{\tau_{\text{stg}}} \mathbf{1}\{Q(t) = q\} \, dt \right]$$

be the expected time spent in CTMC state q *during one stage started from* x*, where* τ_{stg} *is the stage-end time. Let* $A(x, y)$ *denote the probability that a stage started from* $x \in \mathcal{X}$ *reaches the next admitted arrival in stage-start state* $y \in \mathcal{X}$ *before emptying, and set* $\nu_1 = \delta_{\mathbf{1}}$*,* $\nu_{n+1} = \nu_n A$*. Then*

$$\zeta(q) = \sum_{n \geq 1} \sum_{x \in \mathcal{X}} \nu_n(x) H(x, q)$$

is the exact expected busy-period occupation measure of nonzero CTMC states. Consequently,

$$\pi(\mathbf{0}) = \frac{1/\lambda}{1/\lambda + \sum_{q \neq \mathbf{0}} \zeta(q)}, \qquad \pi(q) = \frac{\zeta(q)}{1/\lambda + \sum_{q' \neq \mathbf{0}} \zeta(q')}, \quad q \neq \mathbf{0}. \tag{14}$$

6 Deterministic Staged-Particle Transport

The stage decomposition of Sect. 5 reduces the computation of π to iterating a within-stage transport operator and accumulating occupation measures; this section describes the algorithm that carries out the iteration. It is a deterministic particle method adapted to the discrete state space of a finite-state CTMC. Integer particle counts across queue-length states replace weighted Dirac masses, the embedded jump chain of Definition 4 replaces continuous advection, and inter-stage requantization to exactly P particles plays the role of remeshing. The key ingredient is systematic deterministic quantization, which conserves mass exactly and introduces less than one unit of local count error in every child bin. For more information on related approaches and methods, see [2,5,9].

6.1 Particle Representation and Systematic Quantization

At stage n, the exact stage-start distribution is approximated by P integer particles across states ($n(\mathbf{q})$ particles at $\mathbf{q}$ means empirical probability $n(\mathbf{q})/P$). When a state with count c has transition probabilities $(p_0, \ldots, p_{m-1})$, integer child counts are allocated by deterministic systematic quantization.

Definition 5 (Systematic deterministic quantization). *Given an integer budget c, probabilities $p_0, \ldots, p_{m-1}$, cumulative sums $F_0 = 0$, $F_j = \sum_{r<j} p_r$, and phase $\phi \in [0, 1)$, define*

$$c_j = \lfloor cF_{j+1} - \phi \rfloor - \lfloor cF_j - \phi \rfloor, \qquad j = 0, \ldots, m-1. \tag{15}$$

After each quantization call, update the phase by

$$\phi \leftarrow \{\phi + \phi_g\}, \qquad \phi_g = \frac{\sqrt{5}-1}{2}.$$

In words, (15) lays a uniform grid of c points with spacing $1/c$ and offset ϕ/c along the cumulative-probability axis and assigns child j the points landing in $(F_j, F_{j+1}]$—the deterministic analogue of systematic resampling in particle filtering. Each call commits a signed rounding residual set by ϕ; advancing the phase by the golden-ratio increment ϕ_g makes the offsets $\{\phi_0 + n\phi_g\}$ a Kronecker–Weyl sequence (the golden ratio being the most badly approximable irrational), so successive residuals are equidistributed and cancel rather than drift, as in low-discrepancy quantization [11]. The following guarantees hold for every call, regardless of the phase schedule.

Proposition 4 (Local quantization facts). *For every call of* (15), *the child counts satisfy $\sum_j c_j = c$, and each local rounding error obeys $|c_j - cp_j| < 1$.*

Algorithm 1. Deterministic staged-particle transport with occupation accounting

Require: model $(N, \lambda, \{\mu_i\}, \{K_i\})$, particle population P, stage limit S, stop threshold ε

$p \leftarrow 1$, $D \leftarrow \{\mathbf{1} : 1\}$, $\widehat{\zeta} \leftarrow 0$
for $n = 1$ to S **do**
    if $p < \varepsilon$ **then break**
    end if
    resample D to exactly P integer particles and store the counts in F
    initialize next-arrival count map $A \leftarrow 0$ and absorbed count $a_0 \leftarrow 0$
    while $F \neq \emptyset$ **do**
        initialize empty next-frontier map $F_{\text{new}} \leftarrow 0$
        for all $(\mathbf{q}, c) \in F$ **do**
            $\widehat{\zeta}(\mathbf{q}) \leftarrow \widehat{\zeta}(\mathbf{q}) + p\,(c/P)/R(\mathbf{q})$
            compute $R(\mathbf{q})$ and the EDTC probabilities from Definition 4
            route the c particles with Definition 5
                send admitted-arrival children to A
                send nonzero departure children to F_{new}
                send empty departures to a_0
        end for
        $F \leftarrow F_{\text{new}}$
    end while
    $\widehat{\alpha} \leftarrow \sum_{y \in \mathcal{X}} A(y)/P$
    if $\widehat{\alpha} = 0$ **then break**
    end if
    $D(y) \leftarrow A(y)/(P\widehat{\alpha})$ for $y \in \mathcal{X}$
    $p \leftarrow p\,\widehat{\alpha}$
end for
reconstruct $\widehat{\pi}_{P,S}$ from $\widehat{\zeta}$ by (14)

6.2 Algorithmic Skeleton

Algorithm 1 assembles the quantization primitive into a complete solver: each outer-loop iteration transports the particles of one busy-period stage through the within-stage EDTC until every particle has either triggered a new admitted arrival or been absorbed into $\mathbf{0}$, then passes the accumulated arrival distribution to the next stage.

Remark 2 (Implemented regimes and reported metrics). The implementation uses two exact state representations—an exchangeable histogram representation when all μ_i and K_i are common, and a general labeled-state representation otherwise—with the same deterministic transport mechanism in either case. In full accounting mode it reports P_{blk}, P_{adm}, λ_{eff}, per-server queue-length means, $\mathbb{E}[W_{\max}]$, $\mathbb{E}[R_{\max}]$, and latency CDF grids, the latter by applying the exact mixture formulas of Theorem 1 to the approximated admitted-arrival distribution.

7 Error Analysis and Particle-Limit Consistency

The central feature of the staged solver is that its target is the true stationary distribution of the finite-capacity CTMC: the exact stage kernel is uniformly absorbing, one stage of deterministic transport incurs only $O(P^{-1})$ error, and the stationary approximation converges in ℓ_1 to π as P grows and the stage tail is driven to zero. Full proofs of every result in this section are given in the extended version [6]; for the three central convergence results—Theorems 7 and 8 and Corollary 9—we include short proof sketches inline below.

7.1 Exact Stage Kernel and Geometric Decay

Every stage has a uniformly positive probability of emptying the system, so the mass carried from one stage to the next contracts geometrically. Let $\mathcal{X} = \{\mathbf{q} \in \mathcal{S} : q_i \geq 1 \ \forall i\}$ be the finite set of stage-start states. For $x \in \mathcal{X}$ and $y \in \mathcal{X}$, write $A(x, y)$ for the exact probability that a stage started from x ends at the next admitted arrival in state y rather than emptying first. Thus A is substochastic. Let

$$\Gamma(x, q) = H(x, q), \qquad H_* = \max_{x \in \mathcal{X}} \sum_{q \neq \mathbf{0}} H(x, q),$$

so H_* is the largest expected stage duration over all stage-start states.

Proposition 5 (Uniform stage absorption). *Assume $\lambda > 0$ and $\mu_i > 0$ for all i. There exists $\delta > 0$ such that every stage-start state $x \in \mathcal{X}$ empties before the next admitted arrival with probability at least δ. Equivalently,*

$$\rho := \max_{x \in \mathcal{X}} \sum_{y \in \mathcal{X}} A(x, y) \leq 1 - \delta < 1. \tag{16}$$

Hence, for every nonnegative row vector m on $\mathcal{X}$,

$$\|mA\|_1 \leq \rho \, \|m\|_1. \tag{17}$$

7.2 One-Stage Consistency

For a probability vector d on $\mathcal{X}$, let $T(d) = dA$ be the exact substochastic next-stage distribution and let $\Gamma(d) = d\Gamma$ be the exact within-stage occupation measure on $\mathcal{S} \setminus \{\mathbf{0}\}$, where the matrix Γ has rows $\Gamma(x, \cdot) = H(x, \cdot)$. The deterministic solver induces corresponding approximate maps $\widehat{T}_P(d)$ and $\widehat{\Gamma}_P(d)$ when a particle population of P is used: the input law is first quantized to exactly P particles, particles are propagated through the within-stage EDTC, arrival exits define $\widehat{T}_P(d)$, and accumulated holding-time rewards define $\widehat{\Gamma}_P(d)$.

Theorem 6 (One-stage consistency). *For every finite-capacity model with $\lambda > 0$ and $\mu_i > 0$, there exist finite constants C_T and C_Γ, depending only on the* *model and on the fixed state ordering used by systematic quantization, such that for all integers* $P \geq 1$,

$$\sup_{d \in \Delta(\mathcal{X})} \|\widehat{T}_P(d) - T(d)\|_1 \leq \frac{C_T}{P}, \tag{18}$$

$$\sup_{d \in \Delta(\mathcal{X})} \|\widehat{\Gamma}_P(d) - \Gamma(d)\|_1 \leq \frac{C_\Gamma}{P}. \tag{19}$$

7.3 Global Stationary Error

Define the exact unnormalized stage sequence by

$$m_1 = \delta_1, \qquad m_{n+1} = m_n A,$$

and the deterministic approximation by

$$\widehat{m}_1^{(P)} = \delta_1, \qquad \widehat{m}_{n+1}^{(P)} = \widehat{U}_P(\widehat{m}_n^{(P)}),$$

where, for a nonnegative row vector m on $\mathcal{X}$,

$$\widehat{U}_P(m) = \begin{cases} \|m\|_1 \, \widehat{T}_P(m/\|m\|_1), & m \neq 0, \\ 0, & m = 0, \end{cases} \qquad \widehat{V}_P(m) = \begin{cases} \|m\|_1 \, \widehat{\Gamma}_P(m/\|m\|_1), & m \neq 0, \\ 0, & m = 0. \end{cases}$$

The map $\widehat{U}_P$ is the mass-homogeneous lift of approximate next-stage transport, and $\widehat{V}_P$ is the corresponding mass-homogeneous lift of approximate occupation accounting. Set

$$\varepsilon_P = \frac{C_T}{P}, \qquad \beta_P = \frac{C_\Gamma}{P}.$$

Theorem 7 (Global error bounds and stationary-distribution approximation). *Assume* $\varepsilon_P < 1 - \rho$. *Then, for every* $n \geq 1$,

$$\left\|\widehat{m}_n^{(P)}\right\|_1 \leq (\rho + \varepsilon_P)^{n-1}, \tag{20}$$

$$\sum_{n \geq 1} \left\|\widehat{m}_n^{(P)} - m_n\right\|_1 \leq \frac{\varepsilon_P}{(1-\rho)(1-\rho-\varepsilon_P)}. \tag{21}$$

Let the exact and approximate busy-period occupation measures of nonzero CTMC states be

$$\zeta = \sum_{n \geq 1} m_n \Gamma, \qquad \widehat{\zeta}_P = \sum_{n \geq 1} \widehat{V}_P\left(\widehat{m}_n^{(P)}\right),$$

and let $\widehat{\zeta}_{P,S} = \sum_{n=1}^{S} \widehat{V}_P\left(\widehat{m}_n^{(P)}\right)$ *be the stage-truncated approximation. Then*

$$\|\widehat{\zeta}_P - \zeta\|_1 \leq \frac{\beta_P}{1-\rho-\varepsilon_P} + H_* \frac{\varepsilon_P}{(1-\rho)(1-\rho-\varepsilon_P)}, \tag{22}$$

and

$$\|\widehat{\zeta}_{P,S}-\zeta\|_1 \le \frac{\beta_P}{1-\rho-\varepsilon_P}+H_*\frac{\varepsilon_P}{(1-\rho)(1-\rho-\varepsilon_P)}+(H_*+\beta_P)\frac{(\rho+\varepsilon_P)^S}{1-\rho-\varepsilon_P}. \quad (23)$$

Let π be the exact stationary distribution obtained from (14), *and let $\widehat{\pi}_{P,S}$ be the corresponding particle approximation obtained by replacing ζ with $\widehat{\zeta}_{P,S}$. Then*

$$\|\widehat{\pi}_{P,S}-\pi\|_1 \le 2\lambda\left[\frac{\beta_P}{1-\rho-\varepsilon_P}+H_*\frac{\varepsilon_P}{(1-\rho)(1-\rho-\varepsilon_P)}+(H_*+\beta_P)\frac{(\rho+\varepsilon_P)^S}{1-\rho-\varepsilon_P}\right]. \quad (24)$$

Proof of Theorem 7 (sketch). Write $U(m)=mA$. By (17), $\|U(m)\|_1 \le \rho\|m\|_1$ and U is ρ-Lipschitz in ℓ_1, while the one-stage bounds (18)–(19) lift to $\|\widehat{U}_P(m)-U(m)\|_1 \le \varepsilon_P\|m\|_1$ and $\|\widehat{V}_P(m)-m\Gamma\|_1 \le \beta_P\|m\|_1$. The mass bound (20) follows by induction from $\|\widehat{m}_{n+1}^{(P)}\|_1 \le (\rho+\varepsilon_P)\|\widehat{m}_n^{(P)}\|_1$. Setting $e_n = \|\widehat{m}_n^{(P)} - m_n\|_1$, the triangle inequality gives $e_{n+1} \le \varepsilon_P(\rho+\varepsilon_P)^{n-1} + \rho e_n$ with $e_1 = 0$, hence $e_n \le (\rho+\varepsilon_P)^{n-1} - \rho^{n-1}$; summing the geometric series yields (21). Splitting the occupation error into a per-stage quantization part (weighted by β_P) and a stage-mass propagation part (weighted by H_*) gives (22), and adding the geometric stage-tail term gives (23). Finally the normalization $F(x) = (1/\lambda, x)/(1/\lambda + \|x\|_1)$ is 2λ-Lipschitz in ℓ_1 on nonnegative occupation vectors, so applying it to $x=\widehat{\zeta}_{P,S}$ and $y=\zeta$ produces (24). The full computation appears in [6]. □

The bound (24) has three terms: the first two are $O(P^{-1})$ for fixed S and capture per-stage quantization error; the third is exponentially small in S and captures the truncated stage tail. The following theorem confirms that both sources of error vanish simultaneously.

Theorem 8 (Particle-limit consistency of the staged solver). *Assume the hypotheses of Theorem 7. If $S=S(P)\to\infty$ and $(\rho+\varepsilon_P)^{S(P)}\to 0$ as $P\to\infty$, then*

$$\|\widehat{\pi}_{P,S(P)}-\pi\|_1 \to 0.$$

Hence the staged-particle solver converges to the true stationary distribution of the finite-capacity fork–join CTMC, and every bounded stationary functional converges with it.

Proof of Theorem 8 (sketch). Apply (24). Since $\varepsilon_P = C_T/P$ and $\beta_P = C_\Gamma/P$ vanish as $P\to\infty$, the two particle-resolution terms tend to 0, and the hypothesis $(\rho+\varepsilon_P)^{S(P)}\to 0$ removes the geometric stage-tail term. Hence $\|\widehat{\pi}_{P,S(P)}-\pi\|_1 \to 0$, and convergence of every bounded stationary functional follows from Corollary 9. □

Corollary 9 (Reported stationary functionals)**.** *Let $f:\mathcal{S}\to\mathbb{R}$ be bounded and define $\Pi_f(\pi)=\sum_{\mathbf{q}\in\mathcal{S}}\pi(\mathbf{q})f(\mathbf{q})$. Then*

$$\left|\Pi_f(\widehat{\pi}_{P,S})-\Pi_f(\pi)\right| \le \|f\|_\infty\,\|\widehat{\pi}_{P,S}-\pi\|_1. \quad (25)$$

If $\mathcal{A} = \{\mathbf{q} \in \mathcal{S} : q_i < K_i \ \forall i\}$ *and*

$$\Theta_\phi(\pi) = \frac{\sum_{\mathbf{q}\in\mathcal{A}} \pi(\mathbf{q})\phi(\mathbf{q})}{P_{\mathrm{adm}}}$$

for some kernel $\phi : \mathcal{A} \to [0, M_\phi]$*, then whenever* $\|\widehat{\pi}_{P,S} - \pi\|_1 \le P_{\mathrm{adm}}/2$*,*

$$\left|\Theta_\phi(\widehat{\pi}_{P,S}) - \Theta_\phi(\pi)\right| \le \frac{4M_\phi}{P_{\mathrm{adm}}} \|\widehat{\pi}_{P,S} - \pi\|_1. \tag{26}$$

Hence blocking probability, effective throughput, queue-length moments, admitted-arrival CDFs, $\mathbb{E}[W_{\max}]$*, and* $\mathbb{E}[R_{\max}]$ *all inherit the same order of convergence.*

Proof of Corollary 9 *(sketch).* The linear bound (25) is Hölder duality between ℓ_1 and ℓ_∞. For the admitted-arrival ratio functional, set $a = \sum_{\mathbf{q}\in\mathcal{A}} \widehat{\pi}_{P,S}(\mathbf{q})$, $b = P_{\mathrm{adm}}$, $x = \sum_{\mathbf{q}\in\mathcal{A}} \widehat{\pi}_{P,S}(\mathbf{q})\phi(\mathbf{q})$, and $y = \sum_{\mathbf{q}\in\mathcal{A}} \pi(\mathbf{q})\phi(\mathbf{q})$. Then $|x - y| \le M_\phi\|\widehat{\pi}_{P,S} - \pi\|_1$, $|a - b| \le \|\widehat{\pi}_{P,S} - \pi\|_1$, and $y \le M_\phi b$, so $|x/a - y/b| \le (2M_\phi/a)\|\widehat{\pi}_{P,S} - \pi\|_1$; the hypothesis $\|\widehat{\pi}_{P,S} - \pi\|_1 \le P_{\mathrm{adm}}/2$ forces $a \ge P_{\mathrm{adm}}/2$ and yields (26). Choosing $f = \mathbf{1}\{\mathbf{q} \in \mathcal{B}\}$, $f(\mathbf{q}) = q_i$, or the Erlang-mixture CDF kernels of Theorem 1 specializes the bound to blocking probability, queue-length moments, and the latency metrics; the $\mathbb{E}[W_{\max}]$ and $\mathbb{E}[R_{\max}]$ kernels are bounded because $\mathcal{A}$ and all capacities are finite. Full details are in [6]. □

Remark 3 (Interpretation of the constants). The constants ρ, C_T, C_Γ, and H_* are structural rather than sharp; they separate geometric decay of stage reachability, one-stage quantization error, and the conversion from occupation error to stationary error. The practical message is the qualitative fact of Theorem 8: increasing the particle population drives the solver toward the true stationary distribution.

8 Exact Homogeneous Histogram Reduction

If all rates and capacities are common ($\mu_i \equiv \mu$, $K_i \equiv K$), the model is exchangeable under permutation of server labels.

Definition 6 (Histogram state). *For a fully homogeneous system, define the histogram state*

$$\mathbf{h} = (c_0, c_1, \ldots, c_K), \qquad c_v = |\{i : q_i = v\}|, \qquad \sum_{v=0}^{K} c_v = N.$$

Proposition 10 (Exact homogeneous state-count formula)**.** *The number of distinct histogram states is*

$$|\mathcal{S}_{\mathrm{hist}}| = \binom{N+K}{K}. \tag{27}$$

This is exact, not asymptotic.

Remark 4 (Magnitude of the reduced chain). The reduction is dramatic but not trivializing: at $N = 50$, $K = 10$ the histogram chain still has 75,394,027,566 states, and at $N = 1000$ it grows to 2.91×10^{23}.

9 Numerical Results

On small models where the CTMC can still be solved exactly, this section checks the regenerative accounting and latency-mixture post-processing against direct linear algebra; at branch counts where exact construction is impossible, it compares the staged solver to independent DES.

9.1 Small-State Exact Baselines

Two small models are chosen so that the stationary CTMC can still be solved directly: **Case H2** (homogeneous, $N = 2$, $\lambda = 0.5$, $\mu = (1.0, 1.0)$, $K = (5, 5)$) and **Case X3** (heterogeneous, $N = 3$, $\lambda = 0.4$, $\mu = (1.0, 0.9, 1.1)$, $K = (4, 5, 4)$). For Case H2 the staged solver used $P = 50{,}000$ particles, and for Case X3 it used $P = 30{,}000$ particles. The DES runs used 40,000 measurement arrivals after 5,000 warm-up arrivals in each case.

Table 1. Validated small-state benchmarks. The exact baseline solves $\pi Q = 0$ directly; DES values carry 95% confidence intervals where reported.

Case	Metric	Exact CTMC	Staged	DES
H2	$\mathbb{E}[R_{\max}]$	2.609499	2.609500	2.582268 ± 0.018571
	P_{blk}	0.026999	0.027001	0.024750 ± 0.001523
	λ_{eff}	0.486501	0.486500	0.487625
X3	$\mathbb{E}[R_{\max}]$	2.779452	2.779468	2.778699 ± 0.018130
	P_{blk}	0.028238	0.028232	0.027200 ± 0.001594
	λ_{eff}	0.388705	0.388707	0.389120

Table 1 shows close agreement between the staged solver and the exact CTMC baseline on both examples; the DES values give an independent Monte Carlo check, while the exact comparison is the primary validation of the regenerative accounting, stage transport, and latency-mixture evaluation.

9.2 Homogeneous Finite Buffer Fork-Join Queue Examples up to $N = 1000$

In this example, we use the following parameters

$$\lambda = 0.5, \qquad \mu_i \equiv 1.0, \qquad K_i \equiv 10,$$

with

$$N \in \{2, 5, 10, 20, 50, 100, 500, 1000\}.$$

We vary N at fixed $K_i \equiv 10$ to isolate the effect of branch count on synchronization and cost; per-branch capacity is varied separately in the heterogeneous study of Sect. 9.3. The particle cardinalities are reduced as N grows:

$$P \in \{100{,}000,\ 5{,}000,\ 5{,}000,\ 1{,}000,\ 1{,}000,\ 200,\ 200,\ 200\}.$$

These budgets are set by a wall-clock cap matching the fixed DES budget at each N (so $N = 100$, 500, and 1000 share one population); by Theorem 8, reducing P at large N trades accuracy for runtime only. The DES runs use 5,000,000 measurement arrivals after 1,000 warm-up arrivals; even after exact homogeneous reduction, the chain grows from 66 states at $N = 2$ to $\approx 2.91 \times 10^{23}$ at $N = 1000$.

Table 2. Exact reduced state count $\binom{N+10}{10}$, staged and DES estimates of $\mathbb{E}[R_{\max}]$, relative difference, and wall-clock time.

N	particles	$\binom{N+10}{10}$	staged $\mathbb{E}[R_{\max}]$	DES $\mathbb{E}[R_{\max}]$	$\lvert\Delta\rvert$/DES (%)	solver s	DES s
2	100,000	66	2.857105	2.855441 ± 0.001930	0.058	0.463	378.397
5	5,000	3,003	4.125238	4.126043 ± 0.002123	0.020	9.155	387.086
10	5,000	184,756	5.115245	5.110881 ± 0.002232	0.085	61.540	407.657
20	1,000	30,045,015	6.086543	6.094716 ± 0.002320	0.134	67.800	390.087
50	1,000	7.54×10^{10}	7.343063	7.357331 ± 0.002404	0.194	411.762	408.556
100	200	4.69×10^{13}	8.266130	8.284041 ± 0.002454	0.216	315.647	421.303
500	200	3.00×10^{20}	10.315760	10.305802 ± 0.002521	0.097	4893.450	594.042
1000	200	2.91×10^{23}	11.168633	11.125835 ± 0.002536	0.385	14996.373	842.298

Table 2 reports the results: the staged and DES means stay within 0.4% of each other at all eight branch counts, with some larger-N differences exceeding the DES confidence intervals—visible deterministic particle and truncation error at the chosen budgets. The example illustrates the main computational point: exact reduced-state construction becomes hopelessly large long before $N = 1000$, whereas the staged solver continues to return stationary analytic results (Fig. 1).

Remark 5 (Theoretical advantages vs. DES speed). DES can outrun the staged solver at large N (Table 2), but unlike simulation it yields an explicit error decomposition (Theorem 7), particle-limit convergence to the true π (Theorem 8), RNG-free reproducibility, and exact admitted-arrival mixtures.

9.3 Heterogeneous Robustness, Load, and Convergence Speed

Heterogeneity removes exchangeability, so the solver runs on the labeled state space, and the natural concern is that the populated frontier F of Algorithm 1 grows toward the labeled cardinality. Table 3 addresses this with three configurations of increasing size, varied capacities K_i, rate spreads up to $\mu_{\max}/\mu_{\min} = 2.5$,

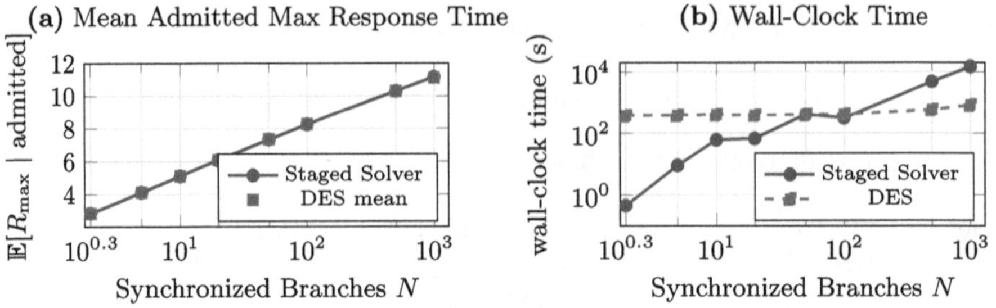

Fig. 1. Homogeneous benchmark from Table 2 as N grows. **(a)** Mean admitted-job maximum response time: staged and DES remain close from $N = 2$ through $N = 1000$. **(b)** Wall-clock times: the particle population varies with N, so the staged runtime is not monotone; it is faster than DES at several moderate N and slower at $N = 500, 1000$.

and bottleneck loads $\rho_{\max} = \lambda/\mu_{\min}$ up to 0.95: **A**: N=3, $\mu = (1.0, 0.9, 1.1)$, $K = (4, 5, 4)$; **B**: N=5, $\mu = (1.0, 1.5, 0.8, 2.0, 1.2)$, $K = (5, 10, 5, 15, 8)$; **C**: N=8, $\mu = (1.0, 1.3, 0.7, 1.6, 0.9, 1.1, 1.4, 0.8)$, $K = (6, 8, 4, 10, 5, 7, 9, 4)$. The reference for A (150 labeled states) is an exact CTMC solve (explicit enumeration of $\mathcal{S}$, assembly of Q from (1)–(2), and solution of $\pi Q = 0$; see [6]); for B and C it is independent DES with 400,000 measurement arrivals after 20,000 warm-up.

Table 3. Heterogeneous benchmarks across load. The reference is the exact CTMC for configuration A and DES (mean ± 95% CI) for B and C. $|F|_{\max}$ is the peak populated frontier and S the number of stages run to $\varepsilon = 10^{-12}$.

cfg	λ	$\rho_{\max}$	$\lvert\mathcal{S}\rvert$	P	staged $\mathbb{E}[R_{\max}]$	reference	$\lvert F\rvert_{\max}$	S	solver s
A	0.40	0.44	150	5000	2.780238	2.779452 (exact)	142	80	1.2
A	0.72	0.80	150	5000	3.586395	3.586244 (exact)	142	208	3.5
A	0.855	0.95	150	5000	3.854596	3.854406 (exact)	142	295	4.7
B	0.50	0.625	57,024	3000	3.531845	3.527451 ± 0.007006	1239	142	24.6
B	0.70	0.875	57,024	3000	4.228166	4.233873 ± 0.008184	1404	283	53.5
C	0.55	0.786	8,316,000	1500	4.682213	4.688632 ± 0.008073	1280	207	73.0

First, accuracy survives the loss of symmetry and high load: configuration A matches the exact baseline to relative error at most 2.8×10^{-4} at all three loads (including $\rho_{\max} = 0.95$), and every B and C estimate lies inside its DES interval. Second, the frontier does not explode—it is bounded by the smaller of the reachable state count and the states P particles can occupy: A saturates at 142 states, while C peaks at 1,280 populated states against 8.3 million labeled ones, so per-stage cost scales with the frontier and stage count, not with $|\mathcal{S}|$. Third, load enters through the stage count S, which roughly quadruples from $\rho_{\max} = 0.44$ to 0.95 in configuration A.

The load effect is isolated in Table 4 on the homogeneous N=10, K=10, μ=1 model: the stages needed to reach $\varepsilon = 10^{-10}$ grow from 77 at $\rho = 0.3$ to 7,000 at $\rho = 0.95$. By Proposition 5 the stage masses decay geometrically at the per-stage survival rate α, so $S(\varepsilon) \approx \lceil \ln(1/\varepsilon)/\ln(1/\alpha) \rceil$; the implied $\widehat{\alpha} = \exp(-\ln(10^{10})/S)$ in the table quantifies its degradation with load. Since finite buffers make Proposition 5 hold for every $\lambda > 0$ with no stability assumption, α stays bounded away from 1 even beyond saturation: convergence slows near $\rho = 1$ but never fails.

Table 4. Stages to threshold $\varepsilon = 10^{-10}$ versus load (homogeneous N=10, K=10, μ=1).

$\rho = \lambda/\mu$	0.30	0.50	0.70	0.80	0.90	0.95
stages S	77	247	897	1,879	4,149	7,000
implied $\widehat{\alpha}$	0.741	0.911	0.975	0.988	0.994	0.997

Finally, we checked whether systematic quantization biases the *tail* of the latency CDF: on configuration B at $\lambda = 0.5$, the staged response-time CDF at its 0.90–0.999 quantile abscissae, compared against a DES with 1.9 million admitted samples, showed signed deviations of -1.5×10^{-4} to -1.7×10^{-5}, each at most 0.42 of the DES CI half-width—no detectable systematic bias. The small negative sign is below Monte Carlo resolution and covered by Corollary 9; near saturation the dominant tail error is stage truncation, controlled by ε via Theorem 7.

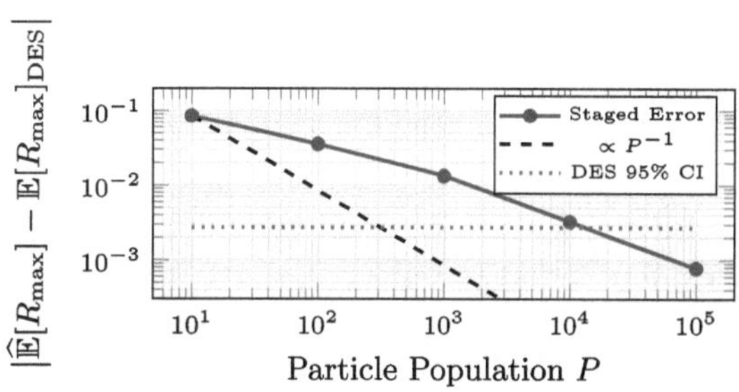

P	$\widehat{\mathbb{E}}[R_{max}]$	$\lvert\Delta\rvert$	rel. err.	solver s
10^1	4.291102	0.085891	2.042%	1.337
10^2	4.169705	0.035506	0.844%	9.179
10^3	4.191849	0.013362	0.318%	63.449
10^4	4.201935	0.003276	0.078%	396.734
10^5	4.204426	0.000785	0.019%	1610.510

Fig. 2. Particle-limit convergence on the heterogeneous N=5 example with $\lambda = 0.5$, $\mu = (1.0, 1.5, 0.8, 2.0, 1.2)$, and $K_i \equiv 20$.

9.4 Particle-Limit Convergence on a Heterogeneous Model

We validate Theorem 8 on the heterogeneous five-server example ($\lambda = 0.5$, $\mu = (1.0, 1.5, 0.8, 2.0, 1.2)$, $K_i \equiv 20$) at five particle ensembles $P \in \{10^1, \ldots, 10^5\}$; the labeled space has $21^5 = 4{,}084{,}101$ states, so an independent DES is the reference

(95% CI half-width 0.002737 on $\mathbb{E}[R_{\max}]$). Figure 2 shows the predicted pattern: the absolute error falls monotonically from 0.086 at $P = 10^1$ to 7.85×10^{-4} at $P = 10^5$, entering the DES CI, consistent with Theorems 7 and 8 and—combined with Tables 1 to 3—supporting controllable-error convergence to the true stationary distribution, as Corollary 9 guarantees for every bounded stationary functional.

10 Conclusion

Fork–join queues are analytically hard exactly where modern parallel systems care most: many synchronized branches, finite buffers, heterogeneity, and maxima-driven job metrics. This paper shows the finite-capacity $M/M/1/K_i$ model nevertheless admits an exact-plus-controlled-approximation decomposition: regenerative busy-period occupation measures determine π and yield exact Erlang-mixture latency CDFs, while the deterministic staged-particle solver converges in ℓ_1 to the true stationary distribution as P grows (Theorem 8), with the explicit error decomposition of Theorem 7 that simulation lacks. Benchmarks reach $N = 1000$ homogeneous branches (2.91×10^{23} reduced states) and heterogeneous systems with millions of labeled states at loads up to 0.95. Next steps: sharper constants, broader heterogeneous suites, and extensions to phase-type services, non-Poisson arrivals, and k-of-n or networked models.

Data Availability Statement. The artifact—solver, exact CTMC and DES baselines, and the scripts reproducing all tables and figures—is openly available at https://doi.org/10.5281/zenodo.19686245.

References

1. Baccelli, F., Makowski, A.M., Shwartz, A.: The fork-join queue and related systems with synchronization constraints: stochastic ordering and computable bounds. Adv. Appl. Probab. **21**(3), 629–660 (1989)
2. Chertock, A.: A practical guide to deterministic particle methods. Handb. Numer. Anal. **18**, 177–202 (2017). https://doi.org/10.1016/bs.hna.2016.11.004
3. Cottet, G.-H., Koumoutsakos, P.D.: Vortex Methods. Cambridge University Press, Cambridge (2000)
4. Dean, J., Ghemawat, S.: MapReduce: simplified data processing on large clusters. In: Proceedings of the 6th Symposium on Operating Systems Design and Implementation (OSDI), pp. 137–150 (2004)
5. Degond, P., Mas-Gallic, S.: The weighted particle method for convection-diffusion equations. Math. Comput. **53**(188), 485–507 (1989)
6. Fiorini, P.M.: Stationary analysis of finite-capacity M/M/1 fork-join queues (extended version with full proofs and appendices). Preprint, ResearchGate (2026). https://www.researchgate.net/publication/406063723_Stationary_Analysis_of_Finite-Capacity_MM1_Fork-Join_Queues
7. Flatto, L., Hahn, S.: Two parallel queues created by arrivals with two demands I. SIAM J. Appl. Math. **44**(5), 1041–1053 (1984)
8. Flatto, L.: Two parallel queues created by arrivals with two demands II. SIAM J. Appl. Math. **45**(5), 861–878 (1985)

9. Li, S., Mehrotra, S.: A New Method for Computing Stationary Distribution and Steady-State Performance Measures of a Continuous-State Markov Chain with a Queuing Application. arXiv preprint arXiv:2112.14315 (2021)
10. Nelson, R., Tantawi, A.N.: Approximate analysis of fork/join synchronization in parallel queues. IEEE Trans. Comput. **37**(6), 739–743 (1988)
11. Pagès, G., Pham, H.: Optimal quantization methods for nonlinear filtering with discrete-time observations. Bernoulli **11**(5), 893–932 (2005). https://doi.org/10.3150/bj/1130077599
12. Stewart, W.J.: Introduction to the Numerical Solution of Markov Chains. Princeton University Press, Princeton (1994)
13. Thomasian, A.: Analysis of fork/join and related queueing systems. ACM Comput. Surv. **47**(2), Article 17 (2014)
14. Varki, E.: Mean value technique for closed fork-join networks. ACM SIGMETRICS Perform. Eval. Rev. **27**(1), 103–112 (1999)
15. Alomari, F., Menascé, D.A.: Efficient response time approximations for multiclass fork and join queues in open and closed queuing networks. IEEE Trans. Parallel Distrib. Syst. **25**(6), 1437–1446 (2014)
16. Atar, R., Mandelbaum, A., Zviran, A.: Control of fork-join networks in heavy traffic. In: 50th Annual Allerton Conference on Communication, Control, and Computing, pp. 823–830 (2012)
17. Ko, S.-S., Serfozo, R.F.: Response times in M/M/s fork-join networks. Adv. Appl. Probab. **36**(3), 854–871 (2004)
18. Ko, S.-S., Serfozo, R.F.: Sojourn times in G/M/1 fork-join networks. Nav. Res. Logist. **55**(5), 432–443 (2008)
19. Konstantopoulos, P., Walrand, J.: Stationary and stability of fork-join networks. J. Appl. Probab. **26**(3), 604–614 (1989)
20. Pinotsi, D., Zazanis, M.A.: Synchronized queues with deterministic arrivals. Oper. Res. Lett. **33**(6), 560–566 (2005)
21. Qiu, Z., Pérez, J.F., Harrison, P.G.: Beyond the mean in fork-join queues: efficient approximation for response-time tails. Perform. Eval. **91**, 99–116 (2015)
22. Rizk, A., Poloczek, F., Ciucu, F.: Computable bounds in fork-join queueing systems. In: Proceedings of ACM SIGMETRICS 2015, pp. 335–346 (2015)
23. Tan, X., Knessl, C.: A fork-join queueing model: diffusion approximation, integral representations and asymptotics. Queueing Syst. **22**(3–4), 287–322 (1996)
24. Varki, E., Merchant, A., Chen, H.: The M/M/1 fork-join queue with variable subtasks. Unpublished manuscript available online (2002)
25. Wright, P.E.: Two parallel processors with coupled inputs. Adv. Appl. Probab. **24**(4), 986–1007 (1992)

Learning Alternating Real-Time Automata

Kazuki Kinoshita[1] and Masaki Waga[1,2](✉)

[1] Kyoto University, Kyoto, Japan
[2] National Institute of Informatics, Tokyo, Japan
mwaga@fos.kuis.kyoto-u.ac.jp

QEST + FORMATS Artifacts Evaluation Available

QEST + FORMATS Artifacts Evaluation Functional

Abstract. We present $\mathrm{AL}^*_{\mathrm{RTA}}$, the first active learning algorithm for alternating real-time automata (ARTAs) via membership and equivalence queries. The algorithm combines ideas from AL^* for alternating finite automata and $\mathrm{NL}^*_{\mathrm{RTA}}$ for nondeterministic real-time automata. We first define ARTAs and show that alternation improves succinctness, although it does not increase expressive power. We then present $\mathrm{AL}^*_{\mathrm{RTA}}$ and analyze its query complexity. Our empirical evaluation suggests that $\mathrm{AL}^*_{\mathrm{RTA}}$ generally learns smaller automata than $\mathrm{NL}^*_{\mathrm{RTA}}$ at the cost of more queries.

Keywords: active automata learning · alternating real-time automata

1 Introduction

Active automata learning, pioneered by Angluin's L^* algorithm [6], aims to identify an unknown language exactly through membership and equivalence queries. In a membership query, the learner asks whether a word belongs to the target language $\mathfrak{L}_{\mathrm{tgt}}$. In an equivalence query, the learner asks whether a hypothesis automaton $\mathcal{A}_{\mathrm{hyp}}$ recognizes the target language $\mathfrak{L}_{\mathrm{tgt}}$; if $\mathfrak{L}_{\mathrm{tgt}} \neq \mathcal{L}(\mathcal{A}_{\mathrm{hyp}})$, the teacher returns a concrete witness $w \in \mathfrak{L}_{\mathrm{tgt}} \triangle \mathcal{L}(\mathcal{A}_{\mathrm{hyp}})$ of deviation, where $\mathfrak{L}_{\mathrm{tgt}} \triangle \mathcal{L}(\mathcal{A}_{\mathrm{hyp}})$ is the symmetric difference of $\mathfrak{L}_{\mathrm{tgt}}$ and $\mathcal{L}(\mathcal{A}_{\mathrm{hyp}})$. For any regular language, L^* identifies the minimum deterministic finite automaton (DFA) recognizing it with a polynomial number of queries.

L^* has been extended in various directions. One direction is to learn more succinct automata by allowing nondeterministic or alternating branching. NL^* [10] learns a class of *nondeterministic* finite automata (NFAs) called *residual* NFAs. AL^* [7] and AL^{**} [9] learn *alternating* finite automata (AFAs). Due to the succinctness of NFAs and AFAs with respect to equivalent DFAs, these algorithms can identify a language in a more succinct form both in theory and in practice.

Another direction is to extend the class of target languages, e. g., to *timed* languages, where each letter is equipped with the delay since the previous letter. Many L^*-style learning algorithms have been proposed for various subclasses of *timed automata (TAs)* [2], such as *deterministic* TAs [23–25], *event-recording automata* [17], and *real-time automata (RTAs)* [4,5]. Despite recent advances

M. Chen and M. Tschaikowski (Eds.): QEST+FORMATS 2026, LNCS 16913, pp. 152–170, 2027.
https://doi.org/10.1007/978-3-032-35298-9_9

in active learning of timed languages, most methods are limited to *deterministic* branching, even when nondeterminism does not add expressive power. One notable exception is NL^*_{RTA} [5], which learns *nondeterministic* RTAs, but no algorithms have been proposed for any subclasses of TAs with *alternating* branching.

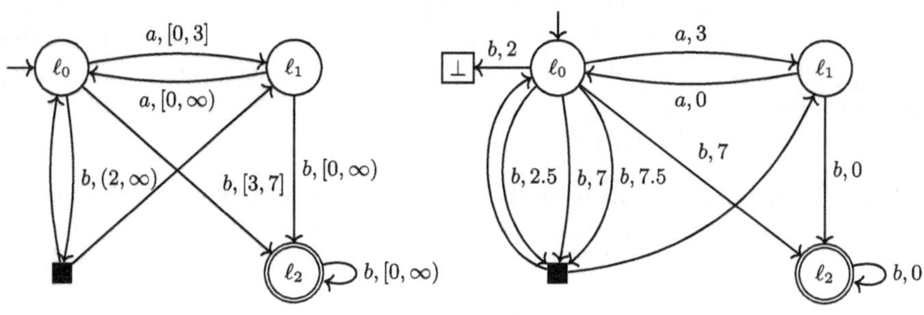

(a) An ARTA $\mathcal{A}$. (b) Corresponding evidence AFA $\mathcal{A}^e$.

Fig. 1. An ARTA and its corresponding evidence AFA. The initial location formula is ℓ_0. The black square represents a universal branching. Some transitions are simplified for illustration, e. g., $\delta(\ell_0, (b, 7)) = (\ell_0 \wedge \ell_1) \vee \ell_2$ in $\mathcal{A}^e$ is split into two edges; $(\ell_0, b, [3, 7], (\ell_0 \wedge \ell_1) \vee \ell_2) \in \Delta$ in $\mathcal{A}$ is similarly split into two edges and merged with others.

Contributions. We propose an active learning algorithm AL^*_{RTA} that learns *alternating RTAs (ARTAs).* Figure 1a depicts an ARTA over $\Sigma = \{a, b\}$. Each transition of an ARTA is labeled with a letter and an interval representing the elapsed time, like RTAs [14]. The target of each transition of an ARTA is a positive Boolean expression over the locations, like alternating finite automata (AFAs) [12]. For example, in the ARTA in Fig. 1a, if we read a letter b with a delay $d \in (2, 3)$ at ℓ_0, we may simultaneously jump to both ℓ_0 and ℓ_1 and this execution is accepted when both spawned branches are accepted at the same time.

We first show that alternating branching improves the succinctness of RTAs, although it does not increase the expressive power. Namely, ARTAs are equi-expressive with RTAs but are *doubly* exponentially more succinct than deterministic RTAs, whereas nondeterministic RTAs are only *singly* exponentially more succinct than deterministic RTAs [5]. This result generalizes the classical succinctness results for AFAs [20,21] to real-time languages, the class of timed languages recognizable by RTAs.

We then present AL^*_{RTA}, which is based on AL^* and NL^*_{RTA}. At a high level, AL^*_{RTA} follows NL^*_{RTA}: it initially constructs an evidence AFA (Fig. 1b) following the idea of AL^* and then generalizes concrete delays on transitions to obtain an ARTA (Fig. 1a). Because delays are drawn from an infinite domain, evidence AFAs necessarily have incomplete transitions. We adapt the conditions in AL^* to accommodate incomplete transitions. We show that AL^*_{RTA} terminates for any real-time language with an exponential query bound in the size of the minimal deterministic RTA with a polynomial dependence on the alphabet size, counterexample length, and timing parameters.

We implemented a prototype library LEARNARTA for AL^*_{RTA} and empirically compared AL^*_{RTA} against NL^*_{RTA}. The results suggest that AL^*_{RTA} generally learns smaller automata than NL^*_{RTA}, suggesting that the aforementioned succinctness result is also helpful in practice. However, this gain comes at a higher learning cost, particularly in terms of the number of queries and the learning time, as reported in [7].

Overall, our contributions are summarized as follows.

- We formulate ARTAs and show that alternation improves the succinctness of real-time languages, although it does not increase expressive power.
- We propose the AL^*_{RTA} algorithm for the active learning of ARTAs and analyze its query complexity.
- We empirically show that AL^*_{RTA} usually learns a smaller automaton than NL^*_{RTA}, at the cost of more queries.

Organization. The remainder of this paper is organized as follows. After reviewing real-time automata and AFAs in Sect. 2, we define ARTAs and discuss their expressive features in Sect. 3. Then, we present the learning algorithm and its termination in Sect. 4. Finally, we empirically evaluate AL^*_{RTA} in Sect. 5, discuss related work in Sect. 6, and conclude in Sect. 7.

2 Preliminaries

Let Σ be a finite alphabet, let $\mathbb{R}_{\geq 0}$ be the set of non-negative reals, and let $\mathbb{N}$ be the set of natural numbers. For sets X, Y, we denote their symmetric difference (i.e., $(X \cup Y) \setminus (X \cap Y)$) as $X \triangle Y$. For a set X, a word is a finite sequence of X. A language is a set of words. We let X^* be the set of words over X and ϵ be the empty word. For $w, w' \in X^*$, we denote their concatenation by $w \cdot w'$. For a word w, we let $\mathit{suff}(w)$ be the set of suffixes of w. We let $\mathcal{I}$ be the set of intervals over $\mathbb{R}_{\geq 0}$ whose endpoints are in $\mathbb{N} \cup \{\infty\}$.

2.1 Timed Words and Real-Time Automata

A *timed word* $w = (a_1, d_1), (a_2, d_2), \ldots, (a_n, d_n) \in (\Sigma \times \mathbb{R}_{\geq 0})^*$ over Σ is a word over $\Sigma \times \mathbb{R}_{\geq 0}$, where each d_i represents the delay between a_{i-1} and a_i[1]. We let $\mathcal{T}(\Sigma)$ be the set of timed words over Σ. A *timed language* is a set of timed words. For a timed language $\mathfrak{L} \subseteq \mathcal{T}(\Sigma)$, we let $\Phi(\mathfrak{L}) \subseteq \Sigma \times \mathbb{R}_{\geq 0}$ be the set of letters with delays in $\mathfrak{L}$, i.e., $\Phi(\mathfrak{L}) = \{(a, d) \mid \exists w, w' \in \mathcal{T}(\Sigma). w \cdot (a, d) \cdot w' \in \mathfrak{L}\}$.

[1] We use a sequence of letters with (relative) delays rather than (absolute) timestamps for the sake of presentation; this choice is not essential.

Definition 1 (real-time automata). *A* real-time automaton (RTA) *[14] is a 5-tuple $\mathcal{A} = (L, \Sigma, \ell_0, F, \Delta)$, where L is a finite set of locations, Σ is the alphabet, $\ell_0 \in L$ is the initial location, $F \subseteq L$ is the set of accepting locations, and $\Delta \subseteq L \times \Sigma \times \mathcal{I} \times L$ is a finite transition relation.*

A *run* of an RTA $\mathcal{A}$ over a timed word $w = (a_1, d_1), (a_2, d_2), \ldots, (a_n, d_n)$ is a sequence of locations $\ell_0, \ell_1, \ldots, \ell_n$ such that for each $i \in \{1, 2, \ldots, n\}$ there is an interval $I_i \in \mathcal{I}$ satisfying $(\ell_{i-1}, a_i, I_i, \ell_i) \in \Delta$ and $d_i \in I_i$. A run is *accepting* if its last location is in F. A timed word w is accepted by an RTA $\mathcal{A}$ if there is an accepting run of $\mathcal{A}$ over w. The timed language recognized by $\mathcal{A}$ is the set of timed words accepted by $\mathcal{A}$. We let $\mathcal{L}(\mathcal{A})$ be the timed language recognized by $\mathcal{A}$. A *real-time language* is a timed language recognized by an RTA.

An RTA is *deterministic* if, for any location $\ell \in L$, letter $a \in \Sigma$, and $(\ell, a, I_1, \ell_1), (\ell, a, I_2, \ell_2) \in \Delta$, $(I_1, \ell_1) \neq (I_2, \ell_2)$ implies $I_1 \cap I_2 = \emptyset$.

Definition 2 (regions of delays [2,4]). *Let $\mathcal{A}$ be an RTA and let $K \in \mathbb{N}$ be the largest integer appearing in the intervals on transitions of $\mathcal{A}$. For each $t \in \mathbb{R}_{\geq 0}$, we define the* region *of t, denoted by $[\![t]\!]$, as $[\![t]\!] = [t, t]$ if we have $t \in \mathbb{N}$ and $t \leq K$, $[\![t]\!] = (\lfloor t \rfloor, \lfloor t \rfloor + 1)$ if we have $t \notin \mathbb{N}$ and $t \leq K$, and $[\![t]\!] = (K, \infty)$ if we have $t > K$.*

The following Myhill-Nerode-style theorem holds for real-time languages.

Theorem 3 ([4]). *For a timed language $\mathfrak{L} \subseteq \mathcal{T}(\Sigma)$, we define an equivalence relation $\equiv_{\mathfrak{L}} \subseteq \mathcal{T}(\Sigma) \times \mathcal{T}(\Sigma)$ by $w_1 \equiv_{\mathfrak{L}} w_2$ if and only if $w_1 \cdot w \in \mathfrak{L} \iff w_2 \cdot w \in \mathfrak{L}$ holds for any $w \in \mathcal{T}(\Sigma)$. $\mathfrak{L}$ is a real-time language if and only if the number of equivalence classes of $\equiv_{\mathfrak{L}}$ is finite and $\equiv_{\mathfrak{L}}$ satisfies the following for some $K \in \mathbb{N}$:*

- *For any $w \in \mathcal{T}(\Sigma)$, $a \in \Sigma$, and $t_1, t_2 \in \mathbb{R}_{\geq 0}$, $[\![t_1]\!] = [\![t_2]\!]$ implies $w \cdot (a, t_1) \equiv_{\mathfrak{L}} w \cdot (a, t_2)$.*
- *For any $w \in \mathcal{T}(\Sigma)$, $a \in \Sigma$, and $t_1, t_2 \in \mathbb{R}_{\geq 0}$, $t_1 > K \wedge t_2 > K$ implies $w \cdot (a, t_1) \equiv_{\mathfrak{L}} w \cdot (a, t_2)$.* □

Moreover, for any real-time language $\mathfrak{L}$, there is a *unique* minimal DRTA $\mathcal{A}$ such that $\mathcal{L}(\mathcal{A}) = \mathfrak{L}$, and the number of locations of $\mathcal{A}$ equals the number of equivalence classes of $\equiv_{\mathfrak{L}}$.

2.2 Alternating Finite Automata

An *alternating finite automaton (AFA)* [12] generalizes DFAs by allowing both existential and universal branching. For a finite set X, we let $\mathbb{B}_{+}(X)$ be the set of positive Boolean expressions over X, i. e., the set defined as follows:

$$\varphi ::= \top \mid \bot \mid x \in X \mid \varphi \wedge \varphi \mid \varphi \vee \varphi.$$

For a set L of locations, we call $\varphi \in \mathbb{B}_{+}(L)$ a *location formula*.

Definition 4 (alternating finite automata). *An* alternating finite automaton (AFA) *is a 5-tuple* $\mathcal{A} = (L, \Sigma, L_0, F, \delta)$, *where* L *is a finite set of locations,* Σ *is a finite alphabet,* $L_0 \in \mathbb{B}_+(L)$ *is the initial location formula,* $F \subseteq L$ *is the set of accepting locations, and* $\delta \colon L \times \Sigma \to \mathbb{B}_+(L)$ *is the transition function.*

We generalize δ to $\delta^* \colon \mathbb{B}_+(L) \times \Sigma^* \to \mathbb{B}_+(L)$ as follows:

$$\delta^*(\varphi, \epsilon) = \varphi \qquad \delta^*(\top, w) = \top \qquad \delta^*(\bot, w) = \bot \qquad \delta^*(\ell, a \cdot w) = \delta^*(\delta(\ell, a), w)$$
$$\delta^*(\varphi \vee \varphi', w) = \delta^*(\varphi, w) \vee \delta^*(\varphi', w) \qquad \delta^*(\varphi \wedge \varphi', w) = \delta^*(\varphi, w) \wedge \delta^*(\varphi', w).$$

For a location formula φ of an AFA $\mathcal{A}$, the evaluation function $E_{\mathcal{A}} \colon \mathbb{B}_+(L) \to \{\top, \bot\}$ is defined as follows:

$$E_{\mathcal{A}}(\top) = \top \qquad E_{\mathcal{A}}(\bot) = \bot \qquad E_{\mathcal{A}}(\ell) = \top \text{ if } \ell \in F \qquad E_{\mathcal{A}}(\ell) = \bot \text{ if } \ell \notin F$$
$$E_{\mathcal{A}}(\varphi \wedge \varphi') = E_{\mathcal{A}}(\varphi) \wedge E_{\mathcal{A}}(\varphi') \qquad E_{\mathcal{A}}(\varphi \vee \varphi') = E_{\mathcal{A}}(\varphi) \vee E_{\mathcal{A}}(\varphi')$$

A word $w \in \Sigma^*$ is *accepted* by an AFA $\mathcal{A}$ if we have $E_{\mathcal{A}}(\delta^*(L_0, w)) = \top$. The language $\mathcal{L}(\mathcal{A}) \subseteq \Sigma^*$ recognized by $\mathcal{A}$ is the set of words accepted by $\mathcal{A}$.

2.3 The AL* algorithm for active learning of AFAs

The AL* *algorithm* [7] learns an AFA $\mathcal{A}_{\mathrm{hyp}}$ recognizing the target language $\mathfrak{L}_{\mathrm{tgt}} \subseteq \Sigma^*$ using finitely many *membership* and *equivalence* queries to the teacher. In a membership query, the learner asks if $w \in \Sigma^*$ belongs to the target language $\mathfrak{L}_{\mathrm{tgt}}$, i.e., $w \in \mathfrak{L}_{\mathrm{tgt}}$. In an equivalence query, the learner asks if the hypothesis automaton $\mathcal{A}_{\mathrm{hyp}}$ recognizes the target language $\mathfrak{L}_{\mathrm{tgt}}$, i.e., $\mathcal{L}(\mathcal{A}_{\mathrm{hyp}}) = \mathfrak{L}_{\mathrm{tgt}}$. When we have $\mathcal{L}(\mathcal{A}_{\mathrm{hyp}}) \neq \mathfrak{L}_{\mathrm{tgt}}$, the teacher returns a counterexample $cex \in \mathcal{L}(\mathcal{A}_{\mathrm{hyp}}) \triangle \mathfrak{L}_{\mathrm{tgt}}$.

	ϵ	a	b	ab	ba
ϵ	$\bot$	$\top$	$\bot$	$\top$	$\top$
a	$\top$	$\top$	$\top$	$\bot$	$\top$
b	$\bot$	$\top$	$\bot$	$\bot$	$\bot$
aa	$\top$	$\top$	$\bot$	$\bot$	$\bot$
ab	$\top$	$\top$	$\bot$	$\bot$	$\bot$
ba	$\top$	$\top$	$\bot$	$\bot$	$\bot$
bb	$\bot$	$\top$	$\bot$	$\bot$	$\bot$
baa	$\top$	$\top$	$\bot$	$\bot$	$\bot$
bab	$\bot$	$\bot$	$\bot$	$\bot$	$\bot$

(a) An observation table. Shaded rows show a minimum-cardinality monotone basis.

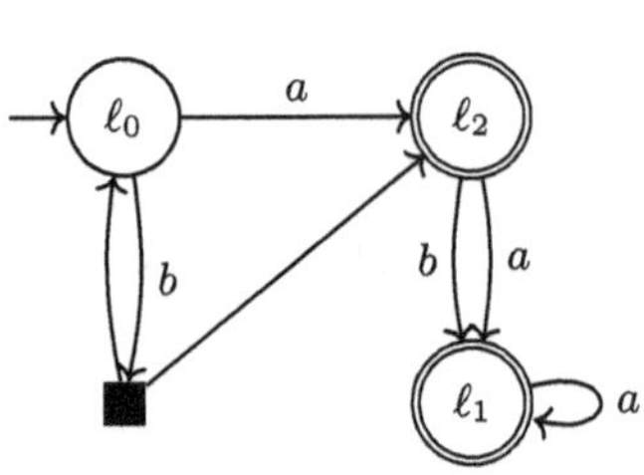

(b) An AFA $\mathcal{A}_{\mathrm{hyp}}$.

Fig. 2. An observation table for AL* and the corresponding AFA. The initial location formula of $\mathcal{A}_{\mathrm{hyp}}$ is ℓ_0. The black square represents a universal branching.

AL* uses a 2-dimensional array T called an *observation table* to maintain the information obtained during learning. See Fig. 2a for an illustration. For finite index sets $S, E \subseteq \Sigma^*$, for each pair $(s, e) \in S \times E$, the observation table stores

Algorithm 1: A variant of the AL* algorithm [7] for AFA learning.

Input : Alphabet Σ and access to membership and equivalence queries for the target language $\mathfrak{L}_{\mathrm{tgt}} \subseteq \Sigma^*$
Output : An AFA $\mathcal{A}_{\mathrm{hyp}}$ satisfying $\mathcal{L}(\mathcal{A}_{\mathrm{hyp}}) = \mathfrak{L}_{\mathrm{tgt}}$

```
S ← {ε}; E ← {ε}; P ← ∅
while true do
    S ← S ∪ {p · a | p ∈ P, a ∈ Σ}
    while the observation table is not P-closed do
        add s ∈ S such that T[s] ∉ 𝔹+(Rows(P)) to P
        S ← S ∪ {p · a | p ∈ P, a ∈ Σ}
    P ← ComputeBasis(S, E, P, T)          // Compute a minimal monotone basis
    A_hyp ← ConstructAFA(S, E, P, T)
    switch eq_𝔏tgt(A_hyp) do
        case ⊤ do
            return A_hyp
        case cex do
            E ← E ∪ suff(cex)
```

whether $s \cdot e \in \mathfrak{L}_{\mathrm{tgt}}$. Since S and E are finite, one can fill the observation table using finite membership queries.

We write $T(s, e) \in \{\top, \bot\}$ for the table entry at row $s \in S$ and column $e \in E$. For each $s \in S$, we define its *row vector* $T[s] \in \{\top, \bot\}^E$ by $T[s](e) = T(s, e)$ for each $e \in E$. For $U \subseteq S$, we write $\mathrm{Rows}(U) = \{T[u] \mid u \in U\}$. For two row vectors $r_1, r_2 \in \{\top, \bot\}^E$, we define their conjunction and disjunction $r_1 \wedge r_2, r_1 \vee r_2 \in \{\top, \bot\}^E$ in a pointwise manner. Namely, $r_1 \wedge r_2$ and $r_1 \vee r_2$ satisfy the following for each $e \in E$: $(r_1 \wedge r_2)(e) = r_1(e) \wedge r_2(e)$ and $(r_1 \vee r_2)(e) = r_1(e) \vee r_2(e)$.

In AL*, one central notion is the *monotone basis* of a set of row vectors. For a set $U \subseteq \{\top, \bot\}^E$ of row vectors, we overload $\mathbb{B}_+(U)$ to denote the set of row vectors obtained by evaluating positive Boolean expressions over U, where $\top$ and $\bot$ denote constant row vectors.

Definition 5 (monotone basis, P-closedness, minimality of P). *Let T be an observation table with finite index sets $S, E \subseteq \Sigma^*$. For $R \subseteq \{\top, \bot\}^E$, $U \subseteq R$ is a* monotone basis *of R if we have $R \subseteq \mathbb{B}_+(U)$. Moreover, U is a* minimal monotone basis *of R if $R \subseteq \mathbb{B}_+(U)$ holds and no strict subset $U' \subsetneq U$ satisfies $R \subseteq \mathbb{B}_+(U')$. $P \subseteq S$ is a* monotone basis *of T if* $\mathrm{Rows}(P)$ *is a monotone basis of* $\mathrm{Rows}(S)$ *and for any $p, p' \in P$, $p \neq p'$ implies $T[p] \neq T[p']$. For $P \subseteq S$, T is* P-closed *if P is a monotone basis of T*[2]. *A monotone basis P is* minimal *if any $P' \subsetneq P$ is not a monotone basis of T.*

Algorithm 1 outlines a variant of the AL* algorithm, partially inspired by the AL** algorithm [9]. We start from $S = E = \{\epsilon\}$ and gradually increase them. In each iteration, the learner first increases the candidate basis P and S to ensure the observation table is P-closed and $P \cdot \Sigma \subseteq S$ holds. The learner then reduces P to be minimal. Once these conditions are satisfied, the learner constructs a hypothesis AFA $\mathcal{A}_{\mathrm{hyp}}$ and asks an equivalence query. The constructed AFA is such that *i)* one location is constructed for each $p \in P$, *ii)* the initial location formula corresponds to the formula representing $T[\epsilon]$, *iii)* the accepting locations

[2] We relax the definition of P-closedness for consistency with $\mathrm{AL}^*_{\mathrm{RTA}}$ in Sect. 4.

correspond to $p \in P$ such that $T(p, \epsilon) = \top$, and *iv)* the successor from a location corresponding to $p \in P$ with $a \in \Sigma$ is the formula representing $T[p \cdot a]$. Figure 2b shows a concrete example. If $\mathcal{L}(\mathcal{A}_{\mathrm{hyp}}) = \mathfrak{L}_{\mathrm{tgt}}$ holds, the algorithm terminates and returns $\mathcal{A}_{\mathrm{hyp}}$. Otherwise, the teacher returns $cex \in \mathcal{L}(\mathcal{A}_{\mathrm{hyp}}) \triangle \mathfrak{L}_{\mathrm{tgt}}$, and the learner adds all suffixes of cex to E to refine the observation table.

3 Alternating Real-Time Automata

We introduce *alternating real-time automata (ARTAs)* by combining notions of RTAs and AFAs. Informally, an ARTA extends an RTA by allowing transition targets to be location formulas in $\mathbb{B}_+(L)$, which enables existential and universal branching over timed transitions.

Definition 6 (alternating real-time automata). *An* alternating real-time automaton (ARTA) *is a 5-tuple $\mathcal{A} = (L, \Sigma, L_0, F, \Delta)$, where L is a finite set of locations, Σ is a finite alphabet, $L_0 \in \mathbb{B}_+(L)$ is the initial location formula, $F \subseteq L$ is the set of accepting locations, and $\Delta \subseteq L \times \Sigma \times \mathcal{I} \times \mathbb{B}_+(L)$ is a finite transition relation satisfying the following condition on determinism: for any $(\ell, a, I_1, \varphi_1), (\ell, a, I_2, \varphi_2) \in \Delta$, $(\ell, a, I_1, \varphi_1) \neq (\ell, a, I_2, \varphi_2)$ implies $I_1 \cap I_2 = \emptyset$.*

Thanks to the condition above on the determinism of the transition relation Δ, for an ARTA $\mathcal{A} = (L, \Sigma, L_0, F, \Delta)$, one can define a transition function $\delta \colon L \times \Sigma \times \mathbb{R}_{\geq 0} \to \mathbb{B}_+(L)$ as follows:

$$\delta(\ell, (a, d)) = \phi \qquad \text{if there exists } (\ell, a, I, \phi) \in \Delta \text{ with } d \in I,$$
$$\delta(\ell, (a, d)) = \bot \qquad \text{if no such edge exists.}$$

We generalize δ to $\delta^* \colon \mathbb{B}_+(L) \times (\Sigma \times \mathbb{R}_{\geq 0})^* \to \mathbb{B}_+(L)$ as follows:

$$\delta^*(\varphi, \epsilon) = \varphi \quad \delta^*(\top, w) = \top \quad \delta^*(\bot, w) = \bot \quad \delta^*(\ell, (a, d) \cdot w) = \delta^*(\delta(\ell, (a, d)), w)$$
$$\delta^*(\varphi \vee \varphi', w) = \delta^*(\varphi, w) \vee \delta^*(\varphi', w) \qquad \delta^*(\varphi \wedge \varphi', w) = \delta^*(\varphi, w) \wedge \delta^*(\varphi', w).$$

For a location formula φ of an ARTA $\mathcal{A} = (L, \Sigma, L_0, F, \Delta)$, the evaluation function $E_{\mathcal{A}} \colon \mathbb{B}_+(L) \to \{\top, \bot\}$ is defined as follows:

$$E_{\mathcal{A}}(\top) = \top \qquad E_{\mathcal{A}}(\bot) = \bot \qquad E_{\mathcal{A}}(\ell) = \top \text{ if } \ell \in F \qquad E_{\mathcal{A}}(\ell) = \bot \text{ if } \ell \notin F$$
$$E_{\mathcal{A}}(\varphi \wedge \varphi') = E_{\mathcal{A}}(\varphi) \wedge E_{\mathcal{A}}(\varphi') \qquad E_{\mathcal{A}}(\varphi \vee \varphi') = E_{\mathcal{A}}(\varphi) \vee E_{\mathcal{A}}(\varphi')$$

An ARTA $\mathcal{A} = (L, \Sigma, L_0, F, \Delta)$ *accepts* a timed word $w \in \mathcal{T}(\Sigma)$ if we have $E_{\mathcal{A}}(\delta^*(L_0, w)) = \top$. The timed language $\mathcal{L}(\mathcal{A})$ recognized by an ARTA $\mathcal{A}$ is the set of timed words accepted by $\mathcal{A}$.

Example 7. Figure 1a depicts an ARTA $\mathcal{A}$ over $\Sigma = \{a, b\}$. $\mathcal{A}$ accepts $(a, 1), (b, 4)$ because we have $E_{\mathcal{A}}(\delta^*(\ell_0, (a, 1), (b, 4))) = E_{\mathcal{A}}(\delta^*(\ell_1, (b, 4))) = E_{\mathcal{A}}(\ell_2) = \top$. $\mathcal{A}$ also accepts $(b, 2.7), (b, 3.5)$ because we have $\delta(\ell_0, (b, 2.7)) = \ell_0 \wedge \ell_1$, $\delta(\ell_0, (b, 3.5)) = (\ell_0 \wedge \ell_1) \vee \ell_2$, and $\delta(\ell_1, (b, 3.5)) = \ell_2$, and thus, $E_{\mathcal{A}}(\delta^*(\ell_0, (b, 2.7), (b, 3.5))) = E_{\mathcal{A}}(\delta^*(\ell_0 \wedge \ell_1, (b, 3.5))) = E_{\mathcal{A}}(((\ell_0 \wedge \ell_1) \vee \ell_2) \wedge \ell_2) = \top$ holds. In contrast, $\mathcal{A}$ does not accept $(b, 2.7), (a, 1)$ because we have $E_{\mathcal{A}}(\delta^*(\ell_0, (b, 2.7), (a, 1))) = E_{\mathcal{A}}(\delta^*(\ell_0 \wedge \ell_1, (a, 1))) = E_{\mathcal{A}}(\ell_1 \wedge \ell_0) = \bot$.

Algorithm 2: $\mathrm{AL}^*_{\mathrm{RTA}}$ for active learning of ARTAs

Input : Alphabet Σ and access to membership and equivalence queries for the target real-time language $\mathfrak{L}_{\mathrm{tgt}} \subseteq \mathcal{T}(\Sigma)$
Output : An ARTA $\mathcal{A}_{\mathrm{hyp}}$ satisfying $\mathcal{L}(\mathcal{A}_{\mathrm{hyp}}) = \mathfrak{L}_{\mathrm{tgt}}$

```
S ← {ε}; E ← {ε} ; P ← ∅
while true do
    if the observation table is not P-closed then
        add s ∈ S such that T[s] ∉ B+(Rows(P)) to P
    else if ∃p ∈ P, σ ∈ Φ(S ∪ E) such that p · σ ∉ S then
        add p · σ to S                          // Ensure evidence-closedness
    else
        P' ← ComputeBasis(S, E, P, T)           // Minimize the monotone basis
        if |P'| < |P| then P ← P'
        else
            A_hyp ← ConstructARTA(S, E, P, T)
            switch eq_{L_tgt}(A_hyp) do
                case ⊤ do
                    return A_hyp
                case cex do
                    cex_α ← G_α(cex)            // Normalize the counterexample
                    E ← E ∪ suff(cex_α)
```

As in the classical setting, where AFAs accept exactly the regular languages [12], alternation does not increase the expressive power of RTAs, i. e., ARTAs recognize real-time languages.

Theorem 8 (expressive power of ARTAs). *DRTAs and ARTAs have the same expressive power. Moreover, for any ARTA with n locations, there is a DRTA recognizing the same timed language with at most 2^{2^n} locations.* □

This double-exponential blow-up in determinization is unavoidable in general.

Theorem 9 (succinctness). *There exist a constant $c > 0$ and a family of ARTAs $(\mathcal{A}_k)_{k \geq 1}$ with at most k locations such that for any sufficiently large k, any DRTA $\mathcal{A}_{\mathcal{D}}$ satisfying $\mathcal{L}(\mathcal{A}_{\mathcal{D}}) = \mathcal{L}(\mathcal{A}_k)$ has at least $2^{2^{ck}}$ locations.* □

4 Active Learning Algorithm for ARTAs

We present our algorithm $\mathrm{AL}^*_{\mathrm{RTA}}$ for the active learning of ARTAs, extending AL^* with ideas used in learning real-time automata [4] to handle infinite domain inputs, i. e., $\Sigma \times \mathbb{R}_{\geq 0}$ rather than Σ. As in AL^*, the learner interacts with a teacher through membership and equivalence queries. In $\mathrm{AL}^*_{\mathrm{RTA}}$, membership queries are asked on timed words and equivalence queries are asked on ARTAs.

Algorithm 2 outlines $\mathrm{AL}^*_{\mathrm{RTA}}$. Compared to AL^*, the key new ingredients are to lift the observation table from words to timed words, and to impose evidence-closedness on observation tables to handle real-valued timestamps.

4.1 Observation Tables for $\mathbf{AL}^*_{\mathbf{RTA}}$

As in AL^*, we use an observation table to store the answers to membership queries for concatenations of a row element and a column element.

Definition 10 (observation tables for $\mathrm{AL}^*_{\mathrm{RTA}}$). *Let $\mathfrak{L}_{\mathrm{tgt}} \subseteq \mathcal{T}(\Sigma)$ be the target real-time language. An observation table for $\mathrm{AL}^*_{\mathrm{RTA}}$ is a 2-dimensional array $T\colon S \times E \to \{\top, \bot\}$ with finite index sets $S, E \subseteq \mathcal{T}(\Sigma)$ such that i) $T(s,e) = \top$ if and only if $s \cdot e \in \mathfrak{L}_{\mathrm{tgt}}$, ii) S is prefix-closed, and iii) E is suffix-closed.*

For each $s \in S$, we write $T[s] \in \{\top, \bot\}^E$ for its row vector, defined by $T[s](e) = T(s,e)$. For $U \subseteq S$, we write $\mathrm{Rows}(U) = \{T[u] \mid u \in U\}$. Row vectors are combined by pointwise conjunction and disjunction as in Sect. 2.3.

Definition 11 (cohesiveness). *Let T be an observation table for $\mathrm{AL}^*_{\mathrm{RTA}}$ with index sets $S, E \subseteq \mathcal{T}(\Sigma)$. For $P \subseteq S$, T is* P-closed *if we have* $\mathrm{Rows}(S) \subseteq \mathbb{B}_+(\mathrm{Rows}(P))$. *For $P \subseteq S$, T is* evidence-closed *if $p \cdot \sigma \in S$ holds for any $p \in P$ and $\sigma \in \Phi(S \cup E)$. T is* floor-distinct *if $(a, d_1), (a, d_2) \in \Phi(S \cup E)$ and $d_1 \neq d_2$ and both $d_1, d_2 \notin \mathbb{N}$ implies $\lfloor d_1 \rfloor \neq \lfloor d_2 \rfloor$. For $P \subseteq S$, T is* cohesive *if it is P-closed, evidence-closed, and floor-distinct.*

Intuitively, P-closedness enables AL^*-style representability via a monotone basis and evidence-closedness ensures that we have a successor row for any $p \in P$ and letter with delay in $S \cup E$. We require floor-distinctness to ensure that the inferred guard intervals are well-defined in $\mathrm{AL}^*_{\mathrm{RTA}}$.

4.2 ARTA Construction from a Cohesive Observation Table

Once we obtain a cohesive observation table T, one can construct a hypothesis ARTA $\mathcal{A}_{\mathrm{hyp}}$. The construction is two-fold: first, we construct an evidence AFA, where each transition is labeled with a letter with a delay, and then we generalize concrete delays on transitions to obtain an ARTA. The construction of an evidence AFA is essentially the same as the AFA construction in AL^*.

Let T be an observation table with index sets $S, E \subseteq \mathcal{T}(\Sigma)$ and let $P \subseteq S$. Let $R_P = \mathrm{Rows}(P) \subseteq \{\top, \bot\}^E$. For each $e \in E$, we define a basis formula $M_{R_P}(e) = \bigwedge_{r \in R_P, r(e) = \top} r$, and for each row vector $r \in \{\top, \bot\}^E$ we define its decomposition $b_{R_P}(r) = \bigvee_{e \in E, r(e) = \top} M_{R_P}(e)$, where empty conjunctions and disjunctions are interpreted as $\top$ and $\bot$, respectively.

Definition 12 (evidence AFA). *Assume the observation table T with index sets $S, E \subseteq \mathcal{T}(\Sigma)$ and monotone basis $P \subseteq S$ is cohesive. The evidence AFA is the AFA $\mathcal{A}^{\mathrm{e}}_{\mathrm{hyp}} = (R_P, \Phi(S \cup E), b_{R_P}(T[\epsilon]), F_{\mathcal{A}^{\mathrm{e}}_{\mathrm{hyp}}}, \delta_{\mathcal{A}^{\mathrm{e}}_{\mathrm{hyp}}})$, where the accepting locations are $F_{\mathcal{A}^{\mathrm{e}}_{\mathrm{hyp}}} = \{r \in R_P \mid r(\epsilon) = \top\}$ and the transition function $\delta_{\mathcal{A}^{\mathrm{e}}_{\mathrm{hyp}}} \colon R_P \times \Phi(S \cup E) \to \mathbb{B}_+(R_P)$ is defined by $\delta_{\mathcal{A}^{\mathrm{e}}_{\mathrm{hyp}}}(r, \sigma) = b_{R_P}(T[p \cdot \sigma])$ for $p \in P$ with $T[p] = r$. Such $p \in P$ is unique by the definition of monotone bases.*

Algorithm 3: ARTA construction from an evidence AFA

Input : Evidence AFA $\mathcal{A}^e_{\text{hyp}} = (R_P, \Phi(S \cup E), L_0, F, \delta_{\mathcal{A}^e_{\text{hyp}}})$
Output : Hypothesis ARTA

$\Delta_{\mathcal{A}_{\text{hyp}}} \leftarrow \emptyset$
for $r \in R_P$ **and** $a \in \Sigma$ **do**
  let $l = d_1, \ldots, d_n$ **be** the increasing sequence of all delays d such that $(a, d) \in \Phi(S \cup E)$
  for $i \leftarrow 2$ **to** n **do**
    if $\delta_{\mathcal{A}^e_{\text{hyp}}}(r, (a, d_{i-1})) = \delta_{\mathcal{A}^e_{\text{hyp}}}(r, (a, d_i))$ **then**
      remove d_{i-1} **from** l
  $d'_1, \ldots, d'_{n'} \leftarrow l$; $I_1, \ldots, I_{n'} \leftarrow \Pi(l)$
  for $i \in \{1, 2, \ldots, n'\}$ **do**
    $\Delta_{\mathcal{A}_{\text{hyp}}} \leftarrow \Delta_{\mathcal{A}_{\text{hyp}}} \cup \{(r, a, I_i, \delta_{\mathcal{A}^e_{\text{hyp}}}(r, (a, d'_i)))\}$
return $(R_P, \Sigma, L_0, F, \Delta_{\mathcal{A}_{\text{hyp}}})$

The evidence AFA $\mathcal{A}^e_{\text{hyp}}$ is defined over the *finite* set $\Phi(S \cup E) \subseteq \Sigma \times \mathbb{R}_{\geq 0}$ of letters with delays appearing in T. The transition function $\delta_{\mathcal{A}^e_{\text{hyp}}}$ is well-defined by evidence-closedness. Following [4], we obtain an ARTA over Σ by inferring guard intervals from these finitely many delays using the *partition function.*

Definition 13 (partition function [4]). *Let $l = d_1, \ldots, d_n$ be a sequence of delays satisfying i) $i < j$ implies $d_i < d_j$ and ii) $\lfloor d_i \rfloor \neq \lfloor d_j \rfloor$ for all $i \neq j$ with $d_i, d_j \notin \mathbb{N}$. We let $d_0 = -0.5$. We define the partition function $\Pi(l) = I_1, \ldots, I_n$, where intervals $I_1, \ldots, I_n \in \mathcal{I}$ are given as follows. Note that we have $\Pi(\epsilon) = \epsilon$.*

$$I_i = \begin{cases} (d_{i-1}, d_i] & \text{if } i < n \text{ and } d_{i-1}, d_i \in \mathbb{N} \\ [\lceil d_{i-1} \rceil, d_i] & \text{if } i < n,\ d_{i-1} \notin \mathbb{N}, \text{ and } d_i \in \mathbb{N} \\ (d_{i-1}, \lceil d_i \rceil) & \text{if } i < n,\ d_{i-1} \in \mathbb{N}, \text{ and } d_i \notin \mathbb{N} \\ [\lceil d_{i-1} \rceil, \lceil d_i \rceil) & \text{if } i < n \text{ and } d_{i-1}, d_i \notin \mathbb{N} \\ (d_{i-1}, \infty) & \text{if } i = n \text{ and } d_{i-1} \in \mathbb{N} \\ [\lceil d_{i-1} \rceil, \infty) & \text{if } i = n \text{ and } d_{i-1} \notin \mathbb{N} \end{cases}$$

Algorithm 3 shows the construction of an ARTA from an evidence AFA. The ARTA returned by Algorithm 3 shares locations, initial formula, and accepting locations with $\mathcal{A}^e_{\text{hyp}}$, while only the transition guards are generalized from finitely many observed delays to intervals via Π.

Example 14. One can obtain the ARTA in Fig. 1a from the evidence AFA in Fig. 1b by using Algorithm 3 and simplifying some transitions. Concretely, the partition function Π generalizes the delays 2, 2.5, 7, and 7.5 on the transitions from ℓ_0 with b into $[0, 2]$, $(2, 3)$, $[3, 7]$, and $(7, \infty)$.

The following theorem shows that this construction is faithful to T. This is essential for the termination of AL^*_{RTA}.

Theorem 15 (faithfulness). *Assume T with index sets $S, E \subseteq \mathcal{T}(\Sigma)$ and monotone basis $P \subseteq S$ is cohesive. Let $\mathcal{A}_{\mathrm{hyp}}$ be the ARTA constructed from T. For any $s \in S$ and $s \in E$, we have $T(s, e) = \top \iff s \cdot e \in \mathcal{L}(\mathcal{A}_{\mathrm{hyp}})$.* □

Since the endpoints of the intervals produced by Definition 13 are independent of d_n, the constants in the ARTA constructed by Algorithm 3 are bounded.

Theorem 16. *Fix $K \in \mathbb{N}$ as in Theorem 3 for $\mathfrak{L}_{\mathrm{tgt}}$. Assume T with index sets $S, E \subseteq \mathcal{T}(\Sigma)$ and monotone basis $P \subseteq S$ is cohesive. Let $\mathcal{A}_{\mathrm{hyp}}$ be the ARTA constructed from T. For any transition (ℓ, a, I, φ) of $\mathcal{A}_{\mathrm{hyp}}$, every finite integer endpoint in I is at most K.* □

4.3 The $\mathbf{AL}^*_{\mathbf{RTA}}$ algorithm

Before presenting the $\mathrm{AL}^*_{\mathrm{RTA}}$ algorithm, we introduce auxiliary functions g_α and G_α to normalize counterexamples returned by equivalence queries. This technique is commonly used in RTA learning [4,5].

Definition 17. *For $\alpha \in (0,1)$, we let $g_\alpha \colon \Sigma \times \mathbb{R}_{\geq 0} \to \Sigma \times \mathbb{R}_{\geq 0}$ by $g_\alpha((a,d)) = (a,d)$ if $d \in \mathbb{N}$, and $g_\alpha((a,d)) = (a, \lfloor d \rfloor + \alpha)$ otherwise. We extend it to timed words $G_\alpha \colon \mathcal{T}(\Sigma) \to \mathcal{T}(\Sigma)$ by applying g_α to each letter with delay.*

It is easy to see that G_α preserves counterexamples during learning.

Theorem 18. *For any real-time language $\mathfrak{L}$, for any $\alpha \in (0,1)$, for any ARTA $\mathcal{A}$, and for any $w \in \mathfrak{L} \triangle \mathcal{L}(\mathcal{A})$, $G_\alpha(w) \in \mathfrak{L} \triangle \mathcal{L}(\mathcal{A})$ holds.* □

Algorithm 2 outlines $\mathrm{AL}^*_{\mathrm{RTA}}$. It maintains an observation table T with finite index sets S, E together with a candidate monotone basis P, which are initialized at line 1 and updated in the main loop line 2 to 17. When S or E is extended, the newly introduced cells in T are filled by using membership queries.

The learner first increases P or S to ensure P-closedness (line 4) and evidence-closedness (line 6). When neither extension step applies, it computes P' by solving a binary integer program (BIP) (line 8). If $|P'| < |P|$, it updates P to P'. The details of the BIP encoding are shown in 4.5.

Otherwise, the learner constructs a hypothesis $\mathcal{A}_{\mathrm{hyp}}$ (line 11) and asks an equivalence query (line 12). If the query succeeds, the algorithm returns the hypothesis (line 14); otherwise the teacher provides a counterexample (line 15). On a counterexample, the learner first normalizes it (line 16) and then refines the table by adding suffixes of the normalized counterexample to E (line 17). This normalization step maintains the observation table in line Algorithm 2 as floor-distinct because *i)* $\Phi(S \cup E)$ increases only at line 17 and *ii)* this normalization ensures that for any $(a, d_1), (a, d_2) \in \Phi(S \cup E)$, $g_\alpha((a, d_1)) = g_\alpha((a, d_2))$ implies $d_1 = d_2$, which ensures the floor-distinctness.

	ϵ
ϵ	$\bot$

(a) Initial observation table T_1.

	ϵ	$(b,7)$
ϵ	$\bot$	$\top$
$(b,7)$	$\top$	$\top$

(b) Second observation table T_2.

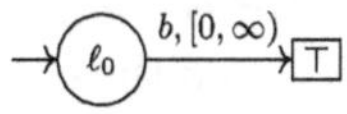

(c) Second hypothesis $\mathcal{A}^2_{\mathrm{hyp}}$.

	ϵ	$(b,7)$	$(b,7.5)$
ϵ	$\bot$	$\top$	$\bot$
$(b,7)$	$\top$	$\top$	$\top$
$(b,7.5)$	$\bot$	$\top$	$\bot$

(d) Third observation table T_3.

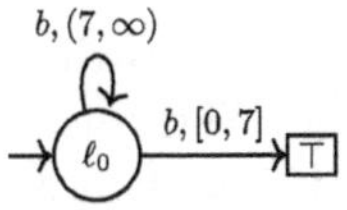

(e) Third hypothesis $\mathcal{A}^3_{\mathrm{hyp}}$.

	ϵ	$(b,2.5)$	$(b,7)$	$(b,7.5)$
ϵ	$\bot$	$\bot$	$\top$	$\bot$
$(b,2.5)$	$\bot$	$\bot$	$\top$	$\bot$
$(b,7)$	$\top$	$\top$	$\top$	$\top$
$(b,7.5)$	$\bot$	$\bot$	$\top$	$\bot$

(f) Fourth observation table T_4.

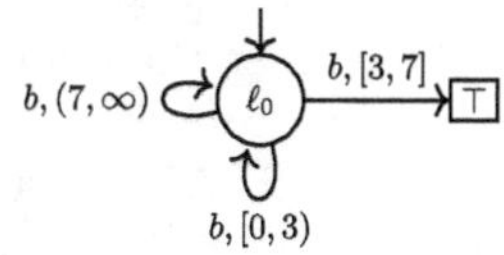

(g) Fourth hypothesis $\mathcal{A}^4_{\mathrm{hyp}}$.

	ϵ	$(b,0)$	$(b,2.5)$	$(b,7)$	$(b,7.5)$	$(a,3)(b,0)$
ϵ	$\bot$	$\bot$	$\bot$	$\top$	$\bot$	$\top$
$(a,3)$	$\bot$	$\top$	$\top$	$\top$	$\top$	$\bot$
$(b,0)$	$\bot$	$\bot$	$\bot$	$\bot$	$\bot$	$\bot$
$(b,2.5)$	$\bot$	$\bot$	$\bot$	$\top$	$\bot$	$\bot$
$(b,7)$	$\top$	$\top$	$\top$	$\top$	$\top$	$\bot$
$(b,7.5)$	$\bot$	$\bot$	$\bot$	$\top$	$\bot$	$\bot$
$(a,3)(a,3)$	$\bot$	$\bot$	$\bot$	$\top$	$\bot$	$\top$
$(a,3)(b,0)$	$\top$	$\top$	$\top$	$\top$	$\top$	$\bot$
$(a,3)(b,2.5)$	$\top$	$\top$	$\top$	$\top$	$\top$	$\bot$
$(a,3)(b,7)$	$\top$	$\top$	$\top$	$\top$	$\top$	$\bot$
$(a,3)(b,7.5)$	$\top$	$\top$	$\top$	$\top$	$\top$	$\bot$
$(b,7)(a,3)$	$\bot$	$\bot$	$\bot$	$\top$	$\bot$	$\bot$
$(b,7)(b,0)$	$\top$	$\top$	$\top$	$\top$	$\top$	$\bot$
$(b,7)(b,2.5)$	$\top$	$\top$	$\top$	$\top$	$\top$	$\bot$
$(b,7)(b,7)$	$\top$	$\top$	$\top$	$\top$	$\top$	$\bot$
$(b,7)(b,7.5)$	$\top$	$\top$	$\top$	$\top$	$\top$	$\bot$

(h) Fifth observation table T_5.

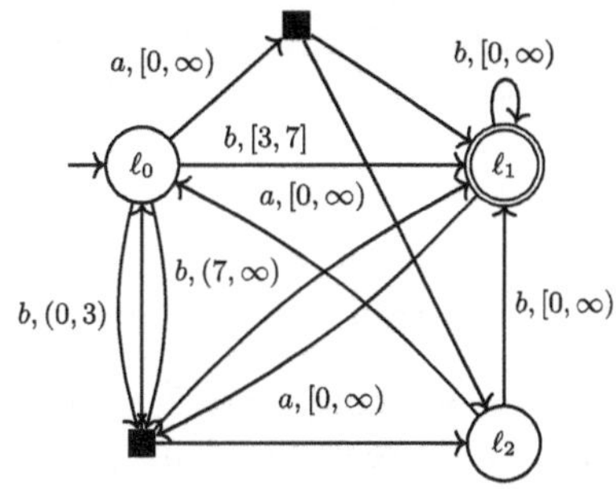

(i) Fifth hypothesis $\mathcal{A}^5_{\mathrm{hyp}}$.

Fig. 3. Observation tables and ARTAs in the example in Sect. 4.4. Shaded rows in the observation tables show a minimum-cardinality monotone basis.

4.4 Worked Example of $\mathrm{AL}^*_{\mathrm{RTA}}$

We illustrate $\mathrm{AL}^*_{\mathrm{RTA}}$ on the target language $\mathfrak{L}_{\mathrm{tgt}}$ recognized by the ARTA in Fig. 1a. We fix $\alpha = \frac{1}{2}$.

Figure 3a shows the initial observation table T_1, where we have $S = E = \{\epsilon\}$ and $P = \emptyset$. Note that $T[\epsilon]$ is represented by $\bigvee_{r\in\emptyset} r$. The corresponding hypothesis ARTA $\mathcal{A}^1_{\mathrm{hyp}}$ is such that $L = F = \emptyset$ and $L_0 = \bot$, which does not accept any timed word. The learner asks an equivalence query, and the teacher returns a counterexample $(b,7)$, which belongs to $\mathfrak{L}_{\mathrm{tgt}}$ but is rejected by $\mathcal{A}^1_{\mathrm{hyp}}$.

The learner then adds $(b,7)$ to E and makes the observation table cohesive. Figures 3b and 3c show the resulting table T_2 and the corresponding hypothesis ARTA $\mathcal{A}^2_{\mathrm{hyp}}$, respectively. Here, we have $P = \{\epsilon\}$; $T[(b,7)]$ is represented by $\bigwedge_{r\in\emptyset} r$. The learner asks an equivalence query, and the teacher returns a counterexample $(b,7.5)$, which does not belong to $\mathfrak{L}_{\mathrm{tgt}}$ but is accepted by $\mathcal{A}^2_{\mathrm{hyp}}$.

The learner then adds $(b,7.5)$ to E and makes the observation table cohesive. The resulting observation table T_3 and the corresponding hypothesis ARTA $\mathcal{A}^3_{\mathrm{hyp}}$ are shown in Figs. 3d and 3e. We still have $P = \{\epsilon\}$; the row of the new prefix timed word $(b,7.5)$ is the same as that of ϵ. The learner asks an equivalence query, and the teacher returns a counterexample $(b,2.5)$.

The learner then adds $(b, 2.5)$ to E and makes the observation table cohesive. The resulting observation table T_4 and the corresponding hypothesis ARTA $\mathcal{A}^4_{\mathrm{hyp}}$ are shown in Figs. 3f and 3g. We again have $P = \{\epsilon\}$; the new timed word $(b, 2.5)$ also has the same row as ϵ. The learner asks an equivalence query, and the teacher returns a counterexample $(a, 3)(b, 0)$.

The learner then adds the suffixes of $(a, 3)(b, 0)$ to E and makes the observation table cohesive. The resulting observation table T_5 and the corresponding hypothesis ARTA $\mathcal{A}^5_{\mathrm{hyp}}$ are shown in Figs. 3h 3i. Here, the monotone basis changes to $P = \{\epsilon, (a, 3), (b, 7)\}$, and the corresponding hypothesis has three locations. The learner asks an equivalence query, and the teacher returns a counterexample $(a, 3.5)(b, 7.5)$.

The fifth hypothesis ARTA already has three locations that match the target ARTA. The learner makes four additional equivalence queries, mainly to refine the transitions rather than to introduce qualitatively new behavior. After the ninth equivalence query, the teacher returns no counterexample, and the final hypothesis is equivalent to the target ARTA in Fig. 1a.

4.5 Monotone Basis Identification via BIP

In $\mathrm{AL}^*_{\mathrm{RTA}}$, we reduce the identification of a minimum-cardinality monotone basis to a binary integer program (BIP) over the distinct row vectors $\mathrm{Rows}(S)$. For each row vector $x \in \mathrm{Rows}(S)$, we let $\mathrm{Pos}(x) = \{e \in E \mid x(e) = \top\}$ and $\mathrm{Neg}(x) = \{e \in E \mid x(e) = \bot\}$. For each triple (x, e^+, e^-) with $x \in \mathrm{Rows}(S)$, $e^+ \in \mathrm{Pos}(x)$, and $e^- \in \mathrm{Neg}(x)$, we identify one *separator obligation*. A row $r \in \mathrm{Rows}(S)$ *covers* (x, e^+, e^-) if $r(e^+) = \top$ and $r(e^-) = \bot$. We then introduce a binary variable $y_r \in \{0, 1\}$ for each $r \in \mathrm{Rows}(S)$ and solve the following BIP:

$$\min \sum_{r \in \mathrm{Rows}(S)} y_r$$

$$\text{s.t.} \sum_{r \in \mathrm{Rows}(S), r(e^+) = \top, r(e^-) = \bot} y_r \geq 1 \quad \forall x \in \mathrm{Rows}(S), e^+ \in \mathrm{Pos}(x), e^- \in \mathrm{Neg}(x).$$

Let $U = \{r \in \mathrm{Rows}(v) \mid y_r = 1\}$ be the selected row set. We map each $r \in U$ back to $s_r \in S$ satisfying $T[s_r] = r$ and set $P = \{s_r \mid r \in U\}$.

Proposition 19. *For $R \subseteq \{\top, \bot\}^E$ and $U \subseteq R$, U is a monotone basis of R if and only if for any $x \in R$, $e^+ \in \mathrm{Pos}(x)$, and $e^- \in \mathrm{Neg}(x)$, there is $r \in U$ covering (x, e^+, e^-).* □

By Proposition 19, the feasible solutions of the BIP are exactly the monotone bases of R, and solving it yields a minimum-cardinality monotone basis. In our implementation, this optimization is solved *approximately* using a relative gap tolerance and a time limit, for the sake of efficiency.

4.6 Query Complexity of $\mathrm{AL}^*_{\mathrm{RTA}}$

Finally, we show the query complexity of $\mathrm{AL}^*_{\mathrm{RTA}}$ using the following bound on the number of distinct rows in observation tables. In what follows, we assume that the BIP from Sect. 4.5 is exactly solved, and we always obtain a minimum-cardinality monotone basis P. We also assume that the mapping from $\mathrm{Rows}(S)$ to S is deterministic; once $s \in S$ for a row vector $r \in \{\top, \bot\}^E$ is selected, the same s is always chosen for r until E changes.

Lemma 20. *Let n be the number of locations of the minimal DRTA recognizing $\mathfrak{L}_{\mathrm{tgt}}$. Every observation table T has at most n distinct row vectors.* □

Theorem 21 (query complexity). *Let $\mathfrak{L}_{\mathrm{tgt}} \subseteq \mathcal{T}(\Sigma)$ be a real-time language, let n be the number of locations of the minimal DRTA recognizing $\mathfrak{L}_{\mathrm{tgt}}$, and let h be the maximum length of counterexamples returned by the teacher. Fix $K \in \mathbb{N}$ as in Theorem 3 for $\mathfrak{L}_{\mathrm{tgt}}$. Algorithm 2 terminates and returns an ARTA recognizing $\mathfrak{L}_{\mathrm{tgt}}$ using at most $M + 1$ equivalence queries and $M^3h^2n + M^2hn + Mh + 1$ membership queries, where $M = 2^n + |\Sigma|(2K + 2)$.*

Proof (sketch). **Number of equivalence queries.** For each suffix $e \in E$, we let $t(e) \in \{\bot, \top\}^n$ as follows. Let $w_1, w_2, \dots, w_n \in \mathcal{T}(\Sigma)$ such that for each $i \neq j$, $w_i \not\equiv w_j$. Then, we let $t(e)(i) = \top \iff w_i \cdot e \in \mathfrak{L}_{\mathrm{tgt}}$. Since there are at most 2^n distinct $t(e)$, discovery of a new $t(e)$ occurs at most 2^n times in Algorithm 2.

Assume the counterexample cex returned by an equivalence query does not reveal a new $t(e)$. Still, processing $G_\alpha(cex)$ must change the observation table so that $G_\alpha(cex)$ is not a counterexample to the next hypothesis ARTA because we have $G_\alpha(cex)$ as a suffix, and thus, the next hypothesis must classify $G_\alpha(cex)$ correctly by Theorems 15 and 18. In such a case, we must discover a new (a, d) in $\Phi(S \cup E)$ such that either $d \leq K$ or $d > K$ and there is no $(a, d') \in \Phi(S \cup E)$ with $d' > K$; otherwise, cex may still be a counterexample to the hypothesis ARTA after processing it. The number of such refinements of $\Phi(S \cup E)$ by equivalence queries is at most $|\Sigma|(2K + 2)$. Overall, the number of equivalence queries is bounded by $2^n + |\Sigma|(2K + 2) + 1$.

Number of membership queries. After each failed equivalence query, at most h new suffixes are added. Thus, we have $|E| \leq Mh + 1$. Moreover, each failed equivalence query increases $\Phi(S \cup E)$ at most by h, and we have $|\Phi(S \cup E)| \leq Mh$. Since P is a monotone basis of T, $|P| = |\mathrm{Rows}(P)| \leq |\mathrm{Rows}(S)| \leq n$ follows from Lemma 20. S increases only to ensure evidence-closedness at line 6. By Lemma 20, there are at most n distinct row vectors in the observation tables appearing in this computation. By determinism of the mapping from $\mathrm{Rows}(S)$ to S in the monotone basis identification, there are at most n distinct $s \in S$ in the monotone bases between each equivalence query. Thus, between each equivalence query, S increases at most by $|\Phi(S \cup E)| \times n \leq Mhn$, and we have $|S| \leq M^2hn + 1$. Overall, the number of membership queries is bounded by $|S| \times |E| \leq (M^2hn + 1) \times (Mh + 1) = M^3h^2n + M^2hn + Mh + 1$. □

In Theorem 21, we have an exponential query bound for $\mathrm{AL}^*_{\mathrm{RTA}}$, while it is polynomial for $\mathrm{NL}^*_{\mathrm{RTA}}$. Nevertheless, in Sect. 5, we empirically observe that the difference in the number of equivalence queries is moderate in practice.

5 Empirical Evaluation

We empirically compare our $\mathrm{AL}^*_{\mathrm{RTA}}$ algorithm with the $\mathrm{NL}^*_{\mathrm{RTA}}$ algorithm. We use our prototype library LEARNARTA[3] as an implementation of $\mathrm{AL}^*_{\mathrm{RTA}}$ and NRTALEARNING [1] as an implementation of $\mathrm{NL}^*_{\mathrm{RTA}}$. In LEARNARTA, monotone basis identification is implemented by approximately solving the BIP from Sect. 4.5. In both implementations, each equivalence query is answered by an exhaustive analysis, such as a reachability analysis.

We used the 190 randomly generated NRTAs taken from [5]. The benchmark consists of 17 groups, each with different numbers of locations and different alphabet sizes. The constant appearing in the guards is at most 20.

For each execution, we measured the number of membership and equivalence queries and the total execution time, including the time to answer the queries. For membership queries, we count the number of unique queried timed words. We conducted all the experiments on a computing server running Ubuntu 24.04 LTS with an Intel Xeon w5-3435X and 125 GiB of RAM.

Results and Discussion. Table 1 summarizes the results. We observe that $\mathrm{AL}^*_{\mathrm{RTA}}$ returned a smaller automaton than $\mathrm{NL}^*_{\mathrm{RTA}}$. This is consistent with the better worst-case succinctness of ARTAs (Theorem 9) than NRTAs [5].

In contrast, $\mathrm{AL}^*_{\mathrm{RTA}}$ almost always used more queries than $\mathrm{NL}^*_{\mathrm{RTA}}$. This is consistent with the theoretical analysis in Theorem 21. The increase in the number of equivalence queries is also reported in untimed cases [7], where AL^* required more equivalence queries than NL^* and L^*. This increase is particularly evident when the alphabet is large. This is likely because in $\mathrm{AL}^*_{\mathrm{RTA}}$, the set of prefixes is initialized with $\{\epsilon\}$ rather than $\{\epsilon\} \cup \{(a, 0) \mid a \in \Sigma\}$, which prevents irrelevant prefixes but often requires more equivalence queries.

The increase in the number of membership queries is partly due to the different counterexample handling strategies: $\mathrm{NL}^*_{\mathrm{RTA}}$ uses an efficient counterexample handling based on Rivest and Schapire's algorithm [22], which reduces the size of observation tables and reduces the number of membership queries. In contrast, $\mathrm{AL}^*_{\mathrm{RTA}}$ uses a simple counterexample handling, much like NL^*.

$\mathrm{AL}^*_{\mathrm{RTA}}$ often took longer time than $\mathrm{NL}^*_{\mathrm{RTA}}$. This is likely due to the difficulty of basis minimization in $\mathrm{AL}^*_{\mathrm{RTA}}$, in addition to the larger number of queries.

[3] LEARNARTA is publicly available at https://github.com/MasWag/LearnARTA. The artifact is available at https://doi.org/10.5281/zenodo.19650471.

Table 1. Summary of the results of experiments. We show the mean value for each group. The columns "# EqQ" and "# MemQ" show the number of equivalence and membership queries, respectively. The columns "$|L_{\mathcal{A}_{\text{hyp}}}|$" show the number of locations of the learned automaton. The columns "Total Time" show the total execution time including the time to answer queries.

$(\lvert L\rvert, \lvert\Sigma\rvert)$	AL^*_{RTA}				NL^*_{RTA}			
	# EqQ	# MemQ	$\lvert L_{\mathcal{A}_{\text{hyp}}}\rvert$	Total Time	# EqQ	# MemQ	$\lvert L_{\mathcal{A}_{\text{hyp}}}\rvert$	Total Time
(3,2)	13.1	975.05	**3**	**0:00.01**	**10.9**	**303.75**	4	0:00.07
(4,2)	17.15	2449.9	**4.05**	**0:00.02**	**15.65**	**810.4**	5.05	0:00.12
(5,2)	22.6	6280.5	**5**	**0:00.06**	**20.7**	**1279.1**	6	0:00.17
(6,2)	**22.3**	10440.1	**6.1**	**0:00.18**	22.4	**1737.5**	7.2	0:00.31
(8,2)	31.3	28058.3	**8**	0:00.77	**27.6**	**3093.6**	9.1	**0:00.45**
(8,4)	48.4	55362.2	**8**	**0:00.66**	**46.4**	**8449.5**	9.1	0:02.05
(10,2)	38.2	50558.9	**10**	0:01.61	**36.4**	**5665.3**	11.1	**0:01.02**
(10,4)	52.4	106987.2	**10**	**0:02.28**	**49.2**	**9664.4**	11.1	0:02.42
(10,6)	71.7	171537.9	**10**	**0:02.92**	**62.8**	**17346.3**	11	0:06.16
(10,8)	97	443801.3	**10**	0:20.75	**83.1**	**20298.7**	11	**0:07.55**
(10,10)	114.2	723024.4	**10**	0:33.37	**92**	**26558.5**	11	**0:13.48**
(12,2)	56	144836.1	**12.3**	0:08.06	**48.4**	**9245.6**	13.3	**0:02.13**
(12,4)	70	295512.5	**12**	0:06.59	**66**	**14968.3**	13	**0:04.39**
(14,4)	81.7	590253.4	**14.1**	0:40.19	**74**	**20697.2**	15.1	**0:07.69**
(16,4)	91.4	812425.1	**16**	1:39.27	**85.5**	**41004.4**	17.1	**0:28.44**
(18,4)	100.7	1136254	**18.3**	2:08.25	**97.5**	**35404**	19.4	**0:18.07**
(20,4)	105.7	1122431.9	**19.8**	1:01.14	**103.9**	**56806.9**	21	**0:46.75**

6 Related Work

Learning Nondeterministic or Alternating Automata. NL* [10] is an L*-style algorithm for learning residual NFAs, which is a subclass of NFAs with some desired properties of DFAs. Thanks to the succinctness of residual NFAs, NL* can learn smaller automata than L*. AL* [7] is an extension of NL* to learn an AFA rather than an NFA. Although the authors conjectured that AL* produces residual AFAs, it is later shown that AL* does not always produce a residual AFA [9], and AL**, a modification of AL* to learn residual AFAs, is proposed. Our AL^*_{RTA} is mainly based on AL* rather than AL** because we aim at identifying smaller ARTAs that are not necessarily residual.

Learning Timed Languages. L*-style learning algorithms have been proposed for various variants of timed automata, including one-clock deterministic timed automata [3,26], (multi-clock) deterministic timed automata [23–25], Mealy machines with timers [11,18,19], event-recording automata [17], and RTAs [4,5].

As we discussed in Sect. 1, most of these algorithms focus on learning deterministic automata. Extending these algorithms to learn automata other than RTAs with nondeterministic or alternating branching is a future direction.

Learning Symbolic Automata. Symbolic automata [13] generalize classical automata so that each transition is labeled with a predicate rather than a letter. Notably, symbolic automata capture real-time automata. So far, no algorithm has been proposed for learning nondeterministic or alternating symbolic automata. Extending learning algorithms for symbolic automata, such as Λ^* [15] and MAT^* [8], or generalizing the theoretical analysis in [16] to support nondeterministic or alternating branching is another future direction.

7 Conclusions and Future Directions

We studied ARTAs, especially focusing on an L*-style learning algorithm for them. Although the introduction of alternating branching does not increase expressive power, it improves succinctness. Our $\mathrm{AL}^*_{\mathrm{RTA}}$ algorithm can learn ARTAs with a termination guarantee. Our empirical evaluation suggests that $\mathrm{AL}^*_{\mathrm{RTA}}$ generally learns smaller automata than $\mathrm{NL}^*_{\mathrm{RTA}}$ at the cost of more queries.

Investigating efficient counterexample handling for ARTAs, particularly along the lines of Rivest and Schapire, is a future direction. Generalizing alternating automata learning for other subclasses of timed automata, e. g., event-recording automata, is another future direction.

Acknowledgements. This work is partially supported by JST BOOST Grant No. JPMJBY24H8, JST PRESTO Grant No. JPMJPR22CA, and JSPS KAKENHI Grant No. 22K17873.

References

1. GitHub: Leslieaj/NRTALearning. https://github.com/Leslieaj/NRTALearning, Accessed 08 Mar 2026
2. Alur, R., Dill, D.L.: A theory of timed automata. Theor. Comput. Sci. **126**(2), 183–235 (1994). https://doi.org/10.1016/0304-3975(94)90010-8
3. An, J., Chen, M., Zhan, B., Zhan, N., Zhang, M.: Learning One-Clock Timed Automata. In: TACAS 2020. LNCS, vol. 12078, pp. 444–462. Springer, Cham (2020). https://doi.org/10.1007/978-3-030-45190-5_25
4. An, J., Wang, L., Zhan, B., Zhan, N., Zhang, M.: Learning real-time automata. SCIENCE CHINA Inf. Sci. **64**(9), 1–17 (2021). https://doi.org/10.1007/s11432-019-2767-4
5. An, J., Zhan, B., Zhan, N., Zhang, M.: Learning nondeterministic real-time automata. ACM Trans. Embed. Comput. Syst. **20**(5s), 99:1–99:26 (2021)
6. Angluin, D.: Learning regular sets from queries and counterexamples. Inf. Comput. **75**(2), 87–106 (1987). https://doi.org/10.1016/0890-5401(87)90052-6

7. Angluin, D., Eisenstat, S., Fisman, D.: Learning regular languages via alternating automata. In: Yang, Q., Wooldridge, M.J. (eds.) Proceedings of the Twenty-Fourth International Joint Conference on Artificial Intelligence, IJCAI 2015, Buenos Aires, Argentina, July 25-31, 2015. pp. 3308–3314. AAAI Press (2015). http://ijcai.org/Abstract/15/466
8. Argyros, G., D'Antoni, L.: The Learnability of Symbolic Automata. In: Chockler, H., Weissenbacher, G. (eds.) CAV 2018. LNCS, vol. 10981, pp. 427–445. Springer, Cham (2018). https://doi.org/10.1007/978-3-319-96145-3_23
9. Berndt, S., Liskiewicz, M., Lutter, M., Reischuk, R.: Learning residual alternating automata. Inf. Comput. **289**(Part), 104981 (2022). https://doi.org/10.1016/J.IC.2022.104981
10. Bollig, B., Habermehl, P., Kern, C., Leucker, M.: Angluin-style learning of NFA. In: Boutilier, C. (ed.) IJCAI 2009, Proceedings of the 21st International Joint Conference on Artificial Intelligence, Pasadena, California, USA, July 11-17, 2009, pp. 1004–1009 (2009). http://ijcai.org/Proceedings/09/Papers/170.pdf
11. Bruyère, V., Garhewal, B., Pérez, G.A., Staquet, G., Vaandrager, F.W.: Active learning of mealy machines with timers. In: QEST+FORMATS, pp. 42–61. Lecture Notes in Computer Science. Springer, Cham (2025)
12. Chandra, A.K., Kozen, D., Stockmeyer, L.J.: Alternation. J. ACM **28**(1), 114–133 (1981). https://doi.org/10.1145/322234.322243
13. D'Antoni, L., Veanes, M.: Minimization of symbolic automata. In: POPL, pp. 541–554. ACM (2014)
14. Dima, C.: Real-time automata. J. Autom. Lang. Comb. **6**(1), 3–23 (2001)
15. Drews, S., D'Antoni, L.: Learning Symbolic Automata. In: Legay, A., Margaria, T. (eds.) TACAS 2017. LNCS, vol. 10205, pp. 173–189. Springer, Heidelberg (2017). https://doi.org/10.1007/978-3-662-54577-5_10
16. Fisman, D., Frenkel, H., Zilles, S.: Inferring symbolic automata. Log. Methods Comput. Sci. **19**(2), (2023). https://doi.org/10.46298/lmcs-19(2:5)2023
17. Grinchtein, O., Jonsson, B., Leucker, M.: Learning of event-recording automata. Theor. Comput. Sci. **411**(47), 4029–4054 (2010)
18. Kogel, P., Klös, V., Glesner, S.: Learning mealy machines with local timers. In: Li, Y., Tahar, S. (eds.) ICFEM, pp. 47–64. Lecture Notes in Computer Science, Springer, Cham (2023). https://doi.org/10.1007/978-981-99-7584-6_4
19. Kogel, P., Schwabe, W., Glesner, S.: Mmlt/ik: Efficiently learning mealy machines with local timers by using imprecise symbol filters. In: Hillston, J., Soudjani, S., Waga, M. (eds.) QEST+FORMATS, pp. 143–159. Lecture Notes in Computer Science. Springer, Cham (2024). https://doi.org/10.1007/978-3-031-68416-6_9
20. Leiss, E.L.: Succinct representation of regular languages by boolean automata. Theor. Comput. Sci. **13**, 323–330 (1981). https://doi.org/10.1016/S0304-3975(81)80005-9
21. Leiss, E.L.: Succinct representation of regular languages by boolean automata II. Theor. Comput. Sci. **38**, 133–136 (1985). https://doi.org/10.1016/0304-3975(85)90215-4
22. Rivest, R.L., Schapire, R.E.: Inference of finite automata using homing sequences. Inf. Comput. **103**(2), 299–347 (1993). https://doi.org/10.1006/INCO.1993.1021
23. Teng, Y., Chen, H., Mi, J., Zhang, M., An, J., Zhan, N.: Active learning of deterministic timed automata via timed classification tree. Sci. China Inf. Sci. **68**(12) (2025)
24. Teng, Y., Zhang, M., An, J.: Learning deterministic multi-clock timed automata. In: HSCC, pp. 6:1–6:11. ACM (2024)

25. Waga, M.: Active learning of deterministic timed automata with myhill-nerode style characterization. In: Enea, C., Lal, A. (eds.) CAV (1), pp. 3–26. Lecture Notes in Computer Science. Springer, Cham (2023). https://doi.org/10.1007/978-3-031-37706-8_1
26. Xu, R., An, J., Zhan, B.: Active learning of one-clock timed automata using constraint solving. In: Bouajjani, A., Holík, L., Wu, Z. (eds.) ATVA, pp. 249–265. Lecture Notes in Computer Science, Springer, Cham (2022). https://doi.org/10.1007/978-3-031-19992-9_16

Multi-objective Statistical Model Checking Using Lightweight Strategy Sampling

Pedro R. D'Argenio[1], Arnd Hartmanns[2], Patrick Wienhöft[3,4], and Mark van Wijk[2](✉)

[1] Universidad Nacional de Córdoba and CONICET, Córdoba, Argentina
[2] University of Twente, Enschede, The Netherlands
mark.vanwijk@utwente.nl
[3] Dresden University of Technology, Dresden, Germany
[4] Centre for Tactile Internet with Human-in-the-Loop (CeTI), Dresden, Germany

Abstract. Statistical model checking delivers quantitative verification results with statistical guarantees. It scales to model sizes and model types that are out of reach for exhaustive, analytical techniques. So far, it has been used to evaluate one property value at a time only. Many practical problems, however, require finding the Pareto front of optimal tradeoffs between multiple objectives. In this paper, we present the first statistical model checking approach for such multi-objective Pareto queries, based on lightweight strategy sampling. We introduce an incremental scheme that almost surely converges to a statistically sound confidence band around the true Pareto front in the long run. To obtain a close underapproximation of the true front in finite time, we propose three heuristic approaches that try to make the best of an a-priori fixed sampling budget. We implement our new techniques in the MODES simulator, and show their effectiveness on benchmarks from the literature.

1 Introduction

Statistical model checking (SMC) [2,55,85] is a popular alternative to probabilistic model checking (PMC) [7,8] for applications where the latter runs into the state space explosion problem due to its exhaustive nature, or where no effective PMC algorithm is available for the model class at hand. SMC circumvents these limitations by applying Monte Carlo simulation: Given an effectively executable probabilistic model M and property of interest ϕ, it performs random executions (or *simulation runs*) of M to obtain samples for the satisfaction or numeric value of ϕ. Given a desired confidence level, SMC then computes a confidence interval around the satisfaction probability or expected value of ϕ, providing a statistical correctness guarantee. The core types of properties handled by PMC and

Authors are sorted in alphabetical order. This work was supported by the EU's H2020 R & I programme under MSCA grant agreement 101008233 (MISSION), by the Interreg North Sea project STORM_SAFE, by SeCyT-UNC grant 33620230100384CB (MECANO), and by NWO VIDI grant VI.Vidi.223.110 (TruSTy).

M. Chen and M. Tschaikowski (Eds.): QEST+FORMATS 2026, LNCS 16913, pp. 171–195, 2027.
https://doi.org/10.1007/978-3-032-35298-9_10

SMC are reachability probabilities and expected accumulated rewards [17,42], and both can be extended to compute probabilities for ω-regular objectives in e.g. LTL [4,49,64]. PMC and more recently SMC [18] have also been extended to other quantities like quantiles [50,76] and conditional value at risk [51].

In practice, we often face tradeoffs between multiple properties, such as maximising a satellite's expected utility without depleting its battery [13], or balancing cost and comfort in heat-pump control [45]. That is, for a probabilistic model with controllable nondeterministic choices, we want to choose among the set of *Pareto-optimal* control strategies: those where no change can be made that improves the value for at least one property without worsening any other. While PMC algorithms for such *multi-objective* problems are more than a decade old [34,35] and cover various models [66] and properties [68], we are not aware of any work that brings the advantages of SMC to the multi-objective setting.

In this paper, we present the **first multi-objective SMC approach** for Pareto queries. Its **memory usage is constant** in the size of the model's state space, preserving the distinguishing advantage of SMC. It uses lightweight strategy sampling (LSS) [56] to evaluate many randomly chosen control strategies w.r.t. all objectives at hand, and from the resulting data constructs a **statistically sound** approximation of the (unknown) true Pareto front.

We first propose an incremental scheme that keeps sampling batches of new strategies, continuously refining an under- and overapproximation of the Pareto front. Assuming ideal LSS and taking care to properly handle statistical error accumulation, the two approximations almost surely converge to a *simultaneous confidence band* for the true Pareto front *in the long run.* Second, we introduce three fixed-budget algorithms aiming to find the best approximation of the Pareto front in *finite time.* They deliver statistically sound *simultaneous lower bounds*, since obtaining guarantees on upper bounds is impossible in finite time.

We implement our approach in the MODEST TOOLSET [40]'s MODES simulator [16] to experimentally evaluate the effectiveness of SMC-based multi-objective model checking and compare the three fixed-budget algorithms, using benchmarks from the quantitative verification and reinforcement learning (RL) [75] literature. Our SMC approaches deliver nontrivial fronts for instances too large (in terms of state space size and dimensionality of the multi-objective query) to be checked by STORM's [47] multi-objective PMC engine. We compare the quality of our approximation with the Pareto fronts computed by STORM where possible. We use Markov decision process (MDP) [12,48,65] models in this paper, but our work directly extends to any model class supported by LSS, e.g. Markov automata [15,30] and probabilistic timed automata [22,44,53].

Related Work. Multi-objective PMC uses linear programming [31,34] or value iteration [35,66]. It is implemented in STORM, PRISM [52], and EPMC [36]; see [3, Section 6] for a summary of the state of the art. While PMC can provide hard (non-statistical) guarantees, it requires exploration of the model's state space. SMC, in contrast, samples instead of exploring, avoiding the state space explosion problem and easily handling models with billions of states [54].

UPPAAL SMC [28] can compare the values of two properties via SMC on a fully stochastic model. UPPAAL STRATEGO [27] brings together SMC and nondeterminism: It first finds a *most permissive* strategy for a set of hard constraints via real-time model checking [11], then among the choices allowed by this strategy, uses RL to optimise a "soft" quantity [26], which is finally evaluated via SMC. STRATEGO thus solves a *lexicographic objectives* problem [39,82] and performs one-dimensional quantitative optimisation only. We, in contrast, determine the trade-offs between multiple equal-priority probabilistic properties. We do so in constant memory, while both STRATEGO's initial model checking and RL store per-state information, potentially limiting scalability. The latter equally applies to other RL-based multi-objective analysis approaches [5,46].

The general field of *multi-objective optimisation* [61,73] uses many different techniques such as evolutionary algorithms [29,69,79], particle swarm optimisation [63], or ant colony optimisation [6]. These are typically applied to functions of real numbers but can be extended to MDPs [80,87]. Like our methods, in finite time, these approaches provide lower bounds only. They however require more than constant memory and typically do not offer statistical guarantees.

2 Background

Definition 1. *A* Markov decision process *(MDP) is a tuple* $\mathcal{M} = \langle \mathcal{S}, s_{init}, \mathcal{A}, \delta \rangle$ *consisting of finite sets of states* $\mathcal{S}$ *(its state space) and actions* $\mathcal{A}$*, an initial state* $s_{init} \in \mathcal{S}$*, and a partial transition probability function* $\delta \colon \mathcal{S} \times \mathcal{A} \rightharpoonup (\mathcal{S} \to [0,1])$*. Define* $\mathcal{A}(s)$ *as the set of all actions for which* $\delta(s,a)$ *is defined. For all* $s \in \mathcal{S}$*, we require* $|\mathcal{A}(s)| \geq 1$ *and* $\forall a \in \mathcal{A}(s) \colon \sum_{s' \in \mathcal{S}} \delta(s,a)(s') = 1$*. A* reward structure *for an MDP is a function* $\mathcal{R} \colon \mathcal{S} \times \mathcal{A} \times \mathcal{S} \to \mathbb{R}$*.*

A reward structure provides a *reward* when moving from one state to another. We call $\delta(s,a)$ for $a \in \mathcal{A}(s)$ a *transition*; its *branches* are $\{\, s' \mid \delta(s,a)(s') > 0 \,\}$.

Example 1. MDP M_R shown in Fig. 1a models the process of writing a paper (action *write*) or not (*stop*), and if successfully written (probability 0.85), choosing to *submit* it to a conference or to *arch*ive it on arXiv. The conference's acceptance rate is 20 %; if rejected, we can try again the next year or go to arXiv. Reward structure $\mathcal{R}_{rec}$ represents the recognition our results get, and $\mathcal{R}_{eff}$ our effort. We annotate branch s' of $\delta(s,a)$ with $(\mathcal{R}_{rec}(s,a,s') \,/\, \mathcal{R}_{eff}(s,a,s'))$, omitting (0/0)s and writing rewards that are the same for all branches on the transition instead.

A path is a sequence $s_0\, a_0 \ldots \in (\mathcal{S} \times \mathcal{A})^\omega$ s.t. $\forall i \colon a_i \in \mathcal{A}(s_i) \wedge \delta(s,a)(s_{i+1}) > 0$. In an MDP, the choice of action in each state is nondeterministic. A *probabilistic memoryless strategy* $\sigma \colon \mathcal{S} \to (\mathcal{A} \to [0,1])$ with $\sum_{a \in \mathcal{A}(s)} \sigma(s)(a) = 1$ makes this choice in all states s. If $\forall s \, \exists a \colon \sigma(s)(a) = 1$, the strategy is *deterministic memoryless* (DM). When removing all state-actions not chosen by σ, we get the induced discrete-time Markov chain $M|_\sigma$, on which we can construct probability measures $\mathbb{P}_\sigma^{M,s}$ on the measurable sets of paths starting in s [33, Sect. 2.2].

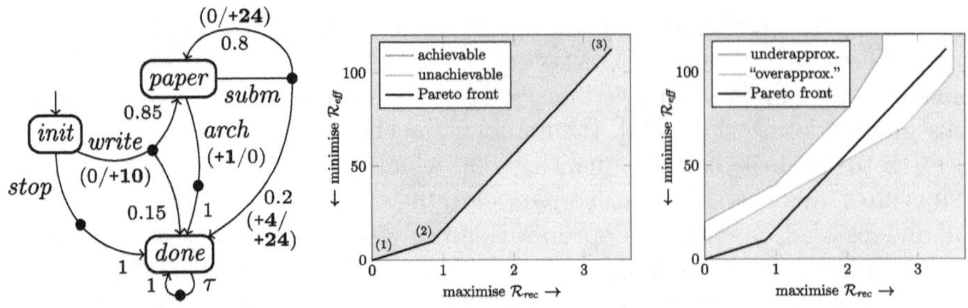

(a) Example MDP M_R (b) The true Pareto front (c) SMC approximations

Fig. 1. An example MDP and the corresponding expected-reward Pareto fronts

Single-Objective Properties. The optimal *reachability probabilities* of a set of goal states $G \subset \mathcal{S}$ are $\mathrm{P}_{\min}(\diamond G) \overset{\text{def}}{=} \inf_\sigma \mathrm{P}^\sigma(\diamond G)$ and $\mathrm{P}_{\max}(\diamond G) \overset{\text{def}}{=} \sup_\sigma \mathrm{P}^\sigma(\diamond G)$ where $\mathrm{P}^\sigma(\diamond G) \overset{\text{def}}{=} \mathbb{P}_\sigma^{M,s_{init}}(\varPi_G)$ with $\varPi_G$ the set of paths that contain a state in G. The optimal *expected rewards* to reach a goal are $\mathrm{E}_{\min}^{\mathcal{R}}(\diamond G) \overset{\text{def}}{=} \inf_\sigma \mathrm{E}^{\mathcal{R},\sigma}(\diamond G)$ and $\mathrm{E}_{\max}^{\mathcal{R}}(\diamond G) \overset{\text{def}}{=} \sup_\sigma \mathrm{E}^{\mathcal{R},\sigma}(\diamond G)$ where $\mathrm{E}^{\mathcal{R},\sigma}(\diamond G) \overset{\text{def}}{=} \mathbb{E}_\sigma^{M,s_{init}}(R_G^{\mathcal{R}})$, $\mathbb{E}_\sigma^{M,s_{init}}(r)$ is the expectation of random variable r under $\mathbb{P}_\sigma^{M,s_{init}}$, and $R_G^{\mathcal{R}}$ maps each path to the sum of the rewards incurred up to the first state in G, or to ∞ if the path does not contain such a state. Note that we consider undiscounted *reachability rewards*, not total-reward or discounted properties. We restrict to *contracting* MDPs, i.e. MDPs that do not contain nontrivial end components. This is common in the literature [9,38], suffices for simulation runs to almost surely terminate—otherwise, we would need more powerful checks that require more than constant memory [4]—and avoids semantic issues which cause current multi-objective *PMC* tools to return inconsistent results on non-contracting MDPs [43]. In this setting, for the properties we consider, DM strategies are optimal, even if the model has both positive and negative rewards [9, p. 164]. For a property $\phi = \mathrm{P}_{opt}(\diamond G)$ or $\phi = \mathrm{E}_{opt}^{\mathcal{R}}(\diamond G)$, we write $\phi(\sigma)$ for $\mathrm{P}^\sigma(\diamond G)$ or $\mathrm{E}^{\mathcal{R},\sigma}(\diamond G)$, respectively.

Multi-objective Properties. Given $\phi_1, \ldots, \phi_d$, the *multi-objective property* $\phi = \mathrm{multi}(\phi_1, \ldots, \phi_d)$ asks for the Pareto front over all the properties. The *achievable set* $Ach(\phi)$ contains all points $\langle v_1, \ldots, v_d \rangle \in \mathbb{R}^d$ for which $\exists\, \sigma\colon \phi_1(\sigma) \sim_1 v_1 \wedge \ldots \wedge \phi_d(\sigma) \sim_d v_d$ where $\sim_i \,=\, \leq$ if ϕ_i is a $\cdot_{\max}$ property and $\sim_i \,=\, \geq$ else. A point $v = \langle v_1, \ldots, v_d \rangle \in Ach(\phi)$ is *Pareto-optimal* if no *other* point $\langle v_1', \ldots, v_d' \rangle \neq v$ with $\forall i\colon v_i \sim_i v_i'$ is achievable; these points form the *Pareto front.*

MDP with multi-objective combinations of reachability probabilities and expected rewards can be transformed into MDP with combinations of only total rewards [35, Prop. 2], for which memoryless strategies are Pareto-optimal [66, Sect. 3.4]. The Pareto front is the convex hull of a finite set of corner points resulting from DM strategies; all its other points are convex combinations of the corners resulting from probabilistic strategies. Note that, for contracting MDPs, [66] also removes [35]'s restriction to non-negative rewards.

An *underapproximation* of the achievable set (and thus the Pareto front) is a convex set of achievable but not necessarily Pareto-optimal points; an *overapproximation*, conversely, is a concave set of unachievable or Pareto-optimal points.

Example 2. Figure 1b shows the achievable set and Pareto front for M_R and $\phi = \text{multi}(\mathrm{E}_{\max}^{\mathcal{R}_{rec}}(\diamond\{\, done \,\}), \mathrm{E}_{\min}^{\mathcal{R}_{eff}}(\diamond\{\, done \,\}))$, visualising the tradeoff between maximising recognition and minimising effort. The corner points correspond to (1) not writing a paper, (2) writing a paper and, if written successfully, archiving it on arXiv, and (3) (re)submitting it to the conference until it is accepted.

Example 3. In Fig. 1c, the blue (top-left) area is an underapproximation of the achievable set of Fig. 1b. The red (bottom-left) area is neither an under- nor an overapproximation. Our methods can statistically approximate these two areas.

Statistical Model Checking. Given a fully stochastic executable formal model (e.g. a large Markov chain compactly specified in a high-level modelling language), a property ϕ that can be evaluated to a real value on a finite path, and user-specified number of runs n and confidence level $\gamma \in (0, 1)$, we can perform *statistical model checking* (SMC): Randomly—using a pseudo-random number generator (PRNG) to implement the model's probabilistic choices—perform n simulation runs, i.e. generate n random paths, evaluate ϕ on each path to obtain samples $X_1, \ldots, X_n$, compute the sample mean $\hat{x} = \frac{1}{n}\sum_{i=1}^{n} X_i$ and a confidence interval (CI) $I = [l, u] \ni \hat{x}$, and return the pair $\langle \hat{x}, I \rangle$. Various methods exist to compute *sound* CIs [17] which guarantee that, when repeating the SMC procedure many times, in the limit $(100 \cdot \gamma)\,\%$ or more of the intervals contain the property's (unknown) true value. The interval's *precision* is $\varepsilon = \frac{1}{2}(u - l)$; for many methods, we can precompute the n needed to achieve a given ε.

Lightweight Strategy Sampling. To apply SMC to nondeterministic models like MDPs, we need to simulate under a given strategy. Many SMC tools implicitly use the uniform random strategy. *Lightweight strategy sampling* (LSS) allows simulating many different strategies using a constant-memory representation for each, e.g. a 32-bit integer. When simulating with such a *strategy identifier* σ and encountering a state s with $|\mathcal{A}(s)| = k > 1$, LSS chooses the $(\mathcal{H}(\sigma.s) \bmod k + 1)$-th action (for some fixed ordering of $\mathcal{A}(s)$) where $\mathcal{H}$ is a hash function and $\sigma.s$ is the concatenation of the bitstring representations of σ and s. In this way, LSS implements a DM strategy. By feeding different information into $\mathcal{H}$, LSS can easily provide history-dependent or partially-informed [15,20,21] strategies as well. Good $\mathcal{H}$ ensure that small changes in the input bits, in particular in σ, have large effects on the output, to cover as much of the strategy space as uniformly as possible. Thus strategy identifiers are "opaque", i.e. their values are not connected to their action choices in any systematic way.[1]

[1] Given user input/domain knowledge, biasing to certain choices may increase the probability of sampling near-optimal strategies; in an uninformed black-box setting, however, no fixed bias can improve results for all models, so we sample uniformly.

An MDP's set of DM strategies—its *strategy space*—is vast but finite. We use 32-bit strategy identifiers, covering up to 2^{32} strategies; while a large space to sample from, it will be smaller than the strategy space of many interesting MDPs. We consider this a *practical* limitation (that could be mitigated by moving to larger identifiers); in the *theoretical* arguments we make in the following sections, we assume *ideal* LSS where by uniformly sampling identifiers we will almost surely eventually encounter every DM strategy. We use a PRNG $\mathfrak{P}$ to sample strategy identifiers; by initialising $\mathfrak{P}$ with a fixed *strategy seed*, we can repeatedly sample the same sequence of strategies to make our experiments more repeatable.

3 Multi-objective SMC

We present an incremental scheme and three fixed simulation budget algorithms. The latter use heuristics to discard seemingly suboptimal strategies. Fix an MDP as in Definition 1, $\phi = \text{multi}(\phi_1, \ldots, \phi_d)$, and confidence level $\gamma = 1 - \alpha$. Recall that the Pareto front's corner points arise from DM strategies after the MDP transformation of [35, Prop. 2], which essentially adds one Boolean state variable per property ϕ_i to track whether its goal set has been reached.

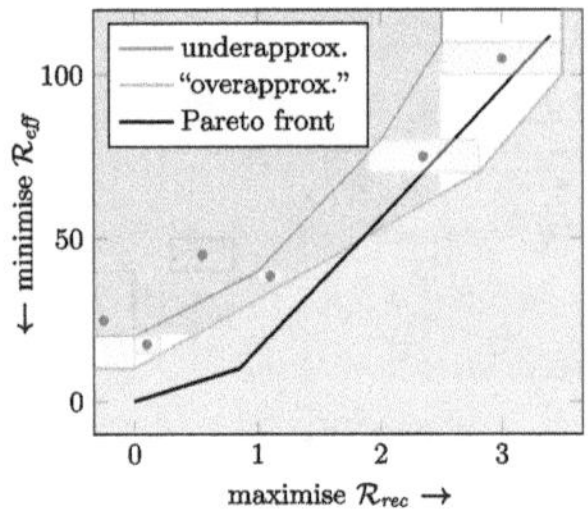

Fig. 2. CI boxes and fronts

Approximating the Pareto Front. All our approaches use LSS to sample deterministic strategies with d bits of memory, which track the visited goal sets to implement the transformation of [35, Prop. 2] on-the-fly, and evaluate them via SMC. For each strategy σ, we thus obtain d pairs $\langle \hat{x}_i^\sigma, I_i^\sigma \rangle$ of a sample mean and a CI with precision ε_i^σ for each dimension, $i \in \{1, \ldots, d\}$; together, the one-dimensional CIs describe a d-dimensional confidence box B_σ around the multi-objective sample mean $\hat{x}^\sigma = \langle \hat{x}_1^\sigma, \ldots, \hat{x}_d^\sigma \rangle$. We ensure that the boxes for the evaluated strategies are *all* simultaneously correct with a-priori probability γ (requirement 1). We write $\mathit{SMC}(\sigma, \gamma, \ldots)$ for the invocation of SMC with LSS for strategy identifier σ and statistical parameters $\gamma, \ldots$ that returns $\langle \hat{x}_\sigma, B_\sigma \rangle$ satisfying requirement 1. The convex hull of the "most pessimistic" corners of all boxes then bounds a candidate underapproximation $\underline{C}$ of the Pareto front; using the "most optimistic" corners gives a candidate overapproximation $\overline{C}$. Given a function stat mapping strategy identifiers to $\langle \hat{x}^\sigma, B_\sigma \rangle$ pairs, we also write $\underline{C}(\mathit{stat})$ and $\overline{C}(\mathit{stat})$ to refer to the respective convex-hull computations.

Example 4. Assume M_R' is a larger version of MDP M_R with the same Pareto-optimal strategies but also many other non-optimal DM ones. Fig. 2 plots a possible result of randomly sampling and evaluating six strategies for M_R'. We draw the sample means (grey dots) and confidence boxes for each, and take the top-left corners (because we want to maximise on the horizontal axis and minimise on the vertical axis) to form $\underline{C}$, which is an underapproximation of the true Pareto front. The most optimistic (bottom-right) corners do not characterise a valid overapproximation because we did not sample good strategies on the bottom left.

Lemma 1. *$\underline{C}$ is an underapproximation with probability γ.*

Proof (sketch). If a most pessimistic corner is unachievable, then its entire box lies in the unachievable region. As the true means for all sampled strategies are by definition achievable points, the box cannot contain the true mean. The probability that this happens for any box is $1 - \gamma$ by requirement 1.

Remark 1. For $\overline{C}$, a similar statement is impossible because we may always have missed a Pareto-optimal strategy. This is a fundamental limitation of a constant-memory sampling-based approach. Only with deeper insights into the model (beyond our assumed black-box view) could we say more, e.g. if we knew how likely it is to sample a Pareto-optimal strategy—but finding that probability essentially solves the multi-objective model checking problem already.

Multiple Comparisons. We perform SMC (1) for each strategy (2) for each dimension, but want all confidence *boxes* to be correct *simultaneously* with probability γ. If we build a CI for each strategy and dimension with confidence γ, then *at least one* CI is wrong with probability $\gg 1 - \gamma$. There are two common ways to counteract this *multiple comparisons problem*: (1) In case of statistical independence, e.g. for the different strategies (since we perform separate simulation runs for each), Šidák correction [70] can be applied. That is, when sampling m strategies, request confidence $\gamma_\sigma = \gamma^{\frac{1}{m}}$ for each strategy. (2) Within a call to $SMC(\sigma, \gamma_\sigma, \ldots)$, all dimensions are estimated from the same simulation runs, so their CIs are not statistically independent. Then Bonferroni correction [59] resp. the union bound can be used, which requires $\sum_{i=1}^{k} \alpha_i \leq \alpha$ for k CIs. Thus the error budget for each dimension can be set to e.g. α_σ / d with $\alpha_\sigma = 1 - \gamma_\sigma$. Šidák correction is harder to work with if we want to distribute the error budget unevenly, and only slightly more efficient than Bonferroni/the union bound beyond very small numbers of comparisons [1]. We thus use the latter throughout.

3.1 Incremental Sampling with Convergence

Our incremental scheme indefinitely samples batches of m strategies as shown in Algorithm 1, updating $\underline{C}$ and $\overline{C}$ whenever a strategy has been evaluated. When interrupted, it returns the current candidate approximations $\underline{C}$ and $\overline{C}$. For Lemma 1 to hold at any interruption point, we need to divide the error budget α over an a priori unknown number of batches. We use the union bound and a user-specified factor $f \in (0,1)$: The i-th batch gets error budget $(1-f)^{i-1} \cdot f \cdot \alpha$. Each batch's budget is then divided equally among the m strategies. We parametrise the SMC invocations in line 6 with precision ε, not the number of runs n; in this way, as α decreases from batch to batch, the number of runs performed per strategy increases. If we fixed n instead, then ε would increase, resulting in larger and larger boxes for later strategies.

1 **function** $\mathit{IncSamp}(\mathit{PRNG}\ \mathfrak{P}, m, \alpha, \varepsilon, f)$
2     $\mathit{stat} := \varnothing$, $\underline{C} := \varnothing$, $\overline{C} := \varnothing$
3   **while** not interrupted **do**
4   **for** $i = 1$ to m **do**
5             $\sigma :=$ next sample from $\mathfrak{P}$
6             $\mathit{stat}(\sigma) := \mathit{SMC}(\sigma, 1 - \frac{f \cdot \alpha}{m}, \varepsilon)$
7             $\underline{C} := \underline{C}(\mathit{stat})$, $\overline{C} := \overline{C}(\mathit{stat})$
8         $\alpha := (1-f) \cdot \alpha$
9   **return** $\langle \underline{C}, \overline{C} \rangle$

Alg. 1: Incremental Sampling

Lemma 2. *With probability* $\gamma = 1 - \alpha$*:*

(1) When interrupted, $\underline{C}$ *returned by Algorithm 1 is an underapproximation.*
(2) When not interrupted, $\underline{C}$ *and* $\overline{C}$ *inside Algorithm 1 almost surely converge to an under- and overapproximation with precision* $2\sqrt{d\varepsilon^2}$*, respectively.*

Proof (sketch). (1) follows directly from Lemma 1 and Algorithm 1's construction (i.e. the use of the union bound to distribute α). For (2), observe that, when sampling indefinitely with ideal LSS on finite MDPs, almost surely all Pareto-optimal strategies σ_i^{opt} will be sampled eventually. At that (unknown) point, with probability γ (by Algorithm 1's construction), all of the $B_{\sigma_i^{opt}}$ contain the strategy's true mean, and thus $\overline{C}$ is an overapproximation. In the worst case, the true mean is at the most pessimistic corner, so at a (d-dimensional Euclidean) distance of $2\sqrt{d\varepsilon^2}$ from the most optimistic corner, or vice-versa.

Thus, in the limit, we are γ-confident that the true Pareto front is fully enveloped by $\underline{C}$ and $\overline{C}$: they form a *simultaneous* γ *confidence band* for the true front.

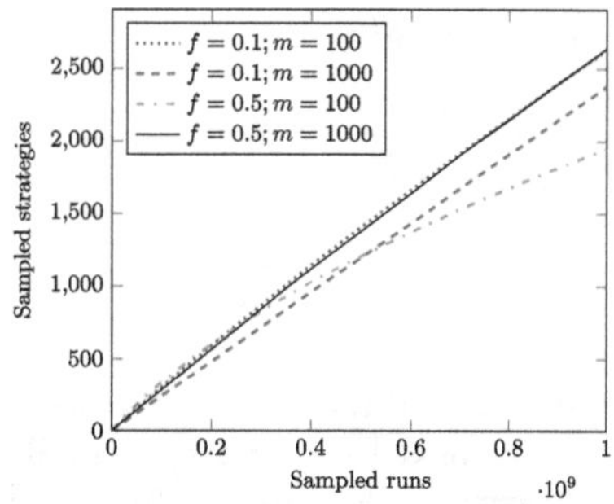

Fig. 3. Sampled strategies vs. runs

Lemma 2 holds independently of the values chosen for m and f. The error budget assigned to a strategy determines the number of simulation runs needed in $SMC(\sigma, \gamma', \varepsilon)$. A larger f results in a higher budget for earlier batches, requiring more runs in later batches to achieve the fixed precision ε. A larger batch size m distributes each batch's budget over more strategies, lowering the budget for earlier strategies. An example of these tradeoffs is visualised in Fig. 3, using the Clopper-Pearson interval [19] and $d = 2, \varepsilon = 0.01, \alpha = 0.1$. On the horizontal axis, the number of runs performed is a proxy for the runtime needed; completing the evaluation of more strategies (higher values on the vertical axis) with fewer runs is "better". As expected, the combination of high $f = 0.5$ and low $m = 100$ starts fast but is overtaken by more conservative approaches in the long run. For large batch sizes, however, the influence of dividing α is small in practice.

3.2 Fixed-Budget Sampling

In practice, we only have finite time. We propose three algorithms—*weight vector refinement* (WVR), *fixed iteration budget* (FIB), and *fixed strategy budget* (FSB)—that try to statistically approximate the achievable set as closely as possible in a fixed simulation budget. They use heuristics to select "promising" strategies, performing fewer simulation runs for seemingly suboptimal ones so that promising ones can be estimated more precisely or more strategies can be sampled. We describe our heuristics in Sect. 3.3; for now, assume one is given as a function select_candidates: $2^{\Sigma} \to 2^{\Sigma}$ that, given a candidate set of strategies, returns a subset of promising strategies. WVR adapts multi-objective PMC [35,66] to the SMC setting, while FIB transfers smart sampling [25] from single-objective to multi-objective SMC. FSB adds dynamic addition of strategies to FIB. While many more schemes are imaginable, these are fundamental corner points.

Since only finitely many strategies can be sampled in finite time, any $\overline{C}$ we compute may not be an overapproximation (as in Ex. 4); and by Remark 1, we cannot quantify the chance for this happening either. The fixed-budget algorithms thus focus on $\underline{C}$, which they guarantee to be an underapproximation with probability γ again. They consist of a heuristic phase that determines a set of candidate strategies, followed by an evaluation phase that returns $\underline{C}$. The latter samples new runs to ensure statistical independence between the phases to avoid bias (see [23, App. A] for an intuitive explanation). Algorithm 2, the evaluation phase, is the same for WVR, FIB, and FSB; the heuristic phases differ, but all perform $I \cdot m \cdot n$ simulation runs (up to rounding) for I a number

function *Eval*(*strategy set* Σ, *run budget* n, *statistical error budget* α)
  $stat := \varnothing$
  foreach $\sigma \in \Sigma$ **do** $stat(\sigma) := SMC(\sigma, 1 - \frac{\alpha}{|\Sigma|}, \lfloor \frac{n}{|\Sigma|} \rfloor)$
  return $\underline{C}(stat)$

Alg. 2: Fixed-budget SMC for multiple objectives, evaluation phase.

1 **function** *WVR*(*PRNG* $\mathfrak{P}$, *heuristic* ρ, m, n, I, α)
2   $\Sigma :=$ next $\lfloor \frac{m}{2} \rfloor$ samples from $\mathfrak{P}$, $stat := \varnothing$
3   **foreach** $\sigma \in \Sigma$ **do** $stat(\sigma) := SMC(\sigma, 1 - \frac{\alpha}{m}, n)$
4   **for** $i = 2$ to I **do**
5   select $w \in \mathbb{R}^d$ in direction of max. distance between $\underline{C}(stat), \overline{C}(stat)$
6     $\Sigma' :=$ select $\lfloor \frac{|\Sigma|}{2} \rfloor$ best strategies in Σ according to $\lambda\sigma.\ w \cdot \hat{x}^\sigma$
7   **do**
8   **foreach** $\sigma \in \Sigma'$ **do** $stat(\sigma) := SMC(\sigma, 1 - \frac{\alpha}{m}, n)$
9       $\Sigma' :=$ select $\lfloor \frac{|\Sigma'|}{2} \rfloor$ best strategies in Σ according to $\lambda\sigma.\ w \cdot \hat{x}^\sigma$
10   **while** $|\Sigma'| > 0$
11   **return** *Eval*(select_candidates$_\rho(\Sigma), I \cdot m \cdot n, \alpha$)

Alg. 3: The weight vector refinement (WVR) algorithm.

of internal iterations. Our implementation then uses $I \cdot m \cdot n$ more runs for the evaluation phase.

Weight Vector Refinement. Listed in Algorithm 3, our WVR algorithm implements the heuristic phase by *linear scalarisation*: it uses a specified number I of weight vectors to convert the multi-objective problem into I single-objective problems, as is done in multi-objective PMC, RL [60,72,83], and optimisation [67,71,81]. When determining the next weight vector, we pick the one with the largest distance between $\underline{C}$ and $\overline{C}$ (line 5) as in [66]. For each vector, we apply a round of smart strategy sampling [25] in lines 7-10, with one exception: The worst-performing strategies are discarded from the local Σ' only, so that they will (temporarily) not receive more runs under the current weight vector. Every **do** loop iteration performs more runs for the remaining strategies to estimate the most promising ones (those with the highest value in the weight vector's direction) most precisely.

Linear scalarisation is a necessity in PMC, to turn ϕ into single-objective properties for which PMC algorithms are available. SMC, however, can estimate the values of multiple objectives at once, enabling more direct approaches:

Fixed Iteration Budget. The FIB algorithm shown in Algorithm 4 samples an initial set of strategies Σ_1, then I times performs simulation runs for the current strategy set Σ_i (line 4) and heuristically discards unpromising strategies to form

1 **function** $\mathit{FIB}(\mathit{PRNG}\ \mathfrak{P}, \mathit{heuristic}\ \rho, m, n, I, \alpha)$
2 $\quad \Sigma_1 :=$ next m samples from $\mathfrak{P}$, $\mathit{stat} := \varnothing$
3 $\quad$ **for** $i = 1$ to I **do**
4 $\quad\quad$ **foreach** $\sigma \in \Sigma_i$ **do** $\mathit{stat}(\sigma) := \mathit{SMC}(\sigma, 1 - \frac{\alpha}{m}, \lfloor \frac{n \cdot m}{|\Sigma_i|} \rfloor)$
5 $\quad\quad \Sigma_{i+1} = \mathsf{select\_candidates}_\rho(\Sigma_i)$
6 $\quad$ **return** $\mathit{Eval}(\mathsf{select\_candidates}_\rho(\Sigma_{I+1}), I \cdot m \cdot n, \alpha)$

Alg. 4: The fixed iteration budget (FIB) algorithm.

1 **function** $\mathit{FSB}(\mathit{PRNG}\ \mathfrak{P}, \mathit{heuristic}\ \rho, m, n, I, \alpha)$
2 $\quad \Sigma_1 := \varnothing$, $\Sigma'_1 :=$ next m samples from $\mathfrak{P}$, $\mathit{stat} := \varnothing$
3 $\quad$ **for** $i = 1$ to I **do**
4 $\quad\quad$ **foreach** $\sigma \in \Sigma_i$ **do** $\mathit{stat}(\sigma) := \mathit{SMC}(\sigma, 1 - \frac{\alpha}{m}, n)$
5 $\quad\quad$ **foreach** $\sigma \in \Sigma'_i$ **do** $\mathit{stat}(\sigma) := \mathit{SMC}(\sigma, 1 - \frac{\alpha}{m}, i \cdot n)$
6 $\quad\quad \Sigma_{i+1} := \mathsf{select\_candidates}_\rho(\Sigma_i \cup \Sigma'_i)$
7 $\quad\quad \Sigma'_{i+1} :=$ next $\lfloor \frac{m - |\Sigma_{i+1}|}{i+1} \rfloor$ samples from $\mathfrak{P}$
8 $\quad$ **return** $\mathit{Eval}(\mathsf{select\_candidates}_\rho(\Sigma_{I+1}), I \cdot m \cdot n, \alpha)$

Alg. 5: The fixed strategy budget (FSB) algorithm.

the next set Σ_{i+1} (line 5). Each iteration evenly distributes $m \cdot n$ runs over the strategies in Σ_i, so that promising strategies are estimated ever more precisely.

Fixed Strategy Budget. Instead of redistributing the budget for discarded strategies over the remaining strategies, FSB (shown in Algorithm 5) samples new strategies so that, in the end, $I \cdot n$ runs were performed for each ultimately surviving strategy. For multi-objective properties with many suboptimal strategies, this increases the probability of finding a Pareto-optimal strategy.

3.3 Strategy Selection Heuristics

We now describe our strategy selection heuristics, beginning with a simple rule:

Definition 2 (a simple rule). *$\sigma \notin \mathsf{select\_candidates}_{simple}(\Sigma)$ iff there is a witness strategy $\varsigma \in \Sigma$ such that, for all $i \in \{1, \ldots, d\}$, $\hat{x}_i^\sigma \leq \hat{x}_i^\varsigma$ if the i-th objective is maximising, and $\hat{x}_i^\sigma \geq \hat{x}_i^\varsigma$ if the i-th objective is minimising.*

This rule effectively constructs a (non-convex) Pareto front from the sample means of the deterministic strategies in Σ as illustrated in Fig. 4, discarding all strategies with suboptimal sample means. In Fig. 4, under the simple rule, σ will be discarded because witness ς has a better sample mean.

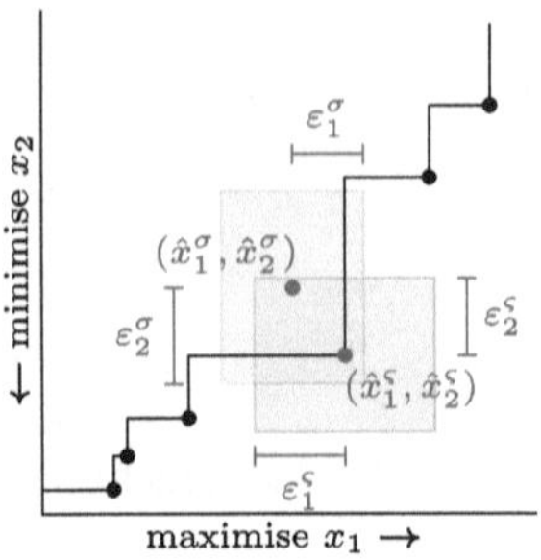

Fig. 4. The simple rule

The simple rule only compares sample means, not taking any statistical considerations into account. In Fig. 4, the confidence boxes for σ and ς overlap, so with the confidence level used, we cannot exclude that σ is in fact the better strategy. A more "precise" rule may decide to perform a probabilistic analysis and reject σ if it is "far enough" from the witness ς, where "far enough" means that the inequalities of Definition 2 hold with a sufficiently large probability. To simplify the process, we only consider CIs to define the criteria for rejection, not the full sample distributions, using the precisions of the CIs to establish "safe" distances.

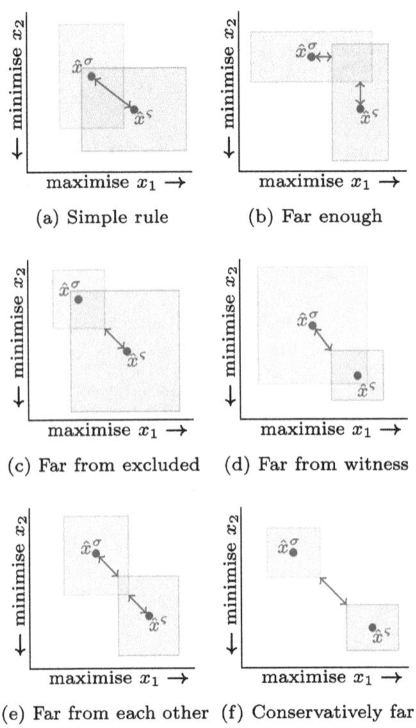

(a) Simple rule (b) Far enough
(c) Far from excluded (d) Far from witness
(e) Far from each other (f) Conservatively far

Fig. 5. Strategy discarding rules.

Figure 5 sketches all criteria we consider based on the sample means (dots) and confidence boxes for σ and witness ς; the double-ended arrows indicate the exclusion rules. Following the simple rule, all situations in Fig. 5 exclude σ as a consequence of the presence of ς. The more precise rules are defined as follows for maximising objectives; the criterion when minimising is the same with + and − as well as $\leq$ and $\geq$ switched:

Definition 3. $\mathsf{select\_candidates}_{rule}(\Sigma)$ *does not contain σ iff there is a witness strategy $\varsigma \in \Sigma$ s.t. $\forall i \in \{1, \ldots, d\}$,*

$$\hat{x}_i^\sigma \leq \hat{x}_i^\varsigma - \min(\varepsilon_i^\varsigma, \varepsilon_i^\sigma) \qquad \textit{(FE)}$$

$$\hat{x}_i^\sigma + \varepsilon_i^\sigma \leq \hat{x}_i^\varsigma \qquad \textit{(FFE)}$$

$$\hat{x}_i^\sigma \leq \hat{x}_i^\varsigma - \varepsilon_i^\varsigma \qquad \textit{(FFW)}$$

$$\hat{x}_i^\sigma \leq \hat{x}_i^\varsigma - \max(\varepsilon_i^\varsigma, \varepsilon_i^\sigma) \qquad \textit{(FFEO)}$$

$$\hat{x}_i^\sigma + \varepsilon_i^\sigma \leq \hat{x}_i^\varsigma - \varepsilon_i^\varsigma \qquad \textit{(CF)}$$

The rules involving ε assume symmetric CIs, but extend analogously to asymmetric ones. Rule "far enough" (FE, Fig. 5b) ensures that, for every objective, at least one of the values estimated with one strategy is outside the other strategy's CIs. Strategy ς cannot exclude σ in Fig. 5a by this rule since $\hat{x}_2^\sigma < \hat{x}_2^\varsigma + \min(\varepsilon_2^\varsigma, \varepsilon_2^\sigma)$, but it can do so in Fig. 5b since $\hat{x}_1^\sigma \leq \hat{x}_1^\varsigma - \varepsilon_1^\varsigma$ and $\hat{x}_2^\sigma \geq \hat{x}_2^\varsigma + \varepsilon_2^\sigma$, and in all of Figs. 5c to 5f.

The next rules focus on the confidence of *one* of the strategies: Rule "far from excluded" (FFE) requires the estimated values for witness ς to be outside the CIs for the excluded strategy σ, shown in Fig. 5c. Dual to this is "far from witness" (FFW), shown in Fig. 5d. These two rules are incomparable but implied by "simple" and "far enough". The intersection of both rules is "far from each other" (FFEO), visualised in Fig. 5e. This rule also excludes σ in the situation of Fig. 5f, but will not exclude it in the situations of Fig. 5a to 5d. The final and most conservative rule is "conservatively far" (CF), which ensures that the CIs obtained with the excluded strategy σ do not overlap with the CIs obtained with the witness ς as shown in Fig. 5f.

4 Experimental Evaluation

We implemented the algorithms and heuristics of Sect. 3 in MODES to experimentally study their effectiveness and performance. Our implementation follows the given pseudocode, except that in the fixed-budget algorithms, any budget unused due to rounding effects in one iteration is carried over to the next.

4.1 Benchmark Set

We compiled a representative benchmark set consisting of benchmarks from PMC and RL. For the former, we use all multi-objective models from [3, Sect. 6.2] that MODES supports (in terms of modelling features) and that have nontrivial Pareto fronts (i.e. of more than one Pareto-optimal point, as computed via PMC). These models are small enough to be checked with PMC. To demonstrate that our approaches keep working when PMC starts to struggle with state space explosion and higher dimensions, we also consider models commonly used in RL literature [10,32,58,74]: (1) the breakable bottles model [78] with 3 objectives, (2) the fruit tree model [84] with both 2 and 6 objectives, (3) deterministic and probabilistic versions of the deep sea treasure case [77], and (4) racetrack benchmarks [37], which we make multi-objective via the puddle world [14] reward scheme. For details, see [23, App. C]. Altogether, we end up with 15 models. Most are parametrised; we consider up to three different parameter valuations per model and one multi-objective property, giving us a final set of 40 benchmark instances. The benchmark selection was done prior to running any experiments, and was not based on any expectations of the performance of our approaches.

4.2 Experimental Setup

Table 1. Configurations

Conf	m	α	n	I
1	100	0.1	1000	10
2	1000	0.1	100	10
3	3333	0.1	30	10

To be able to assess the impact of the number of initial strategies m and the run count parameter n on the fixed-budget algorithms, we use three different configurations as listed in Table 1, and evaluate each algorithm on each instance with all heuristics in all configurations, leading to $3 \cdot 40 \cdot 6 \cdot 3 = 2160$ different experiments. Observe that WVR samples $\frac{m}{2}$, FIB m, and FSB at most $\approx m \cdot \ln I$ strategies. We thus sample < 10000, i.e. *much* fewer than 2^{32}, strategies in each experiment.

We judge each setting by its result's *hypervolume* (HV) [89]: the area under the Pareto front with respect to a "worst possible" reference point. In Fig. 1c, a reasonable reference point is the top left corner; then the HV of the underapproximation is the extent of the visible blue (top left) area. The HV is the only known metric that is strictly monotonic [88], meaning that if one approximation strictly contains another, its HV will be larger. This makes it popular in multi-objective research [57,62,86,90], but a disadvantage is that the reference point is arbitrary. Thus the HV is useful for a qualitative comparison of results on the same model, but not across different models or for a quantitative evaluation.

To quantify the quality of an approximation w.r.t. a reference Pareto front/achievable set, we compute the multiplicative ϵ [88] of the underapproximation. This metric is the least factor by which each point in an approximation should be scalar-multiplied (or divided in case of a minimising objective) in order to be a superset of the reference. A multiplicative ϵ of 1 means that the approximation already fully contains the reference set and is thus optimal. Infinite multiplicative ϵ values can arise when a minimising property needs to approximate the actual value 0: for a value $x \neq 0$, observe that only $\lim_{a\to\infty} \frac{x}{a} = 0$. Since the multiplicative ϵ only considers the worst-performing point of the approximation, it can—and in our benchmarks does—happen that two approximations have equal multiplicative ϵ but one is a superset of the other. Improvements in the rest of the approximation beyond the worst point can only be captured using the HV.

We run each experiment three times, each with a different strategy seed (of 1, 2, and 3). We report the average of the three HVs, and of the multiplicative ϵs if a reference front/set is known (either from insights into the model, or obtained via PMC). We ran our experiments on a 64-bit Ubuntu Linux 22.04 system with AMD Ryzen 9 7950X3D CPU (4.2-5.7 GHz) and 128 GB of RAM.

4.3 Comparison with PMC

Table 2 first lists the number of objectives in the property and the number of states of the MDP for each benchmark instance. The latter has been computed via PMC (using Storm in exact mode where possible, otherwise approximated via mcsta's [41] new sound multi-objective PMC implementation [43] and marked *) or, if that failed, estimated by the authors from model-specific insights (marked **). It then compares the Pareto fronts—obtained via PMC (assumed

Table 2. Benchmarks instances and comparison to PMC.

benchmark instance	objectives	states	PMC time	our time	multiplicative ϵ
care_home.1-160	2	3.46 M*	ERR	TO	–
energy_aware_scheduling.3-2	2	409 k	7.6 s	[5.7 s, 13 s]	[6.84, 20.3]
energy_aware_scheduling.3-3	2	1.86 M	1.4 min	[5.7 s, 13 s]	[7.01, 19.9]
energy_aware_scheduling.3-4	2	10.7 M	15 min	[5.6 s, 13 s]	[7.30, 18.7]
mars_rover.10-1	2	151 k	6.0 s	[1.6 s, 11 s]	[2.15, 3.55]
mars_rover.20-1	2	1.38 M	1.3 min	[1.5 s, 9.7 s]	[15.4, 43.9]
resource_gathering.2-1-1-1	2	3.46 k	0.1 s*	TO	–
breakable_bottles.5-2	3	1.16 k	0.1 s	[1.8 s, 4.4 s]	[1.29, 3.09]
breakable_bottles.10-100	3	36 k	21 s	[7.3 s, 15 s]	[10.1, 32.0]
breakable_bottles.10-1000	3	40.6 M	TO	[9.8 s, 28 s]	[43.5, 117]
deep_sea.5-100	2	2.40 k	0.4 s	[1.7 s, 3.5 s]	[1.02, 1.34]
deep_sea.30-1000	2	532 k	3.8 min	[2.4 s, 22 s]	[1.35, 1.71]
deep_sea.100-5000	2	26.1 M	TO	[2.5 s, 28 s]	[1.71, 3.08]
deep_sea_probabilistic.5-100	2	2.21 k	0.5 s	[1.7 s, 3.5 s]	[1.00, 1.07]
deep_sea_probabilistic.30-1000	2	1.04 M	6.0 min	[2.7 s, 12 s]	[1.67, 2.26]
deep_sea_probabilistic.100-5000	2	126 M	TO	[2.7 s, 12 s]	–
fruit_tree_2.5	2	77	0.1 s	[1.2 s, 17 s]	[1.01, 1.07]
fruit_tree_2.10	2	2.56 k	1.0 s	[1.5 s, 13 s]	[1.01, 1.08]
fruit_tree_2.30	2	2.7 B**	TO	[2.7 s, 11 s]	–
fruit_tree_6.7	6	317	TO	[10 s, 8.4 min]	[1.00, 1.42]
fruit_tree_6.14	6	41.0 k	TO	[5.4 min, TO]	[1.31, 1.53]
fruit_tree_6.30	6	2.7 B**	TO	[4.4 min, TO]	–
puddle_world_barto_big.30	2	411 k*	10 s*	[4.2 s, 7.9 s]	[1.19, ∞)
puddle_world_barto_big.50	2	1.82 M*	46 s*	[4.8 s, 8.8 s]	∞
puddle_world_barto_big.500	2	59.0 M*	TO*	[6.0 s, 20 s]	–
puddle_world_barto_small.50	2	841 k*	36 s*	[3.0 s, 5.4 s]	∞
puddle_world_barto_small.100	2	2.68 M*	1.8 min*	[3.6 s, 6.6 s]	∞
puddle_world_barto_small.500	2	22.4 M*	20 min*	[4.9 s, 11 s]	∞
puddle_world_hansen.80	2	9.15 M*	7.8 min*	[46 s, 53 s]	∞
puddle_world_hansen.500	2	179 M*	TO*	[52 s, 67 s]	–
puddle_world_hansen.1000	2	720 M**	TO*	[56 s, 77 s]	–
puddle_world_ring.60	2	5.41 M*	5.0 min*	[7.0 s, 12 s]	∞
puddle_world_ring.150	2	28.3 M*	52 min*	[8.2 s, 16 s]	∞
puddle_world_ring.500	2	128 M*	TO*	[9.2 s, 24 s]	–
puddle_world_river.30	2	674 k*	27 s*	[44 s, 49 s]	∞
puddle_world_river.50	2	3.39 M*	2.8 min*	[45 s, 50 s]	∞
puddle_world_river.500	2	177 M*	TO*	[55 s, 69 s]	–
puddle_world_tiny.5	2	1.27 k*	0.4 s*	[1.2 s, 4.4 s]	[1.00, ∞)
puddle_world_tiny.10	2	4.55 k*	0.6 s*	[1.2 s, 3.4 s]	[1.00, ∞)
puddle_world_tiny.100	2	131 k*	2.3 s*	[1.3 s, 7.1 s]	[2.26, ∞)

to be correct and taken as the reference) or manually from model insights—with those approximated by our approaches, and the corresponding runtimes: "PMC time" is the runtime of Storm or mcsta, or a timeout "TO" (> 1 hour). Column "our time" gives the interval from lowest to highest runtime of our approaches over all $3 \cdot 6 \cdot 3$ experiments performed per instance (TO > 10 min), while the last column contains analogous intervals over the multiplicative ϵs.

We see that the PMC tools run out time on the larger RL benchmarks, while all our fixed-budget experiments succeed, often in seconds. The SMC-based Pareto front underapproximations can get fairly close to the true Pareto fronts in this time, as observed for the deep_sea and fruit_tree models, but may also be quite far off as observed for the breakable_bottles model.

4.4 Fixed-Budget Algorithms Comparison

We compare the fixed-budget algorithms and heuristics in terms of HV. Each experiment performs a similar number of *simulation runs*, but *wall-clock runtimes* may differ due to per-strategy overhead in MODES and because some strategies induce longer runs.

Best Algorithm, Heuristic, and Configuration. No one combination of configuration, heuristic, and algorithm always wins, as shown in Table 3, where we list the number of models on which each combination outperformed all other combinations (i.e. delivered the highest HV, over the whole table). FSB performs best overall. The total count is 37 instead of 40 since the experiments timed out on care_home and resource_gathering, while there is no unique best combination for the deep_sea_probabilistic.5-100 instance.

Table 3. Config.-heuristic-alg. wins over all.

	FIB			FSB			WVR		
Heuristic	C1	C2	C3	C1	C2	C3	C1	C2	C3
Simple	0	1	1	2	2	5	0	0	4
FE	0	1	0	0	0	8	0	0	0
FFE	0	0	0	0	0	0	0	0	0
FFW	0	1	0	0	0	**9**	0	0	0
FFEO	1	0	0	0	0	0	0	0	0
CF	0	0	0	0	0	2	0	0	0
Total		5			**28**			4	

Table 4. Heuristics.

Heuristic	FIB	FSB	WVR
Simple	6	**16**	**26**
FE	**9**	8	1
FFE	5	1	1
FFW	4	9	3
FFEO	8	1	1
CF	3	2	0

Table 5. Config. wins per heuristic-alg. combination.

	FIB			FSB			WVR		
Heuristic	C1	C2	C3	C1	C2	C3	C1	C2	C3
Simple	5	9	**24**	6	11	**21**	3	2	**31**
FE	7	7	**24**	9	4	**25**	5	5	**26**
FFE	7	5	**26**	9	6	**23**	6	8	**22**
FFW	7	6	**25**	10	3	**25**	5	4	**27**
FFEO	7	6	**25**	11	5	**22**	5	9	**22**
CF	**21**	5	12	14	**15**	9	8	7	**21**

The heuristics significantly influence the number of strategies surviving each iteration. To evaluate their effect on each of the fixed-budget algorithms, we show in Table 4 the number of models for which each heuristic outperformed all others per algorithm (i.e. within a column). The simple strategy obtains the highest HV for most algorithms. It thus appears that it is often beneficial (in our benchmark set) to discard strategies that do not look promising as early as possible and distribute the remaining budget over other strategies, although as observed in Table 3, other heuristics might lead to better overall results.

Table 5 shows the number of models for which each configuration from Table 1 performs best for each heuristic-algorithm pair (i.e. in a 3-column block within a row). On these benchmarks, configuration 3 performs best in almost every case, with the exception of the conservatively-far heuristic. This heuristic needs the intervals to be smaller to discard suboptimal strategies and thus requires more simulation runs. For the other approaches, it appears that having more, and thus presumably a wider variety of, strategies is more important than evaluating each strategy more precisely.

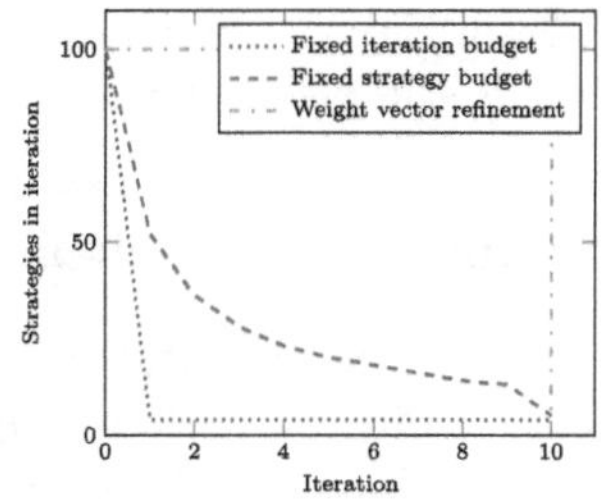

(a) Deep sea treasure

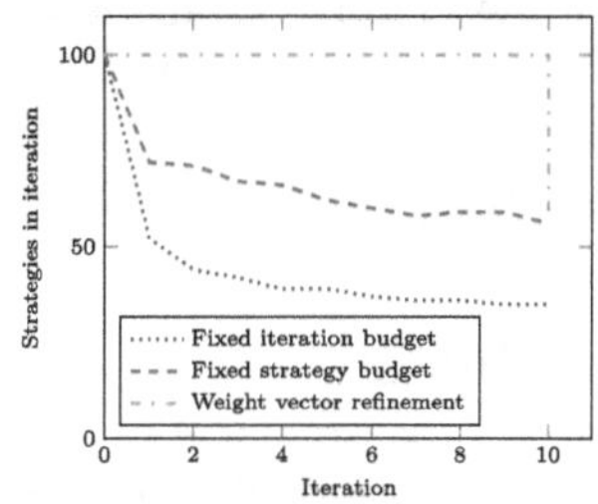

(b) Fruit tree model

Fig. 6. Strategies/iteration

Strategies Per Iteration. The number of strategies remaining in each iteration naturally differs between the algorithms as illustrated for a `deep_sea` and a `fruit_tree` instance using the simple heuristic in Fig. 6. In the `deep_sea` model, most sampled strategies are suboptimal, whereas in the `fruit_tree` model, most strategies are close to Pareto-optimal. When most strategies are determined to be suboptimal, the FIB algorithm instantly discards most strategies. FSB does the same, but uses the remaining budget to sample new strategies. WVR, in contrast, only drops strategies after the last iteration.

Strategy Seeds. As stated previously, each experiment was run with three different strategy seeds. The reason for doing so can be seen in the underapproximations for the three different seeds for one of the racetrack-puddle world models

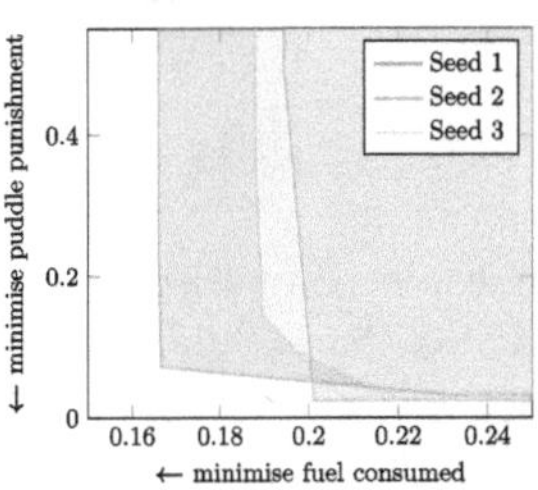

Fig. 7. Strategy seeds.

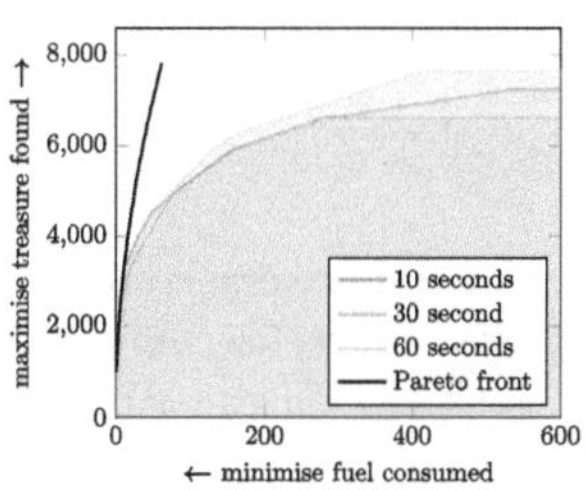

Fig. 8. Incremental scheme.

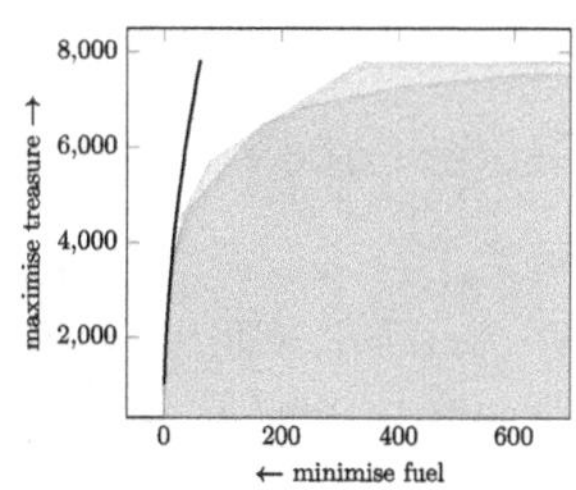

Fig. 9. `deep_sea.10-100`

in Fig. 7: Even though the model, algorithm, heuristic, and configuration are the same, there is still a significant difference in the computed underapproximations.

4.5 The Incremental Scheme

The effect of the incremental scheme's parameters can be seen without running experiments, as in Fig. 3. Nevertheless, we experimentally inspect the underapproximations it finds, which we show for the deep sea treasure model and different timeouts in Fig. 8. As expected, better strategies are sampled over time, and the approximation closes in on the true Pareto front.

Many more plots from our experiments, in particular visualisations like Fig. 9 of the Pareto fronts found via our SMC-based approaches, are in [23, App. D].

5 Conclusion

We introduced the first multi-objective SMC approach that (i) addresses the general case of d equal-priority probabilistic objectives and (ii) is constant-memory w.r.t. the model's state space size. Our algorithms approximate the achievable set and Pareto front. They provide statistical guarantees on the under-approximation, and the incremental approach also comes with a limit guarantee for the overapproximation. Our experiments indicate that, in finite time, the fixed strategy budget algorithm with the simple heuristic, a high number of initial strategies, and a small number of runs for each tends to perform best.

Our approach directly applies to models beyond MDP for which single-objective LSS is available. It also extends to infinite-state models in principle, as long as runs almost surely terminate; however, the overapproximation in the incremental approach would no longer converge because we could no longer assume every strategy to eventually be sampled given fixed-size strategy identifiers.

Data Availability Statement. The exercised versions of MODES and STORM, benchmarks, produced results, and a docker environment are available on ZENODO [24].

References

1. Abdi, H.: The Bonferonni and Šidák corrections for multiple comparisons. In: Salkind, N.J. (ed.) Encyclopedia of measurement and statistics. Sage Publications (2007). https://personal.utdallas.edu/~herve/Abdi-Bonferroni2007-pretty.pdf
2. Agha, G., Palmskog, K.: A survey of statistical model checking. ACM Trans. Model. Comput. Simul. **28**(1), 6:1–6:39 (2018). https://doi.org/10.1145/3158668
3. Andriushchenko, R., et al.: Tools at the frontiers of quantitative verification (QComp 2023 competition report). In: Beyer, D., Hartmanns, A., Kordon, F. (eds.) TOOLympics Challenge 2023. Lecture Notes in Computer Science, vol. 14550, pp. 90–146. Springer, Cham (2024). https://doi.org/10.1007/978-3-031-67695-6_4

4. Ashok, P., Daca, P., Křetínský, J., Weininger, M.: Statistical Model Checking: Black or White? In: Margaria, T., Steffen, B. (eds.) ISoLA 2020. LNCS, vol. 12476, pp. 331–349. Springer, Cham (2020). https://doi.org/10.1007/978-3-030-61362-4_19
5. Auer, P., Chiang, C.K., Ortner, R., Drugan, M.M.: Pareto front identification from stochastic bandit feedback. In: Gretton, A., Robert, C.C. (eds.) 19th International Conference on Artificial Intelligence and Statistics (AISTATS 2016). JMLR Workshop and Conference Proceedings, vol. 51, pp. 939–947. JMLR.org (2016). http://proceedings.mlr.press/v51/auer16.html
6. Awadallah, M.A., Makhadmeh, S.N., Al-Betar, M.A., Dalbah, L.M., Al-Redhaei, A., Kouka, S., Enshassi, O.S.: Multi-objective ant colony optimization: Review. Arch. Comput. Methods Eng. **32**, 995–1037 (2025). https://doi.org/10.1007/s11831-024-10178-4
7. Baier, C.: Probabilistic model checking. In: Esparza, J., Grumberg, O., Sickert, S. (eds.) Dependable Software Systems Engineering, NATO Science for Peace and Security Series - D: Information and Communication Security, vol. 45, pp. 1–23. IOS Press (2016). https://doi.org/10.3233/978-1-61499-627-9-1
8. Baier, C., de Alfaro, L., Forejt, V., Kwiatkowska, M.: Model Checking Probabilistic Systems. In: Handbook of Model Checking, pp. 963–999. Springer, Cham (2018). https://doi.org/10.1007/978-3-319-10575-8_28
9. Baier, C., Klein, J., Leuschner, L., Parker, D., Wunderlich, S.: Ensuring the Reliability of Your Model Checker: Interval Iteration for Markov Decision Processes. In: Majumdar, R., Kunčak, V. (eds.) CAV 2017. LNCS, vol. 10426, pp. 160–180. Springer, Cham (2017). https://doi.org/10.1007/978-3-319-63387-9_8
10. Barto, A.G., Bradtke, S.J., Singh, S.P.: Learning to act using real-time dynamic programming. Artif. Intell. **72**(1–2), 81–138 (1995). https://doi.org/10.1016/0004-3702(94)00011-O
11. Behrmann, G., Cougnard, A., David, A., Fleury, E., Larsen, K.G., Lime, D.: UPPAAL-Tiga: Time for Playing Games! In: Damm, W., Hermanns, H. (eds.) CAV 2007. LNCS, vol. 4590, pp. 121–125. Springer, Heidelberg (2007). https://doi.org/10.1007/978-3-540-73368-3_14
12. Bellman, R.: A Markovian decision process. J. Math. Mech. **6**(5), 679–684 (1957)
13. Bisgaard, M., Gerhardt, D., Hermanns, H., Krcál, J., Nies, G., Stenger, M.: Battery-aware scheduling in low orbit: the GomX-3 case. Formal Aspects Comput. **31**(2), 261–285 (2019). https://doi.org/10.1007/S00165-018-0458-2
14. Boyan, J.A., Moore, A.W.: Generalization in reinforcement learning: Safely approximating the value function. In: Tesauro, G., Touretzky, D.S., Leen, T.K. (eds.) Advances in Neural Information Processing Systems 7 (NIPS 1994). pp. 369–376. MIT Press (1994). https://proceedings.neurips.cc/paper_files/paper/1994/hash/ef50c335cca9f340bde656363ebd02fd-Abstract.html
15. Budde, C.E., D'Argenio, P.R., Hartmanns, A.: Digging for decision trees: A case study in strategy sampling and learning. In: Steffen, B. (ed.) 2nd International Conference on Bridging the Gap Between AI and Reality (AISoLA 2024). Lecture Notes in Computer Science, vol. 15217, pp. 354–378. Springer, Cham (2024). https://doi.org/10.1007/978-3-031-75434-0_24
16. Budde, C.E., D'Argenio, P.R., Hartmanns, A., Sedwards, S.: An efficient statistical model checker for nondeterminism and rare events. Int. J. Softw. Tools Technol. Transf. **22**(6), 759–780 (2020). https://doi.org/10.1007/S10009-020-00563-2

17. Budde, C.E., Hartmanns, A., Meggendorfer, T., Weininger, M., Wienhöft, P.: Sound statistical model checking for probabilities and expected rewards. In: Gurfinkel, A., Heule, M. (eds.) 31st International Conference on Tools and Algorithms for the Construction and Analysis of Systems (TACAS 2025). Lecture Notes in Computer Science, vol. 15696, pp. 167–190. Springer, Cham (2025). https://doi.org/10.1007/978-3-031-90643-5_9
18. Budde, C.E., Hartmanns, A., Meggendorfer, T., Weininger, M., Wienhöft, P.: Statistical model checking beyond means: Quantiles, CVaR, and the DKW inequality. In: 2nd International Joint Conference on Quantitative Evaluation of Systems and Formal Modeling and Analysis of Timed Systems (QEST+FORMATS 2025). Lecture Notes in Computer Science, vol. 16143, pp. 83–94. Springer, Cham (2025). https://doi.org/10.1007/978-3-032-05792-1_5
19. Clopper, C., Pearson, E.S.: The use of confidence or fiducial limits illustrated in the case of the binomial. Biometrika **26**(4), 404–413 (1934). https://doi.org/10.1093/biomet/26.4.404
20. D'Argenio, P.R., Fraire, J.A., Hartmanns, A.: Sampling Distributed Schedulers for Resilient Space Communication. In: Lee, R., Jha, S., Mavridou, A., Giannakopoulou, D. (eds.) NFM 2020. LNCS, vol. 12229, pp. 291–310. Springer, Cham (2020). https://doi.org/10.1007/978-3-030-55754-6_17
21. D'Argenio, P.R., Gerhold, M., Hartmanns, A., Sedwards, S.: A Hierarchy of Scheduler Classes for Stochastic Automata. In: Baier, C., Dal Lago, U. (eds.) FoSSaCS 2018. LNCS, vol. 10803, pp. 384–402. Springer, Cham (2018). https://doi.org/10.1007/978-3-319-89366-2_21
22. D'Argenio, P.R., Hartmanns, A., Legay, A., Sedwards, S.: Statistical Approximation of Optimal Schedulers for Probabilistic Timed Automata. In: Ábrahám, E., Huisman, M. (eds.) IFM 2016. LNCS, vol. 9681, pp. 99–114. Springer, Cham (2016). https://doi.org/10.1007/978-3-319-33693-0_7
23. D'Argenio, P.R., Hartmanns, A., Wienhöft, P., van Wijk, M.: Multi-objective statistical model checking using lightweight strategy sampling (extended version). CoRR **abs/2511.13460** (2025). https://doi.org/10.48550/ARXIV.2511.13460
24. D'Argenio, P.R., Hartmanns, A., Wienhöft, P., van Wijk, M.: Multi-objective statistical model checking using lightweight strategy sampling (artifact) (2026). https://doi.org/10.5281/zenodo.19660159
25. D'Argenio, P.R., Legay, A., Sedwards, S., Traonouez, L.M.: Smart sampling for lightweight verification of Markov decision processes. Int. J. Softw. Tools Technol. Transf. **17**(4), 469–484 (2015). https://doi.org/10.1007/S10009-015-0383-0
26. David, A., et al.: On Time with Minimal Expected Cost! In: Cassez, F., Raskin, J.-F. (eds.) ATVA 2014. LNCS, vol. 8837, pp. 129–145. Springer, Cham (2014). https://doi.org/10.1007/978-3-319-11936-6_10
27. David, A., Jensen, P.G., Larsen, K.G., Mikučionis, M., Taankvist, J.H.: Uppaal Stratego. In: Baier, C., Tinelli, C. (eds.) TACAS 2015. LNCS, vol. 9035, pp. 206–211. Springer, Heidelberg (2015). https://doi.org/10.1007/978-3-662-46681-0_16
28. David, A., Larsen, K.G., Legay, A., Mikučionis, M., Poulsen, D.B.: Uppaal SMC tutorial. Int. J. Softw. Tools Technol. Transfer **17**(4), 397–415 (2015). https://doi.org/10.1007/s10009-014-0361-y
29. Deb, K., Agrawal, S., Pratap, A., Meyarivan, T.: A fast and elitist multiobjective genetic algorithm: NSGA-II. IEEE Trans. Evol. Comput. **6**(2), 182–197 (2002). https://doi.org/10.1109/4235.996017

30. Eisentraut, C., Hermanns, H., Zhang, L.: On probabilistic automata in continuous time. In: 25th Annual IEEE Symposium on Logic in Computer Science (LICS 2010), pp. 342–351. IEEE Computer Society (2010). https://doi.org/10.1109/LICS.2010.41
31. Etessami, K., Kwiatkowska, M.Z., Vardi, M.Y., Yannakakis, M.: Multi-objective model checking of Markov decision processes. Log. Methods Comput. Sci. **4**(4) (2008). https://doi.org/10.2168/LMCS-4(4:8)2008
32. Felten, F., et alC.: A toolkit for reliable benchmarking and research in multi-objective reinforcement learning. In: Oh, A., Naumann, T., Globerson, A., Saenko, K., Hardt, M., Levine, S. (eds.) 37th Annual Conference on Advances in Neural Information Processing Systems (NeurIPS 2023) (2023). http://papers.nips.cc/paper_files/paper/2023/hash/4aa8891583f07ae200ba07843954caeb-Abstract-Datasets_and_Benchmarks.html
33. Forejt, V., Kwiatkowska, M., Norman, G., Parker, D.: Automated Verification Techniques for Probabilistic Systems. In: Bernardo, M., Issarny, V. (eds.) SFM 2011. LNCS, vol. 6659, pp. 53–113. Springer, Heidelberg (2011). https://doi.org/10.1007/978-3-642-21455-4_3
34. Forejt, V., Kwiatkowska, M., Norman, G., Parker, D., Qu, H.: Quantitative Multi-objective Verification for Probabilistic Systems. In: Abdulla, P.A., Leino, K.R.M. (eds.) TACAS 2011. LNCS, vol. 6605, pp. 112–127. Springer, Heidelberg (2011). https://doi.org/10.1007/978-3-642-19835-9_11
35. Forejt, V., Kwiatkowska, M., Parker, D.: Pareto Curves for Probabilistic Model Checking. In: Chakraborty, S., Mukund, M. (eds.) ATVA 2012. LNCS, pp. 317–332. Springer, Heidelberg (2012). https://doi.org/10.1007/978-3-642-33386-6_25
36. Fu, C., et al.: EPMC Gets Knowledge in Multi-agent Systems. In: Finkbeiner, B., Wies, T. (eds.) VMCAI 2022. LNCS, vol. 13182, pp. 93–107. Springer, Cham (2022). https://doi.org/10.1007/978-3-030-94583-1_5
37. Gardner, M.: Mathematical games. Sci. Am. **228**(1), 108–115 (1973). https://doi.org/10.1038/scientificamerican0173-108
38. Haddad, S., Monmege, B.: Reachability in MDPs: Refining Convergence of Value Iteration. In: Ouaknine, J., Potapov, I., Worrell, J. (eds.) RP 2014. LNCS, vol. 8762, pp. 125–137. Springer, Cham (2014). https://doi.org/10.1007/978-3-319-11439-2_10
39. Hahn, E.M., Perez, M., Schewe, S., Somenzi, F., Trivedi, A., Wojtczak, D.: Model-Free Reinforcement Learning for Lexicographic Omega-Regular Objectives. In: Huisman, M., Păsăreanu, C., Zhan, N. (eds.) FM 2021. LNCS, vol. 13047, pp. 142–159. Springer, Cham (2021). https://doi.org/10.1007/978-3-030-90870-6_8
40. Hartmanns, A., Hermanns, H.: The Modest Toolset: An Integrated Environment for Quantitative Modelling and Verification. In: Ábrahám, E., Havelund, K. (eds.) TACAS 2014. LNCS, vol. 8413, pp. 593–598. Springer, Heidelberg (2014). https://doi.org/10.1007/978-3-642-54862-8_51
41. Hartmanns, A., Hermanns, H.: Explicit Model Checking of Very Large MDP Using Partitioning and Secondary Storage. In: Finkbeiner, B., Pu, G., Zhang, L. (eds.) ATVA 2015. LNCS, vol. 9364, pp. 131–147. Springer, Cham (2015). https://doi.org/10.1007/978-3-319-24953-7_10
42. Hartmanns, A., Junges, S., Quatmann, T., Weininger, M.: A practitioner's guide to MDP model checking algorithms. In: Sankaranarayanan, S., Sharygina, N. (eds.) 29th International Conference on Tools and Algorithms for the Construction and Analysis of Systems (TACAS 2023). Lecture Notes in Computer Science, vol. 13993, pp. 469–488. Springer, Cham (2023). https://doi.org/10.1007/978-3-031-30823-9_24

43. Hartmanns, A., Quatmann, T., van Wijk, M.: Tools and algorithms for sound multi-objective probabilistic model checking. In: Sampaio, A., Stoelinga, M. (eds.) 27th International Symposium on Formal Methods (FM 2026). Lecture Notes in Computer Science, vol. 16556, pp. 189–210. Springer (2026). https://doi.org/10.1007/978-3-032-26204-2_10
44. Hartmanns, A., Sedwards, S., D'Argenio, P.R.: Efficient simulation-based verification of probabilistic timed automata. In: 2017 Winter Simulation Conference (WSC 2017), pp. 1419–1430. IEEE (2017). https://doi.org/10.1109/WSC.2017.8247885
45. Hasrat, I.R., Jensen, P.G., Larsen, K.G., Srba, J.: A toolchain for domestic heat-pump control using Uppaal Stratego. Sci. Comput. Program. **230**, 102987 (2023). https://doi.org/10.1016/J.SCICO.2023.102987
46. Hayes, C.F., et al.: A practical guide to multi-objective reinforcement learning and planning. Auton. Agent. Multi-Agent Syst. **36**(1), 1–59 (2022). https://doi.org/10.1007/s10458-022-09552-y
47. Hensel, C., Junges, S., Katoen, J.P., Quatmann, T., Volk, M.: The probabilistic model checker Storm. Int. J. Softw. Tools Technol. Transf. **24**(4), 589–610 (2022). https://doi.org/10.1007/S10009-021-00633-Z
48. Howard, R.A.: Dynamic Programming and Markov Processes, MIT Press (1960)
49. Katoen, J.P.: The probabilistic model checking landscape. In: Grohe, M., Koskinen, E., Shankar, N. (eds.) 31st Annual ACM/IEEE Symposium on Logic in Computer Science (LICS 2016), pp. 31–45. ACM (2016). https://doi.org/10.1145/2933575.2934574
50. Krähmann, D., Schubert, J., Baier, C., Dubslaff, C.: Ratio and Weight Quantiles. In: Italiano, G.F., Pighizzini, G., Sannella, D.T. (eds.) MFCS 2015. LNCS, vol. 9234, pp. 344–356. Springer, Heidelberg (2015). https://doi.org/10.1007/978-3-662-48057-1_27
51. Kretínský, J., Meggendorfer, T.: Conditional value-at-risk for reachability and mean payoff in Markov decision processes. In: Dawar, A., Grädel, E. (eds.) 33rd Annual ACM/IEEE Symposium on Logic in Computer Science (LICS 2018), pp. 609–618. ACM (2018). https://doi.org/10.1145/3209108.3209176
52. Kwiatkowska, M., Norman, G., Parker, D.: PRISM 4.0: Verification of Probabilistic Real-Time Systems. In: Gopalakrishnan, G., Qadeer, S. (eds.) CAV 2011. LNCS, vol. 6806, pp. 585–591. Springer, Heidelberg (2011). https://doi.org/10.1007/978-3-642-22110-1_47
53. Kwiatkowska, M.Z., Norman, G., Segala, R., Sproston, J.: Automatic verification of real-time systems with discrete probability distributions. Theor. Comput. Sci. **282**(1), 101–150 (2002). https://doi.org/10.1016/S0304-3975(01)00046-9
54. Larsen, K.G., Legay, A.: Statistical Model Checking: Past, Present, and Future. In: Margaria, T., Steffen, B. (eds.) ISoLA 2016. LNCS, vol. 9952, pp. 3–15. Springer, Cham (2016). https://doi.org/10.1007/978-3-319-47166-2_1
55. Legay, A., Lukina, A., Traonouez, L.M., Yang, J., Smolka, S.A., Grosu, R.: Statistical Model Checking. In: Steffen, B., Woeginger, G. (eds.) Computing and Software Science. LNCS, vol. 10000, pp. 478–504. Springer, Cham (2019). https://doi.org/10.1007/978-3-319-91908-9_23
56. Legay, A., Sedwards, S., Traonouez, L.-M.: Scalable Verification of Markov Decision Processes. In: Canal, C., Idani, A. (eds.) SEFM 2014. LNCS, vol. 8938, pp. 350–362. Springer, Cham (2015). https://doi.org/10.1007/978-3-319-15201-1_23
57. Li, L., Li, G., Cai, G.: A local Pareto front estimation framework for multi-objective optimization. In: Shen, W., et al. (eds.) 28th International Conference on Computer Supported Cooperative Work in Design (CSCWD 2025), pp. 2103–2108. IEEE (2025). https://doi.org/10.1109/CSCWD64889.2025.11033349

58. Li, Z., Tang, J., Zhao, H., Chen, C., Xie, S.: Dictionary learning-structured reinforcement learning with adaptive-sparsity regularizer. IEEE Trans. Aerosp. Electron. Syst. **60**(2), 1753–1769 (2024). https://doi.org/10.1109/TAES.2023.3342794
59. Miller, R.G.: Simultaneous Statistical Inference. Springer, Cham (1966). https://doi.org/10.1007/978-1-4613-8122-8
60. Moffaert, K.V., Drugan, M.M., Nowé, A.: Scalarized multi-objective reinforcement learning: Novel design techniques. In: 2013 IEEE Symposium on Adaptive Dynamic Programming and Reinforcement Learning (ADPRL 2013), pp. 191–199. IEEE (2013). https://doi.org/10.1109/ADPRL.2013.6615007
61. Nguyen, A.T., Reiter, S., Rigo, P.: A review on simulation-based optimization methods applied to building performance analysis. Appl. Energy **113**, 1043–1058 (2014). https://doi.org/10.1016/j.apenergy.2013.08.061
62. Okumura, N., Takagi, T., Ohta, Y., Sato, H.: Pareto front upconvert on multi-objective building facility control optimization. In: Silva, S., Paquete, L. (eds.) Genetic and Evolutionary Computation Conference (GECCO 2023), pp. 1963–1971. ACM (2023). https://doi.org/10.1145/3583133.3596339
63. Parsopoulos, K.E., Vrahatis, M.N.: Particle swarm optimization method in multi-objective problems. In: Lamont, G.B., Haddad, H., Papadopoulos, G.A., Panda, B. (eds.) 2002 ACM Symposium on Applied Computing (SAC 2002), pp. 603–607. ACM (2002). https://doi.org/10.1145/508791.508907
64. Pnueli, A.: The temporal logic of programs. In: 18th Annual Symposium on Foundations of Computer Science (FOCS 1977), pp. 46–57. IEEE Computer Society (1977). https://doi.org/10.1109/SFCS.1977.32
65. Puterman, M.L.: Markov Decision Processes: Discrete Stochastic Dynamic Programming. Wiley Series in Probability and Statistics, Wiley (1994). https://doi.org/10.1002/9780470316887
66. Quatmann, T.: Verification of multi-objective Markov models Ph.D. thesis, RWTH Aachen University (2023). https://doi.org/10.18154/RWTH-2023-09669
67. Quiroz, E.A.P., Apolinário, H.C.F., Villacorta, K.D.V., Oliveira, P.R.: A linear scalarization proximal point method for quasiconvex multiobjective minimization. J. Optim. Theory Appl. **183**(3), 1028–1052 (2019). https://doi.org/10.1007/S10957-019-01582-Z
68. Randour, M., Raskin, J.F., Sankur, O.: Percentile queries in multi-dimensional Markov decision processes. Formal Methods Syst. Des. **50**(2–3), 207–248 (2017). https://doi.org/10.1007/S10703-016-0262-7
69. Sarker, R.A., Liang, K.H., Newton, C.S.: A new multiobjective evolutionary algorithm. Eur. J. Oper. Res. **140**(1), 12–23 (2002). https://doi.org/10.1016/S0377-2217(01)00190-4
70. Šidák, Z.: Rectangular confidence regions for the means of multivariate normal distributions. J. Am. Stat. Assoc. **62**(318), 626–633 (1967). https://doi.org/10.1080/01621459.1967.10482935
71. Singh, C.: Optimality conditions in multiobjective differentiable programming. J. Optim. Theory Appl. **53**(1), 115–123 (1987). https://doi.org/10.1007/BF00938820
72. Song, F., Xing, H., Wang, X., Luo, S., Dai, P., Li, K.: Offloading dependent tasks in multi-access edge computing: A multi-objective reinforcement learning approach. Future Gener. Comput. Syst. **128**, 333–348 (2022). https://doi.org/10.1016/J.FUTURE.2021.10.013
73. Suman, B., Kumar, P.: A survey of simulated annealing as a tool for single and multiobjective optimization. J. Oper. Res. Soc. **57**(10), 1143–1160 (2006). https://doi.org/10.1057/PALGRAVE.JORS.2602068

74. Sutton, R.S.: Generalization in reinforcement learning: Successful examples using sparse coarse coding. In: Touretzky, D.S., Mozer, M., Hasselmo, M.E. (eds.) Advances in Neural Information Processing Systems 8 (NIPS 1995), pp. 1038–1044. MIT Press (1995). http://papers.nips.cc/paper/1109-generalization-in-reinforcement-learning-successful-examples-using-sparse-coarse-coding
75. Sutton, R.S., Barto, A.G.: Reinforcement learning - an introduction, 2nd edn. MIT Press (2018)
76. Ummels, M., Baier, C.: Computing Quantiles in Markov Reward Models. In: Pfenning, F. (ed.) FoSSaCS 2013. LNCS, vol. 7794, pp. 353–368. Springer, Heidelberg (2013). https://doi.org/10.1007/978-3-642-37075-5_23
77. Vamplew, P., Dazeley, R., Berry, A., Issabekov, R., Dekker, E.: Empirical evaluation methods for multiobjective reinforcement learning algorithms. Mach. Learn. **84**(1–2), 51–80 (2011). https://doi.org/10.1007/S10994-010-5232-5
78. Vamplew, P., Foale, C., Dazeley, R., Bignold, A.: Potential-based multiobjective reinforcement learning approaches to low-impact agents for AI safety. Eng. Appl. Artif. Intell. **100**, 104186 (2021). https://doi.org/10.1016/J.ENGAPPAI.2021.104186
79. Vargas, D.V., Murata, J., Takano, H., Delbem, A.C.B.: General subpopulation framework and taming the conflict inside populations. Evol. Comput. **23**(1), 1–36 (2015). https://doi.org/10.1162/EVCO_A_00118
80. Wan, C., Chen, X., Liu, D.: A multi-objective-driven placement technique for digital microfluidic biochips. J. Circuits Syst. Comput. **28**(5), 1950076:1–1950076:15 (2019). https://doi.org/10.1142/S0218126619500762
81. Wang, C., Hou, Y., Qiu, F., Lei, S., Liu, K.: Resilience enhancement with sequentially proactive operation strategies. IEEE Trans. Power Syst. **32**(4), 2847–2857 (2017). https://doi.org/10.1109/TPWRS.2016.2622858
82. Wray, K.H., Zilberstein, S., Mouaddib, A.I.: Multi-objective MDPs with conditional lexicographic reward preferences. In: Bonet, B., Koenig, S. (eds.) 29th AAAI Conference on Artificial Intelligence (AAAI 2015), pp. 3418–3424. AAAI Press (2015). https://doi.org/10.1609/AAAI.V29I1.9647
83. Yamaguchi, T., Nagahama, S., Ichikawa, Y., Takadama, K.: Model-Based Multi-objective Reinforcement Learning with Unknown Weights. In: Yamamoto, S., Mori, H. (eds.) HCII 2019. LNCS, vol. 11570, pp. 311–321. Springer, Cham (2019). https://doi.org/10.1007/978-3-030-22649-7_25
84. Yang, R., Sun, X., Narasimhan, K.: A generalized algorithm for multi-objective reinforcement learning and policy adaptation. In: Wallach, H.M., Larochelle, H., Beygelzimer, A., d'Alché Buc, F., Fox, E.B., Garnett, R. (eds.) 33rd Annual Conference on Advances in Neural Information Processing Systems (NeurIPS 2019), pp. 14610–14621 (2019). https://proceedings.neurips.cc/paper/2019/hash/4a46fbfca3f1465a27b210f4bdfe6ab3-Abstract.html
85. Younes, H.L.S., Simmons, R.G.: Probabilistic Verification of Discrete Event Systems Using Acceptance Sampling. In: Brinksma, E., Larsen, K.G. (eds.) CAV 2002. LNCS, vol. 2404, pp. 223–235. Springer, Heidelberg (2002). https://doi.org/10.1007/3-540-45657-0_17
86. Zhang, X., Tian, Y., Jin, Y.: A knee point-driven evolutionary algorithm for many-objective optimization. IEEE Trans. Evol. Comput. **19**(6), 761–776 (2015). https://doi.org/10.1109/TEVC.2014.2378512
87. Zhu, Y., Liang, S., Xue, G., Yang, R., Wu, X.: An efficient multi-objective optimization approach for sensor management via multi-Bernoulli filtering. EURASIP J. Adv. Signal Process. **2022**(1), 62 (2022). https://doi.org/10.1186/S13634-022-00881-4

88. Zitzler, E., Knowles, J., Thiele, L.: Quality Assessment of Pareto Set Approximations. In: Branke, J., Deb, K., Miettinen, K., Słowiński, R. (eds.) Multiobjective Optimization. LNCS, vol. 5252, pp. 373–404. Springer, Heidelberg (2008). https://doi.org/10.1007/978-3-540-88908-3_14
89. Zitzler, E., Thiele, L.: Multiobjective optimization using evolutionary algorithms – a comparative case study. In: Eiben, A.E., Bäck, T., Schoenauer, M., Schwefel, H.P. (eds.) 5th International Conference on Parallel Problem Solving from Nature (PPSN 1998). Lecture Notes in Computer Science, vol. 1498, pp. 292–304. Springer, Cham (1998).https://doi.org/10.1007/BFB0056872
90. Zitzler, E., Thiele, L., Laumanns, M., Fonseca, C.M., da Fonseca, V.G.: Performance assessment of multiobjective optimizers: an analysis and review. IEEE Trans. Evol. Comput. **7**(2), 117–132 (2003). https://doi.org/10.1109/TEVC.2003.810758

Adaptive Importance Sampling for Static Fault Trees with Two-State Markovian Components

Leonardo Paroli(✉), Laura Carnevali, and Enrico Vicario

Department of Information Engineering, University of Florence, Florence, Italy
{leonardo.paroli,laura.carnevali,enrico.vicario}@unifi.it

Abstract. In the simulation of complex systems, quantities of interest are often limited to specific events with extremely low probability. This becomes particularly relevant in various processes of reliability engineering. Rare event simulation (RES) addresses this challenge by focusing the evaluation on a subset of the space of outcomes, reducing estimator variance to achieve an efficient trade-off between estimator accuracy and sample size. We propose an adaptive importance sampling (AIS) technique to estimate top-event failures in static fault trees (SFTs) with self-healing leaf components that alternate two states, operational and failed, with exponential inter-times. To this end, we leverage cross-entropy (CE) adaptation to iteratively refine a parametric importance function by tilting failure and repair rates of individual components in a balanced way. In particular, at each time instant, we decompose a system-level stochastic optimization problem into a set of component-level subproblems, jointly solved by recursive updates. Experimental results on a set of randomly generated models with increasing event rarity show better estimator accuracy and variance reduction with respect to alternative baselines.

Keywords: static fault trees · two-state continuous time Markov chains · adaptive importance sampling · cross-entropy adaptation

1 Introduction

In the development of highly reliable systems, quantitative evaluation is often aimed to estimate threats to dependability [1] that, by design, occur as rare events with very low probability [2] [3]. The issue is further exacerbated in the case of threats occurring over time in systems with autonomous components [4]. A common approach to face this challenge is the analysis of models that specify the probability space of the system through some stochastic formalism that captures the uncertainty arising from the intertwined effects of concurrency and timing [5–11]. To this end, fault trees with stochastic parameters provide a simple but well grounded model [12–15], able to bridge theoretical foundation and industrial practice [16] and open to model-to-model transformation back and forth from several different domain-specific or general-purpose formalisms [17,18].

M. Chen and M. Tschaikowski (Eds.): QEST+FORMATS 2026, LNCS 16913, pp. 196–215, 2027.
https://doi.org/10.1007/978-3-032-35298-9_11

Resulting models often break limits for practical feasibility of numerical analysis [19,20], due to the size or class of the underlying stochastic process [21]. In this case, stochastic Monte Carlo simulation becomes the first choice analysis approach. However, highly reliable systems exhibit statistically elusive failures [4], making crude Monte Carlo (CMC) simulation prohibitively expensive, as the required number of trials scales inversely with the failure event probability [3,22].

Rare event simulation (RES) [2,23] mitigates this issue through two main methods that make the target event more frequent. On the one hand, in importance splitting (ISPLIT) [24], favorable realizations are replicated while pruning less promising ones, estimating the target probability as the product of the conditional probabilities of the surviving realizations. In particular, ISPLIT methods have been often applied to dynamic fault trees (DFTs) [25–29], which naturally capture order-dependent failures rather than Boolean combinations of failures.

Conversely, in importance sampling (IS) [3,30], estimates are produced with affordable effort by drawing samples from an importance distribution and weighting them to compensate for the deviation from the target distribution. Forcing and failure biasing [31–33] estimates the unreliability of highly dependable systems by including the distance to the target states into the importance function. The path zero variance approximation (Path-ZVA) [34] uses path-based measures to quantify the distance to the target, estimating rare-event probabilities in large discrete time Markov chains and in DFTs [35]. The IS scheme in [36] evaluates the failure probability of static fault trees (SFTs) without repeated events, assuming non-repairable components with time-invariant failure probabilities.

In IS methods, identification of the optimal distribution remains challenging, especially for high-dimensional systems. Adaptive importance sampling (AIS) [37] mitigates the issue by iteratively refining a parametric importance distribution, learning its parameters from previously sampled values. The approach presented in [38] for Markovian queueing networks efficiently estimates the probability of a full buffer within a given time interval, splitting simulation into regenerative cycles if the time horizon is long. A notable class of AIS methods leverages cross-entropy (CE) adaptation [39] to derive the importance distribution by minimizing its Kullback–Leibler divergence from the optimal distribution. CE-based approaches have been successfully applied in a variety of domains, including reliability evaluation of series and parallel systems [40] and power systems [41], leveraging Gaussian distributions to define proposal distributions. The approach of [42] has explored extensions based on exponential tilting and advanced sampling strategies for rare realizations in discrete-time Markov processes. In [43], a CE-based AIS algorithm is presented to evaluate reliability properties of hybrid systems specified as continuous-time piecewise deterministic Markov processes, exploiting the system decomposition into minimal cut sets and path sets.

To the best of our knowledge, AIS schemes specifically tailored to SFTs with repeated events and components characterized by continuous-time Markovian behavior remain underexplored. In particular, existing methods do not exploit the SFT structure, or they rely on global parameterizations that struggle to scale as system complexity (e.g., number of components) and event rarity increase.

In this paper, we introduce an AIS approach for estimation of rare-event probabilities in SFTs with repeated events, where individual components are modeled by two-state continuous-time Markov chains (CTMCs). By using exponential-tilting CE adaptation, we iteratively refine failure and repair rates of components to optimize a parametric importance function, decomposing a system-wide stochastic optimization problem into manageable component-level subproblems, jointly solved through recursive updates. As our approach exploits the SFT structure, we improve unconstrained CE in terms of stability, by defining closed-form coordinate-wise maximizers for failure and repair rate tilting parameters, and by providing a provably convergent smoothed update tailored to the specific decomposition. Experiments on randomly generated models show that our method outperforms alternative CE and CMC baselines in both estimator accuracy and variance reduction, even as the target event becomes increasingly rare.

In the following, we recall IS (Sect. 2) and formulate our problem (Sect. 3). Then, we present our method (Sect. 4) and discuss experimental results (Sect. 5), threats to validity (Sect. 6), and final conclusions (Sect. 7).

2 Importance Sampling

Let $(\Omega, \mathcal{F}, P)$ be a probability space, where Ω is a sample space, $\mathcal{F}$ is a σ-algebra of events, and P is a probability measure over $\mathcal{F}$. We consider a stochastic process $\{X(t), t \in \mathcal{T}\}$ characterizing the system evolution over the time domain $\mathcal{T}$. Let $x \in \Omega$ denote a realization (trajectory) of $X(t)$ and $f_0(x)$ denote the PDF of realizations. The probability of a set A of realizations is $P_{f_0}(A) = \int \mathbb{I}_A(x)\, f_0(x)\, dx$ where $\mathbb{I}_A(x)$ is the indicator function. The CMC estimator of $P_{f_0}(A)$ is

$$\widehat{P}_{\text{MC}}(A) = \frac{1}{M} \sum_{m=1}^{M} \mathbb{I}_A(x^{(m)}), \qquad x^{(m)} \sim f_0(x) \tag{1}$$

where M denotes the number of sampled realizations $x^{(1)}, \ldots, x^{(M)}$. This estimator exhibits high variance and inefficiency for a rare event A, as it requires a huge sample size to observe a sufficient number of realizations in A.

IS [3] reduces variance by drawing realizations from a proposal (importance) PDF f^* that increases the likelihood of A, preserving unbiasedness by weighting each realization x by the likelihood ratio $w(x) = f_0(x)/f^*(x)$, assuming that $f^*(x) > 0$ for each realization x such that $f_0(x) > 0$. The resulting estimator is

$$\widehat{P}_{\text{IS}}(A) = \frac{1}{M} \sum_{m=1}^{M} w(x^{(m)})\, \mathbb{I}_A(x^{(m)}), \qquad x^{(m)} \sim f^*(x) \tag{2}$$

The optimal importance yielding a *zero variance* estimator is:

$$f^*(x) = \frac{\mathbb{I}_A(x)\, f_0(x)}{P_{f_0}(A)} \tag{3}$$

Since $P_{f_0}(A)$ is unknown, a proposal PDF is selected within a parametric family $\{f(x,\theta)\}$ by determining θ so that $f(x,\theta) \approx f^*(x)$. AIS iteratively updates θ using previously generated realizations, progressively steering $f(x,\theta)$ toward f^*.

The CE method builds $f(x,\theta)$ solving $\theta^\star = \arg\min_\theta \mathrm{KL}(f^* \parallel f(x;\theta))$, where $\mathrm{KL}(f^* \parallel f(x;\theta)) = \int f^*(x)\log(f^*(x)/f(x,\theta))\,dx$ is the Kullback-Leibler (KL) divergence. Since $\mathrm{KL}(f^* \parallel f(x;\theta)) = \int f^*(x)\log(f^*(x))\,dx - \int f^*(x)\log(f(x,\theta))\,dx$, where the minuend is independent of θ, minimizing $\mathrm{KL}(f^* \parallel f(x,\theta))$ is equivalent to solving $\theta^\star = \arg\max_\theta \int f^*(x)\log(f(x,\theta))\,dx$. Expressing the integral as an expectation given f^* yields $\theta^\star = \arg\max_\theta J(\theta)$ where $J(\theta) = \mathbb{E}_{f^*}[\log(f(x,\theta))]$. Since f^* is expressed as in Eq. (3) and $P_{f_0}(A)$ is independent of θ, we obtain $J(\theta) \propto \mathbb{E}_{f(x,\theta)}[\mathbb{I}_A(x)\,(f_0(x)/f(x,\theta))\log(f(x,\theta))]$. We approximate $J(\theta)$ by M simulated realizations $x^{(m)} \sim f(x,\theta)$, solving the stochastic optimization problem:

$$\theta^\star = \arg\max_\theta \sum_{m=1}^{M} \mathbb{I}_A(x^{(m)})\frac{f_0(x^{(m)})}{f(x^{(m)},\theta)}\log f(x^{(m)},\theta) \tag{4}$$

3 Problem Formulation

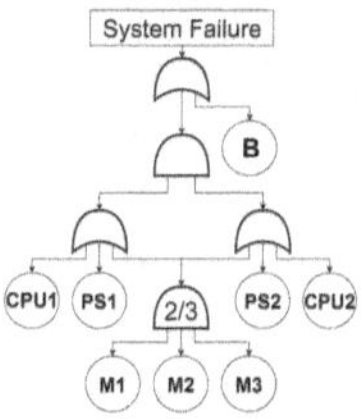

Fig. 1. SFT for a partially redundant computer system (from [12]).

Table 1. MTTF and MTTR values for the components of the SFT in Fig. 1.

Component	MTTF (yrs)	MTTR (yrs)
PS1	2.28×10^1	6.84×10^{-4}
PS2	2.28×10^1	6.84×10^{-4}
CPU1	9.43×10^1	9.13×10^{-4}
CPU2	9.43×10^1	9.13×10^{-4}
M1	4.57×10^1	9.13×10^{-4}
M2	4.57×10^1	9.13×10^{-4}
M3	4.57×10^1	9.13×10^{-4}
B	9.26×10^1	1.38×10^{-3}

The system consists of components with *independent* failure and repair cycles characterized by mean time to failure (MTTF) and mean time to repair (MTTR), fit by exponential distributions, and it fails only if specific *combinations* of components concurrently fail. While structurally straightforward, this model closely reflects the usual engineering practice [44,45], where component MTTFs and MTTRs are often the only available metrics, and the SFT formalism assumes component independence. Formally, the system is specified by a tuple $\langle C, Y, F\rangle$:

- $C = \{c_1, \ldots, c_N\}$ is the set of components.
- Y specifies the behavior of each component $c_i \in C$ as a two-state CTMC $\mathcal{C}_i$ modeling the time-to-failure and the time-to-repair by exponential random variables whose rates λ_i (failure rate) and μ_i (repair rate), respectively, fit

the MTTF and the MTTR of component c_i, respectively, i.e., $C_i = \langle S, Q_i, s_i^0 \rangle$, where $S = \{\text{OK}, \text{KO}\}$ is the set of states, $Q_i = [-\lambda_i, \lambda_i; \mu_i, -\mu_i]$ is the infinitesimal generator matrix, and $s_i^0 \in S$ is the initial state.

- F is the considered system failure mode, specified by an SFT with *repeated events*, i.e., $F = \langle B, G, T, I \rangle$ where: B is the set of basic events (one per component); G is the set of gates ($B \cap G = \emptyset$); $T : G \to \{\text{AND}, \text{OR}, \text{VOT}(k/N)\}$ defines the type of each gate; $I : G \to \mathcal{P}(B \cup G)$ defines the inputs of each gate ($I(g) \neq \emptyset$ and $|I(g)| = N$ if $T(g) = \text{VOT}(k/N)\ \forall\, g \in G$).
  The semantics $\pi : (B \cup G) \times \mathcal{P}(B) \to \{0, 1\}$ of F indicates if node $e \in \{B \cup G\}$ fails given the set $S \in \mathcal{P}(B)$ of failed basic events. In particular, an AND, OR, VOT(k/N) gate fails if all, one, k over N of its inputs fail, respectively.

As running example, we consider a partially redundant computer system from [12], with failure logic modeled by the SFT in Fig. 1, and with component MTTFs and MTTRs in Table 1, aligned in magnitude with those in [46–48]. The system consists of eight components: a bus (B), two CPUs (CPU1, CPU2), three memory units (M1, M2, M3), and two power supplies (PS1, PS2). In particular, the memory subsystem is shared between the two redundant computing units: such repeated events introduce non-trivial dependencies at the fault-tree level, preventing direct application of Boolean formulas over the disjoint states of the components. A system failure is caused by a failure of the bus or the computing subsystem which, in turn, fails if its two redundant units fail. The failure of each such unit is caused by a failure of its CPU, or its power supply, or the shared memory subsystem which, in turn, fails if two of its three memory units fail.

Without loss of generality, we evaluate the *system unreliability* $P_A(t)$, defined as the probability that a system failure event A occurs within the interval $[0, t]$. Note that our method can be applied as well to evaluate any transient probability.

4 Solution Method

This section presents the novel contribution of the paper. Specifically, we solve the stochastic optimization problem of Eq. (4) by deriving a component-wise decomposition of the likelihood function under exponential tilting (Sect. 4.1), enabling independent parameter updates via coordinate maximization (Sect. 4.2).

4.1 Cross–Entropy Update Via Exponential Tilting

To make the system failures more frequent during simulation while preserving unbiasedness, our IS approach increases the failure rates and decreases the repair rates of components through multiplicative tilting factors, exploiting component independence to derive separate update equations. Specifically, exponential tilting [2,23,49] *jointly* scales the failure rate and repair rates of all the components, biasing the stochastic process $X(t)$ toward behaviors that increase the probability of sampling realizations in the rare event A. Note that tilting increases the likelihood of observing rare events, but may also increase the estimator variance due to more extreme likelihood ratios: therefore, the key challenge is to balance biasing across the components, which motivates our adaptive scheme.

In the following, for our system model, we scale the original rates to $\lambda_i' = \alpha_i \lambda_i$ and $\mu_i' = \beta_i \mu_i$ with $\alpha_i, \beta_i > 0 \forall\ i \in \{1, \ldots, N\}$ and then we decompose the system-level optimization problem of Eq. (4) over $\theta = (\bar{\alpha}, \bar{\beta})$ where $\bar{\alpha} = [\alpha_1, \ldots, \alpha_N]$ and $\bar{\beta} = [\beta_1, \ldots, \beta_N]$ into a set of (jointly solved) component-level optimization problems over (α_i, β_i). We then derive the closed-form expressions of the transition rates λ_i' and μ_i' of each component c_i.

We consider the system behavior in a time interval $[0, T]$. For each component c_i and each realization x_i, let $N_{\text{fail},i}(x_i)$ and $N_{\text{repair},i}(x_i)$ denote the number of OK→KO and KO→OK transitions, respectively, and let $S_{\text{OK}}(x_i)$ and $S_{\text{KO}}(x_i)$ denote the total time spent by c_i in the OK and KO states, respectively, and let $q_u = \sum_{v \neq u} q_{uv}$ be the total exit rate from state u. For a realization x_i visiting the sequence of states $\{u_k\}_{k=0,\ldots,K_i}$ over the interval $[0, T]$, where K_i is the number of transitions, the path probability under fixed rates is

$$\mathbb{P}(x_i) = \prod_{k=0}^{K_i} q_{u_k u_{k+1}} e^{(-\int_0^T q_{X_i(t)}\, dt)}$$

When rates are tilted, the exit rates become $q_{\text{OK}} = \lambda_i'$ and $q_{\text{KO}} = \mu_i'$, and the integrated hazard term decomposes as $\int_0^T q_{X_i(t)}\, dt = \lambda_i' S_{\text{OK}}(x_i) + \mu_i' S_{\text{KO}}(x_i)$. Given that each OK → KO and KO → OK transition contributes a factor λ_i' and μ_i' to the product term, respectively, the path probability can be written as

$$\mathbb{P}(x_i) = (\lambda_i')^{N_{\text{fail},i}(x_i)} (\mu_i')^{N_{\text{repair},i}(x_i)} e^{(-\lambda_i' S_{\text{OK}}(x_i) - \mu_i' S_{\text{KO}}(x_i))}$$

Thus, the likelihood function for $\lambda_i' = \alpha_i \lambda_i$ and $\mu_i' = \beta_i \mu_i$ can be defined as:

$$f_i(x_i; \alpha_i, \beta_i) = (\alpha_i \lambda_i)^{N_{\text{fail},i}(x_i)} (\beta_i \mu_i)^{N_{\text{repair},i}(x_i)} e^{(-(\alpha_i \lambda_i) S_{\text{OK}}(x_i) - (\beta_i \mu_i) S_{\text{KO}}(x_i))}$$

Finally, applying the logarithm yields the log-likelihood:

$$\begin{aligned} \ell_i(x_i, \alpha_i, \beta_i) &:= \log f_i(x_i; \alpha_i, \beta_i) \\ &= N_{\text{fail},i}(x_i) \log(\alpha_i \lambda_i) - \alpha_i \lambda_i S_{\text{OK}}(x_i) \\ &+ N_{\text{repair},i}(x_i) \log(\beta_i \mu_i) - \beta_i \mu_i S_{\text{KO}}(x_i) \end{aligned} \tag{5}$$

Given that components are independent of each other, the system-level likelihood function $f(x, \theta) = f(x, \bar{\alpha}, \bar{\beta})$ decomposes as the product of the component-level likelihood functions $f(x_i, \theta_i) = f(x_i, \alpha_i, \beta_i) \forall i \in \{1, \ldots, N\}$ or, equivalently, the system-level log likelihood $\log f(x, \theta)$ decomposes as the sum of the component-level log-likelihood functions $\ell_i(x_i, \alpha_i, \beta_i)$. According to this, we decompose the system-level optimization problem in Eq. (4) into a set of component-level optimization problems, each defined as:

$$(\alpha_i^*, \beta_i^*) = \arg\max_{(\alpha_i, \beta_i)} \widetilde{J}_i(\alpha_i, \beta_i) = \arg\max_{(\alpha_i, \beta_i)} \sum_{m=1}^{M} w_m\, \mathbb{I}_A(x^{(m)})\, \ell_i(x_i^{(m)}, \alpha_i, \beta_i) \tag{6}$$

where $w_m := f_0(x^{(m)})/f(x^{(m)}, \bar{\alpha}, \bar{\beta})$, $x^{(m)} \sim f(x, \bar{\alpha}, \bar{\beta})$ is a system-level realization, and $x_i^{(m)}$ is the corresponding component-level realization for c_i.

The derivative of $\widetilde{J}_i(\alpha_i, \beta_i)$ w.r.t. α_i is:

$$\begin{aligned}\frac{\partial \widetilde{J}_i}{\partial \alpha_i} &= \sum_{m=1}^{M} w_m \, \mathbb{I}_A(x^{(m)}) \frac{\partial}{\partial \alpha_i} (N_{\text{fail},i}^{(m)} \log(\alpha_i \lambda_i) - \alpha_i \lambda_i S_{\text{OK},i}^{(m)}) \\ &= \sum_{m=1}^{M} w_m \, \mathbb{I}_A(x^{(m)}) \left(\frac{N_{\text{fail},i}^{(m)}}{\alpha_i} - \lambda_i S_{\text{OK},i}^{(m)} \right) \end{aligned} \tag{7}$$

where $N_{\text{fail},i}^{(m)} := N_{\text{fail},i}(x^{(m)})$ and $S_{\text{OK},i}^{(m)} := S_{\text{OK}}(x^{(m)})$. Solving $\partial \widetilde{J}_i / \partial \alpha_i = 0$ w.r.t. α_i and, similarly, $\partial \widetilde{J}_i / \partial \beta_i = 0$ w.r.t. β_i yields the conditional maximizers α_i^* and β_i^*, respectively, as proved by Lemma 1:

$$\boxed{\alpha_i^* = \frac{\sum_{m=1}^{M} w_m \, \mathbb{I}_A(x^{(m)}) \, N_{\text{fail},i}^{(m)}}{\lambda_i \sum_{m=1}^{M} w_m \, \mathbb{I}_A(x^{(m)}) \, S_{\text{OK},i}^{(m)}}} \tag{8}$$

$$\boxed{\beta_i^* = \frac{\sum_{m=1}^{M} w_m \, \mathbb{I}_A(x^{(m)}) \, N_{\text{repair},i}^{(m)}}{\mu_i \sum_{m=1}^{M} w_m \, \mathbb{I}_A(x^{(m)}) \, S_{\text{KO},i}^{(m)}}} \tag{9}$$

For independent two-state CTMCs, the expectation of the random variables $S_{\text{KO},i}$, $S_{\text{KO},i}$, $N_{\text{fail},i}$ and $N_{\text{repair},i}$ of Eqs. (8) and 9 could, in principle, be derived analytically. However, the optimization objective would remain conditioned on the occurrence of the system-level rare event, whose dependence on the SFT structure couples the component realizations in a non-trivial way. Investigating hybrid analytical-simulative updates is an interesting direction for future work.

Lemma 1. *For each component c_i, the objective $\widetilde{J}_i(\alpha_i, \beta_i)$ is concave in both α_i and β_i $\forall\, \alpha_i, \beta_i > 0$. Thus, any α_i satisfying $\partial \widetilde{J}_i / \partial \alpha_i = 0$ and any β_i satisfying $\partial \widetilde{J}_i / \partial \beta_i = 0$ comprise the unique global maximizers in their respective coordinate.*

Proof. Consider the first derivative of $\widetilde{J}_i$ w.r.t α_i in Eq. (7). Its second derivative is $\partial^2 \widetilde{J}_i / \partial^2 \alpha_i = -\sum_{s=1}^{M} w_s \, \mathbb{I}_A(x^{(s)}) \, N_{\text{fail},i}^{(s)} / \alpha_i^2 \leq 0$, Hence $\widetilde{J}_i$ is concave in α_i, and its stationarity equation has a unique solution, which is necessarily the global maximizer along that coordinate (analogous result for β_i).

4.2 Cross–Entropy Update in Adaptive Importance Sampling

The solutions in Eqs. (8) and 9 for the stochastic optimization problem in Eq. (6) are numerically unstable, as they require the observation of a sufficient number of realizations $x^{(m)}$ yielding the rare event A to accurately estimate $w_m, N_{\text{repair},i}^{(m)} N_{\text{fail},i}^{(m)}$, $S_{\text{OK},i}^{(m)}$ and $S_{\text{KO},i}^{(m)}$. If A is extremely rare, these quantities may be dominated by sampling noise or may be undefined in early iterations.

In principle, we could use *gradient ascent* to iteratively estimate the solutions in Eqs. (8) and 9 from the partial derivatives $\partial \widetilde{J}_i / \partial \alpha_i$ and $\partial \widetilde{J}_i / \partial \beta_i$ with learning rate $\eta \in (0, 1]$, yielding the parameters $\alpha_i^{(k)}$ and $\beta_i^{(k)}$ at iteration $k > 0$:

$$\alpha_i^{(k)} = \alpha_i^{(k-1)} + \eta \left. \frac{\partial \widetilde{J}_i}{\partial \alpha_i} \right|_{\alpha_i = \alpha_i^{(k-1)}}, \quad \beta_i^{(k)} = \beta_i^{(k-1)} + \eta \left. \frac{\partial \widetilde{J}_i}{\partial \beta_i} \right|_{\beta_i = \beta_i^{(k-1)}}$$

However, gradient ascent performs updates that move the parameters *locally* in the direction of the objective, thus slowly approaching the optimum through a sequence of incremental steps whose number depends on the learning rate.

Therefore, we iteratively estimate the *coordinate-wise maximizers* α_i^* and β_i^* through a recursion with smoothing $\eta \in (0,1]$:

$$\alpha_i^{(k)} = (1-\eta)\,\alpha_i^{(k-1)} + \eta\,\alpha_i^{\star(k-1)}, \qquad \beta_i^{(k)} = (1-\eta)\,\beta_i^{(k-1)} + \eta\,\beta_i^{\star(k-1)} \quad (10)$$

where $\alpha_i^{*(k)}$ and $\beta_i^{*(k)}$ are computed according to Eqs. (8) and 9, respectively, from M realizations observed at iteration k. Note that larger values of η accelerate convergence but increase sensitivity to sampling variability. In the following, we prove the convergence of $\alpha_i^{(k)}$ and $\beta_i^{(k)}$ in Eq. (10).

Lemma 2 (Consistency of $\alpha_i^{\star(k)}$ and $\beta_i^{\star(k)}$). *If realizations $\{x^{(m)}\}_{m=1}^M$ are i.i.d. and the denominators in Eqs. 8 and 9 are strictly positive, then $\alpha_i^{*(k)}$ and $\beta_i^{\star(k)}$ satisfy $\alpha_i^{\star(k)} \xrightarrow[k\to\infty]{} \alpha_i^\star$ and $\beta_i^{\star(k)} \xrightarrow[k\to\infty]{} \beta_i^\star$ almost surely.*

Proof. Both the numerator and denominator of $\alpha_i^{\star(k)}$ are weighted sample averages of integrable random variables. By the Strong Law of Large Numbers, they converge almost surely to their corresponding expectations. Since the limiting denominator is strictly positive, the ratio converges almost surely to $\alpha_i^\star$.

Lemma 3 (Stability of linear contraction). *Let $e_k := \alpha_i^{(k)} - \alpha_i^\star$ be the estimation error at iteration $k > 0$. If $e_k = (1-\eta)e_{k-1} + \eta(\alpha_i^{\star(k-1)} - \alpha_i^\star)$ with smoothing $\eta \in (0,1]$, then $e_k \to 0$.*

Proof. By unrolling $e_k = (1-\eta)^k e_0 + \eta \sum_{j=0}^{k-1}(1-\eta)^{k-1-j}\left(\alpha_i^{\star(j)} - \alpha_i^\star\right)$, we note that the first term converges to zero since $0 < 1-\eta < 1$; for the second term, the geometric weights ensure that early values of $(\alpha_i^{\star(j)} - \alpha_i^\star)$ are damped, while $(\alpha_i^{\star(j)} - \alpha_i^\star) \to 0$ suppresses the contribution of later terms. Hence $e_k \to 0$.

Theorem 1 (Convergence of the smoothed CE update). *Under the assumptions of Lemmas 2 and 3, $\alpha_i^{(k)}$ satisfies $\lim_{k\to\infty} \alpha_i^{(k)} = \alpha_i^\star$ almost surely.*

Proof. By Lemma 2 $(\alpha_i^{\star(k)} - \alpha_i^\star)$ converges almost surely to zero. Lemma 3 then implies $e_k \to 0$, completing the proof.

The recursion in Eq. (10) terminates once the relative error $\widehat{\mathrm{RE}}_k$ of the estimator $P_{IS}^{(k)}(A)$ in Eq. (2) derived at iteration k is not larger than a threshold $\epsilon \in (0,1]$:

$$\widehat{\mathrm{RE}}_k = \frac{\sqrt{\widehat{\mathrm{Var}}[P_{IS}^{(k)}(A)]}}{P_{IS}^{(k)}(A)} \leq \epsilon \quad (11)$$

where $\widehat{\mathrm{Var}}[P_{IS}^{(k)}(A)] = (1/M)\sum_{m=1}^M (w_m\,\mathbb{I}_A,(x^{(m)}) - P_{IS}^{(k)}(A))^2$ is the sample variance estimated from M realizations $\{x^{(m)}\}$.

The rarity of the event A can vary significantly over a time interval $[0, T]$, as, e.g., system components may degrade, operational conditions may evolve, and mission phases may have different failure modes. Thus, a single global set of tilting parameters $(\bar{\alpha}, \bar{\beta})$ cannot simultaneously provide an effective biasing distribution for all times $\{t_z\}_{z=0}^{Z}$, $0 = t_0 < t_1 < \cdots < t_Z = T$. Therefore, we estimate a distinct set of tilting parameters $(\bar{\alpha}(t_z), \bar{\beta}(t_z))$ for each time t_z, using the solution $(\alpha_i^*(t_{z-1}), \beta_i^*(t_{z-1}))$ computed at time t_{z-1} as the initial guess:

$$\alpha_i^{(k)}(t_z) = (1-\eta)\,\alpha_i^{(k-1)}(t_z) + \eta\,\alpha_i^*(t_z), \qquad \alpha_i^{(0)}(t_z) = \alpha_i^*(t_{z-1}) \tag{12}$$

$$\beta_i^{(k)}(t_z) = (1-\eta)\,\beta_i^{(k-1)}(t_z) + \eta\,\beta_i^*(t_z), \qquad \beta_i^{(0)}(t_z) = \beta_i^*(t_{z-1}) \tag{13}$$

5 Experimentation

In this section, we present our experimental setup (Sect. 5.1) and results (Sect. 5.2). Experiments were executed on a dedicated virtual machine configured with 8 virtual CPU cores and 32 GB of RAM, running on an x86_64 architecture hosted on an Intel(R) Xeon(R) Gold 5120 processor operating at 2.20 GHz.

5.1 Experimental Setup

Research Questions. We define research questions evaluating the feasibility and effectiveness of our AIS method at estimating the system unreliability $P_A(t)$:

- **RQ1**: is our method feasible in terms of sample size and run time?
- **RQ2**: is our method accurate in terms of relative error and sample variance?
- **RQ3**: is our method feasible and accurate compared to alternative baselines?

Alternative Baselines. CMC serves as a reference estimator: for each model instance and time point t, M independent realizations of the underlying system CTMC are simulated, estimating the failure probability as an empirical mean.

The base cross-entropy importance sampling (CE) method is considered as an adaptive variance reduction baseline: similarly to AIS, CE estimates the failure probability by simulating realizations under a biased probability density $f'(x, \hat{\lambda}, \hat{\mu})$; however, component rates λ_i, μ_i are directly optimized to $\lambda_i', \mu_i' \in \mathbb{R}$, without any constraint or proportionality to tilting parameters.

We consider an upper bound $\hat{P}_A^{UB}(t)$ on $P_A(t)$ based on the union bound (UB) over the set of minimal cut sets $\mathcal{U} = \{U_1, \ldots, U_C\}$ of the SFT. By Boole's inequality, $P_A(t) = P(\bigcup_{U_j \in \mathcal{U}} A_j(t)) \leq \hat{P}_A^{UB}(t) := \sum_{U_j \in \mathcal{U}} P(A_j(t))$, where $A_j(t)$ is the event that all components in U_j have failed before t. We compute the impact of each cut set by integrating over all failure times of a triggering component, conditioned on the failure of the remaining components in the cut set:

$$\hat{P}_A^{UB}(t) = \int_0^t \sum_{U_j \in \mathcal{U}} \sum_{i \in U_j} \left(h_i(\tau) \prod_{k \in U_j \setminus i} p_k(\tau) \right) d\tau \tag{14}$$

where $p_k(\tau)$ denotes the probability that component k has failed by time τ, and $h_i(\tau)$ represents the instantaneous failure rate of component c_i.

For our running example we also take into account Adaptive Multilevel Splitting (AMS) [50], an ISPLIT variant that dynamically adapts the scoring functions and survival thresholds (for larger models, we omit AMS as its computational cost and variance, comparable with those of CMC, are due to its specialization in capturing order-dependent failures rather than combinations of failures). For each SFT and time t, we simulate M realizations, each assigned a score measuring its proximity to failure. At each iteration, a fixed fraction $k = \lfloor 0.2M \rfloor$ of the lowest-scoring realizations is discarded and replaced by clones of higher-scoring ones. This resampling scheme continues until all realizations reach the failure event or the time horizon t. The probability estimate is computed as the product of conditional survival probabilities across the induced splitting levels. Let L be the number of levels and $\hat{p}_\ell$ the survival probability at level ℓ. With fixed k, we have $\hat{p}_\ell = 1 - \frac{k}{M}$. Thus, the AMS estimator is:

$$\hat{P}_A^{\mathrm{AMS}}(t) = \left(1 - \frac{k}{M}\right)^L \cdot \frac{1}{M} \sum_{m=1}^{M} \mathbb{I}_A(x^{(m)}) \tag{15}$$

Test Suite Generation. We adopt a rejection-based generation procedure that targets predefined probability ranges, yielding synthetic models with controlled unreliability values. We consider three intervals, defining three benchmark suites: $[10^{-5}, 10^{-3}]$ (Suite A), $[10^{-7}, 10^{-5}]$ (Suite B), and $[10^{-11}, 10^{-9}]$ (Suite C). We generate SFTs by randomly combining component failures by AND, OR, and VOT(k/n) gates, possibly with repeated events. Component failure and repair rates (λ_i, μ_i) are independently sampled within $[10^{-5}, 10^{-3}]$ and $[10, 10^3]$, respectively. The system unreliability $P_A(t)$ is approximated by $\hat{P}_A^{UB}(t)$, discarding the model if $\hat{P}_A^{UB}(\Delta t)$ and $\hat{P}_A^{UB}(T)$ are not within the considered range, where Δt is the first time instant (also the time step) and T is the time limit. For each Suite A, B, and C and for each number $N \in \{20, 35, 50\}$ of components, we repeat the procedure until 10 valid SFTs are generated, yielding 90 models.

Performance Measures. For each method and time point t, we compute the system unreliability estimate $\hat{P}_A(t)$ and its variance over R independent repetitions, estimating the relative error as in Eq. (11). To quantify improvements over CMC (when available), we compute the variance reduction factor $\mathrm{VRF}(t) = \mathrm{Var}_{\mathrm{CMC}}(t)/\mathrm{Var}_{\mathrm{method}}(t)$, where values greater than 1 indicate variance reduction w.r.t. CMC. Along the whole system unreliability, we report 95% confidence intervals (CI) obtained via bootstrap resampling. Specifically, let $\{P_A^1(t), \ldots, P_A^R(t)\}$ be a set of probability estimates and $\hat{P}_A(t) = \frac{1}{R}\sum_{i=1}^{R} P_A^i(t)$ be the estimated mean over R repetitions. To estimate the uncertainty of $\hat{P}_A(t)$ we generate B bootstrap estimates by sampling with replacement from the original R samples. For each bootstrap sample $b \in \{1, \ldots, B\}$, we produce a set of bootstrap estimates $\{\hat{P}_A^{(1)}(t), \ldots, \hat{P}_A^{(B)}(t)\}$ where $\hat{P}_A^{(b)}(t) = \frac{1}{R}\sum_{i=1}^{R} P_A^{(b)}(t)$.

We provide a $(1-\xi)$ confidence interval by using the percentile method, yielding the lower bound $\mathrm{CI}_{\mathrm{low}} = \mathrm{quantile}\big(\{\hat{P}_A^{(b)}(t)\}, \xi/2\big)$ and upper bound $\mathrm{CI}_{\mathrm{high}} = \mathrm{quantile}\big(\{\hat{P}_A^{(b)}(t)\}, 1-\xi/2\big)$. We use $B = 1000$ and $\xi = 0.05$. Finally, we measure the average wall-clock time required to compute estimates for each method.

5.2 Experimental Results

Running Example. We evaluate all methods (i.e. AIS, CMC, CE and AMS) on a time grid $t \in \{\Delta t, 2\Delta t, \ldots, T\}$ with $\Delta t = 0.1$ and $T = 10$, repeating each experiment $R = 30$ times to assess statistical variability. All methods use sample size $M = 10^4$ samples for each time t. AIS is configured with a maximum of 100 adaptation steps, with initial tilting parameters set to $\alpha_i^0 = 100$ and $\beta_i^0 = 0.5$ for all components c_i, with bounds $\alpha_i \in [1, 10^3]$ and $\beta_i \in [0.01, 10^2]$. CE is configured with the same parameters: 100 adaptation steps, initial rates set at $\lambda_i^0 = 100\lambda_i$ and $\mu_i^0 = 0.5\mu_i$, and bounds $\lambda_i' \in [\lambda_i, 10^3\lambda_i]$ and $\mu_i' \in [0.01\mu_i, 10^2\mu_i]$. A smoothing factor η of 0.2 is used to stabilize parameter updates for both AIS and CE. AMS is configured with a kill fraction of 20%.

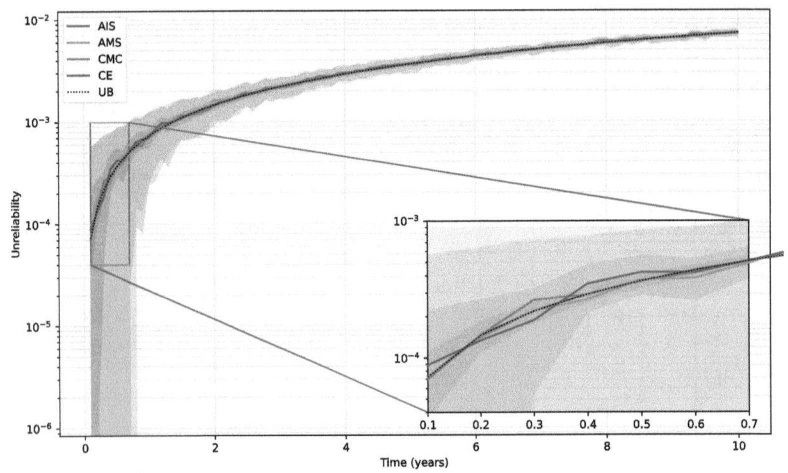

Fig. 2. For the model of Fig. 1 and Table 1, UB and system unreliability with band-wise 95% CI computed by AIS and alternative baselines (CMC, CE, AMS).

Figure 2 shows the system unreliability and bootstrap band-wise 95% CI computed by all methods on the model introduced in Fig. 1 and Table 1. All methods correctly estimate the system unreliability, while exhibiting significantly different variance. In particular, AIS yields the tightest CIs, CMC and AMS show substantially larger variability, and CE lies in between. AIS provides lower-variance estimates, while also consistently remaining below the upper bound UB (**RQ2**), whereas alternative methods clearly deviate (**RQ3**). For all methods, the

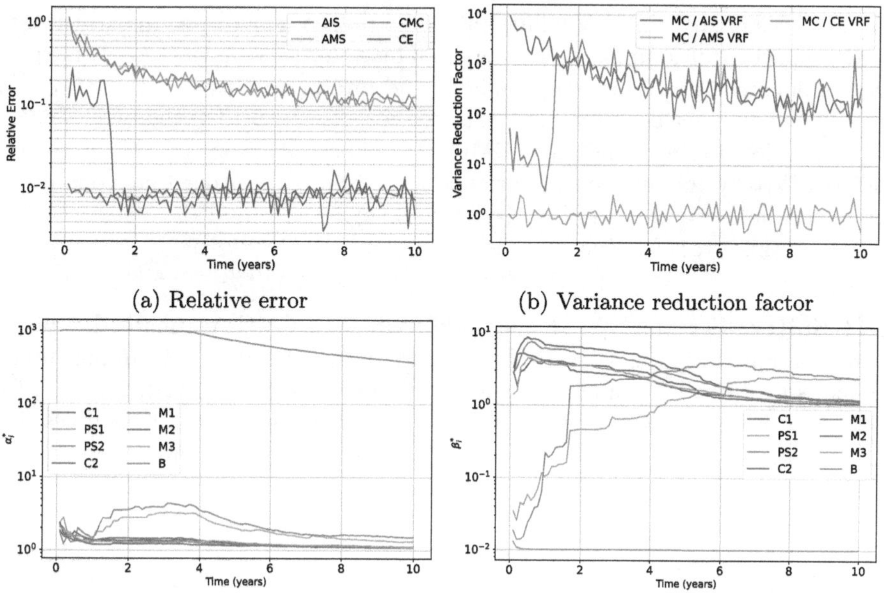

(a) Relative error

(b) Variance reduction factor

(c) $\alpha_i^*(t)$ mean values

(d) $\beta_i^*(t)$ mean values

Fig. 3. For the model of Fig. 1 and Table 1: (a) relative error and (b) variance reduction achieved by AIS and alternative baselines (CMC, CE and AMS), averaged over 30 repetitions; (c) $\alpha_i^*(t)$ and (d) $\beta_i^*(t)$ of AIS for each component c_i.

observed decrease in variance over time is expected, since the failure probability increases over time while the sample size remains constant.

Figure 3a shows the relative error over time (computed as in Eq. (11)): CMC and AMS exhibit similar behaviors, with values consistently above those of AIS and CE, while AIS maintains relative error at or below the threshold $\epsilon = 0.01$ across the full time interval $[0, T]$. Figure 3b reports VRF values over time (computed as previously described in this section) for AIS, CE and AMS w.r.t. CMC, showing that AIS achieves VRF at least two orders of magnitude larger than that of AMS (**RQ2**, **RQ3**). CE performance in Figs. 3a and 3b varies over time, with higher variance and relative error at early times, where the failure probability is low, and improving as the probability increases. In contrast, AIS remains stable over time; its tilting parameters α_i, β_i guide the update of transition rates and mitigate the numerical instability typically observed in IS schemes (**RQ2**, **RQ3**).

Figures 3c and 3d present the AIS-optimized tilting rates $\alpha_i^*(t)$ and $\beta_i^*(t)$ for each component c_i. AIS correctly identifies the BUS B, which can directly trigger the system failure, as the most critical component, substantially increasing its failure rate while decreasing its repair rate with respect to the other components. AIS further identifies PS1 and PS2 as critical due to their low reliability in the SFT, with their joint failures also leading to the system failure. Their tilting parameters similarly amplify failure rates and reduce repair rates in early stages (with respect to all components except B). This adaptive behavior closely mirrors the system structure, supporting validity and interpretability of AIS (**RQ2**).

AIS has a higher per-run computational cost (7.72 ± 0.23 s) than CE (3.00 ± 0.02 s), AMS (0.71 ± 0.03 s), and CMC (0.68 ± 0.01 s). However, this increased cost is offset by its substantial variance reduction, suggesting that competing methods would require significantly larger sample sizes (and thus, significantly longer runtime) to achieve comparable accuracy (**RQ1, RQ3**).

Test Suites. For Suite A, all methods (AIS, CE, CMC) estimate the whole system unreliability $\hat{P}_A(t)$ in probability regimes $[10^{-5}, 10^{-3}]$, enabling direct comparison of feasibility and accuracy. They consider a grid $t \in \{\Delta t, 2\Delta t, \ldots, T\}$ with $\Delta t = 0.1$ and $T = 10$, using $R = 30$ repetitions and $M = 10^4$ samples per time point. AIS and CE follow the configuration used for the running example.

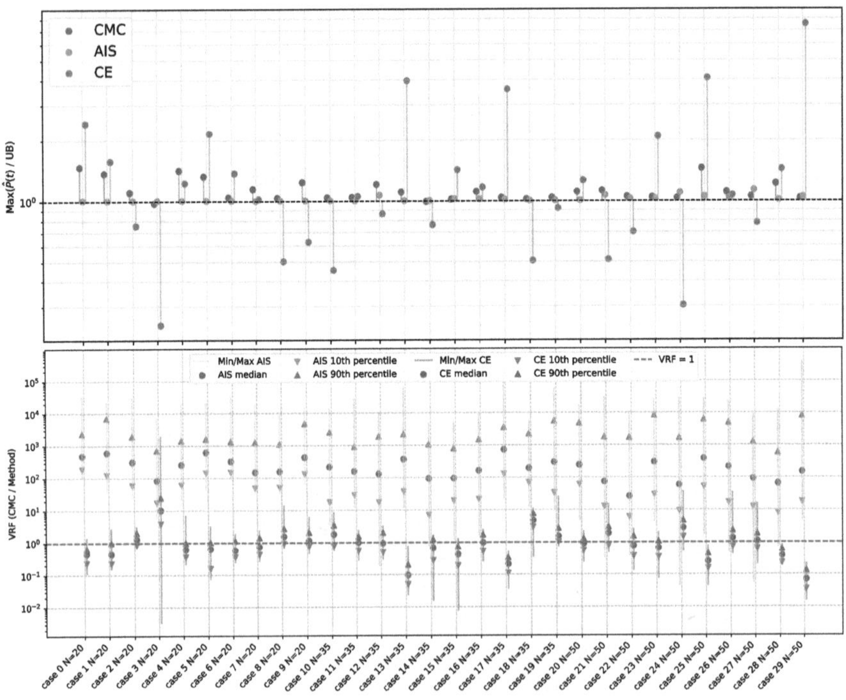

Fig. 4. Suite A: above, $\max_t \hat{P}_A(t)/\text{UB}$ for AIS, CE and CMC; below, VRF (min, max and percentiles) of AIS and CE w.r.t. CMC, over the whole computed CDFs.

The top plot of Fig. 4 shows $\max_t \hat{P}_A(t)/\text{UB}$ for each method, representing the maximum divergence of the estimated system unreliability from the upper bound UB. CMC and CE frequently exceed the upper bound, while AIS yields values consistently below it, preserving validity with respect to this theoretical constraint (**RQ2**, **RQ3**). The bottom plot of Fig. 4 reports VRF values for AIS and CE with respect to CMC: AIS significantly outperforms both baselines, achieving median VRF values between 10^2 and 10^3, and peaks up to 10^4, whereas CE provides only marginal improvements and often behaves similarly to CMC. These results demonstrate that AIS achieves substantially lower variance (**RQ2**).

Suite B consists of models with lower failure probabilities within $[10^{-7}, 10^{-5}]$. The evaluation is focused on the earliest time point $t_{min} = 0.1$, where the failure probability is smallest. In this setting, we consider $R = 30$ repetitions, with CMC using $M = 10^6$ samples and AIS and CE using $M = 10^4$, maintaining the same configuration used for the running example test.

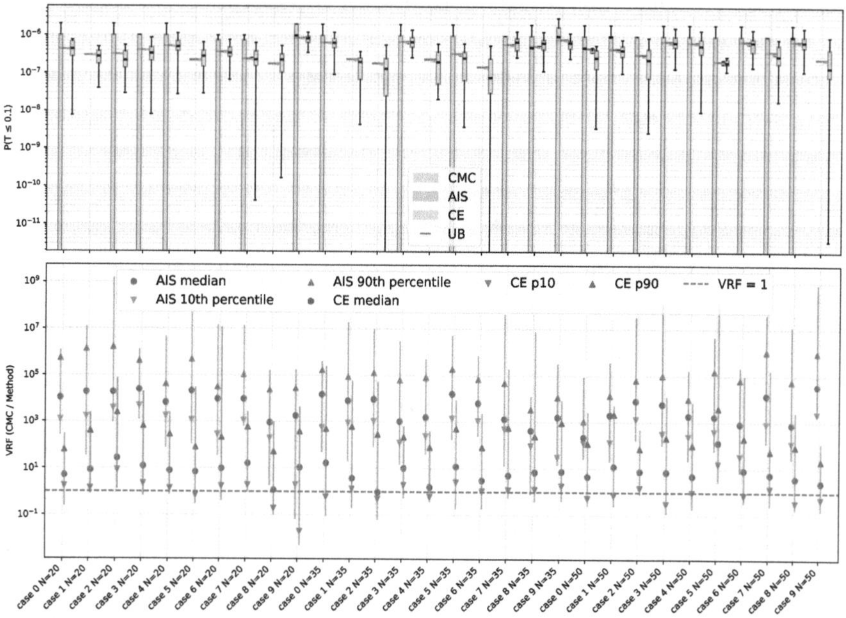

Fig. 5. Suite B: above, system unreliability $\hat{P}_A(t_{min})$ for AIS, CE and CMC (min and max, as vertical black lines; quartiles, as colored rectangles; median, as horizontal black lines), and upper bound UB (as horizontal green lines); below, VRF (min, max and percentiles) of AIS and CE w.r.t. CMC, for time t_{min}.

The top plot of Fig. 5 presents the results on $\hat{P}_A(t_{min})$, summarized via box-plots, highlighting clear differences in stability and validity across methods. CMC exhibits the largest variability, with high variance across all models, also consistently exceeding UB. CE shows lower but still non-negligible variance, and occasional proximity to UB. In contrast, AIS presents tight interquartile ranges and low variance, systematically below UB across all models (**RQ2**).

The bottom plot of Fig. 5 shows the VRFs for AIS and CE w.r.t. CMC, confirming the trends observed for Suite A. In particular, AIS exhibits from three up to five orders of magnitude lower variance than both CMC and CE, and it notably maintains high accuracy even with a limited sample size, making it suitable for scenarios in which CMC becomes impractical (**RQ3**).

Suite C pushes the evaluation to extreme rare-event regimes $[10^{-11}, 10^{-9}]$, again focusing on $t_{min} = 0.1$ with $R = 30$ repetitions. Due to memory limitations, CMC is capped at $M = 10^7$ samples, while AIS and CE are configured as in the previous suites. In this setting, as shown in Fig. 6, the limitations of the baseline methods become evident: CMC is unable to observe the rare event, while CE often fails to produce non-zero estimates. In particular, CE is strongly dependent on its initial configuration, requiring a specific setting for each model in order to observe the event. In contrast, AIS remains stable across all models, providing estimates with controlled variance, below the upper bound UB, while using the same initial configuration for all models (**RQ2**, **RQ3**).

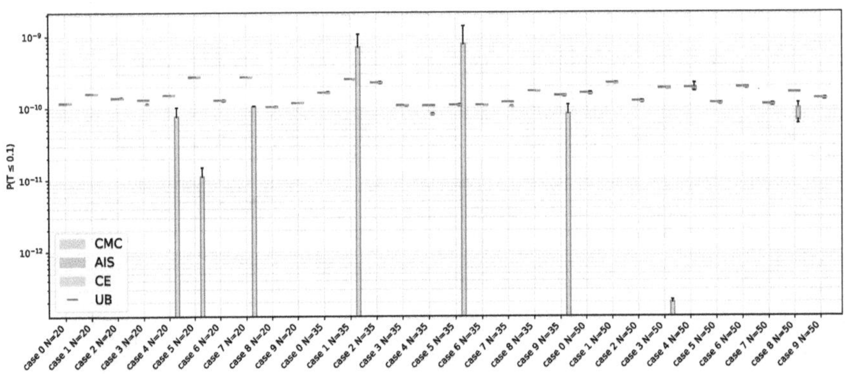

Fig. 6. Suite C: above, system unreliability $\hat{P}_A(t_{min})$ for AIS, CE and CMC (min and max, as vertical black lines; quartiles, as colored rectangles; median, as horizontal black lines), and upper bound UB (as horizontal green lines).

Table 2. Average runtime (mean ± standard deviation, in seconds) per repetition for AIS, CE, and CMC across the three generated experimental suites.

Suite	**AIS (s)**	**CE (s)**	**CMC (s)**
A	5.01 ± 1.87	5.38 ± 0.39	0.49 ± 0.01
B	18.36 ± 5.52	4.18 ± 1.16	4.65 ± 2.53
C	20.35 ± 7.56	4.68 ± 2.43	6.68 ± 3.29

The observed runtimes across all the three test suites are summarized in Table 2. Consistently with the running example, AIS incurs higher per-run computational cost than CMC, and it is comparable to or slower than CE, while offering substantially higher variance reduction (**RQ1**, **RQ3**). Due to space limits, we leave as future work the evaluation of the estimator accuracy achieved by each method within the same fixed computational budget.

In summary, across all the three test suites, AIS provides valid estimates under a fixed initial configuration of parameters (**RQ1**) and maintains accuracy in increasingly rare-event regimes (**RQ2**). Moreover, w.r.t alternative baselines CMC and CE, AIS consistently achieves substantial variance reduction (up to five orders of magnitude) whereas both CMC and CE either require significantly larger sample sizes or fail to produce reliable results (**RQ3**).

6 Threats to Validity

We discuss potential threats to validity [51]. As for *internal* validity, we mitigate the inherent variance of stochastic simulation (especially in extremely rare-event regimes like Suite C), by repeating each experiment $R = 30$ times and reporting bootstrap CIs. The consistent performance achieved for all suites and system sizes suggests that the observed gains are robust. Notably, while CE is sensitive to initial parameter settings and requires case-specific tuning, our approach uses the same initial setting for all suites to better mirror realistic usage scenarios.

Regarding *external* validity, we mitigate the potential limitations of synthetic models in capturing real-world dependencies (despite controlling event rarity and system size) by considering a realistic case study, obtaining consistent variance reduction. Moreover, our approach assumes independent two-state CTMC components, and thus its effectiveness is not yet guaranteed for other system models, e.g., with inter-component dependencies or more complex degradation behaviors.

As for *construct* validity, we mitigate potential measurement bias by evaluating accuracy through standard RES metrics (e.g., relative error, variance). To address numerical instability in extremely rare-event regimes (e.g., Suite C), we further mitigate estimation risks by complementing variance with CIs and consistency checks against the upper bound UB. To isolate algorithmic performance, all methods are executed in a strictly controlled environment.

Finally, regarding *conclusion* validity, we reduce model selection bias by considering diverse systems in terms of size, topology, and failure rarity, and we compare methods under identical sampling budgets. While our AIS approach consistently outperforms CMC and CE and shows significant advantages over AMS in the considered settings, we mitigate the risk of over-generalization by acknowledging that results are conditioned on the selected model class, parameter ranges, and implementation choices, and should therefore be interpreted as a robust validation within the domain investigated in this paper.

7 Conclusion

We introduced a novel AIS method to estimate rare-event probabilities in SFTs where components behave as two-state CTMCs, outperforming alternative baselines in terms of estimator accuracy and variance reduction. A natural direction for future work is the extension to richer stochastic component models, namely semi-Markov and non-Markovian models [20,26,28]. In these settings, new likelihood decompositions and parameter-update rules would be required, while the underlying cross-entropy adaptation principle would remain applicable.

In our implementation, the tilting parameters are initialized identically for every component based on an initial guess. Although the CE adaptation progressively corrects poor initializations, the quality of the initial guess can significantly affect both convergence speed and numerical stability. A solution could be learning initializations of the tilting parameters by exploiting structural information of the model, in particular through supervised learning (using previously solved instances as training data) or through graph-based models such as graph neural networks, which could exploit topological features of the SFT (e.g., depth, gate types, minimal cut sets, component importance, or redundancy patterns).

Another direction would be the integration of AIS with complementary RES techniques like AMS to perform both realization selection and distribution tilting. In particular, rather than considering only the realizations $x^{(m)}$ that lead to the event A representing the system failure (as defined by $\mathbb{I}_A(x^{(m)})$ in Eq. (4)), we could consider a scoring function $\mathbb{S}_A(x^{(m)})$ to include near-failure realizations, which did not lead to the system failure while being critically close to it.

Data availability statement. An artifact for experimental reproduction of the results in Sect. 5 is publicly available at DOI 10.5281/zenodo.19628909.

References

1. Avizienis, A., Laprie, J.-C., Randell, B., Landwehr, C.: Basic concepts and taxonomy of dependable and secure computing. IEEE Trans. Dependable Secure Comput. **1**(1), 11–33 (2004)
2. Rubino, G., Tuffin, B. (eds.): Rare Event Simulation using Monte Carlo Methods. Wiley, Chichester (2009). ISBN 978-0-470-57717-3
3. Rubinstein, R.Y., Kroese, D.P.: Simulation and the Monte Carlo Method, 3rd edn. Wiley, Hoboken (2016)
4. Zio, E.: Some challenges and opportunities in reliability engineering. IEEE Trans. Reliab. **65**(4), 1769–1782 (2016)
5. Sahner, R.A., Trivedi, K.S.: Reliability modeling using sharpe. IEEE Trans. Reliab. **36**(2), 186–193 (2009)
6. Amparore, E.G., Balbo, G., Beccuti, M., Donatelli, S., Franceschinis, G.: 30 years of GreatSPN. In: Principles of Performance and Reliability Modeling and Evaluation, pp. 227–254 (2016)
7. Kwiatkowska, M., Norman, G., Parker, D.: PRISM: probabilistic symbolic model checker. In: TOOLS, pp. 200–204. Springer, Cham (2002)
8. Zimmermann, A.: Modelling and performance evaluation with TimeNET 4.4. In: QEST, pp. 300–303. Springer, Cham (2017)
9. Ciardo, G., Miner, A.S.: Smart: the stochastic model checking analyzer for reliability and timing. In: International Conference on the Quantitative Evaluation of Systems, pp. 338–339. IEEE (2004)

10. Paolieri, M., Biagi, M., Carnevali, L., Vicario, E.: The ORIS tool: quantitative evaluation of non-Markovian systems. IEEE Trans. Softw. Eng. **47**(6), 1211–1225 (2019)
11. Carnevali, L., Cerboni, S., Montecchi, L., Vicario, E.: Faultflow: an MDE library for dependability evaluation of component-based systems. IEEE Trans. Dependable Secure Comput. **22**(4), 3431–3448 (2025)
12. Ruijters, E., Stoelinga, M.: Fault tree analysis: a survey of the state-of-the-art in modeling, analysis and tools. Comput. Sci. Rev. **15**, 29–62 (2014). https://api.semanticscholar.org/CorpusID:206990190
13. Meshkat, L., Dugan, J.B., Andrews, J.D.: Dependability analysis of systems with on-demand and active failure modes, using dynamic fault trees. IEEE Trans. Reliab. **51**(2), 240–251 (2002)
14. Bobbio, A., Franceschinis, G., Gaeta, R., Portinale, L.: Parametric fault tree for the dependability analysis of redundant systems and its high-level petri net semantics. IEEE Trans. Softw. Eng. **29**(3), 270–287 (2003)
15. Veeraraghavan, M., Kishor, S.: Trivedi: an improved algorithm for symbolic reliability analysis. IEEE Trans. Reliab. **40**(3), 347–358 (2002)
16. Katoen, J.-P., Stoelinga, M.: Boosting fault tree analysis by formal methods. In: ModelEd, TestEd, TrustEd: Essays Dedicated to Ed Brinksma on the Occasion of His 60th Birthday, pp. 368–389. Springer, Cham (2017)
17. Silva, A.R.D.: Model-driven engineering: a survey supported by the unified conceptual model. Comput. Lang. Syst. Struct. **43**, 139–155 (2015)
18. Douglas, C.: Schmidt: model-driven engineering. Comput. IEEE Comput. Soc. **39**(2), 25 (2006)
19. Reibman, A., Trivedi, K.: Numerical transient analysis of Markov models. Comput. Oper. Res. **15**(1), 19–36 (1988)
20. Choi, H., Kulkarni, V.G., Trivedi, K.S.: Markov regenerative stochastic petri nets. Perform. Eval. **20**(1–3), 337–357 (1994)
21. Ciardo, G., German, R., Lindemann, C.: A characterization of the stochastic process underlying a stochastic petri net. IEEE Trans. Softw. Eng. **20**(7), 506–515 (1994)
22. Asmussen, S., Glynn, P.W.: Stochastic Simulation: Algorithms and Analysis. Springer, New York (2007)
23. Heidelberger, P.: Fast simulation of rare events in queueing and reliability models. ACM Trans. Model. Comput. Simul. **5**(1), 43–85 (1995). https://doi.org/10.1145/203011.203017
24. L'Ecuyer, P., Le Gland, F., Lezaud, P., Tuffin, B.: Splitting Techniques, chapter 3, pp. 39–61. https://doi.org/10.1002/9780470745403.ch3
25. Budde, C.E., Biagi, M., Monti, R.E., D'Argenio, P.R., Stoelinga, M.: Rare event simulation for non-Markovian repairable fault trees. In: Biere, A., Parker, D. (eds.) TACAS 2020. LNCS, vol. 12078, pp. 463–482. Springer, Cham (2020). https://doi.org/10.1007/978-3-030-45190-5_26
26. Budde, C.E., Ruijters, E., Stoelinga, M.: The dynamic fault tree rare event simulator. In: Gribaudo, M., Jansen, D.N., Remke, A. (eds.) QEST 2020. LNCS, vol. 12289, pp. 233–238. Springer, Cham (2020). https://doi.org/10.1007/978-3-030-59854-9_17
27. Budde, C.E., D'Argenio, P.R., Monti, R.E., Stoelinga, M.: Analysis of non-Markovian repairable fault trees through rare event simulation. Int. J. Softw. Tools Technol. Transf. **24**(5), 821–841 (2022). https://doi.org/10.1007/S10009-022-00675-X

28. Dengler, G., Carnevali, L., Budde, C.E., Vicario, E.: Transient evaluation of non-Markovian models by stochastic state classes and simulation. In: International Conference on Quantitative Evaluation of Systems and Formal Modeling and Analysis of Timed Systems, pp. 213–232. Springer, Cham (2024)
29. Dengler, G., Budde, C.E., Carnevali, L., Hartmanns, A.: Time-sensitive importance splitting. In: International Conference on Quantitative Evaluation of Systems and Formal Modeling and Analysis of Timed Systems, pp. 21–41. Springer, Cham (2025)
30. Tabandeh, A., Jia, G., Gardoni, P.: A review and assessment of importance sampling methods for reliability analysis. Struct. Saf. **97**, 102216 (2022)
31. Shahabuddin, P.: Fast transient simulation of Markovian models of highly dependable systems. Perform. Eval. **20**(1–3), 267–286 (1994)
32. Carrasco, J.A., et al.: Failure distance-based simulation of repairable fault-tolerant systems. In: Proceedings of the 5th International Conference on Modeling Techniques and Tools for Computer Performance Evaluation, pp. 351–365 (1992)
33. Juan, A.: Carrasco: failure transition distance-based importance sampling schemes for the simulation of repairable fault-tolerant computer systems. IEEE Trans. Reliab. **55**(2), 207–236 (2006)
34. Reijsbergen, D., De Boer, P.-T., Scheinhardt, W., Juneja, S.: Path-ZVA: general, efficient, and automated importance sampling for highly reliable Markovian systems. ACM Trans. Model. Comput. Simul. (TOMACS) **28**(3), 1–25 (2018)
35. Ruijters, E., Reijsbergen, D., de Boer, P.-T., Stoelinga, M.: Rare event simulation for dynamic fault trees. Reliab. Eng. Syst. Saf. **186**, 220–231 (2019)
36. Zhao, Y., O'Leary, J.: Solving large fault trees with importance sampling and tree search. In: 2024 Annual Reliability and Maintainability Symposium (RAMS), pp. 1–6. IEEE (2024)
37. Monica, F., Bugallo, V., Elvira, L., Martino, D., Luengo, J., Miguez, P.M.: Djuric: adaptive importance sampling: the past, the present, and the future. IEEE Signal Process. Mag. **34**(4), 60–79 (2017)
38. Buchholz, P.: An adaptive importance sampling approach for the transient analysis of Markovian queueing networks. Eur. Trans. Telecommun. **13**(4), 317–329 (2002)
39. Rubinstein, R.Y., Kroese, D.P.: The Cross-Entropy Method. Springer, New York (2004)
40. Kurtz, T.G., Song, J.: Cross-entropy-based adaptive importance sampling using gaussian mixture. Stochast. Process. Appl. **123**(11), 3968–3994 (2013)
41. Bermúdez, E., Wehenkel, L.: Cross-entropy based rare-event simulation for the identification of dangerous events in power systems. IEEE Trans. Power Syst. **23**(2), 823–835 (2008)
42. Aguilar, J., Gatto, R.: A unified perspective on exponential tilt and bridge algorithms for rare trajectories of discrete Markov processes. arXiv preprint (2023). Provides a modern analysis of exponential tilting and its relation to other rare-event sampling methods
43. Chennetier, J., Chraibi, M., Dutfoy, A., Garnier, J.: Adaptive importance sampling based on fault tree analysis for PDMPs. Reliab. Eng. Syst. Saf. **244**, 109823 (2024)
44. Vesely, W.E., Goldberg, F.F., Roberts, N.H., Haasl, D.F.: Fault tree handbook. Technical report (1981)
45. Rausand, M., Hoyland, A.: System Reliability Theory: Models, Statistical Methods, and Applications, vol. 396. Wiley, Hoboken (2003)
46. Department of Defense. Reliability prediction of electronic equipment. Military Handbook MIL-HDBK-217F, Department of Defense, Washington, DC, USA (1991)

47. Schroeder, B., Pinheiro, E., Weber, W.-D.: Dram errors in the wild: a large-scale field study. SIGMETRICS Perform. Eval. Rev. **37**(1), 193–204 (2009). https://doi.org/10.1145/2492101.1555372. ISSN 0163-5999
48. Falck, J., Felgemacher, C., Rojko, A., Liserre, M., Zacharias, P.: Reliability of power electronic systems: an industry perspective. IEEE Ind. Electron. Mag. **12**(2), 24–35 (2018). https://doi.org/10.1109/MIE.2018.2825481
49. Blanchet, J., Glynn, P.W.: Efficient rare-event simulation for the maximum of heavy-tailed random walks. Ann. Appl. Probab. **18**(4), 1351–1378 (2008). https://doi.org/10.1214/07-AAP494
50. Cérou, F., Guyader, A.: Adaptive multilevel splitting for rare event analysis. Stoch. Anal. Appl. **25**(2), 417–443 (2007)
51. Runeson, P., Höst, M.: Guidelines for conducting and reporting case study research in software engineering. Empir. Softw. Eng. **14**(2), 131–164 (2009)

MightyPPL: Towards Model Checking MTL

Hsi-Ming Ho[1](✉), Shankara Narayanan Krishna[2](✉), Khushraj Madnani[4](✉), Rupak Majumdar[3](✉), and Paritosh Pandya[2](✉)

[1] University of Sussex, Brighton, UK
hsimho@gmail.com
[2] Indian Institute of Technology Bombay, Mumbai, India
[3] Max Planck Institute for Software Systems (MPI-SWS), Kaiserslautern, Germany
[4] Indian Institute of Technology Guwahati, Guwahati, India

Abstract. The theoretical foundation for model checking timed systems against *Metric Interval Temporal Logic* (MITL) was established in the early 1990s, yet the first practical tool supporting future MITL (MightyL) did not emerge until 2017. Recently, there has been growing interest in extending this toolchain to support more expressive logical operators, including *past modalities*, *Pnueli modalities*, and limited use of *singular intervals*. MightyPPL is one such toolchain. We introduce an upgraded version of MightyPPL that enables—for the first time—the model checking of *Metric Temporal Logic* (MTL) properties of the form $\mathbf{G}(p \Rightarrow \mathbf{F}_{=5}\, q)$ in addition to Pnueli and Past modalities. We discuss the tool's underlying architecture and implementation, and present a performance evaluation against the Tempora tool across diverse satisfiability and model checking benchmarks, demonstrating that MightyPPL delivers significantly better performance.

1 Introduction

Timed Logics and Automata. *Metric Temporal Logic* (MTL) [14] extends *Linear Temporal Logic* (LTL) [19] with temporal operators decorated by *time intervals*; e.g., 'each request is followed by an acknowledgement after *exactly* 5 *seconds*' is written $\mathbf{G}(req \Rightarrow \mathbf{F}_{[5,5]}\, ack)$. MTL and its signal variant STL [17] are widely used in cyber-physical systems. *Timed automata* (TAs) [3] are the *de facto* model for timed systems and benefit from decades of mature zone-based tooling (Uppaal [5], LTSmin [15], TChecker [11]). Solving satisfiability and model checking for MTL therefore reduces, in principle, to translating formulae into TAs. Such translations are, however, only possible when the logic is restricted to *Metric Interval Temporal Logic* (MITL), i.e., MTL without *singular ('punctual') intervals* [4]. Until recently, MightyL [10] was the only tool translating future MITL formulae into TAs; its successor MightyPPL [12] added *past modalities* and *Pnueli modalities* but still disallowed punctual intervals.

M. Chen and M. Tschaikowski (Eds.): QEST+FORMATS 2026, LNCS 16913, pp. 216–225, 2027.
https://doi.org/10.1007/978-3-032-35298-9_12

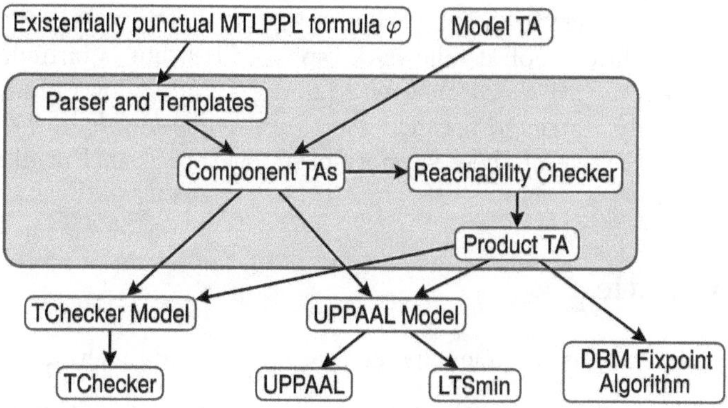

Fig. 1. MIGHTYPPL architecture.

Table 1. Feature coverage of MIGHTYPPL vs. prior tools. ✓= supported, ✗= not supported, (✓) = partially supported.

Feature	MIGHTYL	TEMPORA	MIGHTYPPL
Future MITL	✓	✓	✓
Past modalities	✗	✓	✓
Pnueli modalities	✗	✗	✓
Punctual intervals (one-sided MTL)	✗	(✓)	✓
Symbolic alphabets	✗	✗	✓
Multiple backend support	(✓)	✗	✓
Built-in fixpoint algorithm	✗	✗	✓

Contributions. We present an updated version of MIGHTYPPL [12] that, for the first time in a practical tool, handles a useful class of formulae with punctual intervals—including 'one-sided' MTL [9] with past and Pnueli modalities [12]. Our contributions, relative to [12] and to prior tools, are: (i) an end-to-end implementation with multiple interchangeable backends (UPPAAL, LTSMIN, TCHECKER, and a built-in symbolic fixpoint algorithm); (ii) a reachability-driven product construction (flattening) that keeps monolithic TAs small enough for explicit analysis; and (iii) an extensive empirical comparison against TEMPORA [1], the only other tool supporting MITL with past modalities, on satisfiability and model-checking benchmarks. On the benchmarks in Sect. 4, MIGHTYPPL outperforms TEMPORA by orders of magnitude on the majority of instances, scales to multi-core model checking via LTSMIN, and produces correct answers on all instances whereas TEMPORA reports wrong results on several instances.

Related Work. A fragment of MTL with a decidable satisfiability problem—Flat-MTL, which subsumes the 'one-sided' MTL of [9]—was studied in [7,8]. To the best of our knowledge, however, no tool supporting Flat-MTL has been implemented. Beyond MIGHTYL and the original MIGHTYPPL of [12], TEMPORA is the most closely related tool; it follows the spirit of [18] but is formulated in the framework of *generalized timed automata* [2]. TEMPORA allows punctual intervals, but in a highly restricted manner: *they are permitted only at the outermost level of a formula.* None of these support past together with Pnueli modalities (Table 1).

2 Preliminaries

Due to space limitations, we refer the reader to [12] for more detailed definitions.

Timed Words. Let $\mathbb{R}_{\geq 0}, \mathbb{N}$ denote the non-negative reals and naturals; let $\mathbb{I}$ (resp. $\mathbb{I}_0$) denote intervals $\langle l, u \rangle$ with $l \leq u \in \mathbb{N} \cup \{\infty\}$ (resp. with $l = 0$). Fix a finite set AP of atomic propositions and let $\Sigma_{\mathsf{AP}} = 2^{\mathsf{AP}}$. An infinite (resp. finite) *timed word* over Σ_{AP} is an infinite (resp. finite) non-Zeno (weakly) monotonic sequence $\rho = (\sigma_1, \tau_1)(\sigma_2, \tau_2) \ldots$ with $\sigma_i \in \Sigma_{\mathsf{AP}}$ and $\tau_i \in \mathbb{R}_{\geq 0}$.

MTL **with Past and Pnueli modalities** [12]. MTLPPL formulae over AP are

$$\varphi ::= p \mid \neg\varphi \mid \varphi \wedge \varphi \mid \varphi\, \mathbf{U}_I\, \varphi \mid \varphi\, \mathbf{S}_I\, \varphi \mid \mathbf{Pn}_J(\varphi_1, \ldots, \varphi_k) \mid \overleftarrow{\mathbf{Pn}}_J(\varphi_1, \ldots, \varphi_k),$$

where $a \in \mathsf{AP}$, $I \in \mathbb{I}$, $J \in \mathbb{I}_0$. We also write psuedo-arithmetic expressions for intervals, e.g., '$= 5$' for $[5, 5]$. The pointwise semantics is standard on Booleans; $\mathbf{U}_I$ and $\mathbf{S}_I$ are the (strict) future and past metric versions of 'Until' and 'Since' of LTL. The Pnueli modalities $\mathbf{Pn}_J$ and $\overleftarrow{\mathbf{Pn}}_J$ assert the existence of n strictly increasing (resp. decreasing) positions $i_n > \cdots > i_1 > i$ with $\tau_{i_k} - \tau_i \in J$ and $\rho, i_k \models \varphi_k$. Derived operators like $\mathbf{F}_I, \overleftarrow{\mathbf{F}}_I, \mathbf{G}_I, \overleftarrow{\mathbf{G}}_I, \mathbf{X}, \mathbf{Y}$ are defined as usual.
Existentially Punctual MTLPPL. An MTLPPL formula is *existentially punctual* if, when written in negation normal form, all of its punctual subformulae only appear in (i) φ_2 in larger subformulae $\varphi_1\, \mathbf{U}_I\, \varphi_2$; (ii) φ_1 in larger $\varphi_1\, \mathbf{R}_I\, \varphi_2$ ($\mathbf{R}_I$ is the dual of $\mathbf{U}_I$); (iii) φ_k ($1 \leq k \leq n$) in larger $\mathbf{Pn}(\varphi_1, \ldots, \varphi_n)$ or $\overleftarrow{\mathbf{Pn}}(\varphi_1, \ldots, \varphi_n)$.

For example, $\mathbf{F}(p \wedge \mathbf{F}_{=5}\, q)$ and $\neg\, \mathbf{G}_{[0,20]}(\mathit{req} \Rightarrow ((\neg\, \overleftarrow{\mathbf{F}}_{[0,2]}(\mathit{ack}_2 \wedge \overleftarrow{\mathbf{F}}_{=1}\, \mathit{ack}_1))\, \mathbf{U}_{=10}\, \mathit{grant}))$ are existentially punctual. 'One-sided' MTL of [9] is the fragment of existentially punctual MTLPPL without past and Pnueli modalities.

Timed Automata and Model Checking. A *timed automaton* (TA) $\mathcal{A}$ over Σ_{AP} is a finite automaton equipped with real-valued clocks, clock constraints on transitions, and clock resets [3]; $\mathcal{A}$ accepts a timed language $[\![\mathcal{A}]\!]$ under Büchi or finite-word acceptance. Given a formula $\varphi \in$ MTLPPL and a model TA $\mathcal{M}$, the *satisfiability* problem asks whether $[\![\varphi]\!] \neq \emptyset$, and the *model-checking* problem asks whether $[\![\mathcal{M}]\!] \subseteq [\![\varphi]\!]$, i.e., whether every behaviour of $\mathcal{M}$ satisfies φ.

3 Tool Design and Implementation

Overview. The tool MightyPPL is a command-line program written in C++17. Given an existentially punctual MTLPPL formula, MightyPPL can output component TAs individually (the '`compflat`' mode), or generate the synchronous product of component TAs as a single monolithic TA (the '`flat`' mode). The system is depicted in Fig. 1. The source code repository is hosted on GitHub at

https://github.com/hsimho/MightyPPL

Symbolic Encoding of Atomic Propositions. In MightyL and Tempora, atomic propositions are represented by Boolean variables which are set to **true** or **false** non-deterministically. MightyPPL works similarly in the '`compflat`' mode, but it also uses wildcard '$*$' values to effectively force the back-end tools to perform 'symbolic execution' (rather than 'exhaustive testing').

Forward and Backward Reachability Analysis. MightyPPL uses BDDs to synchronise transitions of component TAs into joint transitions in the '`flat`' mode. It enumerates only the combinations of component transitions whose symbolic label conjunctions are satisfiable, and generates only locations and transitions forward-reachable from the initial product location; a backward reachability pass further rules out locations with no accepting run.

Using the Tool. MightyPPL is invoked as

```
mitppl <in_spec_file> --{fin|inf} [out_file --{tck|xml}]
```

where `in_spec_file` contains the MITPPL formula and `fin`/`inf` selects finite-/infinite-word semantics; the optional output file format is `tck`/`xml`. The flag `--compflatten` selects the '`compflat`' mode; For model checking, the model $\mathcal{M}$ can be specified by editing `MightyPPL.cpp` or the generated files.

4 Experiments

All experiments were conducted on a desktop machine with an Intel i9-13900K CPU and 64GB memory. We used commit d711ace of TChecker, commit 4bbf3cc of MoniTAal, commit 1c8f7c2 of PARDIBAAL, commit 5aca063 of BuDDy. The main baseline is Tempora [1] (commit 1ba0bad), the only other tool supporting MITL with past modalities.

Simple MTL **Formulae.** We consider the finite- and infinite-word satisfiability of the handcrafted MTL formulae below:

$$\varphi_1 = \mathbf{F}\left((p\,\mathbf{U}_{[2,3]}\,q) \wedge \neg(p\,\mathbf{U}_{[0,3]}\,q)\right), \qquad \varphi_2 = \mathbf{F}\left((p\,\mathbf{S}_{[2,3]}\,q) \wedge \neg(p\,\mathbf{S}_{[0,3]}\,q)\right),$$
$$\varphi_3 = \mathbf{G}(\neg p \vee \mathbf{F}_{[1,\infty)}\,q) \wedge \mathbf{F}_{[2,3]}\,(p \wedge \neg(\mathbf{F}\,q)), \qquad \varphi_4 = \mathbf{F}_{[5,\infty)}\,(p \wedge (\top\,\mathbf{S}_{[2,3]}\,q)) \wedge \mathbf{G}(\neg q),$$
$$\varphi_5 = (\mathbf{F}_{[1,2]}\,p) \wedge \mathbf{G}_{[1,3]}\,(\neg p \vee (\top\,\mathbf{S}_{[1,2]}\,q)), \qquad \varphi_6 = (\mathbf{F}_{[1,2]}\,p) \wedge \mathbf{G}_{[1,3]}\,(\neg p \vee (\top\,\mathbf{S}_{[3,6]}\,q)),$$
$$\varphi_7 = \mathbf{F}\left((p\,\mathbf{S}_{[1,3]}\,(\mathbf{F}_{[2,5]}\,q)) \wedge \mathbf{G}(\neg q)\right), \qquad \varphi_8 = \mathbf{F}\left((p\,\mathbf{S}_{[1,3]}\,(\mathbf{F}_{=2}\,q)) \wedge \mathbf{G}(\neg q)\right),$$
$$\varphi_9 = \mathbf{F}\left((p\,\mathbf{S}_{[1,3]}\,(\mathbf{F}_{=4}\,q)) \wedge \mathbf{G}(\neg q)\right), \qquad \varphi_{10} = \mathbf{F}\left((p\,\mathbf{S}_{[1,5]}\,(\mathbf{F}_{[2,3]}\,q)) \wedge \neg(p\,\mathbf{S}\,q)\right),$$
$$\varphi_{11} = \mathbf{F}\left((p\,\mathbf{S}_{=2}\,(\mathbf{F}_{[2,3]}\,q)) \wedge \neg(p\,\mathbf{S}\,q)\right), \qquad \varphi_{12} = \mathbf{F}\left((p\,\mathbf{S}_{=4}\,(\mathbf{F}_{[2,3]}\,q)) \wedge \neg(p\,\mathbf{S}\,q)\right),$$
$$\varphi_{13} = \mathbf{F}\left((\mathbf{F}_{[2,5]}(p\,\mathbf{S}_{[1,3]}\,q)) \wedge \mathbf{G}(\neg q)\right), \qquad \varphi_{14} = \mathbf{F}\left((\mathbf{F}_{[2,5]}(p\,\mathbf{S}_{=3}\,q)) \wedge \mathbf{G}(\neg q)\right),$$
$$\varphi_{15} = \mathbf{F}\left((\mathbf{F}_{[2,5]}(p\,\mathbf{S}_{=1}\,q)) \wedge \mathbf{G}(\neg q)\right), \qquad \varphi_{16} = \mathbf{F}\left((\mathbf{F}_{[2,3]}(p\,\mathbf{S}_{[1,5]}\,q)) \wedge \neg(p\,\mathbf{S}\,q)\right),$$
$$\varphi_{17} = \mathbf{F}\left((\mathbf{F}_{=3}(p\,\mathbf{S}_{[2,5]}\,q)) \wedge \neg(p\,\mathbf{S}\,q)\right), \qquad \varphi_{18} = \mathbf{F}\left((\mathbf{F}_{=1}(p\,\mathbf{S}_{[2,5]}\,q)) \wedge \neg(p\,\mathbf{S}\,q)\right),$$

$$\varphi_{19} = \mathbf{F}\left((p\,\mathbf{S}_{[1,2]}\,q) \wedge (p\,\mathbf{S}_{[3,4]}\,r) \wedge \neg\left(\top\,\mathbf{S}\,(q \wedge (p\,\mathbf{S}_{[1,2]}\,r))\right)\right),$$
$$\varphi_{20} = \mathbf{F}\left((p\,\mathbf{S}_{[1,2]}\,q) \wedge (p\,\mathbf{S}_{[3,4]}\,r) \wedge \neg\left(\top\,\mathbf{S}\,(q \wedge (p\,\mathbf{S}_{[1,3]}\,r))\right)\right),$$
$$\varphi'_i = \mathbf{F}\left(\mathbf{G}_{[0,5]}(p_1 \wedge \cdots \wedge p_i) \wedge \mathbf{F}_{[0,3]}(\neg p_i)\right) \quad \text{for } 10 \le i \le 15.$$

The results are listed in Table 2. TEMPORA performs better on some testcases (e.g., φ_5, φ_6) but much worse on others (e.g., $\varphi_2, \varphi_{19}, \varphi_{20}$). The advantange of MIGHTYPPL's symbolic encoding is highlighted by the results on $\varphi'_{13}, \varphi'_{14}, \varphi'_{15}$. We also note that TEMPORA outputs wrong results on $\varphi_7, \varphi_{10}, \varphi_{13}, \varphi_{16}$.

Timed Lamp. This is a simple case study adapted from [6]. We consider a lamp with a single button controlled by the user. Whenever the user *pushes* the button, the lamp *blinks* at the same instant and enters a state, where it blinks at exactly every 1 time unit for three more times. The simple system can be modelled by the TA $\mathcal{M}_{lamp}$ of Fig. 2, which is model-checked against the following formulae (over finite or infinite timed words):

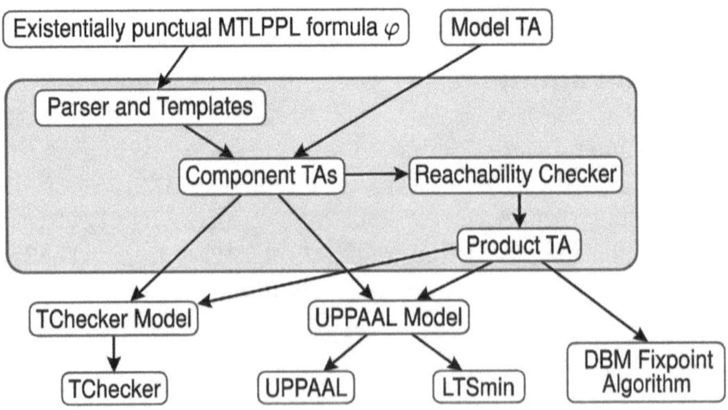

Fig. 2. The model TA $\mathcal{M}_{lamp}$.

Table 2. Execution times on the simple MTL benchmarks. Times are in seconds. 'TO' indicates timeouts (600 s). '-' means unsupported. Incorrect results are marked by '*'.

φ	Sat?	Tempora infinite	Tempora finite	MightyPPL tck --inf compflat	MightyPPL tck --fin compflat	MightyPPL tck --inf flat	MightyPPL tck --fin flat
φ_1	✗	3.675	0.013	0.008	0.008	0.000	0.000
φ_2	✗	TO	0.010	0.013	0.009	0.000	0.000
φ_5	✓	0.004	0.004	0.084	0.088	22.410	2.829
φ_6	✗	0.021	0.005	0.185	0.094	22.038	2.934
φ_7	✓	TO	0.191*	0.008	0.009	0.013	0.008
φ_9	✗	-	-	0.009	0.010	0.013	0.008
φ_{10}	✓	201.135*	0.049*	0.009	0.008	0.022	0.012
φ_{12}	✗	-	-	0.009	0.012	0.022	0.012
φ_{13}	✓	TO	0.203*	0.007	0.010	0.012	0.008
φ_{16}	✓	113.864*	0.055*	0.008	0.010	0.016	0.011
φ_{19}	✓	TO	0.120	23.002	0.787	TO	231.522
φ_{20}	✗	TO	3.520	565.107	4.187	TO	227.865
φ'_{13}	✗	9.274	0.191	0.008	0.008	0.009	0.007
φ'_{14}	✗	29.443	0.546	0.007	0.007	0.010	0.008
φ'_{15}	✗	TO	1.983	0.007	0.007	0.010	0.008

$$
\begin{aligned}
\alpha_1 &= \mathbf{G}(push \Rightarrow \mathbf{F}_{\geq 3}\, blink),\\
\alpha_2 &= \mathbf{G}((push \wedge \mathbf{G}_{(0,\infty)}(\neg push)) \Rightarrow \mathbf{F}_{=3}\, blink),\\
\alpha_3 &= \mathbf{G}((push \wedge \mathbf{G}_{(0,\infty)}(\neg push)) \Rightarrow \neg(\mathbf{F}_{(2,3)}\, blink)),\\
\alpha_4 &= \mathbf{G}((push \wedge (\mathbf{X}_{(0,\infty)}\, push) \wedge \mathbf{F}_{[0,1]}(push \wedge \mathbf{G}_{(0,\infty)}(\neg push))) \Rightarrow \neg(\mathbf{F}_{=3}\, blink)),\\
\alpha_5 &= \mathbf{G}((push \wedge (\mathbf{X}_{(0,\infty)}\, push) \wedge \mathbf{F}_{[0,1]}(push \wedge \mathbf{G}_{(0,\infty)}(\neg push))) \Rightarrow \neg(\mathbf{F}_{[4,5]}\, blink)),
\end{aligned}
$$

$$
\begin{aligned}
\alpha_6 &= \mathbf{G}((blink \wedge \mathbf{G}_{(0,\infty)}\, blink) \Rightarrow \overleftarrow{\mathbf{F}}_{=3}(push)),\\
\alpha_7 &= \mathbf{G}((blink \wedge \mathbf{G}_{(0,\infty)}\, blink) \Rightarrow \neg\overleftarrow{\mathbf{F}}_{=2}(push)),\\
\alpha_8 &= \mathbf{G}((push \wedge blink) \Rightarrow \mathbf{F}_{\leq 5}(\neg blink \wedge \mathbf{G}_{[0,2]}(\neg blink))),\\
\alpha_9 &= \mathbf{G}((blink \wedge \mathbf{F}_{\leq 1}\, blink \wedge \mathbf{G}_{\leq 9}(\mathbf{F}_{\leq 1}\, blink)) \Rightarrow \mathbf{Pn}_{\leq 9}(push, push)),\\
\alpha_{10} &= \mathbf{G}((blink \wedge \mathbf{F}_{\leq 1}\, blink \wedge \mathbf{G}_{\leq 9}(\mathbf{F}_{\leq 1}\, blink)) \Rightarrow \mathbf{Pn}_{\leq 9}(push, push, push)).
\end{aligned}
$$

The results are in Table 3. MightyPPL finishes model checking each of the 10 properties in < 0.4s in every configuration; Tempora supports 4 of them but performs worse on α_3, α_5, and produces a wrong answer on α_8.

Table 3. Execution times on the timed lamp benchmarks. Times are in seconds. 'TO' indicates timeouts (600 s). '-' means unsupported. Incorrect results are marked by '*'.

φ	Hold?	TEMPORA `infinite`	TEMPORA `finite`	MIGHTYPPL `tck --inf compflat`	MIGHTYPPL `tck --fin compflat`	MIGHTYPPL `tck --inf flat`	MIGHTYPPL `tck --fin flat`
α_1	✓	0.002	0.000	0.009	0.009	0.014	0.009
α_2	✓	-	-	0.009	0.010	0.033	0.016
α_3	✓	6.834	0.037	0.008	0.009	0.027	0.014
α_4	✓	-	-	0.013	0.013	0.055	0.023
α_5	✓	TO	2.572	0.013	0.014	0.055	0.022
α_6	✓	-	-	0.027	0.016	0.061	0.025
α_7	✓	-	-	0.012	0.011	0.027	0.014
α_8	✗	0.003	0.000*	0.009	0.011	0.027	0.013
α_9	✓	-	-	0.023	0.022	0.259	0.079
α_{10}	✗	-	-	0.014	0.031	0.370	0.109

Fischer's Mutual Exclusion Protocol. This is a classic mutual exclusion algorithm (from [16]) that has been used as a standard example for UPPAAL. We use the file `fischer.sh` from [11] to generate the system models (asynchronous networks of TAs—one TA $\mathcal{P}_i$ for each of the N processes) with a fixed delay $K = 5$; then we add the atomic propositions $idle_i$, req_i, $wait_i$, cs_i to indicate which location the process $\mathcal{P}_i$ is in. The system models are model-checked against the following formulae (over infinite timed words):

$$\beta_1 = \bigwedge_i \left(\mathbf{G}(req_i \Rightarrow \mathbf{F}_{[0,10]}(cs_i \wedge \mathbf{Y}(\neg cs_i)))\right),$$

$$\beta_2 = \bigwedge_i \left(\mathbf{G}((cs_i \wedge \mathbf{Y}(\neg cs_i)) \Rightarrow \overleftarrow{\mathbf{F}}_{[5,10]} req_i)\right),$$

$$\beta_3 = \bigwedge_i \left(\mathbf{G}((cs_i \wedge \mathbf{Y}(\neg cs_i)) \Rightarrow \neg\mathbf{Pn}_{[0,11]}(cs_i \wedge \mathbf{Y}(\neg cs_i), cs_i \wedge \mathbf{Y}(\neg cs_i)))\right),$$

$$\beta_4 = \bigwedge_i \left(\mathbf{G}((cs_i \wedge \mathbf{Y}(\neg cs_i)) \Rightarrow \neg\mathbf{Pn}_{[0,10]}(cs_i \wedge \mathbf{Y}(\neg cs_i), cs_i \wedge \mathbf{Y}(\neg cs_i)))\right),$$

$$\beta_5 = \bigwedge_i \left(\mathbf{G}((cs_i \wedge \mathbf{Y}(\neg cs_i)) \Rightarrow \overleftarrow{\mathbf{F}}(wait_i \wedge \overleftarrow{\mathbf{F}}_{[0,5]}(req_i \wedge \neg\mathbf{Y}(wait_i))))\right),$$

$$\beta_6 = \bigwedge_i \left(\mathbf{G}((cs_i \wedge \mathbf{Y}(\neg cs_i)) \Rightarrow \overleftarrow{\mathbf{F}}_{[5,20]}(wait_i \wedge \overleftarrow{\mathbf{F}}_{[0,5]}(req_i \wedge \neg\mathbf{Y}(wait_i))))\right).$$

The results are in Table 4. While TEMPORA demonstrates noticeable performance gains on β_5, it produces incorrect results for the closely related β_6.

Pinwheel Scheduling. The *pinwheel scheduling* problem [13] asks whether there is an *infinite* schedule for tasks $\{i \mid 1 \leq i \leq k\}$ (each with a *relative deadline* $a_i \geq 2$), e.g., on each day we can schedule a task, and each task is scheduled at least once in every a_i days. We use the model TA $\mathcal{M}^k_{pin}$ to enforce that exactly one of $\{p_1, \ldots, p_k\}$ is scheduled in each slot, and any two scheduled tasks must be separated by at least 1 time unit. We then model-check $\mathcal{M}^k_{pin}$ against the corresponding formula for an instance $(a_1, \ldots, a_k)$ of the problem:

$$\varphi_{(a_1,\ldots,a_k)} = \Big(\bigwedge_{i\in\{1,\ldots,k\}} \mathbf{F}\, p_i\Big) \Rightarrow \neg \bigwedge_{i\in\{1,\ldots,k\}} \mathbf{G}(p_i \Rightarrow \mathbf{X}\,\mathbf{F}_{[0,a_i]}\, p_i)\,.$$

The results are in Table 5. To highlight the effect of using multiple cores, here we omit the comparison with TEMPORA (which times out on all the instances tested) and exclude the time used for flattening and compiling (which on the most difficult instances take ~ 110s). The single-threaded performance of TCHECKER in the '`flat`' mode is roughly comparable to that of LTSMIN. The latter, however, can run on multiple threads and we can see that a speedup is generally observed as the number of threads increases, and this improvement is sometimes linear with respect to the thread count. Furthermore, we notice that using more than 16 threads does not always yield significant performance gains, likely due to hardware limitations and increased synchronisation overhead.

Table 4. Execution times on the Fischer benchmarks. Times are in seconds. 'TO' indicates timeouts (600 s). '-' means unsupported. Incorrect results are marked by '*'.

φ	N	Hold?	TEMPORA `infinite`	MIGHTYPPL `tck --inf compflat`
β_1	2	✗	0.141	0.010
β_1	3	✗	484.967	0.029
β_1	4	✗	TO	0.087
β_2	2	✗	0.062*	0.266
β_2	3	✗	112.774*	TO
β_3	2	✗	-	0.156
β_3	3	✗	-	6.504
β_3	4	✗	-	TO
β_5	2	✓	0.056	3.841
β_5	3	✓	33.281	TO
β_6	2	✗	0.031*	2.957
β_6	3	✗	8.136*	TO

Table 5. Execution times on the pinwheel scheduling benchmarks. Times are in seconds. Numbers in the heading are the numbers of threads. 'TO' indicates timeouts (600 s).

		MightyPPL tck --inf compflat	MightyPPL tck --inf flat	MightyPPL opaal2lts-mc flat					
φ	Hold?			1	2	4	8	16	32
$(3, 4, 5, 7)$	✓	1.899	0.068	0.028	0.027	0.036	0.063	0.145	0.428
$(3, 4, 5, 8)$	✗	0.291	0.059	0.013	0.033	0.036	0.063	0.145	0.305
$(2, 7, 9, 9, 16)$	✓	TO	1.087	0.912	0.519	0.319	0.286	0.452	0.748
$(2, 7, 9, 10, 16)$	✗	11.154	0.331	0.063	0.043	0.050	0.075	0.156	0.333
$(3, 7, 8, 8, 9, 10)$	✓	TO	10.881	9.106	4.789	2.814	2.122	3.254	3.890
$(3, 7, 8, 8, 9, 11)$	✗	TO	2.484	0.474	0.356	0.242	0.232	0.336	0.637
$(4, 6, 8, 9, 9, 9, 15)$	✓	TO	202.238	TO	TO	375.863	312.715	392.217	475.140
$(4, 6, 8, 9, 9, 9, 16)$	✗	TO	16.072	4.467	3.426	0.269	0.486	1.971	1.937
$(5, 5, 8, 9, 9, 9, 11)$	✓	TO	125.591	491.325	260.836	153.255	120.500	174.310	201.452
$(5, 5, 8, 9, 9, 9, 12)$	✗	TO	15.704	3.014	1.121	1.799	0.480	3.263	9.223
$(6, 7, 7, 7, 8, 8, 10)$	✓	TO	78.520	210.064	113.902	64.913	48.763	70.275	77.565
$(6, 7, 7, 7, 8, 8, 11)$	✗	TO	16.579	8.268	2.543	2.471	2.895	5.219	3.431

Acknowledgments. The work of Khushraj Madnani was supported by the ANRF ARG-MATRICS Grant (File No. ANRF/ARGM/2025/003152/MTR) under the project titled "Approaching the Decidability Purlieu of Timed Languages (ADePT)."

Data Availability Statement. The tool implementation, benchmark models, and datasets generated during the current study are available in the Zenodo repository, https://doi.org/10.5281/zenodo.19852060.

References

1. Akshay, S., Contractor, P., Gastin, P., Govind, R., Srivathsan, B.: TEMPORA: efficient verification of metric temporal properties with past in pointwise semantics. In: TACAS, LNCS, vol. 16505, pp. 640–659, Springer (2026)
2. Akshay, S., Gastin, P., Govind, R., Joshi, A.R., Srivathsan, B.: A unified model for real-time systems: symbolic techniques and implementation. In: International Conference on Computer Aided Verification, pp. 266–288, Springer (2023)
3. Alur, R., Dill, D.L.: A theory of timed automata. Theor. Comput. Sci. **126**(2), 183–235 (1994)
4. Alur, R., Feder, T., Henzinger, T.: The benefits of relaxing punctuality. J.ACM **43**(1), 116–146 (1996)
5. Behrmann, G., Cougnard, A., David, A., Fleury, E., Larsen, K.G., Lime, D.: Uppaaltiga: time for playing games! In: International Conference on Computer Aided Verification, pp. 121–125, Springer (2007)
6. Bersani, M.M., Rossi, M., San Pietro, P.: A tool for deciding the satisfiability of continuous-time metric temporal logic. Acta Informatica **53**(2), 171–206 (2016)

7. Bouyer, P., Markey, N., Ouaknine, J., Worrell, J.: The cost of punctuality. In: LICS, pp. 109–120 (2007). https://doi.org/10.1109/LICS.2007.49
8. Bouyer, P., Markey, N., Ouaknine, J., Worrell, J.: On expressiveness and complexity in real-time model checking. In: Aceto, L., Damgård, I., Goldberg, L.A., Halldórsson, M.M., Ingólfsdóttir, A., Walukiewicz, I. (eds.) ICALP 2008. LNCS, vol. 5126, pp. 124–135. Springer, Heidelberg (2008). https://doi.org/10.1007/978-3-540-70583-3_11
9. Bouyer, P., Srivathsan, B., Vishwanath, V.: Model-checking real-time systems: revisiting the alternating automaton route. In: FoSSaCS, LNCS, vol. 15691, pp. 399–421, Springer (2025)
10. Brihaye, T., Geeraerts, G., Ho, H., Monmege, B.: Mightyl: a compositional translation from MITL to timed automata. In: Majumdar, R., Kuncak, V. (eds.) Computer Aided Verification - 29th International Conference, CAV 2017, Heidelberg, Germany, 24–28 July 2017, Proceedings, Part I, Lecture Notes in Computer Science, vol. 10426, pp. 421–440, Springer (2017). https://doi.org/10.1007/978-3-319-63387-9_21
11. Herbreteau, F., Point, G.: The tchecker tool and libraries. https://github.com/ticktac-project/tchecker (2019). Accessed 20 Apr 2026
12. Ho, H.M., Krishna, S.N., Madnani, K., Majumdar, R., Pandya, P.: MightyPPL: model checking MITL with past and PNUELI modalities. In: TACAS, LNCS, vol. 16505, pp. 457–479, Springer (2026)
13. Holte, R., Mok, A., Rosier, L., Tulchinsky, I., Varvel, D.: The pinwheel: a real-time scheduling problem. In: Proceedings of the 22nd Hawaii International Conference of System Science, pp. 693–702 (1989)
14. Koymans, R.: Specifying real-time properties with metric temporal logic. Real Time Syst. **2**(4), 255–299 (1990)
15. Laarman, A., Olesen, M.C., Dalsgaard, A.E., Larsen, K.G., van de Pol, J.: Multi-core Emptiness Checking of Timed Büchi Automata Using Inclusion Abstraction. In: Sharygina, N., Veith, H. (eds.) CAV 2013. LNCS, vol. 8044, pp. 968–983. Springer, Heidelberg (2013). https://doi.org/10.1007/978-3-642-39799-8_69
16. Lamport, L.: A fast mutual exclusion algorithm. ACM Trans. Comput. Syst. (TOCS) **5**(1), 1–11 (1987)
17. Maler, O., Nickovic, D.: Monitoring temporal properties of continuous signals. In: International Symposium on Formal Techniques in Real-Time and Fault-Tolerant Systems, pp. 152–166. Springer (2004)
18. Maler, O., Nickovic, D., Pnueli, A.: From MITL to timed automata. In: Asarin, E., Bouyer, P. (eds.) FORMATS 2006. LNCS, vol. 4202, pp. 274–289. Springer, Heidelberg (2006). https://doi.org/10.1007/11867340_20
19. Pnueli, A.: The temporal logic of programs. In: 18th Annual Symposium on Foundations of Computer Science (SFCS 1977), pp. 46–57 (1977). https://doi.org/10.1109/SFCS.1977.32

An Algebraic Framework for Quantitative Semantics of Spatio-Temporal Logic with Graph Operators

Sheryl Paul[1(✉)], Vidisha Kudalkar[1], Anand Balakrishnan[2], Tianhao Wu[1], Lars Lindemann[3], and Jyotirmoy V. Deshmukh[1]

[1] University of Southern California, Los Angeles, CA, USA
{sherylpa,kudalkar,wutianha,jdeshmuk}@usc.edu
[2] University of Texas at Austin, Austin, TX, USA
anandbal@utexas.edu
[3] ETH Zürich, Zürich, Switzerland
llindemann@ethz.ch

Abstract. Spatio-Temporal Logic with Graph Operators (STL-GO) extends Signal Temporal Logic (STL) to multi-agent systems by introducing graph operators that reason about the number of neighboring agents satisfying a property, and multi-agent quantifiers. While Boolean semantics for STL-GO are well-defined, quantitative semantics have not been developed. Existing quantitative semantics for spatio-temporal logics such as STREL are insufficient for the counting constraints in STL-GO's graph operators. Thus, we develop quantitative semantics for STL-GO, as a layered algebraic construction that separates temporal aggregation from graph-operator aggregation (governed by an abstract accumulator with a monotone fold and readout). We prove that soundness and completeness of the full semantics reduce to monotonicity conditions on these components, and show that standard min-max and Boolean instantiations satisfy them automatically. We implement the framework and evaluate it on two multi-agent environments: a 2D bounded region with stochastic Dubins-car dynamics and a 3D Earth-satellite system, under four semantic instantiations (Boolean, min-max, signed-deficit, and hybrid of min-max and signed-deficit), demonstrating the tradeoffs between accumulator choices and reporting scalability in the number of agents and time horizon.

Keywords: Signal Temporal Logic · multi-agent systems · robustness semantics · graph operators · semirings

1 Introduction

Signal Temporal Logic [23] is widely used to describe objectives in many applications like in cyber-physical systems (CPS). Its applications span robot task

S. Paul, V. Kudalkar and A. Balakrishnan—Equal contribution.

M. Chen and M. Tschaikowski (Eds.): QEST+FORMATS 2026, LNCS 16913, pp. 226–245, 2027.
https://doi.org/10.1007/978-3-032-35298-9_13

objectives [17,33], control synthesis [31], scenario generation [18] and CPS monitoring [8]. A key enabler of this success is the notion of *quantitative semantics*: which assign a real-valued robustness score to each trajectory, based on *how well* the trajectory satisfies the specification. A substantial body of work has explored such semantics, including classical robustness measures [11], spatial and temporal robustness [1,10,32], stochastic robustness distributions [7], and smooth approximations [15,24,28]. These quantitative semantics have enabled a wide range of applications, including robustness-aware monitoring [16,27], falsification [34], model predictive control [31], reinforcement learning [3,4,21,29,30], and signal classification [35].

Modern systems, however, are generally multi-agent. Applications such as robot swarms, networked CPS, distributed sensing, and social systems require reasoning not only about temporal evolution but also about interactions between agents. These interactions are naturally captured by time-varying graphs encoding communication, sensing, or influence relationships. As a result, specifying multi-agent behavior requires combining temporal logic with relational and spatial reasoning over graphs.

Several spatio-temporal logics have been proposed to address this need, including SSTL [26], SpaTeL [14], SaSTL [22], STREL [6,25], STL-GO [36], and HyperLTL [12]. These logics have enabled monitoring [5,27] and control synthesis [2,19] of spatio-temporal cyber-physical systems. These logics differ in their treatment of spatial structure, expressiveness of quantification, and ability to handle dynamic relations. While Boolean semantics for these logics are generally well understood, quantitative semantics are less developed. Notably, STREL admits an elegant robustness interpretation based on min-max algebra [5], while other frameworks either lack quantitative semantics or rely on constructions that are difficult to interpret or extend. Among these, of particular interest to us is STL-GO, which significantly extends the expressiveness of prior frameworks by offering support for directed graph relations, multiple simultaneous graph layers, count-based constraints over neighbors, and quantification over both graphs and agents. However, this expressiveness introduces new challenges for quantitative semantics. In particular, graph operators in STL-GO require aggregating values over sets of neighbors subject to cardinality constraints. Such aggregation cannot be captured by standard semiring operations (e.g., min or max) alone.

To illustrate the challenges, consider a team of agents subject to the requirement that each agent must have at least 3 neighbors within communication range R satisfying a property φ (Fig. 1). In STL-GO, this is expressed as $\mathbf{FA}_{\mathcal{V}}\ \mathbf{In}^{\exists}_{\mathcal{G},[3,\infty)}\ \varphi$. Plain STL lacks graph operators and multi-agent quantifiers; expressing this would require expanding the counting constraint into a conjunction over agents of disjunctions: $\bigwedge_{i \in V} \bigvee_{\{j_1,j_2,j_3\} \subseteq V \setminus \{i\}} \bigwedge_{k=1}^{3} \left(\|x^{j_k} - x^i\| \leq R \wedge \varphi(x^{j_k}) \right)$, across all $\binom{N-1}{3}$ neighbor subsets with explicit distance predicates. Applying min-max robustness to this expansion yields nested min/max operations that collapse to the margin of the 3rd-best neighbor of the worst-off agent, with the following limitations: (i) *Count-margin insensitivity*: whether a

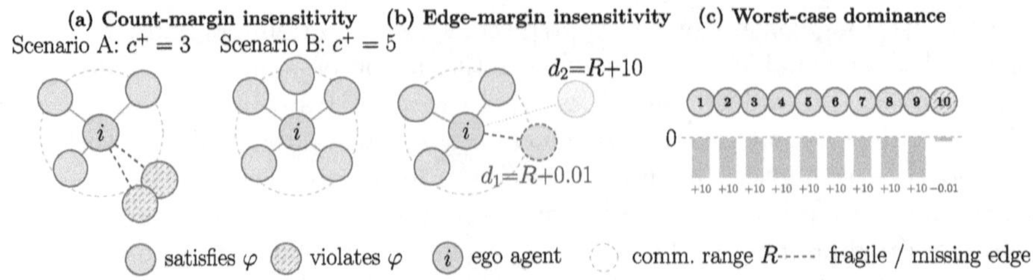

Fig. 1. Limitations of existing spatio-temporal robustness semantics. **(a)** Both scenarios satisfy "at least 3 neighbors satisfy φ," but A has zero spare neighbors and B has two; min-max robustness assigns both the same value. STREL cannot express this counting requirement. **(b)** The dashed-border agent satisfies φ but lies just outside range ($d_1 = R+0.01$); the faded agent is far beyond $d_2 = R + 10$. Both are treated the same way by the robustness semantics. **(c)** The all-agent STL-GO quantifier $\mathbf{FA}_\mathcal{V}$ assigns robustness according to the worst agent (-0.01), discarding 9 agents at $+10$.

drone has exactly three qualifying neighbors or five, the robustness is identical so long as the third-best neighbor's margin is the same, even though the latter configuration is far more resilient. (ii) *Edge-margin insensitivity*: a drone just outside communication range that robustly satisfies the predicate is treated similarly to a drone that might be far away from the range, regardless of how close it is to qualifying as a neighbor. (iii) *Worst-case dominance*: at the system level, the all-agent quantifier assigns robustness based on the single worst agent, discarding the fact that the vast majority may be performing well. Moreover, STREL cannot express the counting requirement at all; its spatial operators reason about path existence, not neighbor counts. These limitations motivate the development of richer quantitative semantics for STL-GO that separately account for individual signal margins, the number of satisfying neighbors, and the robustness of the graph structure itself.

Contributions. Our contributions are: (i) a layered algebraic semantics for STL-GO that separates temporal, graph, and multi-agent operators into distinct algebraic components, with temporal and Boolean operators functioning as in standard STL robustness; (ii) an abstract accumulator for graph operators that factors into a monoid fold and a readout map, isolating counting-specific aggregation and admitting product spaces that propagate robustness values alongside qualifying agent sets; (iii) multi-agent semantics in which agent quantifiers aggregate over an agent-indexed valuation space via semiring operations, completing the semantics for full STL-GO; (iv) soundness and completeness results showing that monotone folds and readouts preserve threshold semantics, with compositionality across all layers by structural induction; and (v) experimental evaluation of four semantic instantiations (Boolean, min-max, signed-deficit, and hybrid) on two multi-agent environments, demonstrating accumulator tradeoffs and scalability in agents and time horizon.

2 Preliminaries

We first describe the multi-agent system and the spatio-temporal logic STL-GO.

2.1 Multi-agent System

We consider a multi-agent system consisting of a finite set of agents denoted by $\mathcal{V} = \{1, \ldots, N\}$. Each agent $i \in \mathcal{V}$ has a state space $\mathcal{X}$, and the joint system state at time t is denoted by $\mathbf{X}_t = (x_t^1, \ldots, x_t^N) \in \mathcal{X}^{|\mathcal{V}|}$, where $x_t^i \in \mathcal{X}$ is the state of agent i at time t. In addition to agent states, the system may include an external environment or world state $w_t \in \mathcal{W}$. Here, $\mathcal{W}$ represents the environment space. Agent interactions are modeled by a collection of time-varying, weighted directed graphs $\mathbf{G}_t = \{\mathcal{G}_t^\tau \mid \tau \in \mathcal{T}\}$, where $\mathcal{T}$ denotes the set of all graph types, and each graph $\mathcal{G}_t^\tau = (\mathcal{V}, \mathcal{E}_t^\tau, w_t^\tau)$ consists of: (i) a set of directed edges $\mathcal{E}_t^\tau \subseteq \mathcal{V} \times \mathcal{V}$, (ii) a weight function $w_t^\tau : \mathcal{E}_t^\tau \to \mathbb{R}$. Each graph type $\tau \in \mathcal{T}$ represents a distinct interaction modality. A *multi-agent execution trace* is a sequence $\mathcal{M} = \{(\mathbf{X}_0, w_0, \mathbf{G}_0), (\mathbf{X}_1, w_1, \mathbf{G}_1), \ldots\}$ over discrete time steps $t \in \mathbb{N}$. We write $(\mathcal{M}, i, t)$ to denote evaluation of agent-local formulas at agent $i \in \mathcal{V}$ and time t, and $(\mathcal{M}, t)$ for multi-agent formulas evaluated over the joint system at time t. We write $(\mathcal{M}, i, t) \models \varphi$ to denote that an agent-local formula φ is *Boolean satisfied* at agent i and time t and $(\mathcal{M}, t) \models \phi$ to denote that a multi-agent formula ϕ is *satisfied* over the system.[1]

2.2 Signal Temporal Logic with Graph Operators (STL-GO)

Spatio-Temporal Logic with Graph Operators (STL-GO) [36] extends STL [23] to reason simultaneously about spatio-temporal agent behaviour and the *topological* relationships between agents encoded in interaction graphs. Its syntax and semantics are stratified into agent-local formulas φ and multi-agent formulas ϕ.

Agent-Local Formulas. The syntax for agent-local STL-GO formulas is:

$$\varphi ::= \top \mid \mu_x \mid \neg\varphi \mid \varphi \wedge \varphi \mid \varphi\, \mathbf{U}_{[t_1,t_2]}\, \varphi \mid \mathbf{In}_{\mathcal{G},E}^{W,\#}\, \varphi \mid \mathbf{Out}_{\mathcal{G},E}^{W,\#}\, \varphi,$$

where $\mu_x : \mathcal{X} \to \mathbb{B}$ is an atomic predicate over a single agent's state; $\neg$ and $\wedge$ are the standard Boolean connectives; $\mathbf{U}_{[t_1,t_2]}$ is the bounded Until operator from STL; and $\mathbf{In}_{\mathcal{G},E}^{W,\#}$, $\mathbf{Out}_{\mathcal{G},E}^{W,\#}$ are the *incoming* and *outgoing graph operators* introduced by STL-GO, where $W = [w_1, w_2] \subseteq \mathbb{R}^\infty$ is an edge-weight interval, $E = [e_1, e_2]$ with $e_1 \in \mathbb{N}$, $e_2 \in \mathbb{N} \cup \{+\infty\}$ constrains the count of qualifying edges, and $\# \in \{\exists, \forall\}$ quantifies over the graph collection $\mathbf{G}_t$. The semantics $(\mathcal{M}, i, t) \models \varphi$ are defined inductively; Boolean and temporal clauses follow STL:

[1] We only consider traces over discrete-time in this work.

$$\begin{aligned}
(\mathcal{M},i,t) &\models \mu_x && \text{iff } \mu_x(\mathbf{x}_t^i),\\
(\mathcal{M},i,t) &\models \neg\varphi && \text{iff } (\mathcal{M},i,t) \not\models \varphi,\\
(\mathcal{M},i,t) &\models \varphi_1 \wedge \varphi_2 && \text{iff } (\mathcal{M},i,t) \models \varphi_1 \text{ and } (\mathcal{M},i,t) \models \varphi_2,\\
(\mathcal{M},i,t) &\models \varphi_1 \mathbf{U}_{[t_1,t_2]}\, \varphi_2 && \text{iff } \exists t' \in t \oplus [t_1,t_2].\ (\mathcal{M},i,t') \models \varphi_2\\
& && \quad \wedge\ \forall t'' \in [t,t'].\ (\mathcal{M},i,t'') \models \varphi_1.
\end{aligned}$$

The graph operators count the neighbours of agent i that satisfy φ and lie within the weight interval W:

$$(\mathcal{M},i,t) \models \mathbf{In}^{W,\#}_{\mathcal{G},E}\varphi \quad \text{iff} \quad \#\mathcal{G}^\tau \in \mathbf{G}_t.\ \left|\{(j,i) \in \mathcal{E}_t^\tau \mid w_t^\tau(j,i) \in W,\ (\mathcal{M},j,t) \models \varphi\}\right| \in E$$

The **Out** operator is similarly defined except using $(i,j) \in \mathcal{E}_t^\tau$ instead. Here $\# = \exists$ requires the count condition to hold in *some* graph $\mathcal{G}^\tau \in \mathbf{G}_t$; $\# = \forall$ requires it in *every* graph. When edge weights are unconstrained, we abbreviate $\mathbf{In}^{\#}_{\mathcal{G},E}\varphi$ and $\mathbf{Out}^{\#}_{\mathcal{G},E}\varphi$ (setting $W = (-\infty,\infty)$).

Multi-agent Formulas. Multi-agent STL-GO formulas ϕ compose agent-local formulas and predicates over the joint system state:

$$\phi ::= \top \mid \mu \mid i.\varphi \mid \neg\phi \mid \phi \wedge \phi \mid \phi\, \mathbf{U}_{[t_1,t_2]}\, \phi,$$

where $\mu : \mathcal{X}^{|\mathcal{V}|} \times \mathcal{W} \to \mathbb{B}$ is an atomic predicate over the joint agent and world state, and $i.\varphi$ embeds the agent-local formula φ for agent i into the multi-agent context. The semantics $(\mathcal{M},t) \models \phi$ follow the same structure as the agent-local case, with: $(\mathcal{M},t) \models \mu$ iff $\mu(\mathbf{X}_t, w_t)$, and $(\mathcal{M},t) \models i.\varphi$ iff $(\mathcal{M},i,t) \models \varphi$. All remaining clauses ($\neg$, $\wedge$, $\mathbf{U}_{[t_1,t_2]}$) are identical in structure to the agent-local case. The derived operators $\mathbf{F}_{[t_1,t_2]}$ and $\mathbf{G}_{[t_1,t_2]}$ follow from Until as in STL. For convenience, the agent-universal and agent-existential shorthands are: $\mathbf{FA}_{\mathcal{V}}\,\varphi \triangleq \bigwedge_{i\in\mathcal{V}} i.\varphi$, and $\mathbf{EX}_{\mathcal{V}}\,\varphi \triangleq \bigvee_{i\in\mathcal{V}} i.\varphi$.

The semantics of STL-GO presented so far assume the standard Boolean satisfaction semantics; i.e., for a given agent-centric STL-GO formula φ, or for a multi-agent STL-GO formula ϕ, a given multi-agent execution trace either satisfies the given formulas or violates them. *Quantitative semantics* generalize the notion of Boolean semantics by defining a function that maps a formula and a multi-agent trace to some (signed) number that indicates a quantitative degree of satisfaction. Many different quantitative semantics for logics such as STL and STREL have been defined in the literature [5,10,11,32]. Of particular relevance to this paper are generalization of quantitative semantics as operations defined using suitable algebras. We also define the quantitative semantics of STL-GO more generally using specific algebraic operations. Before we define them in the next section, we review some key ideas from algebra.

2.3 Background on Algebraic Structures

Definition 1 (Monoid). *A* monoid *is a triple $(M,\oplus,e)$ consisting of a set M, a binary operation $\oplus : M \times M \to M$, and an element $e \in M$, such that for all*

$a, b, c \in M$: $(a \oplus b) \oplus c = a \oplus (b \oplus c)$ and $a \oplus e = e \oplus a = a$, i.e., all the elements satisfy the property of associativity and e acts as an identity element. Further, a commutative monoid *is a monoid $(M, \oplus, e)$ that additionally satisfies, for all $a, b \in M$: $a \oplus b = b \oplus a$.*

Definition 2 (Semiring [13,20]**).** *A tuple $\mathbb{K} = (K, \oplus, \otimes, \bot, \top)$ is a* semiring *with underlying set K if $(K, \oplus, \bot)$ is a commutative monoid with identity $\bot$; $(K, \otimes, \top)$ is a monoid with identity $\top$; $\otimes$ distributes over $\oplus$; and $\bot$ is an annihilator for $\otimes$ (i.e. $k \otimes \bot = \bot \otimes k = \bot$ for all $k \in K$).*

A semiring $\mathbb{K}$ is *commutative* if $\otimes$ is also commutative, and *simple* if $k \oplus \top = \top$ for all $k \in K$. It is *additively* (resp. *multiplicatively*) *idempotent* if $k \oplus k = k$ (resp. $k \otimes k = k$) for all $k \in K$. A semiring that is commutative, additively idempotent, and simple is called a *constraint semiring* (c-semiring). A c-semiring that is also multiplicatively idempotent induces a bounded distributive lattice, with $\top$ as supremum and $\bot$ as infimum.

Example 1. Consider the semiring $(\mathbb{R} \cup \{-\infty, +\infty\}, \max, \min, -\infty, +\infty)$. Then $(\mathbb{R}, \max, -\infty)$ is a commutative monoid with identity $-\infty$, and $(\mathbb{R}, \min, +\infty)$ is a monoid with identity $+\infty$. By the distributivity of min over max, we have $\min(a, \max(b, c)) = \max(\min(a, b), \min(a, c))$. The element $-\infty$ is an annihilator for min, since $\min(a, -\infty) = -\infty$. Both operations are idempotent, i.e., $\max(a, a) = a$ and $\min(a, a) = a$. The semiring is commutative, and simple since $\max(k, +\infty) = +\infty$ for all $k \in K$.

When considering a unary operator, $\ominus : K \to K$, on a bounded distributive lattice K, these operators give rise to a special case of a useful algebra:

Definition 3 (De Morgan Algebra [9]**).** *A* De Morgan algebra *is a structure $(K, \oplus, \otimes, \ominus, \bot, \top)$ such that $(K, \oplus, \otimes, \bot, \top)$ is a bounded distributive lattice and $\ominus$ is a negation on K satisfying, for all $a, b \in K$: $\ominus(a \oplus b) = \ominus a \otimes \ominus b$, $\ominus(a \otimes b) = \ominus a \oplus \ominus b$, $\ominus \ominus a = a$.*

The two De Morgan algebras of primary interest here are: (i) The *Boolean algebra* $(\mathbb{B}, \vee, \wedge, \neg, \mathbf{0}, \mathbf{1})$, giving qualitative (Boolean) satisfaction semantics. (ii) The *min-max algebra* $(\mathbb{R}^{\infty}, \max, \min, -, -\infty, +\infty)$ over the extended reals, which recovers the quantitative (robustness) semantics of STL [11].

Since we are interested in the semantics of multi-agent systems, we will consider how such semirings can be extended to multiple components, and how we can generalize operations on multiple semirings. Let semirings be $\{(K_\ell, \oplus_\ell, \otimes_\ell, \bot_\ell, \top_\ell)\}_{\ell=1}^{m}$ and their product semiring is $K := \prod_{\ell=1}^{m} K_\ell$, with componentwise operations: $(x \oplus y)_\ell := x_\ell \oplus_\ell y_\ell$, $(x \otimes y)_\ell := x_\ell \otimes_\ell y_\ell$ and identities: $\bot := (\bot_\ell)_\ell$, $\top := (\top_\ell)_\ell$. From this, we can algebraically define the notion of a general *commutative monoid* operating on this product semiring as a *semimodule*.

Definition 4 (Semimodule). *Let $(K, \oplus, \otimes, \bot, \top)$ be a semiring. A semimodule over K is a commutative monoid $(S, \oplus_S, \bot_S)$ together with a scalar multiplication $\otimes : K \times S \to S$, such that for all $k, k' \in K$ and $s, s' \in S$:*

$$k \otimes (s \oplus_S s') = (k \otimes s) \oplus_S (k \otimes s'); \qquad (k \oplus k') \otimes s = (k \otimes s) \oplus_S (k' \otimes s)$$
$$(k \otimes k') \otimes s = k \otimes (k' \otimes s); \qquad \top \otimes s = s; \quad k \otimes \bot_S = \bot_S$$

Definition 5 ($\mathbb{K}$-Labeled Trace). *Let $(K, \oplus, \otimes, \ominus, \bot, \top)$ be a De Morgan algebra and L a space universe. A* $\mathbb{K}$-labeled trace *is a pair (σ, ν) where $\sigma = \mathcal{S}_0\mathcal{S}_1 \cdots$ is a finite sequence of spatial models and $\nu(\mathcal{S}_i, l, \mu) \in K$ is the value of atomic predicate μ at location $l \in L$ in model $\mathcal{S}_i$.*

3 Layered Algebraic Semantics for STL-GO

In addition to the operators introduced by STL, STL-GO also introduces *graph operators* and *multi-agent operators* which require mechanisms beyond those required for defining standard STL robustness. While the latter includes only Boolean and temporal operators that can be governed by a single algebraic structure such as a DeMorgan Algebra or semiring, STL-GO requires multiple forms of algebraic structures. To capture this uniformly, we present STL-GO semantics as a layered algebraic construction. We begin with the temporal semantics, which provide the base algebraic structure.

3.1 Semirings for Temporal Operators

We begin with ego-centric temporal operators for STL-GO which generalize robustness semantics of STL. Let $(K, \oplus, \otimes, \ominus, \bot, \top)$ be an algebra, where $(K, \oplus, \bot)$ is a commutative monoid, $(K, \otimes, \top)$ is a monoid, $\ominus$ is an involutive negation satisfying De Morgan laws when present. The set K represents the space of semantic values assigned to formulas.

Ego-Centric Boolean and Temporal Semantics. To evaluate a formula φ for agent i at time t, the ego-centric semantics $\rho_i(\sigma, \varphi, t) \in K$ can be defined as:

$$\rho_i(\sigma, \mu, t) = \nu(\mathcal{S}_t, i, \mu), \quad \rho_i(\sigma, \varphi_1 \wedge \varphi_2, t) = \rho_i(\sigma, \varphi_1, t) \otimes \rho_i(\sigma, \varphi_2, t),$$
$$\rho_i(\sigma, \neg\varphi, t) = \ominus\, \rho_i(\sigma, \varphi, t), \quad \rho_i(\sigma, \varphi_1 \vee \varphi_2, t) = \rho_i(\sigma, \varphi_1, t) \oplus \rho_i(\sigma, \varphi_2, t),$$
$$\rho_i(\sigma, \varphi_1 \mathbf{U}_{[t_1,t_2]} \varphi_2, t) = \bigoplus_{t' \in t+[t_1,t_2]} \left(\rho_i(\sigma, \varphi_2, t') \otimes \bigotimes_{t'' \in [t, t']} \rho_i(\sigma, \varphi_1, t'') \right)$$

Derived operators such as **F** and **G** follow in the usual way.

Example 2. Different choices of $(K, \oplus, \otimes)$ yield different interpretations:
(i) Boolean Semantics: $K = \{0, 1\}$, $\oplus = \vee$, $\otimes = \wedge$ recovers the classical qualitative satisfaction. *(ii) Min-max Semantics:* $K = \mathbb{R}^{\infty}$, $\oplus = \max$, $\otimes = \min$ recovers the standard STL robustness, where values represent perturbation margins.

Recall the definition of product semirings. This construction allows formulas to simultaneously produce multiple semantic quantities. E.g., we may combine a robustness value with quantities such as counts, or relational data. The temporal semantics may be extended componentwise to this setting, ensuring consistency across outputs. This multi-output structure will be essential for defining graph operators which we describe below.

3.2 Multi-agent and Graph Operators

Graph Operators. We now define the semantics of graph operators. Unlike temporal operators, which are induced directly by the semiring structure on K, graph operators require aggregation over dynamically defined neighborhoods of agents. This aggregation is not, in general, expressible purely in terms of the semiring operations $(\oplus, \otimes)$.

Let $\mathcal{V}$ denote the set of agents, and define $S := K^{\mathcal{V}}$. Each element $s \in S$ assigns to every agent $i \in \mathcal{V}$ a semantic value $s(i) \in K$. The space S inherits algebraic structure from K: $(S, \oplus, \bot)$ is a commutative monoid under addition: $(s \oplus s')(i) := s(i) \oplus s'(i)$, S is a semimodule over K with scalar multiplication defined: $(k \otimes s)(i) := k \otimes s(i)$. For a fixed trace σ, formula φ, and time t, the ego-centric semantics induce an element $s_{\varphi,t} \in S$, $s_{\varphi,t}(i) := \rho_i(\sigma, \varphi, t)$, which collects the local evaluations across all agents.

Eligible Neighbors. For an agent $i \in \mathcal{V}$, time t, graph type $\tau \in \mathcal{T}$, and weight interval W, we define the sets of eligible neighbors: $\mathcal{N}_i^{\tau,\mathbf{In}}(t, W) = \{j \in \mathcal{V} \mid (j,i) \in \mathcal{E}_t^{\tau},\ w_t^{\tau}(j,i) \in W\}$, $\mathcal{N}_i^{\tau,\mathbf{Out}}(t, W)$ is defined similarly except with $(i,j) \in \mathcal{E}_t^{\tau}$. The graph operators act by restricting the global valuation $s_{\varphi,t} \in S$ to the neighborhood of i, yielding a multiset of values: $\{s_{\varphi,t}(j) \mid j \in \mathcal{N}_i^{\tau,\circ}(t, W)\} \subseteq K$ where $\circ \in \{\mathbf{In}, \mathbf{Out}\}$.

Aggregation Monoid. To combine neighbor values we introduce an aggregation space $(M, \boxplus, e)$, a commutative monoid. It serves as a domain for aggregating multisets of semantic values. Choices may include: $M = K$ with $\boxplus = \oplus$ (idempotent reduction), $M = \mathbb{Z}$ with $\boxplus = +$ (counting), $M = \mathbb{R}$ with $\boxplus = +$ (averaging), product spaces $M = K \times M'$ for hybrid semantics.

Let $\mathcal{A}_E : K^* \to M$ be a fold map that aggregates a multiset of neighbor values subject to the count constraint E, where K^* denotes the set of finite multisets over K. Then for each graph type τ define: $\mathcal{A}_E(\{\, s_{\varphi,t}(j) \mid j \in \mathcal{N}_i^{\tau,\circ}(t, W) \,\}) \in M$ with $\circ \in \{\mathbf{In}, \mathbf{Out}\}$. This folding operation summarizes the multiset of neighbor values into a single aggregate element of M. Different choices of $\mathcal{A}_E$ induce different interpretations of the count constraint.

Readout. To obtain a value compatible with the temporal semantics, we map the aggregated result back into K via a readout map $h : M \to K$. This yields, for each graph type τ: $h(\mathcal{A}_E(\{\, s_{\varphi,t}(j) \mid j \in \mathcal{N}_i^{\tau,\circ}(t, W) \,\})) \in$ K with $\circ \in \{\mathbf{In}, \mathbf{Out}\}$.

Quantification over Graphs. Finally, we aggregate across graph types using semiring operations: $\mathcal{Q}_{\exists}^{K} = \bigoplus$, and $\mathcal{Q}_{\forall}^{K} = \bigotimes$. The graph operators are then defined as: $\rho_i(\sigma, \mathbf{In}_{\mathcal{G},E}^{W,\#}\varphi, t) = \mathcal{Q}_{\#}^{K}\Big(h\Big(\mathcal{A}_E(\{\, s_{\varphi,t}(j) \mid j \in \mathcal{N}_i^{\tau,\mathbf{In}}(t, W)\,\})\Big)\Big)_{\tau\in\mathcal{T}}$ and similarly for the **Out** operator. For a fixed trace σ, formula φ, time t, and quantifier $\# \in \{\exists, \forall\}$, we define the *induced local graph operator* $T_i^{\#} : S \to K$ by $T_i^{\#}(s) = \mathcal{Q}_{\#}^{K}\Big(\, h(\mathcal{A}_E(\{\, s(j) \mid j \in \mathcal{N}_i^{\tau,\circ}(t, W)\,\}))\Big)_{\tau\in\mathcal{T}}$. The graph operator semantics are then $\rho_i(\sigma, \mathbf{In}_{\mathcal{G},E}^{W,\#}\varphi, t) = T_i^{\#}(s_{\varphi,t})$, and analogously for **Out**.

In this setting, graph aggregation is fully internal to the semiring, and the resulting operator preserves the algebraic structure of K.

Multi-agent Aggregation and Quantifiers. In addition to neighborhood-based aggregation, STL-GO supports aggregation over the entire agent set. This yields the semantics of multi-agent quantifiers. Let $s_{\varphi,t} \in S = K^{\mathcal{V}}$ denote the agent-indexed valuation induced by a formula φ at time t, i.e., $s_{\varphi,t}(i) := \rho_i(\sigma, \varphi, t)$. We define a global aggregation map $\mathsf{Acc} : S \to K$, which combines the values across all agents. The existential and universal quantifiers are obtained as special cases: $\mathsf{Acc}_{\oplus}(s) = \bigoplus_{i\in\mathcal{V}} s(i)$, $\mathsf{Acc}_{\otimes}(s) = \bigotimes_{i\in\mathcal{V}} s(i)$. The semantics of multi-agent quantifiers are then: $\rho(\sigma, \mathbf{EX}_{\mathcal{V}}\,\varphi, t) = \mathsf{Acc}_{\oplus}(s_{\varphi,t})$, $\rho(\sigma, \mathbf{FA}_{\mathcal{V}}\,\varphi, t) = \mathsf{Acc}_{\otimes}(s_{\varphi,t})$. More generally, Acc may be instantiated as an arbitrary aggregation map, potentially factoring through an intermediate space M_A as in the graph-operator construction: $S \to M_A \to K$.

3.3 Spatio-Temporal Algebras

The construction of Sect. 3.2 separates aggregation into an intermediate space M and a readout map $h : M \to K$. This enables the use of richer aggregation domains that carry additional structural information beyond scalar robustness. We now show how to construct such domains via product algebras, yielding *'spatio-temporal implicants'* that generalize classical temporal semantics.

Product Aggregation Spaces. Let $(K, \oplus, \otimes)$ be the base semiring used for temporal semantics. We extend the aggregation space by introducing an auxiliary structure that tracks the *origin* of satisfaction. Let $R := \mathcal{P}(\mathcal{V} \times \mathbb{T})$ where $\mathbb{T}$ denotes the time domain. Elements of R represent sets of agent-time pairs witnessing the satisfaction of a formula. We define the product aggregation space $M := K \times R$ with componentwise aggregation $(r_1, R_1) \boxplus (r_2, R_2) := (r_1 \oplus r_2,\ R_1 \cup R_2)$. Thus, M simultaneously carries: (i) a scalar semantic value $r \in K$, (ii) a relational component $R \subseteq \mathcal{V} \times \mathbb{T}$.

Spatio-Temporal Implicants. In classical temporal logic, implicants correspond to time indices at which a formula is satisfied. We extend this by interpreting the relational component R as a spatio-temporal implicant: a set that acts as a witness to the satisfaction of the formula. Our construction generalizes classic implicants by lifting them to the spatio-temporal domain: $R \subseteq \mathcal{V} \times \mathbb{T}$, allowing formulas to track both *which agents* and *when* they contribute to satisfaction.

For e.g., for the ego agent i fixed by the enclosing graph operator, writing $r_j = \rho_j(\sigma, \varphi, t)$, we may define the fold as the ordered pair in $M = K \times R$: $\mathcal{A}_E(\{r_j\}) = \Big(\bigoplus_j r_j,\ \{(j,t) \mid j \in \mathcal{N}_i^{\tau,\circ}(t, W)\} \Big)$, where the first component is the scalar aggregate in K and the second is the witness set in R. More generally, R may track only those neighbours that contribute positively (e.g. $r_j \succ \bot_K$).

Readout and Coupling. The readout map $h : M \to K$ allows the relational component to influence the final semantic value. A general form is: $h(r, R) = \Phi\big(r,\ \pi_{\mathcal{V}}(R)\big)$, where $\pi_{\mathcal{V}}(R) = \{i \mid \exists t,\ (i,t) \in R\}$ projects onto the set of participating agents, and Φ combines signal-level and structural information. For e.g., a simple example is a hybrid robustness measure which augments the scalar robustness r with a count-based correction term: $h(r, R) = r + \alpha\big(|\pi_{\mathcal{V}}(R)|\big)$, where $\alpha \in \mathbb{R}$ is some constant.

General Semiring Compositions. More generally, the product construction extends to arbitrary families of semirings: $K = \prod_{\ell=1}^{m} K_\ell,\ M = \prod_{\ell=1}^{m} M_\ell$, with layer-specific aggregation and readout maps. This enables the simultaneous propagation of multiple semantic quantities, such as: (a) robustness values, (b) counts or densities, (c) witness sets, (d) probabilistic weights. Thus, STL-GO admits a unified algebraic semantics in which temporal, spatial, and structural information are propagated compositionally through product semiring constructions. For some examples of algebras to pick for different semantic objectives, the reader is encouraged to look at the Supplementary Material.

4 Structural Guarantees and Complexity

We now characterize when the generalized aggregation framework described yields semantics that are consistent with Boolean satisfaction.

Definition 6 (Soundness and Completeness). *We write $\sigma, i, t \models \varphi$ to denote that agent i satisfies φ at time t along trace σ. In addition, we associate to each formula a quantitative semantics (robustness value) $\rho_i(\sigma, \varphi, t) \in K$, which measures the degree of satisfaction of the formula in the underlying algebra K. The quantitative semantics defined by the function $\rho_i(\sigma, \varphi, t) \in K$ is* sound *if $\rho_i(\sigma, \varphi, t) \succ \bot_K \Rightarrow \sigma, i, t \models \varphi$, and* complete *if $\sigma, i, t \models \varphi \Rightarrow \rho_i(\sigma, \varphi, t) \succ \bot_K$, i.e., soundness ensures that positive robustness implies satisfaction, while completeness ensures all satisfying configurations are assigned positive robustness.*

The temporal semantics are sound and complete by classical results [11]; our goal is to identify conditions the graph-operator pipeline $K^* \xrightarrow{\mathcal{A}_E} M \xrightarrow{h} K$ preserves this property. Throughout, we assume that K is equipped with a partial order $\preceq$ compatible with the semiring operations, and write $\bot_K$ for the threshold separating satisfaction from violation (e.g., $0 \in \mathbb{R}^\infty$ for the min-max algebra).

Before turning to soundness and completeness, we first record a weaker but very general structural property of the layered semantics. Namely, if each stage of the graph-aggregation pipeline is monotone with respect to the natural orders on the semantic and aggregation spaces, then the overall operator preserves the pointwise ordering of agent-wise valuations. This result is independent of any particular interpretation of the threshold $\bot_K$ and applies uniformly to scalar, product, and hybrid semantics.

Definition 7 (Monotone map). *A function $f : X \to Y$ between partially ordered sets is monotone if $x \preceq x' \Rightarrow f(x) \preceq f(x')$. Monotonicity implies that increasing the robustness of inputs does not decrease the value of the output.*

Theorem 1 (Order preservation of layered aggregation). *Let $(K, \preceq_K)$ be a partially ordered semiring, and let $S := K^{\mathcal{V}}$ be the induced semimodule of agent-indexed valuations equipped with the pointwise order $s \preceq_S s' \iff s(i) \preceq_K s'(i)\ \forall i \in \mathcal{V}$. Let $(M, \preceq_M)$ be a partially ordered aggregation space. Fix an agent $i \in \mathcal{V}$, time t, graph type $\tau \in \mathcal{T}$, and consider the graph-aggregation pipeline $S \xrightarrow{\mathrm{res}_i^{\tau}} K^{\mathcal{N}_i^{\tau,\circ}(t,W)} \xrightarrow{\mathcal{A}_E} M \xrightarrow{h} K$, where res_i^{τ} denotes restriction of an agent-wise valuation to the eligible neighborhood $\mathcal{N}_i^{\tau,\circ}(t, W)$. Assume: (1) the restriction map res_i^{τ} is order-preserving; (2) the fold map $\mathcal{A}_E$ is monotone with respect to the product order on $K^{\mathcal{N}_i^{\tau,\circ}(t,W)}$; (3) the readout map $h : M \to K$ is monotone; (4) the graph-type quantifier $\mathcal{Q}_{\#}^{K}$ is monotone with respect to the pointwise order on $K^{|\mathcal{T}|}$. Then the induced local graph operator $T_i^{\#} : S \to K$ is monotone, i.e., $s \preceq_S s' \implies T_i^{\#}(s) \preceq_K T_i^{\#}(s')$. If this holds for every agent $i \in \mathcal{V}$, then the induced global operator $T : S \to S$, $(T(s))(i) := T_i^{\#}(s)$, is monotone with respect to the pointwise order on S.*

Proof. The proof is in the Supplementary Material. □

Theorem 1 is the structural basis for the results that follow.

Proposition 1 (Soundness and Completeness of Graph Aggregation). *Let $(K, \oplus, \otimes, \ominus, \bot_K, \top_K)$ be a partially ordered De Morgan algebra with order $\preceq$. Let $(M, \boxplus, e)$ be a commutative monoid with order $\preceq_M$, and let $\mathcal{A}_E : K^* \to M$, $h : M \to K$ define the aggregation pipeline for a count constraint $E = [e_1, e_2]$. Define the composed map $F_E := h \circ \mathcal{A}_E : K^* \to K$. Assume: (1) For each input r_j, $r_j \succ \bot_K \iff$ neighbor j satisfies φ; (2) For every configuration $\{r_j\}$, if $c^+ = |\{j : r_j \succ \bot_K\}|$, $c^+ \in [e_1, e_2] \Rightarrow F_E(\{r_j\}) \succ \bot_K$, and $c^+ \notin [e_1, e_2] \Rightarrow F_E(\{r_j\}) \preceq \bot_K$.*

Then the induced graph operator is sound and complete. Consequently, since by Assumption 1 $c^+ = \left|\{j : \sigma, j, t \models \varphi\}\right|$, the quantitative sign agrees with Boolean satisfaction: $F_E(\{r_j\}) \succ \bot_K \iff (\mathcal{M}, i, t) \models \mathbf{In}_{\mathcal{G},E}^{W,\#}\varphi$.

Proof. The proof can be found in the Supplementary Material.

□

Lemma 1 (Quantifiers preserve threshold semantics). *Let $(K, \oplus, \otimes, \ominus, \bot_K, \top_K)$ be an ordered De Morgan algebra, and let $\mathcal{V}$ be finite. Assume that for all finite families $\{x_i\}_{i\in I} \subseteq K$, $\bigoplus_{i\in I} x_i \succ \bot_K \iff \exists i \in I : x_i \succ$*

Algorithm 1. Quantitative monitoring of STL-GO

Input: Trace σ; formula ϕ; algebra K; accumulator $\mathcal{A}_E$, readout h.

for all subformulae ψ of ϕ, bottom-up; agents i; times t **do**
    switch root operator of ψ **do**
        case μ (predicate): $\rho_i(\sigma, \psi, t) \leftarrow \nu(\mathcal{S}_t, i, \mu)$
        end case
        case $\neg, \wedge, \mathbf{U}_{[t_1,t_2]}$ (Boolean/temporal): apply $\ominus, \otimes$, and the $\mathbf{U}_{[t_1,t_2]}$ recurrence over K
        end case
        case $\mathbf{EX}_{\mathcal{V}}, \mathbf{FA}_{\mathcal{V}}$ (quantifier): apply $\bigoplus_i, \bigotimes_i$ over K
        end case
        case $\mathbf{In}^{W,\#}_{\mathcal{G},E}\psi_1$ / $\mathbf{Out}^{W,\#}_{\mathcal{G},E}\psi_1$ (graph):
            $\rho_i(\sigma, \psi, t) \leftarrow \mathcal{Q}^K_{\#}\big(h(\mathcal{A}_E(\{\rho_j(\sigma, \psi_1, t) \mid j \in \mathcal{N}_i^{\tau,\circ}(t, W)\}))\big)_{\tau \in \mathcal{T}}$
        end case
    end switch
end for
return $\rho(\sigma, \phi, \cdot)$

$\bot_K$, *and* $\bigotimes_{i\in I} x_i \succ \bot_K \iff \forall i \in I : x_i \succ \bot_K$. *Then the multi-agent quantifiers satisfy:* $\rho(\sigma, \mathbf{EX}_{\mathcal{V}}\,\varphi, t) \succ \bot_K \iff \exists i \in \mathcal{V} : \rho_i(\sigma, \varphi, t) \succ \bot_K$, *and* $\rho(\sigma, \mathbf{FA}_{\mathcal{V}}\,\varphi, t) \succ \bot_K \iff \forall i \in \mathcal{V} : \rho_i(\sigma, \varphi, t) \succ \bot_K$. *Consequently, if the agent-local semantics of* φ *are sound and complete, then so are the multi-agent quantifiers* $\mathbf{EX}_{\mathcal{V}}\,\varphi$ *and* $\mathbf{FA}_{\mathcal{V}}\,\varphi$.

Proof. Proof can be found in the Supplementary Material. □

Theorem 2 (End-to-end soundness and completeness). *Assume that: (1) atomic predicate valuations are sound and complete, i.e., for every atomic predicate* μ, $\rho_i(\sigma, \mu, t) \succ \bot_K \Leftrightarrow \sigma, i, t \models \mu$, *(2) the temporal/Boolean operators admit an interpretation over* $(K, \oplus, \otimes, \ominus, \bot_K, \top_K)$ *that is sound and complete with respect to the Boolean STL semantics; (3) each graph operator satisfies Proposition 1; (4) multi-agent and graph-type quantifiers satisfy Lemma 1; then for every STL-GO formula* ϕ, $\rho(\sigma, \phi, t) \succ \bot_K \iff \sigma, t \models \phi$, *that is, the quantitative semantics are sound and complete with respect to the Boolean semantics.*

Proof. The proof follows by structural induction on the formula ϕ, and is detailed in the Supplementary Material. □

The preceding results establish soundness and completeness by verifying that each syntactic construct (temporal and graph operators, and quantifiers) preserves correspondence between quantitative robustness and Boolean satisfaction.

Monitoring Algorithm and Runtime Complexity. In Algorithm 1 we describe a bottom-up procedure (with respect to the specification syntax) that computes the robustness of a given finite multi-agent trace, generalizing the Boolean STL-GO monitor of [36].

Theorem 3 (Monitoring complexity). *Let φ be an STL-GO formula of size $|\varphi|$, evaluated over N agents, horizon T, and M graph types. Let $\Delta := \max_{i,t,\tau} \left|\mathcal{N}_i^{\tau,\circ}(t,W)\right|$ denote the largest number of eligible neighbours any agent i has in any graph type τ at any time t. Assume that (i) all temporal operators use fixed-size bounded windows, (ii) the eligible-neighbour sets $\mathcal{N}_i^{\tau,\circ}(t,W)$ are precomputed, and (iii) each accumulator update, readout, and semiring operation takes $O(1)$ time. Then Algorithm 1 computes the robustness signal $\rho(\sigma,\varphi,\cdot)$ in $O(|\varphi|\, N\, T\, M\, \Delta)$ time.*

Proof. The proof can be found in the Supplementary Material. □

5 Experiments

We evaluate the STL-GO robustness semantics on two multi-agent simulation environments with distinct dynamics and interaction structures: The two environments are chosen to stress different aspects of the semantics.

Environment 1: 2D Bounded Region. We simulate $N \in \{10, 20, 50, 100\}$ agents over $T = 50$ discrete timesteps moving in a bounded planar region $[-50, 50]^2 \subset \mathbb{R}^2$. Each agent's state is $\mathbf{x}_t^i = (x_t^i, y_t^i, \theta_t^i) \in \mathbb{R}^2 \times [0, 2\pi)$, comprising position and heading angle. The dynamics follow a stochastic Dubins-car model: $x_{t+1}^i = x_t^i + v_t^i \cos\theta_{t+1}^i \Delta t$, $y_{t+1}^i = y_t^i + v_t^i \sin\theta_{t+1}^i \Delta t$, $\theta_{t+1}^i = (\theta_t^i + \omega_t^i \Delta t) \bmod 2\pi$, where $v_t^i \sim \text{Unif}[0, 10]$ and $\omega_t^i \sim \text{Unif}[-\pi/4, \pi/4]$ are linear and angular velocities. (See Fig. 2a). The 2D Dubins-car region models a planar multi-robot team (e.g. ground search-and-rescue or distributed sensing), where the sensing/communication graphs are dense and time-varying, stressing the *graph-operator* aggregation over large neighbourhoods.

Environment 2: 3D Sphere and Free Space. This environment is inspired by Earth-satellite systems, comprising two agent classes: ground stations constrained to a spherical surface and satellites moving freely in 3D space. Both are simulated for $T = 50$ timesteps (see Fig. 2b).

Sphere-constrained agents (N_s). Each agent's state is parameterized by spherical angles (θ_t^i, ϕ_t^i) on a sphere of radius R_s, with Cartesian position given by $(x_t^i, y_t^i, z_t^i) = R_s(\sin\theta_t^i \cos\phi_t^i,\ \sin\theta_t^i \sin\phi_t^i,\ \cos\theta_t^i)$. The dynamics follow a random angular walk with $\Delta\theta_t^i, \Delta\phi_t^i \sim \text{Unif}[-\pi/8, \pi/8]$ applied additively.

Free-space agents (N_f). These agents move freely in $\mathbb{R}^3$, initialized at some radius R_f from the origin. At each timestep, small perturbations are applied in spherical coordinates: $\Delta\phi_t^i, \Delta\theta_t^i \sim \text{Unif}[-\pi/100, \pi/100]$ and $\Delta r_t^i \sim \text{Unif}[-0.1, 0.1]$, producing slow drifting trajectories near their initial radii. This system models a heterogeneous space-asset network (ground stations on a sphere, free-flying satellites), where connectivity is sparse and geometry-driven, stressing the *multi-graph* quantification and edge-weight constraints spanning the dense/sparse and homogeneous/heterogeneous regimes that arise in real distributed CPS.

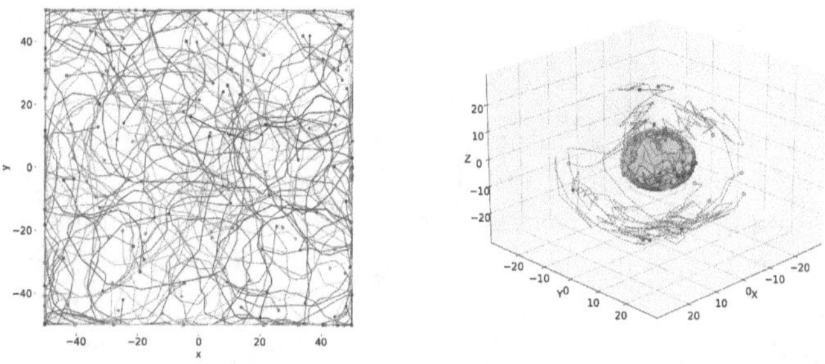

(a) $|\mathcal{V}| = 100$ (b) $|\mathcal{V}| = 20, N_s = 14, N_f = 6$

Fig. 2. (a) Environment 1: 50×50 space with 100 agents, (b) Environment 2: 3D environment with 14 agents on the sphere, and 6 agents in space.

Interaction Graphs. At each timestep, three interaction graphs are computed from the pairwise Euclidean distances $d_{ij}^t = \|\mathbf{p}_t^i - \mathbf{p}_t^j\|_2$: *(i) Distance graph* $\mathcal{G}_t^{\text{dist}}$: a weighted complete graph with edge weight $w_{ij}^t = d_{ij}^t$ for $i \neq j$. *(ii) Sensing graph* $\mathcal{G}_t^{\text{sense}}$: an unweighted graph with edge (i, j) iff $d_{ij}^t \leq R_{\text{sense}}$ and j is within the field of view of i, i.e. the angle between the relative position $\mathbf{p}_t^j - \mathbf{p}_t^i$ and i's heading vector $(\cos\theta_t^i, \sin\theta_t^i)$ is at most a half-angle β. *(iii) Communication graph* $\mathcal{G}_t^{\text{comm}}$: an unweighted graph with edge (i, j) iff $d_{ij}^t \leq R_{\text{comm}}$. The topology of all the graphs changes at every timestep as agents move.

Specifications. We consider the following STL-GO specifications of increasing complexity, parameterized by a nesting depth $\ell \geq 0$.
Incoming Reachability ($\varphi_\ell^{r\text{-}in}$): The specifications $\varphi_\ell^{r\text{-}in}$ (Eq. 1) is defined by recursively nesting the **In** operator over the sensing graph $\mathcal{G}_t^{\text{sense}}$, with edge-count bound $E = [1, 4]$ and unconstrained weights $W = [-\infty, \infty]$. It asserts that there exists an agent which, at some time within $[0, 10]$, can be reached from a goal-satisfying agent via ℓ successive incoming edges over the sensing graph.

$$\psi_0 = \mathbf{In}^{W,\exists}_{\mathcal{G}_t^{\text{sense}},E}\,\texttt{goal},\ \psi_{\ell+1} = \mathbf{In}^{W,\exists}_{\mathcal{G}_t^{\text{sense}},E}\,\psi_\ell,\ \varphi_\ell^{r\text{-}in} = \mathbf{EX}_{\mathcal{V}}\,\mathbf{F}_{[0,10]}\,\psi_\ell, \tag{1}$$

Bi-Directional Reachability($\varphi_\ell^{bi\text{-}dir}$): The formulas $\varphi_\ell^{bi\text{-}dir}$ (Eq. 2) necessitates bi-directional reachability over $\mathcal{G}_t^{\text{sense}}, \mathcal{G}_t^{\text{comm}}$, with $E = [1, 2]$ and $W = [-\infty, \infty]$. The recursively defined subformulas ψ_ℓ^{out} and ψ_ℓ^{in} check reachability to a goal-satisfying agent via ℓ outgoing and incoming hops, respectively. Thus, $\varphi_\ell^{bi\text{-}dir}$ asserts that, within $[0, 10]$, there exists an agent that can both reach and be reached by a goal-satisfying agent in ℓ hops.

$$\begin{aligned}
\psi_0^{\text{out}} &= \mathbf{Out}^{W,\exists}_{\{\mathcal{G}_t^{\text{sense}},\mathcal{G}_t^{\text{comm}}\},E}\,\texttt{goal}, \quad \psi_{\ell+1}^{\text{out}} = \mathbf{Out}^{W,\exists}_{\{\mathcal{G}_t^{\text{sense}},\mathcal{G}_t^{\text{comm}}\},E}\,\psi_\ell^{\text{out}} \\
\psi_0^{\text{in}} &= \mathbf{In}^{W,\exists}_{\{\mathcal{G}_t^{\text{sense}},\mathcal{G}_t^{\text{comm}}\},E}\,\texttt{goal}, \quad \psi_{\ell+1}^{\text{in}} = \mathbf{In}^{W,\exists}_{\{\mathcal{G}_t^{\text{sense}},\mathcal{G}_t^{\text{comm}}\},E}\,\psi_\ell^{\text{in}}, \\
\varphi_\ell^{bi\text{-}dir} &= \mathbf{EX}_{\mathcal{V}}\,\mathbf{F}_{[0,10]}\left(\psi_\ell^{\text{out}} \wedge \psi_\ell^{\text{in}}\right), \quad \ell = 0, 1, \ldots
\end{aligned} \tag{2}$$

Alternating Spatio-Temporal Operators ($\varphi_\ell^{r\text{-}sp}$): These specifications (Eq. 3) interleaves spatial and temporal operators in alternation over $\mathcal{G}_t^{\text{dist}}, \mathcal{G}_t^{\text{sense}}, \mathcal{G}_t^{\text{comm}}$ graphs with edge-count bound $E = [1, 20]$ and weight interval $W = [0, 20]$. At each nesting level ℓ, a temporal operator $\mathsf{T} = \{\mathbf{F}, \mathbf{G}\}$ is applied first, followed by a spatial operator $\mathsf{S} = \{\mathbf{In}, \mathbf{Out}\}$. This allows the specification to encode diverse spatio-temporal patterns. The atomic predicate `goal` holds at agent i iff its position lies within the goal region $\mathcal{G}_{\text{goal}} \subset \mathbb{R}^d$. At each nesting level ℓ, the temporal operator $\mathsf{T}_\ell \in \{\mathbf{F}, \mathbf{G}\}$ and the spatial operator $\mathsf{S}_\ell \in \{\mathbf{In}_{\mathcal{G},E}^{W,\exists}, \mathbf{Out}_{\mathcal{G},E}^{W,\exists}\}$ alternate, and the graph type is drawn cyclically from $\{\mathcal{G}_t^{\text{dist}}, \mathcal{G}_t^{\text{sense}}, \mathcal{G}_t^{\text{comm}}\}$.

$$\psi_0^{sp} = \texttt{goal}, \quad \psi_{\ell+1}^{sp} = \mathsf{T}_\ell\, \mathsf{S}_\ell\, \psi_\ell^{sp}, \quad \varphi_\ell^{r\text{-}sp} = \mathbf{EX}_{\mathcal{V}}\mathbf{F}_{[0,10]}\, \psi_\ell^{sp}, \quad \ell = 0, 1, \dots \tag{3}$$

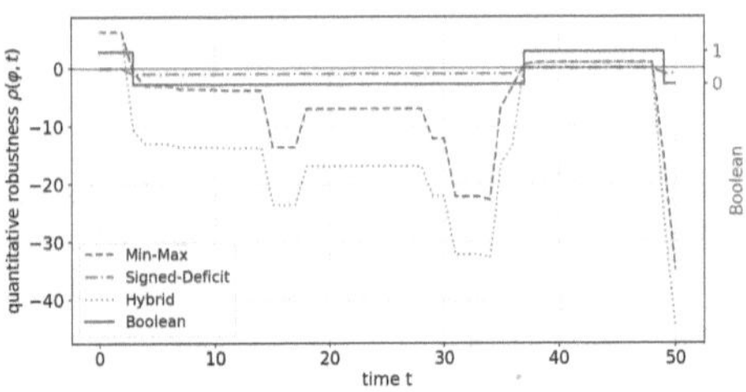

Fig. 3. Robustness values across semantics for a sampled trajectory.

These three formulas isolate the semantic target of the operators: $\varphi_\ell^{r\text{-}in}$ stresses nested *incoming* count-constrained reachability; $\varphi_\ell^{bi\text{-}dir}$ adds multi-graph bidirectional reachability; and $\varphi_\ell^{r\text{-}sp}$ interleaves temporal and spatial operators to exercise the full layered composition. The recursive nesting depth ℓ controls the scaling complexity.

Evaluation Metrics and Methodology. For each environment, we evaluate STL-GO specifications under four semantic instantiations: (i) *Boolean*: classical qualitative satisfaction ($K = \mathbb{B}$), yielding a binary verdict; (ii) *Min-max*: standard STL robustness with the accumulator $\mathcal{A}_E^{\text{mm}} = \min(r_{(e_1)}, -r_{(e_2+1)})$; (iii) *Signed-deficit*: the counting accumulator $\mathcal{A}_E^{\text{cd}}(\{r_j\}) = \min(c^+ - e_1,\, e_2 - c^+)$ clipped to 0 when $c^+ \in [e_1, e_2]$, where $c^+ = |\{j : r_j \succ \bot_K\}|$ counts satisfying neighbours; (iv) *Hybrid*: the deficit-corrected accumulator $\mathcal{A}_E^{\text{h}} = r_{(k)} + \alpha \mathcal{A}_E^{\text{cd}}(\{r_j\})$, combining min-max and signed-deficit with exchange-rate parameter $\alpha > 0$. All share the same temporal De Morgan algebra for temporal operators and differ in the graph-operator accumulator. We measure the wall-clock T(s) to evaluate the full STL-GO specification for all agents at all timesteps, averaged over 100 independently generated trajectories and graph sequences. To assess scalability, we vary two parameters independently: (i) the number of agents $|\mathcal{V}| \in \{10, 20, 50, 100\}$ with fixed $\mathcal{H} = 50$, which affects neighbor set sizes and the cost of the accumulator and multi-agent quantifiers; and (ii) the time horizon $\mathcal{H} \in \{10, 20, 50, 100\}$ with fixed $|\mathcal{V}| = 20$, which affects the cost of temporal operators. For each configuration, we report average monitoring time per trajectory across 100 runs, broken down by semantic instantiation.

Results and Discussion. From Table 1 we see that across all specifications the four semantics produce identical Boolean outcomes (Sat./Vio.), confirming that

Table 1. Results across $\varphi_\ell^{r\text{-}in}$, $\varphi_\ell^{bi\text{-}dir}$ (denoted $\varphi_\ell^{b\text{-}d}$), and $\varphi_\ell^{r\text{-}sp}$ over increasing recursive depths. For each, we report the percentage of satisfying trajectories (Sat.%), the robustness range [min, max] over the 100 trajectories, and computation time (T, seconds) for 100 trajectories, at fixed horizon $\mathcal{H} = 50$. For Boolean semantics this range lies in $\{0,1\}$, so $[0,1]$ indicates mixed verdicts and $[1,1]$ universal satisfaction. $\varphi_\ell^{r\text{-}in}$ and $\varphi_\ell^{b\text{-}d}$ use the 2D environment ($|\mathcal{V}| = 100$); $\varphi_\ell^{r\text{-}sp}$ uses the 3D environment ($N_s = 14$, $N_f = 6$). Robustness and time are rounded to the nearest integer.

Spec	Sat.%	Bool.		Min-Max		Sign.-Def.		Hybrid	
		[min, max]	T(s)	[min, max]	T(s)	[min, max]	T(s)	[min, max]	T(s)
$\varphi_0^{r\text{-}in}$	94	[0, 1]	3	[−7, 9]	3	[−1, 1]	3	[−17, 25]	3
$\varphi_1^{r\text{-}in}$	67	[0, 1]	3	[−16, 9]	3	[−1, 1]	3	[−36, 19]	3
$\varphi_2^{r\text{-}in}$	55	[0, 1]	3	[−16, 9]	3	[−1, 1]	3	[−46, 26]	3
$\varphi_3^{r\text{-}in}$	54	[0, 1]	3	[−16, 9]	3	[−1, 1]	3	[−56, 28]	3
$\varphi_4^{r\text{-}in}$	54	[0, 1]	3	[−16, 9]	3	[−1, 1]	4	[−66, 36]	3
$\varphi_0^{b\text{-}d}$	100	[1, 1]	12	[3, 10]	11	[0, 0]	11	[4, 20]	11
$\varphi_1^{b\text{-}d}$	100	[1, 1]	30	[3, 10]	28	[0, 0]	27	[7, 30]	27
$\varphi_2^{b\text{-}d}$	100	[1, 1]	81	[3, 10]	73	[0, 0]	71	[8, 40]	76
$\varphi_0^{r\text{-}sp}$	65	[0, 1]	1	[−18, 10]	1	[−1, 1]	1	[−28, 20]	1
$\varphi_1^{r\text{-}sp}$	10	[0, 1]	1	[−14, 9]	1	[−1, 0]	1	[−34, 19]	1
$\varphi_2^{r\text{-}sp}$	26	[0, 1]	6	[−19, 10]	6	[−1, 1]	6	[−49, 37]	6
$\varphi_3^{r\text{-}sp}$	6	[0, 1]	71	[−6, 10]	69	[−1, 0]	68	[−46, 30]	75

the quantitative semantics preserve correctness with respect to Boolean satisfaction; Fig. 3 visualizes this for a sampled trajectory.[2] Runtimes are similar for the simpler $\varphi_\ell^{r\text{-}in}$, while differences grow with complexity in $\varphi_\ell^{bi\text{-}dir}$ and $\varphi_\ell^{r\text{-}sp}$, signed-deficit typically being fastest due to its simpler aggregation. The hybrid semantics is slightly more expensive but provides richer information: its larger range reflects the count-based term, which grows with the number of participating agents rather than being bounded by extrema. We emphasize that the four semantics compute different quantities and are not meant to compete on a single axis; the timing comparison quantifies the *overhead* of the richer accumulators relative to the Boolean baseline, establishing acceptable cost.

Table 2 reports the effect of time horizon $\mathcal{H}$ and number of agents $|\mathcal{V}|$ on runtime for the spec $\varphi_2^{r\text{-}in}$. We fix $\varphi_2^{r\text{-}in}$ as the ablation baseline because it exercises nested graph operators without the large constant factors of the bidirectional family, making it ideal for isolating the effects of $\mathcal{H}$ and $|\mathcal{V}|$. Across $|\mathcal{V}|$, all four semantics scale near-identically, with differences within run-to-run variation. The clearest separation appears in the time horizon $\mathcal{H}$, where the

[2] Boolean is drawn as a post-step switching only at integer samples, while the quantitative curves are linearly interpolated; a Min-Max zero-crossing and the Boolean transition may thus appear shifted by up to one timestep, even though they agree in sign at each sampled t.

Table 2. Ablation: wall-clock time (s) for $\varphi_2^{r\text{-}in}$ over 100 trajectories, varying time horizon $\mathcal{H}$ (left, fixed $|\mathcal{V}| = 20$) and number of agents $|\mathcal{V}|$ (right, fixed $\mathcal{H} = 50$). Semantics: Boolean (Bool), Min-max (MM), Signed-Deficit (SD), Hybrid (Hyb.).

(a) Ablation on time horizon $\mathcal{H}$.

$\mathcal{H}$	Bool	MM	SD	Hyb.
10	0.29	0.31	0.32	0.30
20	0.62	0.71	0.73	0.74
50	2.31	3.24	2.84	3.07
100	6.70	10.65	9.58	9.68

(b) Ablation on number of agents.

$\lvert\mathcal{V}\rvert$	Bool	MM	SD	Hyb.
10	1.21	1.25	1.27	1.51
20	2.16	3.09	2.86	2.87
50	9.31	10.10	9.90	10.38
100	25.35	26.68	27.61	26.19

Boolean semantics scales best and the three quantitative accumulators track one another closely.

Conclusions. We introduced a unified algebraic framework for quantitative semantics of STL-GO, separating temporal reasoning, graph aggregation, and multi-agent quantification. A key insight is that graph-based aggregation cannot, in general, be captured by standard semiring operations alone; instead, the structure of the aggregator determines whether the quantitative semantics align with Boolean satisfaction, and under appropriate conditions we obtain compositional soundness and completeness. Our experiments show that these semantics preserve Boolean correctness while incurring only modest overhead and scaling predictably with problem size. The proposed framework relies on the choice of aggregation operators, which must be threshold-aligned with Boolean semantics. While we identify suitable classes (e.g., order-statistic and hybrid), systematic construction of expressive and well-behaved (e.g., smooth) aggregators remains open. Future work includes developing efficient algorithms, designing smooth aggregators, and extending the framework to stochastic multi-agent settings.

References

1. Akazaki, T., Hasuo, I.: Time robustness in mtl and expressivity in hybrid system falsification. In: International Conference on Computer Aided Verification, pp. 356–374. Springer (2015)
2. Alsalehi, S., Mehdipour, N., Bartocci, E., Belta, C.: Neural network-based control for multi-agent systems from spatio-temporal specifications. In: 2021 60th IEEE Conference on Decision and Control (CDC), pp. 5110–5115. IEEE (2021)
3. Balakrishnan, A., Deshmukh, J.V.: Structured reward shaping using signal temporal logic specifications. In: 2019 IEEE/RSJ International Conference on Intelligent Robots and Systems (IROS), pp. 3481–3486 (2019). https://doi.org/10.1109/IROS40897.2019.8968254

4. Balakrishnan, A., Jaksic, S., Aguilar, E., Nickovic, D., Deshmukh, J.: Poster abstract: model-free reinforcement learning for symbolic automata-encoded objectives. In: Proceedings of the 25th ACM International Conference on Hybrid Systems: Computation and Control. HSCC 2022, ACM, New York, NY, USA (2022). https://doi.org/10.1145/3501710.3524734
5. Balakrishnan, A., Paul, S., Silvetti, S., Nenzi, L., Deshmukh, J.V.: Monitoring spatially distributed cyber-physical systems with alternating finite automata. In: Proceedings of the 28th ACM International Conference on Hybrid Systems: Computation and Control. HSCC 2025, ACM, New York, NY, USA (2025). https://doi.org/10.1145/3716863.3718033
6. Bartocci, E., Bortolussi, L., Loreti, M., Nenzi, L.: Monitoring mobile and spatially distributed cyber-physical systems. In: Proceedings of the 15th ACM-IEEE International Conference on Formal Methods and Models for System Design, MEMOCODE 2017, pp. 146–155. ACM, New York, NY, USA (2017). https://doi.org/10.1145/3127041.3127050
7. Bartocci, E., Bortolussi, L., Nenzi, L., Sanguinetti, G.: On the robustness of temporal properties for stochastic models. arXiv preprint arXiv:1309.0866 (2013)
8. Chen, H., Lin, S., Smolka, S.A., Paoletti, N.: An stl-based formulation of resilience in cyber-physical systems. In: International Conference on Formal Modeling and Analysis of Timed Systems, pp. 117–135. Springer (2022)
9. Cignoli, R., Gallego, M.S.D.: Dualities for some de morgan algebras with operators and lukasiewicz algebras. J. Aust. Math. Soc. **34**(3), 377–393 (1983). https://doi.org/10.1017/S1446788700023806
10. Donzé, A., Maler, O.: Robust satisfaction of temporal logic over real-valued signals. In: Chatterjee, K., Henzinger, T.A. (eds.) FORMATS 2010. LNCS, vol. 6246, pp. 92–106. Springer, Heidelberg (2010). https://doi.org/10.1007/978-3-642-15297-9_9
11. Fainekos, G.E., Pappas, G.J.: Robustness of temporal logic specifications for continuous-time signals. Theor. Comput. Sci. **410**(42), 4262–4291 (2009). https://doi.org/10.1016/j.tcs.2009.06.021
12. Finkbeiner, B., Zimmermann, M.: The first-order logic of hyperproperties. arXiv preprint arXiv:1610.04388 (2016)
13. Golan, J.S.: Semirings and Their Applications. Springer Netherlands, Dordrecht (1999). https://doi.org/10.1007/978-94-015-9333-5
14. Haghighi, I., Jones, A., Kong, Z., Bartocci, E., Gros, R., Belta, C.: Spatel: a novel spatial-temporal logic and its applications to networked systems. In: Proceedings of the 18th International Conference on Hybrid Systems: Computation and Control, pp. 189–198 (2015)
15. Haghighi, I., Mehdipour, N., Bartocci, E., Belta, C.: Control from signal temporal logic specifications with smooth cumulative quantitative semantics. In: 2019 IEEE 58th Conference on Decision and Control (CDC), pp. 4361–4366. IEEE (2019)
16. Jakšić, S., Bartocci, E., Grosu, R., Nguyen, T., Ničković, D.: Quantitative monitoring of stl with edit distance. Formal Methods Syst. Des. **53**(1), 83–112 (2018). https://doi.org/10.1007/s10703-018-0319-x
17. Karagulle, R., Vasile, C.I., Ozay, N.: Safe and optimal learning from preferences via weighted temporal logic with applications in robotics and formula 1. arXiv preprint arXiv:2511.08502 (2025)
18. Kudalkar, V., et al.: Sampling-based and gradient-based efficient scenario generation. In: International Conference on Runtime Verification, pp. 70–88. Springer (2024)

19. Kudalkar, V., Ponguluri, S., Balakrishnan, A., Deshmukh, J.V.: Sampling-based multi-agent path planning guided by spatio-temporal logic mission objectives. In: Proceedings of the International Conference on Automated Planning and Scheduling, vol. 36, pp. 133–141 (2026)
20. Kuich, W., Salomaa, A.: Semirings, Automata. Languages. Springer, Berlin, Heidelberg (1986)
21. Li, X., Vasile, C.I., Belta, C.: Reinforcement learning with temporal logic rewards. In: 2017 IEEE/RSJ International Conference on Intelligent Robots and Systems (IROS), pp. 3834–3839. IEEE (2017)
22. Ma, M., Bartocci, E., Lifland, E., Stankovic, J., Feng, L.: Sastl: spatial aggregation signal temporal logic for runtime monitoring in smart cities. In: 2020 ACM/IEEE 11th International Conference on Cyber-Physical Systems (ICCPS), pp. 51–62. IEEE (2020)
23. Maler, O., Nickovic, D.: Monitoring temporal properties of continuous signals. In: Lakhnech, Y., Yovine, S. (eds.) Formal Techniques, Modelling and Analysis of Timed and Fault-Tolerant Systems, pp. 152–166. Springer, Berlin Heidelberg, Berlin, Heidelberg (2004)
24. Mehdipour, N., Vasile, C.I., Belta, C.: Arithmetic-geometric mean robustness for control from signal temporal logic specifications. In: 2019 American Control Conference (ACC), pp. 1690–1695. IEEE (2019)
25. Nenzi, L., Bartocci, E., Bortolussi, L., Loreti, M.: A logic for monitoring dynamic networks of spatially-distributed cyber-physical systems. Logical Methods in Computer Science **18**(1) (2022). https://doi.org/10.46298/lmcs-18(1:4)2022
26. Nenzi, L., Bortolussi, L., Ciancia, V., Loreti, M., Massink, M.: Qualitative and quantitative monitoring of spatio-temporal properties with sstl. Log. Methods Comput. Sci. **14**(4) (2018). https://doi.org/10.23638/LMCS-14(4:2)2018, https://lmcs.episciences.org/4913
27. Nenzi, L., Bortolussi, L., Ciancia, V., Loreti, M., Massink, M.: Qualitative and quantitative monitoring of spatio-temporal properties with sstl. Log. Methods Comput. Sci. (Modal and temporal logics) **14** (2018)
28. Pant, Y.V., Abbas, H., Mangharam, R.: Smooth operator: Control using the smooth robustness of temporal logic. In: 2017 IEEE Conference on Control Technology and Applications (CCTA), pp. 1235–1240. IEEE (2017)
29. Paul, S., Balakrishnan, A., Qin, X., Deshmukh, J.V.: Multi-agent path finding for timed tasks using evolutionary games. In: Hillston, J., Soudjani, S., Waga, M. (eds.) Quantitative Evaluation of Systems and Formal Modeling and Analysis of Timed Systems, pp. 302–321. Springer Nature Switzerland, Cham (2024)
30. Paul, S., Deshmukh, J.V.: Survival of the Fittest: Evolutionary Adaptation of Policies for Environmental Shifts. IOS Press (2024). https://doi.org/10.3233/faia240874
31. Raman, V., Donzé, A., Maasoumy, M., Murray, R.M., Sangiovanni-Vincentelli, A., Seshia, S.A.: Model predictive control with signal temporal logic specifications. In: 53rd IEEE Conference on Decision and Control, pp. 81–87. IEEE (2014)
32. Rodionova, A., Bartocci, E., Nickovic, D., Grosu, R.: Temporal logic as filtering. In: Proceedings of the 19th International Conference on Hybrid Systems: Computation and Control, pp. 11–20 (2016)
33. Silano, G., Afifi, A., Saska, M., Franchi, A.: Stl-based motion planning and uncertainty-aware risk analysis for human-robot collaboration with a multi-rotor aerial vehicle. arXiv preprint arXiv:2509.10692 (2025)

34. Waga, M.: Falsification of cyber-physical systems with robustness-guided black-box checking. In: Proceedings of the 23rd International Conference on Hybrid Systems: Computation and Control, pp. 1–13 (2020)
35. Yoo, C., Belta, C.: Rich time series classification using temporal logic. In: Robotics: Science and Systems (2017)
36. Zhao, Y., Yu, X., Hoxha, B., Fainekos, G., Deshmukh, J., Lindemann, L.: Stl-go: spatio-temporal logic with graph operators for distributed systems with multiple network topologies. ACM Trans. Embed. Comput. Syst. **24**(5s), 1–23 (2025)

Wasserstein Error Bounds for Aggregations of Continuous-Time Markov Chains

Fabian Michel(✉)

Universität der Bundeswehr München, Werner-Heisenberg-Weg 39, 85579 Neubiberg, Germany
fabian.michel@unibw.de

Abstract. We study approximations of a finite continuous-time Markov chain by a Markov chain on a reduced state space, and we provide formal error bounds for the approximated transient distributions in the Wasserstein distance. These bounds extend previous work on error bounds in the total variation distance. A Wasserstein matrix norm is used to bound the error caused by the lower-dimensional approximation of the dynamics. To control the propagation of the accumulated error, we rely on the concept of coarse Ricci curvature of a Markov chain. The usefulness of the bounds depends strongly on the curvature of the chain. A running example demonstrates that positive curvature results in better bounds than those considered in previous works. In contrast, negative curvature results in exponentially exploding bounds. Computing the bounds exactly requires solving many linear programs which currently limits the practical applicability.

Keywords: Markov chains · State space reduction · Formal error bounds · Wasserstein distance · Aggregation · Coarse Ricci curvature

1 Introduction

State aggregation in dynamic systems has been studied extensively (see [14] for one of the earliest works). Due to the curse of dimensionality, continuous-time Markov chains (CTMCs) with large state spaces can quickly become computationally intractable without state space reduction. One way to reduce computation time – or to obtain a model that is more amenable to manual or automated analysis – is to approximate the original model with an aggregated model on a lower-dimensional state space. Various cases where exact model reductions are possible have been analysed in the literature (see, e.g. [4]).

Formal error bounds for the approximation error when exact aggregation is impossible have received less attention. [4] derived upper and lower bounds for the transient distribution of a Markov chain from an aggregation. [1] improved these bounds, and they were later extended to a more general setting in [11].

M. Chen and M. Tschaikowski (Eds.): QEST+FORMATS 2026, LNCS 16913, pp. 246–264, 2027.
https://doi.org/10.1007/978-3-032-35298-9_14

Both [1,11] measured the error (the difference between approximated and actual transient distributions) in the total variation distance. However, depending on the model properties of interest, which are usually not the transient distributions themselves, other error metrics might be more natural. In addition, if the state space has an associated distance between states, it is more natural to work with the Wasserstein distance. The latter is also much more suitable for potential generalizations to continuous-time and -state Markov processes.

Measuring the error in the Wasserstein distance instead of the total variation distance introduces additional complications compared to [11]. While the error caused by the approximation of the dynamics of a Markov chain on a lower-dimensional state space can be controlled in a similar way to [11], it is no longer true that the error accumulated in the calculation up to a given time point cannot blow up at a later stage. To deal with the accumulated error propagation, the concept of coarse Ricci curvature of a Markov chain [12] turns out to be the right tool. Essentially, the Ricci curvature measures the rate at which two transient distributions of a given Markov chain move toward or away from one another.

1.1 Our Contribution

Our main contribution is a theory for calculating formal error bounds for the difference between approximated and actual transient distributions of a CTMC in the Wasserstein distance. Our central result is the following: Consider a CTMC on the state space $S = \{1, \ldots, n\}$ equipped with some metric. Let $p_0 \in \mathbb{R}^n$ be its initial distribution, denote the generator by $Q \in \mathbb{R}^{n \times n}$, such that the transient distribution is $p_t^{\mathsf{T}} = p_0^{\mathsf{T}} e^{tQ}$, and consider the approximation $\widetilde{p}_t^{\mathsf{T}} = \pi_0^{\mathsf{T}} e^{t\Theta} A$ with aggregated initial distribution $\pi_0 \in \mathbb{R}^m$, aggregated generator $\Theta \in \mathbb{R}^{m \times m}$, and disaggregation matrix $A \in \mathbb{R}^{m \times n}$ (details in Sect. 2.1). We do not consider how to find π_0, Θ, A, but only how to compute error bounds for a given approximation. Similarly to [11, Theorem 5], we can prove:

$$\frac{\mathrm{d}}{\mathrm{d}t^+} \mathrm{W}_1\left(\widetilde{p}_t, p_t\right) \leq \|\Theta A - AQ\|_{\mathrm{W}_1} + \mathrm{W}_1\left(\widetilde{p}_t, p_t\right) \cdot (-K) \tag{1}$$

where $\mathrm{W}_1(\cdot,\cdot)$ is the 1-Wasserstein distance, $\|\cdot\|_{\mathrm{W}_1}$ is a Wasserstein matrix norm on matrices with rows summing to 0 (see Definition 3), and where K is a lower bound on the coarse Ricci curvature $\underline{\kappa}(Q)$ of the Markov chain (see [12] and Definition 4). Error bounds for the transient distribution at a given time point can be obtained by integrating (1). The exact result is stated in Theorem 1.

1.2 Paper Structure

Section 2 introduces the notation, the concept of aggregation, a running example, the Wasserstein distance and coarse Ricci curvature. The proof of Theorem 1 is given in Sect. 3.1 and Sect. 3.2. Section 3.4 applies the derived bounds to an aggregation of the running example. A short conclusion is provided in Sect. 4.

2 Preliminaries

2.1 Markov Chains and State Space Reduction

We consider time-homogeneous CTMCs on the finite state space $S = \{1, \ldots, n\}$ with generator matrix $Q \in \mathbb{R}^{n \times n}$. $Q(i,j)$ is the transition rate from state i to j, and $Q(i,i) = -\sum_{j \neq i} Q(i,j)$. For the initial distribution $p_0 \in \mathbb{R}^n$, the transient distribution of the CTMC is given by $p_t^\intercal = p_0^\intercal e^{tQ}$.

We refer to state space reduction as aggregation, even though there are not necessarily groups of states which are aggregated into one single macro state. Instead, we consider the following form of CTMC aggregation onto a lower-dimensional state space with m states (where $m \leq n$): the original chain is approximated by setting $\widetilde{p}_t^\intercal := \pi_t^\intercal A := \pi_0^\intercal e^{t\Theta} A$ where $\pi_0 \in \mathbb{R}^m$ can be an arbitrary probability distribution, $\Theta \in \mathbb{R}^{m \times m}$ can be an arbitrary CTMC generator matrix, and $A \in \mathbb{R}^{m \times n}$ can be an arbitrary row-stochastic matrix ($A(i,j) \geq 0$, rows summing to 1; not necessarily quadratic). Here, $\widetilde{p}_t$ is intended to approximate the transient distributions of the original CTMC, i.e., p_t. We do not consider how to obtain appropriate π_0, Θ, A and assume they are already given. All results in this paper hold for an arbitrary choice of π_0, Θ, A within the above limits. We call A disaggregation matrix since A describes how to blow up the aggregated transient distribution π_t to the full-state-space approximation $\widetilde{p}_t$ (via $\widetilde{p}_t^\intercal = \pi_t^\intercal A$). The generator Θ is intended to approximate the dynamics on the lower-dimensional state space of size m. Note that, in this paper, in contrast to [11], we assume that the reduced model is again a Markov chain. An example of an aggregation is provided in the next section.

2.2 Running Example

We use a simple reliability model for illustration: consider 6 servers in a network where every server has 3 neighbors (see Fig. 1, left). Assume that a server fails with rate λ, is repaired with rate μ, and has an additional failure rate of η per failed neighboring server (due to load redistribution). We set $\lambda = 0.01, \mu = 1, \eta = 0.2$. The arising CTMC has $n = 2^6 = 64$ states, one state for every combination of running and failed servers. This model is not big enough to merit approximation by aggregation, but it is well suited for illustrating the main points of the paper. One potential aggregation is to put all states where the number of failed servers is identical into one macro state, resulting in $m = 7$ aggregates (separated by the dashed lines in Fig. 1, right). Numbering the original states from 1 to 64 and the aggregates from 1 to 7, a natural choice for π_0, A, and Θ is then the following (see [11, Section 2.2]), with the additional aggregation matrix $\Lambda \in \mathbb{R}^{n \times m}$:

$$\Lambda(j,i) := \begin{cases} 1 & \text{if original CTMC state } j \text{ is part of aggregate } i \\ 0 & \text{otherwise} \end{cases}$$

$$A(i,j) = \begin{cases} \frac{1}{\text{size of aggregate } i} & \text{if original CTMC state } j \text{ is part of aggregate } i \\ 0 & \text{otherwise} \end{cases}$$

$$\pi_0^\top = p_0^\top \Lambda, \qquad \Theta = AQ\Lambda$$

The aggregated model does not represent the original model exactly because the neighbor relations among the failed servers are ignored. Our assumptions on A from the previous section permit a more general form of aggregation where we do not simply group states together and where no aggregation matrix Λ exists.

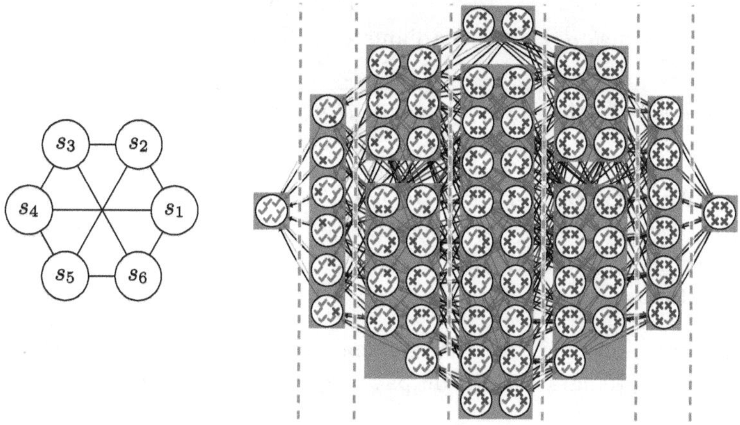

Fig. 1. Left: a network of 6 servers, each of which is connected to 3 neighbors. Right: arising CTMC on 64 states, containing each possible configuration of running and failed servers (indicated by tickmarks and crosses). Arrows represent possible transitions, the shade represents the transition rate, using a hybrid linear-log scaling of the rates (darker arrow = higher rate). States which are equivalent under labelled graph automorphisms of the server network are clustered together in the blue rectangles. The proposed aggregates are separated by the orange dashed lines. (Color figure online)

2.3 Wasserstein Distance

We will measure the error caused by the aggregation in the Wasserstein distance [6,16], sometimes also called Kantorovich-Rubinstein distance [7,9].

Definition 1. *Given a metric d on the state space $S = \{1, \ldots, n\}$, we define the 1-Wasserstein distance of two probability distributions $p, q \in \mathbb{R}^n$ on S as:*

$$\mathrm{W}_1(p,q) = \min_{\gamma \in \Gamma(p,q)} \sum_{r,s \in S} d(r,s) \cdot \gamma(r,s) \tag{2}$$

$$= \max_{\substack{f \in \mathbb{R}^n \text{ is 1-Lipschitz w.r.t. } d \\ \forall s \in S: 0 \le f(s) \le d_{\max}}} \left(\sum_{s \in S} f(s) \cdot p(s) - \sum_{s \in S} f(s) \cdot q(s) \right) \tag{3}$$

$$\text{with } \Gamma(p,q) = \{\gamma \text{ probability measure on } S \times S : \\ \forall A \subseteq S: \ \gamma(A \times S) = p(A), \gamma(S \times A) = q(A)\}$$

$$\text{and } d_{\max} = \max_{r,s \in S} d(r,s) \quad \text{the diameter of } S$$

$\Gamma(p,q)$ is the set of couplings of p and q. Note: $f \in \mathbb{R}^n$ being 1-Lipschitz means $|f(r) - f(s)| \le d(r,s)$ for $r, s \in \{1, \ldots, n\}$, with $f(s)$ the s-th component of f.

Remark 1. We can interpret the right hand side of (2) as $\mathbb{E}[d(X,Y)]$ where X and Y are random variables with laws p and q, and where the expectation is taken according to the optimal coupling of X and Y (which minimizes their expected distance).

See [2, Theorem 2.10] for the existence of the minimum in (2), and [18, Theorem 1.14 & Remark 1.15 (i) on page 34] for the equivalence of (2) and (3).

Remark 2. The restriction $\forall s \in S : 0 \le f(s) \le d_{\max}$ does not change the maximum in (3). This is due to two reasons: on the one hand, adding a constant to a function f leaves the objective value over which we maximize unchanged. On the other hand, because f needs to be 1-Lipschitz, the difference between the maximum and the minimum of f can be at most $d_{\max}$. Therefore, we can shift any 1-Lipschitz f (by adding the appropriate constant) such that it falls within the range $[0, d_{\max}]$ while keeping the objective value unchanged.

Remark 3. If d is the discrete metric, i.e., $d(r,s) = 1$ if $r \ne s$ and $d(s,s) = 0$, then we have the equality (see [10, page 6] for details)

$$\mathrm{W}_1(p,q) = \frac{1}{2} \sum_{s \in S} |p(s) - q(s)| = \frac{1}{2} \|p - q\|_1$$

$$= \text{total variation distance between } p \text{ and } q$$

We can also derive a relation between the total variation distance and the Wasserstein distance for a general metric (not necessarily the discrete one):

$$\mathrm{W}_1(p,q) \overset{[10,\text{ equation }(2.6)]}{\le} \frac{d_{\max}}{2} \cdot \|p - q\|_1 \tag{4}$$

Wasserstein Norm for Matrices. For the error bounds developed later for the aggregation, it will be helpful to define a Wasserstein norm for matrices.

Definition 2. *Let $D \in \mathbb{R}^{m \times n}$ with rows summing to 0 and assume that d is a metric on $S = \{1, \ldots, n\}$ with $d_{\max} = \max_{r,s \in S} d(r, s)$. We define the vector*

$$\vdots D \vdots_{\mathrm{W}_1} := \begin{pmatrix} \max_{f \in \mathbb{R}^n \text{ is 1-Lip. w.r.t. } d,\ \forall s \in S:\, 0 \leq f(s) \leq d_{\max}} D_1 f \\ \max_{f \in \mathbb{R}^n \text{ is 1-Lip. w.r.t. } d,\ \forall s \in S:\, 0 \leq f(s) \leq d_{\max}} D_2 f \\ \vdots \\ \max_{f \in \mathbb{R}^n \text{ is 1-Lip. w.r.t. } d,\ \forall s \in S:\, 0 \leq f(s) \leq d_{\max}} D_m f \end{pmatrix} \in \mathbb{R}^m$$

Here, D_i denotes the i-th row of D.

$\vdots D \vdots_{\mathrm{W}_1}$ measures the Wasserstein distance between the positive and the negative part of each row D_i, if the Wasserstein distance is generalized to any two measures of equal mass (due to the rows summing to 0, the positive and negative parts have equal mass). The generalization is straightforward given (3). For a CTMC aggregation, the rows of $\Theta A - AQ$ sum to 0 so that Definition 2 is applicable to this matrix. Indeed, $(\Theta A - AQ)\mathbf{1}_n = \Theta \mathbf{1}_m - A\mathbf{0}_n = \mathbf{0}_m$ because Θ, Q are generators and A is row-stochastic ($\mathbf{1}_n, \mathbf{0}_n$ are the vectors of all ones/zeros).

Definition 3. *Let $D \in \mathbb{R}^{m \times n}$. We define*

$$\|D\|_{\mathrm{W}_1} := \begin{cases} \max\limits_{i \in \{1,\ldots,m\}} \max\limits_{\substack{f \in \mathbb{R}^n \text{ is 1-Lip. w.r.t. } d, \\ \forall s: 0 \leq f(s) \leq d_{\max}}} D_i f & \text{if all rows of } D \text{ sum to } 0 \\ \infty & \text{otherwise} \end{cases}$$

$\|\cdot\|_{\mathrm{W}_1}$ is a norm on the space of matrices with rows summing to 0. This can be seen by noting that $\|\cdot\|_{\mathrm{W}_1}$ is the maximum of the row-wise Kantorovich-Rubinstein norm (see [8, Chapter VIII, §4, 4.3] or [7]), and therefore inherits the norm properties directly. $\|\cdot\|_{\mathrm{W}_1}$ is not sub-multiplicative in general. Furthermore, if d is the discrete metric, then (if the rows of D sum to 0) $\|D\|_{\mathrm{W}_1} = \frac{1}{2}\|D\|_{\infty}$, where $\|D\|_{\infty}$ is the matrix norm given by $\|D\|_{\infty} = \max_{1 \leq i \leq m} \sum_{j=1}^{n} |D(i,j)|$.

2.4 Ricci Curvature

The so-called Ricci curvature, which was originally defined from a geometric point of view for a metric space [13], has been extended by [12] to the setting of discrete-time Markov chains. In [12, after Example 4 on page 814], Ricci curvature is also defined for CTMCs, although without further discussion:

Definition 4. *Given a CTMC with generator $Q \in \mathbb{R}^{n \times n}$, two states $r, s \in S = \{1, \ldots, n\}$, and a metric d on S, we define the coarse Ricci curvature of the CTMC along the states r and s, with $r \neq s$, as*

$$\kappa(r, s) := -\frac{1}{d(r, s)} \cdot \left.\frac{\mathrm{d}}{\mathrm{d}t^{+}}\right|_{t=0} \mathrm{W}_1\left(\delta_r^{\mathsf{T}} e^{tQ}, \delta_s^{\mathsf{T}} e^{tQ}\right)$$

with $\delta_r, \delta_s \in \mathbb{R}^n$ being the Dirac measures concentrated on r and s, respectively. The derivative exists by Lemma 1 and Corollary 1 below. We also set $\underline{\kappa}(Q) := \min_{r \neq s} \kappa(r, s)$. (Note that $\kappa(r, s)$ and $\underline{\kappa}(Q)$ can be both positive or negative.)

The Ricci curvature will help us bound the error caused by aggregation.

Consider the running example from Sect. 2.2, where we equip the state space with the Manhattan distance, i.e., we define the distance d of two CTMC states as the number of servers which are in a different state (working vs. failed). Using Lemma 2 below, we can calculate $\underline{\kappa}(Q) = 0.41$. This Ricci curvature is achieved along the states $r =$ and $s =$, for example. In Fig. 2, these states and their outgoing transitions are highlighted in orange and blue. The outgoing transitions from $r =$ (in orange) all have rate $\lambda = 0.01$, and the outgoing transitions from $s =$ (in blue) either have rate $\mu = 1$ (for the transitions to the left in Fig. 2; these are repairs) or rate $\lambda + 3\eta = 0.61$ (for the transitions to the right; these are failures). We further have $d(\cdot, \cdot) = 3$. Let $X_t^{(u)}$ be the CTMC started in $X_0^{(u)} = u$ for $u \in \{r, s\}$. Then, $\mathrm{W}_1\left(\delta_r^\intercal e^{tQ}, \delta_s^\intercal e^{tQ}\right) = \mathbb{E}[d(X_t^{(r)}, X_t^{(s)})]$ if we consider the expectation under the optimal coupling of $X_t^{(r)}$ and $X_t^{(s)}$. In the following paragraph, we will give an informal account of how the optimal coupling looks like in this particular case (only near $t = 0$, which is sufficient to derive $\kappa(r, s)$). Lemma 2 below can be used to formally verify that $\kappa(r, s) = 0.41$.

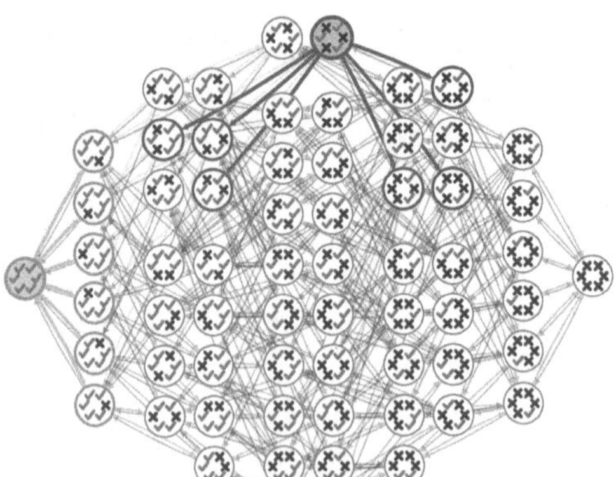

Fig. 2. Two highlighted states $r =$ (orange) and $s =$ (blue) with their outgoing transitions (Color figure online)

Coupling the failure transitions from $r =$ with the repair transitions from $s =$ in an optimal way, we see that, near $t = 0$, $d(X_t^{(r)}, X_t^{(s)})$ jumps down to 1 at rate $3\lambda = 0.03$ (we couple three of the failure transitions from r, each with rate λ, with an equally sized part of the three repair transitions from s), and down to 2 at rate $3\mu - 3\lambda = 2.97$ (we couple the remaining part of the repair transitions from s with the part where the chain stays in r). We also find that it jumps up to 4 at rate $3(\lambda + 3\eta) - 3\lambda = 9\eta = 1.8$ under an optimal coupling (we couple the three remaining failure transitions from r with an equally sized part of the failure transitions from s, leaving the CTMC state distance unchanged at 3, and we couple the remaining part of the failure transitions from s with the part where the chain stays in r). This results in

$$\begin{aligned}\kappa(r,s) &= -\frac{1}{d\,(r,s)} \cdot \frac{\mathrm{d}}{\mathrm{d}t^{+}}\bigg|_{t=0} \mathbb{E}\left[d(X_t^{(r)}, X_t^{(s)})\right] \\ &= -\frac{1}{3} \cdot (0.03 \cdot (1-3) + 2.97 \cdot (2-3) + 1.8 \cdot (4-3)) = 0.41\end{aligned}$$

Note: if we had used the discrete metric instead ($d(r,s) = 1$ for $r \neq s$), we would have $\kappa(r,s) = 0$ for $r =$  and $s =$ , which is easy to see as the immediate successors of r and s do not intersect, i.e., their distance is still 1, just as $d(r,s) = 1$. In fact, $\underline{\kappa}(Q) = 0$ for the discrete metric.

3 Wasserstein Error Bounds

Recall that we approximate the CTMC's transient distribution $p_t^{\mathsf{T}} = p_0^{\mathsf{T}} e^{tQ}$ by $\widetilde{p}_t$, defined as $\widetilde{p}_t^{\mathsf{T}} = \pi_0^{\mathsf{T}} e^{t\Theta} A$ with $\pi_0 \in \mathbb{R}^m, \Theta \in \mathbb{R}^{m \times m}, A \in \mathbb{R}^{m \times n}$.

Our goal is to prove Theorem 1 below, which is a generalization of [11, Theorem 5] (at least in the setting where our reduced model on the lower-dimensional state space is also a Markov chain). For the proof of Theorem 1, we will split the error growth into two classes, which are treated in Sect. 3.1 and Sect. 3.2: first, we consider how the error accumulated up to a given time point will propagate in the remaining calculation, and then, we look at the error caused by the approximation of the dynamics on a lower-dimensional state space.

We start with a general result which gives us a way to calculate the derivative of the Wasserstein distance between two probability distributions which depend on a time parameter.

Lemma 1. *Let $p_u, q_u \in \mathbb{R}^n$ be probability measures depending on a parameter $u \geq 0$. Further assume that p_u and q_u are continuous for $u \geq 0$, that p_u and q_u have one-sided right derivatives for $u \geq 0$ which are locally bounded near 0^+, and denote the (one-sided) derivatives in u by $\dot{p}_u$ and $\dot{q}_u$. Then, the one-sided derivative of* $\mathrm{W}_1(p_u, q_u)$ *at $u = 0$ exists and*

$$\frac{\mathrm{d}}{\mathrm{d}u^{+}}\bigg|_{u=0} \mathrm{W}_1(p_u, q_u) = \max_{f \in M} \left(\dot{p}_0^{\mathsf{T}} - \dot{q}_0^{\mathsf{T}}\right) f$$

$$\text{where } M := \operatorname*{arg\,max}_{\substack{f \in \mathbb{R}^n \text{ is 1-Lipschitz w.r.t. } d \\ \forall s \in S: 0 \leq f(s) \leq d_{\max}}} \left(p_0^{\mathsf{T}} - q_0^{\mathsf{T}}\right) f$$

Proof. By (3), we have

$$\mathrm{W}_1(p_u, q_u) = \max_{\substack{f \in \mathbb{R}^n \text{ is 1-Lipschitz w.r.t. } d \\ \forall s \in S: 0 \leq f(s) \leq d_{\max}}} \left(p_u^{\mathsf{T}} f - q_u^{\mathsf{T}} f\right) \tag{5}$$

We will use Danskin's Theorem [5, Theorem I on page 22]. In particular, we use the version proven in [3], which requires that the maximization in (5) is over a compact subset of a Banach space. Indeed, the set $\mathcal{V}$ of all $f \in \mathbb{R}^n$ which are 1-Lipschitz w.r.t. d and which satisfy $0 \leq f(s) \leq d_{\max}$ for all s is clearly a compact subset of the vector space $\mathbb{R}^n$ with the Euclidean norm, which is a Banach space. We further have to verify the three hypotheses from [3]:

- **H1**. The map $\mathbb{R}^n \ni f \mapsto (p_u^\intercal f - q_u^\intercal f) \in \mathbb{R}$ is clearly continuous with respect to the Euclidean topology.
- **H2**. For all $f \in \mathcal{V}$ and for all $u \in [0, \varepsilon)$ (for some $\varepsilon > 0$), the one-sided derivative $\frac{\mathrm{d}}{\mathrm{d}u^+}(p_u^\intercal f - q_u^\intercal f) = \frac{\mathrm{d}}{\mathrm{d}u^+}(p_u^\intercal - q_u^\intercal)f$ clearly exists: it is equal to the linear combination with weights $f(s)$ of $\dot{p}_u - \dot{q}_u$, which we assumed to exist. The derivatives are locally bounded by assumption.
- **H3**. The map $(u, f) \mapsto (p_u^\intercal f - q_u^\intercal f)$ is clearly continuous: it is linear in f, continuous in u because p_u and q_u are continuous in u by assumption, and $(a, b) \mapsto a^\intercal b$ is a continuous map as well.

Therefore, by [3, Theorem 10.1], the right derivative of $\mathrm{W}_1(p_u, q_u)$ exists at $u = 0$ and is given by

$$\left.\frac{\mathrm{d}}{\mathrm{d}u^+}\right|_{u=0} \mathrm{W}_1(p_u, q_u) = \max_{f \in M} \left.\frac{\mathrm{d}}{\mathrm{d}u^+}\right|_{u=0} (p_u^\intercal f - q_u^\intercal f) = \max_{f \in M} (\dot{p}_0^\intercal - \dot{q}_0^\intercal) f$$

$$\text{where } M := \underset{\substack{f \in \mathbb{R}^n \text{ is 1-Lipschitz w.r.t. } d \\ \forall s \in S: 0 \leq f(s) \leq d_{\max}}}{\arg\max} (p_0^\intercal f - q_0^\intercal f) \qquad \square$$

In the subsequent applications of Lemma 1, we will not mention all of the assumptions explicitly whenever these assumptions are straightforward to show.

3.1 Bounding the Growth of the Accumulated Errror

In this section, we consider how to bound the growth of the error which has already accumulated up to the current time point of the calculation in the aggregated CTMC. Given the actual transient distribution p_t and its approximation $\widetilde{p}_t$, we will look at how the Wasserstein distance between the two would develop if both distributions were to evolve according to the original generator Q from time t onwards. That is, we ignore the approximation of the dynamics and just look at the way in which the accumulated error propagates.

A first step in that direction is the following direct corollary of Lemma 1:

Corollary 1. *Let $p, q \in \mathbb{R}^n$ be probability measures, and let $Q \in \mathbb{R}^{n \times n}$ be a generator matrix. Then,*

$$\begin{aligned}\left.\frac{\mathrm{d}}{\mathrm{d}u^+}\right|_{u=0} \mathrm{W}_1\left(p^\intercal e^{uQ}, q^\intercal e^{uQ}\right) &= \max_{f \in M} (p^\intercal - q^\intercal) Q f \\ &\leq \max_{\substack{f \in \mathbb{R}^n \text{ is 1-Lipschitz w.r.t. } d \\ \forall s \in S: 0 \leq f(s) \leq d_{\max}}} (p^\intercal - q^\intercal) Q f \\ \text{with } M &= \operatorname*{arg\,max}_{\substack{f \in \mathbb{R}^n \text{ is 1-Lipschitz w.r.t. } d \\ \forall s \in S: 0 \leq f(s) \leq d_{\max}}} (p^\intercal - q^\intercal) f\end{aligned}$$

Corollary 1 gives us a way to calculate the coarse Ricci curvature from Definition 4 with a linear program, which is helpful for applications, but also for the subsequent theory:

Lemma 2. *For $r \neq s$, we have that*

$$\begin{aligned}\kappa(r, s) &= -\frac{1}{d(r, s)} \cdot V \quad \text{where } V \text{ is the solution of} \\ \max_{f \in \mathbb{R}^n, f \geq 0} (Q_r - Q_s) f \quad &\text{s.t.} \quad f(r) - f(s) = d(r, s) \\ &\text{and } \forall \widetilde{r}, \widetilde{s} : f(\widetilde{r}) - f(\widetilde{s}) \leq d(\widetilde{r}, \widetilde{s})\end{aligned}$$

where Q_r and Q_s are the r-th and s-th row of Q.

Proof. Let $\delta_r, \delta_s \in \mathbb{R}^n$ be the Dirac probability measures concentrated on r and s, respectively, and let Q be a generator matrix. By Corollary 1, we have

$$\left.\frac{\mathrm{d}}{\mathrm{d}u^+}\right|_{u=0} \mathrm{W}_1\left(\delta_r^\intercal e^{uQ}, \delta_s^\intercal e^{uQ}\right) = \max_{f \in M} (\delta_r^\intercal - \delta_s^\intercal) Q f = \max_{f \in M} (Q_r - Q_s) f$$

where Q_r and Q_s are the r-th and s-th row of Q, and where

$$\begin{aligned}M &= \operatorname*{arg\,max}_{\substack{f \in \mathbb{R}^n \text{ is 1-Lipschitz w.r.t. } d \\ \forall \widetilde{s} \in S: 0 \leq f(\widetilde{s}) \leq d_{\max}}} \underbrace{(\delta_r^\intercal - \delta_s^\intercal) f}_{= f(r) - f(s)} \\ &= \{f \in \mathbb{R}^n : f \text{ is } 1\text{–Lip. w.r.t. } d, \ \forall \widetilde{s} \in S : 0 \leq f(\widetilde{s}) \leq d_{\max}, \\ &\qquad f(r) - f(s) = d(r, s)\}\end{aligned}$$

Dropping the restriction to f with $f(\widetilde{s}) \leq d_{\max}$ does not change anything by Remark 2. □

The next lemma shows how the coarse Ricci curvature from Definition 4 can be used to bound the rate at which the Wasserstein distance between two transient distributions of a CTMC grows. This will later help us to bound the rate at which the accumulated error continues to grow for the aggregated CTMC.

Lemma 3. *Let $p, q \in \mathbb{R}^n$ be probability measures, and let Q be a generator matrix. Then,*

$$\left.\frac{\mathrm{d}}{\mathrm{d}u^+}\right|_{u=0} \mathrm{W}_1\left(p^\mathsf{T} e^{uQ}, q^\mathsf{T} e^{uQ}\right) \leq -\underline{\kappa}(Q) \cdot \mathrm{W}_1(p, q)$$

where $\underline{\kappa}(Q)$ was defined in Definition 4.

Proof. The derivative exists by Corollary 1. The claimed upper bound follows from [17, Theorem 1.9], but as this is not immediately obvious, we give a self-contained proof for our specific setting.

Let γ be a coupling achieving the Wasserstein distance between p and q, that is, $\mathrm{W}_1(p, q) = \sum_{r,s} d(r, s)\gamma(r, s)$. For every $u \geq 0$ and all state pairs r, s, let $\eta_u^{(r,s)}$ be a coupling of $\delta_r^\mathsf{T} e^{uQ}$ and $\delta_s^\mathsf{T} e^{uQ}$ achieving the Wasserstein distance between the two distributions. Then, $\beta_u := \sum_{r,s} \gamma(r, s) \cdot \eta_u^{(r,s)}$ is a coupling between $p^\mathsf{T} e^{uQ}$ and $q^\mathsf{T} e^{uQ}$. Thus,

$$\mathrm{W}_1\left(p^\mathsf{T} e^{uQ}, q^\mathsf{T} e^{uQ}\right) \leq \sum_{i,j} \beta_u(i, j) \cdot d(i, j) = \sum_{i,j} \sum_{r,s} \gamma(r, s) \cdot \eta_u^{(r,s)}(i, j) \cdot d(i, j)$$
$$= \sum_{r,s} \gamma(r, s) \cdot \sum_{i,j} \eta_u^{(r,s)}(i, j) \cdot d(i, j) = \sum_{r,s} \gamma(r, s) \cdot \mathrm{W}_1\left(\delta_r^\mathsf{T} e^{uQ}, \delta_s^\mathsf{T} e^{uQ}\right)$$

Since the inequality above is an equality for $u = 0$, we can bound the derivative of the left-hand side at $u = 0$ by the derivative of the right-hand side:

$$\left.\frac{\mathrm{d}}{\mathrm{d}u^+}\right|_{u=0} \mathrm{W}_1\left(p^\mathsf{T} e^{uQ}, q^\mathsf{T} e^{uQ}\right) \leq \left.\frac{\mathrm{d}}{\mathrm{d}u^+}\right|_{u=0} \sum_{r,s} \gamma(r, s) \cdot \mathrm{W}_1\left(\delta_r^\mathsf{T} e^{uQ}, \delta_s^\mathsf{T} e^{uQ}\right)$$
$$= \sum_{r \neq s} \gamma(r, s) \cdot \underbrace{\left.\frac{\mathrm{d}}{\mathrm{d}u^+}\right|_{u=0} \mathrm{W}_1\left(\delta_r^\mathsf{T} e^{uQ}, \delta_s^\mathsf{T} e^{uQ}\right)}_{d(r,s) \cdot (-\kappa(r,s))}$$

The existence of the derivatives follows from Lemma 1. We conclude

$$\left.\frac{\mathrm{d}}{\mathrm{d}u^+}\right|_{u=0} \mathrm{W}_1\left(p^\mathsf{T} e^{uQ}, q^\mathsf{T} e^{uQ}\right) \leq \sum_{r \neq s} \gamma(r, s) d(r, s) \cdot \big(- \kappa(r, s)\big)$$
$$\leq \sum_{r \neq s} \gamma(r, s) d(r, s) \cdot \left(\max_{r \neq s} -\kappa(r, s)\right) = \mathrm{W}_1(p, q) \cdot (-\underline{\kappa}(Q)) \qquad \square$$

The next lemma provides a lower bound for κ (i.e., an upper bound for the derivative of the Wasserstein distance between two transient distributions). The bound is computationally less expensive than solving the linear program from Lemma 2, since it only requires simple matrix-vector multiplications.

Lemma 4. *For $r \neq s$, it holds that*

$$\kappa(r,s) \geq -\frac{\min\{Q_r d(r,\cdot),\ Q_r d(s,\cdot)\} + \min\{Q_s d(s,\cdot),\ Q_s d(r,\cdot)\}}{d(r,s)} =: k(r,s)$$

where Q_r is the r-th row of Q and $d(r,\cdot) := \big(d(r,1),\ldots,d(r,n)\big)^{\top}$.

Proof. By Lemma 2, we have

$$\kappa(r,s) = -\frac{\max_{f\in M}(Q_r - Q_s)f}{d(r,s)}$$
$$\text{with } M = \{f \in \mathbb{R}^n :\ f \geq 0 \text{ is } 1-\text{Lip. w.r.t. } d,\ f(r) - f(s) = d(r,s)\}$$

Hence, it suffices to show that

$$\max_{f\in M}(Q_r - Q_s)f \leq \min\{Q_r d(r,\cdot),\ Q_r d(s,\cdot)\} + \min\{Q_s d(s,\cdot),\ Q_s d(r,\cdot)\}$$

Indeed, we have, for arbitrary $\widetilde{r}, \widetilde{s} \in S$,

$$\begin{aligned}
(Q_r - Q_s)f &= (Qf)(r) - (Qf)(s) = \sum_k Q(r,k)f(k) - \sum_k Q(s,k)f(k) \\
&\overset{\circledast}{=} \sum_k Q(r,k)\big(f(k) - f(\widetilde{r})\big) + \sum_k Q(s,k)\big(f(\widetilde{s}) - f(k)\big) \\
&= Q(r,r)\big(f(r) - f(\widetilde{r})\big) + Q(s,s)\big(f(\widetilde{s}) - f(s)\big) \\
&\quad + \sum_{k\neq r} \underbrace{Q(r,k)}_{\geq 0}\big(f(k) - f(\widetilde{r})\big) + \sum_{k\neq s} \underbrace{Q(s,k)}_{\geq 0}\big(f(\widetilde{s}) - f(k)\big) \\
&\overset{\text{for } f \text{ 1-Lip.}}{\leq} Q(r,r)\big(f(r) - f(\widetilde{r})\big) + Q(s,s)\big(f(\widetilde{s}) - f(s)\big) \\
&\quad + \sum_{k\neq r} Q(r,k)d(k,\widetilde{r}) + \sum_{k\neq s} Q(s,k)d(\widetilde{s},k) \\
&= Q_r d(\widetilde{r},\cdot) + Q_s d(\widetilde{s},\cdot) \\
&\quad + \underbrace{Q(r,r)\big(f(r) - f(\widetilde{r}) - d(\widetilde{r},r)\big) + Q(s,s)\big(f(\widetilde{s}) - f(s) - d(\widetilde{s},s)\big)}_{\circledcirc}
\end{aligned}$$

where $\circledast$ holds because each row of Q sums to 0. We can now finish the proof by considering all four possible combinations of $\widetilde{r} \in \{r,s\}$ and $\widetilde{s} \in \{r,s\}$ and by noting that the extra term $\circledcirc$ always disappears if $f \in M$: for $\widetilde{r} = r$, we have $f(r) - f(\widetilde{r}) - d(\widetilde{r},r) = f(r) - f(r) - d(r,r) = 0$; for $\widetilde{r} = s$, we have $f(r) - f(\widetilde{r}) - d(\widetilde{r},r) = f(r) - f(s) - d(r,s) = 0$ because $f(r) - f(s) = d(r,s)$ for all $f \in M$; and the same applies to $Q(s,s)\big(f(\widetilde{s}) - f(s) - d(\widetilde{s},s)\big)$. □

Corollary 2. *Let $p, q \in \mathbb{R}^n$ be probability measures, and let $Q \in \mathbb{R}^{n\times n}$ be a generator matrix. Then,*

$$\left.\frac{\mathrm{d}}{\mathrm{d}u^+}\right|_{u=0} \mathrm{W}_1\left(p^{\top}e^{uQ}, q^{\top}e^{uQ}\right) \leq -\underline{k}(Q)\cdot \mathrm{W}_1(p,q) \qquad \text{with } \underline{k}(Q) := \min_{r\neq s} k(r,s)$$

with $k(r,s)$ defined in Lemma 4.

Proof. By Lemma 4, $\kappa(r,s) \geq k(r,s)$ for $r \neq s$. Hence, $\underline{\kappa}(Q) = \min_{r \neq s} \kappa(r,s) \geq \min_{r \neq s} k(r,s) = \underline{k}(Q)$ and the result follows by Lemma 3. □

If d is the discrete metric, we can simplify the expression for $k(r,s)$ to $k(r,s) = Q(r,s) + Q(s,r)$ and derive that

$$\left.\frac{\mathrm{d}}{\mathrm{d}u^+}\right|_{u=0} \mathrm{W}_1\left(p^\intercal e^{uQ}, q^\intercal e^{uQ}\right) \leq 0$$

See [10, Lemma 13 and Corollary 14] for details.

3.2 Bounding the Approximation Error

In the previous section, we have seen how to bound the rate of growth of the Wasserstein distance between two transient distributions of the same CTMC (i.e., with the same generator). This helps us to analyze how the error caused by aggregation which has already accumulated up to a given time might blow up (or even decrease over time). To provide complete error bounds, we also need to consider the error caused by approximating the generator of the original CTMC on a lower-dimensional state space. The following corollary of Lemma 1 fills that gap:

Corollary 3. *Let $\pi \in \mathbb{R}^m$ be a probability measure, and let $\Theta \in \mathbb{R}^{m \times m}, Q \in \mathbb{R}^{n \times n}$ be generator matrices. Further, let $A \in \mathbb{R}^{m \times n}$ be a matrix with non-negative entries and with each row summing to 1. Then,*

$$\left.\frac{\mathrm{d}}{\mathrm{d}u^+}\right|_{u=0} \mathrm{W}_1\left(\pi^\intercal e^{u\Theta} A, \pi^\intercal A e^{uQ}\right) = \max_{\substack{f \in \mathbb{R}^n \text{ is 1-Lipschitz w.r.t. } d \\ \forall s \in S: 0 \leq f(s) \leq d_{\max}}} \pi^\intercal(\Theta A - AQ) f$$

Proof. Note that the set M from Lemma 1 is the set of all 1-Lipschitz functions in this case because the Wasserstein distance between the two probability measures $\pi^\intercal e^{u\Theta} A$ and $\pi^\intercal A e^{uQ}$ is 0 for $u = 0$. □

We can immediately derive an upper bound for the derivative in Corollary 3 which is easier to compute ($¦\Theta A - AQ¦_{\mathrm{W}_1}$ needs to be computed only once):

Corollary 4. *Assume that $\Theta \in \mathbb{R}^{m \times m}, Q \in \mathbb{R}^{n \times n}, A \in \mathbb{R}^{m \times n}$ are fixed. Then,*

$$\forall \pi \in \mathbb{R}^m \text{ prob. measure}: \left.\frac{\mathrm{d}}{\mathrm{d}u^+}\right|_{u=0} \mathrm{W}_1\left(\pi^\intercal e^{u\Theta} A, \pi^\intercal A e^{uQ}\right) \leq \pi^\intercal ¦\Theta A - AQ¦_{\mathrm{W}_1}$$

where $¦\cdot¦_{\mathrm{W}_1}$ was defined in Definition 2.

3.3 Overall Error Bound

Theorem 1. *Consider a CTMC with initial distribution $p_0 \in \mathbb{R}^n$, generator $Q \in \mathbb{R}^{n\times n}$ and transient distribution $p_t^\top = p_0^\top e^{tQ}$, and consider the approximation $\widetilde{p}_t$, defined as*

$$\widetilde{p}_t^\top = \pi_0^\top e^{t\Theta} A \quad \text{with}\, \pi_0 \in \mathbb{R}^m, \Theta \in \mathbb{R}^{m\times m}, A \in \mathbb{R}^{m\times n}$$

where π_0 is an arbitrary probability distribution (intended as an approximation of p_0 on a lower-dimensional state space), Θ is an arbitrary CTMC generator matrix (intended as an approximation of Q), and A is an arbitrary row-stochastic matrix (non-negative entries, each row sums to 1; the disaggregation matrix).

Then, $\mathrm{W}_1(\widetilde{p}_t, p_t)$ is continuous in t, and

$$\frac{\mathrm{d}}{\mathrm{d}t^+}\mathrm{W}_1(\widetilde{p}_t, p_t) \leq \pi_t^\top \,¦\, \Theta A - AQ \,¦_{\mathrm{W}_1} + \mathrm{W}_1(\widetilde{p}_t, p_t) \cdot \big(- \underline{\kappa}(Q)\big) \tag{6}$$

where $\pi_t^\top = \pi_0^\top e^{t\Theta}$, and where t^+ indicates that we consider the one-sided derivative into the positive t-direction (the right derivative). This derivative exists for all $t \geq 0$. $\underline{\kappa}(Q)$ was defined in Definition 4. The bound remains valid when replacing $\underline{\kappa}(Q)$ with $\underline{k}(Q)$ from Corollary 2 (the latter being easier to compute).

Proof. $\mathrm{W}_1(\widetilde{p}_t, p_t)$ is continuous in t because $\widetilde{p}_t$ and p_t are continuous in t. Indeed, by the triangle inequality for the Wasserstein distance,

$$\begin{aligned} \mathrm{W}_1(\widetilde{p}_{t+u}, p_{t+u}) &\leq \mathrm{W}_1(\widetilde{p}_{t+u}, \widetilde{p}_t) + \mathrm{W}_1(\widetilde{p}_t, p_t) + \mathrm{W}_1(p_t, p_{t+u}) \\ \mathrm{W}_1(\widetilde{p}_t, p_t) &\leq \mathrm{W}_1(\widetilde{p}_t, \widetilde{p}_{t+u}) + \mathrm{W}_1(\widetilde{p}_{t+u}, p_{t+u}) + \mathrm{W}_1(p_{t+u}, p_t) \end{aligned}$$

Subtracting $\mathrm{W}_1(\widetilde{p}_t, p_t)$ from both sides in the first equation, and subtracting $\mathrm{W}_1(\widetilde{p}_{t+u}, p_{t+u})$ in the second equation, we get

$$|\mathrm{W}_1(\widetilde{p}_{t+u}, p_{t+u}) - \mathrm{W}_1(\widetilde{p}_t, p_t)| \leq \mathrm{W}_1(\widetilde{p}_t, \widetilde{p}_{t+u}) + \mathrm{W}_1(p_t, p_{t+u})$$

Furthermore, by continuity of p_t,

$$\mathrm{W}_1(p_t, p_{t+u}) \overset{(4)}{\leq} \frac{d_{\max}}{2} \cdot \|p_t - p_{t+u}\|_1 \to 0 \quad \text{for} \quad u \to 0$$

The same holds for $\mathrm{W}_1(\widetilde{p}_t, \widetilde{p}_{t+u})$, showing that $\mathrm{W}_1(\widetilde{p}_t, p_t)$ is continuous in t.

For the proof of the main statement, note that

$$\begin{aligned} \mathrm{W}_1(\widetilde{p}_{t+u}, p_{t+u}) &\overset{\triangle\text{-inequ.}}{\leq} \mathrm{W}_1(\widetilde{p}_{t+u}^\top, \widetilde{p}_t^\top e^{uQ}) + \mathrm{W}_1(\widetilde{p}_t^\top e^{uQ}, p_{t+u}^\top) \\ &= \mathrm{W}_1(\pi_t^\top e^{u\Theta} A, \pi_t^\top A e^{uQ}) + \mathrm{W}_1(\widetilde{p}_t^\top e^{uQ}, p_t^\top e^{uQ}) \end{aligned}$$

and apply Corollary 4 as well as Lemma 3 (or Corollary 2 when $\underline{\kappa}(Q)$ is replaced by $\underline{k}(Q)$). Also note that for $u = 0$, the inequality in the equation above is actually an equality, which is why we can bound the one-sided derivative of the left-hand side at $u = 0$ by the one-sided derivative of the right-hand side. The existence of the derivative was shown in Lemma 1. □

Since $\pi_t^\top \,\vdots\, \Theta A - AQ \,\vdots_{\mathrm{W}_1} \le \|\Theta A - AQ\|_{\mathrm{W}_1}$, we can also deduce from Theorem 1 by integration that

$$\mathrm{W}_1(\widetilde{p}_t, p_t) \le \begin{cases} \left(\mathrm{W}_1(\widetilde{p}_0, p_0) - \frac{\|\Theta A - AQ\|_{\mathrm{W}_1}}{\underline{\kappa}(Q)}\right) \cdot e^{-\underline{\kappa}(Q)\cdot t} + \frac{\|\Theta A - AQ\|_{\mathrm{W}_1}}{\underline{\kappa}(Q)} & \text{if } \underline{\kappa}(Q) \neq 0 \\ \mathrm{W}_1(\widetilde{p}_0, p_0) + t \cdot \|\Theta A - AQ\|_{\mathrm{W}_1} & \text{otherwise} \end{cases} \quad (7)$$

Note that (7) provides an upper bound for $\mathrm{W}_1(\widetilde{p}_t, p_t)$, but the derivative of the bound in (7) need not be an upper bound for $\frac{\mathrm{d}}{\mathrm{d}t^+}\mathrm{W}_1(\widetilde{p}_t, p_t)$ when $\underline{\kappa}(Q) > 0$: $\mathrm{W}_1(\widetilde{p}_t, p_t)$ might be strictly smaller than its bound on the right-hand side of (7), and we derived (7) from (6) which uses $\mathrm{W}_1(\widetilde{p}_t, p_t) \cdot (-\underline{\kappa}(Q))$ on the right-hand side. Indeed, to derive (7) from Theorem 1, it is crucial that we have shown continuity of $\mathrm{W}_1(\widetilde{p}_t, p_t)$. If we only knew that the right derivatives of $\mathrm{W}_1(\widetilde{p}_t, p_t)$ existed at every t, $\mathrm{W}_1(\widetilde{p}_t, p_t)$ could have upwards jumps (while staying right-continuous and still having right derivatives) crossing the bound from (7).

Remark 4. A result similar to Theorem 1 can be shown for discrete-time Markov chains. This case is actually simpler to analyze and covered in [10, Section 3.2].

3.4 Error Bound for the Running Example

Consider again the running example with the Manhattan distance, as discussed in Sect. 2.4, and with the aggregation by the number of failed servers from Sect. 2.2. In Sect. 2.4, we saw that $\underline{\kappa}(Q) = 0.41$. The easier-to-calculate lower bound $\underline{k}(Q) \approx -8.4$ is too coarse to be helpful here (but [10] provides examples where it can be used to speed up the computation of $\underline{\kappa}(Q)$ considerably). We can further calculate that $\|\Theta A - AQ\|_{\mathrm{W}_1} = 0.48$. Figure 3 compares the actual error and the bounds derived in this paper. The bound derived by direct integration of the inequality from Theorem 1 is of a similar order of magnitude as the actual error, and is even tight near $t = 0$. The bound from (7) quickly grows out of the pictured range, but, if the initial distribution is chosen uniformly over the states where one server has failed, (7) is also tight near $t = 0$. This suggests that [11, Theorem 11] can be generalized to the Wasserstein error bounds considered in this paper, i.e., the presented bounds are tight in a certain sense.

An application of the bounds would be the following: As the number of failed servers is 1-Lipschitz w.r.t. the Manhattan distance d, we can derive an approximation with explicit bounds on the expected number of failed servers at any time using the error bounds from Fig. 3. With initial state , the bound from Theorem 1 is ≈ 0.062 at time 5. Combining the approximated expected number of failed servers obtained from $\widetilde{p}_5$ and this bound tells us that the real value must be in the range 0.110 ± 0.062 (the actual expectation is ≈ 0.107).

Note: a positive Ricci curvature $\underline{\kappa}(Q)$ results in an error bound which does, in the long run, stabilize in the range $[0,\ \underline{\kappa}(Q)^{-1}\,\|\Theta A - AQ\|_{\mathrm{W}_1}]$ by (7). If $\|\Theta A - AQ\|_{\mathrm{W}_1} = 0$, then the error bound disappears as $t \to \infty$ if $\underline{\kappa}(Q) > 0$. The bounds from [11] cannot exhibit such a behavior as they are non-decreasing.

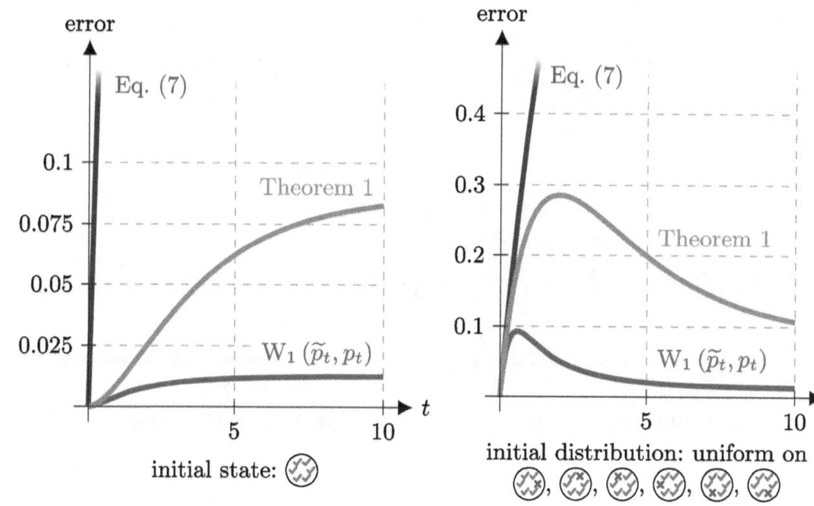

Fig. 3. Comparison of actual error, the bound derived by integration of Theorem 1, and the bound given in (7) for the running example with the aggregation from Sect. 2.2 and with the Manhattan distance. (Color figure online)

Such a case is shown in Fig. 4, where an aggregation with $\|\Theta A - AQ\|_{W_1} = 0$ is considered: When d is the Manhattan distance and if the initial distribution is such that the initial error is not zero, the bound in Theorem 1 captures that the approximated distributions $\widetilde{p}_t$ converge to the correct stationary distribution. When d is the discrete metric instead (which basically results in the bounds considered in [11]), we saw that $\underline{\kappa}(Q) = 0$ in Sect. 2.4, which results in the bound from Theorem 1 not being able to capture the convergence of the $\widetilde{p}_t$ to the actual stationary distribution of the model.

A negative Ricci curvature $\underline{\kappa}(Q)$ can result in exponentially exploding error bounds. It depends strongly on the CTMC structure and the chosen metric whether or not the error bound from Theorem 1 explodes. The running example showed that, if the CTMC converges quickly to a stationary distribution and if the chosen metric can quantitatively capture that convergence in the Wasserstein distance, then the bounds do not explode and are an improvement over the bounds from [11]. Examples where the bounds do explode are provided in [10].

3.5 Computational Aspects

In practice, if we want to use Theorem 1 to bound the error, we need to solve the m linear programs (with n variables and $\mathcal{O}(n^2)$ inequalities) given in Definition 2 to determine $\|\Theta A - AQ\|_{W_1}$ and $\mathcal{O}(n^2)$ linear programs (with n variables, $\mathcal{O}(n^2)$ inequalities and 1 equality) as given in Lemma 2 to find $\underline{\kappa}(Q)$. Solving these programs exactly can easily lead to a prohibitive runtime exceeding the time required to compute transient distributions of the original model directly.

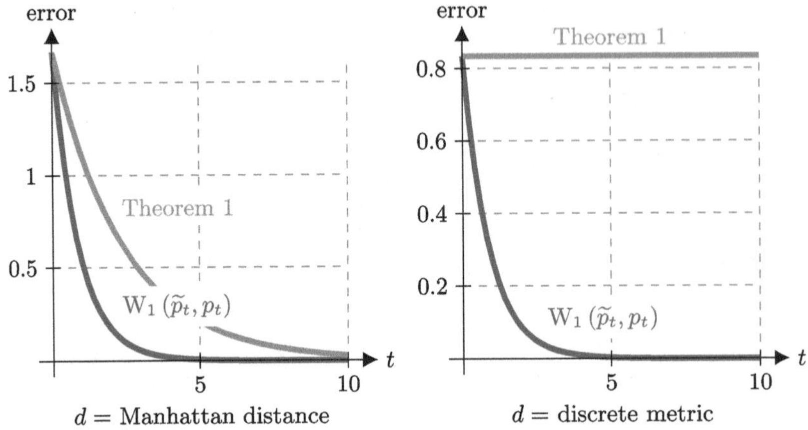

Fig. 4. Actual error and the bound derived by integration of Theorem 1 for the running example when states which are equivalent under labelled graph automorphisms are aggregated (marked by the blue rectangles in Fig. 1, right), resulting in $\Theta A = AQ$. The initial distribution was chosen to be a Dirac measure on ⚙. (Color figure online)

Simpler bounds, like the one in Corollary 2, or hybrid approaches, which e.g. calculate $\kappa(r, s)$ only for state pairs with low $k(r, s)$, are necessary. [10, Section 3.1.6 to 3.1.8] discusses this in more detail, but also shows that further work is required to improve the applicability of the bounds in realistic settings, both from the computational side, but sometimes also from the qualitative side: (6) in Theorem 1 can be too crude to obtain useful bounds for some models. With the theory presented here, it is usually necessary to manually identify model properties which lead to bounds on the Ricci curvature (see [10, Section 3.1.4]).

4 Conclusion

This paper has shown that the error bounds for approximate transient distributions derived for a CTMC aggregation in [11] for the total variation distance can be extended to the Wasserstein distance, employing coarse Ricci curvature to bound the propagation of the accumulated error. Using Wasserstein distances can be advantageous, especially if the Ricci curvature is positive, in which case the bounds can decrease over time, which was impossible for the bounds in [11]. A negative curvature, however, results in exponentially exploding bounds. A big obstacle for the practical applicability is the required computation time for the error bounds (see Sect. 3.5). At this point, the bounds are most attractive when transient distributions have to be calculated for many different initial distributions and time points, if an expensive analysis is run on the reduced model, or if manual analysis yields lower bounds on the Ricci curvature.

Future Work. For wider practical applicability, it is necessary to design efficient algorithms finding good lower bounds on the Ricci curvature of a CTMC. Also, as shown in [10], there are examples where the Ricci curvature is negative, but only a small part of the state pairs has a negative curvature. Local notions of Ricci curvature combined with the aggregated transient distributions might be exploited to improve the bounds. It would also be interesting to consider a modified Wasserstein distance which can be used if the reduced model is no longer a Markov chain, such as the reductions found in [15]. Another avenue for future work would be to develop algorithms which find metrics under which a given CTMC has a high curvature.

Acknowledgments. The author would like to thank Markus Siegle and the anonymous reviewers for offering helpful comments on draft versions of this manuscript.

Data Availability Statement. The models, scripts, and tools to reproduce the reported results are archived and publicly available at DOI 10.5281/zenodo.19567948.

Disclosure of Interests. The author has no competing interests to declare that are relevant to the content of this article.

References

1. Abate, A., Andriushchenko, R., Češka, M., Kwiatkowska, M.: Adaptive formal approximations of Markov chains. Perform. Eval. **148**(102207) (2021). https://www.sciencedirect.com/science/article/pii/S0166531621000249
2. Ambrosio, L., Brué, E., Semola, D.: Lectures on Optimal Transport, 1st edn. Springer (2021). https://doi.org/10.1007/978-3-030-72162-6
3. Başar, T., Bernhard, P.: Appendix B: Danskin's Theorem. In: H-Infinity-Optimal Control and Related Minimax Design Problems: A Dynamic Game Approach, pp. 383–389. Birkhäuser, Boston, MA (2008). https://doi.org/10.1007/978-0-8176-4757-5_10
4. Buchholz, P.: Exact and ordinary lumpability in finite Markov chains. J. Appl. Probab. **31**(1), 59–75 (1994). https://www.jstor.org/stable/3215235
5. Danskin, J.M.: The Theory of Max-Min and its Application to Weapons Allocation Problems, Econometrics and Operations Research, vol. 5. Springer, 1st edn. (1967). https://doi.org/10.1007/978-3-642-46092-0
6. Dobrushin, R.L.: Gibbsian random fields for particles without hard core. Theor. Math. Phys. **4**(1), 705–719 (1970). https://doi.org/10.1007/BF01246670
7. Hanin, L.G.: Kantorovich-Rubinstein norm and its application in the theory of Lipschitz spaces. Proc. Am. Math. Soc. **115**(2), 345–352 (1992). http://www.jstor.org/stable/2159251
8. Kantorovich, L.V., Akilov, G.P.: Functional Analysis, 2nd edn. Elsevier (1982). https://doi.org/10.1016/C2013-0-03044-7
9. Kantorovich, L.V., Rubinstein, G.S.: On a space of completely additive functions. Vestnik Leningrad. Univ. **13**(7), 52–59 (1958)
10. Michel, F.: Wasserstein error bounds for aggregations of continuous-time Markov chains (2025). https://doi.org/10.48550/arXiv.2512.16634, https://arxiv.org/abs/2512.16634, arXiv preprint arXiv:2512.16634

11. Michel, F., Siegle, M.: Formal error bounds for the state space reduction of Markov chains. Perform. Eval. **167**, 102464 (2025). https://doi.org/10.1016/j.peva.2024.102464, https://www.sciencedirect.com/science/article/pii/S0166531624000695
12. Ollivier, Y.: Ricci curvature of Markov chains on metric spaces. J. Funct. Anal. **256**(3), 810–864 (2009). https://doi.org/10.1016/j.jfa.2008.11.001, https://www.sciencedirect.com/science/article/pii/S002212360800493X
13. Ricci, M.M.G., Levi-Civita, T.: Méthodes de calcul différentiel absolu et leurs applications. Mathematische Annalen **54**, 125–201 (1901). http://eudml.org/doc/157997
14. Simon, H.A., Ando, A.: Aggregation of variables in dynamic systems. Econometric **29**(2), 111–138 (1961). https://www.jstor.org/stable/1909285
15. Sonnentag, P., Michel, F., Siegle, M.: The Arnoldi aggregation for approximate transient distributions of Markov chains (2025). https://doi.org/10.48550/arXiv.2508.02078, https://arxiv.org/abs/2508.02078, arXiv preprint arXiv:2508.02078, to appear in *Lecture Notes of the Institute for Computer Sciences, Social Informatics and Telecommunications Engineering*
16. Vaserstein, L.N.: Markov processes over denumerable products of spaces, describing large systems of automata. Probl. Peredachi Inf. **5**(3), 64–72 (1969). https://www.mathnet.ru/links/546d2602f023a6beab855597b91139d0/ppi1811.pdf
17. Veysseire, L.: Courbure de Ricci grossière de processus markoviens. Ph.D. thesis, Ecole normale supérieure de Lyon - ENS LYON (2012). https://theses.hal.science/tel-00737099
18. Villani, C.: Topics in Optimal Transportation, Graduate Studies in Mathematics, vol. 58. American Mathematical Society, 1st edn. (2003)

The Component Queues Method for Quasi-Birth-Death Process Fitting and Analysis

Julianna Bor[1](✉), Giuliano Casale[1], and Evgenia Smirni[2]

[1] Department of Computing, Imperial College London, London, UK
{jb1316,gcasale}@imperial.ac.uk

[2] Department of Computer Science, William & Mary, Williamsburg, USA
exsmir@wm.edu

Abstract. Markovian queueing analysis requires fitting models of arrival and service processes, such as phase-type (PH) distributions, to accurately represent system measurements. However, fitting real-world data is challenging in practice due to model over-parameterization, the large order of the fitting equations, and the limited understanding of the impact of fitting decision on queueing. Motivated by these issues, we present a methodology that uses quasi-birth-death (QBD) processes to fit PH distributions directly from measured queueing data. Core to the proposed method is a new result that reduces the solution of a broad class of QBD processes to the analysis of a collection of simpler queueing systems featuring PH distributions of small order, leading to exact or approximate solutions depending on the chosen arrival process. The proposed method reduces the computational complexity of solving a QBD process from cubic to quadratic for this special class of QBD processes. Numerical results on real traces demonstrate that this feature enables accurate PH fitting from real-world system data.

1 Introduction

An important branch of stochastic modeling focuses on fitting real-world data to the family of distributions compatible with a given queueing model. In particular, phase-type (PH) distributions are routinely used to parameterize queueing systems, as they give rise to an underlying stochastic process that is a continuous-time Markov chain (CTMC) [14]. PH distribution fitting assumes that the empirical distribution obtained from the data can be well approximated by the distribution of passage times in absorbing CTMCs. Unfortunately, this may not hold in practice unless the number of states of the CTMC is large, which introduces difficulties in the fitting process. In fact, existing algorithms struggle to fit general PH models with more than a handful of states, as they incur significant computational and numerical issues, especially when the model is over-parameterized, i.e., it has more parameters than degrees of freedom [3]. However, working with small or medium-sized PH distributions can considerably constrain the range of

M. Chen and M. Tschaikowski (Eds.): QEST+FORMATS 2026, LNCS 16913, pp. 265–283, 2027.
https://doi.org/10.1007/978-3-032-35298-9_15

distributions that can be fitted, as the number of states is linearly proportional to the degrees of freedom available for fitting [16].

Motivated by the above challenges, this paper proposes a novel fitting algorithm to estimate input PH distributions for a class of Markovian queueing systems, focusing on those that can be modeled as quasi-birth-death (QBD) processes. Surveys such as [2] indicate a gap in the literature to estimate arrival and service distributions tailored to QBD processes. Unlike the direct fitting of a PH distribution from traces of inter-arrival or service times, inverse problems involving QBDs fitted from queueing data have rarely been considered [22]. Intuitively, such a development would allow the PH distribution fitting process to capture the salient aspects that drive the observed QBD behavior, filtering out less influential characteristics of the distributions from a queueing analysis perspective. The present work seeks to address this gap by focusing on fitting the QBD service process from the arrival process and the marginal distribution of queue-length probabilities, whether measured or simulated. This is particularly relevant when service times are unavailable, as is often the case with cloud providers.

To make our technique effective, we focus on PH distributions represented as compositional PH models obtained via the Kronecker Product Composition (KPC) operator [16]. We prove the surprising result that parameterizing QBD processes with PH service distributions defined through the KPC operator allows the solution of the QBD process to be *decomposed* into the solutions of independent, but much smaller, QBD processes. Depending on the characteristics of the arrival process, this approach can be exact or approximate. Although divide-and-conquer approaches already exist to explore the space of PH distributions for fitting purposes (e.g., [16]), to the best of our knowledge, this is the first work to develop a PH distribution fitting method of this kind that efficiently searches a space of QBD processes.

Our KPC-based approach reduces the time complexity of calculating marginal queue-length probabilities in a QBD from cubic to quadratic. Leveraging this fast computation of queue-length probabilities alongside numerical search methods, we develop a fitting algorithm that can efficiently explore PH distributions that result in QBD processes that accurately fit the empirical queueing data. Validation experiments on real traces reveal that our method can learn service distributions from queueing data more accurately than existing methods that directly fit the service distributions. This improvement in accuracy also extends to the response time distributions computed from the resulting queueing models. Furthermore, the fitted PH models exhibit low prediction errors when used to parameterize QBD models with different arrival processes, showing low generalization errors.

The remainder of the paper is organized as follows. Section 2 summarizes related work in the area of PH distribution fitting. Section 3 provides basic definitions and relevant background. Section 4 develops the decomposition result for solving QBD processes. Section 5 applies the results of the previous section to the QBD process fitting. In Sect. 6, we report and discuss our numerical results. Section 7 concludes the paper and outlines future work.

2 Related Work

Most fitting algorithms use PH distributions or Markovian arrival processes (MAPs) to fit trace data. PH distributions and MAPs can be readily plugged into various families of performance models, including quasi-birth-death (QBD) processes, queueing networks, stochastic Petri nets and Markovian process algebras [5,10,13,29,32,33]. PH distribution fitting, in particular, has received considerable attention over the years due to its importance in modeling and simulation studies, software reliability, web server modeling, load balancing, and risk theory [1,6,8,18,24,26,27,30,35]. Established fitting approaches include the expectation maximization (EM) algorithm, the cumulative distribution function (CDF) fitting, and moment matching algorithms [9,19,34]. EM algorithms, for which many solutions are available [3,7,31,34], are understood to perform well on shorter traces; however, they are computationally costly and require long-running executions to find accurate PH distributions. Therefore, EM algorithms are overly demanding for fitting the long traces necessary for heavy-tailed distributions. The high computational costs of EM algorithms are overcome by direct fitting of the cumulative distribution function (CDF) [19] or moment matching algorithms [9,16,21]. CDF and moment-based fitting typically require inverting analytical expressions that describe certain statistics of the PH distribution, such as high-order polynomials used to compute the moments of PH distributions. The computational complexity of such expressions is directly proportional to the size of the corresponding PH distribution. Therefore, methods relying on quantities such as moments tend to induce significant restrictions on the size of the PH distribution they attempt to fit. The result is either the fitting of small PH distributions or compositional models constructed based on Kronecker algebra, such as process superposition or Kronecker product composition, which define large distributions in terms of smaller PH distributions. An interesting recent approach uses Bernstein exponentials [20], which form a subclass of acyclic PH distributions, to fit a distribution by sampling its density function. The method allows for tunable accuracy, notably enabling even the fitting of bounded uniform distributions; however, it can converge slowly and require very large PH distributions that may make the following queueing analysis intractable.

3 Preliminaries

In this paper, we consider QBD processes where both inter-arrival and service times follow a PH distribution. A PH distribution is completely determined by a matrix $\boldsymbol{A}_0$ of transition rates between phases and a probability vector $\boldsymbol{\alpha}$ that defines the initial probabilities of starting in each phase. An arrival process with PH inter-arrival times is a Markovian Arrival Process (MAP). Instead of the vector-matrix pair $(\boldsymbol{\alpha}, \boldsymbol{A}_0)$, a PH distribution can also be represented using

MAP notation as $\boldsymbol{A} = (\boldsymbol{A}_0, \boldsymbol{A}_1)$ where $\boldsymbol{A}_0 \in \mathbb{R}^{a \times a}$ defines the transition rates between phases and $\boldsymbol{A}_1 \in \mathbb{R}^{a \times a}$ describes external transitions corresponding to arrivals to the system. $\boldsymbol{A}_1$ is a rank one matrix given by $\boldsymbol{A}_1 := -\boldsymbol{A}_0 \mathbf{1} \boldsymbol{\alpha}$ where $\boldsymbol{\alpha}$ is the row vector of initial probabilities and $\mathbf{1}$ is a column vector of ones. The infinitesimal generator of the above PH arrival process is $\boldsymbol{Q}_A = \boldsymbol{A}_0 + \boldsymbol{A}_1$. Keeping the matrix notation, the service process is represented by $\boldsymbol{S} = (\boldsymbol{S}_0, \boldsymbol{S}_1)$ where $\boldsymbol{S}_0, \boldsymbol{S}_1 \in \mathbb{R}^{s \times s}$ and the corresponding infinitesimal generator is $\boldsymbol{Q}_S = \boldsymbol{S}_0 + \boldsymbol{S}_1$.

Let matrix $\boldsymbol{B}$ denote the service rates (backward transitions), matrix $\boldsymbol{L}$ the local transitions, and matrix $\boldsymbol{F}$ the arrival rates (forward transitions). We focus on QBD processes with the following infinitesimal generator

$$\boldsymbol{Q} := \begin{bmatrix} \boldsymbol{L}_0 & \boldsymbol{F} & \mathbf{0} & \dots \\ \boldsymbol{B} & \boldsymbol{L} & \boldsymbol{F} & \dots \\ \mathbf{0} & \boldsymbol{B} & \boldsymbol{L} & \dots \\ \vdots & \vdots & \vdots & \ddots \end{bmatrix} \qquad \begin{array}{l} \boldsymbol{B} := \boldsymbol{I}_a \otimes \boldsymbol{S}_1 \\ \boldsymbol{L} := \boldsymbol{A}_0 \oplus \boldsymbol{S}_0 = \boldsymbol{A}_0 \otimes \boldsymbol{I}_s + \boldsymbol{I}_a \otimes \boldsymbol{S}_0 \\ \boldsymbol{F} := \boldsymbol{A}_1 \otimes \boldsymbol{I}_s \end{array} \tag{1}$$

where $\boldsymbol{B}, \boldsymbol{L}, \boldsymbol{F} \in \mathbb{R}^{as \times as}$ and $\boldsymbol{I}_a, \boldsymbol{I}_s$ are identity matrices of dimension $a \times a$ and $s \times s$, respectively, and $\otimes$ denotes the Kronecker product operation. Matrix $\boldsymbol{L}_0 \in \mathbb{R}^{as \times as}$ is different to $\boldsymbol{L}$ as only the arrival process has internal phase transition. Hence, $\boldsymbol{L}_0$ is defined as

$$\boldsymbol{L}_0 := \boldsymbol{A}_0 \otimes \boldsymbol{I}_s \tag{2}$$

The state space of the QBD process can be described with triplets (A, S, L), where $A \in \{1, ..., a\}$ and $S \in \{1, ..., s\}$ denote the internal phase of the arrival and service processes, respectively, and $L \in \mathbb{N}_0$ denotes the level of the process. If the stationary distribution exists, it is the unique solution to the equation

$$\boldsymbol{\pi}_l = \boldsymbol{\pi}_0 \boldsymbol{R}^l \tag{3}$$

where $\boldsymbol{\pi}_l = (\pi_{l1}, \pi_{l2}, ..., \pi_{lm})$ denotes the steady state probabilities corresponding to level l and $m = as$ is the number of internal phases within a given level. Therefore, $|\boldsymbol{\pi}_0| = (1 - \rho)$ where ρ is the utilization of the queue, that is, the probability of a busy server, which occurs whenever the queue is non-empty. Furthermore, $\boldsymbol{R} \in \mathbb{R}^{as \times as}$ is defined as

$$\boldsymbol{R} = -\boldsymbol{F}(\boldsymbol{L} + \boldsymbol{F}\boldsymbol{G})^{-1} \tag{4}$$

where $\boldsymbol{G} \in \mathbb{R}^{as \times as}$ is the minimal non-negative solution to the equation

$$\boldsymbol{B} + \boldsymbol{L}\boldsymbol{G} + \boldsymbol{F}\boldsymbol{G}^2 = 0 \tag{5}$$

The matrix $\boldsymbol{G}$ above has a probabilistic interpretation and plays a central role in our proposed fitting algorithm. Given the state space defined above, the elements of $\boldsymbol{G}$ correspond to the following downward first-passage times. Let $X_t \in (A, S, L), t \geq 0$ denote a QBD process and let $i = (i_0, i_1, l)$ and

$j = (j_0, j_1, l-1)$. Denoting the elements of $\boldsymbol{G}$ as $\boldsymbol{G}_{ij}$, the elements of $\boldsymbol{G}$ correspond to the following conditional probabilities:

$$\boldsymbol{G}_{ij} = P(T_{l-1} < \infty \wedge X_{T_{l-1}} = j | X_0 = i) \tag{6}$$

where T_{l-1} is the first passage time to level $(l-1)$. Assuming that the QBD process is positive recurrent, the process reaches any level in a finite time with probability 1, regardless of the starting state. Therefore, the rows of $\boldsymbol{G}$ correspond to the conditional probabilities that, given the process starts in state i of level l, the first passage to level $l-1$ hits the state j of level $l-1$. Since the columns of $\boldsymbol{G}$ correspond to all possible states of level $l-1$, they sum to 1, making $\boldsymbol{G}$ a stochastic matrix.

We now introduce a novel characterization result for the PH/PH/1 queue used throughout the paper. A similar result is established for the MAP/G/1 queue in [12, Theorem 12.10]; unlike this earlier result, our characterization allows the service phase to be explicitly included in the definition of the state space, which is essential for the models considered in this paper.

Theorem 1. *Let* $\boldsymbol{\pi}_0 = (\pi_{01}, \ldots, \pi_{0m})$, *then*

$$\boldsymbol{\pi}_0 = (1-\rho)\boldsymbol{x} \tag{7}$$

where $\boldsymbol{x}$ *is the solution to* $\boldsymbol{x}\boldsymbol{G} = \boldsymbol{x}$, $\boldsymbol{x}\boldsymbol{1}_m = 1$.

The proof of Theorem 1 can be found in [11, Section 5.3.4, Chap. 5]. Despite the tractability of the above formulas, it is well-known that a queue with a large PH service distribution, i.e., with many phases, is expensive to solve. However, limiting the size of the PH service distribution would significantly affect the type of data the model could fit. In this paper, we overcome this by focusing on KPC-PH distributions, composed of small PH building blocks [16]. Here, we recall the key KPC-PH definition used throughout the paper.

Definition 1. *Let* $\boldsymbol{N} = (\boldsymbol{N}_0, \boldsymbol{N}_1)$ *represent a general PH distribution and* $\boldsymbol{H} = (\boldsymbol{H}_0, \boldsymbol{H}_1)$ *a hyperexponential distribution. The KPC-PH distribution denoted by* $\boldsymbol{S} = \boldsymbol{N} \otimes \boldsymbol{H} = (\boldsymbol{S}_0, \boldsymbol{S}_1)$ *is defined as* $\boldsymbol{S}_0 = -\boldsymbol{N}_0 \otimes \boldsymbol{H}_0$ *and* $\boldsymbol{S}_1 = \boldsymbol{N}_1 \otimes \boldsymbol{H}_1$.

The fact that $\boldsymbol{H}$ corresponds to a hyperexponential distribution ensures that $\boldsymbol{S}$, as defined above, produces a valid PH distribution. Let $X_{\boldsymbol{N}}$ be the random variable denoting the inter-arrival time of process $\boldsymbol{N}$ in Definition 1 and let $X_{\boldsymbol{H}}$ and $X_{\boldsymbol{S}}$ be the corresponding random variables for $\boldsymbol{H}$ and $\boldsymbol{S}$. It can be seen that the following relation holds [17]

$$E[X_{\boldsymbol{S}}^k] = \frac{1}{k!} E[X_{\boldsymbol{N}}^k] E[X_{\boldsymbol{H}}^k] \tag{8}$$

Since the composed PH distributions have independent degrees of freedom, the overall degrees of freedom of the KPC-PH distribution grow linearly with the number of composed PH distributions, allowing larger compositions to be used for fitting more characteristics of the empirical dataset.

4 Obtaining the Queue-Length Distribution

4.1 Introducing Component Queues

Let $\boldsymbol{A}/\boldsymbol{S}/1$ denote, analogously to Kendall's notation, a KPC-queue with a PH renewal arrival process $\boldsymbol{A}$ and KPC-PH service time distribution $\boldsymbol{S} = \boldsymbol{N} \otimes \boldsymbol{H}_1 \otimes \ldots \otimes \boldsymbol{H}_J$. We follow the convention that scalar multiplication of a process applies to all representation matrices, that is, $c\boldsymbol{A} = (c\boldsymbol{A}_0, c\boldsymbol{A}_1)$ for real c. Our main result is a strategy to decompose the KPC-PH queue into a collection of smaller queueing systems, called *component queues*, enabling a divide-and-conquer solution strategy for the QBD model. We begin by formally defining component queues.

Definition 2 (Component queues). *Consider a KPC queue $\boldsymbol{A}/\boldsymbol{S}/1$ where $\boldsymbol{S} = \boldsymbol{N} \otimes \boldsymbol{H}_1 \otimes \ldots \otimes \boldsymbol{H}_J$ is the Kronecker product composition of a general PH and J hyperexponential distributions, its decomposition into component queues is given by the following set of queueing processes*

1. *$CQ_N := \boldsymbol{A}/\boldsymbol{N}^*/1$ where $\boldsymbol{N}^* := \mu^*_{H_1} \cdots \mu^*_{H_J} \boldsymbol{N}$, and $\mu^*_{H_k}$ is the mean departure rate of process $\boldsymbol{H}_k, k = 1, \ldots, J$.*
2. *$CQ_{H_k} = M/\boldsymbol{H}^*_k/1$, where M is a Poisson process with unit rate and*

$$\boldsymbol{H}^*_k := (\mu^*_N/\lambda^*_A) \left(\prod_{1 \le i \le J, i \ne k} \mu^*_{H_i} \right) \boldsymbol{H}_k$$

where μ^*_N is the mean departure rate of process $\boldsymbol{N}$, $\mu^*_{H_i}$ are the mean departure rates of processes $\boldsymbol{H}_i$, and λ^*_A is the mean arrival rate of process $\boldsymbol{A}$.

The component queues CQ_N and CQ_{H_k}, $1 \le k \le J$ defined above are themselves QBD processes. Moreover, each component queue has the same utilization as the original $\boldsymbol{A}/\boldsymbol{S}/1$ queue. The latter property can be readily verified by the definitions, using the recursive expression of the Kronecker product composition moments in (8) for $k = 1$, as the service rates are reciprocals of the mean service times. Thus, if the $\boldsymbol{A}/\boldsymbol{S}/1$ queue is stable, so are its component queues.

4.2 Approximating the Matrix $\boldsymbol{G}$

We approximate the matrix $\boldsymbol{G}$ of the original queue via the matrices $\boldsymbol{G}_N$ and $\boldsymbol{G}_{H_k}, k = 1, \ldots, J$, which are analogously defined as $\boldsymbol{G}$, but for the component queues CQ_N and CQ_{H_k}. Let $\widetilde{\boldsymbol{G}}$ denote the approximation of the matrix $\boldsymbol{G}$. In the case of Poisson arrivals, $\widetilde{\boldsymbol{G}}$ is the same as $\boldsymbol{G}$, so the approximation holds without error. Formally, we have

Proposition 1. *Let $M/\boldsymbol{N} \otimes \boldsymbol{H}_1 \otimes \ldots \otimes \boldsymbol{H}_J/1$ be a KPC-PH queue with Poisson arrivals and let $\widetilde{\boldsymbol{G}} = \boldsymbol{G}_N \otimes \boldsymbol{G}_{H_1} \otimes \ldots \otimes \boldsymbol{G}_{H_J}$. Then $\widetilde{\boldsymbol{G}} = \boldsymbol{G} = \boldsymbol{1}\boldsymbol{\alpha}$, where $\boldsymbol{\alpha}$ is the embedded distribution of the service process phase at departure instants.*

The proof of Proposition 1 is given in [11, Section 5.3.2, Chap. 5]. While it gives an exact decomposition of the $\boldsymbol{G}$ matrix for queues with Poisson inputs, if the arrival process is an arbitrary PH then $\boldsymbol{G}$ and $\widetilde{\boldsymbol{G}}$ are no longer identical and our method provides an approximation. The following result outlines for this case a relationship between $\boldsymbol{G}$ and its approximation $\widetilde{\boldsymbol{G}}$.

Theorem 2. *Consider a queue with PH arrival process $\boldsymbol{A}$ and KPC-PH service process $\boldsymbol{S} = \boldsymbol{N} \otimes \boldsymbol{H}_1 \otimes \ldots \otimes \boldsymbol{H}_J$, where $\boldsymbol{H}_k, k = 1, \ldots, J$ are hyperexponential. Defining $\widetilde{\boldsymbol{G}} = \boldsymbol{G}_N \otimes \boldsymbol{G}_{H_1} \otimes \ldots \otimes \boldsymbol{G}_{H_J}$, we have the following relationship between $\widetilde{\boldsymbol{G}}$ and $\boldsymbol{G}$*

$$\widetilde{\boldsymbol{G}} \approx (\boldsymbol{I}_{an} \otimes \boldsymbol{G}_{H_1} \otimes \ldots \otimes \boldsymbol{G}_{H_J})\boldsymbol{G} \tag{9}$$

where a and n denote the number of phases of process $\boldsymbol{A}$ and $\boldsymbol{N}$, respectively, and $\boldsymbol{I}_{an}$ is the identity matrix with dimension $an \times an$.

An approximate lumping argument supporting Theorem 2 is given in [11, Section 5.3.2, Chap. 5]. As we explain in the following sections, the above result is fundamental, as it allows us to estimate steady-state performance metrics for the QBD process in terms of the $\boldsymbol{G}_N$ and $\boldsymbol{G}_{H_k}$ matrices of the individual component queues CQ_N and CQ_{H_k}. As these processes are significantly smaller than the original queue, the net result is a large reduction in computational cost of the QBD analysis.

The key result in Theorem 2 is interpreted as follows. Matrix $\widetilde{\boldsymbol{G}} = \boldsymbol{G}_N \otimes \boldsymbol{G}_{H_1} \otimes \ldots \otimes \boldsymbol{G}_{H_J}$, which is efficiently constructed with a Kronecker product of matrices, each of which is obtained by solving the smaller component queues, may be seen as an approximation of $\boldsymbol{G}$ by means of aggregation. The aggregation is applied by the premultiplying factors on the right-hand side of (9). Compared to the exact model, relying on $\widetilde{\boldsymbol{G}}$ thus loses some information about the internal phases of process $\boldsymbol{H}$, but its computation can be done much more efficiently than computing $\boldsymbol{G}$. This provides a novel approximation for the solution of the QBD, which becomes exact in the case of Poisson arrivals.

4.3 Example

To illustrate (9), we now show it in a small example. Let the number of hyperexponential components in the KPC service be $J = 1$, and the degree of the arrival process $\boldsymbol{A}$ and the service components $\boldsymbol{N}, \boldsymbol{H}$ be 2, so that $an = 4$ and let $\boldsymbol{\alpha}_H = [\alpha_1, \alpha_2]$ be the embedded distribution of the hyperexponential component. From Proposition 1, we know that $\boldsymbol{G}_H = \boldsymbol{1}\boldsymbol{\alpha}_H$. Multiplying $\boldsymbol{I}_4 \otimes \boldsymbol{G}_H$ with $\boldsymbol{G} = (g_{ij})_{1 \leq i,j, \leq m}$ implies that the $(2i-1)$th and $2i$th rows of $\widetilde{\boldsymbol{G}}$ are identical. Furthermore, they are computed from $\boldsymbol{G}$ by averaging these specific rows, weighted according to the embedded distribution $\boldsymbol{\alpha}_H$ of $\boldsymbol{H}$. That is,

$$\widetilde{\boldsymbol{G}} = \boldsymbol{G}_N \otimes \boldsymbol{G}_H = (\boldsymbol{I}_4 \otimes \boldsymbol{G}_H)\boldsymbol{G} = \begin{bmatrix} \alpha_1 g_{11} + \alpha_2 g_{21} & \alpha_1 g_{12} + \alpha_2 g_{22} & \cdots & \alpha_1 g_{18} + \alpha_2 g_{28} \\ \alpha_1 g_{11} + \alpha_2 g_{21} & \alpha_1 g_{12} + \alpha_2 g_{22} & \cdots & \alpha_1 g_{18} + \alpha_2 g_{28} \\ \alpha_1 g_{31} + \alpha_2 g_{41} & \alpha_1 g_{32} + \alpha_2 g_{42} & \cdots & \alpha_1 g_{38} + \alpha_2 g_{48} \\ \alpha_1 g_{31} + \alpha_2 g_{41} & \alpha_1 g_{32} + \alpha_2 g_{42} & \cdots & \alpha_1 g_{38} + \alpha_2 g_{48} \\ \vdots & \vdots & \ddots & \vdots \\ \alpha_1 g_{71} + \alpha_2 g_{81} & \alpha_1 g_{72} + \alpha_2 g_{82} & \cdots & \alpha_1 g_{78} + \alpha_2 g_{88} \end{bmatrix}$$

which shows the aforementioned aggregation property. We also provide a numerical example to further illustrate the aggregation process. Consider a queue with the following hyperexponential arrival process with mean 3 and squared coefficient of variation (SCV) 11/9

$$\boldsymbol{A}_0 = \begin{bmatrix} -0.25 & 0 \\ 0 & -0.5 \end{bmatrix} \qquad \boldsymbol{A}_1 = \begin{bmatrix} 0.125 & 0.125 \\ 0.25 & 0.25 \end{bmatrix}$$

and a KPC service process where the first component is Erlang-2 with unit mean and the second component is a 2-phase hyperexponential distribution with the following parameters

$$\boldsymbol{H}_0 = \begin{bmatrix} -10 & 0 \\ 0 & -20 \end{bmatrix} \qquad \boldsymbol{H}_1 = \begin{bmatrix} 5 & 5 \\ 5 & 15 \end{bmatrix}$$

corresponding to mean 1/15 and SCV 5/4. Below, we can see how closely $\boldsymbol{G}$ aligns with the aggregated matrix $\widetilde{\boldsymbol{G}} = \boldsymbol{G}_N \otimes \boldsymbol{G}_H$,

$$\boldsymbol{G} = \begin{bmatrix} 0.488 & 0.488 & 0 & 0 & 0.012 & 0.012 & 0 & 0 \\ 0.494 & 0.494 & 0 & 0 & 0.006 & 0.006 & 0 & 0 \\ 0.494 & 0.494 & 0 & 0 & 0.006 & 0.006 & 0 & 0 \\ 0.497 & 0.497 & 0 & 0 & 0.003 & 0.003 & 0 & 0 \\ 0.024 & 0.024 & 0 & 0 & 0.476 & 0.476 & 0 & 0 \\ 0.012 & 0.012 & 0 & 0 & 0.488 & 0.488 & 0 & 0 \\ 0.012 & 0.012 & 0 & 0 & 0.488 & 0.488 & 0 & 0 \\ 0.006 & 0.006 & 0 & 0 & 0.494 & 0.494 & 0 & 0 \end{bmatrix} \quad \widetilde{\boldsymbol{G}} = \begin{bmatrix} 0.491 & 0.491 & 0 & 0 & 0.009 & 0.009 & 0 & 0 \\ 0.491 & 0.491 & 0 & 0 & 0.009 & 0.009 & 0 & 0 \\ 0.496 & 0.496 & 0 & 0 & 0.005 & 0.005 & 0 & 0 \\ 0.496 & 0.496 & 0 & 0 & 0.005 & 0.005 & 0 & 0 \\ 0.018 & 0.018 & 0 & 0 & 0.482 & 0.482 & 0 & 0 \\ 0.018 & 0.018 & 0 & 0 & 0.482 & 0.482 & 0 & 0 \\ 0.009 & 0.009 & 0 & 0 & 0.491 & 0.491 & 0 & 0 \\ 0.009 & 0.009 & 0 & 0 & 0.491 & 0.491 & 0 & 0 \end{bmatrix}$$

As visible, the proposed approximation based on component queues effectively provides a local weighted sum of neighboring elements of $\boldsymbol{G}$. The benefit is that the expression on the right can be computed efficiently via Kronecker products, whereas the original $\boldsymbol{G}$ matrix on the left cannot.

4.4 Approximating the Matrix $\boldsymbol{R}$

The formula (3), which is used for calculating the steady-state queue-length probability vectors, requires the calculation of matrix $\boldsymbol{R}$. In this subsection, $\widetilde{\boldsymbol{G}}$, previously derived, is used to obtain an estimate for $\boldsymbol{R}$. Using (4) and substituting the approximation $\widetilde{\boldsymbol{G}}$ for $\boldsymbol{G}$, we propose the approximation

$$\boldsymbol{R} \approx -\boldsymbol{F}(\boldsymbol{L} + \boldsymbol{F}\widetilde{\boldsymbol{G}})^{-1} \tag{10}$$

In the above formula, the computational bottleneck is the inverse calculation, which, as we now show, exhibits special properties for the class of QBDs we consider. First, we have found through numerical experimentation that the inversion of matrix $\boldsymbol{L}$ is very efficient due to its sparsity. Furthermore, we prove in the following result that $\boldsymbol{F}\widetilde{\boldsymbol{G}}$ is a low-rank matrix.

Theorem 3. *Given a KPC queue with PH arrival process* $\boldsymbol{A} = (\boldsymbol{A}_0, \boldsymbol{A}_1)$ *and service process* $\boldsymbol{S} = \boldsymbol{N} \otimes \boldsymbol{H}_1 \otimes \ldots \otimes \boldsymbol{H}_J$*, where* $\boldsymbol{H}_k, k = 1, \ldots, J$ *are hyperexponential and* $\widetilde{\boldsymbol{G}} = \boldsymbol{G}_N \otimes \boldsymbol{G}_{H_1} \otimes \ldots \otimes \boldsymbol{G}_{H_J}$ *as before,* $rank(\boldsymbol{F}\widetilde{\boldsymbol{G}}) \leq n$*, where* n *is the number of phases of process* $\boldsymbol{N}$*.*

The proof of Theorem 3 is given in [11, Section 5.3.3, Chap. 5]. The low-rank of $\boldsymbol{F}\widetilde{\boldsymbol{G}}$ is an important property of our method, which allows us to efficiently compute the inverse of the sum in (10) using the Woodbury matrix identity:

$$\begin{aligned} \boldsymbol{R} &\approx -\boldsymbol{F}(\boldsymbol{L} + \boldsymbol{F}\widetilde{\boldsymbol{G}})^{-1} = -\boldsymbol{F}(\boldsymbol{L} + \boldsymbol{U}\boldsymbol{\Sigma}\boldsymbol{V})^{-1} \\ &\qquad = -\boldsymbol{F}(\boldsymbol{L}^{-1} - \boldsymbol{L}^{-1}\boldsymbol{U}(\boldsymbol{\Sigma}^{-1} + \boldsymbol{V}\boldsymbol{L}^{-1}\boldsymbol{U})^{-1}\boldsymbol{V}\boldsymbol{L}^{-1}) \end{aligned} \quad (11)$$

where $\boldsymbol{F}, \boldsymbol{L}, \widetilde{\boldsymbol{G}} \in \mathbb{R}^{m \times m}, \boldsymbol{U} \in \mathbb{R}^{m \times k}, \boldsymbol{\Sigma} \in \mathbb{R}^{k \times k}, \boldsymbol{V} \in \mathbb{R}^{k \times m}$, and $k = rank(\boldsymbol{F}\widetilde{\boldsymbol{G}})$. The Woodbury matrix identity relies on the decomposition of $\boldsymbol{F}\widetilde{\boldsymbol{G}} = \boldsymbol{U}\boldsymbol{\Sigma}\boldsymbol{V}$, where $\boldsymbol{\Sigma}$ is invertible and hence $rank(\boldsymbol{\Sigma}) = rank(\boldsymbol{F}\widetilde{\boldsymbol{G}})$. A straightforward approach to obtain $\boldsymbol{U}, \boldsymbol{\Sigma}, \boldsymbol{V}$ is to use the Singular Value Decomposition (SVD) of $\boldsymbol{F}\widetilde{\boldsymbol{G}}$. The computational complexity of this operation for a matrix with dimension $m \times m$ is typically $\mathcal{O}(m^3)$ in general [4]. However, we can improve that by exploiting the special Kronecker-form structure of $\boldsymbol{F}\widetilde{\boldsymbol{G}}$. We know

$$\boldsymbol{F}_{H_k}\boldsymbol{G}_{H_k} = \boldsymbol{I}_{h_k}(\mathbf{1}_{h_k}\boldsymbol{\alpha}_{H_k}) = \mathbf{1}_{h_k}\boldsymbol{\alpha}_{H_k}, k = 1, \ldots, J$$

from Proposition 1. Let $\boldsymbol{F}_N\boldsymbol{G}_N = \boldsymbol{U}_N\boldsymbol{\Sigma}\boldsymbol{V}_N$ be the SVD of matrix $\boldsymbol{F}_N\boldsymbol{G}_N$. Define $\boldsymbol{U}$ and $\boldsymbol{V}$ as follows

$$\begin{aligned} \boldsymbol{F}\widetilde{\boldsymbol{G}} &= \boldsymbol{F}_N\boldsymbol{G}_N \otimes \boldsymbol{F}_{H_1}\boldsymbol{G}_{H_1} \otimes \ldots \otimes \boldsymbol{F}_{H_J}\boldsymbol{G}_{H_J} \\ &= (\boldsymbol{U}_N\boldsymbol{\Sigma}\boldsymbol{V}_N) \otimes (\mathbf{1}_{h_1}\boldsymbol{\alpha}_{H_1}) \otimes \ldots \otimes (\mathbf{1}_{h_J}\boldsymbol{\alpha}_{H_J}) \\ &= (\boldsymbol{U}_N \otimes \mathbf{1}_{h_1} \otimes \ldots \otimes \mathbf{1}_{h_J})\boldsymbol{\Sigma}(\boldsymbol{V}_N \otimes \boldsymbol{\alpha}_{H_1} \ldots \otimes \boldsymbol{\alpha}_{H_J}) = \boldsymbol{U}\boldsymbol{\Sigma}\boldsymbol{V} \end{aligned} \quad (12)$$

Since $rank(\boldsymbol{F}\widetilde{\boldsymbol{G}}) = rank(\boldsymbol{F}_N\boldsymbol{G}_N) = rank(\boldsymbol{\Sigma})$, this decomposition has the required properties. Moreover, it is significantly faster to calculate as it only requires the SVD of a much smaller matrix $\boldsymbol{F}_N\boldsymbol{G}_N$. The SVD of $\boldsymbol{F}_N\boldsymbol{G}_N$ takes $\mathcal{O}(n^3)$ time and calculating the Kronecker products to obtain $\boldsymbol{U}$ and $\boldsymbol{V}$ takes $\mathcal{O}(m^2)$ time. Therefore, the entire decomposition runs in $\mathcal{O}(n^3+m^2)$ time. Recall that the size of $\boldsymbol{F}_N\boldsymbol{G}_N$ is $n \times n$ and $\boldsymbol{F}\widetilde{\boldsymbol{G}}$ is a $m \times m$ matrix. Under the assumption that $n \ll m$, $\mathcal{O}(n^3 + m^2)$ is dominated by the m^2 factor.

4.5 Approximating the Queue-Length Probabilities

With the efficient approximation of matrix $\boldsymbol{R}$ in (3), we can now compute the steady-state probability vectors $\widetilde{\boldsymbol{\pi}}_l = (\pi_{l1}, \ldots, \pi_{lm})$ for $l \in \mathbb{N}_0$, where m is the number of phases in each level. As stated in the background section, to calculate $\boldsymbol{\pi}_0$, one must first solve the associated system of equations for $\boldsymbol{x}$. It is known that if a solution $\boldsymbol{x}$ exists, then it is unique for the system. Using our approximate $\widetilde{\boldsymbol{G}} = \boldsymbol{G}_N \otimes \boldsymbol{G}_{H_1} \otimes \cdots \otimes \boldsymbol{G}_{H_1}$, we have similarly to (7) that $\boldsymbol{x}\widetilde{\boldsymbol{G}} = \boldsymbol{x}$, $\boldsymbol{x}\mathbf{1}_s = 1$.

Let $\boldsymbol{x}_N$ and $\boldsymbol{x}_{H_k}, k = 1, \ldots, J$ be the solutions of the above system of equation where $\widetilde{\boldsymbol{G}}$ is replaced by $\boldsymbol{G}_N$ and $\boldsymbol{G}_{H_k}, k = 1, \ldots, J$, respectively.

$$\begin{aligned}(\boldsymbol{x}_N \otimes \boldsymbol{x}_{H_1} \otimes \cdots \otimes \boldsymbol{x}_{H_J})\widetilde{\boldsymbol{G}} &= (\boldsymbol{x}_N \otimes \boldsymbol{x}_{H_1} \otimes \cdots \otimes \boldsymbol{x}_{H_J})(\boldsymbol{G}_N \otimes \boldsymbol{G}_{H_1} \otimes \cdots \otimes \boldsymbol{G}_{H_J}) \\ &= \boldsymbol{x}_N\boldsymbol{G}_N \otimes \boldsymbol{x}_{H_1}\boldsymbol{G}_{H_1} \otimes \cdots \otimes \boldsymbol{x}_{H_J}\boldsymbol{G}_{H_J} \\ &= \boldsymbol{x}_N \otimes \boldsymbol{x}_{H_1} \otimes \cdots \otimes \boldsymbol{x}_{H_J}\end{aligned}$$

where the last equation holds because of the definition of the vectors $\boldsymbol{x}_i$. Furthermore,

$$\begin{aligned}(\boldsymbol{x}_N \otimes \boldsymbol{x}_{H_1} \otimes \cdots \otimes \boldsymbol{x}_{H_J})\mathbf{1}_s &= (\boldsymbol{x}_N \otimes \boldsymbol{x}_{H_1} \otimes \cdots \otimes \boldsymbol{x}_{H_J})(\mathbf{1}_n \otimes \mathbf{1}_{h_1} \otimes \cdots \otimes \mathbf{1}_{h_J}) \\ &= \boldsymbol{x}_N\mathbf{1}_n \otimes \boldsymbol{x}_{H_1}\mathbf{1}_{h_1} \otimes \cdots \otimes \boldsymbol{x}_{H_J}\mathbf{1}_{h_J} \\ &= 1 \otimes 1 \otimes \cdots \otimes 1 = 1\end{aligned}$$

Therefore, $\boldsymbol{x}_N \otimes \boldsymbol{x}_{H_1} \otimes \cdots \otimes \boldsymbol{x}_{H_k}$ is the unique solution to the approximate system of equations $\boldsymbol{x}\widetilde{\boldsymbol{G}} = \boldsymbol{x}$, $\boldsymbol{x}\mathbf{1}_s = 1$. Furthermore, Proposition 1 states that $\boldsymbol{G}_{H_k} = \mathbf{1}_{h_k}\boldsymbol{\alpha}_{H_k}$. This implies that $\boldsymbol{x}_{H_k} = \boldsymbol{\alpha}_{H_k}$, since $\boldsymbol{\alpha}_{H_k}$ is a probability vector and $\boldsymbol{\alpha}_{H_k}\boldsymbol{G}_{H_k} = \boldsymbol{\alpha}_{H_k}\mathbf{1}_{h_k}\boldsymbol{\alpha}_{H_k} = 1\boldsymbol{\alpha}_{H_k} = \boldsymbol{\alpha}_{H_k}$.
Let $\widetilde{\boldsymbol{R}}$ be the approximation of $\boldsymbol{R}$ given by (11). Based on the previous results, $\widetilde{\boldsymbol{\pi}}_l$ can be computed as

$$\widetilde{\boldsymbol{\pi}}_0 = (1-\rho)(\boldsymbol{x}_N \otimes \boldsymbol{\alpha}_{H_1} \otimes \ldots \otimes \boldsymbol{\alpha}_{H_J}) \tag{13}$$

$$\widetilde{\boldsymbol{\pi}}_l = \widetilde{\boldsymbol{\pi}}_0\widetilde{\boldsymbol{R}}^l \qquad \text{for} \quad l > 0 \tag{14}$$

5 Fitting Algorithm

In this section, the theoretical results established in the previous section are combined with a search algorithm to find a QBD process that matches a given set of queue-length probabilities. Our main contribution is the optimization problem formulation. The user may choose different numerical solvers based on their preferred cost and convergence properties.

5.1 Decision Variables

The decision variables are a set of vector parameters used to construct the components of the KPC-PH distribution. Initially, we describe the construction of the hyperexponential PH components, denoted as $\boldsymbol{H}_i = (\boldsymbol{H}_{i0}, \boldsymbol{H}_{i1}), i = 1, \ldots, J$, specified by the parameter vectors $\boldsymbol{\alpha}_{H_i} \in \mathbb{R}^{h_i - 1}$ and $\boldsymbol{\lambda}_{H_i} \in \mathbb{R}^{h_i}$. Note, that $\boldsymbol{\alpha}_{H_i}$ has one dimension less than the number of phases in process $\boldsymbol{H}$, with the last element chosen to satisfy the constraint that the elements sum to 1. Therefore, the actual probability vector is formulated as $\boldsymbol{\alpha}'_{H_i} = (\boldsymbol{\alpha}_{H_i}, 1 - \boldsymbol{\alpha}_{H_i}\mathbf{1}_{h_i - 1})$.

To ensure $\boldsymbol{\alpha}'_{H_i}$ is a valid probability vector, the following constraints apply to the decision variables: $\mathbf{0} \leq \boldsymbol{\alpha}_{H_i}$ and $\boldsymbol{\alpha}_{H_i}\mathbf{1} \leq 1$ for $i = 1, \ldots, J$. The diagonal

Algorithm 1. Calculating the approximate queue-length probabilities

Input: $\boldsymbol{\alpha}_N, \boldsymbol{\lambda}_N, \boldsymbol{\alpha}_{H_1}, \boldsymbol{\lambda}_{H_1} \ldots, \boldsymbol{\alpha}_{H_J}, \boldsymbol{\lambda}_{H_J}, \boldsymbol{A}$

Construct the PH distributions from $\boldsymbol{\alpha}_i, \boldsymbol{\lambda}_i$ parameters as shown in Section 5.1
Calculate $\boldsymbol{F}$ and $\boldsymbol{L}$ from the components (1)
Solve CQ_N with cyclic reduction to get $\boldsymbol{G}_N$ $\triangleright\ \mathcal{O}(L_{CR}n^3)$
Calculate the SVD decomposition of $\boldsymbol{F}_N\boldsymbol{G}_N = \boldsymbol{U}_N\boldsymbol{\Sigma}\boldsymbol{V}_N$ $\triangleright\ \mathcal{O}(n^3)$
Calculate $\boldsymbol{U}$ and $\boldsymbol{V}$ for decomposition $\boldsymbol{F}\widetilde{\boldsymbol{G}} = \boldsymbol{U}\boldsymbol{\Sigma}\boldsymbol{V}$ (12) $\triangleright\ \mathcal{O}(m^2)$
Calculate $\widetilde{\boldsymbol{\pi}}_0$ from the components (4.5)
Run Algorithm 2 with input $\boldsymbol{F}, \boldsymbol{L}, \widetilde{\boldsymbol{\pi}}_0, \boldsymbol{U}, \boldsymbol{\Sigma}, \boldsymbol{V}$ to get $\widetilde{\pi}_1, \ldots, \widetilde{\pi}_t$ $\triangleright\ \mathcal{O}(m^2 t)$
Calculate marginal probability for level 0 as $\widetilde{\pi}_0 = \widetilde{\boldsymbol{\pi}}_0 \mathbf{1}_m$
return $\widetilde{\pi} = (\widetilde{\pi}_0, \ldots, \widetilde{\pi}_t)$

Algorithm 2. Calculating $\widetilde{\pi}_1, \ldots, \widetilde{\pi}_t$

Input: $\boldsymbol{F}, \boldsymbol{L}, \widetilde{\boldsymbol{\pi}}_0, \boldsymbol{U}, \boldsymbol{\Sigma}, \boldsymbol{V},$

$\boldsymbol{\Sigma}^{-1} = \mathrm{inv}(\boldsymbol{\Sigma})$ $\triangleright\ \mathcal{O}(k)$ as $\boldsymbol{\Sigma}$ is diagonal
$\boldsymbol{L}^{-1} = \mathrm{inv}(\boldsymbol{L})$ $\triangleright\ \mathcal{O}(m^2)$ when $\boldsymbol{L}$ is upper triangular
$\boldsymbol{A} = \boldsymbol{V}\boldsymbol{L}^{-1}$ $\triangleright\ \mathcal{O}(km^2)$ as $\boldsymbol{V}$ is rectangular
$\boldsymbol{B} = \boldsymbol{\Sigma}^{-1} + \boldsymbol{A}\boldsymbol{U}$ $\triangleright\ \mathcal{O}(k^2 m)$ as $\boldsymbol{U}$ is rectangular
$\widetilde{\pi}_0 = (1-\rho), \quad \widetilde{\boldsymbol{\pi}}_{l-1} = \widetilde{\boldsymbol{\pi}}_0$
for $l = 1, \ldots, t$ **do**
  $\boldsymbol{u}_0 = \widetilde{\boldsymbol{\pi}}_{l-1}(-\boldsymbol{F})\boldsymbol{L}^{-1}$ $\triangleright\ \mathcal{O}(m^2)$
  $\boldsymbol{u}_1 = \boldsymbol{u}_0\boldsymbol{U}$ $\triangleright\ \mathcal{O}(km)$ as $\boldsymbol{U}$ is rectangular
  $\boldsymbol{u}_2 = (\boldsymbol{B}^\top \backslash \boldsymbol{u}_1^\top)^\top$ $\triangleright\ \mathcal{O}(k^3)$ to solve system of linear equations
  $\boldsymbol{u}_2 = \boldsymbol{u}_2\boldsymbol{V}$ $\triangleright\ \mathcal{O}(km)$ as $\boldsymbol{V}$ is rectangular
  $\boldsymbol{u}_1 = \boldsymbol{u}_2\boldsymbol{L}^{-1}$ $\triangleright\ \mathcal{O}(m^2)$
  $\widetilde{\boldsymbol{\pi}}_l = \boldsymbol{u}_0 - \boldsymbol{u}_1, \quad \widetilde{\pi}_l = \widetilde{\boldsymbol{\pi}}_l\mathbf{1}_m, \quad \widetilde{\boldsymbol{\pi}}_{l-1} = \widetilde{\boldsymbol{\pi}}_l$
end for $\triangleright\ \mathcal{O}((m^2+k^3)t)$
return $\widetilde{\pi}_1, \ldots, \widetilde{\pi}_t$

elements of $\boldsymbol{H}_{i0}$ are stored in the decision vector $\mathbf{0} \le \boldsymbol{\lambda}_{H_i}, i = 1, \ldots, J$, so that $\boldsymbol{H}_{i0} = \mathrm{diag}(-\boldsymbol{\lambda}_{H_i})$ and $\boldsymbol{H}_{i1} = \boldsymbol{\lambda}_{H_i}^\top \boldsymbol{\alpha}'_{H_i}$.

Constructing $\boldsymbol{N} = (\boldsymbol{N}_0, \boldsymbol{N}_1)$ requires additional effort as the matrices $\boldsymbol{N}_0, \boldsymbol{N}_1$ are no longer diagonal. The decision vector $\boldsymbol{\alpha}_N \in \mathbb{R}^{n-1}$ is analogous to $\boldsymbol{\alpha}_{H_k}$ and thus subject to the same constraints: $\mathbf{0} \le \boldsymbol{\alpha}_N$ and $\boldsymbol{\alpha}_N\mathbf{1} \le 1$. Finally, $\mathbf{0} \le \boldsymbol{\lambda}_N \in \mathbb{R}^{n^2}$ records both the off-diagonal elements of $\boldsymbol{N}_0$ and the exit rates $\boldsymbol{t}$. Here, $\boldsymbol{N}_1 = \boldsymbol{t}\boldsymbol{\alpha}_N$, and the diagonal elements of $\boldsymbol{N}_0$ are set to ensure the row sums of $\boldsymbol{Q}_N = \boldsymbol{N}_0 + \boldsymbol{N}_1$ sum to 0 as required.

5.2 Outline of the Fitting Algorithm

During the fitting process, the assumption is that we have information on the arrival process $\boldsymbol{A}$ and a finite number t of marginal queue-length probabilities $\pi = (\pi_0, \pi_1, \ldots, \pi_t)$. The marginal queue length probabilities are calculated from the queue-length probabilities by summing the probability vectors at each level, that is $\pi_0 = \boldsymbol{\pi}_0\mathbf{1}_m$ and $\pi_1 = \boldsymbol{\pi}_1\mathbf{1}_m$. We start by implementing the equations of Sect. 4 in Algorithm 1, to calculate the approximate marginal queue-length distribution from the decision variables $\boldsymbol{\alpha}_N, \boldsymbol{\lambda}_N, \boldsymbol{\alpha}_{H_1}, \boldsymbol{\lambda}_{H_1} \ldots, \boldsymbol{\alpha}_{H_J}, \boldsymbol{\lambda}_{H_J}$, and the arrival process $\boldsymbol{A}$. The time complexity of the main steps is noted inline and discussed in detail in [11, Section 5.4.5, Chap. 5]. Once the relevant matrices and $\widetilde{\boldsymbol{\pi}}_0$ are calculated, Algorithm 1 calls Algorithm 2 to calculate the remaining marginal probabilities iteratively. To reduce the time complexity of the computation, we do not compute matrix $\boldsymbol{R}$ directly. Instead, the Woodbury identity is applied at every iteration, and matrix-matrix multiplication is entirely avoided. We point to [11, Section 5.4.5, Chap. 5] for a detailed analysis of the time complexity corresponding to the steps in Algorithm 2.

The optimization problem to find a QBD process with a KPC-PH service distribution, that has the given arrival process ($\boldsymbol{A}$), and that when solved has the given marginal queue length probabilities (π_l for $l = 0, \ldots, t$), can be written as follows

$$\begin{aligned} &\min \textstyle\sum_{l=0}^{t} |\widetilde{\pi}_l - \pi_l|/\pi_l \\ &\text{s.t. } \boldsymbol{\alpha}_i \mathbf{1} \leq 1, \; i \in I := \{N, H_1, \ldots, H_J\} \\ &\qquad \boldsymbol{\alpha}_i, \boldsymbol{\lambda}_i \geq \mathbf{0}, \; i \in I \end{aligned} \tag{15}$$

where $\boldsymbol{\alpha}_N \in \mathbb{R}^{n-1}, \boldsymbol{\alpha}_{H_i} \in \mathbb{R}^{h_i-1}, \boldsymbol{\lambda}_N \in \mathbb{R}^{n \times n}, \boldsymbol{\lambda}_{H_i} \in \mathbb{R}^{h_i}$ with $i = 1, \ldots, J$ are the decision variables, and $\widetilde{\boldsymbol{\pi}} = f(\boldsymbol{\alpha}_N, \boldsymbol{\lambda}_N, \boldsymbol{\alpha}_{H_1}, \boldsymbol{\lambda}_{H_1} \ldots, \boldsymbol{\alpha}_{H_J}, \boldsymbol{\lambda}_{H_J} | \boldsymbol{A})$ where function f is the implementation of Algorithm 1.

The vector $\boldsymbol{\pi}$ and the arrival process $\boldsymbol{A}$ are inputs to the problem and remain unchanged throughout. In the case of an irreducible Markov chain, $\boldsymbol{\pi} > 0$; hence, division by π_l is always allowed up to machine precision. Thus, we have $V = n - 1 + n^2 + \sum_{i=1}^{J}(2h_i - 1)$ variables and $C = V + n - 1 + \sum_{i=1}^{J}(h_i - 1)$ constraints. The objective function calculates the error between the input $\boldsymbol{\pi}$ and approximate $\widetilde{\boldsymbol{\pi}} = (\widetilde{\pi}_0, \ldots, \widetilde{\pi}_t)$. Normalizing the difference forces the fitting algorithm to correctly capture the tail of the distribution. This is necessary because the probabilities corresponding to higher levels are very close to zero; therefore, otherwise, the difference between the approximate and the input probabilities would be negligible.

5.3 Using a Gamma Distribution as N

As expected, the general PH distribution $\boldsymbol{N}$ requires more parameters than the hyperexponential components $\boldsymbol{H}_k$. To reduce the number of parameters and therefore the dimension of the optimization problem, we note the following. The SCV of $\boldsymbol{N}$ can be chosen to be $SCV_N \leq 1$ without loss of generality. This is because a PH distribution with $SCV > 1$ is hyperexponential. Setting $SCV_N > 1$ would mean the KPC-PH distribution is hyperexponential [25]. However, setting $SCV_N = 1$ gives a hyperexponential KPC-PH as well. In this case, the first component is lost, but the number of hyperexponential components can be increased from J to $J + 1$ to counteract the lost component.
During the optimization, we reduce the dimension of the search problem by using a gamma distribution, which requires only two parameters. At each iteration, the gamma distribution is approximated with an acyclic phase-type (APH) distribution. This approach is natural since the gamma distribution generalizes the

Erlang distribution by allowing a non-integer shape parameter.

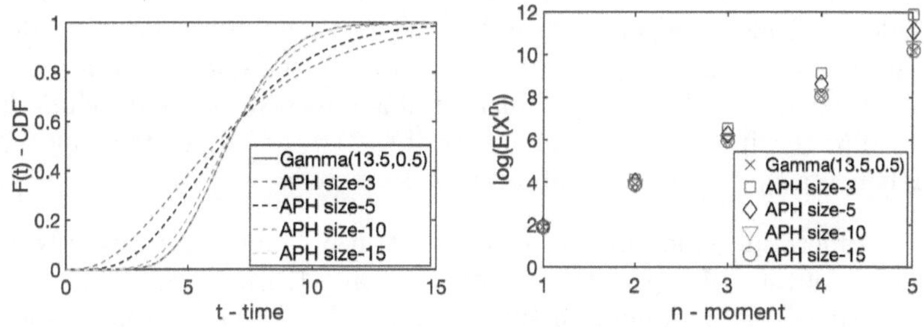

(a) CDF (b) Logmoments

Fig. 1. Gamma distribution with $SCV = 0.074$ approximated by APH distributions of increasing size.

Figure 1 shows that a gamma distribution with an SCV as low as .074 is closely matched by an APH distribution of size 10, and is matched perfectly by an APH distribution of size 15. The updated version of (15) that uses a two-parameter gamma distribution is as follows:

$$\begin{aligned} &\min \textstyle\sum_{l=1}^{t} |\widetilde{\pi}_l - \pi_l|/\pi_l \\ &\text{s.t. } \boldsymbol{\alpha}_i \mathbf{1} \leq 1, \ i \in I := \{H_1, \ldots, H_J\} \\ &\quad\ k, \theta, \boldsymbol{\alpha}_i, \boldsymbol{\lambda}_i \geq \mathbf{0}, \ i \in I \end{aligned} \tag{16}$$

where $\widetilde{\boldsymbol{\pi}} = f(k, \theta, \boldsymbol{\alpha}_{H_1}, \boldsymbol{\lambda}_{H_1} \ldots, \boldsymbol{\alpha}_{H_J}, \boldsymbol{\lambda}_{H_J} | \boldsymbol{A})$ and function f is updated to include the conversion of the gamma distribution into an APH distribution before the queue-length probabilities are computed. The parameters (k, θ) correspond to the gamma distribution.

6 Numerical Results

We now conduct numerical tests to evaluate the proposed divide-and-conquer fitting algorithm. In our implementation, we used MATLAB and its built-in solver fmincon [23] to solve the optimization Problem 16. The numerical tests were run on an Ubuntu machine with an Intel Xeon E5540 CPU having 16 virtual cores and 32GB of RAM.

6.1 Evaluating the Fitting Algorithm

In this section, we use two real-life Microsoft production server traces, `Radius Authentication` and `BuildServer00`, collected from the RADIUS authentication server and a Microsoft build server, respectively, to evaluate the fitting algorithm outlined in (15) and implemented in MATLAB. In the upcoming figures, we refer to our method as *KPC-QBD*. We first compare *KPC-QBD* to an alternative queue-based fitting method, which we term *exact QBD*. The alternative approach involves using optimization Problem 15; however, instead of the divide-and-conquer algorithm we developed, it employs the cyclic-reduction algorithm implemented in QMAM [28] to compute the queue-length probabilities. We also employ other well-established fitting algorithms as benchmarks, all of which directly fit traces using moments of the trace. These fitting algorithms are implemented in the KPC toolbox [16] and include:

- Prony's method, which utilizes characteristic polynomials to exactly match moments of the trace (see, [17, Theorem 1], and [15]).
- Hyperexponential parameter fitting, an approximate moment-matching algorithm that uses a non-linear optimization problem to find the hyperexponential distribution that best approximates the first three moments of the empirical trace.
- KPC-PH algorithm [16] that uses a divide-and-conquer approach to directly match moments of a trace.

For all Kronecker-composition methods (KPC-PH, KPC-QBD, and exact QBD), the number J of hyperexponential building blocks was experimentally selected by evaluating multiple values of J.

Initially, examining the marginal queue-length probabilities in Fig. 2, the difference between *KPC-QBD* and *exact QBD* appears small. Fitting is performed at the 25% utilization level where some improvement is noticeable. We have observed similar result at 50% utilization. However, a significant difference in fitting accuracy becomes evident upon inspecting the empirical Cumulative Distribution Function (CDF) of the traces directly in Fig. 3. *KPC-QBD* not only outperforms the exact QBD method but also surpasses all other benchmark methods that directly fit the traces. Noticeably, the sizes of the PH distributions fitted by the *exact QBD* method are significantly smaller compared to those produced by *KPC-QBD*. This size difference is an anticipated limitation of *exact QBD*, attributable to the slow computation of cyclic reduction at each iteration of the optimization problem while evaluating its objective function.

6.2 Response Time Distributions

To further validate the proposed *KPC-QBD* algorithm, we calculate the response time CDFs at 25% and 50% utilization for both the `Buildserver00` and the `Radius Authentication` traces. There are toolboxes that provide solvers for response time calculations such as [28]. In Figs. 4 and 5, the response time CDFs are computed, except for *KPC-QBD* distributions, which are simulated due to their large number of states. It is apparent from these figures that the *KPC-QBD* distributions offer a much closer fit compared to the other methods, including the *exact QBD* distributions.

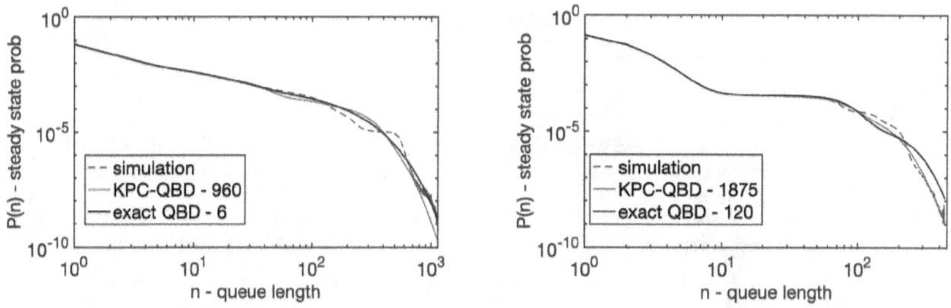

(a) **`Radius Authentication`** (b) **`Buildserver00`**

Fig. 2. Queueing results at 25% utilization when using KPC-QBD and exact QBD methods to fit queueing probabilities.

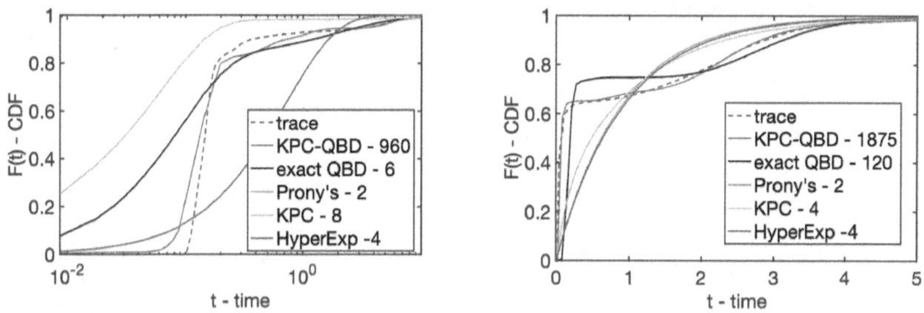

(a) **`Radius Authentication`** (b) **`Buildserver00`**

Fig. 3. Empirical CDF of traces and the corresponding fitted distributions.

Motivated by the accuracy of the *KPC-QBD* method, we next reduce the size of its distributions to make response time calculations numerically feasible while preserving accuracy. The reduction is performed using the algorithm described in [11, Algorithm 3, Chap. 5]. A reduced PH distribution size ensures a smoother fit. This smoothing effect occurs because the reduced distribution has fewer degrees of freedom, mitigating any overfitting that may have occurred due to overparameterization. We run simulations of varying lengths to compute the CCDF of response times for the `Buildserver00` and `Radius Authentication` traces. Achieving 99% convergence requires 10^6 iterations, taking about 38 s. These results are compared with the original and reduced KPC-QBD distributions in Fig. 6. The CCDF emphasizes tail behavior, where reduction errors are most pronounced. For `Buildserver00`, a reduced size of 110 maintains high accuracy and computes the PH response time distribution in 10 s, a four times speedup over simulation. For `Radius Authentication`, a size of 130 provides a reasonable estimate in about 27 s.

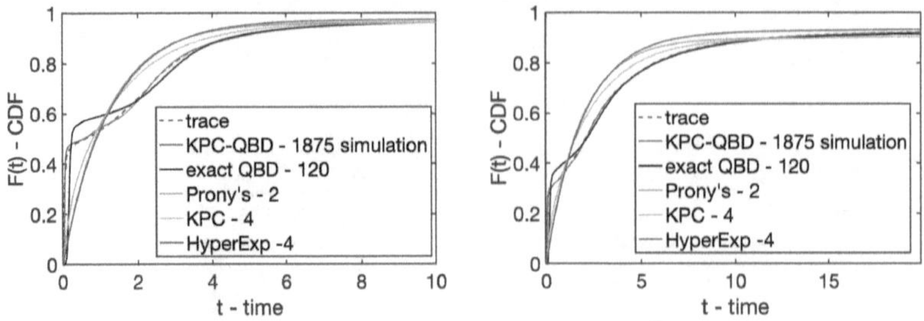

(a) $\rho = 0.25$ (b) $\rho = 0.50$

Fig. 4. `Buildserver00` trace CDF of response time, KPC-QBD and benchmark methods.

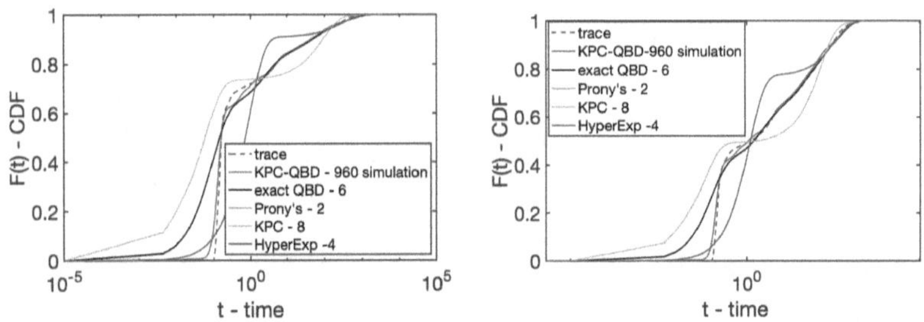

(a) $\rho = 0.25$ (b) $\rho = 0.50$

Fig. 5. `Radius Authentication` trace CDF of response time, KPC-QBD and benchmark methods.

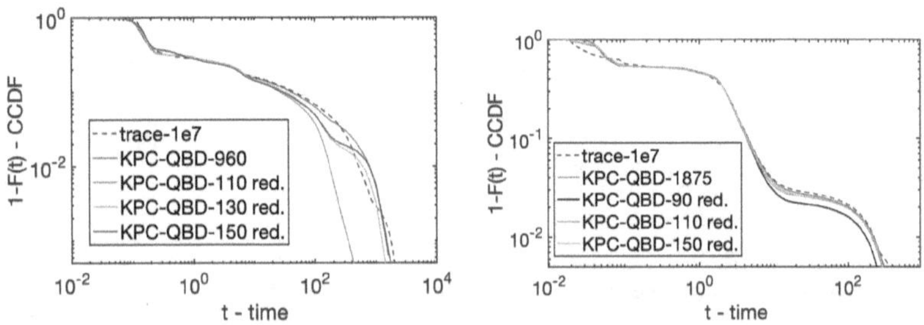

(a) `Radius Authentication` (b) `Buildserver00`

Fig. 6. Response time CCDF at 25% utilization, comparing empirical simulation with original and reduced KPC-QBD distributions.

7 Conclusion

In this paper, we present theoretical results for the efficient approximation of marginal queue-length probabilities for a QBD process with PH-distributed arrival times and compositional PH-distributed service times. We formulate an optimization problem that utilizes this approximation of queue-length probabilities in its objective function and aims to find a QBD process with a KPC-PH service distribution, that closely matches a given set of marginal queue-length probabilities.

Future work could examine the extension of the presented results to a broader class of Markovian processes, such as matrix exponential distributions, Markovian arrival processes, or rational arrival processes. Queueing systems associated with these processes can be studied using matrix analytic methods. Therefore, our results may also be used to fit and analyze these processes.

Acknowledgement. This work has received funding from the European Union's Horizon Europe research and innovation programme under grant agreement No 101299128. Evgenia Smirni is partially supported by the U.S. National Science Foundation (NSF) grant #2402942. The work was done in part during a visit of the first author to William & Mary, VA, US. We are grateful to Andreas Stathopoulos for constructive comments.

References

1. Ali, A., Pinciroli, R., Yan, F., Smirni, E.: Opt. inference serving on serverless platforms. Proc. VLDB Endow. **15**(10) (2022)
2. Asanjarani, A., Nazarathy, Y., Taylor, P.: A survey of param. and state estimation in queues. Queueing Syst. **97** (2021)
3. Asmussen, S., Nerman, O., Olsson, M.: Fitting phase-type distributions via the EM algorithm. Scan. J. Stat. (1996)
4. Baglama, J., Reichel, L.: Augmented implicitly restarted lanczos bidiagonalization methods. SIAM J. Sci. Comput. **27**(1), 19–42 (2005). https://doi.org/10.1137/04060593X
5. Banik, A.D., Chaudhry, M.L.: Efficient computational analysis of stationary probabilities for the queueing system BMAP/G/1/N with or without vacation(s). INFORMS J. Comput. **29**(1), 140–151 (2017)
6. Bladt, M.: A review on PH dist. and their use in risk theory. ASTIN Bull. J. IAA **35**(1), 145–161 (2005)
7. Bladt, M., Esparza, L.J.R., Nielsen, B.F.: Fisher information and statistical inference for phase-type distributions. J. Appl. Probab. **48**(A), 277–293 (2011)
8. Bladt, M., Nielsen, B.F.: Matrix-Exponential Distributions in Applied Probability, vol. 81. Springer (2017)
9. Bobbio, A., Horváth, A., Telek, M.: Matching three moments with minimal acyclic phase type distributions. Stoch. Model. **21**(2–3), 303–326 (2005)
10. Bolch, G., Greiner, S., De Meer, H., Trivedi, K.S.: Queueing Networks and Markov Chains: Modeling and Performance Evaluation with Computer Science Applications. Wiley (2006)
11. Bor, J.: Analytical Methods and Distribution Fitting for Response Times in Markovian Queueing Systems. Ph.D. thesis, Imperial College London, London, UK (2025)
12. Breuer, L., Baum, D.: An Introduction to Queueing Theory and Matrix-Analytic Methods. Springer, Heidelberg (2006)
13. Buchholz, P.: On a Markovian process algebra. Citeseer (1994)
14. Buchholz, P., Kriege, J., Felko, I.: Input Modeling with Phase-Type Distributions and Markov Models: Theory and Applications. Springer (2014)

15. Casale, G., Smirni, E.: KPC-toolbox: fitting Markovian arrival processes and phase-type distributions with matlab. SIGMETRICS Perform. Eval. Rev. **39**(4), 47 (2012)
16. Casale, G., Zhang, E.Z., Smirni, E.: KPC-toolbox: best recipes for automatic trace fitting using Markovian arrival processes. Perform. Eval. **67**(9), 873–896 (2010)
17. Casale, G., Zhang, E.Z., Smirni, E.: Trace data characterization and fitting for Markov modeling. Perform. Eval. **67**(2), 61–79 (2010)
18. Ciardo, G., Riska, A., Smirni, E.: Equiload: a load balancing policy for clustered web servers. PEVA **46**(2-3) (2001)
19. Feldmann, A., Whitt, W.: Fitting mixtures of exponentials to long-tail distributions to analyze network performance models. Perform. Eval. **31**(3–4), 245–279 (1998)
20. Horváth, A., Vicario, E.: Computer Performance Engineering and Stochastic Modelling, pp. 201–215. Springer (2023). Construction of PH dist. by Bernstein Exponentials
21. Horváth, G.: Moment matching-based distribution fitting with generalized hyper-erlang distributions. In: Dudin, A., De Turck, K. (eds.) ASMTA 2013. LNCS, vol. 7984, pp. 232–246. Springer, Heidelberg (2013). https://doi.org/10.1007/978-3-642-39408-9_17
22. Kumaran, J., Mitchell, K., Liefvoort, A.: The waiting time distribution of an MEP/MEP/1 queue. In: Proceedings of the 19th International Teletraffic Congress (ITC19), pp. 687–696 (2005)
23. MathWorks: fmincon Documentation Home. MathWorks (2024). https://uk.mathworks.com/help/optim/ug/fmincon.html
24. Nasr, W.W., Taaffe, M.R.: Fitting the $\mathrm{Ph_t}/\mathrm{M_t}/s/c$ time-dependent departure process for use in tandem queueing networks. INFORMS J. Comput. **25**(4), 758–773 (2013)
25. Neuts, M.F.: Matrix-geometric solutions in stochastic models: an algorithmic approach. Courier Corporation (1994)
26. Okamura, H., Dohi, T., Trivedi, K.S.: Improvement of expectation-maximization algorithm for phase-type distributions with grouped and truncated data. Appl. Stoch. Model. Bus. Ind. **29**(2), 141–156 (2013)
27. Pacheco-Sanchez, S., Casale, G., Scotney, B., McClean, S., Parr, G., Dawson, S.: Markovian workload characterization for QoS prediction in the cloud. In: IEEE CLOUD 2011, pp. 147–154. IEEE (2011)
28. Pérez, J.F., Van Velthoven, J., Van Houdt, B.: Q-mam: a tool for solving infinite queues using matrix-analytic methods. In: Proceedings of the 3rd International Conference on Performance Evaluation Methodologies and Tools, p. 9. Brussels, BEL (2008)
29. Rastpour, A., Ingolfsson, A., Sandikçi, B.: Algorithms for queueing systems with reneging and priorities modeled as quasi-birth-death processes. INFORMS J. Comput. **34**(3), 1693–1710 (2022)
30. Reinecke, P., Horváth, G.: PH dists for realistic modelling in discrete-event sim. In: SimuTools, pp. 283–290 (2012)
31. Riska, A., Diev, V., Smirni, E.: An EM-based technique for approximating long-tailed data sets with PH distributions. Perform. Eval. **55**(1), 147–164 (2004)
32. Riska, A., Smirni, E.: M/G/1-type Markov processes: a tutorial. In: Calzarossa, M.C., Tucci, S. (eds.) Performance 2002. LNCS, vol. 2459, pp. 36–63. Springer, Heidelberg (2002). https://doi.org/10.1007/3-540-45798-4_3
33. Riska, A., Smirni, E.: ETAQA solutions for infinite Markov processes with repetitive structure. IJOC **19**(2) (2007)

34. Thümmler, A., Buchholz, P., Telek, M.: A novel approach for PH fitting with the EM algorithm. IEEE Trans. Dependable Secur. Comput. **3**(3), 245–258 (2006)
35. Yu, K., Huang, M., Brill, P.H.: An algorithm for fitting heavy-tailed distributions via generalized hyperexponentials. INFORMS J. Comput. **24**(1), 42–52 (2012). https://doi.org/10.1287/IJOC.1100.0443

Stability Checking of Markov Jump Linear Systems via Probabilistic Temporal Logic

Lena Becker(✉) and Holger Hermanns

Saarland University, Saarland Informatics Campus, 66123 Saarbrücken, Germany
{becker,hermanns}@depend.uni-saarland.de

Abstract. Markov jump linear systems (MJLSs) model dynamical phenomena subject to random switching among multiple linear modes, driven by an underlying Markov chain. Classical notions such as mean and mean-square stability characterize the long-term asymptotic behaviour of the first and second moments of an MJLS, but they can be overly conservative or even misleading when only a specific subset of initial conditions is of interest. We tackle this challenge through the lens of model checking, where reasoning about specific sets of initial conditions is intrinsic to the approach. We begin by formalizing probabilistic computation tree logic (PCTL) on MJLSs, enabling the specification of state-based temporal properties for these systems. Building on this foundation, we extend the logic to capture moment-based stability properties relative to a prescribed set of initial states. While we ultimately do not obtain a decision procedure for the entire base logic, the logical extensions can be handled – albeit with some technical subtleties – by exploiting linear-algebraic techniques.

Keywords: PCTL model checking · Dynamical systems · Stability

1 Introduction

Dynamical systems are a fundamental tool for modelling the evolution of real-world processes over time, typically with the objective of designing controllers that ensure a desired system behaviour. In many applications, however, the underlying dynamics are subject to uncertainty and may exhibit abrupt changes, e.g. in case of system failures or volatile environmental circumstances, inducing switches between different modes of dynamics, each covering the behaviour in a specific situation. We here focus on models where these switches are probabilistic in nature, and thus governed by a Markov chain. In such Markov jump linear systems (MJLSs) [9,10], modes are mapped to distinct system dynamics, adding discrete behaviour to the continuous-state system. These system dynamics are typically assumed to be linear, as many systems can be locally approximated by linear dynamics near operating points. MJLSs have been used to model a wide range of application domains including flight control [13], power systems analysis [22], and robotic systems such as lower-limb exoskeletons [23].

In settings where the system cannot be controlled, or where the controller is already fixed, one must rely on analysis techniques to assess its behaviour.

M. Chen and M. Tschaikowski (Eds.): QEST+FORMATS 2026, LNCS 16913, pp. 284–301, 2027.
https://doi.org/10.1007/978-3-032-35298-9_16

A commonly analysed property of dynamical systems in this context is their stability, i.e. whether the system's state converges in the long run, independently of the initial conditions. The stability analysis of MJLSs not only needs to take into account the dynamical behaviour of all system modes, but also the long-run behaviour of the underlying Markov chain. It may yield counterintuitive results: even if each individual dynamics considered in isolation is stable, the MJLS as a whole may exhibit unstable behaviour, and vice versa.

In the literature, different notions of stability for MJLSs have been explored, prominently among them *mean stability* [21] – analysing the evolution of the first moment – and *mean-square stability* [8,9,12] – accounting for cancellation effects due to opposing outliers by considering the second moment as the stability criterion of choice. Both stability notions are typically characterized via eigenvalue analysis of specific operators that capture the combined continuous and discrete probabilistic dynamics of the MJLS. In doing so, it is standard to aim for global guarantees covering all initial states. This perspective may however be overly conservative in practice. In particular, instability might arise only for a small fraction of initial states that may not even be of practical relevance, e.g. if corresponding to physically unreachable configurations, such as physically impossible operating regions of a drone.

Despite its intuitive importance, the MJLS stability problem for a given set of initial states has not been tackled in the literature, at least to the best of our knowledge. It is a genuine model-checking problem, so we attack it by embedding appropriate logical operators into a typical temporal-logic framework. Probabilistic model checking, based on logics such as probabilistic computation tree logic (PCTL), enables the specification of temporal properties with respect to specific sets of initial states [15]. PCTL for MJLSs arises naturally by conservatively extending work on PCTL for discrete-time Markov chains (DTMCs) [3], as well as by specializing work for general Markov processes (GMPs) [24]. When turning to the model-checking problem for this logic, however, it turns out that the latter work must necessarily be incomplete, because of a Skolem-hardness result [7] that applies to a specific fragment of PCTL model checking on MJLSs (and thus equally on GMPs from [24]). We proceed by extending PCTL with propositions to capture moment-based stability properties and discuss the intricacies when checking properties that include such propositions. Altogether, we arrive at a logic that enables the specification of refined stability-like properties as well as their combination with classical reachability or branching-time objectives. A computational example enriches the exposition.

The remainder of this paper is structured as follows: Sect. 2 develops relevant background on MJLSs. In Sect. 3, we define a semantics for classical PCTL for MJLSs and give an algorithm to verify such PCTL properties. Furthermore, we discuss limitations in model checking reachability properties due to the uncountable state space of MJLSs. Finally, in Sect. 4, we extend PCTL with the aforementioned new propositions that capture the first and second moments of MJLSs and describe an algorithm to verify these properties.

Related Work. *Discrete-time stochastic systems* that are given by the equation $x(k+1) = \mathcal{F}_a(x(k), w(k))$, where a ranges over the available modes, generalize MJLSs in multiple aspects: the functions $\mathcal{F}_a$ may be non-linear, the switching is not necessarily governed by a Markov chain, and a stochastic noise $w(k)$ is incorporated. Contrasting our approach, existing work on the verification of such systems is based on abstractions into finite-state models [20]. A PCTL semantics is then defined on these abstractions and model checking is carried out similarly to standard algorithms for Markov chains or Markov decision processes.

Markov reward models (MRMs) extend DTMCs by associating rewards with the states and accumulating them additively [1]. The use of multiplicative rewards instead [2] yields a modelling formalism that closely resembles MJLSs with a single continuous dimension. This prevents modelling dependencies across multiple continuous dimensions. Further, since only positive rewards are supported, behaviours that rely on sign changes or cancellations are excluded. The authors consider a notion of expectation that is similar in spirit to our treatment of continuous state variables, but oscillatory behaviour is handled by distinguishing between limit superior and limit inferior. In contrast, we consider a notion based on limits of Cesàro-averaged expected values.

Linear (discrete-time) dynamical systems (LDSs) given by $x(k+1) = Ax(k)$ have been subject to a long line of research investigating the decidability of reachability problems [17].[1] While deciding reachability of a specific point is possible in polynomial time [16], reachability of more complex sets, such as a four-dimensional affine subspace in a five-dimensional system, can be encoded as longstanding open problems like the Skolem problem [7]. Furthermore, model checking linear temporal logic (LTL) is known to be decidable for LDSs up to dimension three [18]. Since LDSs can be viewed as a deterministic special case of MJLSs, these results provide important insight into the inherent difficulty of verification problems in our setting. In particular, they indicate that even in the absence of stochastic switching, reachability and model-checking problems quickly approach the frontier of decidability, thereby setting clear limits on what can be expected for the more general case of MJLSs.

General Markov processes (GMPs) extend finite-state DTMCs to an uncountable state space by replacing the transition probability matrix with a stochastic kernel defined over a measurable state space. Existing work explores a PCTL semantics for GMPs and shows that reach-avoid properties can be characterized via Bellman fixed-point equations [24]. As we show in Eq. 1, MJLSs can be expressed in the framework of GMPs and, thus, these results are directly relevant to our setting. However, just like MJLSs, GMPs should also be affected by the aforementioned difficulties in deciding reachability problems.

[1] The term "reachability" here refers to the set of states reachable without considering the option to choose controller inputs in a stepwise way, i.e. the orbit problem.

$$A_1 = \begin{bmatrix} 0.35 & 13.45 \\ 0.35 & -7.65 \end{bmatrix} \quad A_3 = \begin{bmatrix} 0.35 & 2.65 \\ 0.35 & -0.35 \end{bmatrix}$$

$$A_2 = \begin{bmatrix} -17.66 & 0.35 \\ 13.44 & 0.35 \end{bmatrix} \quad P = \begin{bmatrix} 1/7 & 3/7 & 3/7 \\ 1/4 & 1/2 & 1/4 \\ 1/3 & 1/3 & 1/3 \end{bmatrix}$$

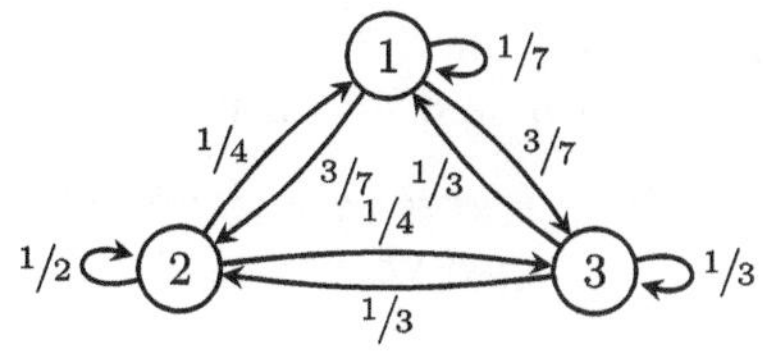

Fig. 1. An MJLS $(P, \{A_1, A_2, A_3\})$ with the DTMC visualized on the right.

2 Background

Notation. The indicator function that evaluates to 1 if $x \in S$ and 0 otherwise is denoted $\mathbf{1}_S(x)$. For $i \in \{1, \ldots, m\}$, let $\vec{e}_i$ be the i-th canonical base vector for the m-dimensional space. Let I_n and 0_n be the identity and all-zeros matrix of size $n \times n$, respectively. $A^\top$ denotes the transpose of matrix A. Further, $\rho(A) = \max |\lambda_i|$, where λ_i ranges over the eigenvalues of A, denotes the spectral radius of matrix A, which is the maximum of the absolute values of A's eigenvalues.

The block diagonal matrix blkdiag$(A_1, \ldots, A_m)$ is defined as the matrix that puts A_1 to A_m on the main diagonal and fills all other entries with 0. We further define the Kronecker product $A \otimes B$ of matrices A and B as usual.

We will make use of a function VEC that vectorizes an $m \times n$ matrix to a vector of length mn by stacking the matrix columns underneath each other, i.e. $\text{VEC}\left((a_{ij})\right) = \left[\, a_{11} \cdots a_{m1} \cdots a_{1n} \cdots a_{mn} \,\right]^\top$. Its inverse function $\text{VEC}^{-1}_{m \times n}$ transforms a vector of length mn into an $m \times n$ matrix. The function VECH carries this concept to symmetric matrices and only vectorizes the lower triangular elements. In particular, VECH "half-vectorizes" a symmetric $n \times n$ matrix to a vector of length $\frac{1}{2}n(n+1)$ and is given by $\text{VECH}\left((a_{ij})\right) = \left[\, a_{11} \cdots a_{n1}\, a_{22} \cdots a_{n2} \cdots a_{nn} \,\right]^\top$.

Finally, we define a function SUM^{mn}_n that reshapes a vector of length mn into an $n \times m$ matrix (via $\text{VEC}^{-1}_{n \times m}$) and then sums its columns, resulting in a vector of n entries. It is given by $\text{SUM}^{mn}_n\left(\left[\, a_1 \cdots a_{mn} \,\right]^\top\right) = (1^\top_m \otimes I_n) \cdot \left[\, a_1 \cdots a_{mn} \,\right]^\top$, where 1_m is the all-ones vector of length m.

LDSs. A linear (discrete-time) dynamical system (LDS) is described by a state matrix $A \in \mathbb{R}^{n \times n}$, and, given an initial state $x(0) \in \mathbb{R}^n$, evolves according to the equation $x(k+1) = Ax(k)$. Properties of LDSs, in particular stability [5,6] and reachability (here: orbit problem) [7,16], have been studied extensively.

MJLSs. Markov jump linear systems (MJLSs) generalize LDSs by allowing for switching between a finite set of state matrices. The switching is governed by an underlying discrete-time Markov chain (DTMC), whose modes are associated with the corresponding state matrices. Formally, an MJLS $\mathfrak{J}$ with m modes and n continuous state variables is a tuple $(P, \{A_1, \ldots, A_m\})$, where $P \in \mathbb{R}^{m \times m}$ is the transition probability matrix of Markov chain $\{\theta(k) \mid k \in \mathbb{N}\}$ that assumes values from 1 to m and A_1 to $A_m \in \mathbb{R}^{n \times n}$ are the state matrices. $P(\theta, \theta')$ denotes the transition probability from mode θ to θ', i.e. $Pr\left(\theta(k+1) = \theta' \mid \theta(k) = \theta\right)$ for $k \in \mathbb{N}$. Let $x(k) \in \mathbb{R}^n$ denote the continuous state at time $k \in \mathbb{N}$. Then, this continuous state evolves according to the equation $x(k+1) = A_{\theta(k)}x(k)$ [9].

For the remainder of this paper, let $\mathfrak{J} = (P, \{A_1, \ldots, A_m\})$ be an MJLS with m modes and n continuous state variables.

Example 1. Figure 1 shows an MJLS with $m = 3$ and $n = 2$ (values are rounded, see the extended version [4] for the precise values).

MJLS Measurability. The state space $S_\mathfrak{J}$ of MJLS $\mathfrak{J}$ is of the form $\mathbb{R}^n \times \{1, \ldots, m\}$. An infinite path of $\mathfrak{J}$ is a sequence $(x_0, \theta_0)(x_1, \theta_1)(x_2, \theta_2) \cdots \in S_\mathfrak{J}^\omega$, where for all $i \geq 0$, $P(\theta_i, \theta_{i+1}) > 0$ and $x_{i+1} = A_{\theta_i} x_i$. Similarly, finite paths of length $k \in \mathbb{N}$ are sequences $(x_0, \theta_0) \cdots (x_k, \theta_k) \in S_\mathfrak{J}^{k+1}$. Given a state $(x, \theta) \in S_\mathfrak{J}$, $\mathit{Paths}((x, \theta))$ and $\mathit{Paths}_k((x, \theta))$ describe the set of all infinite and finite paths of length k, respectively, beginning in (x, θ).

The cylinder set $\mathit{Cyl}(\pi)$ of a finite path π is the set of all infinite extensions of π. The probability of such a cylinder set is defined via the corresponding cylinder set in the Markov chain [3]:

$$\mathit{Pr}(\mathit{Cyl}((x_0, \theta_0) \cdots (x_k, \theta_k))) = \mathit{Pr}(\mathit{Cyl}(\theta_0 \cdots \theta_k)) = \prod_{i=0}^{k-1} P(\theta_i, \theta_{i+1})$$

The σ-algebra $\Sigma^\mathfrak{J}$ associated with MJLS $\mathfrak{J}$ is the smallest σ-algebra that contains all cylinder sets $\mathit{Cyl}(\hat{\pi})$ where $\hat{\pi}$ ranges over all finite paths in $\mathfrak{J}$.

GMPs. A general Markov process (GMP) [24] is a discrete-time stochastic process on a Borel set X, whose evolution is characterized by a stochastic kernel $Q \colon X \times \mathfrak{B}(X) \to [0, 1]$, where $\mathfrak{B}(X)$ is the Borel σ-algebra on X. GMPs generalize MJLSs, as we can lift MJLS $\mathfrak{J}$ to the GMP setting by choosing $X = S_\mathfrak{J}$ and

$$Q((x, \theta), B) = \sum_{\theta'=1}^{m} \mathbf{1}_B((A_\theta x, \theta')) \cdot P(\theta, \theta') \tag{1}$$

2.1 Stability of MJLSs

There are different notions of stability for MJLSs. In the following, we summarize well-established results on *mean* and *mean-square stability,* which analyse the system state's expected value and uncentred variance, respectively, on the long run [9,21]. All results in the remainder of this section stem from [9], notationally adjusted to our setting. The first and second (uncentred) moment at time $k \in \mathbb{N}$ of MJLS $\mathfrak{J}$ with initial state $(x, \theta) \in S_\mathfrak{J}$ are defined as follows:

$$\mathbb{E}_{(x,\theta)}[x(k)] = \sum_{(x_0,\theta_0)\cdots(x_k,\theta_k) \in \mathit{Paths}_k((x,\theta))} x_k \cdot \mathit{Pr}(\mathit{Cyl}((x_0, \theta_0) \cdots (x_k, \theta_k)))$$

$$\mathbb{V}_{(x,\theta)}[x(k)] = \mathbb{E}_{(x,\theta)}[x(k)x(k)^\intercal] = \sum_{(x_0,\theta_0)\cdots(x_k,\theta_k) \in \mathit{Paths}_k((x,\theta))} x_k x_k^\intercal \cdot \mathit{Pr}(\mathit{Cyl}((x_0, \theta_0) \cdots (x_k, \theta_k)))$$

Note that, as it is standard for multivariate random variables, the first moment is a vector of the first moments of each random variable, whereas the second (uncentred) moment is an $n \times n$ symmetric matrix. In this matrix, position (i, j) represents the covariance between random variable x_i and x_j, which are the respective continuous state variables in our case. Further note that while variance is classically defined as the centred second moment $\mathbb{E}\left[(X - \mathbb{E}[X])^2\right]$, this definition uses the uncentred moment. However, $\mathbb{E}\left[(X - \mathbb{E}[X])^2\right] = \mathbb{V}(X) - (\mathbb{E}[X])^2$ shows the direct connection between centred and uncentred second moment.

Definition 1. *MJLS* $\mathfrak{J}$ *is* mean stable (MS) *iff* $\forall (x, \theta) \in S_{\mathfrak{J}}. \mathbb{E}_{(x,\theta)}[x(k)] \to 0$ *for* $k \to \infty$*, and it is* mean-square stable (MSS) *iff* $\forall (x, \theta) \in S_{\mathfrak{J}}. \mathbb{V}_{(x,\theta)}[x(k)] \to 0$ *for* $k \to \infty$*.*

Note that if control inputs and stochastic noise were added to the system, convergence points different from the origin might arise. In this case, the definition would require a unique convergence point across all initial states. We remark further that mean-square stability obviously implies mean stability. To arrive at an algorithmic check for stability of $\mathfrak{J}$, some further definitions are required upfront. For some initial state $(x, \theta) \in S_{\mathfrak{J}}$ and Markov chain mode $\theta' \in \{1, \ldots, m\}$,

$$\mathbb{E}_{(x,\theta)}[x(k) \mid \theta(k) = \theta'] = \sum_{(x_0,\theta_0)\cdots(x_k,\theta') \in Paths_k((x,\theta))} x_k \cdot Pr\left(Cyl\left(\theta_0 \cdots \theta'\right)\right)$$

$$\mathbb{V}_{(x,\theta)}[x(k) \mid \theta(k) = \theta'] = \sum_{(x_0,\theta_0)\cdots(x_k,\theta') \in Paths_k((x,\theta))} x_k x_k^{\mathsf{T}} \cdot Pr\left(Cyl\left(\theta_0 \cdots \theta'\right)\right)$$

denote the MJLS state's expected value and uncentred variance at time k with initial state (x, θ), restricted on the Markov chain mode θ' at time k. Note that the sum of $\mathbb{E}_{(x,\theta)}[x(k) \mid \theta(k) = i]$ over all $i \in \{1, \ldots, m\}$ is the overall expected state value at time k starting in (x, θ), i.e. $\mathbb{E}_{(x,\theta)}[x(k)]$. The analogous property for $\mathbb{V}_{(x,\theta)}[x(k)]$ holds as well. Furthermore, the following two equalities hold:

$$\mathbb{E}_{(x,\theta)}[x(k+1) \mid \theta(k+1) = j] = \sum_{i=1}^{m} P(i,j) \cdot A_i \cdot \mathbb{E}_{(x,\theta)}[x(k) \mid \theta(k) = i]$$

$$\mathbb{V}_{(x,\theta)}[x(k+1) \mid \theta(k+1) = j] = \sum_{i=1}^{m} P(i,j) \cdot A_i \cdot \mathbb{V}_{(x,\theta)}[x(k) \mid \theta(k) = i] \cdot A_i^{\mathsf{T}}$$

which can be exploited by an operator $\mathcal{B}_{\mathfrak{J}}$ that propagates the conditional expected value to the next time step, where $\mathcal{B}_{\mathfrak{J}} = (P^{\mathsf{T}} \otimes I_n) \cdot \mathrm{blkdiag}(A_1, \ldots, A_m)$. This operator satisfies:

$$\begin{bmatrix} \mathbb{E}_{(x,\theta)}[x(k+1) \mid \theta(k+1) = 1] \\ \vdots \\ \mathbb{E}_{(x,\theta)}[x(k+1) \mid \theta(k+1) = m] \end{bmatrix} = \mathcal{B}_{\mathfrak{J}} \cdot \begin{bmatrix} \mathbb{E}_{(x,\theta)}[x(k) \mid \theta(k) = 1] \\ \vdots \\ \mathbb{E}_{(x,\theta)}[x(k) \mid \theta(k) = m] \end{bmatrix}$$

Similarly, the operator $\mathcal{T}_{\mathfrak{J}} = (P^{\top} \otimes I_{n^2}) \cdot \text{blkdiag}(A_1 \otimes A_1, \ldots, A_m \otimes A_m)$ propagates the conditional uncentred variance to the next time step. Since both operators characterize the stepwise evolution of the respective moments, it is their properties that determine the respective limiting behaviour. Thus, we can resort to eigenvalue analysis to determine stability with respect to mean and uncentred variance:

Theorem 1. (cf. [9]). *MJLS $\mathfrak{J}$ is MS iff $\rho(\mathcal{B}_{\mathfrak{J}}) < 1$. It is MSS iff $\rho(\mathcal{T}_{\mathfrak{J}}) < 1$.*

Example 2. The MJLS from Fig. 1 is neither MSS nor MS, because $\rho(\mathcal{B}_{\mathfrak{J}}) \approx 11.4$.

3 Probabilistic Computation Tree Logic for MJLSs

Originally established to specify properties on probabilistic systems such as DTMCs [15], PCTL is a temporal logic that suits well our modelling framework of MJLSs.

We define its syntax as usual:

$$\begin{aligned} \Phi &::= \mathit{true} \;\mid\; a \;\mid\; \Phi_1 \wedge \Phi_2 \;\mid\; \neg\Phi \;\mid\; \mathcal{P}_J(\varphi) \\ \varphi &::= \mathcal{X}\,\Phi \;\mid\; \Phi_1\,\mathcal{U}^{\leq k}\,\Phi_2 \;\mid\; \Phi_1\,\mathcal{U}\,\Phi_2 \end{aligned}$$

where $a \in AP_x \sqcup AP_\theta$, $J \subseteq [0,1]$, and $k \in \mathbb{N}$. We denote the step-bounded fragment of PCTL with PCTL$^-$, i.e. all constructs except the unbounded until. We define AP_x and AP_θ as the set of atomic propositions on the continuous and discrete state, respectively, with $\sqcup$ denoting disjoint union. States $(x,\theta) \in S_{\mathfrak{J}}$ can be labelled with atomic propositions:

$$L((x,\theta)) = \{a \in AP_x \mid x \in \Pi_a\} \sqcup \{a \in AP_\theta \mid a \in L_{\text{MC}}(\theta)\}$$

where L_{MC} is the labelling function for the discrete states and Π_a are convex polytopes, which are bounded convex subsets of $\mathbb{R}^n$ that can be represented equivalently either as the intersection of finitely many half-spaces or as the convex hull of finitely many vertices [25]. We restrict the labelling of the continuous state with atomic propositions to convex polytopes, since this class of sets provides a suitable balance between expressiveness and computational tractability. In particular, membership of a state in a convex polytope can be checked efficiently via its halfspace representation [14]. Further, the model-checking algorithm presented below requires repeated set operations and transformations. Convex polytopes are well suited as a basis for this, since operations such as affine (and thus linear) transformations and intersection preserve the polytopic structure [25]. Thus, the sets arising during the recursive evaluation of formulas remain computationally manageable. By contrast, more general classes of sets may fail to retain such a tractable representation under these operations, making them less suitable for algorithmic verification.

Let $a \in AP_x \sqcup AP_\theta$, and $(x, \theta) \in S_{\mathfrak{J}}$. We define the satisfaction relation $\models$ of PCTL state formulas analogously to DTMCs [3]:

$$
\begin{array}{llll}
(x,\theta) \models \mathit{true} & & (x,\theta) \models a & \text{iff } a \in L((x,\theta)) \\
(x,\theta) \models \neg\Phi & \text{iff } (x,\theta) \not\models \Phi & (x,\theta) \models \Phi_1 \wedge \Phi_2 & \text{iff } (x,\theta) \models \Phi_1 \wedge (x,\theta) \models \Phi_2 \\
(x,\theta) \models \mathcal{P}_J(\varphi) & \multicolumn{3}{l}{\text{iff } \mathit{Pr}\left(\{\pi \in \mathit{Paths}\left((x,\theta)\right) \mid \pi \models \varphi\}\right) \in J}
\end{array}
$$

and for an infinite path $\pi = (x_0, \theta_0)(x_1, \theta_1) \cdots \in S_{\mathfrak{J}}^\omega$, where $\pi[i]$ denotes (x_i, θ_i):

$$
\begin{array}{ll}
\pi \models \mathcal{X}\,\Phi & \text{iff } \pi[1] \models \Phi \\
\pi \models \Phi_1\,\mathcal{U}^{\leq k}\,\Phi_2 & \text{iff } \exists j \leq k.\, \pi[j] \models \Phi_2 \wedge \forall i < j.\, \pi[i] \models \Phi_1 \\
\pi \models \Phi_1\,\mathcal{U}\,\Phi_2 & \text{iff } \exists j \geq 0.\, \pi[j] \models \Phi_2 \wedge \forall i < j.\, \pi[i] \models \Phi_1
\end{array}
$$

Theorem 2. *For each PCTL path formula φ and state $(x, \theta) \in S_{\mathfrak{J}}$ of MJLS $\mathfrak{J}$, the set $\{\pi \in \mathit{Paths}((x,\theta)) \mid \pi \models \varphi\}$ is measurable, i.e. it is an element of $\Sigma^{\mathfrak{J}}$.*

This entails that the probabilities of the relevant sets can be constructed from the available operations on the cylinder sets. Due to space limitations, we omit the proofs in this paper. They can be found in the extended version [4].

Relation to GMPs. As per the discussion above, MJLS can be embedded into general Markov processes (GMPs) [24] with the lifting of Eq. 1. This is consistent with the respective semantics of PCTL:

Lemma 1. *Given MJLS $\mathfrak{J}$, let Borel set $X = S_{\mathfrak{J}}$ with the associated Borel σ-algebra and stochastic kernel Q from Eq. 1. Let $\models_{\mathsf{GMP}}$ denote the PCTL satisfaction relation for GMPs as given in [24]. Then, for any PCTL formula Φ and state $(x, \theta) \in S_{\mathfrak{J}}$, $(x, \theta) \models_{\mathsf{GMP}} \Phi$ iff $(x, \theta) \models \Phi$.*

Model Checking PCTL$^-$. Just like PCTL model checking on Markov chains (or CTL on Kripke structures), its model checking on MJLSs is based on a bottom-up traversal of the parse tree of the formula in question. Along the way, the algorithm computes the satisfaction set $\mathrm{Sat}(\Phi) = \{(x,\theta) \in S_{\mathfrak{J}} \mid (x,\theta) \models \Phi\}$ of the respective subformula Φ as described in the following, where we first restrict to PCTL$^-$ for the sake of exposition.

Theorem 3. *Let Φ be a PCTL$^-$ formula. For every state $(x, \theta) \in S_{\mathfrak{J}}$ of MJLS $\mathfrak{J}$, it holds that:*

$$
\mathit{Sat}(\mathit{true}) = S_{\mathfrak{J}} \qquad \mathit{Sat}(a) = \begin{cases} \Pi_a \times \{1, \ldots, m\} & \text{if } a \in AP_x \\ \mathbb{R}^n \times L_{MC}(\theta) & \text{if } a \in AP_\theta \end{cases}
$$

$$
\mathit{Sat}(\neg\Phi) = S_{\mathfrak{J}} \setminus \mathit{Sat}(\Phi) \qquad \mathit{Sat}(\Phi_1 \wedge \Phi_2) = \mathit{Sat}(\Phi_1) \cap \mathit{Sat}(\Phi_2)
$$

$$
\mathit{Sat}(\mathcal{P}_J(\mathcal{X}\,\Phi)) = \left\{ (x,\theta) \in S_{\mathfrak{J}} \mid \sum_{\theta'=1}^{m} \mathbf{1}_{\mathit{Sat}(\Phi)}\left((A_\theta x, \theta')\right) \cdot P(\theta, \theta') \in J \right\}
$$

$$
\mathit{Sat}\left(\mathcal{P}_J\left(\Phi_1\,\mathcal{U}^{\leq k}\,\Phi_2\right)\right) = \{(x,\theta) \in S_{\mathfrak{J}} \mid U_k\left((x,\theta)\right) \in J\}
$$

where

$$U_0\left((x,\theta)\right) = \mathbf{1}_{Sat(\Phi_2)}\left((x,\theta)\right)$$

$$U_{k+1}\left((x,\theta)\right) = U_0\left((x,\theta)\right) + \mathbf{1}_{Sat(\Phi_1)\setminus Sat(\Phi_2)}\left((x,\theta)\right) \sum_{\theta'=1}^{m} P\left(\theta,\theta'\right) \cdot U_k\left((A_\theta x,\theta')\right)$$

Model Checking $\Phi_1 \,\mathcal{U}\, \Phi_2$. We now turn to the model checking of the full logic. The only missing case is the computation of Sat $(\Phi_1 \,\mathcal{U}\, \Phi_2)$, where we can assume that Sat (Φ_1) and Sat (Φ_2) are already computed. In principle, the work of [24] provides a guideline, since it establishes a connection between PCTL and dynamic programming (albeit not formulated for satisfaction sets, but for individual states) for the GMP setting. More concretely, it views the until $\mathcal{U}$ as the limit of the bounded-until $\mathcal{U}^{\leq k}$ for $k \to \infty$, which is quite natural. However, the existence of this limit appears not to be guaranteed. The reason is as follows: Since LDSs are a special case of MJLSs (with only a single mode and $P = [\,1\,]$), LDSs are also a special case of GMPs. As an LDS only exhibits a single path given an initial state, reachability in the path-based setting (such as LTL and MSO) coincides with reachability in the branching-time view, such as in PCTL.

Thus, the negative results with respect to decidability for LDSs reachability carry over to both MJLSs and GMPs. To the best of our knowledge, this implication for GMPs has not been observed before. Notably: if Sat (Φ_2) (i.e. the space that is supposed to be reached eventually) is at least four-dimensional, LDS reachability is an open problem [7] linked to the Skolem problem. (On the other hand, point-to-point reachability is decidable for LDSs in polynomial time, and so is point-to-hyperplane reachability for system dimensions up to four [17].)

This contrast is striking. While the fixed-point characterization for GMPs [24] suggests a conceptually simple recursive computation, even basic reachability questions for the strictly simpler class of LDSs are unresolved in general. This indicates that the apparent simplicity of the recursion does not translate into algorithmic tractability. We conclude this section with the observation that the model-checking problem for PCTL beyond PCTL$^-$ on MJLSs is open.

4 Adding Stability Properties to PCTL

This section first discusses how stability properties can be added to the base logic, then turns to model checking algorithmics, and finally discusses computability aspects.

4.1 Logical Extension

Our main motivation, as stated earlier, lies in the fact that a negative answer to the stability analysis of an MJLS may be overly conservative. Two scenarios immediately come to mind in which a more refined answer may be useful:

1. Even though the moment does converge for all initial states, the limit value is not necessarily the origin for all initial states. Thus, the MJLS is not considered stable. However, in this scenario, convergence itself might be considered of higher importance than convergence to the origin.

2. For some set of initial states, the moment diverges. Again, the MJLS is not considered stable. However, these initial states may be insignificant, e.g. because they correspond to impossible configurations in the real-world process. In this case, one would like to restrict the stability analysis to the set of relevant initial states.

Example 3. A combination of the two scenarios above is well possible. Recall that the system illustrated in Fig. 1 is not MS. In fact, for most initial states, the first moment diverges. However, there exist initial states whose first moment stabilizes in the long run, though not to the origin. For instance, the long-run expected value of initial state $\left[\,16.047\ -0.254\,\right]^{\mathsf{T}}$ with $\theta = 1$ is $\left[\,17.377\ 0.8\,\right]^{\mathsf{T}}$, while for $\left[\,0.324\ 31.814\,\right]^{\mathsf{T}}$ with $\theta = 2$ it is $\left[\,-2.69\ -0.124\,\right]^{\mathsf{T}}$.

These observations motivate a logical framework capable of specifying convergence properties relative to sets of initial states of interest. We address this with an extension of PCTL for MJLSs to express properties on the system's first and second moments, given a specific initial state. To arrive there, we extend the atomic propositions of PCTL with corresponding propositions in both step-bounded and step-unbounded variants. Just like the original atomic propositions, these are state formulas and can be nested within other PCTL formulas. The converse, i.e. nesting other PCTL formulas within the new operators as it is possible for e.g. next $\mathcal{X}$, is not supported in our choice of extension. Other than that, the semantics as defined in Sect. 3 remains unchanged. We stipulate

$$a \in AP_x \sqcup AP_\theta \sqcup \{\mathcal{E}_\Pi^k, \mathcal{E}_\Pi, \mathcal{V}_\Xi^k, \mathcal{V}_\Xi \mid k \in \mathbb{N}\}$$

where Π and Ξ are convex polytopes in $\mathbb{R}^n$ and $\mathbb{R}^{\frac{1}{2}n(n+1)}$, respectively. For some convex polytope Π, $\mathcal{E}_\Pi^k$ labels all states in which the first moment up to step k on average lies in Π and, analogously, $\mathcal{E}_\Pi$ considers the long-run first moment value. The operators $\mathcal{V}_\Xi^k$ and $\mathcal{V}_\Xi$ are defined in the same way for the second moment. We restrict Π and Ξ to convex polytopes again for the possibility of efficient representation, while still providing an intuitive modelling framework. Since the second moment is an $n \times n$ matrix, the bounding polytope Ξ has to reflect this and put bounds on n^2 variables. We can reduce this to $\frac{1}{2}n(n+1)$ as the second moment matrix is symmetric. For this, we stack the lower triangular part of a matrix into a vector with the VECH operator. With this, we define:

$$\mathcal{E}_\Pi^k \in L((x,\theta)) \quad \text{iff} \quad \frac{1}{k+1}\sum_{i=0}^{k} \mathbb{E}_{(x,\theta)}\left[x(i)\right] \quad \in \Pi \tag{2}$$

$$\mathcal{E}_\Pi \in L((x,\theta)) \quad \text{iff} \quad \lim_{k\to\infty}\frac{1}{k+1}\sum_{i=0}^{k} \mathbb{E}_{(x,\theta)}\left[x(i)\right] \quad \in \Pi \tag{3}$$

$$\mathcal{V}_\Xi^k \in L((x,\theta)) \quad \text{iff} \quad \text{VECH}\left(\frac{1}{k+1}\sum_{i=0}^{k} \mathbb{V}_{(x,\theta)}[x(i)]\right) \quad \in \Xi \tag{4}$$

$$\mathcal{V}_\Xi \in L((x,\theta)) \quad \text{iff} \quad \text{VECH}\left(\lim_{k\to\infty}\frac{1}{k+1}\sum_{i=0}^{k} \mathbb{V}_{(x,\theta)}[x(i)]\right) \quad \in \Xi \tag{5}$$

The above definitions use time-averaging summation, yielding the Cesàro limit in (3) and (5). This is a design choice, and e.g. (3) could have been replaced by the simpler $\lim_{k\to\infty} \mathbb{E}_{(x,\theta)}[x(k)]$, which corresponds to the traditional steady-state distribution on Markov chains.[2] Ensuring existence of the latter however requires a handful of additional definitions (recurrence, aperiodicity, ergodicity) [19] that we jump across by instead using Cesàro limits. We do this for brevity of the exposition paired with simplicity of the subsequent algorithmisation: the Cesàro limit always exists (so no check needs to be implemented) and agrees with the steady-state distribution whenever that exists. The connection to mean and mean-square stability is thus immediate: in an MJLS that is MS or MSS, respectively, the entire state space is labelled with $\mathcal{E}_\Pi$ or $\mathcal{V}_\Xi$ for any polytope Π or Ξ that contains the origin. However, the converse implication does not hold, due to the use of Cesàro limits. Even if the Cesàro-averaged moment converges, the corresponding moment sequence itself may fail to converge. Hence, the proposed operators capture a slightly different notion of convergence than classical MS and MSS, focusing on long-run average behaviour rather than pointwise limits and remaining well-defined even in the presence of oscillations. Further, note that variance is typically defined as the centred second moment. However, using the uncentred second moment aligns more closely with classical mean-square stability. Centring can be incorporated directly into the specification of the polytope Ξ if desired.

Although the step-unbounded operators describe limit behaviour, and thus finite path prefixes do not affect the limit, nesting them within other PCTL formulas like next $\mathcal{X}$ is still meaningful. Even a single transition may change the system dynamics to the extent that a state not satisfying $\mathcal{E}_\Pi$ (or $\mathcal{V}_\Xi$) transitions to one that does, such that $\mathcal{P}_J(\mathcal{X}\,\mathcal{E}_\Pi)$ for some interval J can still be satisfied.

Example 4. Consider again the MJLS from Fig. 1. Let $\Pi \in \mathbb{R}^2$ be a convex polytope with vertices $(30, 14)$, $(22, 35)$, $(-32, -5)$, and $(36, -27)$. We are interested in the set of states that satisfy $\mathcal{E}_\Pi$. Furthermore, we would like to allow for a five-step grace period to reach a state with the desired long-run expected value, while still avoiding unsafe states. We define another convex polytope Λ with vertices $(-15, -10)$, $(50, -29)$, $(45, 50)$, and $(0, 21)$ of safe states. With a slight abuse of notation, atomic proposition Λ labels all states in polytope Λ. Requiring the reach-avoid probability to be at least 0.3, we desire to check the PCTL formula $\mathcal{P}_{\geq 0.3}(\Lambda\,\mathcal{U}^{\leq 5}\,\mathcal{E}_\Pi)$. For an illustration of the polytopes Π and Λ, see Fig. 2. We shall return to the evaluation of the formulas in Example 5.

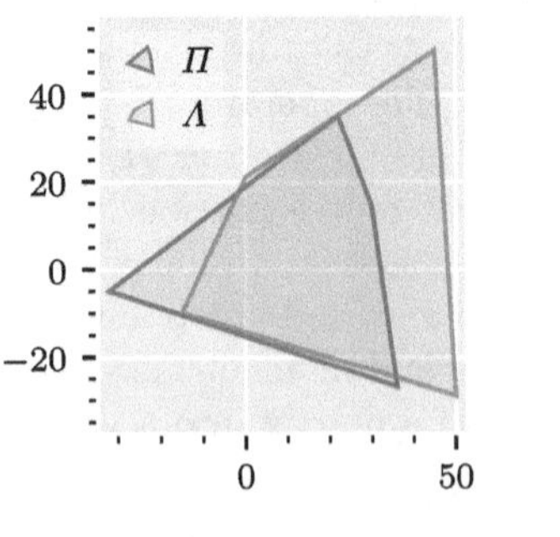

Fig. 2. A visualization of Π and Λ from Example 4.

[2] Another possible characterization would be via limit superior and inferior (cf. [2]).

4.2 Model Checking

We now present algorithms for computing, for a given polytope Π, the satisfaction sets of $\mathcal{E}_{\Pi}^{k}$, followed by $\mathcal{E}_{\Pi}$, and then describe the needed adaptations to also cover the variance operators.

Model Checking $\mathcal{E}_{\Pi}^{k}$. The satisfaction set of $\mathcal{E}_{\Pi}^{k}$ for an MJLS $\mathfrak{J}$ relies on propagating the operator $\mathcal{B}_{\mathfrak{J}}$ for k steps. Since $\mathcal{B}_{\mathfrak{J}}$ propagates the conditional expected value, we initialize with the corresponding values at time 0, i.e. for a given state $(x, \theta) \in S_{\mathfrak{J}}$:

$$\left[\mathbb{E}_{(x,\theta)}\left[x(0) \mid \theta(0) = 1\right] \cdots \mathbb{E}_{(x,\theta)}\left[x(0) \mid \theta(0) = \theta\right] \cdots \mathbb{E}_{(x,\theta)}\left[x(0) \mid \theta(0) = m\right]\right]^{\intercal}$$

which reduces to $\left[0 \cdots 0 \; x \; 0 \cdots 0\right]^{\intercal} = \vec{e}_{\theta} \otimes x$. Finally, observe that:

$$\mathbb{E}_{(x,\theta)}\left[x(k)\right] = \textsc{sum}_{n}^{mn}\left(\mathcal{B}_{\mathfrak{J}}^{k}(\vec{e}_{\theta} \otimes x)\right) \tag{6}$$

The satisfaction set is then obtained as described in the following theorem.

Theorem 4. *Let $k \in \mathbb{N}$ and let Π be a convex polytope in $\mathbb{R}^n$. For every state $(x, \theta) \in S_{\mathfrak{J}}$ of MJLS $\mathfrak{J}$, it holds that:*

$$Sat\left(\mathcal{E}_{\Pi}^{k}\right) = \left\{(x, \theta) \in S_{\mathfrak{J}} \,\middle|\, \textsc{sum}_{n}^{mn}\left(\left(\frac{1}{k+1}\sum_{i=0}^{k}\mathcal{B}_{\mathfrak{J}}^{i}\right)(\vec{e}_{\theta} \otimes x)\right) \in \Pi\right\}$$

While the computation of the satisfaction seems to iterate over the uncountable state space, this is actually not necessary. First, note that propagating $\mathcal{B}$ and computing the Cesàro average is independent of the initial states. The remainder is then computed for each mode of the Markov chain separately (and forming the union of the resulting sets): Given a mode θ, the multiplication of the Cesàro average and $(\vec{e}_{\theta} \otimes x)$ amounts to mn linear expressions over only n variables. $\textsc{sum}_{n}^{mn}$ and the constraint of membership in Π reduces this to an n-dimensional equation system of n variables. The correctness for this part of the model-checking algorithm follows immediately from the above derivation.

Again, since the algorithm is based on solving linear equation systems and computing set membership of a convex polytope, the resulting satisfaction set remains representable.

Computing the Cesàro Limit. Before turning to the computation of Sat $(\mathcal{E}_{\Pi})$ we first describe how to compute the Cesàro-averaged long-term first moment for one specific initial state $(x, \theta) \in S_{\mathfrak{J}}$, i.e. $\lim_{k\to\infty} \frac{1}{k+1} \sum_{i=0}^{k} \mathbb{E}_{(x,\theta)}\left[x(i)\right]$.

Obviously, if the initial state vector is the zero vector or the MJLS is MS, the limit trivially is zero. Otherwise, the analysis relies on linear-algebraic computations revolving around Eq. 6. However, the approach differs in technical aspects if $\mathcal{B}_{\mathfrak{J}}$ is not diagonalizable.

If $\mathcal{B}_{\mathfrak{J}}$ is diagonalizable, we can represent the vector $\vec{e}_{\theta} \otimes x$ of conditional expected values at time 0 as a linear combination of the (complete set of potentially complex and linearly-independent) eigenvectors $v_1, \ldots, v_{mn}$ of $\mathcal{B}_{\mathfrak{J}}$, i.e.

$\vec{e}_\theta \otimes x = \sum_{j=1}^{mn} a_j v_j$. With Eq. 6 and since $\mathcal{B}_\mathfrak{J} v_j = \lambda_j v_j$ for all eigenpairs (λ_j, v_j), we get:

$$\lim_{k\to\infty} \frac{1}{k+1} \sum_{i=0}^{k} \mathbb{E}_{(x,\theta)}[x(i)] = \lim_{k\to\infty} \frac{1}{k+1} \sum_{i=0}^{k} \text{SUM}_n^{mn} \left(\sum_{j=1}^{mn} a_j \lambda_j^i v_j \right)$$

For those eigenvalues λ_j that lie on or within the unit circle, i.e. $|\lambda_j| \le 1$, the limit can be determined per summand. It is $\text{SUM}_n^{mn}(a_j v_j)$ for $\lambda_j = 1$ and 0 otherwise. If there are summands with $|\lambda_j| > 1$, the limit only exists if those summands are cancelled out, i.e. $\sum_{j|\lambda_j=\lambda} \text{SUM}_n^{mn}(a_j v_j) = 0$ for every $|\lambda| > 1$.

If $\mathcal{B}_\mathfrak{J}$ is not diagonalizable, it is not possible to perform the above computations as there is no complete set of eigenvectors. Instead, we use the Jordan normal form of $\mathcal{B}_\mathfrak{J} = M \cdot \text{blkdiag}(J_{m_1}(\lambda_1), \ldots, J_{m_b}(\lambda_b)) \cdot M^{-1}$, where λ_j are eigenvalues with multiplicity m_j and $J_{m_j}(\lambda_j)$ are the corresponding Jordan blocks [11]. A Jordan block $J_{m_j}(\lambda_j)$ is an $m_j \times m_j$ square matrix with all-λ_j on its main diagonal and all-ones on the superdiagonal. Substituting the Jordan normal form into the limit to be computed, we obtain:

$$\lim_{k\to\infty} \frac{1}{k+1} \sum_{i=0}^{k} \mathbb{E}_{(x,\theta)}\left[x(i)\right] = \tag{7}$$

$$\lim_{k\to\infty} \frac{1}{k+1} \sum_{i=0}^{k} \text{SUM}_n^{mn}\left(M\left(\text{blkdiag}(J_{m_1}^i(\lambda_1), \ldots, J_{m_b}^i(\lambda_b))\right) M^{-1}(\vec{e}_\theta \otimes x)\right)$$

For some $i \ge m$, when zooming into Jordan block $J_m^i(\lambda)$, the k-th diagonal ($k \ge 0$) is constant and equal to $\binom{i}{k}\lambda^{i-k}$. In particular, the main diagonal (i.e. 0-th diagonal) is equal to λ^i.

The Cesàro limit of the above term can then be computed for every Jordan block separately and depends on the corresponding eigenvalue λ. For eigenvalues λ_j inside the unit circle, all limits of the entries of the Jordan block are 0. For eigenvalues λ_j on the unit circle, the limits on the main diagonal exist (1 for $\lambda = 1$ and 0 otherwise) but not for the superdiagonals. For eigenvalues λ_j outside the unit circle, all limits of the upper triangular matrix do not exist. We compute the limit of each block symbolically: For a Jordan block $J_{m_j}(\lambda_j)$ where $|\lambda_j| < 1$, we simply create a zero matrix. If $|\lambda_j| \ge 1$, we create a matrix K_j and fill the diagonals corresponding to non-converging entries with symbolic values ∞_{ja_j}. Each non-converging diagonal is filled with a different symbolic value, denoted by the second index of ∞_{ja_j}, since the different non-converging sequences cannot cancel each other out. Finally, we combine all K_j to a block matrix and perform the outer multiplications of Eq. 7. If all symbolic values vanish, the limit can be computed by applying SUM_n^{mn} to the multiplication result.

Model Checking $\mathcal{E}_\Pi$. Algorithm 1 computes $\text{Sat}\,(\mathcal{E}_\Pi)$ in a fashion similar to what the above algorithm does for a single initial state. However, starting with the entire state space, this algorithm removes all initial states that do not satisfy the formula step by step via set intersections.

Algorithm 1. Satisfaction set of $\mathcal{E}_{\Pi}$

procedure SATE(Π)
  $\mathcal{B}_{\mathfrak{J}} \leftarrow (P^{\mathsf{T}} \otimes I_n) \cdot \textsc{BlkDiag}(A_1, \dots, A_m)$
  if $\rho(\mathcal{B}_{\mathfrak{J}}) < 1$ **then** ▷ *if MS, convergence to origin for every state*
    if $[0 \cdots 0]^{\mathsf{T}} \in \Pi$ **then return** $S_{\mathfrak{J}}$ **else return** $\emptyset$
  else if $\mathcal{B}_{\mathfrak{J}}$ is diagonalizable **then**
    $(\lambda_1, v_1), \dots, (\lambda_{mn}, v_{mn}) \leftarrow \textsc{EigenPairs}(\mathcal{B}_{\mathfrak{J}})$
    ▷ *move to coefficient domain*
    $U \leftarrow \{(a_1, \dots, a_{mn}) \in \mathbb{R}^{mn} \mid \exists (x, \theta) \in S_{\mathfrak{J}}.\, \vec{e}_\theta \otimes x = \sum_{j=1}^{mn} a_j v_j\}$
    ▷ *exclude all that diverge*
    $V \leftarrow U \cap \{(a_1, \dots, a_{mn}) \in \mathbb{R}^{mn} \mid \forall |\lambda| > 1.\, \sum_{j \mid \lambda_j = \lambda} \mathrm{SUM}_n^{mn}(a_j v_j) = 0\}$
    ▷ *exclude all with result* $\notin \Pi$
    $W \leftarrow V \cap \{(a_1, \dots, a_{mn}) \in \mathbb{R}^{mn} \mid \sum_{j \mid \lambda_j = 1} \mathrm{SUM}_n^{mn}(a_j v_j) \in \Pi\}$
    ▷ *back to state domain*
    return $\{(x, \theta) \in S_{\mathfrak{J}} \mid \exists (a_1, \dots, a_{mn}) \in W.\, \vec{e}_\theta \otimes x = \sum_{j=1}^{mn} a_j v_j\}$
  else
    $M, (\lambda_1, a_1), \dots, (\lambda_\ell, a_\ell) \leftarrow \textsc{JordanDecomp}(\mathcal{B}_{\mathfrak{J}})$
    for all $j \leftarrow 1$ to ℓ **do**
      $K_j \leftarrow 0_{a_j}$ ▷ *start off with zero matrix*
      ▷ FILLDIAGS *fills j-th diag. with j-th arg., starting with main diag.*
      if $|\lambda_j| > 1$ **then** $K_j \leftarrow \textsc{FillDiags}(\infty_{j1}, \infty_{j2}, \dots, \infty_{ja_j})$
      else if $\lambda_j = 1$ **then** $K_j \leftarrow \textsc{FillDiags}(1, \infty_{j2}, \dots, \infty_{ja_j})$
      else if $|\lambda_j| = 1$ **then** $K_j \leftarrow \textsc{FillDiags}(0, \infty_{j2}, \dots, \infty_{ja_j})$
    $r \leftarrow \{(x, \theta) \in U \mid \mathrm{SUM}_n^{mn}(M \cdot \textsc{BlkDiag}(K_1, \dots, K_\ell) \cdot M^{-1}(\vec{e}_\theta \otimes x)) \in \Pi\}$
    return r ▷ $\in \Pi$ *implies that no* ∞ *is contained in result*

The first step is again to check for mean stability. If the MJLS is MS, Sat $(\mathcal{E}_{\Pi})$ encompasses the entire state space provided the polytope Π contains the origin and evaluates to the empty set otherwise. If the system is not MS, we again distinguish between diagonalizable and non-diagonalizable $\mathcal{B}_{\mathfrak{J}}$.

For diagonalizable $\mathcal{B}_{\mathfrak{J}}$, we proceed via linear combinations of the eigenvectors of $\mathcal{B}_{\mathfrak{J}}$. In particular, we first compute the set of coefficients a_1 to a_{mn} for which the result of $\sum_{j=1}^{mn} a_j v_j$ is a vector of shape $\vec{e}_\theta \otimes x$ for some $(x, \theta) \in S_{\mathfrak{J}}$. In other words, we want to compute the set of coefficients such that all rows but $n(\theta-1)+1$ to $n(\theta-1)+n$ are set to zero, for every mode θ. Algorithmically, this can be realized by computing the nullspace of the matrix obtained by removing those rows from the modal matrix of $\mathcal{B}_{\mathfrak{J}}$. Then, from the resulting set, filter out all initial states whose first moment diverges, i.e. all those coefficients where $|\lambda| > 1$ and the contribution of this eigenvalue is not cancelled out. Finally, we remove all elements for which the limit value lies outside the desired polytope Π and translate the resulting set back to the state domain.

If $\mathcal{B}_{\mathfrak{J}}$ is not diagonalizable, we compute the same symbolic matrices K_i as above, since they are independent of the initial state. Then, compute the set of all initial states for which the multiplication with this Jordan decomposition does not contain any of the symbolic values as those represent a non-converging term and return all those initial states that are included in polytope Π.

Theorem 5. *Let Π be some convex polytope in $\mathbb{R}^n$. For every state $(x, \theta) \in S_{\mathfrak{J}}$ of MJLS $\mathfrak{J}$, it holds that $Sat(\mathcal{E}_\Pi) = \text{SATE}(\Pi)$.*

Again, we must ask whether the resulting satisfaction set is representable. For the diagonalizable case, we first compute the nullspace of the modal matrix, which results in a convex set. In lines 10 and 12, we intersect with sets that are convex themselves. Thus, the resulting sets are also convex. The approach of the non-diagonalizable case boils down to solving linear equation systems and computing set membership of a convex polytope, similar to the step-bounded algorithm. Thus, the satisfaction set remains representable as well.

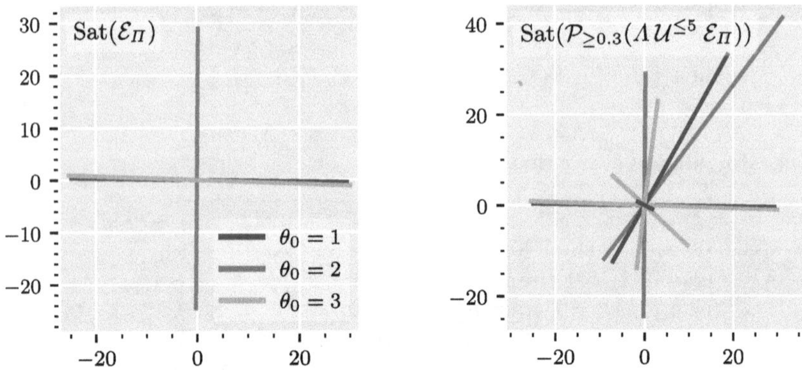

Fig. 3. A visualization of the satisfaction sets of $\mathcal{E}_\Pi$ and $\mathcal{P}_{\geq 0.3}\left(\Lambda\,\mathcal{U}^{\leq 5}\,\mathcal{E}_\Pi\right)$.

Example 5. The satisfaction sets of the formulas $\mathcal{E}_\Pi$ and $\mathcal{P}_{\geq 0.3}\left(\Lambda\,\mathcal{U}^{\leq 5}\,\mathcal{E}_\Pi\right)$ as defined in Example 4 are visualized in Fig. 3. The colour represents the corresponding mode of the Markov chain.[3]

Model Checking $\mathcal{V}^k_\Xi$ and $\mathcal{V}_\Xi$. The satisfaction sets of the variance operators are computed mostly analogously to the algorithms described above. However, some technical adaptions are necessary, which we describe in the following.

Example 6. There is no initial state for which the second moment converges, except trivially the origin. Consequently, depending on whether $\left[\begin{smallmatrix}0 & 0 & 0\end{smallmatrix}\right]^\top \in \Xi$, Sat$\left(\mathcal{V}_\Xi\right)$ is either $\{([\begin{smallmatrix}0\\0\end{smallmatrix}], 1), ([\begin{smallmatrix}0\\0\end{smallmatrix}], 2), ([\begin{smallmatrix}0\\0\end{smallmatrix}], 3)\}$ or $\emptyset$.

[3] This example has been implemented and analysed using Wolfram Mathematica with version 14.3. While Mathematica is able to compute exact results for $\mathcal{E}_\Pi$, the exact analysis of the until operator was aborted after 120 min. This is likely due to the exponential increase of possible sets per time step that need to be backpropagated (and afterwards intersected). We obtained the final results with a numerical analysis instead, with an (initial) 100-digit precision.

First and foremost, to compute the satisfaction sets, we replace the operator $\mathcal{B}_{\mathfrak{J}}$ with $\mathcal{T}_{\mathfrak{J}}$, which has the following effects: Since it propagates the vectorized conditional second moments, the use of $\mathcal{T}_{\mathfrak{J}}$ lifts the whole computation to a dimension of mn^2 and up to mn^2 eigenvalues are produced. For the same reason, $xx^{\intercal}$ has to be vectorized before combining it with $\vec{e}_\theta$. Similarly, after computing $\text{SUM}_{n^2}^{mn^2}$, the result has to be reshaped to fit the dimension of the polytope Ξ. This can be done by first de-vectorizing with $\text{VEC}_{n\times n}^{-1}$ and then half-vectorizing the result again with VECH. Analogously to Theorem 4, the satisfaction set of $\mathcal{V}_{\Xi}^{k}$ is obtained as follows, with Algorithm 1 being adapted in a similar fashion:

$$\begin{aligned}\text{Sat}\left(\mathcal{V}_{\Xi}^{k}\right) = \{(x,\theta) \in S_{\mathfrak{J}} \mid \\ \text{VECH}\left(\text{VEC}_{n\times n}^{-1}\left(\text{SUM}_{n^2}^{mn^2}\left(\frac{1}{k+1}\sum_{i=0}^{k}\mathcal{T}_{\mathfrak{J}}^{i}\left(\vec{e}_\theta \otimes \text{VEC}\left(xx^{\intercal}\right)\right)\right)\right)\right) \in \Xi\right\}\end{aligned}$$

4.3 Computational Issues and Numerical Approximation

The procedures developed above for the analysis of stability-related PCTL properties face limitations when it comes to effective implementation. The issue is not one of logical decidability, but of exact computability. In particular, the algorithms rely on eigenvalue computations for the operators $\mathcal{B}$ and $\mathcal{T}$. Even for MJLSs with rational parameters, these eigenvalues are generally irrational, and determining them amounts to solving polynomial equations whose degree grows with the system dimension. For degrees greater than four, no general closed-form solution exists (see Abel-Ruffini theorem [11]). This frontier is exceeded easily, for instance already for $m = 3$ and $n = 2$ in the analysis targetting the first moment ($\text{Sat}\left(\mathcal{E}_{\Pi}\right)$). Consequently, exact symbolic computation is in general not possible, and the methods must instead rely on numerical approximations. The accuracy of the analysis therefore depends on the precision and stability of the underlying numerical routines.

5 Conclusion

In this paper, we have addressed PCTL model checking for MJLSs, enriched with propositions that allow reasoning about stability in a way that can take into consideration specific initial state sets. This enables a novel view on MJLSs that so far has been impossible to express and decide, based on characterizations of the convergence of the first and second moments, relative to a given initial state. For the PCTL fragment excluding the until operator, we have provided a straightforward decision algorithm, while we have highlighted that the general model-checking problem for step-unbounded properties is open and closely related to other longstanding open problems. For the logic enriched with stability operators (until aside), we have succeeded in developing a corresponding model-checking procedure that harvests linear-algebraic techniques. More broadly, our results show that probabilistic model checking provides a natural framework for analysing state-dependent stability properties of MJLSs beyond classical global stability notions. Due to the setup of our logic, stability characteristics can be nested inside ordinary PCTL formulas, but the converse is not supported. While it is possible to define an extended PCTL semantics also covering this variant, the model-checking problem for the resulting logic needs to be considered open. This is an interesting problem for future work.

Acknowledgements. This work has been partially funded by DFG grant 389792660 as part of TRR 248 – CPEC, by the EU under grant agreement No 101008233 – MISSION, and as part of the INTERREG North Sea project STORM_SAFE.

References

1. Andova, S., Hermanns, H., Katoen, J.-P.: Discrete-time rewards model-checked. In: Larsen, K.G., Niebert, P. (eds.) FORMATS 2003. LNCS, vol. 2791, pp. 88–104. Springer, Heidelberg (2004). https://doi.org/10.1007/978-3-540-40903-8_8
2. Baier, C., Chatterjee, K., Meggendorfer, T., Piribauer, J.: Multiplicative rewards in Markovian models. In: 40th Annual ACM/IEEE Symposium on Logic in Computer Science, LICS 2025, pp. 499–512. IEEE (2025). https://doi.org/10.1109/LICS65433.2025.00044
3. Baier, C., Katoen, J.-P.: Principles of Model Checking. MIT Press (2008)
4. Becker, L., Hermanns, H.: Stability checking of Markov jump linear systems via probabilistic temporal logic (extended version). CoRR abs/2507.15574 (2026). https://doi.org/10.48550/arXiv.2507.15574
5. Brogan, W.L.: Modern Control Theory, 3rd edn. Prentice-Hall (1991)
6. Chen, C.T.: Linear System Theory and Design, 4th edn. Oxford University Press (2014)
7. Chonev, V., Ouaknine, J., Worrell, J.: On the complexity of the orbit problem. J. ACM **63**(3), 23:1–23:18 (2016). https://doi.org/10.1145/2857050
8. Costa, O.L.V., Fragoso, M.D.: Stability results for discrete-time linear systems with Markovian jumping parameters. J. Math. Anal. Appl. **179**(1), 154–178 (1993). https://doi.org/10.1006/jmaa.1993.1341
9. Costa, O.L.V., Fragoso, M.D., Marques, R.P.: Discrete-Time Markov Jump Linear Systems. Springer, Cham (2005). https://doi.org/10.1007/b138575
10. Costa, O.L.V., Fragoso, M.D., Todorov, M.G.: Continuous-Time Markov Jump Linear Systems. Springer, Cham (2012). https://doi.org/10.1007/978-3-642-34100-7
11. Dummit, D.S., Foote, R.M.: Abstract Algebra, 3rd edn. Wiley (2003)
12. Feng, X., Loparo, K., Ji, Y., Chizeck, H.: Stochastic stability properties of jump linear systems. IEEE Trans. Autom. Control **37**(1), 38–53 (1992). https://doi.org/10.1109/9.109637
13. Gray, W., Gonzalez, O., Dogan, M.: Stability analysis of digital linear flight controllers subject to electromagnetic disturbances. IEEE Trans. Aerosp. Electron. Syst. **36**(4), 1204–1218 (2000). https://doi.org/10.1109/7.892669
14. Grötschel, M., Lovász, L., Schrijver, A.: Geometric Algorithms and Combinatorial Optimization, Algorithms and Combinatorics, vol. 2, 2nd edn. Springer (1993). https://doi.org/10.1007/978-3-642-78240-4

15. Hansson, H., Jonsson, B.: A logic for reasoning about time and reliability. Formal Aspects Comput. **6**(5), 512–535 (1994). https://doi.org/10.1007/bf01211866
16. Kannan, R., Lipton, R.J.: Polynomial-time algorithm for the orbit problem. J. ACM **33**(4), 808–821 (1986). https://doi.org/10.1145/6490.6496
17. Karimov, T., Kelmendi, E., Ouaknine, J., Worrell, J.: What's Decidable About Discrete Linear Dynamical Systems?, pp. 21–38. Springer, Cham (2022). https://doi.org/10.1007/978-3-031-22337-2_2
18. Karimov, T., Ouaknine, J., Worrell, J.: On LTL model checking for low-dimensional discrete linear dynamical systems. In: 45th International Symposium on Mathematical Foundations of Computer Science, MFCS 2020. LIPIcs, vol. 170, pp. 54:1–54:14. Schloss Dagstuhl – Leibniz-Zentrum für Informatik (2020). https://doi.org/10.4230/LIPIcs.MFCS.2020.54
19. Kemeny, J.G., Snell, J.L.: Finite Markov Chains. Springer, Cham (1976)
20. Lahijanian, M., Andersson, S.B., Belta, C.: Formal verification and synthesis for discrete-time stochastic systems. IEEE Trans. Autom. Control **60**(8), 2031–2045 (2015). https://doi.org/10.1109/TAC.2015.2398883
21. Lian, J., Liu, J., Zhuang, Y.: Mean stability of positive Markov jump linear systems with homogeneous and switching transition probabilities. IEEE Trans. Circuits Syst. II Express Briefs **62**(8), 801–805 (2015). https://doi.org/10.1109/TCSII.2015.2433371
22. Loparo, K., Abdel-Malek, F.: A probabilistic approach to dynamic power system security. IEEE Trans. Circuits Syst. **37**(6), 787–798 (1990). https://doi.org/10.1109/31.55036
23. Nogueira, S.L., Siqueira, A.A.G., Inoue, R.S., Terra, M.H.: Markov jump linear systems-based position estimation for lower limb exoskeletons. Sensors **14**(1), 1835–1849 (2014). https://doi.org/10.3390/s140101835
24. Ramponi, F., Chatterjee, D., Summers, S., Lygeros, J.: On the connections between PCTL and dynamic programming. In: 13th ACM International Conference on Hybrid Systems: Computation and Control, HSCC 2010, pp. 253–262. ACM (2010). https://doi.org/10.1145/1755952.1755988
25. Ziegler, G.M.: Lectures on Polytopes, Graduate Texts in Mathematics, vol. 152. Springer, Cham (1994). https://doi.org/10.1007/978-1-4613-8431-1

LF-Mc: An Efficient Verifier for Lingua Franca

Mario Reja[1(✉)], Mikheil Rukhaia[2], Kyungmin Bae[3], Mircea Marin[1], and Peter Csaba Ölveczky[4]

[1] West University of Timişsoara, Timişoara, Romania
mario.reja@e-uvt.ro
[2] Institute of Applied Mathematics, TSU, Tbilisi, Georgia
[3] Pohang University of Science and Technology, Pohang, Korea
[4] University of Oslo, Oslo, Norway

Abstract. This tool paper presents the LF-mc verification tool for the Lingua Franca (LF) coordination language for cyber-physical systems. LF-mc is the only such tool which is based on LF's logical-time semantics, and therefore gives correct analysis results. LF-mc is also the only verifier for LF that supports a model of nondeterministic environments. LF-mc is fully integrated into the LF toolchain, and provides system simulation as well as fully automatic unbounded and time-bounded reachability analysis and LTL model checking for both state-based and action-based properties. Extensive benchmarking shows that LF-mc significantly outperforms the LF Verifier and PolyVer analysis tools for LF.

1 Introduction

Lingua Franca (LF) [7] is a recent reactor-oriented "coordination language for efficient, deterministic, multi-threaded, time-sensitive, embedded, and distributed programs" developed by the group of Edward A. Lee at UC Berkeley [5].

An LF program consists of *reactors*—with state variables, input and output ports, actions, timers, and a list of *reactions*—and immediate and delayed connections. A reaction executes *reaction code* when triggered by actions, inputs, or timers. The reaction code is given in a target language such as C or Rust, hence "Lingua Franca," and may update state variables, produce outputs, and/or schedule future actions. Reaction executions are considered to be *instantaneous*, so that a reaction's output traveling along an *immediate* connection may trigger *another* reaction execution *at the same logical time*, and so on. LF has a *logical-time semantics* [4,7,9]: all reactions executed at the same logical time happen in *one* step, where the order of reaction executions satisfies certain constraints to ensure *deterministic* behavior. *Physical actions* model external environments, and are nondeterministic in *when* they take place and *what* "values" they have.

Providing an executable formalization of LF's logical-time semantics is known to be challenging (e.g., [4]). In our paper [12] we provide the first such formalization, for a significant subset of LF, in rewriting logic [10], and show how

M. Chen and M. Tschaikowski (Eds.): QEST+FORMATS 2026, LNCS 16913, pp. 302–312, 2027.
https://doi.org/10.1007/978-3-032-35298-9_17

Maude [3] and Real-Time Maude [11] can be used to formally analyze LF programs.

Other approaches to verifying LF programs [2,4,16] use an *event-based* semantics, in which *each* reaction execution corresponds to a transition. Analyses based on this semantics explore additional "intermediate" states not reachable with a logical-time semantics. In contrast, in our logical-time semantics *all* reactions executed at the same logical time *together* define a single transition.

Unlike other verification approaches for LF, we also support a model of nondeterministic physical actions (while others model check deterministic programs).

The LF-MC tool presented in this paper uses our formalization of the LF semantics to support automatic Maude-based formal analysis of LF programs. LF-MC is fully integrated into the LF toolchain: The user provides an LF program annotated with desired properties; LF-MC then generates the Maude model corresponding to the LF program, invokes Maude to execute analysis commands checking the user's properties, and outputs the analysis in readable form.

LF-MC is distributed as an artifact [15] and a Docker image, for `amd64` and `arm64`, at https://ghcr.io/symbolicsafety/lf-mc:v1.2. The image bundles the main components of LF-MC: a modified LF compiler `lfc` and the Maude run-time. The `.lf` programs used in our evaluation are included in `/opt/lf-maude/examples/src/`. The source code is available at https://github.com/symbolicsafety/lf-maude, which contains Maude-specific components, and https://github.com/symbolicsafety/lf-mc, which hosts the Java-based LF compiler extensions.

2 Lingua Franca

Lingua Franca (LF) [5,7,9] supports *reactor-oriented* programming [6,8]: system components are modeled by *reactors*, which are *deterministic* actors with local state variables, ports, actions, timers, and an ordered list of *reactions*. A reaction *invocation* is triggered by, and may produce, *discrete events*. A triggering event can be (the presence of an event at) an input port, a timer, or an action. A *logical action* is used to schedule future events. A reaction has the form `reaction(`*triggers*`) ->` *effects* `{=` *body* `=}`, where *triggers* is a set of triggers, *effects* are the ports and actions that the reaction *may* write to (resp., schedule), and *body* specifies the reaction's behavior in a *target language* supported by LF.

LF has a *logical-time* semantics: all reaction invocations (or executions) at the same logical time happen in a *single step*. The reactions executed in this step may be executed in different orders, as long as a reaction r_1 whose output triggers a reaction r_2 is executed before r_2, and if reactions in the same reactor are executed at the same logical time, then they are executed in order of declaration.

Physical actions represent external events from the physical environment, the timing and values of which are not controlled by the program. LF distinguishes between events from the physical environment and events produced by the program by considering two time lines: *logical time* and *physical time*.

Example 1. The figure below shows the architecture of the ensuing LF program, which models a driver assistance system (ADAS). A `Camera c` and a `LiDAR l` are connected (with immediate connections) to an `ADASProcessor` `v`, which has a delayed connection to the `Brakes` reactor `b`. The `LiDAR` and the `Camera` are triggered periodically by *timers* with respective periods 34 and 17. Reaction **1** in the `ADASProcessor` schedules a logical ('`L`') action (that will trigger reaction **2**) with delay 50 msec. The driver can press the brake pedal at any time; this is modeled by the physical ('`P`') action `a` in the `Pedal` component.

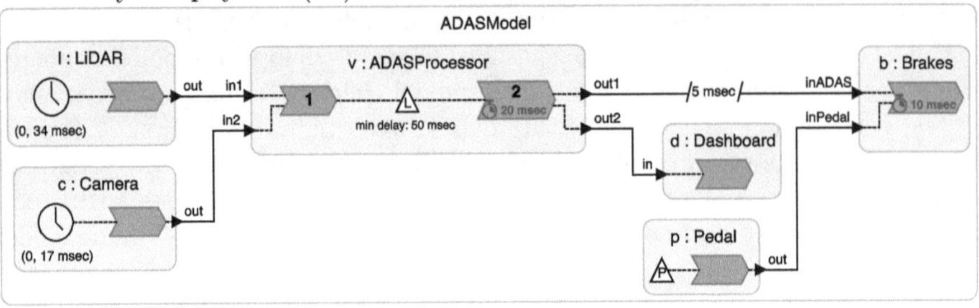


```
target C;
reactor Camera { output out:int;   state  frame:int;   timer  t(0, 17 msec);
                 reaction(t) -> out {= self->frame = 1; lf_set(out, self->frame); =} }
reactor LiDAR { output out:int;   state  frame:int;   timer  t(0, 34 msec);
                reaction(t) -> out {= self->frame = 2;  lf_set(out, self->frame); =} }
reactor Pedal { output out:int;   physical action a;
                reaction(a) -> out {= lf_set(out, 1); =} }
reactor Brakes { input inADAS:int;   input inPedal:int;   state brakesApplied:int = 0;
                 reaction(inADAS, inPedal) {= self->brakesApplied = 1; =} }
reactor ADASProcessor {input in1:int; input in2:int; output out1:int; output out2:int;
                       logical action a(50 msec);    state requestStop:int;
        reaction(in1,in2) -> a {= lf_schedule(a, 0);  self->requestStop = 1; =}
        reaction(a) -> out1, out2 {= if (self->requestStop == 1) lf_set(out1, 1); =} }
reactor Dashboard { input in:int;  state received:int;
                    reaction(in) {= self->received = 1; =} }
main reactor ADASModel { c = new Camera();  l = new LiDAR();  v = new ADASProcessor();
                         b = new Brakes();  d = new Dashboard();  p = new Pedal();
                         l.out -> v.in1;   c.out -> v.in2;  v.out2 -> d.in;
                         v.out1 -> b.inADAS after 5 msec;   p.out -> b.inPedal; }

@maudePhysAct(name="a", inReactor="p", vals="69, 71", period=5, timeNonDet=true)
@maude(analysis="simulation", rewrites=51)
@maude(analysis="reachability", timeBound=5, goal="brakesApplied in b == 1")
@maude(analysis="ltl", timeBound=55, goal="<>(brakesApplied in b == 1)")
@maude(analysis="ltl", timeBound=54, goal="<>(brakesApplied in b == 1)")
@maude(analysis="ltl",  goal="[]((l.1 invoked /\ requestStop in v == 1) ->
                              (<> brakesApplied in b == 1))")
@maude(analysis="reachability", goal="(p.1 invoked) /\ (brakesApplied in b != 1)")
```
□

3 Related Work on Verifying Lingua Franca Programs

Sirjani et al. [16] "hand translate" three LF programs to *an extension* of Timed Rebeca [1]. This yields an *event-based* semantics, and they perform exhaustive state space analysis. Physical actions are translated into deterministic behaviors.

The LF VERIFIER [4] also provides an *event-based* semantics. The tool supports safety metric temporal logic properties, where all temporal operators have upper bounds (e.g., $\Box_{[2,5]}(p \rightarrow \Diamond_{[1,6]} q)$), so that it is sufficient to perform *bounded* model checking (e.g., up to time 11 for the above formula), and uses UCLID5 [14] to compile such *bounded* model checking to an SMT problem to be solved by Z3. LF VERIFIER supports both state-based and action-based ("dataflow") properties, but does *not* support physical actions. The POLYVER tool [2] extends this approach to *polyglot* LF programs, where the reaction code of different reactors can be given in different languages (currently C and Rust).

4 The LF-MC Verification Tool

The LF-MC tool extends the LF compiler (`lfc`) to:

1. generate a Maude model from the LF code, annotated with intuitive queries,
2. generate Maude analysis commands for the user queries,
3. invoke Maude to execute the analysis commands on the Maude model, and
4. output the results of the analysis commands in an understandable form.

LF-MC supports a fragment of LF comparable to that of LF VERIFIER (see Table 1), extended with the following model for physical actions: Each physical action has a user-defined period and a finite range of possible values. At the end of each period, the physical action *may* or *must* take place (depending on the `timeNonDet` flag); if it takes place, its value is selected nondeterministically from the finite set. The user should add to her LF program annotations of the form

```
@maudePhysAct(name="action", inReactor="reactor", vals="values", period=p, timeNonDet=b)
```

to specify these parameters of physical actions, where *values* can be written either as an integer interval (e.g., `2 .. 5`) or as a comma-separated list of (Boolean or integer) values. LF-MC captures all 22 LF VERIFIER benchmarks and all four LF programs in [4,16] (which all have physical actions!).

LF-MC supports unbounded and time-bounded simulation, reachability analysis, and LTL model checking, for both state and action ("dataflow") properties.

Table 1 compares the LF features and analysis capabilities supported by LF-MC, LF VERIFIER [4], the work in [16], and POLYVER [2]. "ub" and "tb" denote that, resp., *unbounded* and *time-bounded* analyses are supported.

4.1 Property Specification Language

Desired properties are specified in LF source files using annotations of the form

```
@maude(analysis="simulation", timeBound=t,  rewrites=n)
@maude(analysis="reachability", timeBound=t,  rewrites=n, goal="stateFormula")
@maude(analysis="ltl", timeBound=t,  goal="LTLformula")
```

Table 1. Comparison of supported features

Feature	LF-MC	LF VERIFIER	[16]	POLYVER
LF Features				
Logical-time semantics	✓	×	×	×
Two timelines	×	×	×	×
Nondeterministic physical actions	✓	×	×	×
Logical actions	✓	✓	×	✓
Minimum spacing of actions	×	×	×	×
Multiple reaction triggers	✓	✓	×	✓
Verification Capabilities				
Reachability analysis	ub+tb	×	ub?	×
LTL model checking	ub+tb	×	×	×
Timed temporal logic model checking	×	tb	×	tb
State properties	✓	✓	✓	✓
Dataflow properties	✓	✓	×	✓
Tool Integration				
Automated model and analysis command generation	✓	✓	×	✓
Native languages supported	C	C	—	C, Rust

`timeBound` and `rewrites` (a bound on the number of rewrite steps) are optional.

LF-MC supports the following analyses: (i) `simulation` performs bounded simulation and prints the resulting state; (ii) `reachability` outputs the path to a reachable state (if any) satisfying a given state formula; and (iii) `ltl` performs LTL model checking, and outputs a readable counterexample path if the property does not hold. The reachability annotation may also include the attribute `type="!"` to restrict the search to deadlocked (final) states.

A *stateFormula* is a Boolean combination (using `!` or `~` for negation, `/\` for conjunction, `\/` for disjunction, `->` for implication, etc.) of *state propositions*. An *LTLformula* extends state formulas to LTL formulas by allowing the temporal operators `[]`, `<>`, `U`, `W` (weak until), and `O` (next). State propositions are:

- r . n `invoked`, which holds if the nth reaction in reactor r was invoked in the big step leading to the current state;
- `event(`r`,` *trigger*`,` *val*`)` `inQueue`, where *val* is optional, which holds if the event queue contains the/an event `event(`r`,` *trigger*`,` *val*`)`;
- a *relational expression* over reactor variables and input ports;
- `remainingTime` t, which holds when the remaining time until the specified time bound is t (only for time-bounded analysis queries).

Relational expressions are built from Boolean or integer constants, *var* `in` r (the value of variable *var* in reactor r), and *port* `in` r (the value of input port *port* in reactor r). These expressions may be combined using arithmetic operators (e.g., `+`, `-`, `*`) and relational operators (e.g., `==`, `!=`, and `>=`, `>`, `<=`, and `<` for integers).

To analyze a `.lf` file containing an annotated LF program, LF-MC can be invoked from the command line as follows: `lfc` *path/fileName*`.lf`

LF-MC prints LTL counterexamples and search paths in an intuitive form. In each state, only the event queue, the values of reactor variables and input ports, and the reactions executed in the last step are retained, as shown in Appendix A.

4.2 Architecture and Implementation

LF-MC extends the LF compiler. This allows us to inherit parsing, name resolution, and attribute validation from the LF compiler. This enables a semantics-based model synthesis rather than ad-hoc source-to-source translation. The tool is composed of two parts: a front-end, written in Java, and a backend, written in Maude, that performs the formal verification. The front-end generates a `.maude` file containing both the translated LF program and the analysis commands, and invokes the Maude backend to execute these commands and return the results.

To translate annotated LF programs into Maude code and commands, we reuse existing LF infrastructure. This includes using ANTLR [13] as the parser generator for LF and reaction code, as well as for parsing analysis annotations.

Our implementation extends the Java classes in the LF implementation used to represent reactors, reactions, actions, timers etc. with naming, typing and other necessary information required by the Maude backend. We follow the way the UCLID5 transformation was implemented and provide the `MaudeGenerator` and `MaudeRunner` classes. We use the `CodeBuilder` object (already implemented inside the LF framework) to generate Maude files.

4.3 Benchmarking

We compare the performance of LF-MC with LF VERIFIER and POLYVER on the LF VERIFIER benchmark suite, which consists of 22 LF programs drawn from real-world applications and other benchmarks suites. LF-MC could successfully verify the desired—or stronger—properties of all programs in this suite.

For various reasons (e.g., LF VERIFIER's Docker file does not build on arm64; wrong reaction numbering; etc.) we could not run all LF VERIFIER experiments from its publicly available artifact. We therefore use the numbers reported in [4] on a platform comparable to ours. For the LF VERIFIER benchmarks we could run, our figures are similar to those in the table. Likewise, we use the runtime results from [2] for POLYVER, since the LLM model o1-mini used in POLYVER is deprecated and inaccessible. All LF-MC experiments were executed using the Docker image `lf-mc:v1.2`, on a laptop running macOS 15.7.4, equipped with a 3.2 GHz Apple M1 Max processor (10 cores: 8 performance and 2 efficiency) and 32GB of RAM. POLYVER ran its benchmarks using a single core of a 3.7 GHz Intel Core i9 machine with 62 GB RAM. LF VERIFIER used a macOS laptop with a 2.3 GHz 8-core Intel Core i9 and 16 GB RAM.

Table 2 summarizes the experimental results. For LF VERIFIER, "Gen." shows the time taken to generate SMT encodings , and "Solving" is the time spent on SMT solving. For LF-MC, "Gen." shows the time needed to parse the annotated `.lf` file, produce a `.maude` file, and invoke Maude; and "Solving" is the time

Maude uses to analyze the properties. For PolyVer, "SOT" presents the synthesis oracle time, "VOT" the verification oracle time and "UT" is the Uclid5 total time.

Table 2. LF Verifier vs. LF-MC vs. PolyVer (times in seconds)

Program	LF Verifier			LF-MC			PolyVer			
	Gen.	Solving	Total	Gen.	Solving	Total	SOT	VOT	UT	Total
ADAS	21.07	1.72	22.79	1.775	0.002	1.777	220.9	3.2	21.5	245.6
AircraftDoor	6.57	0.02	6.59	1.699	<0.001	1.699	82.1	1.2	2.0	85.3
Alarm	5.36	0.01	5.37	1.663	<0.001	1.663	62.3	0.9	2.1	65.3
CoopSchedule	21.1	0.31	21.41	1.686	0.009	1.695	54.5	0.8	13.4	68.7
Elevator	325.14	1038.02	1363.16	1.747	0.002	1.749	413.1	33.4	664.8	1111.3
Election	41.65	3.27	44.92	1.716	<0.001	1.716	124.2	1.4	83.9	209.5
Election2	12.21	0.17	12.38	1.754	<0.001	1.754	142.8	1.5	11.7	156.0
Factorial	21.46	13.15	34.61	1.694	0.01	1.704	138.2	1.9	63.0	203.1
Fibonacci	94.96	153.49	248.45	1.694	0.01	1.704	167.3	2.4	390.9	560.6
PingPong	13.42	0.14	13.56	1.688	<0.001	1.688	131.0	1.8	8.2	141.0
Pipe (I)	142.41	12.15	154.56	1.717	0.001	1.718	147.4	2.4	172.2	322.0
ProcessMsg	13.23	0.74	13.97	1.681	0.008	1.689	116.6	1.8	5.1	123.5
ProcessSync	4.99	0.01	5.00	1.681	<0.001	1.681	35.2	0.4	1.7	37.3
Railroad	158.14	4.33	162.47	1.751	0.01	1.761	678.9	9.0	264.8	952.7
Ring	15.83	0.07	15.90	1.754	<0.001	1.754	105.6	1.8	6.8	114.2
RoadsideUnit	74.07	0.62	74.69	1.713	<0.001	1.713	259.0	3.8	25.3	288.1
SafeSend	6.48	0.03	6.51	1.697	<0.001	1.697	124.3	1.8	2.3	128.4
Subway	29.96	0.31	30.27	1.704	<0.001	1.704	266.5	4.5	67.2	338.2
Thermostat	13.41	0.14	13.55	1.587	<0.001	1.587	192.6	2.6	3.5	198.7
TrafficLight	–	–	–	1.708	<0.001	1.708	462.1	4.4	216.1	682.6
TrainDoor	5.65	0.01	5.66	1.704	<0.001	1.704	74.5	1.2	2.1	77.8
UnsafeSend	6.5	0.02	6.52	1.703	<0.001	1.703	122.1	1.8	2.9	126.8

LF-MC drastically outperforms both LF Verifier and PolyVer, possibly because LF-MC executes many reactions in one step, resulting in fewer states. Since LF is *deterministic*, it is sufficient to execute all reactions in a single "big step" in *some* allowed order, whereas event-based model checkers, such as LF Verifier, explore *all* such sequences of reaction executions, which can be exponential in the number of reactions executed.

5 Concluding Remarks

LF-MC is the only verification tool for Lingua Franca (LF) that is based on LF's *logical-time* semantics. LF-MC is fully automatic and integrated into the

LF toolchain: the user enters an LF program annotated with desired properties, and LF-MC calls Maude to perform the corresponding analysis and displays the analysis results in a readable form. LF-MC is also the only verifier for LF which supports a model of nondeterministic physical actions.

LF-MC supports unbounded and time-bounded simulation, reachability analysis, and LTL model checking for both state- and action-based properties. Using the correct semantics (and more appropriate analysis tools?) gives the correct analysis results *and* much more efficient analysis, as our benchmarking shows.

A key take-away message from this work is the suitability of rewriting logic and Maude to formalize the semantics and provide analysis for sophisticated (real-time) languages. Formalizing the LF semantics was manageable in rewriting logic, which uses rewrite rules (for transitions) and equations (to define functions to, e.g., perform a big step). This task seems impossible using, e.g., timed automata (because of C programs, complex big steps, and unbounded event queues) or actor-based languages such as Timed Rebeca. It also seems very hard to encode model checking for the correct LF semantics as SMT formulas. Finally, Maude comes with purpose-made reachability and LTL model checking implementations, which should be more efficient than encoding model checking as SMT problems.

Acknowledgments. This work was supported by NATO SPS program project G6133 and by the Shota Rustaveli National Science Foundation of Georgia under the project FR-25-5957. Bae has been supported by the Institute of Information & Communications Technology Planning & Evaluation (IITP) grant (No. RS-2024-00439856) and by the National Research Foundation of Korea (NRF) grant (No. RS-2021-NR060080), both funded by the Korea government (MSIT).

Data Availability Statement. The LF-MC tool, the Maude run-time, and the annotated `.lf` programs used in our evaluation are archived as an artifact [15] and distributed as a Docker image at https://ghcr.io/symbolicsafety/lf-mc:v1.2.

A An Example

The LF program in Sect. 2 ends with seven LF-MC annotations. The first one specifies that the physical action `a` in the `Pedal` reactor `p` has period 5, possible values 69 and 71, and may or may not take place when its period ends (`timeNonDet=true`). The other annotations specify the following analysis commands: (1) simulate the system for (up to) 51 rewrite steps; (2) check whether it is possible to reach a state within time 5 in which the state variable `brakesApplied` in the reactor `b` has value 1; (3) check whether one can *always* reach such a state within time 55; (4) the same, but with time bound 54; (5) check whether any state, reached in a big step where the first (and only) reaction in the `LiDAR` reactor `l` was executed, where the value of the variable `requestStop` is 1, must imply that `brakesApplied` sooner or later will be 1; and (6) check whether it is possible to reach a *bad state*, in which the first/only (physical) action in `p` was invoked, but where `brakesApplied` is different from 1. The last two properties have no time bound.

The following shows (parts of) the result of running LF-MC on this annotated LF program:[1]

```
linux> lfc lf-maude/examples/src/ADAS-model.lf
[...]
==========================================
rewrite [51] in SIMULATION-ADASMODEL : initSystem timeBound INF .
rewrites: 3949 in 2ms cpu (2ms real) (1320735 rewrites/second)
result ClockedSystem: {...} in time 75000000 timeBound INF
==========================================
reduce in ANALYSIS-ADASMODEL : search in 'ANALYSIS-ADASMODEL : initSystem timeBound 5000000
    =>* (@ b.sv.brakesApplied in b === @ [1]) .
rewrites: 772 in 11ms cpu (12ms real) (64912 rewrites/second)
result SearchOutput: trace
 { invoked: none
   queue: (event(c, c.t.t, [0]) event(l, l.t.t, [0])) at tag(0, 1)
   c :[ inports: empty   state: c.sv.frame |-> [0]]
   l :[ inports: empty   state: l.sv.frame |-> [0]]
   v :[ inports: (v.in.in1 |-> [0]) ; v.in.in2 |-> [0]    state: v.sv.requestStop |-> [0]]
   b :[ inports: (b.in.inADAS |-> [0]) ; b.in.inPedal |-> [0]   state: b.sv.brakesApplied |-> [0]]
   d :[ inports: d.in.in |-> [0]   state: d.sv.received |-> [0]]
   p :[ inports: empty    state: empty] }
=['step]=>  ...  =['step]=>
 { invoked: (b . 1) ; (p . 1)
   queue: ...
   c :[ inports: empty   state: c.sv.frame |-> [1]]
   l :[ inports: empty   state: l.sv.frame |-> [2]]
   v :[ inports: (v.in.in1 |-> [2]) ; v.in.in2 |-> [1]   state: v.sv.requestStop |-> [1]]
   b :[ inports: (b.in.inADAS |-> [0]) ; b.in.inPedal |-> [1]   state: b.sv.brakesApplied |-> [1]]
   d :[ inports: d.in.in |-> [0]   state: d.sv.received |-> [0]]
   p :[ inports: empty   state: empty] }
==========================================
reduce in MODELCHECKER-ADASMODEL : modelCheck(initSystem timeBound 55000000,
    <> @ b.sv.brakesApplied in b === @ [1]) .
rewrites: 6535 in 4ms cpu (4ms real) (1326095 rewrites/second)
result Bool: true
==========================================
reduce in MODELCHECKER-ADASMODEL : modelCheck(initSystem timeBound 54000000,
    <> @ b.sv.brakesApplied in b === @ [1]) .
rewrites: 3222 in 0ms cpu (1ms real) (3231695 rewrites/second)
result ModelCheckResult:
counterexample
 { invoked: none
   queue: (event(c, c.t.t, [0]) event(l, l.t.t, [0])) at tag(0, 1)  ... }
===[ 'step ]===>
 { invoked: (c . 1) ; (l . 1) ; (v . 1)  ... }
===[ 'tick ]===>  ...  ===[ 'tick ]===>
begin loop
 { invoked: none
   queue: ...
   c :[ inports: empty    state: c.sv.frame |-> [1] ]
   l :[ inports: empty   state: l.sv.frame |-> [2] ]
   v :[ inports: (v.in.in1 |-> [2]) ; v.in.in2 |-> [1]   state: v.sv.requestStop |-> [1] ]
   b :[ inports: (b.in.inADAS |-> [0]) ; b.in.inPedal |-> [0]   state: b.sv.brakesApplied |-> [0] ]
   d :[ inports: d.in.in |-> [0]    state: d.sv.received |-> [0] ]
   p :[ inports: empty    state: empty ] }
===[ deadlock ]===>
end loop
==========================================
reduce in MODELCHECKER-ADASMODEL : modelCheck(initSystem timeBound INF,
   []((l . 1) invoked /\ @ v.sv.requestStop in v === @ [1] -> <> @ b.sv.brakesApplied in b === @ [1])) .
```

[1] LF-MC assumes that time bounds in annotations are given in milliseconds and converts them to nanoseconds, which is the time unit used by the LF compiler.

```
rewrites: 81361 in 53ms cpu (54ms real) (1515158 rewrites/second)
result Bool: true
==========================================
reduce in ANALYSIS-ADASMODEL : search in 'ANALYSIS-ADASMODEL : initSystem timeBound INF
    =>* ((p . 1) invoked /\ @ b.sv.brakesApplied in b ==/= @ [1]) .
rewrites: 89546 in 63ms cpu (63ms real) (1409462 rewrites/second)
result SearchOutput:  No Solution
```

The output shows that a desired state is reachable in query (2) and that a bad state is unreachable in query (6). The LTL properties hold, except for the one where the brakes must be applied within time 54. In search paths and counterexamples, each transition is presented in the following form:

{*sourceSimplifiedState*}
===[*ruleLabel*]===>
{*targetSimplifiedState*}.

In each state, only the event queue, the values of reactor variables and input ports of each reactor, and the reactions executed in the last step are shown. Specifically, each state is presented in the following format:

invoked: *reactionsExecutedInLastStep*
queue: *eventQueue*
r_1 : [inports: p_1^1 |-> [v_1^1] ; ... ; $p_{l_1}^1$ |-> [$v_{l_1}^1$]
state: x_1^1 |-> [w_1^1] ; ... ; $x_{k_1}^1$ |-> [$w_{k_1}^1$]]
...
r_n : [inports: p_1^n |-> [v_1^n] ; ... ; $p_{l_n}^n$ |-> [$v_{l_n}^n$]
state: x_1^n |-> [w_1^n] ; ... ; $x_{k_n}^n$ |-> [$w_{k_n}^n$]]

References

1. Aceto, L., Cimini, M., Ingólfsdóttir, A., Reynisson, A.H., Sigurdarson, S.H., Sirjani, M.: Modelling and simulation of asynchronous real-time systems using Timed Rebeca. In: 10th International Workshop on the Foundations of Coordination Languages and Software Architectures (FOCLASA 2011). EPTCS, vol. 58, pp. 1–19 (2011)
2. Chen, P.W., Lin, S., Godbole, A., Singh, R., Polgreen, E., Lee, E., Seshia, S.: PolyVer: a compositional approach for polyglot system modeling and verification. In: Proceedings of the 25th Conference on Formal Methods in Computer-Aided Design - FMCAD 2025, pp. 1–13. TU Wien Academic Press, Vienna (2025)
3. Clavel, M., Durán, F., Eker, S., Lincoln, P., Martí-Oliet, N., Meseguer, J., Talcott, C.: All About Maude - A High-Performance Logical Framework. LNCS, vol. 4350. Springer, Heidelberg (2007). https://doi.org/10.1007/978-3-540-71999-1
4. Lin, S., et al.: Towards building verifiable CPS using Lingua Franca. ACM Trans. Embed. Comput. Syst. **22**(5s), 1–24 (2023)
5. https://www.lf-lang.org/ . Accessed 20 Mar 2026
6. Lohstroh, M.: Reactors: A Deterministic Model of Concurrent Computation for Reactive Systems, Berkeley. Ph.D. thesis (2020)
7. Lohstroh, M., Menard, C., Bateni, S., Lee, E.A.: Toward a Lingua Franca for deterministic concurrent systems. ACM Trans. Embed. Comput. Syst. (TECS) **20**(4), 1–27 (2021)

8. Lohstroh, M., et al.: Actors revisited for time-critical systems. In: Proceedings of the 56th Design Automation Conference 2019 (DAC'19), pp. 152:1–152:4. ACM, New York (2019)
9. Menard, C., et al.: High-performance deterministic concurrency using Lingua Franca. ACM Trans. Arch. Code Optim. **20**(4), 1–29 (2023)
10. Meseguer, J.: Conditional rewriting logic as a unified model of concurrency. Theor. Comput. Sci. **96**(1), 73–155 (1992)
11. Ölveczky, P.C.: Real-Time Maude and its applications. In: Escobar, S. (ed.) WRLA 2014. LNCS, vol. 8663, pp. 42–79. Springer, Cham (2014). https://doi.org/10.1007/978-3-319-12904-4_3
12. Ölveczky, P.C., Reja, M., Rukhaia, M., Bae, K., Marin, M.: Efficient verification of Lingua Franca programs. In: Proceedings of TACAS'26. LNCS, vol. 16506, pp. 23–41. Springer, Heidelberg (2026). https://doi.org/10.1007/978-3-032-22749-2_2
13. Parr, T.: The definitive ANTLR 4 reference. The Pragmatic Bookshelf, Dallas, Texas (2013). https://www.antlr.org/
14. Polgreen, E., et al.: UCLID5: multi-modal formal modeling, verification, and synthesis. In: Proceedings of CAV'22. LNCS, vol. 13371, pp. 538–551. Springer, Cham (2022). https://doi.org/10.1007/978-3-031-13185-1_27
15. Reja, M., Rukhaia, M., Bae, K., Marin, M., Ölveczky, P.C.: Artifact for "LF-MC: an efficient verifier for Lingua Franca". Submitted to QEST+FORMATS (2026). https://doi.org/10.5281/zenodo.19568831 (2026)
16. Sirjani, M., Lee, E.A., Khamespanah, E.: Verification of cyberphysical systems. Mathematics **8**(7), 1068 (2020)

The Cost of Repetition: A Compositional Scalability Model for Attack Trees

Clemens Fruböse and Eva Hetzel(✉)

KASTEL Security Research Labs, Karlsruhe Institute of Technology, Karlsruhe, Germany
{clemens.fruboese,eva.hetzel}@kit.edu

Abstract. Repeated cyber attacks rarely entail constant attacker costs: Tool reuse, learning effects, detection, and access burn-out can produce economies or diseconomies of scale. Yet existing quantitative attack tree analyses typically treat costs and impacts as static values and therefore miss how attacker incentives change under repetition. In this paper, we add a previously overlooked dimension to attack tree based risk analysis, which is the number of attack executions. We lift cost and damage from static scalars to execution-indexed functions and provide compositional AND/OR propagation rules that yield path-level cost and damage profiles, enabling joint cost–damage analysis under repeated executions. The resulting cost–damage relations can be non-concave due to scalable costs and nonlinear damage effects (e.g., economies of scale, saturation, or threshold behavior).

On illustrative attack trees, we show that scalability reshapes optimal attacker choices across objectives (net benefit, return on investment, budget- and target-constrained): A path optimal for a single execution may be suboptimal for multiple executions, even switching paths across executions can be optimal. This accounts for why attacker strategies evolve with scale. Our framework provides a sound basis to compare such objectives, to anticipate path choices as attack executions increase, and to support quantitative, repeatable security assessments.

Keywords: Attack trees · cumulative cost curves · risk quantification · scalability of attacks · security economics

1 Introduction

The scalability of an attack is the damage-to-cost ratio between the attacker's costs and the damage resulting from the attack subject to multiple attack executions [13]. Considering the scalability of attacks in risk analysis can reveal attacks that have high initial costs, but are profitable in the attacker's eyes if executed multiple times. This can occur because of diminishing marginal cost for an additional attack or increased damage when a certain number of assets is affected. For example, in critical infrastructure (such as energy systems), many

M. Chen and M. Tschaikowski (Eds.): QEST+FORMATS 2026, LNCS 16913, pp. 313–330, 2027.
https://doi.org/10.1007/978-3-032-35298-9_18

components are designed to include redundancies to mitigate the impact of a single attack. An attacker would therefore want to attack multiple times to damage the system. Since this important aspect of attacks has been overlooked in previous risk analysis frameworks, we propose extending the cost–damage analysis by the dimension of the number of attacks as it can explain attacker behavior previously deemed unlikely.

Incorporating this additional dimension into the cost–damage comparison renders the cost–damage analysis three dimensional, as depicted in Fig. 1. This view captures the behavior of the cost–damage properties of attack paths under repeated attack executions. We observe that this view reduces to existing cost–damage analysis in the case of one single attack. However, a projection of these foliations onto the cost–damage plane yields a trajectory showcasing the changing cost–damage behavior over attack repetitions. In particular, intersections of these trajectories indicate a shift in the choice of the optimal attack path.

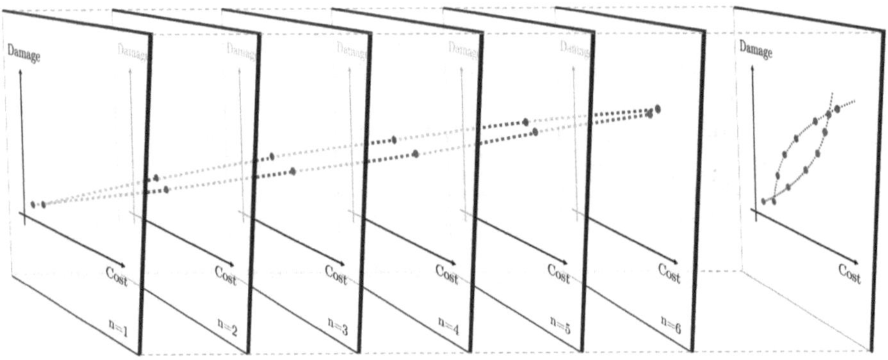

Fig. 1. Three-dimensional foliations of the cost–damage plane, subject to several attack executions. Each surface corresponds to a fixed execution count; big dots mark the discrete cumulative cost–damage points. Connecting the dots across execution counts yields trajectories on the cost–damage plane, revealing how path economics evolve and where optimal choices may switch captured by the notion of scalability of different attacks.

Constructing such trajectories is non-trivial: (i) basic events along a composite path can scale differently; (ii) multi-stage attacks may deliver partial damage before completion; and (iii) assessing costs separately for each n is cumbersome.

To address this, we use attack trees and their natural decomposition to propagate upwards the cost and damage functions of each node. We revisit observations from economics to assess that cost behavior is naturally determined as cost functions with respect to a given number of executions. Inverting these functions enables to obtain the number of possible attacks given a budget. Using the dependency of damage on successful attacks, we can construct the cost–damage trajectories for each path and investigate the scalabilities as depicted in Fig. 2.

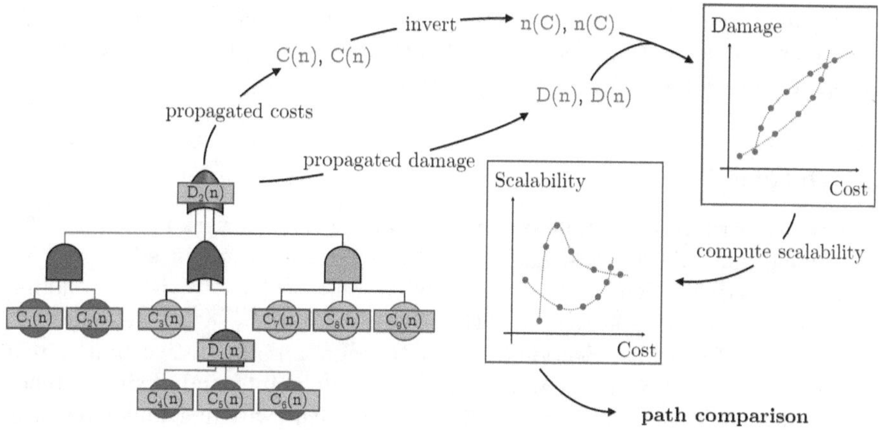

Fig. 2. Procedure overview. Cost and damage functions are propagated along the attack tree to obtain path-level profiles. Their combination yields cost–damage trajectories; the scalability computation highlights efficiency points enabling path comparison.

Our approach formalizes an attacker analyzing the attack space for the most cost-efficient attack: While depending on the goals, the attacker may want to maximize return on investment (ROI) or net profit or optimize with respect to given boundary conditions such as minimal required damage or maximal budget. For example, a company may want to harm a competing company in a cost-efficient way but only if the cost of the attack does not ruin the malicious company, too.

As the notion of scalability relates the damage to the required cost, the choice according to these circumstances can be predicted. As a consequence—based on the expected attacker types—preventive measures to harden the most critical attack paths can be implemented.

1.1 Contributions

We summarize the paper's main contributions as follows:

- **Scalability-aware attack trees.** We lift attack tree quantification from static leaf costs to *execution-indexed cumulative cost functions* $C_{v_i}(n)$ and provide *compositional propagation rules* for AND/OR refinements to obtain path-level cost profiles $C_P(n)$ and their budget-indexed counterparts.
- **Nonlinear attack economics under repetition.** By combining scalable costs with (potentially saturating or threshold) damage, our framework captures *non-concave cost–damage behavior*, implying that attacker optima need not be unique or interior (boundary optima and multiple local optima can arise), in contrast to standard diminishing-returns assumptions.
- **Objective-dependent path reversals.** We connect common attacker objectives (net benefit, ROI, budget/target constraints) to *average vs.*

marginal scalability diagnostics and show that scalability induces *path-preference reversals* as budget grows; we illustrate these effects on a multi-stage example and a real-world case study.

1.2 Outline

The paper is structured as follows: After reviewing related work in Sect. 2, setup and propagation rules for cost and damage functions are introduced in Sect. 3.1. Section 3.2 formally defines attack scalability and Sect. 3.3 revisits economies and diseconomies of scale together with cost function classes as well as exemplary damage functions. Section 4 works through an illustrative and a real-world example. Section 5.1 characterizes efficiency points and their existence, leading to a path reversal. Section 5.2 analyzes path switching under repeated executions—demonstrated in Sect. 5.3. Sections 6 and 7 close with a discussion and conclusions.

2 Related Work

Hetzel, Nemes, and Müller-Quade [13] consider the scalability of attacks with a more theoretical approach, where costs are represented by oracle accesses. This oracle can break any assumption the security of a protocol is based on once per access. The impact of an attack is measured in affected units, e.g., participants in an authentication infrastructure or votes in a voting scheme. In this work, we take a more practical approach using attack trees [25]. Our framework can be used to further analyze existing attack trees to improve risk analysis and highlight so far overlooked threats.

Quantifying security using attack trees is a well-established approach [17,19,23]. Additionally, relating the attacker's costs to the resulting damage has been proposed [2,4,6,20]. A common limitation, however, is the absence of a temporal component, which would reveal how optimal choices evolve under repeated executions. Although Jürgenson and Willemson consider time by allowing the attacker to sequence elementary steps to switch to a different path in case of failure, each attack is only conducted once [15]. André, Lime, Ramparison, and Stoelinga add the duration of an attack as an additional parameter to the attacker's costs and the damage of an attack [1]. Here, the temporal component is reflected by *multiple executions*, i.e., repeating entire paths.

Classical work has long recognized nonlinear effects under repetition: Learning/experience and coordination curves yield economies and diseconomies of scale [5,12,29]. Damage need not be linear either—saturation or redundancies can diminish marginal damage [16,21], whereas in interdependent infrastructures (e.g., power grids) numerous perturbations exceeding a threshold can escalate via cascades [7,24]. These effects have however not been considered in attack tree based risk analysis.

In their seminal article from 2002, Gordon and Loeb propose a defender centric economic model for determining which assets warrant protection investment

[11]. In this sense, our paper studies the complementary attacker-side problem: choosing the most cost-effective attack strategy.

3 Preliminaries and Model of Scalability

In this section, we provide the basis for our framework. After a short repetition of attack trees, we present various definitions of attack scalability. Then we provide a collection of cost and damage functions to be plugged into the scalability definitions.

3.1 Propagation of Cost and Damage Along Attack Trees

Attack trees [25] are a widely used tool in risk analysis. Ideally, they are a representation of all possible attack vectors an attacker could take to harm the system that is analyzed. Attack trees contain the attacker's goal in the root node. The internal nodes of the tree can most commonly be either OR or AND nodes, indicating that either one of the child nodes is sufficient to reach the goal or that all of them have to be fulfilled. The leaf nodes represent the so-called elementary attacks, which are not split up any further and are assumed to be independent of each other. While damage of an attack can be assigned to any node, only leaf nodes carry the attacker's costs as proposed by Lopuhaä-Zwakenberg and Stoelinga [20]. Both impact as well as costs are traditionally fixed values [25], which we lift to functions in this paper. The terms impact and damage are used synonymously here.

The different attack paths can be found by identifying the minimal cut sets of the tree. Each minimal cut set consists of elementary attacks which can lead to the attacker's goal if they are combined. Since the cut sets are minimal, the goal cannot be reached anymore if one of the elementary attacks is removed from the set. Looking at the attack paths resulting from the minimal cut sets, all OR connections have been removed from the tree, since at every OR connection, the attacker can take different routes to reach the root node.

To compute the attacker's costs for each path, at AND connections, the costs of the child nodes can simply be added:

$$C_{\text{total}} = \sum_{i=1}^{k} C_{v_i}, \tag{1}$$

where i is the index of one of the k child vertices. In more complex trees, there might be further connections, e.g., where an inner node can be executed by choosing two out of three elementary attacks [18]. Such connections are treated analogously to OR connections in the minimal cut set analysis. Gates that require the parallel (PAND) or sequential (SAND) execution of attacks are of less interest in our framework, since monetary costs in particular are added at AND gates to compute the costs of the path, independent of the order of execution.

In an attack tree, all paths lead to at least the impact at the root node. Further damage can be caused by internal nodes of the tree, which can be interpreted as collateral damage resulting from intermediate steps leading to the actual goal of the attacker at the root node. Nevertheless, it can also show a nonlinear behavior after multiple attack executions, as discussed in Sect. 3.4.

3.2 Definition of Attack Scalability

Intuitively, the scalability of an attack relates damage and required cost to the number of executions. We quantify scalability through *cost–damage efficiency*, i.e., the ratio of cumulative damage to cumulative cost. We provide both an execution-indexed and a cost-indexed view, since cumulative cost and damage are often given in execution-indexed form, while the cost-indexed view is more natural for economic comparisons.

Setup. Let $C : \mathbb{N}_{\geq 1} \to (0, \infty)$ be the cumulative cost function on execution count n with $C(n+1) > C(n)$ for all $n \geq 1$. Let $D : \mathbb{N}_{\geq 1} \to [0, D_{\max}]$ be the cumulative (monetized) damage function that is non-decreasing and bounded by $D_{\max}$. They assign cost and damage to the of number of executions, respectively.

Definition 1 (Execution-indexed Scalability Ratio[1]**).** *The execution-indexed scalability ratio of the attack is defined as the function*

$$S(n) := \frac{D(n)}{C(n)}, \quad n \geq 1. \tag{2}$$

This definition can be understood as a measure of average cost-efficiency after $n \geq 1$ executions and is the cost-aware relaxation of [13].

As different attacks usually have different costs, attackers are more interested in the impact with respect to costs than with attack executions. Accordingly, we obtain $D\big(n(C)\big)$, damage as function of cost, by inverting the function $C(n)$ to calculate the number of possible attacks given a specific budget:

$$n(C') := C^{-1}(C'), \;\; C' \geq C_{\min} := C(n = 1). \tag{3}$$

Definition 2 (Cost-indexed Scalability Ratio). *The cost-indexed scalability ratio of the attack is defined as the function*

$$S(C') := \frac{D(n(C'))}{C'}, \quad C' \geq C_{\min} := C(n = 1). \tag{4}$$

These definitions of scalability measure average cost efficiency and hence mask the behavior around a state of interest. Accordingly, the notion of marginal utility gives insight into the current behavior of efficiency and can hence answer whether a further investment is beneficial.

[1] This ratio is mathematically equivalent to the classical benefit–cost ratio (BCR) known from economics, but since we analyze the growth behavior of attacks, we use the term scalability to emphasize the technical interpretation.

Definition 3 (Marginal Utility). *The marginal utility is defined as*

$$MU(C') := \frac{\Delta D(n(C'))}{\Delta C'}, \quad C' \geq C_{\min} := C(n = 1). \tag{5}$$

3.3 Attack Costs Under Repeated Attacks

All cumulative cost functions, denoted by $C(n)$, are strictly monotonically increasing, since upon every attack execution, the attacker has to either invest more or can benefit from previous investments with barely any additional costs. On the other hand, attacks can also become more expensive, e.g., when resources are used up or complexity increases. Accordingly, cost functions may be convex, concave, or neither.

The extended version of this paper [10] presents four cost functions derived from economic laws and their corresponding cumulative cost curves: Wright's law, which describes an increase in experience and efficiency with every produced item [29], Brooks's law, which describes diseconomies of scale due to growing communication overhead in large groups [5], non-rival goods which cannot be exhausted [26], and lastly constant costs if no scaling can be observed. The different types of cost functions are summarized in Table 1 in the top rows. The table also includes some examples for attacks that show the respective cost behavior across attack executions.

3.4 Damage Under Repeated Attacks

Similar to the attacker's costs, the damage of an attack (per attack execution) can change with every time the attack is performed. On one hand, many systems rely on redundant components to ensure availability. While the first attack only leads to a small damage (the cost of replacing or repairing the redundant component without any consequences to the rest of the system) a second attack can cause much more harm.

On the other hand, a system might have higher value and lower value targets. Once an attacker has compromised all higher value targets, the additional damage in further attacks decreases. Another reason for the damage per attack execution to decline might be that full destruction of the system is approached.

Just like the cumulative cost function, the cumulative damage function $D(n)$ is monotonically increasing, since further attacks do not reduce damage. Damage functions can also be convex or concave functions and their domain and range are limited, since once the target is fully destroyed, no more damage is possible. In the extended version [10], two types of damage functions are discussed exemplarily [14,27,28]. They can also be found in Table 1 in the bottom rows.

If independence and linearity of the cumulative damage are assumed for illustration, one may write $D(n) = n \cdot d_1$ with constant per-execution damage d_1; in general, interactions, saturation, or synergies imply nonlinearity and are captured by the definitions of Sect. 3.2.

Table 1. Examples of cost and damage functions, at the top and bottom of the table respectively

Type	Reasoning	Characterization, Typical Values	Examples
Sublinear [29]	Gained experience and efficiency	$C(n) = c_1 \cdot \left(1 + \frac{1}{1+\log_2(b)} \cdot (n^{1+\log_2(b)} - 1)\right)$, $b \in [0.75, 0.9]$	Malware development for device class, lock picking
Superlinear [5]	Growing communication overhead and training effort	$C(n) = n \cdot \frac{n^2-n}{2}$	Bypassing layered detection systems, coordinated multi-actor attacks
Zeroth Order [26]	Goods used by multiple parties	$C(n) = \varepsilon \cdot n + b$, ε very small	Self-replicating malware, supply chain attacks
First Order	No cost saving	$C(n) = c_1 \cdot n$, c_1 equal to single attack cost	Physical-layer attack (e. g., GNSS jamming, break open a car)
Saturating [14,27]	Few high value targets	$D(n) = D_{\max} \cdot \frac{n^b}{n^b+a}$, $D_{\max}$ total system value, $a = 1 - \frac{D_{\max}}{D(1)}$, b discriminatory power	Peer-to-peer networks with supernodes
Cascading [28]	Interdependent infrastructures	$D(n) = \frac{D_{\max}}{1+\exp(-b(n-n_0))}$, $D_{\max}$ total system value, b growth rate, n_0 executions to damage half	Redundant systems (e. g., electrical grid)

4 Application of the Framework to Examples

In this section, we present two example attack trees: one to showcase our scalability framework and one real-world example. The latter stems from a commissioned security analysis of the recently introduced German electronic patient record.

4.1 Application to a Generic Tree

To illustrate the framework, we first give an example tree in Fig. 3. Looking at the structure of the tree, we find all minimal sets which lead to a damage and obtain three different paths:

$$P1 = \{v_1, v_2, v_7, v_{10}, v_{11}\}, \quad P2 = \{v_3, v_4, v_8, v_{10}, v_{11}\}, \quad P3 = \{v_5, v_6, v_9, v_{11}\}.$$

When we assign cumulative costs and damages to the nodes as indicated in Fig. 3, we can compute the cumulative costs and damages for each path using the propagation rules from Sect. 3:

$$C_{P1}(n) = \frac{1}{40}\frac{n^3 - n^2}{2} + 0.5 + \varepsilon n \qquad D_{P1} = \frac{12}{1+\exp(-n+17)} + 35 \cdot \frac{n}{n+28}$$

$$C_{P2}(n) = 1.5 + \varepsilon n + n, \qquad D_{P2} = \frac{12}{1+\exp(-n+17)} + 35 \cdot \frac{n}{n+28}$$

$$C_{P3}(n) = \frac{5(n^{1+\log_2(0.6)} - 1)}{1 + \log_2(0.6)} + 5 + 2 + \varepsilon n \qquad D_{P3} = 35 \cdot \frac{n}{n+28} \; ; \varepsilon > 0.$$

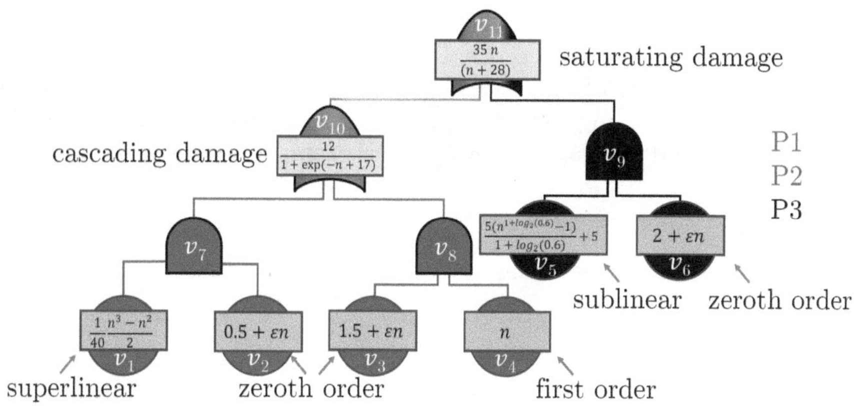

Fig. 3. Example attack tree to illustrate the framework. According to standard notation, AND and OR gates are represented with a round or tapered top, respectively. Each node may carry a damage function (yellow box), while only leaf nodes can carry a cost function (gray box). (Color figure online)

The cumulative costs are plotted in Fig. 4a. One can see that the combination of different basic events amounts to the shape of the curves: While for five iterations $P2$ is more expensive than $P3$, this is reversed at their crossing. We then invert the cumulative cost functions using the variable transformation of Eq. 3 to obtain the number of possible attack executions with respect to budget in Fig. 4b. We only consider cumulative cost greater than the respective $C_{\min} := C(n = 1)$.

By plugging the cumulative cost functions into the cumulative damage functions, we obtain the cost–damage graph shown in Fig. 5a. While useful for some

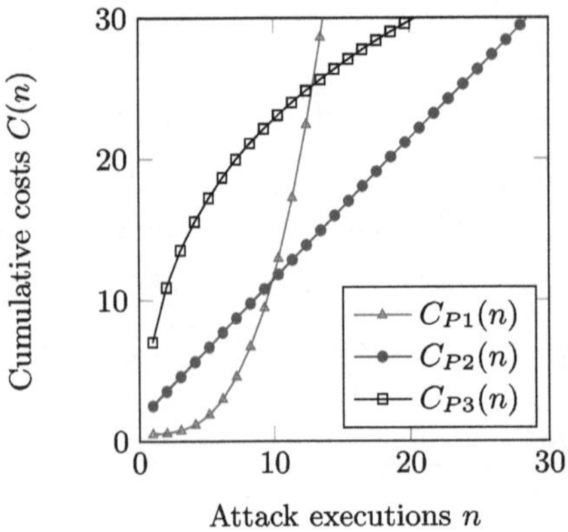

(a) Cumulative cost given attack executions.

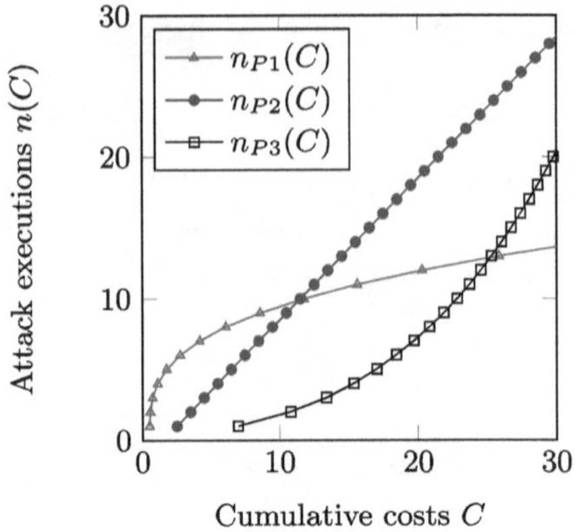

(b) Attainable number of attack executions given cumulative costs.

Fig. 4. Cumulative cost for different attack paths (left) and corresponding possible attack executions given a budget (right). We observe the superlinear, linear and sublinear type. Note their respective axial symmetry along the bisectrix.

decision problems, Sect. 5.1 argues that the scalability graph better captures different attacker goals. Using Definition 2, we therefore derive the scalability plot in Fig. 5b. A detailed scalability analysis of this example tree follows in Sect. 5.3.

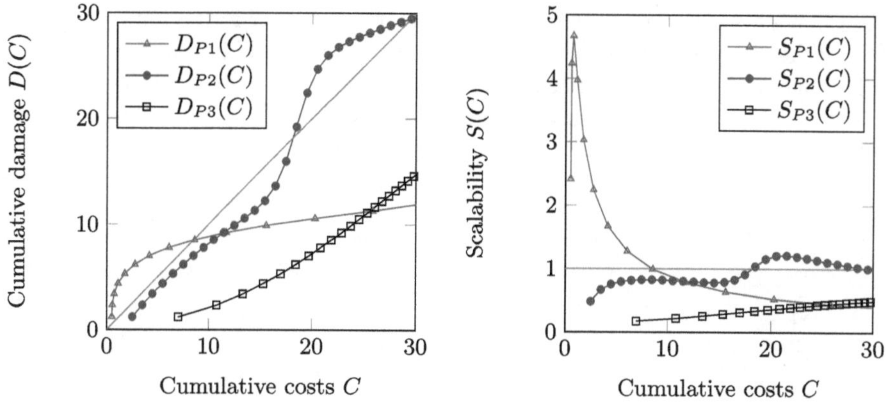

(a) Cost–damage plot.

(b) Cost-indexed scalability plot.

Fig. 5. Cost–damage plot (left) and scalability plot (right) for all paths. The thin gray line represents equality of cost and damage, attacks above it are profitable. (Color figure online)

4.2 Electronic Patient Record (ePA for All)

Before the rollout of the electronic patient record ePA in Germany in January 2025, a security analysis using attack trees has been performed [9]. Five attacker goals were identified, leading to five attack tree root nodes: unauthorized reading, manipulating or deleting of a record, denial of service of the record, and revealing treatment/ diseases/ personal data of the insured person. The report assigns success probabilities to attacks and attacker types.

Looking, e.g., at the attack tree with the attacker goal unauthorized reading of a record (partially shown in the extended version [10]) and the group of external attackers like hackers or hacktivists, there are seven different major paths the attacker could take, which again split up into multiple subpaths. The two groups of paths with the highest success rates for this type of attacker are via the nodes "gain access via the primary system of the healthcare provider" ($p = 24\%$) and "gain access to the front end of the insured person" ($p = 59\%$). Therefore, the latter seems to have a higher risk. However, when the attacker gains access to the front end of one user, they can only read this user's patient record. When gaining access to the primary system of the healthcare provider, all of their patients are affected. The success probability of this attack path is smaller, but it seems to scale much better. The report [9] does note that healthcare providers have access to a large number of patient records, but this is not reflected in the classification merely based on success probabilities.

Figure 6 compares the cost-indexed scalability ratio curves of two attack paths, where the red solid line corresponds to an attack via the healthcare provider and the blue dashed line corresponds to an attack via the insured person. Please note the logarithmic scale. With the analysis in our framework we can confirm the intuition that indeed the attack via the healthcare provider scales much better and might be preferred by the attacker in the long run. Once the attacker has invested two units of cost, the solid scalability curve crosses the dashed one and stays at a far higher scalability value. A more extensive analysis of this part of the attack tree on unauthorized reading of a patient record can be found in the extended version of this paper [10].

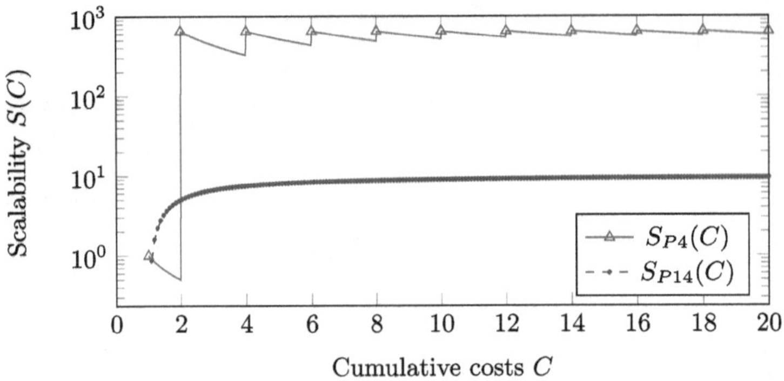

Fig. 6. Selection of two cost-indexed scalability ratio curves of the attack tree on unauthorized reading of a record with $c_1 = 1$ and $D(n) = n$.

5 Choice of Attack Vector Based on Scalability

After identifying the scalability curves for all attack tree paths, the conclusions of the security analysis are still ambiguous. Different attacker budgets and goals might lead to different path preferences.

Continuous relaxation. We now treat cumulative cost and damage as sufficiently smooth to enable closed-form solutions and existence arguments for efficiency points using basic calculus, in the spirit of classical security-economics models. Accordingly, we will then require that $C : [1, \infty) \to (0, \infty)$ be continuously differentiable, $n \in \mathbb{R}$ with $\mathrm{d}/\mathrm{d}n \ C(n) > 0$ for all $n > 0$ and $D : [1, \infty) \to [0, D_{\max}]$ continuously differentiable, non-decreasing and bounded by $D_{\max}$. We require $n > 1, \ C \geq C_{\min} = C(1)$ to exclude poles of high scalability at $C = 0$.

5.1 Objectives and Their Relation to Scalability

Existing work has considered different attacker behavior subject to different economic rationales [6,11,20]. We formalize four natural attacker goals with budget indexing for a path P:

(a) Maximize net profit (NP) as difference between damage and cost. The attacker searches for the best combination of path and cost:

$$\max_{P, C \geq C_{\min,P}} (NP_P(C)) := \max_{P, C \geq C_{\min,P}} \left(D_P(n(C)) - C\right). \tag{6}$$

(b) Maximize return on investment (ROI). The attacker seeks the best return on investment—defined as the ratio of net benefit by cumulative cost:

$$\max_{P,C\geq C_{\text{min},P}} \text{ROI}_P(C) := \max_{P,C\geq C_{\text{min},P}} \left(\frac{D_P(n(C)) - C}{C}\right). \tag{7}$$

(c) Maximize damage subject to a budget. Given a budget $B > 0$, the objective is

$$\max_{P,C_{\text{min},P}\leq C\leq B} D_P(n(C)). \tag{8}$$

(d) Minimize cost subject to a required damage. Given a required minimum damage $D_{\text{min}} > 0$, the objective is

$$\min_{P,C\geq C_{\text{min},P}} (C) \quad \text{s.t.} \quad D_P(n(C)) \geq D_{\text{min}}. \tag{9}$$

We note that $D_P(C)$ is continuous, non-decreasing, and bounded above by some D_{max} as we are using the continuous relaxation to enable closed-form solutions. Hence, both $NP_P(C)$ and $\text{ROI}_P(C)$ reach a maximum in their half-open domain $[C_{\text{min},P}, \infty)$. However, we did not impose a concavity assumption on D_P. In particular, the interplay of $D(n)$ and $n(C)$ can yield a *non-concave* (e.g., locally convex) function $D(n(C))$. Thus, the maxima need not be unique.

For a fixed path P, the four objectives yield the following solution characterizations which again justify the consideration of scalability S and marginal utility MU:

(a) *Profit peak.* Interior stationary points of $NP_P(C)$ satisfy

$$\partial_C \ NP_P(C)\big|_{C=C^\star} = 0 \iff MU_P(C^\star) = 1,$$

i.e., marginal benefit equals marginal cost. Since $D_P(C)$ is not assumed to be concave, such stationary points are only necessary conditions: $C^\star$ may correspond to a local maximum, a local minimum, or a stationary inflection point. Consequently, the global optimum may also occur at the boundary of the feasible domain (e.g., at saturation or hitting D_{max}).

(b) *Efficiency peak.* Noting that $\text{ROI}_P(C) = S_P(C) - 1$, interior extrema of the ROI satisfy

$$\partial_C \ ROI_P(C)\big|_{C=C^\star} = 0 \iff MU_P(C^\star) = S_P(C^\star),$$

i.e., marginal utility equals average scalability. As above, without concavity, such points need not be unique nor correspond to a global maximum.

(c) *Budget frontier.* For objective (c), monotonicity of D_P implies that the maximum attainable damage under budget B is given by $D_P(B)$.

(d) *Damage frontier.* For objective (d), the optimal solution corresponds to the minimal cost at which the damage threshold D_{min} is reached; uniqueness holds if and only if $D_P(C)$ is strictly increasing in a neighborhood of that level.

Implications of Non-concavity of the Cost–Damage Function. Overall, the absence of concavity in $D_P(C)$ is a central feature of scalable attacks: It allows for multiple local optima, boundary solutions, and qualitative changes in optimal behavior as budget or execution scale varies. While our model explicitly permits non-concave damage functions, we note that non-concavity already stems from allowing the costs of attacks to incorporate economies of scale, refining classical economical analysis assuming diminishing returns [11].

5.2 Path Switching Under Scale

The attacker may formulate a ranking of favorable attacks for a single execution due to a cost–damage analysis. This ranking can change based on the budget, as observed in Fig. 5, indicated by the intersection of paths. However, switching paths *within an attack*, i.e., combining their most favorable parts, can be better than following one path only, e.g. for attacker goal (a), allowing path switching trivially cannot decrease the net profit since it also comprises sticking to a path.

Following that argumentation, we immediately observe that goals (c) and (d) can also be fulfilled better if path switching is permitted. Further, for goal (b) switching paths can—counter-intuitively—result in a *better* ROI, as the mediant inequality does not hold here. The intuition behind that is that a threshold damage can be surpassed if several attacks of bad individual scaling are combined. An example of beneficial path switching can be found in Sect. 5.3 below.

Allowing path switching turns the optimization into a resource-allocation problem: The optimal achievable damage as a function of budget is given by the supremal convolution of the per-path cost–damage functions, i.e., by optimally splitting the total cost across paths and summing their contributions[2]—a well-known setting in mathematical economics [3,8]. In a discretized model (integer number of executions), each option $(P, C(n))$ constitutes an item; for any fixed P these options are mutually exclusive (selecting n executions excludes any other execution of that path), which can be encoded as a conflict-constrained knapsack formulation, see [22].

In this paper we do not propose a full algorithmic solver for switching under shared-node nonlinearities; instead, we formalize switching as a resource-allocation problem and use it to demonstrate that scalability can create incentives for mixed strategies that can dominate any single path. Designing efficient heuristics and tool support is important future work.

5.3 Path Reversal and Path Switching Example

Figure 5 shows that multiple attack executions can induce path reversal beyond a certain budget marked by intersections of the graphs. Moreover, we note that scalability analysis fundamentally requires a *comparison of curves*: Although

[2] Allowing path switching also requires keeping track of the attack executions for every node shared by several attack paths, because cost and damage of that node are in general nonlinear.

path $P1$ exhibits high scalability at small costs—three executions of $P1$ are optimal for attacker goal (b)—it does not "scale well" (in intuitive understanding) and does not simultaneously maximize attacker goal (a). As illustrated in Fig. 7a, net profit corresponds to vertical distances to the bisectrix, where path $P2$ attains a higher maximum than $P1$, highlighting that attacker goals may not align and require careful attacker characterization.

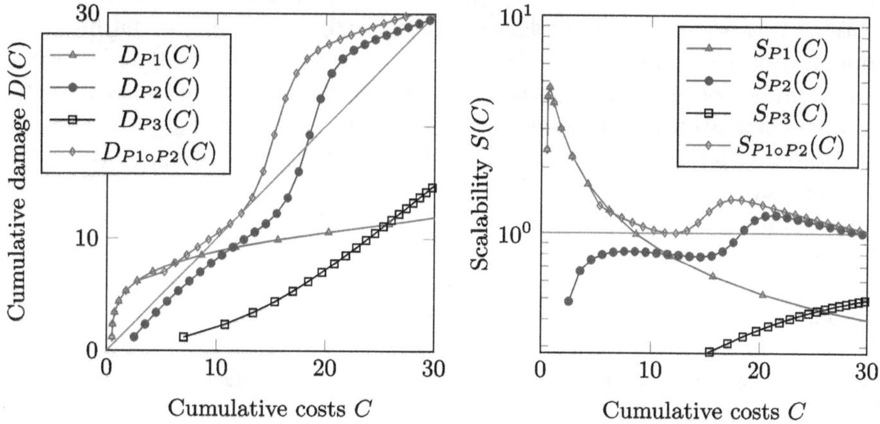

(a) Cost–damage plot with path switch.

(b) Scalability plot.

Fig. 7. Cost–damage plot (left) and scalability plot (right) for all paths. We note that path switching can substantially increase net profit (maximal vertical distance to bisectrix) and the scalability curve.

Now we consider *path switching*, i.e., combining different paths to achieve maximal damage: While attack $P1$ offers an attractive behavior at its beginning, its decreasing scalability makes this attack path less appealing for more attacks. In contrast, $P2$ offers a good behavior at more executions. We hence combine these paths such that for the first six executions we follow path $P1$ and afterwards follow path $P2$[3]. We take care of the initial investment which switching to $P2$ requires and make sure not to double count the executions at their path intersection nodes. The results for the cost–damage plot and the scalability plot are shown in Figs. 7a and 7b respectively.

The corresponding path dominates all single paths in terms of cost–damage plot and scalability plot as depicted in Fig. 7 (except for the switching point). This attack hence yields greater damage at given costs and shows a better scalability than any single attack path. Consequently, extending the analysis from single attacks to multiple executions must also consider path switching, since it can reveal powerful attacks overlooked by standard risk assessment.

[3] We note that finding the optimal switching point is in general a combinatorial problem, see Sect. 5.2.

To assess how sensitive the observed path rankings and intersections are to moderate parameter uncertainty in this illustrative tree, we provide a sensitivity analysis in the extended version [10].

6 Discussion

The central qualitative effect enabled by our model is that attacker preferences can *change with scale.* Because we lift costs to execution-indexed cumulative functions and analyze their budget-indexed view, the ranking of paths under common objectives (net profit, ROI, budget-/target-constrained) may reverse as the number of executions grows. This helps explain why an attack vector that appears unattractive in a one-shot assessment may become preferable when amortized over many executions, and conversely why initially cheap vectors may lose their appeal once their marginal efficiency deteriorates.

A second key implication is that cost–damage relations need not exhibit globally diminishing returns. Many classical cost–benefit analyses rely (implicitly or explicitly) on regularity assumptions (e.g., concavity of a benefit function or unimodality of net benefit) to obtain a unique interior optimum from first-order conditions. In contrast, scalable attacks can naturally yield *non-concave* $D_P(C)$ (and hence non-concave net benefit and ROI) due to sublinear cost growth combined with nonlinear damage.

Limitations

Interdependencies. We treat paths independently and repetition as re-executing one chosen path. In reality, executions can change other paths' feasibility/costs (reuse, exposure, patching, burned access). Additionally, the system dynamics may be heavily altered by defender's reactions. This requires a state-dependent, sequential model; greedy re-selection need not be globally optimal.

Success Ratio. We implicitly assume a static 100% success ratio if the required basic attacks are activated. Probabilistic success can be incorporated by treating the number of successful executions as a random variable $\mathcal{N}$ and replacing $D(n)$ by $\mathbb{E}[D(\mathcal{N})]$; we leave this extension to future work to keep the present contribution focused on compositional propagation and objective-induced reversals.

Solutions to the Path-Switching Problem. While the specialized cost–damage optimization problem (only single executions of each attack and damages independent of the amount of repeated executions) is already NP-hard and has been proven to not be applicable to knapsack heuristics [20], new heuristic approaches for the discrete path-switching optimization can further foster the adaptation of execution-dependent costs and damages.

Limited Application Domain of Economic Models. While it is clear that the range of a damage function is limited to the total value of the system, the applicability of economic cost models is also limited. Attacks becoming infinitely expensive at an infinite number of executions is a simplification of reality. Once an asset is destroyed, there are no more attack executions to be expected.

7 Conclusion and Outlook

We have introduced scalability-aware attack tree analysis by lifting leaf costs to execution-indexed cumulative functions and propagating them compositionally through AND/OR connections to obtain path-level cost profiles and budget-indexed views. This exposes a core phenomenon that is hidden in static cost–damage analyses: Under repetition, cost–damage relations can become non-concave and attacker-optimal choices can reverse with scale. As a result, a path that is optimal for a single execution may become suboptimal over many executions, and switching paths across executions can be economically optimal.

We illustrated these effects on representative attack-tree examples, demonstrating that conclusions drawn from single-execution or static cost–damage views may no longer hold once repetition is considered. Although the extension represents a huge refinement of existing schemes, at the same time, several important aspects remain open, most notably implementable heuristics (and tool support) for computing beneficial path-switching strategies across executions.

Acknowledgments. This work was funded by the Topic Engineering Secure Systems of the Helmholtz Association (HGF) and supported by KASTEL Security Research Labs (structure 46.23.02).

Disclosure of Interests. The authors have no competing interests to declare that are relevant to the content of this article.

References

1. André, É., Lime, D., Ramparison, M., Stoelinga, M.: Parametric analyses of attack-fault trees. Fund. Inform. **182**(1), 69–94 (2021)
2. Aslanyan, Z., Nielson, F.: Pareto efficient solutions of attack-defence trees. In: International Conference on Principles of Security and Trust, pp. 95–114. Springer (2015)
3. Bayón, L., García-Nieto, P., García-Rubio, R., Grau, J.M., Ruiz, M., Suárez, P.: The operation of infimal/supremal convolution in mathematical economics. Int. J. Comput. Math. **93**(5), 735–748 (2016)
4. Bistarelli, S., Fioravanti, F., Peretti, P.: Defense trees for economic evaluation of security investments. In: First International Conference on Availability, Reliability and Security (ARES 2006), pp. 416–423. IEEE (2006)
5. Brooks, F.P.: The mythical man-month. Datamation **20**(12), 44–52 (1974)
6. Buldas, A., Laud, P., Priisalu, J., Saarepera, M., Willemson, J.: Rational choice of security measures via multi-parameter attack trees. In: International Workshop on Critical Information Infrastructures Security, pp. 235–248. Springer (2006)
7. Buldyrev, S.V., Parshani, R., Paul, G., Stanley, H.E., Havlin, S.: Catastrophic cascade of failures in interdependent networks. Nature **464**(7291), 1025–1028 (2010)
8. Burgert, C., Rüschendorf, L.: On the optimal risk allocation problem. Stat. Decisions-Int. J. Stoch. Methods Models **24**(1), 153–172 (2006)
9. Fraunhofer SIT: Abschlussbericht Sicherheitsanalyse des Gesamtsystems ePA für alle (2024). https://www.gematik.de/media/gematik/Medien/ePA_fuer_alle/Abschlussbericht_Sicherheitsanalyse_ePA_fuer_alle_Frauenhofer_SIT.pdf. Accessed 11 Dec 2025

10. Fruböse, C., Hetzel, E.: The cost of repetition: a compositional scalability model for attack trees (2026). https://doi.org/10.5445/IR/1000194155
11. Gordon, L.A., Loeb, M.P.: The economics of information security investment. ACM Trans. Inf. Syst. Secur. (TISSEC) **5**(4), 438–457 (2002)
12. Hax, A.C., Majluf, N.S.: Competitive cost dynamics: the experience curve. Interfaces **12**(5), 50–61 (1982)
13. Hetzel, E., Nemes, M., Müller-Quade, J.: Attack once, compromise all? On the scalability of attacks. In: International Joint Conference on Electronic Voting, pp. 90–106. Springer (2025)
14. Iliaev, D., Oren, S., Segev, E.: A tullock-contest-based approach for cyber security investments. Ann. Oper. Res. **320**(1), 61–84 (2023)
15. Jürgenson, A., Willemson, J.: Serial model for attack tree computations. In: International Conference on Information Security and Cryptology, pp. 118–128. Springer (2009)
16. Kempe, D., Kleinberg, J., Tardos, É.: Maximizing the spread of influence through a social network. In: Proceedings of the Ninth ACM SIGKDD International Conference on Knowledge Discovery and Data Mining, pp. 137–146 (2003)
17. Kordy, B., Kordy, P., Mauw, S., Schweitzer, P.: Adtool: security analysis with attack–defense trees. In: International Conference on Quantitative Evaluation of Systems, pp. 173–176. Springer (2013)
18. Kumar, R., Stoelinga, M.: Quantitative security and safety analysis with attack-fault trees. In: IEEE 18th International Symposium on High Assurance Systems Engineering (HASE), pp. 25–32. IEEE (2017)
19. Lopuhaä-Zwakenberg, M., Budde, C.E., Stoelinga, M.: Efficient and generic algorithms for quantitative attack tree analysis. IEEE Trans. Dependable Secure Comput. **20**(5), 4169–4187 (2022)
20. Lopuhaä-Zwakenberg, M., Stoelinga, M.: Cost-damage analysis of attack trees. In: 2023 53rd Annual IEEE/IFIP International Conference on Dependable Systems and Networks (DSN), pp. 545–558. IEEE (2023)
21. Mossel, E., Roch, S.: Submodularity of influence in social networks: from local to global. SIAM J. Comput. **39**(6), 2176–2188 (2010)
22. Pferschy, U., Schauer, J.: The knapsack problem with conflict graphs. J. Graph Algorithms Appl. **13**(2), 233–249 (2009)
23. Rana, A., Gupta, S., Gupta, B.: A comprehensive framework for quantitative risk assessment of organizational networks using fair-modified attack trees. Front. Comput. Sci. **6**, 1304288 (2024)
24. Schäfer, B., Witthaut, D., Timme, M., Latora, V.: Dynamically induced cascading failures in power grids. Nat. Commun. **9**(1), 1975 (2018)
25. Schneier, B.: Attack trees. Dr. Dobb's J. **24**(12), 21–29 (1999)
26. Shapiro, C., Varian, H.R.: Information Rules: A Strategic Guide to the Network Economy. Harvard Business Press (1999)
27. Tullock, G.: Efficient rent seeking. In: Toward a theory of the rent-seeking society, vol. 97, p. 112 (1980)
28. Verhulst, P.F.: Recherches mathématiques sur la loi d'accroissement de la population, vol. 18. Académie Royale de Bruxelles (1844)
29. Wright, T.P.: Factors affecting the cost of airplanes. J. Aeronaut. Sci. **3**(4), 122–128 (1936)

Witnesses and Counterexamples for Timed Bisimulation

Alexander Lieb[1](✉) and Malte Lochau[2](✉)

[1] Real-Time Systems Lab, TU Darmstadt, Darmstadt, Germany
alexander.lieb@es.tu-darmstadt.de

[2] Model-Based Engineering Group, University of Siegen, Siegen, Germany
malte.lochau@uni-siegen.de

Abstract. Timed automata provide a modeling formalism for time-critical properties of reactive systems with discrete-state/continuous-time behaviors. To handle the infinite state space of timed automata, recent verification tools use zone graphs, a symbolic semantic model that guarantees sound results, at least for properties reducible to reachability problems. If we instead want to compare the behavior of two timed automata, checking for timed trace equivalence is undecidable. Fortunately, timed bisimulation equivalence is decidable, but currently available checks do not provide useful explanations of the results. To overcome this limitation, we use a recently proposed extension of zone graphs by so-called virtual clocks. The extension not only facilitates effective tool support for timed bisimilarity checking but also enables the derivation of useful explanations from the results. If timed bisimilarity holds, all witnesses derivable from the composed symbolic representation of both models are indeed valid for both models. If timed bisimilarity does not hold, we describe how to obtain counterexamples, making explicit the behavioral differences. These witnesses/counterexamples may serve as test cases in later stages of system refinement.

1 Introduction

Timed automata are a successful formalism for reasoning about time-critical behaviors [3]. Application examples include communication protocols [10] and safety-critical software [14]. Timed automata incorporate clock variables to express quantifiable timing constraints on valid execution traces. These variables impose infinite state spaces being computationally intractable. Thus, most analysis tools for timed automata use zone graphs, a symbolic representation which is, however, limited to checking problems reducible to reachability [12]. In contrast, a complete comparison of the timing behavior of two timed automata, based on timed trace equivalence checking is undecidable [4], whereas timed bisimulation has been proven to be decidable [8,15]. But, to make timed bisimilarity checking an effective tool that scales to realistic applications, it must be performed on zone graphs as well. Due to the limited precision of zone graphs, the outcomes of bisimilarity checks are neither reliable [25] nor do they provide

M. Chen and M. Tschaikowski (Eds.): QEST+FORMATS 2026, LNCS 16913, pp. 331–349, 2027.
https://doi.org/10.1007/978-3-032-35298-9_19

any further *explanation* of the outcome [9,20]. The first problem is tackled in [16] by extending zone graphs with virtual clocks, while the second issue is unsolved until now, although being particularly relevant for many applications in which fine-grained insights into behavioral differences are essential. For example, Cortés et al. [11] propose to use timed bisimulation in mutation-based testing to check whether a generated mutant of a TA is equivalent to the original model. In such a setting, a counterexample can be used to generate test cases, distinguishing the original model and the mutant [2]. Conversely, when a mutant turns out to be equivalent by accident, the developer benefits from a witness that explains why the original model and the mutant exhibit the same behavior.

Contribution. In this paper, we use virtual clocks [16] to facilitate the derivation of explanations of timed bisimilarity checks. If timed bisimilarity holds, we describe how to derive a complete and finite witness representation from the composed symbolic representation of both models. If timed bisimilarity does not hold, we describe how to obtain counterexamples, making explicit the behavioral differences. To do so, we introduce the new *Product of TLTS with Virtual Clocks (PTVC)* and *Product of Zone Graphs with Virtual Clocks (PZVC)*. We provide tool support and present evaluation results gained from applying our tool to community benchmarks. The results show that bisimilarity checking with subsequent witness/counterexample generation scales well to larger models.

Related Work. Timed bisimulation originates from Moller and Tofts [18] and Yi [24]. Čerāns [8] provides the first decidability proof using region graphs. Aceto et al. [1] describe a formula that is fulfilled if and only if two states are timed bisimilar. Weise and Lenzkes [25] propose a more space-efficient approach using a variation of zone graphs. However, their descriptions lack crucial details and no tool is available. Tanimoto et al. [23] use timed bisimulation to verify whether behavioral abstractions preserve time-critical properties, but no description of the checking procedure itself is given. To the best of our knowledge, the tool in [16] is the only available tool for effective timed bisimulation checking. However, it only gives a yes/no-answer without any further explanation. Although Wimmer et al. [26] establish a comprehensive theory for explanations in terms of certificates for reachability analysis in timed automata, no comparable approach exists so far for timed bisimulation.

Open Science Policy. The open-source implementation of our approach and all examples are available at https://github.com/Echtzeitsysteme/tchecker. We also provide a Virtual Machine at https://doi.org/10.5281/zenodo.20702440. All proofs can be found at https://arxiv.org/abs/2606.16736.

2 Background

We revisit basic definitions of timed automata [4] and timed bisimulation [8]. We either assume as a time domain $\mathbb{T} = \mathbb{N}^{\geq 0}$ for discrete time, or $\mathbb{T} = \mathbb{R}^{\geq 0}$ for dense time. All results apply in both settings. We assume a finite set C of numerical variables over $\mathbb{T}$, called *clocks*. The values of all clocks in C increase synchronously, but can be reset to zero independently. The set $\mathcal{B}(C)$ of *clock constraints* ϕ over C is defined inductively as: $\phi := \text{true} \mid c \lesseqgtr n \mid c - c' \lesseqgtr n \mid \phi \wedge \phi$, with $n \in \mathbb{N}^{\geq 0}$, $\lesseqgtr \in \{<, \leq, >, \geq\}$, and $c, c' \in C$. A clock constraint is *atomic* if it does not contain the $\wedge$-operator.

Definition 1 (Timed Automaton (TA) [12]). *A TA is a tuple* $(L, l_0, \Sigma, C, I, E)$, *where* L *is a finite set of* locations, $l_0 \in L$ *is an* initial location, Σ *is a finite set of* actions, C *is a finite set of* clocks *such that* $C \cap \Sigma = \emptyset$, $I : L \to \mathcal{B}(C)$ *is a function assigning* invariants *to locations, and* $E \subseteq L \times \mathcal{B}(C) \times \Sigma \times 2^C \times L$ *is a relation containing* switches. *We write* $l_1 \xrightarrow{g,\sigma,R} l_2$ *for* $(l_1, g, \sigma, R, l_2) \in E$.

To simplify proofs, we assume TA to be *diagonal-free* (i.e., no atomic constraint in the TA has the form $c - c' \lesseqgtr n$). For any non-diagonal-free TA, a language-equivalent diagonal-free TA can be constructed [7]. We reconsider so-called difference constraints when we define zone graphs later on.

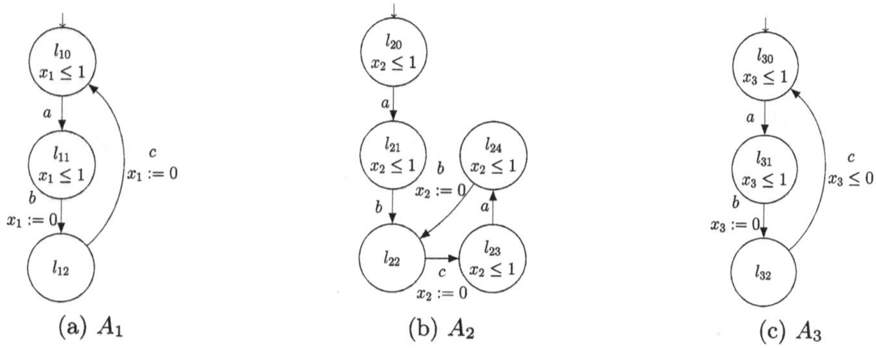

Fig. 1. Examples of Timed Automata

Example 1. Figure 1 shows three TA, each with one clock (x_1, x_2, or x_3), labeled over the same alphabet $\Sigma = \{a, b, c\}$. The initial location of A_1 has invariant $x_1 \leq 1$ and an outgoing switch labeled a to l_{11}, which also has invariant $x_1 \leq 1$. l_{11} has a single outgoing switch labeled b to l_{12}, which resets x_1. Finally, location l_{12} has an outgoing switch to the initial location, labeled c, which also resets x_1. A_2 is similar to A_1 in its first two locations, except that the switch labeled b does not reset x_2. From the third location, the switch labeled c leads to a new location l_{23}, which resembles the initial location with invariant $x_2 \leq 1$ and an outgoing switch labeled a. Finally, l_{24} has an outgoing switch labeled b to l_{22}, which resets x_2. A_3 differs from A_1 primarily in the naming of locations and clocks, as well as in the edge labeled c, which involves a guard instead of a reset.

Clock constraints are evaluated by *clock valuations* $u : C \to \mathbb{T}$, which assign a value $u(c) \in \mathbb{T}$ to each clock $c \in C$. By $u + d$, we denote the clock valuation that assigns to each clock c the value $u(c) + d$. For a set $R \subseteq C$, the clock valuation

$u' = [R \to 0]u$ assigns to each clock $c \in R$ the value $u'(c) = 0$ and for each clock $c \in C \setminus R$ the value $u'(c) = u(c)$. We denote the clock valuation u with $\forall c \in C : u(c) = 0$ with $[C \to 0]$. To check whether a clock valuation u satisfies a $\phi \in \mathcal{B}(C)$, denoted $u \models \phi$, we replace all occurrences of any clock $c \in C$ in ϕ with its value $u(c)$ and check whether the result is true. We next define the operational semantics of TA as a *Timed Labeled Transition Systems (TLTS)*.

Definition 2 (TLTS). *Let A be a TA. The* TLTS A_{TLTS} *of A is a tuple $(V_{TS}, (l_0, [C \to 0]), \Sigma \cup \mathbb{T}, \hookrightarrow)$, where $V_{TS} = L \times (C \to \mathbb{T})$ is a set of states, $(l_0, [C \to 0]) \in V_{TS}$ is the initial state, $\Sigma \cup \mathbb{T}$ is a set of transition labels, where $\Sigma \cap \mathbb{T} = \emptyset$, and $\hookrightarrow \subseteq V_{TS} \times \Sigma \cup \mathbb{T} \times V_{TS}$ (transitions) is the least relation s.t.*

- $(l, u) \overset{d}{\hookrightarrow} (l, u + d)$ *if* $(u + d) \models I(l)$ *for* $d \in \mathbb{T}$, *and*
- $(l_1, u_1) \overset{\sigma}{\hookrightarrow} (l_2, u_2)$ *if* $l_1 \xrightarrow{g,\sigma,R} l_2$ *with* $u_1 \models g$, $u_2 = [R \to 0]u_1$, $u_2 \models I(l_2)$, *and* $\sigma \in \Sigma$.

To compare two TA, we define *timed bisimulation* on their TLTS [8,17].

Definition 3 (Timed Bisimulation). *Let A and B be TA over Σ. We denote $A_{TLTS} = (V_{TS,A}, (l_{A,0}, u_{A,0}), \Sigma \cup \mathbb{T}, \hookrightarrow_A)$ and $B_{TLTS} = (V_{TS,B}, (l_{B,0}, u_{B,0}), \Sigma \cup \mathbb{T}, \hookrightarrow_B)$, respectively. A Relation $R \subseteq V_{TS,A} \times V_{TS,B}$ is a timed bisimulation, iff for all $((l_{A,1}, u_{A,1}), (l_{B,1}, u_{B,1})) \in R$ it holds that*

1. *if* $(l_{A,1}, u_{A,1}) \overset{\mu}{\hookrightarrow}_A (l_{A,2}, u_{A,2})$ *for* $\mu \in \Sigma \cup \mathbb{T}$, *then there exists a* $(l_{B,1}, u_{B,1}) \overset{\mu}{\hookrightarrow}_B (l_{B,2}, u_{B,2})$ *with* $((l_{A,2}, u_{A,2}), (l_{B,2}, u_{B,2})) \in R$, *and*
2. *if* $(l_{B,1}, u_{B,1}) \overset{\mu}{\hookrightarrow}_B (l_{B,2}, u_{B,2})$ *for* $\mu \in \Sigma \cup \mathbb{T}$, *then there exists a* $(l_{A,1}, u_{A,1}) \overset{\mu}{\hookrightarrow}_A (l_{A,2}, u_{A,2})$ *with* $((l_{A,2}, u_{A,2}), (l_{B,2}, u_{B,2})) \in R$.

A is timed bisimilar to B, denoted $A \sim B$, iff there exists a timed bisimulation R with $((l_{A,0}, u_{A,0}), (l_{B,0}, u_{B,0})) \in R$.

Example 2. We compare the TA from Example 1 by considering the excerpt of their respective TLTS, shown in Fig. 2. In $A_{\mathrm{TLTS},1}$, the initial state $(l_{10}, x_1 = 0)$ can delay up to one time unit (due to the invariant of l_{10}) or take the transition labeled a, followed by a delay or b. At location l_{12}, infinite delays are possible, or a transition labeled c can be used to reach the initial state again. Analogously, $A_{\mathrm{TLTS},2}$ has an equivalent beginning, except that there is no reset at the transition labeled b and the state $(l_{21}, x_2 = 1)$ has an outgoing transition to $(l_{22}, x_2 = 1)$. The part concerning l_{23} and l_{24} of $A_{\mathrm{TLTS},2}$ is analogous to $A_{\mathrm{TLTS},1}$. $A_{\mathrm{TLTS},3}$ is equivalent to $A_{\mathrm{TLTS},1}$, but has only one state with an outgoing transition labeled c. To show $A_1 \sim A_2$, we construct a timed bisimulation $R_{1,2}$ with $((l_{10}, x_1 = 0), (l_{20}, x_2 = 0)) \in R_{1,2}$. By Definition 3, this implies $((l_{10}, x_1 = 1), (l_{20}, x_2 = 1)) \in R_{1,2}$, $((l_{11}, x_1 = 0), (l_{21}, x_2 = 0)) \in R_{1,2}$, and $((l_{11}, x_1 = 1), (l_{21}, x_2 = 1)) \in R_{1,2}$. Iteratively applying this procedure

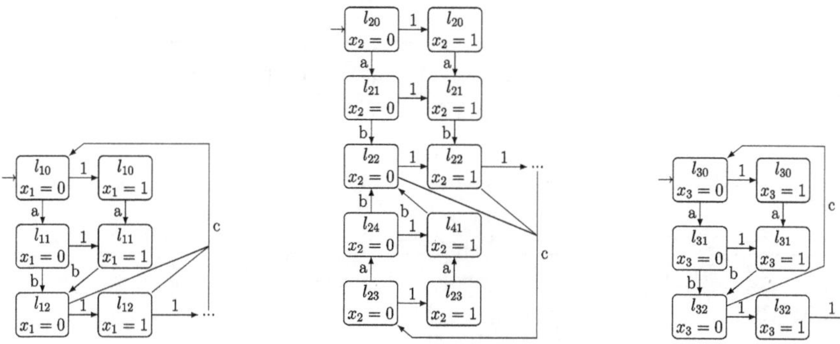

(a) $A_{\text{TLTS},1}$ (b) $A_{\text{TLTS},2}$ (c) $A_{\text{TLTS},3}$

Fig. 2. Extracts of the TLTS of the TA of Fig. 1

confirms that $R_{1,2}$ fulfills the conditions of Definition 3. In contrast, assuming a timed bisimulation $R_{1,3}$ with $((l_{10}, x_1 = 0), (l_{30}, x_3 = 0)) \in R_{1,3}$ implies $((l_{12}, x_1 = 1), (l_{32}, x_3 = 1)) \in R_{1,3}$, which contradicts the assumption that $R_{1,3}$ is a timed bisimulation since $(l_{12}, x_1 = 1)$ has a transition labeled c, which is not the case for $(l_{32}, x_3 = 1)$. Hence, no timed bisimulation that contains $((l_{10}, x_1 = 0), (l_{30}, x_3 = 0))$ exists and $A_1 \not\sim A_3$. The reader might recognizes that the actual TLTS are infinitely large (even uncountable in dense time model).

Since TLTS are infinitely large they are not tractable by analysis tools. This problem is solved by symbolic TA semantics. A *symbolic state* consists of a location and a *zone*. A zone comprises a set of clock valuations, where we denote $2^{C \to \mathbb{T}} = \mathcal{D}(C)$. $(l_A, u_A) \in (l_B, D_B)$ holds if and only if $l_A = l_B$ and $u_A \in D_B$. A zone $D \in \mathcal{D}(C)$ is typically represented by a clock constraint $\phi \in \mathcal{B}(C)$, such that $D = \{u \mid u \in (C \to \mathbb{T}) \land u \models \phi\}$. Assume a zone $D \in \mathcal{D}(C)$, a clock constraint $\phi_{cc} \in \mathcal{B}(C)$, and any $R \subseteq C$. We use the operations $D \land \phi_{cc} = \{u \mid u \in D \land u \models \phi_{cc}\}$ ($\land$-operation), $D^{\uparrow} = \{u + d \mid d \in \mathbb{T} \land u \in D\}$ (future-operation), and $R(D) = \{[R \to 0]u \mid u \in D\}$ (reset-operation). The following definition is a slightly adapted definition of the zone graphs of Bengtsson and Yi [6]. A transition labeled $\varepsilon \notin \Sigma$ is used to represent delay transitions [27].

Definition 4 (Zone Graph). *Let A be a TA. A* zone graph $A_{ZG} = (V_{ZG}, (l_0, D_0), \Sigma \cup \{\varepsilon\}, \hookrightarrow\!\!\!\!\!\rightarrow)$ *of A is a transition system, where $V_{ZG} = L \times \mathcal{D}(C)$ is the set of symbolic states, $(l_0, \{[C \to 0]\}) \in V_{ZG}$ is the initial symbolic state, and $\hookrightarrow\!\!\!\!\!\rightarrow \subseteq V_{ZG} \times \Sigma \cup \{\varepsilon\} \times V_{ZG}$ is the least relation satisfying the rules:*

- *$(l, D) \overset{\varepsilon}{\hookrightarrow\!\!\!\!\!\rightarrow} (l, D^{\uparrow} \land I(l))$, and*
- *$(l_1, D_1) \overset{\sigma}{\hookrightarrow\!\!\!\!\!\rightarrow} (l_2, R(D_1 \land g) \land I(l_2))$, if $l_1 \xrightarrow{g,\sigma,R} l_2$ and $R(D_1 \land g) \land I(l_2) \neq \emptyset$.*

While zone graphs can be used for checking reachability of TA states, they do not preserve enough information of the TA to check for timed bisimulation [25].

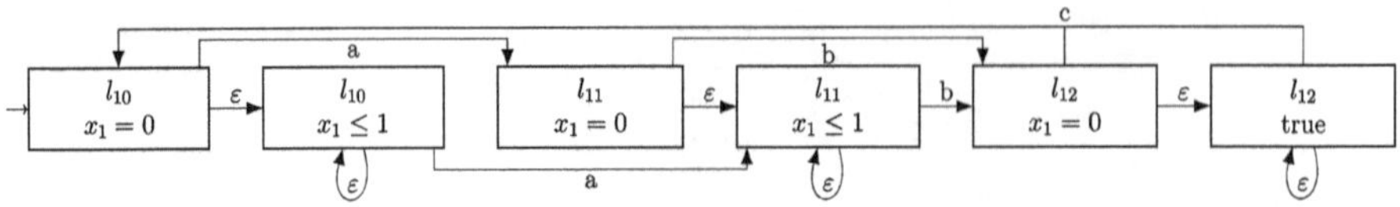

Fig. 3. Zone Graph of A_1.

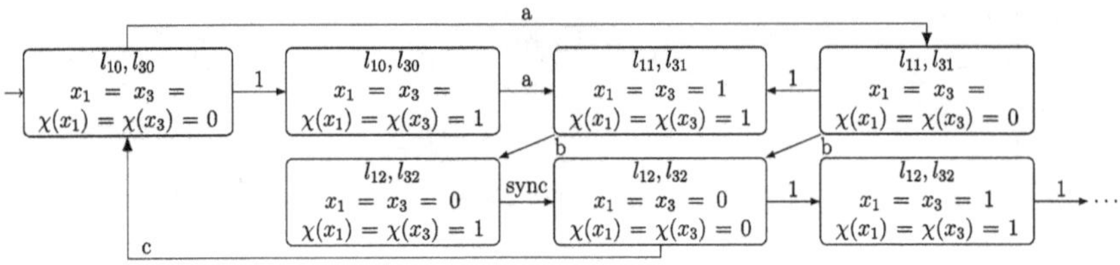

Fig. 4. PTVC of A_1 and A_3.

Example 3. Applying Definition 4 to A_1 yields the zone graph in Fig. 3. A_3 has the exact same zone graph up-to renaming of locations and the clock although A_1 and A_3 not bisimilar.

Zone graphs lack two important properties for bisimulation checking: (1) prone to false positives due to subtle effects of clock resets and (2) they do not contain sufficient information to explain the outcome. Issue (1) is addressed in [16] by adding *virtual clocks* into zone graphs, whereas (2) is addressed in the remainder.

3 Finding Witnesses

We describe how to represent in a finite way a timed bisimulation between two TA. For this, we first describe a set-based construction of bisimulation witnesses in a single unified graph based on *synchronous product of TLTS* [19]. We extend the product by a set of *virtual clocks*, called PTVC. Let C_{AB} be the disjoint union of the set clocks A, B of two TA. Each original clock $c \in C_{AB}$ is paired with exactly one virtual clock $\chi(c) \in \chi(C_{AB})$ assuming $\chi(C_{AB}) \cap C_{AB} = \emptyset$, which initially has the same interpretation as its original clock. We decompose action transitions into two steps. First, only the original clocks, which are reset at action transitions of the TA, are also reset in the PTVC. By leaving the virtual clocks unaffected, they preserve the missing information required to correctly distinguish non-bisimilar states (see Example 3). Second, virtual clocks are reset before the next transition takes place to synchronize with their original clocks.

Example 4. Figure 4 shows the PTVC of A_1 and A_3 (only delay transitions for $d = 1$). In the initial state, all clocks are set to zero. As both TA have enabled a-transitions, the initial state has an a-transition to the state with (l_{11}, l_{31}). A 1-delay is also possible in both TA, resulting in a 1-labeled transition after which both TA, again, have an a-transition. This a-transition leads to a state in which only a b-transition is enabled. In the next state, the original clocks are set to zero due to the clock resets of the switches while the virtual clock values remain unchanged. Next, a sync-transition is added which targets the state in which the values of the virtual clocks are also set to zero. After another delay, we reach the state with location pair (l_{12}, l_{32}) and all clocks are set to 1. Here, A_1 enables a c-transition, but A_3 does not, so the state of the product has no c-transition.

Definition 5 (PTVC). *The PTVC of two TA A and B is a labeled transition system* $(V_{PTVC}, s_{AB,0}, \Sigma \cup \mathbb{T} \cup \{sync\}, \rightsquigarrow)$. $V_{PTVC} \subseteq (L_A \times L_B) \times (C_{AB} \cup \chi(C_{AB}) \to \mathbb{T})$ *is the set of states* $((l_A, l_B), u)$ *with* $\forall c \in C_{AB} : (u(c) = u(\chi(c)) \vee u(c) = 0)$. *We denote* $V_{PTVC,s} \subseteq V_{PTVC}$ *for the set of all states* $((l_A, l_B), u_s) \in V_{PTVC}$ *with* $\forall c \in C_{AB} : u(c) = u(\chi(c))$. *Moreover,* $s_{AB,0} = ((l_{A,0}, l_{B,0}), [C_{AB} \cup \chi(C_{AB}) \to 0]) \in V_{PTVC,s}$ *is the initial state,* $\Sigma \cup \mathbb{T} \cup \{sync\}$ *is the set of transition labels, and* $\rightsquigarrow \subseteq V_{PTVC} \times \Sigma \cup \mathbb{T} \cup \{sync\} \times V_{PTVC}$ *is a set of transitions, which is the least relation satisfying*

- $((l_A, l_B), u) \overset{sync}{\rightsquigarrow} ((l_A, l_B), u_{sync})$, *if* $((l_A, l_B), u) \notin V_{PTVC,s}$, $\forall c \in C_{AB} : u(c) = u_{sync}(c)$, *and* $((l_A, l_B), u_{sync}) \in V_{PTVC,s}$,
- $((l_A, l_B), u) \overset{d}{\rightsquigarrow} ((l_A, l_B), u + d)$ *if* $((l_A, l_B), u) \in V_{PTVC,s}$, $(u + d) \models I(l_A)$, *and* $(u + d) \models I(l_B)$ *for* $d \in \mathbb{T}$, *and*
- $((l_A, l_B), u) \overset{\sigma}{\rightsquigarrow} ((l_{A,1}, l_{B,1}), u_1)$ *if* $((l_A, l_B), u) \in V_{PTVC,s}$, $l_A \xrightarrow{g_A, \sigma, R_A} l_{A,1}$ *with* $u \models g_A$, $l_B \xrightarrow{g_B, \sigma, R_B} l_{B,1}$ *with* $u \models g_B$, $u_1 = [R_A \cup R_B \to 0]u$, $u_1 \models I(l_{A,1})$, $u_1 \models I(l_{B,1})$, *and* $\sigma \in \Sigma$.

We call a state $((l_A, l_B), u) \in V_{\text{PTVC},s}$ *synchronized.* The target of a delay transition of a synchronized state is also synchronized, as delay transitions cannot perform clock resets, while the target of an action transition might not be synchronized due to resets. Based on PTVC, we define *virtual bisimulation.*

Definition 6 (Virtual Bisimulation). *Given two TA A, B, with their PTVC* $(V_{PTVC}, s_{AB,0}, \Sigma \cup \mathbb{T} \cup \{sync\}, \rightsquigarrow)$. *A relation* $R_{virt} \subseteq V_{PTVC,s}$ *is called a virtual bisimulation, if and only if for all elements* $((l_A, l_B), u) \in R_{virt}$ *it holds that*

1. *for any* $d \in \mathbb{T}$ *with* $u + d \models I(l_A)$, *there exists a transition* $((l_A, l_B), u) \overset{d}{\rightsquigarrow} ((l_A, l_B), u + d)$ *and* $((l_A, l_B), u + d) \in R_{virt}$,
2. *if there exists a transition* $l_A \xrightarrow{g_A, \sigma, R_A} l_{A,1}$ *with* $u \models g_A$ *and* $[R_A \to 0]u \models I(l_{A,1})$, *then there exists a transition* $((l_A, l_B), u) \overset{\sigma}{\rightsquigarrow} ((l_{A,1}, l_{B,1}), u_1)$ *with* $\forall c \in C_A : [R_A \to 0]u(c) = u_1(c)$ *and either*
   (a) $((l_{A,1}, l_{B,1}), u_1)$ *is synchronized and* $((l_{A,1}, l_{B,1}), u_1) \in R_{virt}$ *holds, or*
   (b) $((l_{A,1}, l_{B,1}), u_1)$ *is not synchronized and there exists a transition* $((l_{A,1}, l_{B,1}), u_1) \overset{sync}{\rightsquigarrow} ((l_{A,1}, l_{B,1}), u_{1,s})$ *with* $((l_{A,1}, l_{B,1}), u_{1,s}) \in R_{virt}$, *and*
3. *the analog for B.*

TA A and B are *virtually bisimilar* if and only if there exists a virtual bisimulation containing the initial state of their PTVC. Moreover, two TA are virtually bisimilar if and only if they are timed bisimilar according to Definition 3, where the mapping from a virtual bisimulation to a timed bisimulation is straightforward. As a virtual bisimulation may, again, contain infinitely many state pairs, we use a *symbolic* representation. Given a set of symbolic states S, we denote $\text{states}(S) = \{(l, u) \mid \exists (l, D) \in S : u \in D\}$. We next describe how to find a set S of symbolic states such that states(S) forms a virtual bisimulation. To do so, we construct the product of zone graphs with virtual clocks (PZVC).

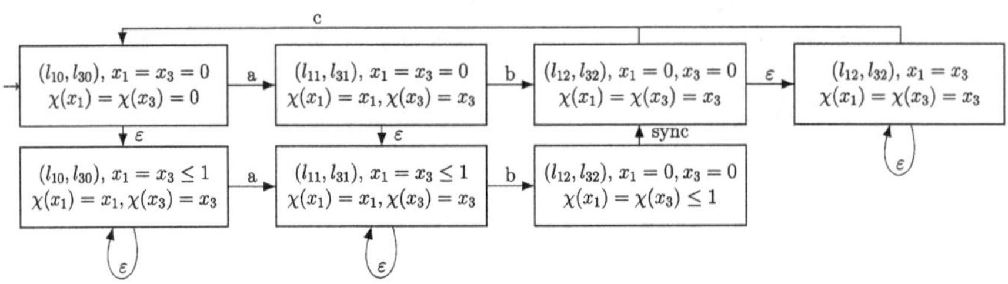

Fig. 5. PZVC of A_1 and A_3.

Example 5. Figure 5 shows the PZVC of A_1 and A_3. In the initial state, all clocks are set to zero. An ε-transition allows time to elapse as long as x_1 and x_3 are smaller or equal 1. Since the next ε-transition is a self-loop, we take the a-transition which is either followed by an ε-transition (self-loop), or the b-transition. As both TA reset their clocks in this step, x_1 and x_3 are also reset in the PZVC, while the virtual clocks remain unchanged. Analogously to Example 4, we synchronize the symbolic state as the values of original and their virtual clocks may differ. After another ε-transition, we reach $((l_{12}, l_{32}), x_1 = x_3 = \chi(x_1) = \chi(x_3))$. Since l_{12} and l_{32} have outgoing c-transitions, this symbolic state also has one. However, since the transition of l_{32} has a guard, we must comply to it such that the target only contains the initial state again. If A_3 would contain a reset instead of a guard (i.e., A_3 would be timed bisimilar to A_1), the zone of the target would be $x_1 = x_3 = 0 \wedge \chi(x_1) = \chi(x_3) \geq 0$ instead of the virtual clock values being equal to 0.

Definition 7 (PZVC). *A PZVC of two TA A and B is a labeled transition system $(V_{PZVC}, s_{AB,0}, \Sigma \cup \{\varepsilon\} \cup \{sync\}, \rightsquigarrow)$. $V_{PZVC} \subseteq (L_A \times L_B) \times \mathcal{D}(C_{AB} \cup \chi(C_{AB}))$ is the set of all symbolic states $((l_A, l_B), D)$ with $\forall c \in C_{AB} : (\forall u \in D : u(c) = u(\chi(c))) \vee (\forall u \in D : u(c) = 0)$. We denote $V_{PZVC,s} \subseteq V_{PZVC}$ for the set of all symbolic states $((l_A, l_B), D) \in V_{PZVC}$ with $\forall c \in C_{AB} : \forall u \in D : u(c) = u(\chi(c))$. Moreover, $s_{AB,0} = ((l_{A,0}, l_{B,0}), [C_{AB} \cup \chi(C_{AB}) \rightarrow 0]) \in V_{PZVC,s}$ is the initial symbolic state, $\Sigma \cup \{\varepsilon\} \cup \{sync\}$ is the set of transition labels, and $\rightsquigarrow \subseteq V_{PZVC} \times \Sigma \cup \{\varepsilon\} \cup \{sync\} \times V_{PZVC}$ is the least transition relation satisfying*

- *$((l_A, l_B), D) \overset{sync}{\rightsquigarrow} ((l_A, l_B), D_{sync})$, if $((l_A, l_B), D) \notin V_{PZVC,s}$, $\forall u \in D : \exists u_{sync} \in D_{sync} : \forall c \in C_{AB} : u(c) = u_{sync}(c)$, $\forall u_{sync} \in D_{sync} : \exists u \in D : \forall c \in C_{AB} : u(c) = u_{sync}(c)$, and $((l_A, l_B), D_{sync}) \in V_{PZVC,s}$,*

- $((l_A, l_B), D) \overset{\varepsilon}{\rightsquigarrow} ((l_A, l_B), D^{\uparrow} \wedge I(l_A) \wedge I(l_B))$, *if* $((l_A, l_B), D) \in V_{PZVC,s}$, *and*
- $((l_{A,1}, l_{B,1}), D) \overset{\sigma}{\rightsquigarrow} ((l_{A,2}, l_{B,2}), R_A(R_B(D \wedge g_A \wedge g_B)) \wedge I(l_{A,2}) \wedge I(l_{B,2}))$, *if* $((l_{A,1}, l_{B,1}), D) \in V_{PZVC,s}$, $l_{A,1} \xrightarrow{g_A, \sigma, R_A} l_{A,2}$, $l_{B,1} \xrightarrow{g_B, \sigma, R_B} l_{B,2}$, *and* $R_A(R_B(D \wedge g_A \wedge g_B)) \wedge I(l_{A,2}) \wedge I(l_{B,2}) \neq \emptyset$.

Again, we call a symbolic state $(l, D) \in V_{\mathrm{PZVC},s}$ *synchronized.* We next define symbolic virtual bisimulation over PZVC.

Definition 8 (Symbolic Virtual Bisimulation). *A relation* $R_{symb,virt} \subseteq V_{PZVC,s}$ *is a symbolic virtual bisimulation, if and only if states*$(R_{symb,virt})$ *is a virtual bisimulation according to Definition 6.*

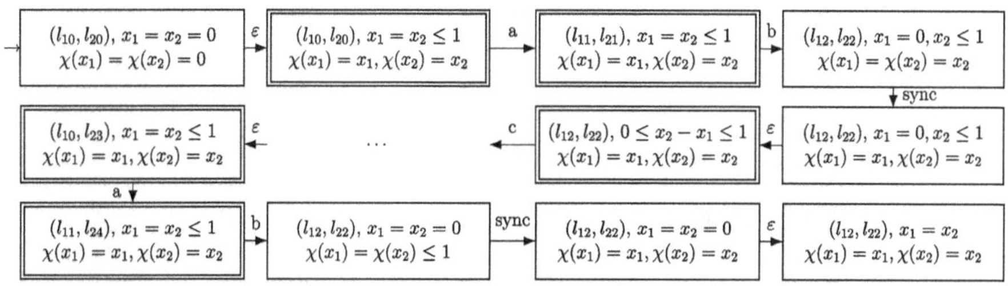

Fig. 6. Finding a Witness for $A_1 \sim A_2$.

Example 6. Figure 6 shows how to create a witness for the bisimulation of A_1 and A_2 (Example 1). We start at the initial symbolic state of the PZVC and add the target of the ε-transition to the set (doubled border). The initial symbolic state is a subset of the target and, therefore, we remove the initial symbolic state and the set now only contains the second symbolic state. This replacement is always feasible as the source of an epsilon transition is always a subset of its target. Since the outgoing ε-transition of the second symbolic state is a self-loop, we ignore it. The target of the outgoing a-transition is added to the set. Since the second symbolic state does not have any other outgoing transitions, we now consider the newly added symbolic state. The ε-transition is a self-loop and, therefore, we consider the outgoing action transition. Since the target of the b-transition is not synchronized (as there are clock valuations in the zone for which x_1 and $\chi(x_1)$ are not equal), we add the symbolic state after the sync-transition, which is replaced by the target of the outgoing ε-transition. We continue until all outgoing transitions of all contained symbolic state have been processed. Finally, the symbolic virtual bisimulation contains those symbolic states marked with doubled border. The last symbolic state is a subset of the symbolic state after the second ε-transition and is, therefore, not added to the set.

In case of non-deterministic choices, this construction becomes more complicated. We call clock constraints over virtual clocks *virtual constraints* and use *contradictions*, which are sets of virtual constraints [16]. The checking if a symbolic state can become part of a symbolic virtual bisimulation returns a contradiction. If the symbolic state can be added, the contradiction is empty. Otherwise, the contradiction consists of virtual constraints such that each constraint can be satisfied by at least one state within the symbolic state. Any state that satisfies a virtual constraint of the contradiction cannot be part of a virtual bisimulation and must be excluded.

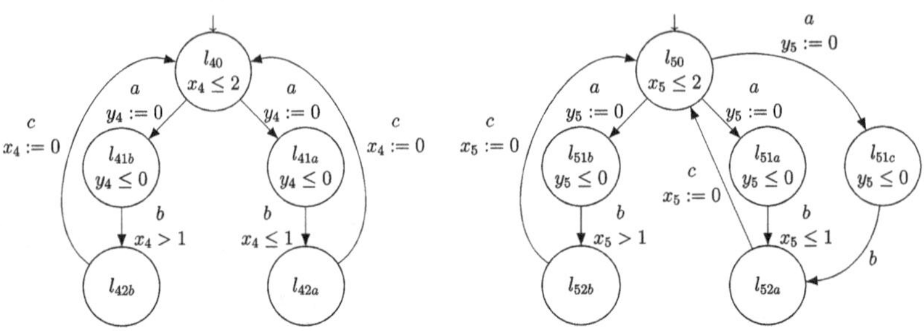

(a) $A_{\text{non},1}$ (b) $A_{\text{non},2}$

Fig. 7. Two timed bisimilar TA with Non-Deterministic Choices

Example 7. Figure 7 shows two bisimilar TA with non-deterministic choices, where $A_{\text{non},2}$ has one additional branch. If action a is used while $x_5 \leq 1$ holds, the additional branch is timed bisimilar to the branch on the right hand side of $A_{\text{non},1}$. Therefore, we can combine these two branches into one symbolic state which leads to the contradiction $\{\chi(x_4) = \chi(y_4) = \chi(x_5) = \chi(y_5) > 1\}$. For $x_5 > 1$, the additional branch is timed bisimilar to the branch on the left hand side such that the contradiction $\{\chi(x_4) = \chi(y_4) = \chi(x_5) = \chi(y_5) \leq 1\}$ holds.

To obtain finite witnesses, we utilize k-normalization [21,22], where k is the largest constant in any clock constraint of the TA thus serving as upper bound for all zones. The image of normalization is finite and there exists a virtual bisimulation covering all states of a symbolic state if and only if one exists for its k-normalized counterpart [8,16]. Since the initial symbolic state of the PZVC contains the initial state of the PTVC, any symbolic virtual bisimulation that includes this symbolic state also yields a virtual bisimulation containing the initial state of the PTVC. To construct the bisimulation, we rely on forward- and backward-stability. Forward-stability guarantees that any transition from a state $((l_{A,1}, l_{B,1}), u_1) \in ((l_{A,1}, l_{B,1}), D_1)$ to a state $((l_{A,2}, l_{B,2}), u_2)$, whether due to action, delay, or synchronization, has a corresponding symbolic transition from $((l_{A,1}, l_{B,1}), D_1)$ to $((l_{A,2}, l_{B,2}), D_2)$ such that $((l_{A,2}, l_{B,2}), u_2) \in ((l_{A,2}, l_{B,2}), D_2)$. Backward-stability guarantees the reverse: If

$((l_{A,2}, l_{B,2}), D_2)$ has an incoming symbolic transition from $((l_{A,1}, l_{B,1}), D_1)$, then every $((l_{A,2}, l_{B,2}), u_2) \in ((l_{A,2}, l_{B,2}), D_2)$ is reached by a transition from some $((l_{A,1}, l_{B,1}), u_1) \in ((l_{A,1}, l_{B,1}), D_1)$.

Proposition 1. *Given two timed bisimilar TA A and B, a finite symbolic virtual bisimulation $R_{symb,virt}$ is constructed starting from the initial symbolic state by iteratively applying the following steps:*

- *For any $((l_A, l_B), D) \in R_{symb,virt}$ with outgoing ε-transition $((l_A, l_B), D) \overset{\varepsilon}{\rightsquigarrow} ((l_A, l_B), D_\varepsilon)$, add the k-normalized $((l_A, l_B), D_\varepsilon)$ to $R_{symb,virt}$ and*
- *for any $((l_A, l_B), D) \in R_{symb,virt}$, if there exists a transition $l_A \xrightarrow{g_A,\sigma,R_A} l_{A,\sigma}$ with $D_{\sigma,A} = R_A(D \wedge g_A) \wedge I(l_A) \neq \emptyset$, then there exists a finite set of symbolic states $\{((l_{A,\sigma}, l_{B,1}), D_{inter,1}), ..., ((l_{A,\sigma}, l_{B,n}), D_{inter,n})\}$ such that*
  1. *for any $((l_{A,\sigma}, l_{B,i}), D_{inter,i})$ there exists a symbolic state $((l_{A,\sigma}, l_{B,i}), D_{\sigma,i})$ with $D_{inter,i} \subseteq D_{\sigma,i}$ such that there exists a transition $((l_A, l_B), D) \overset{\sigma}{\rightsquigarrow} ((l_{A,\sigma}, l_{B,i}), D_{\sigma,i})$,*
  2. *if $((l_{A,\sigma}, l_{B,i}), D_{inter,i})$ is synchronized, it can be element of a symbolic virtual bisimulation,*
  3. *if $((l_{A,\sigma}, l_{B,i}), D_{inter,i})$ is not synchronized, there exists a transition $((l_{A,\sigma}, l_{B,i}), D_{inter,i}) \overset{sync}{\rightsquigarrow} ((l_{A,\sigma}, l_{B,i}), D_{sync,i})$ and $((l_{A,\sigma}, l_{B,i}), D_{sync,i})$ can be element of a symbolic virtual bisimulation, and*
  4. *for any $u_{\sigma,A} \in D_{\sigma,A}$ there exists a $((l_{A,\sigma}, l_{B,i}), D_{inter,i})$ with a $u_{inter,i} \models \phi_{virt}(D_{inter,i})$ with $\forall c \in C_A \cup \chi(C_{AB}) : u_{\sigma,A}(c) = u_{inter,i}(c)$ and vice versa.*

  If $((l_{A,\sigma}, l_{B,i}), D_{inter,i})$ is synchronized, add its k-normalized version; if it is not synchronized, add its k-normalized version of $((l_{A,\sigma}, l_{B,i}), D_{sync,i})$, and
- *do the analog for B.*

To summarize, for two bisimilar TA, A and B, we first apply the algorithm from [16] and then apply the construction steps from Proposition 1 to obtain a witness as shown in Example 7.

4 Finding Counterexamples

We describe the construction of counterexamples showing that two TA are not timed bisimilar. For deterministic TA, a counterexample represents a single finite path through the PVTC whose final state contradicts Definition 6.

Example 8. To compare A_1 and A_3, we consider the sequence of transitions labeled a, b, 1 in the PTVC. The resulting state, $(l_{12}, l_{32}, x_1 = x_3 = \chi(x_1) = \chi(x_3) = 1)$, is synchronized, and there exists a transition $l_{12} \xrightarrow{true,c,\{x\}} l_{10}$. While the preconditions of the second case in Definition 6 are satisfied, the state lacks an outgoing transition labeled c according to Definition 5. This is due to the fact that there is no outgoing transition from l_{32} labeled c whose guard is fulfilled.

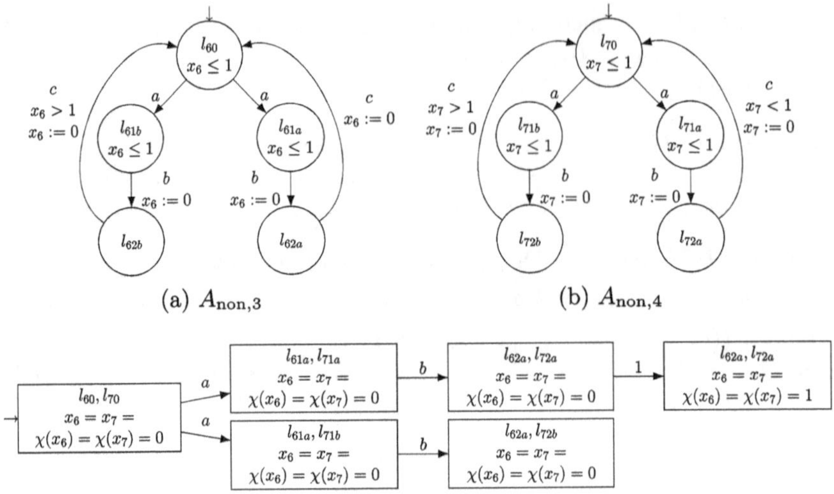

(a) $A_{\text{non},3}$ (b) $A_{\text{non},4}$

(c) Contradiction DAG

Fig. 8. Two TA with Non-Deterministic Choices

However, in case a non-deterministic choice occurs, counterexamples become more complicated in terms of directed acyclic graphs (DAG).

Example 9. The top of Fig. 8 shows two non-deterministic TA. Beside names of clocks and locations, the only difference is the additional guard in one of the transitions labeled c of $A_{\text{non},4}$. To check for timed bisimilarity, we assume a virtual bisimulation R to contain the state $((l_{60}, l_{70}), x_6 = x_7 = \chi(x_6) = \chi(x_7) = 0)$ and consider the transition $l_{60} \xrightarrow{true,a,\emptyset} l_{61a}$. By Definition 6, the existence of this transition implies either $((l_{61a}, l_{71a}), x_6 = x_7 = \chi(x_6) = \chi(x_7) = 0) \in R$ or $((l_{61a}, l_{71b}), x_6 = x_7 = \chi(x_6) = \chi(x_7) = 0) \in R$. Therefore, in case such a non-deterministic choice occurs, we have to consider all possible options, as depicted in Fig. 8c. The first case implies $((l_{62a}, l_{72a}), x_6 = x_7 = \chi(x_6) = \chi(x_7) = 1) \in R$ and the second case implies $((l_{62a}, l_{72b}), x_6 = x_7 = \chi(x_6) = \chi(x_7) = 0) \in R$. In both cases, the part regarding $A_{\text{non},3}$ has an outgoing transition labeled c, while the part regarding $A_{\text{non},4}$ does not. Therefore, in both cases, R is not a virtual bisimulation and $A_{\text{non},3}$ and $A_{\text{non},4}$ are not timed bisimilar.

The following definition generalizes the concept from Examples 8 and 9.

Definition 9 (Contradiction DAGs). *Let A, B be two TA, with corresponding PTVC $(V_{PTVC}, s_{AB,0}, \Sigma \cup \mathbb{T} \cup \{sync\}, \rightsquigarrow)$. A Contradiction DAG (S, E) of A and B is a finite directed acyclic graph where $S \subseteq V_{PTVC}$ is the set of states and $E \subseteq (S \times \Sigma \cup \mathbb{T} \cup \{sync\} \times S)$ is the set of edges with $E \subseteq \rightsquigarrow$. For any $s_1 \in S$, either $s_1 = ((l_{A,1}, l_{B,1}), u_1)$ is a leaf, in which case it is synchronized and*

- *either there exists a $d \in \mathbb{T}$ such that $u_1 + d \models I(l_{A,1})$ but no transition $s_1 \overset{d}{\rightsquigarrow} s_2$, or*

- *there exists a transition* $l_{A,1} \xrightarrow{g_A,\sigma,R_A} l_{A,2}$ *with* $u_1 \models g_A$ *and* $[R_A \to 0]u_1 \models I(l_{A,2})$ *but no transition* $s_1 \overset{\sigma}{\rightsquigarrow} ((l_{A,2}, l_{B,2}), u_2)$ *with* $\forall c \in C_A : [R_A \to 0]u_1(c) = u_2(c)$*, or*
- *an analog condition for B*

holds, or it is not a leaf, in which case the sync-transition is element of E in case s_1 *is not synchronized, or there exists exactly one outgoing transition* $(s_1, d, s_2) \in E$ *with* $d \in \mathbb{T}$*, or there exists a transition* $l_{A,1} \xrightarrow{g_A,\sigma,R_A} l_{A,2}$ *s.t.*

- *for all edges* $(s_1, \sigma_1, s_2) \in E$ *the statement* $\sigma_1 = \sigma$ *holds, and*
- *for any outgoing transition* $l_{B,1} \xrightarrow{g_B,\sigma,R_B} l_{B,2}$*, it holds that if there exists a transition* $(s_1, \sigma, ((l_{A,2}, l_{B,2}), u_2)) \in \rightsquigarrow$ *with* $u_2 = R_A(R_B(u_1))$*, then* $(s_1, \sigma, ((l_{A,2}, l_{B,2}), u_2)) \in E$ *holds,*

or the symmetric conditions hold, with the roles of A and B reversed.

Proposition 2. *Let A, B be two TA. A and B are virtually bisimilar if and only if the initial state of the PTVC cannot be the root of a Contradiction DAG.*

A Contradiction DAG is finite thus no symbolic representation is required and mapping to a Contradiction DAG without virtual clocks is also straightforward. As shown in Example 7, the algorithm from [16] produces contradictions for symbolic states that cannot belong to a symbolic virtual bisimulation. If a delay within a contradiction is possible, we use the highest possible delay that still fulfills the contradiction as it is unclear whether a particular state contributes to a contradiction solely due to a delay transition. Such a state may appear contradictory only because of the presence of that delay transition. Once we reach a state where only delays less than 1 are allowed, we have to use a delay value s.t. any possible constraint of the TA does not change its evaluation if we get closer to the upper bound. In that case, we stay in the same *region*, which is the base element for the decidability proof of Čerāns [8]. Two clock valuations u_1, u_2 are region-equivalent, if and only if

- for any clock c, either $u_1(c) > k \wedge u_2(c) > k$ or the integral part of $u_1(c)$ and $u_2(c)$ are the same and the fractional part of $u_1(c)$ is zero if and only if the fractional part of $u_2(c)$ is zero and
- for any clocks c_1, c_2, if the integral part of $u_1(c_1)$ and $u_1(c_2)$ are lower or equal k, the fractional part of $u_1(c_1)$ is smaller than the fractional part of $u_1(c_2)$ if and only if the fractional part of $u_2(c_1)$ is smaller than the fractional part of $u_2(c_2)$.

Let u be the valuation and $0 < d_{\max} < 1$ be the upper bound, we search for a value $0 \leq d_{\text{step}} < d_{\max}$ s.t. for any delay d between d_{step} and $d_{\max}$, $u + d$ is in the same region as $u + d_{\text{step}}$. If for any $c \in C$, $u(c)$ has a fractional value $\{u(c)\}$ such that $1 - \{u(c)\} < d_{\max}$, we require $1 - \{u(c)\} < d_{\text{step}} < d_{\max}$. Moreover, if for any $c \in C : \{u(c)\} = 0$, we require $0 < d_{\text{step}} < d_{\max}$. In practice, we choose the center of the resulting interval. In case there is no highest possible delay, we choose one such that all clock values are above the highest occurring constant in any of the TA. The following proposition formalizes this approach.

Proposition 3. *Let A and B be two non timed bisimilar TA with maximum used constant k. A Contradiction DAG can be constructed starting from the initial state as the root, by iteratively applying the following steps:*

- *If all leaves of the DAG serve as leaves of a Contradiction DAG, stop.*
- *If any leaf of the DAG is not synchronized, add a sync-transition.*
- *For any synchronized leaf* $((l_A, l_B), u)$ *of the current DAG that cannot be a leaf of a Contradiction DAG, for which a c exists such that* $u(c) \leq k$*, and for which a delay exists such that the target state fulfills the contradiction of its symbolic state and the resulting valuation is not region-equivalent to* u*, proceed as follows. If* $u + k + 1$ *fulfills the contradiction of the symbolic state, add the transition labeled* $k + 1$*, if the maximum delay satisfying the condition is an integer* d*, add the transition labeled* d*, and if the maximum lies strictly between two consecutive integers* d *and* $d + 1$*, add the transition labeled* $d + d_{step}$*.*
- *For any synchronized leaf* $((l_A, l_B), u)$ *of the current DAG that cannot be a leaf of a Contradiction DAG and which does not fulfill the preconditions of the previous case, proceed as follows. There exists either a transition* $l_A \xrightarrow{g_A, \sigma, R_A} l_{A,1}$ *such that* $u \models g_A$*,* $[R_A \to 0]u \models I(l_{A,1})$*, and for any* $l_B \xrightarrow{g_B, \sigma, R_B} l_{B,1}$ *with* $u \models g_B$ *and* $[R_B \to 0]u \models I(l_{B,1})$*, the corresponding transition* $((l_A, l_B), u) \overset{\sigma}{\rightsquigarrow} ((l_{A,1}, l_{B,1}), R_A(R_B(u)))$ *leads to a state whose synchronized clock valuation fulfills the contradiction of its symbolic state and we add all such transitions to the Contradiction DAG, or an analogous transition exists with the roles of A and B swapped.*
- *If a transition introduces a cycle, there exists an alternative that allows the above procedure to continue without forming a cycle. Location-equivalent states with region-equivalent clock valuations are considered equivalent.*

5 Experimental Evaluation

Implementation. Our tool is open-source software[1]. It extends the open-source model-checker TChecker [13]. We also provide a GUI[2].

Research Questions.

(RQ1) What are the sizes of witnesses/counterexamples in comparison to the sizes of the input models?

(RQ2) What is the runtime overhead caused by witness/counterexample generation compared to bare timed bisimulation checking?

Community Models. We use three TA models widely used as community benchmarks for TA analysis techniques: *Root Contention Protocol (RCP)* [10], the

[1] https://github.com/Echtzeitsysteme/tchecker/.

[2] https://github.com/Echtzeitsysteme/tchecker-webapp.

IEEE1394 root contention protocol (ten locations, 26 switches, two clocks); *Audio/Video Components (AVC)* protocol for message transmissions over a shared bus (18 locations, 30 switches, single clock), which includes a non-deterministic choice, and the synchronous product of a variation of the *train-gate model (TG)* [5] with three trains (73 locations, 129 switches, three clocks).

Synthetic Models. For RQ2, we also generated synthetic models. Our generator allows to adjust number of locations, model depth/width, and number of clocks.

Comparison Models. For each community model, we generated three mutants using the approach in [2] such that one mutant is timed bisimilar to the original while the others are not. This ground truth was calculated with our tool from [16] and the witnesses/counterexamples were checked manually using the GUI. Each original model was checked against its one-to-one copy and the mutants twice: once with and once without witness/counterexample generation.

Setup. All runs were performed on an Intel i7-6700K processor with 64 GB of RAM, Linux Mint 21.2 ("Victoria"). All experiments were repeated ten times.

Table 1. Results for the community models

		without		with		Witness/DAG Size
		avg. time [ms]	variance	avg. time [ms]	variance	# nodes
RCP	1-to-1	1.5	0.0	2.0	0.2	10
	Bisim	1.5	0.0	1.8	0.0	10
	Non-Bisim 1	0.4	0.0	1.0	0.3	5
	Non-Bisim 2	0.3	0.0	0.5	0.0	4
AV	1-to-1	0.9	0.0	1.1	0.0	20
	Bisim	0.9	0.0	1.1	0.0	20
	Non-Bisim 1	0.9	0.0	2.2	1.0	21
	Non-Bisim 2	0.3	0.0	0.8	0.1	12
TG	1-to-1	228.4	0.3	376.4	1.0	765
	Bisim	284.8	0.3	461.2	3.6	765
	Non-Bisim 1	233.4	3.1	237.9	0.2	18
	Non-Bisim 2	203096.6	10310.5	203211.1	53037.3	21

Results (Community Models). The results are summarized in Table 1. For the RCP model, the highest average runtime is 2ms with low variances. The bisimilar cases produce witnesses whose size matches the number of TA locations. For the non-bisimilar cases, we have smaller Contradiction DAGs, while runtime is increased by a factor of up to 2.5. For the AVC model, the sizes of the witness are, once again, equivalent to the number of locations, despite having the non-deterministic choice. We can see that witness generation noticeably impacts runtime, especially for non-bisimilar cases. For the TG model, we see for the bisimilar cases that the average runtime is below 0.5 s, still with low variances. However, the size of the witness is ten times the number of locations. In the first non-bisimilar case the average runtime is still quite low, while the Contradiction DAG is small. Since the second non-bisimilar case hits an edge case of the original algorithm, the average runtime increases to 3 min and 23 s.

Results (Synthetic Models). We used model sizes ranging from three up to 781 locations as well as models with four and five clocks. Figure 9 presents the evaluation results. The left-hand side shows the results for models with four clocks, while the right-hand side corresponds to models with five clocks. The first row depicts the evaluation duration with witness or DAG generation enabled, the second row shows the same evaluation without generation, and the last row illustrates the resulting runtime overhead in percent. Each graph includes results for models with depths of three and five, with the x-axis representing different values of the width parameter. There is always one graph for the experiments in which the corresponding model where checked against its 1-to-1 copy (*timed bisimilar, tb*), and one graph in which the original model is checked against a version where the guard of the very last switch was modified (*not tb*). For depth three, the number of locations increases from three to 31, while for depth five, it grows from five to 781. The corresponding runtimes range from below one millisecond up to approximately seven seconds.

The results reveal that, while the overhead is relatively large for smaller models, it decreases as the model size increases.

Answer to RQ1. The number of locations is a possible indicator for the size of the Witness/Contradiction DAG, where the Contradiction DAG is up to twice as large due to additional delay transitions. However, this indicator looses its reliability when the special structure of synchronous products comes into play.

Answer to RQ2. The additional generation of the Witness/Contradiction DAG causes an acceptable runtime overhead which decreases as model sizes increase.

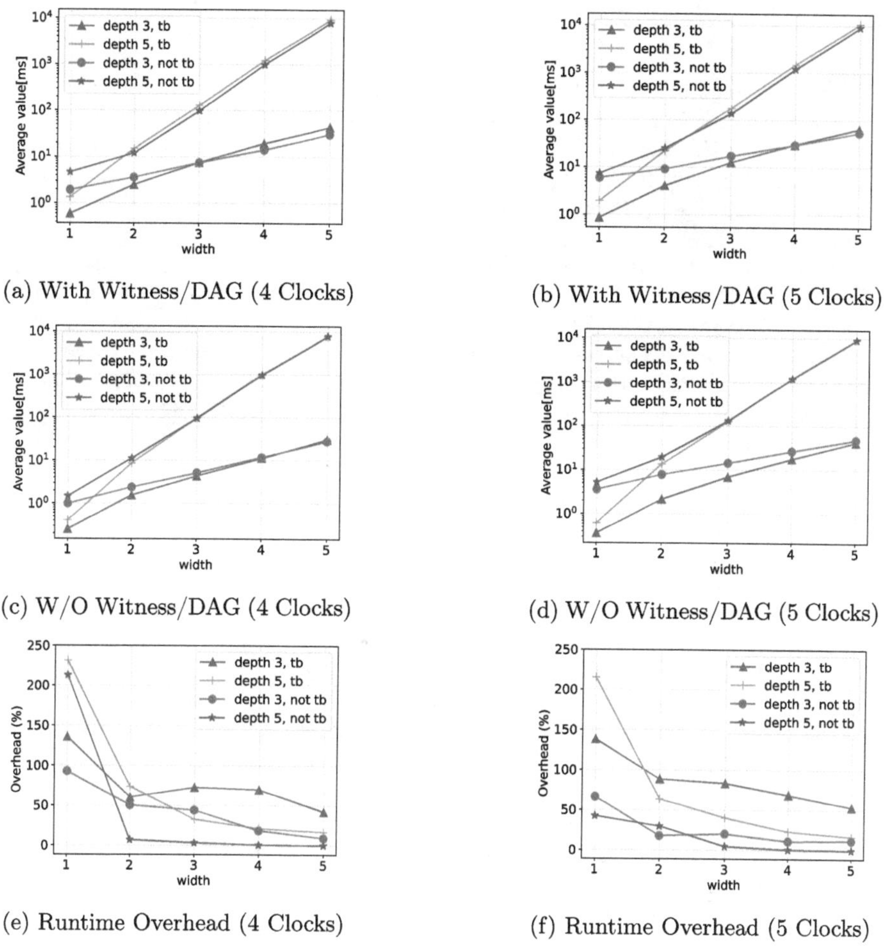

(a) With Witness/DAG (4 Clocks)

(b) With Witness/DAG (5 Clocks)

(c) W/O Witness/DAG (4 Clocks)

(d) W/O Witness/DAG (5 Clocks)

(e) Runtime Overhead (4 Clocks)

(f) Runtime Overhead (5 Clocks)

Fig. 9. Evaluation Results for the Artificial Model

6 Conclusion

We presented a method for generating witnesses to explain that two TA are timed bisimilar, as well as counterexamples to explain why they are not. Our open-source implementation extends the model checker TChecker, and our experimental evaluation shows that the approach is practically applicable and scalable. As future work, we plan a case study to enhance our understanding of developers' needs and to assess the real-world applicability of our approach, for instance in automated test generation using mutation-based testing [2,11].

Data Availability Statement. The implementation and the used models are openly available at https://doi.org/10.5281/zenodo.20702440.

References

1. Aceto, L., Ingólfsdóttir, A., Pedersen, M.L., Poulsen, J.: Characteristic formulae for timed automata. RAIRO-Theor. Inf. Appl. **34**(6), 565–584 (2000)
2. Aichernig, B.K., Hörmaier, K., Lorber, F.: Debugging with timed automata mutations. In: Bondavalli, A., Di Giandomenico, F. (eds.) SAFECOMP 2014. LNCS, vol. 8666, pp. 49–64. Springer, Cham (2014). https://doi.org/10.1007/978-3-319-10506-2_4
3. Alur, R., Dill, D.: Automata for modeling real-time systems. In: Paterson, M.S. (ed.) ICALP 1990. LNCS, vol. 443, pp. 322–335. Springer, Heidelberg (1990). https://doi.org/10.1007/BFb0032042
4. Alur, R., Dill, D.: The theory of timed automata. In: de Bakker, J.W., Huizing, C., de Roever, W.P., Rozenberg, G. (eds.) REX 1991. LNCS, vol. 600, pp. 45–73. Springer, Heidelberg (1992). https://doi.org/10.1007/BFb0031987
5. Alur, R., Dill, D.L.: A theory of timed automata. Theoret. Comput. Sci. **126**(2), 183–235 (1994)
6. Bengtsson, J., Yi, W.: Timed automata: semantics, algorithms and tools. In: Desel, J., Reisig, W., Rozenberg, G. (eds.) ACPN 2003. LNCS, vol. 3098, pp. 87–124. Springer, Heidelberg (2004). https://doi.org/10.1007/978-3-540-27755-2_3
7. Bérard, B., Petit, A., Diekert, V., Gastin, P.: Characterization of the expressive power of silent transitions in timed automata. Fundam. Inf. **36**(2,3), 145–182 (1998)
8. Čerāns, K.: Decidability of bisimulation equivalences for parallel timer processes. In: von Bochmann, G., Probst, D.K. (eds.) CAV 1992. LNCS, vol. 663, pp. 302–315. Springer, Heidelberg (1993). https://doi.org/10.1007/3-540-56496-9_24
9. Clarke, E.M., Grumberg, O., McMillan, K.L., Zhao, X.: Efficient generation of counterexamples and witnesses in symbolic model checking. In: 32nd Design Automation Conference, pp. 427–432 (1995)
10. Collomb-Annichini, A., Sighireanu, M.: Parameterized reachability analysis of the IEEE 1394 root contention protocol using TReX. In: Proceedings of the Real-Time Tools Workshop, pp. 1–20 (2001)
11. Cortés, D., Ortiz, J., Basile, D., Aranda, J., Perrouin, G., Schobbens, P.Y.: Time for networks: mutation testing for timed automata networks. In: Proceedings of the 2024 IEEE/ACM 12th International Conference on Formal Methods in Software Engineering, pp. 44–54. Association for Computing Machinery, New York, NY, USA (2024)
12. Henzinger, T., Nicollin, X., Sifakis, J., Yovine, S.: Symbolic model checking for real-time systems. Inf. Comput. **111**(2), 193–244 (1994)
13. Herbreteau, F.: Tchecker. https://github.com/ticktac-project/tchecker/. Accessed 23 Sept 2025
14. Jensen, H.E., Larsen, K.G., Skou, A.: Modelling and analysis of a collision avoidance protocol using spin and uppaal. In: DIMACS 1996 (1996)
15. Larsen, K.G., Yi, W.: Time abstracted bisimulation: Implicit specifications and decidability. In: Brookes, S., Main, M., Melton, A., Mislove, M., Schmidt, D. (eds.) MFPS 1993. LNCS, vol. 802, pp. 160–176. Springer, Heidelberg (1994). https://doi.org/10.1007/3-540-58027-1_8
16. Lieb, A., Göttmann, H., Luthmann, L., Lochau, M., Schürr, A.: Checking timed bisimilarity with virtual clocks. Fundamenta Informaticae (2026). https://arxiv.org/abs/2412.15799. Accepted for publication, to appear

17. Lynch, N., Vaandrager, F.: Forward and backward simulations for timing-based systems. In: de Bakker, J.W., Huizing, C., de Roever, W.P., Rozenberg, G. (eds.) REX 1991. LNCS, vol. 600, pp. 397–446. Springer, Heidelberg (1992). https://doi.org/10.1007/BFb0032002
18. Moller, F., Tofts, C.: A temporal calculus of communicating systems. In: Baeten, J.C.M., Klop, J.W. (eds.) CONCUR 1990. LNCS, vol. 458, pp. 401–415. Springer, Heidelberg (1990). https://doi.org/10.1007/BFb0039073
19. Nicollin, X., Sifakis, J.: The algebra of timed processes, ATP: theory and application. Inf. Comput. **114**(1), 131–178 (1994). https://www.sciencedirect.com/science/article/pii/S0890540184710832
20. Pearl, J.: Causality, 2 edn. Cambridge University Press (2009)
21. Pettersson, P.: Modelling and Verification of Real-Time Systems Using Timed Automata: Theory and Practice. Ph.D. thesis, Uppsala University (1999)
22. Rokicki, T.G.: Representing and Modeling Digital Circuits. Ph.D. thesis, Stanford University (1993)
23. Tanimoto, T., Sasaki, S., Nakata, A., Higashino, T.: A global timed bisimulation preserving abstraction for parametric time-interval automata. In: Wang, F. (ed.) ATVA 2004. LNCS, vol. 3299, pp. 179–195. Springer, Heidelberg (2004). https://doi.org/10.1007/978-3-540-30476-0_18
24. Wang, Y.: Real-time behaviour of asynchronous agents. In: Baeten, J.C.M., Klop, J.W. (eds.) CONCUR 1990. LNCS, vol. 458, pp. 502–520. Springer, Heidelberg (1990). https://doi.org/10.1007/BFb0039080
25. Weise, C., Lenzkes, D.: Efficient scaling-invariant checking of timed bisimulation. In: Reischuk, R., Morvan, M. (eds.) STACS 1997. LNCS, vol. 1200, pp. 177–188. Springer, Heidelberg (1997). https://doi.org/10.1007/BFb0023458
26. Wimmer, S., Mutius, J.V.: Verified certification of reachability checking for timed automata. In: International Conference on Tools and Algorithms for the Construction and Analysis of Systems, pp. 425–443. Springer, Cham (2020)
27. Yi, W., Pettersson, P., Daniels, M.: Automatic verification of real-time communicating systems by constraint-solving. In: Hogrefe, D., Leue, S. (eds.) Formal Description Techniques VII: Proceedings of the 7th IFIP WG 6.1 International Conference on Formal Description Techniques, pp. 243–258. Springer, Boston (1995)

Collective Decision-Making Under Timing Constraints

Julia Klein[1,2] and Tatjana Petrov[2,3,4(✉)]

[1] Department of Computer and Information Science, University of Konstanz, Konstanz, Germany
[2] Centre for the Advanced Study of Collective Behaviour, University of Konstanz, Konstanz, Germany
[3] Department of Mathematics, Informatics and Geoscience, University of Trieste, Trieste, Italy
[4] Max Planck Institute of Animal Behaviour, Radolfzell, Germany
tatjana.petrov@gmail.com

Abstract. Achieving consensus is essential for coordinated group decisions in collective systems, yet the timescales of consensus dynamics critically depend on system size and interaction mechanisms. In this paper, we study consensus in stochastic swarms of finite size under three timing constraints: time to reach a decision (majority), time to hold a decision, and time to recover from a wrong decision. We propose a formal framework for specifying and analysing consensus dynamics as parameterised temporal properties, enabling systematic comparison of mechanistic models. First, we express consensus and recovery as parameterised properties in bounded linear temporal logic (BLTL). Then, we use statistical model checking to compute the consensus and recovery probabilities for fixed parameter values. Finally, we employ Gaussian process classification to approximate the robustness landscape over continuous parameter domains from a limited number of statistical model checking evaluations, thereby enabling efficient, uncertainty-aware exploration of timing parameters. Unlike existing approaches that abstract away either timing constraints or finite population effects, our framework enables systematic exploration of consensus and recovery probabilities across timing parameters for a fixed group size. We demonstrate our methodology in two application scenarios: in swarm robotics, we use our logical specification of consensus to guide model design; in systems biology, we use our consensus notion to elucidate the timescales governing the cell cycle switch.

Keywords: Collective Decision Making · Robustness · Probabilistic Verification · Swarm Robotics · Systems Biology

1 Introduction

The ability to reach a consensus, or common agreement, is essential in biological, social, and engineered systems. Consensus ensures that collective decisions

M. Chen and M. Tschaikowski (Eds.): QEST+FORMATS 2026, LNCS 16913, pp. 350–368, 2027.
https://doi.org/10.1007/978-3-032-35298-9_20

are made, leading to coherent actions and strategies that benefit the group as a whole. Understanding which local interaction mechanisms lead to robust consensus dynamics can elucidate natural design principles in biological systems and guide the design of engineered systems. The speed, accuracy and robustness of collective decision-making are influenced by multiple factors, including group size, noise, intrinsic characteristics of individuals, interaction and communication between agents. To illustrate, consider the classical voter model, where agents choose between two options, and, upon meeting, one may convince the other one to switch state [11]. Although this model reaches majority efficiently, it is surprisingly fragile: even a few disruptive agents can drive a population of hundred agents into persistent indecision [31]. A simple extension adds an intermediate undecided state, inspired by the honeybee house-hunting process [34] and known as the cross-inhibition model. It maintains efficiency while greatly improving robustness to disruptive individuals [1,31,37]. More broadly, subgroups need not be harmful: a few well-informed or even uninformed agents can help the system to collectively agree on a decision [12], and heterogeneity can generally boost swarm performance [2]. While previous literature quantified accuracy or convergence, these works do not formalize when a decision must be reached, how long it must be held, or how quickly a system should recover from an undesirable consensus. These temporal requirements are intrinsic in biological regulation (slow–fast processes, checkpoints) and in swarm robotics (real-time mission constraints). However, they are usually ignored due to the lack of a unified formal framework.

In this paper, we study consensus in stochastic swarms of finite size under three timing constraints: time to reach a decision, time to hold a decision, and time to recover from the wrong decision. We propose a formal framework which allows us to systematically explore and compare different mechanistic models with respect to their consensus and recovery probability landscape. First, we propose a formal consensus and recovery definition as a parametrised property in bounded linear temporal logic (BLTL), explicitly encoding the three timing requirements. Then, we employ statistical model checking to compute the probability of satisfying the consensus and recovery properties for concrete timing parameters. Finally, we use Gaussian process classification to approximate the probability landscape over continuous parameter domains from a limited number of statistical model checking evaluations, enabling efficient and systematic exploration of timing parameters.

We focus on the scenario of a finite group deciding between two options X and Y, where one option is qualitatively better than the other. We analyse several interaction mechanisms: the voter model, the cross-inhibition model, and several related extensions, relevant for our target application scenarios: swarm robotics and systems biology. For each given mechanistic model, our approach constructs the consensus probability landscape over the specified timing parameters. We demonstrate how the consensus landscape can be used to guide the model design in swarm robotics, and how it can elucidate the natural operating regimes of the cell cycle switch model in systems biology. Our results reveal operational regimes that are difficult to capture using deterministic analyses or direct stochastic simulation, highlighting the broader utility of formal verification tools for studying timing-critical collective decision-making.

Classical notions of consensus focus on eventual agreement and asymptotic convergence. In many natural and engineered systems, however, decisions are only useful if they are reached and maintained within relevant time windows. In swarm robotics, these constraints are often imposed by mission requirements, whereas in systems biology they reflect characteristic timescales of regulatory processes and are typically unknown a priori. Rather than prescribing such timescales, one seeks to identify the temporal regimes under which reliable collective decisions emerge. This motivates our treatment of reaching, holding, and recovery times as explicit parameters of the consensus specification.

Related works. Consensus mechanisms have been extensively studied in opinion dynamics, network science, statistical physics, and multi-agent systems. Vicsek et al. [35] showed how local alignment rules generate large-scale collective order, while Couzin et al. [13] demonstrated that a small number of informed individuals can guide large groups towards a common decision. In control theory, Olfati-Saber et al. [29] established fundamental connections between consensus protocols and graph topology. Across these works, consensus is typically treated as an asymptotic property, without explicit requirements on when agreement is reached, how long it is maintained, or how quickly incorrect decisions are corrected.

Transient phenomena such as indecision, switching, and unstable majorities are often analysed through trajectories or stochastic simulations, but rarely formalised as properties with precise temporal semantics. This limits systematic comparison and quantitative verification across models. Our work is inspired by [31,36], which highlight the role of cross-inhibition in stabilising collective decisions under disruptive behaviour, as well as by our previous studies on robustness in finite stochastic systems [24,25]. Here, we address the temporal dimension by encoding reaching, holding, and recovery requirements directly in a probabilistic logical framework.

The formal-methods workflow proposed here is closely related to logic-based parameter synthesis in systems biology [5,14,18]. Foundational notions of robustness with respect to temporal-logic specifications were introduced for deterministic systems in [15,17,32] and later extended to stochastic models in [3,4]. Similar to our setting, these works study chemical reaction networks with continuous-time Markov chain semantics, but focus on signal temporal logic properties rather than collective decision-making. To the best of our knowledge, consensus dynamics and swarm-robotic applications have not been addressed within this framework.

Timing constraints arise naturally in both biological regulation and swarm robotics, yet remain largely unexplored in finite-size stochastic models. In systems biology, slow–fast dynamics, metastability, and noise-driven switching play a crucial role in processes such as checkpoint enforcement and irreversible transitions, challenging deterministic and mean-field descriptions [6,16,19,21,22].

In swarm robotics, convergence and recovery must satisfy mission constraints despite communication delays, heterogeneous agents, and environmental uncertainty [23]. Existing studies predominantly focus on accuracy, equilibrium selection, or asymptotic behaviour. In contrast, we explicitly formalise consensus in terms of reaching, holding, and recovery times in finite stochastic populations, enabling a principled temporal characterisation of collective decision-making under uncertainty.

Finally, the proposed framework naturally extends to richer settings, including varying decision difficulty, multiple options, spatial interactions, and additional model parameters.

Contributions. The main contributions of this paper are: (i) a temporal-logic characterisation of consensus that explicitly captures reaching, holding, and recovery times; (ii) a scalable verification framework combining statistical model checking and Gaussian process classification to infer probability landscapes over temporal parameters; (iii) applications to swarm robotics and systems biology, revealing mechanistic differences in the temporal reliability of three swarm-decision models and two biochemical switches; (iv) identification of finite-size effects and temporal trade-offs that are difficult to capture using deterministic or mean-field approaches.

2 Methods

2.1 Models of Decision-Making

We model collective decision-making using Chemical reaction networks (CRNs), a classical formalism for stochastic population dynamics. Under the well-stirred assumption, a CRNs induces a continuous-time Markov chain (CTMC) [20,33] on a discrete state space.

In all models studied here, individuals choose between two options of unequal quality, where option X is superior to option Y. Without loss of generality, we focus on the asymmetric scenario where $q_x > q_y$ across all models and application domains.

We consider two application scenarios: (1) swarm robotics, using three decision models called voter model, cross-inhibition model, and combined model. For these models, we analyse consensus and recovery from a wrong decision, with respect to reaching time, holding time, and recovery time; (2) systems biology, using two biochemical switch models called simply catalytic (SC) and cell cycle (CC) model, taken from [10]. Here, we analyse reaching and holding times for reaching a stable consensus for the better option.

Voter Model. The classic voter model assumes that an individual can immediately adopt the opinion of an interacting neighbour with a different opinion (at the respective rate) [11]. The corresponding reaction system is defined as follows:

$$\begin{aligned} x + y &\xrightarrow{q_x} x + x \\ x + y &\xrightarrow{q_y} y + y \end{aligned} \tag{1}$$

To exemplify, from the population state in the underlying CTMC model counting e.g. 10 copies of agents X and 90 copies of agents Y, if the first reaction occurs, the next state will count 11 copies of agent X and 89 copies of agent Y. The propensity of each reaction is proportional to the amounts of reactants, multiplied by the reaction rate (hence, for the example state, the propensities will be $10 \cdot 90 q_x$ and $10 \cdot 90 q_y$, respectively). The time of next reaction will be an exponentially distributed random variable, with parameter being the sum of propensities of all reactions that are enabled in the current state.

Cross-inhibition model. In the cross-inhibition model [31], individuals switch opinion only after two consecutive interaction with individuals of different opinion, modelled via a transition through an undecided state U:

$$\begin{aligned} x + y &\xrightarrow{q_x} x + u \\ x + y &\xrightarrow{q_y} u + y \\ x + u &\xrightarrow{q_x} x + x \\ y + u &\xrightarrow{q_y} y + y \end{aligned} \tag{2}$$

Combined Model. In our previous work [25], we observed that the cross-inhibition model is substantially more robust than the voter model to reach consensus, particularly in the presence of disruptive individuals. However, unlike the voter model, the cross-inhibition model can also stabilise an undesirable consensus on the inferior option. Intuitively, this could indicate that the voter model is inferior in maintaining a consensus but better in recovering from an initial bad decision.

Motivated by these strengths, we aim to find a decision model that supports both convergence to a good consensus and fast recovery from a bad decision. To this end, we propose a combined model that incorporates the direct switching reactions of the voter model together with the indirect transitions of the cross-inhibition model.

This hybrid structure is designed to merge the stabilising effect of cross-inhibition with the responsiveness of voter dynamics. We assume direct switching to be five times more likely, thus reactions in 1 are included with multiplicity five, and those in 2 once.

Zealots. Previous studies have shown that the presence of a few stubborn agents, also known as *zealots*, who maintain their opinion regardless of peer influence, can dramatically disrupt consensus formation in the voter model [28]. Variants such as the cross-inhibition model significantly improve robustness to such disruptions. Consequently, analysing the robustness of consensus with respect to the amount of zealots in the population is an interesting task. In our framework, zealots are modeled as agents that never change their own opinion when interacting with other agents.

The second application scenario is based on a systems biology work [9]. The authors argue that the biochemical network controlling the entry into mitosis in eukaryotic cells shares its structure and dynamics with the so-called Approximate Majority (AM) protocol. In the AM model, agents converge to a consensus by switching through an intermediate 'undecided' state; this makes the AM model mechanistically equivalent to our cross-inhibition model. Additionally, they compare AM to a direct competition (DC) network, which corresponds to our voter model. They then derive two more switch networks – simply catalytic and cell-cycle switch model (denoted later in text by SC and CC) that reproduce the key behaviours of AM but are biologically more plausible. Each of these two biological models differs from DC and AM in mechanistic details, yet share the same computational signature: nonlinearity and double positive feedback through competition and intermediate (undecided) states. From the dynamical systems point of view, all of the studied models feature the noise-induced bimodality. We introduce the two additional models below.

Simply Catalytic (SC) model. The SC model from [10] is a variant derived from AM by replacing the mutually autocatalytic reactions, which are biologically unrealistic, with pairs of mutually catalytic reactions mediated by intermediate catalyst species. This eliminates the need for species that directly replicate themselves. Instead, catalysis occurs only indirectly through helper molecules. The resulting reaction system is:

$$\begin{array}{ll} x+w \xrightarrow{q_y} w+b & z+y \xrightarrow{q} y+u \\ b+w \xrightarrow{q_y} w+y & u+y \xrightarrow{q_y} y+w \\ y+r \xrightarrow{q_x} r+b & p+x \xrightarrow{q} x+q \\ b+r \xrightarrow{q_x} r+x & q+x \xrightarrow{q_x} x+r \\ w+s \xrightarrow{q} s+u & r+t \xrightarrow{q} t+q \\ u+s \xrightarrow{q} s+z & q+t \xrightarrow{q} t+p \end{array} \tag{3}$$

Intuitively, the SC model preserves the decision-making logic of Approximate Majority while replacing biologically unrealistic self-replication mechanisms with catalytically mediated interactions. It therefore serves as an intermediate step between abstract population protocols and biologically plausible regulatory networks.

Cell Cycle (CC) switch. The CC model further transforms the SC model to better reflect the actual biochemical regulatory architecture of the cell-cycle switch [10]. In this version, one of the two competing states, here Y, loses its catalytic activity. Activation of X proceeds through a catalytic feedback loop, while suppression of Y occurs via a separate catalytic pathway. This structure implements both double-positive and double-negative feedback loops which are also found in real cell-cycle regulation. Thus, the CC model provides a biologically plausible biochemical implementation of the computational decision-making algorithm. The reactions are:

$$\begin{aligned} x+z &\xrightarrow{q_y} z+b & z+x &\xrightarrow{q} x+u \\ b+z &\xrightarrow{q_y} z+y & u+x &\xrightarrow{q} x+w \\ y+r &\xrightarrow{q_x} r+b & p+x &\xrightarrow{q} x+q \\ b+r &\xrightarrow{q_x} r+x & q+x &\xrightarrow{q_x} x+r \\ w+s &\xrightarrow{q} s+u & r+t &\xrightarrow{q} t+q \\ u+s &\xrightarrow{q_y} s+z & q+t &\xrightarrow{q} t+p \end{aligned} \tag{4}$$

The CC model can be interpreted as a biologically grounded implementation of the same underlying decision-making principle, where activation and inhibition are realised through feedback loops resembling those found in eukaryotic cell-cycle regulation.

2.2 Bounded Linear Temporal Logic (BLTL)

To formally evaluate whether a group reaches a consensus over time, we express properties using Bounded Linear Temporal Logic (BLTL). BLTL is a variant of Linear Temporal Logic (LTL) with explicit time bounds on operators (F, G, X, U and W), either as a number of steps or a real-time bound. Without these temporal bounds, the two syntaxes are equivalent. BLTL is well-suited for specifying time-critical behavioural constraints (see [8]) and can be efficiently verified using statistical model checking (SMC) [27]. In this work, we use PlasmaLab [8] for probabilistic verification, as it supports BLTL-based statistical model checking of properties which, to the best of our knowledge, is not available in other freely available probabilistic verification tools.

2.3 Formally Analysing Consensus and Recovery with BLTL

Consensus refers to the alignment of a sufficiently large fraction of the population towards the same decision, enabling coordinated behaviour among individuals. A single trace of our population model represents one possible evolution of the population state (counting the amount of individuals of each of the sub-populations) over time. Intuitively, each single trace may reach a majority in favour of one decision or another. Consensus is characterised by five parameters: a majority threshold m, a distance threshold d, a reaching time t, a holding time h, and a recovery time r. Throughout the paper, we assume that option X is preferable to Y.

For readability, we introduce shorthand predicates for consensus states:

$$\begin{aligned} Xcons &:= (x \geq \min_m) \wedge (x - y \geq d), \\ Ycons &:= (y \geq \min_m) \wedge (y - x \geq d). \end{aligned}$$

Good consensus. A *good* consensus, i.e. consensus for the superior option X, is expressed in BLTL as:

$$F_{\leq t}(G_{\leq h}(Xcons)) \tag{5}$$

In words, the formula specifies that a group, before time t, reaches the majority in favour of the decision of better quality, X, so that the total number of agents for the voted decision is greater than the majority parameter $\min_m = (m/100) \cdot N$, the absolute difference between the groups of agents is greater than the distance d, and the group remains in such a state for at least h time units.

Recovering bad consensus. If the group initially chooses the *bad* option Y, we want to analyse its ability to quickly recover from this decision and reach a consensus for the *good* option X. Accordingly, the property of recovery is defined with the following formula:

$$(\neg Ycons)\ U_{\leq t}\ (Ycons\ \wedge\ F_{\leq r}(G_{\leq h}(Xcons))) \tag{6}$$

This formula specifies that a group first reaches a majority for Y before reaching time t, and escapes this majority within reaching time r and subsequently reaches a good consensus for X and stays in such state for at least h time units. This property expresses the ability to recover from an incorrect decision.

Good and recovering consensus. Finally, we define a combined property capturing systems that either directly reach a good consensus, or recover from a bad consensus and subsequently stabilize on the good decision:

$$F_{\leq t}(G_{\leq h}(Xcons))\ \vee\ (\neg Ycons\ U_{\leq t}\ (Ycons \wedge F_{\leq r}(G_{\leq h}Xcons))) \tag{7}$$

This combined specification captures systems that either converge quickly to a good consensus, or initially make a wrong decision but subsequently recover within the allowed recovery window.

Basic properties. The parameterised consensus specifications exhibit natural monotonicity properties. In particular, increasing the reaching time t or recovery time r cannot decrease the satisfaction probability, while decreasing the holding time, or relaxing majority or distance thresholds similarly cannot decrease satisfaction. These properties follow directly from the semantics of BLTL.

Interpretation of temporal parameters. The temporal parameters t, r, and h capture different aspects of decision-making dynamics and are typically associated with distinct time scales. In particular, the reaching time t and recovery time r characterise transient behaviour and are expected to be relatively short, while the holding time h reflects the stability of a consensus and is typically chosen on a longer time scale.

We emphasize that the temporal bounds are not necessarily fixed design requirements. In engineered systems, such as robotic swarms, they may represent explicit mission constraints. In natural systems, however, the relevant timescales are often unknown and constitute an object of study themselves. In this setting, varying the temporal parameters allows us to identify operating regimes in which consensus is achieved reliably and maintained sufficiently long.

2.4 Model Checking and Robustness Analysis

Statistical model checking (SMC) uses random sampling to simulate system behaviour and estimate the probability that a given property is satisfied [27]. By repeatedly generating system trajectories and evaluating the property, SMC provides unbiased statistical estimates with lower computational cost than exhaustive verification. We use PlasmaLab to evaluate the BLTL properties across different models and parameter settings.

While SMC provides accurate estimates for individual parameter values, exploring the full parameter space is computationally prohibitive. To address this, we employ Gaussian process classification (GPC) [7] to approximate the robustness landscape over continuous parameter domains. GPC treats property satisfaction as a binary classification problem and learns a smooth approximation of the satisfaction probability from a limited number of SMC samples. This enables efficient interpolation across the parameter space, significantly reducing the number of required simulations.

In addition, GPC provides uncertainty-aware predictions, allowing us to identify regions where estimates are more or less reliable. This makes it well-suited for systematic exploration of high-dimensional parameter spaces and for comparing models in terms of their robustness. For further details, we refer to [30].

Computational cost. All experiments were performed on a MacBook Pro (M1 Pro, 32 GB RAM, macOS Sequoia). Evaluating one SMC probability point with 4,239 Monte Carlo simulations typically takes 1 s for the swarm robotic models and around 15 s for the systems biology models. For the GPC interpolation, we sample 324 SMC points per landscape, resulting in total runtimes of 5–100 min per model.

2.5 Quantitative Robustness Measures

We introduce quantitative measures to enable systematic comparison between models.

Let $\theta \in \Theta$ denote a vector of parameters, which may include model parameters (e.g., reaction rates) and/or specification parameters (e.g., reaching time t, holding time h, recovery time r, or thresholds such as m and d). The parameter space Θ is assumed to be a predefined domain, typically given as a product of intervals over the parameters of interest (e.g., ranges of reaching time t, recovery time r, or holding time h). In practice, these intervals correspond to the parameter ranges explored during statistical model checking.

Let $P(\theta) \in [0, 1]$ be the satisfaction probability of a given temporal property. In this work, we primarily consider variations in temporal parameters (e.g., t, r, h), although the framework naturally extends to other model or specification parameters.

Definition 1 (Average satisfaction probability). *Average satisfaction probability* is defined by $\overline{P} := \mathbb{E}_{\theta \in \Theta}[P(\theta)]$, representing the overall likelihood of satisfying the specification across the considered parameter ranges.

Definition 2 (Robustness score). We define the robustness score as $\mathcal{R}_\lambda := \mathbb{E}_{\theta\in\Theta}[P(\theta)] - \lambda \cdot \mathrm{Var}_{\theta\in\Theta}[P(\theta)]$, where $\lambda \geq 0$ is a weighting parameter controlling the trade-off between average performance and variability. This score captures that models with high average satisfaction probability and low sensitivity to parameter variations achieve higher scores.

We use the robustness score as a compact scalar summary of the robustness landscape. Since $P(\theta) \in [0,1]$, the variance is bounded. In this work, we set $\lambda = 1$, providing a natural balance without additional scaling.

Definition 3 (α-feasible set). For a fixed probability threshold $\alpha \in (0,1)$, let the *feasible set* be defined as $\mathcal{F}_\alpha := \{\theta \in \Theta \mid P(\theta) \geq \alpha\}$, i.e., the set of parameter values for which the specification is satisfied with probability at least α. This feasible set provides a geometric representation of robustness, from which both global measures (e.g., volume) and structural properties (e.g., trade-offs between parameters) can be derived.

Definition 4 (Robustness measures via α-feasible set). *Robustness volume* is defined by $V_\alpha := \frac{\mu(\mathcal{F}_\alpha)}{\mu(\Theta)}$, where μ denotes the Lebesgue measure over the parameter space Θ, capturing the fraction of the parameter space for which the model satisfies the specification with probability at least α

These measures provide a concise summary of the robustness landscape and allow for principled comparison of different decision-making mechanisms beyond visual inspection.

When focusing on temporal parameters, robustness can also be interpreted in terms of achievable temporal guarantees, such as *minimum reaching time* or *maximum holding time*. For instance, the minimum reaching time can be computed from the α-feasible set by $t_\alpha^{\min} := \min\{t \mid P(\theta) \geq \alpha \text{ for some } \theta \in \Theta \text{ with reaching time } t\}$. This perspective can be useful in e.g. model selection, by detecting desirable models as those that reach consensus quickly (small t), recover rapidly (small r), and maintain consensus for extended periods (large h). Alternatively, computing achievable temporal guarantees for one temporal parameter at a time also provides boundary values, overapproximation of the feasible set $\mathcal{F}_\alpha$. As such it can be used to reveal the time scales providing stable consensus operating regime of a given model.

3 Results

In all experiments, we consider a group size of $N = 100$, and asymmetric option qualities encoded through the reaction rates $q_x = 1.05$ and $q_y = 0.95$. Consensus is defined using the property parameters majority $m = 50$, distance $d = 10$ and holding time $h = 50$, if not stated otherwise. Initially, the system starts from a well-balanced state with an equal amount of individuals voting for X and Y. The chosen baseline parameters are consistent with earlier studies in [24,25,31].

For all model variants, we evaluate the BLTL properties using the statistical model checking tool Plasmalab [8]. We employ the Monte Carlo algorithm with

4,239 simulations per experiment, ensuring an error margin of $\epsilon = 0.025$ and a confidence bound of $\Delta = 0.01$. To obtain smooth robustness landscapes, we further apply Gaussian process classification.

All models and scripts are available at the project repository. and archived under version v1.0.0 at Zenodo.

3.1 Swarm Robotics

First, we analyse how the three models of swarm robotics, the voter model, the cross-inhibition model, and the combined model compare in reaching a good consensus and recovering from an initial bad decision.

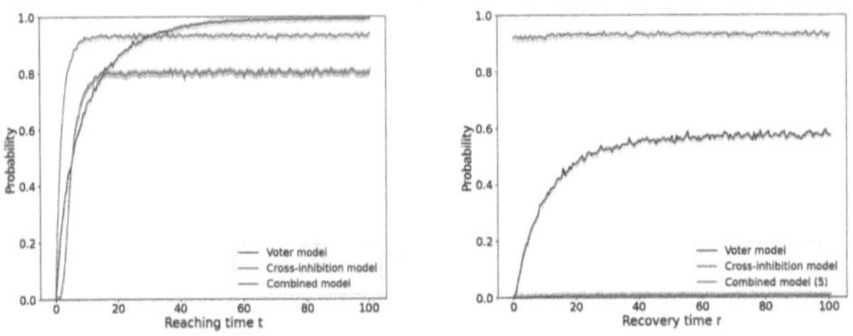

(a) (b)

Fig. 1. Robustness analysis of consensus and recovery for the voter model (blue), cross-inhibition model (red), and combined model (green). Curves show the satisfaction probabilities, with shaded indicating confidence intervals. (a) Robustness of reaching good consensus (see property 5 (Color figure online)) with respect to varying reaching time. (b) Robustness of recovery from an initial bad decision to a good consensus (see property 6 with respect to varying recovery time; reaching time is fixed to $t = 10$.

Figure 1(a) shows the probability of reaching a good consensus, for option X, under increasing reaching time. We see that the combined model initially performs significantly better than both the voter and cross-inhibition models. For short reaching times, it achieves substantially higher consensus probabilities, demonstrating its ability to converge quickly to the superior option. As the reaching time exceeds approximately $t \approx 30$, the voter model surpasses the two other models and ultimately becomes the most reliable at long timescales. The cross-inhibition model, in contrast, rapidly stabilises at a probability of around 0.8.

Figure 1(b) presents the recovery probability after an initial bad consensus. Here, the qualitative differences between the models are more pronounced. The cross-inhibition model exhibits almost no capacity for recovery, with probabilities close to zero across all recovery times. The voter model begins with zero probability as well but gradually improves, reaching probabilities around 0.6 for larger recovery times. This confirms our initial intuition that the voter model is better in recovering from bad decisions than the cross-inhibition model. The combined model performs best and achieves high recovery probabilities around 0.9 already for short recovery times and maintains this throughout.

Additionally, in both analyses we observe that the combined model is more robust with respect to variations in the temporal parameter. It offers the fastest convergence to a good consensus at short timescales and is consistently the most robust model for recovering from an initial bad decision.

Next, we analyse the robustness of the joined probability to not only reaching a good consensus but also recovering from an initial bad decision for the three decision-making models under two different reaching-time constraints.

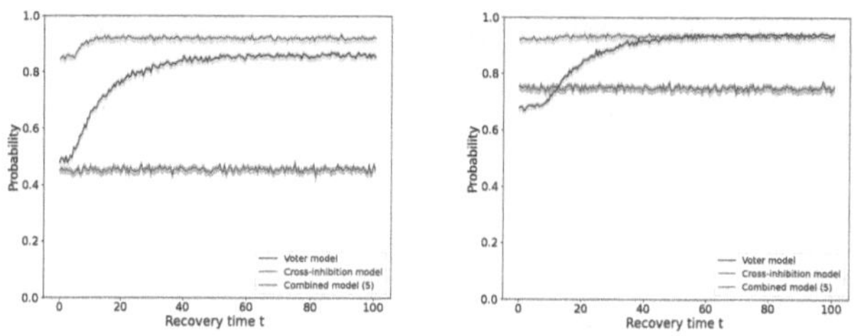

(a) (b)

Fig. 2. Robustness analysis of both reaching a good consensus and recovering from an initial bad decision (see property 7) for the voter model (blue), cross-inhibition model (red), and combined model (green), with respect to varying recovery time. (a) Results with reaching time fixed to $t = 5$. (b) Results with reaching time fixed to $t = 10$. (Color figure online)

In Fig. 2(a), where the reaching time is fixed to $t = 5$, the combined model exhibits the highest satisfaction probability across the full range of recovery times. In particular, it clearly dominates for short recovery times, achieving high consensus and recovery probability even when the allowed recovery window is small. As expected, the combined model merges the ability to reach a good consensus but also to recover from a bad decision. The voter model starts with lower probabilities but improves steadily as the recovery time increases, still with a clear difference to the combined model. In contrast, the cross-inhibition model remains at a probability of around 0.45, reflecting its intrinsic ability to meet the good consensus but inability to escape a bad consensus.

In Fig. 2(b), where the reaching time is fixed to $t = 10$, all models achieve higher satisfaction probabilities. The cross-inhibition model now stabilises around 0.75 but still shows minimal sensitivity to recovery time, consistent with its poor recovery capability. The combined model again performs best for short recovery times, reaching directly high probabilities. Only after a recovery time of approximately $r = 40$, the voter model reaches comparable probabilities, highlighting its slower, but eventually effective, decision-making dynamics.

Overall, these results emphasise that the combined model consistently achieves strong consensus and recovery probabilities on short timescales, while retaining high robustness across the entire parameter range.

After analysing how the three models differ with respect to reaching and recovery time, we next investigated the combined influence of both temporal parameters on the ability to reach a good consensus and recover from an initial bad decision. Therefore, we varied both the reaching time t and the recovery time r simultaneously and used GPC to infer the full robustness landscape over this two-dimensional parameter space. Note that GPC introduces some interpolation noise, which is why some regions in the resulting heatmaps appear patchy.

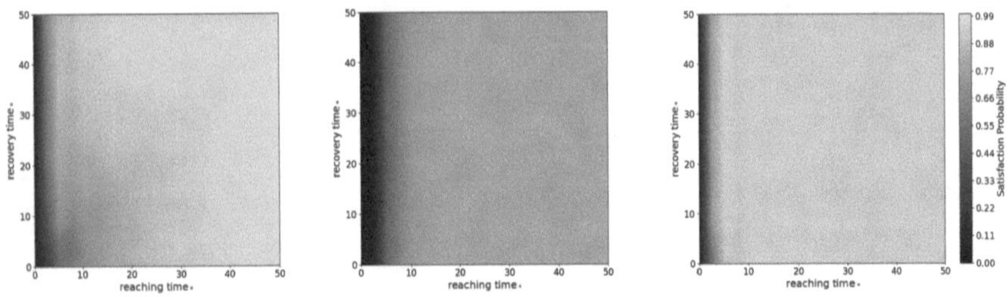

(a) (b) (c)

Fig. 3. Robustness analysis of both reaching a good consensus and recovering from an initial bad decision (see property 7), with respect to varying reaching and recovery time. Probabilities range from 0 (dark blue) to 1 (yellow). (a) Results for voter model. (b) Results cross-inhibition model. (c) Results for combined model. (Color figure online)

Figure 3 shows that the cross-inhibition model performs worst overall, achieving only moderate satisfaction probabilities due to its limited recovery capability. The voter model gradually improves as reaching and recovery times increase, but requires relatively long time horizons to approach high probabilities. In contrast, the combined model consistently achieves high probabilities across most of the parameter space, including short reaching and recovery times, confirming its superior short-timescale performance.

To highlight the low-time regime, Fig. 5 (see Appendix) provides a zoomed-in view. The combined model clearly dominates, maintaining high satisfaction probabilities even for small temporal parameters, whereas the voter model improves more gradually. Overall, the combined model provides the most robust behaviour, combining rapid convergence to a good consensus with efficient recovery from incorrect decisions.

Robustness Evaluation. We compute the robustness measures introduced in Sect. 2.5. Table 1 (see Appendix) summarizes the quantitative robustness measures, as well as the achievable temporal guarantees required to achieve a required satisfaction probability for varying zealot counts and for $\alpha \in \{0.9, 0.5\}$. The quantitative measures confirm the trends observed in the robustness landscapes.

3.2 Systems Biology

For the systems biology scenarios, we analyse the robustness of the Direct Competition (DC, similar to our voter model), Approximate Majority (AM, similar to our cross-inhibition model), Simply Catalytic (SC) and Cell Cycle (CC) model with respect to reaching and holding a good consensus.

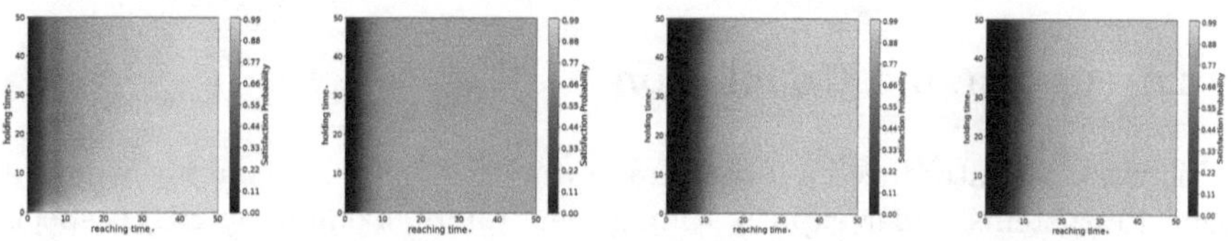

(a) (b) (c) (d)

Fig. 4. Robustness analysis of reaching a good consensus (see property 5) with respect to varying reaching and holding time. Both models start initially with: $x = y = b = w = z = u = p = r = q = s = t = 50$, $N = 150$. (a) Results for DC model, similar to our voter model. (b) Results for AM model, similar to our cross-inhibition model. (c) Results for SC model. (d) Results for CC model.

Figure 4 shows the robustness landscape of reaching a consensus on the superior option X as a function of the reaching and holding times. The DC model displays a distinctive landscape for low values of reaching and holding times: it achieves high probabilities already for very short reaching times, providing the holding time is also small. Then, it gradually increases in reliability as both temporal parameters grow. For sufficiently long reaching and holding times, the DC model becomes robust in reaching a good consensus. In contrast, the AM model performs noticeable worse. Its probability of reaching a good consensus remains limited, typically not exceeding values around 0.75, and it does not benefit as strongly from increases in either reaching or holding time.

Overall, the SC and CC models display remarkably similar behaviour and they both outperform the AM model. Both achieve consistently high probabilities starting from a reaching time of roughly 25. However, the CC model attains slightly lower probabilities than SC.

Our comparison of the SC and CC architectures provides insight into how biological realism shapes consensus dynamics. Cardelli and Csikász-Nagy [10] introduced SC as a computationally faithful refinement of Approximate Majority (AM), whereas CC incorporates additional biochemical constraints from the eukaryotic cell-cycle switch. These constraints introduce stabilising feedback loops and inactive–active forms that render the CC network more realistic, but also less responsive. Consistent with this interpretation, our analysis shows that although both networks behave almost identically in terms of consensus reliability, CC consistently performs marginally worse than SC.

Robustness evaluation. Table 1 reports the robustness measures and temporal guarantees for reaching and holding times at satisfaction thresholds $\alpha \in 0.9, 0.5$. The quantitative results are consistent with the robustness landscapes.

Together, these results demonstrate that explicit timing constraints fundamentally shape consensus reliability and reveal qualitative differences that only emerge in finite-size stochastic systems. Our framework provides a principled way to compare decision-making mechanisms across temporal regimes, making timescale-dependent behaviours explicit and measurable. We next discuss its broader implications, limitations, and future directions.

4 Discussion and Conclusion

In this work, we introduced a formal framework for analysing consensus in finite stochastic systems under explicit timing constraints. Combining BLTL specifications, statistical model checking, and Gaussian process classification, our approach characterises when consensus is reached, maintained, and recovered with statistical guarantees.

Unlike deterministic or mean-field models [26], our framework captures finite-size effects, stochastic fluctuations, and noise-induced transitions that are central to biological and robotic collective systems, revealing operational regimes that are difficult to detect using asymptotic analyses or stochastic simulations alone.

Applications. The framework applies to both biochemical and robotic collective systems, where timing constraints are crucial. In systems biology, it provides probabilistic guarantees on reaching, holding, and recovery times, complementing existing approaches to multiscale stochastic dynamics. In swarm robotics, it enables principled comparison of decision mechanisms under explicit mission requirements. Our results show that biologically motivated switch architectures remain robust despite additional mechanistic constraints, while combining direct competition with cross-inhibition yields particularly strong performance across temporal regimes. The framework naturally extends to richer settings, including varying decision difficulty, multiple options, and spatially structured interactions.

Limitations and future work. GPC becomes increasingly sensitive to hyperparameter choices in high-dimensional parameter spaces and requires more samples for accurate inference, while the computational cost of SMC grows with system size and complexity. Future work will focus on automated model exploration and learning.

Acknowledgments. JK and TP were funded by the Deutsche Forschungsgemeinschaft (DFG, German Research Foundation) under Germany's Excellence Strategy – EXC 2117 – 422037984. The authors would like to thank Heiko Hamann and Andreagiovanni Reina for the inspiring discussions on the topic.

Disclosure of Interests. The authors have no competing interests to declare that are relevant to the content of this article.

A Appendix

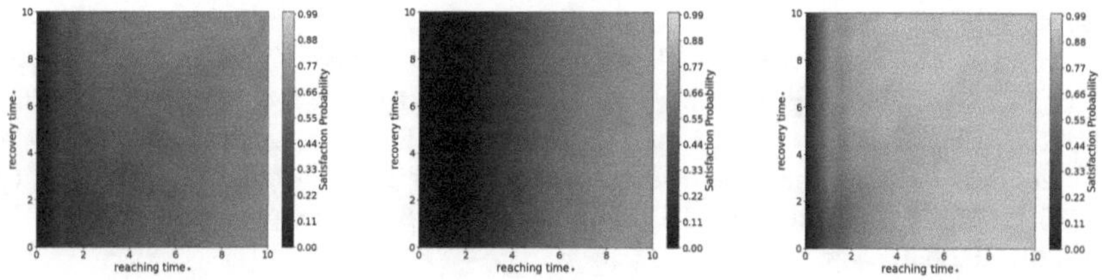

(a) (b) (c)

Fig. 5. Zoomed-in view of robustness analysis in Fig. 3. (a) Results for voter model. (b) Results cross-inhibition model. (c) Results for combined model.

Table 1. Quantitative robustness comparison across case studies: good consensus.

Model	$\mathcal{R}_1$	$\overline{P}$	$V_{0.9}$	$t_{0.9}^{\min}$	$h_{0.9}^{\max}$	$V_{0.5}$	$t_{0.5}^{\min}$	$h_{0.5}^{\max}$
(0 zealots)								
Voter (DC)	0.550	0.806	0.56	5	50	0.903	2	50
Cross-inhibition (AM)	0.471	0.702	0.00	-	-	0.885	5	50
SC	0.334	0.680	0.453	21	75	0.75	10	75
CC	0.325	0.664	0.37	23	75	0.75	10	75
(10 zealots)								
Voter	0.173	0.481	0.155	5	13	0.43	2	35
Cross-inhibition	0.461	0.706	0.00	-	-	0.85	7	50
(20 zealots)								
Voter	0.090	0.232	0.093	5	5	0.195	2	13
Cross-inhibition	0.442	0.711	0.003	50	36	0.835	7	50
(30 zealots)								
Voter	-0.145	0.142	0.07	5	2	0.122	2	7
Cross-inhibition	0.379	0.680	0.27	23	50	0.76	7	50

References

1. Angluin, D., Aspnes, J., Eisenstat, D.: A simple population protocol for fast robust approximate majority. Distrib. Comput. **21**(2), 87–102 (2008). https://doi.org/10.1007/s00446-008-0059-z

2. Antonic, N., Zakir, R., Dorigo, M., Reina, A.: Collective robustness of heterogeneous decision-makers against stubborn individuals. In: Proceedings of the 23rd International Conference on Autonomous Agents and Multiagent Systems. p. 68–77. AAMAS '24, Richland, SC (2024)
3. Bartocci, E., Bortolussi, L., Nenzi, L., Sanguinetti, G.: On the robustness of temporal properties for stochastic models. Electron. Proc. Theor. Comput. Sci. **125**, 3–19 (2013). https://doi.org/10.4204/EPTCS.125.1, http://arxiv.org/abs/1309.0866v1
4. Bartocci, E., Bortolussi, L., Nenzi, L., Sanguinetti, G.: System design of stochastic models using robustness of temporal properties. Theor. Comput. Sci. **587**, 3–25 (2015). https://doi.org/10.1016/j.tcs.2015.02.046, https://doi.org/10.1016/j.tcs.2015.02.046
5. Bartocci, E., Lió, P.: Computational modeling, formal analysis, and tools for systems biology. PLoS Comput. Biol. **12**(1), e1004591 (2016). https://doi.org/10.1371/journal.pcbi.1004591
6. Bokes, P., Klein, J., Petrov, T.: Accelerating reactions at the DNA can slow down transient gene expression. In: International Conference on Computational Methods in Systems Biology, pp. 44–60. Springer (2020)
7. Bortolussi, L., Milios, D., Sanguinetti, G.: Smoothed model checking for uncertain continuous-time markov chains. Inf. Comput. **247**, 235–253 (Apr 2016). https://doi.org/10.1016/j.ic.2016.01.004, https://linkinghub.elsevier.com/retrieve/pii/S0890540116000055
8. Boyer, B., Corre, K., Legay, A., Sedwards, S.: Plasma-lab: a flexible, distributable statistical model checking library. In: International Conference on Quantitative Evaluation of Systems, pp. 160–164. Springer (2013). https://doi.org/10.1007/978-3-642-40196-1_12
9. Cardelli, L., Csikász-Nagy, A.: The cell cycle switch computes approximate majority. Sci. Rep. **2**(1), 656 (2012). https://doi.org/10.1038/srep00656
10. Cardelli, L., Csikász-Nagy, A.: The cell cycle switch computes approximate majority. Sci. Rep. **2**(1), 656 (2012). https://doi.org/10.1038/srep00656, http://www.nature.com/articles/srep00656
11. Clifford, P., Sudbury, A.: A model for spatial conflict. Biometrika **60**, 581–588 (1973). https://doi.org/10.1093/biomet/60.3.581
12. Couzin, I.D., et al.: Uninformed individuals promote democratic consensus in animal groups. Science **334**(6062), 1578–1580 (2011). https://doi.org/10.1126/science.1210280
13. Couzin, I.D., Krause, J., Franks, N.R., Levin, S.A.: Effective leadership and decision-making in animal groups on the move. Nature **433**(7025), 513–516 (2005). https://doi.org/10.1038/nature03236
14. Danos, V., Laneve, C.: Formal molecular biology. Theoret. Comput. Sci. **325**(1), 69–110 (2004). https://doi.org/10.1016/j.tcs.2004.03.065
15. Donzé, A., Fanchon, E., Gattepaille, L.M., Maler, O., Tracqui, P.: Robustness analysis and behavior discrimination in enzymatic reaction networks. PLoS ONE **6**(9), e24246 (2011). https://doi.org/10.1371/journal.pone.0024246
16. Eftimie, R., Dushoff, J., Bridle, B.W., Bramson, J.L., Earn, D.J.: Multi-stability and multi-instability phenomena in a mathematical model of tumor-immune-virus interactions. Bull. Math. Biol. **73**(12), 2932–2961 (2011). https://doi.org/10.1007/s11538-011-9653-6
17. Fainekos, G.E., Pappas, G.J.: Robustness of temporal logic specifications for continuous-time signals. Theoret. Comput. Sci. **410**(42), 4262–4291 (2009). https://doi.org/10.1016/j.tcs.2009.06.021

18. Fisher, J., Henzinger, T.: Executable cell biology. Nat. Biotechnol. **25**(11), 1239–49 (2007)
19. Ghaffarizadeh, A., Flann, N.S., Podgorski, G.J.: Multistable switches and their role in cellular differentiation networks. BMC Bioinform. **15**(7), 1–13 (2014). https://doi.org/10.1186/1471-2105-15-S7-S1
20. Gillespie, D.T.: Exact stochastic simulation of coupled chemical reactions. J. Phys. Chem. **81**(25), 2340–2361 (1977). https://doi.org/10.1021/j100540a008, https://pubs.acs.org/doi/10.1021/j100540a008
21. Guet, C., Henzinger, T.A., Igler, C., Petrov, T., Sezgin, A.: Long Lived Transients in Gene Regulation (2021). https://doi.org/10.1016/j.tcs.2021.05.023
22. Gunawardena, J.: Time-scale separation-Michaelis and Menten's old idea, still bearing fruit. FEBS J. **281**(2), 473–488 (2014). https://doi.org/10.1111/febs.12586
23. Hamann, H.: Swarm Robotics: A Formal Approach, vol. 221. Springer (2018). https://doi.org/10.1007/978-3-319-74528-2
24. Klein, J., d'Onofrio, A., Petrov, T.: Exploring consensus robustness in swarms with disruptive individuals. In: International Symposium on Leveraging Applications of Formal Methods. pp. 33–48. Springer (2024). https://doi.org/10.1007/978-3-031-73606-3_3
25. Klein, J., Petrov, T.: Quantifying consensus in stochastic swarms with disruptive individuals. In: European Control Conference (ECC), pp. 272–277. IEEE (2025)
26. Kurtz, T.G.: Limit theorems and diffusion approximations for density dependent markov chains. Stochast. Syst. Model. Ident. Optim. **I**, 67–78 (1976). https://doi.org/10.1007/BFb0103212
27. Legay, A., Delahaye, B., Bensalem, S.: Statistical model checking: an overview. In: International Conference on Runtime Verification, pp. 122–135. Springer (2010). https://doi.org/10.1007/978-3-642-16612-9_11
28. Mobilia, M., Petersen, A., Redner, S.: On the role of zealotry in the voter model. J. Stat. Mech. Theory Exp. **08**, P08029 (2007). https://doi.org/10.1088/1742-5468/2007/08/P08029
29. Olfati-Saber, R., Fax, J.A., Murray, R.M.: Consensus and cooperation in networked multi-agent systems. Proc. IEEE **95**(1), 215–233 (2007). https://doi.org/10.1109/JPROC.2006.887293
30. Rasmussen, C.E., Williams, C.K.I.: Gaussian Processes for Machine Learning, 3, vol. 2. MIT Press, Cambridge, MA (2006)
31. Reina, A., Zakir, R., De Masi, G., Ferrante, E.: Cross-inhibition leads to group consensus despite the presence of strongly opinionated minorities and asocial behaviour. Commun. Phys. **6**(1), 236 (2023). https://doi.org/10.1038/s42005-023-01345-3
32. Rizk, A., Batt, G., Fages, F., Soliman, S.: On a continuous degree of satisfaction of temporal logic formulae with applications to systems biology. In: International Conference on Computational Methods in Systems Biology, pp. 251–268. Springer (2008). https://doi.org/10.1007/978-3-540-88562-7_19
33. Schnoerr, D., Sanguinetti, G., Grima, R.: Approximation and inference methods for stochastic biochemical kinetics–a tutorial review. J. Phys. A Math. Theor. **50**(9), 093001 (2017). https://doi.org/10.1088/1751-8121/aa54d9
34. Seeley, T.D., Visscher, P.K., Schlegel, T., Hogan, P.M., Franks, N.R., Marshall, J.A.: Stop signals provide cross inhibition in collective decision-making by honeybee swarms. Science **335**(6064), 108–111 (2012). https://doi.org/10.1126/science.1210361

35. Vicsek, T., Czirók, A., Ben-Jacob, E., Cohen, I., Shochet, O.: Novel type of phase transition in a system of self-driven particles. Phys. Rev. Lett. **75**(6), 1226 (1995). https://doi.org/10.1103/PhysRevLett.75.1226
36. Zakir, R., Dorigo, M., Reina, A.: Miscommunication between robots can improve group accuracy in best-of-n decision-making. In: Proceedings of the 2024 IEEE/RSJ International Conference on Intelligent Robots and Systems (IROS 2024), pp. 9014–9021. IEEE Press (2024). https://doi.org/10.1109/IROS58592.2024.10801865
37. Zakir, R., Carletti, T., Dorigo, M., Reina, A.: Bio-inspired decision making in swarms under biases from stubborn robots, corrupted communication, and independent discovery. arXiv preprint arXiv:2509.07561 (2025). https://doi.org/10.48550/arXiv.2509.07561

Effective Stochastic Automata Model Checking by Interval Abstraction

Pedro R. D'Argenio[1], Arnd Hartmanns[2], and Annabell Petri[2](✉)

[1] Universidad Nacional de Córdoba and CONICET, Córdoba, Argentina
[2] University of Twente, Enschede, The Netherlands
annabell.petri@utwente.nl

Abstract. Stochastic automata (SA) are a formal stochastic continuous-time model based on countdown timers whose expiration times follow general probability distributions. SA are particularly useful to faithfully model and analyse dependable systems involving faults, maintenance, and repairs. Effective SA analysis approaches have so far been limited to statistical model checking and thus deterministic SA, while previously proposed model-checking techniques apply to limited subclasses of SA only, or do not scale. In this paper, we present the first dedicated SA model checking approach that is general and effective: It puts few restrictions on the input SA, and we show in our experimental evaluation that it works well for nontrivial examples. It combines a refinable interval abstraction of the continuous distributions with a direct application of the "big time steps" semantics of SA, providing upper/lower bounds on maximum/minimum reachability probabilities. We extend the Modest and Jani modelling formalisms with support for SA, and provide a prototype implementation of our approach in Rust.

1 Introduction

In dependable systems, component faults happen randomly over time, with schemes involving redundancy, regular inspections, maintenance, and repairs attempting to prevent escalation into overall system failure. Similarly, in high-performance systems, random queueing and service times, message loss probabilities, and signal transmission and propagation delays determine the system's key performance properties such as throughput or response times. Formal models for dependability and performance evaluation [4] need to capture these aspects of stochastic time in a realistic way, as close as possible to the real system or the available data. Two widely-used formalisms are continuous-time Markov chains (CTMCs) and generalized stochastic Petri nets (GSPNs) [19,38].

Authors are sorted in alphabetical order. This work was supported by the EU's H2020 R & I programme under MSCA grant agreement 101008233 (MISSION), by the Interreg North Sea project STORM_SAFE, by SeCyT-UNC grant 33620230100384CB (MECANO), and by NWO VIDI grant VI.Vidi.223.110 (TruSTy).

M. Chen and M. Tschaikowski (Eds.): QEST+FORMATS 2026, LNCS 16913, pp. 369–388, 2027.
https://doi.org/10.1007/978-3-032-35298-9_21

In both, all delays must be exponentially distributed. The resulting memoryless nature of these models admits scalable analytical analysis methods such as probabilistic model checking [2,3]. In reality, however, many inter-event times are not exponentially distributed, and hard to approximate by CTMCs/phase-type distributions: time to failure is more realistically modelled by a Weibull distribution in most cases, and inspections typically happen around fixed intervals. We thus need *non-Markovian* formalisms that directly incorporate general continuous probability distributions, such as stochastic automata (SA) [15] or Petri nets with general transitions (e.g. HPnGs [23]). In this paper, we use SA, because they are conceptually simple yet highly expressive. Additionally, their compositionality makes them attractive for modelling complex component-based systems.

SA extend labelled transition systems with stochastic *timers* (that may be *reset* to values sampled from continuous probability distributions to then decrease over time and *expire* when reaching value zero) and *guard sets* that enable an edge when all their timers have expired. As SA are non-Markovian *and* allow nondeterministic choices, their analysis is hard. The only available tool with dedicated SA support, FIG [9], employs statistical model checking [1,37] (i.e. Monte Carlo simulation) and is thus restricted to (weakly) deterministic SA [16]. Analytical approaches for more general models, such as HPnGs or stochastic timed automata (STA) [6,25], do not scale for typical SA that have many discrete locations and several timers possibly reset repeatedly on loops.

We present the first dedicated model checking approach for SA that is effective (i.e. works for typical and nontrivial SA models) and implemented in an available tool. It is based on *interval abstraction* [21]: we replace each clock reset, i.e. each sampling from a continuous distribution μ, by a discrete distribution over a finite set of intervals that partition μ's support. The choice of concrete value from an interval is nondeterministic, and we propagate the intervals symbolically through the semantics of the SA. This turns the SA into a Markov decision process (MDP) [5,32] that overapproximates the SA's semantics. As a result, the max. (min.) probability of eventually reaching a set of MDP states corresponding to a goal location in the SA—which we compute by value iteration [30]—is an upper (lower) bound on the corresponding probability in the SA.

We specify our approach formally and show the overapproximation property (Sect. 3). We extend the MODEST [6,26] and JANI [11] languages with support for timers and thus SA, providing a new user-friendly high-level modelling language for SA and a means to conveniently exchange SA models between tools (Sect. 4). Our new SA model checking approach is implemented in a prototype tool written in Rust, which we use to evaluate its performance and scalability characteristics on example SA from and inspired by the literature (Sect. 5). We find that, despite its prototypical nature, our tool already works well for reasonably-sized SA even when using many intervals to obtain results with good accuracy.

Related Work. FIG's simulation-based approach (mentioned above) is limited to a subclass of *input-output* SA that syntactically guarantees all nondeterminism to be spurious, i.e. minimum and maximum reachability probabilities coincide.

Two SA model checking algorithms were proposed by Bryans et al. in 2003 [7]. They use a clock-based variant of SA with severe restrictions (clocks may only be used out of locations in which they have just been reset, and all distributions must be non-negative and bounded). The first algorithm uses a "region tree" and requires solving nested integrals over the probability density functions of each expired clock, which becomes infeasible as the tree grows in depth. The second algorithm avoids the nested integrals by discretising the distributions. No tool implements either algorithm today. We note that our interval abstraction does not discretise the sampling results, but instead replaces continuous probabilistic choices by discrete probabilistic followed by continuous nondeterministic ones. For nondeterministic SA, both algorithms require a strategy to be user-specified, while our approach includes the optimisation over nondeterministic choices.

Model checking tool support exists for the following stochastic-timed formalisms that are more general than SA: (1) STA extend timed automata by allowing clocks to be compared to real-valued variables whose values can be sampled from general probability distributions. Compared to SA, they also support continuous nondeterministic delays, and are very (unrealistically) flexible in how clocks and the real-valued variables can be combined. Interval abstraction has been implemented for STA in the MODEST TOOLSET's [29] MCSTA model checker [25] in a way that reduces the STA to a probabilistic timed automaton [36], which is turned into an MDP via the digital clocks approach [31,35]. The implementation is limited to abstraction intervals of width 1 (or infinity). The digital clocks semantics enumerates all time steps of duration 1, making the MDP very large, but also preventing refinement to smaller intervals. (2) Stochastic hybrid automata (SHA) [21] further extend STA with continuous variables whose evolution over time is governed by differential equations and inclusions. The PROHVER tool applies interval abstraction to SHA, followed by a hybrid reachability analysis via PHAVER [22], to overapproximate reachability probabilities. Here, the abstraction intervals have to be specified manually by the user, and the hybrid reachability step—unnecessary for SA—is the scalability and performance bottleneck. (3) While PROHVER applies to linear SHA of type I, the REALYST tool [17] focuses on rectangular hybrid automata with random events (RAEs), into which (some) SA can also be encoded. REALYST's approach works for time-bounded reachability only, and requires structurally non-Zeno RAEs, which complicates the analysis of SA with cycles. As every timer reset must be encoded as a separate random variable in the RAE, each of which REALYST turns into one dimension of the model's continuous state, scalability—already limited by the complexity of hybrid reachability—for typical SA is low.

Finally, Buchholz et al. [8] present model checking algorithms for a generalization of CTMCs with clocks that they also call "stochastic automata", but clearly point out that their definition is less general than the one of [15] that we use.

2 Background

Let $\mathbb{R}_{\geqslant 0}$ be $[0, \infty)$, $\mathbb{R}_{>0} \stackrel{\text{def}}{=} (0, \infty)$, and $\mathbb{R}_{\infty} \stackrel{\text{def}}{=} \mathbb{R} \cup \{\infty\}$. A discrete *probability distribution* is a function $\mu \colon S \to [0,1]$ over a countable set S with $\sum_{s \in S} \mu(s) = 1$. Its *support* is $spt(\mu) \stackrel{\text{def}}{=} \{ s \in S | \mu(s) > 0 \}$. Now let S be an arbitrary set. We write $\mathcal{P}(S)$ for its powerset. Given a σ-algebra $\sigma(S)$ over S, subset $A \subseteq S$ is *measurable* if $A \in \sigma(S)$. A *probability measure* over S is a function $\mu \colon \sigma(S) \to [0,1]$ with $\mu(S) = 1$ and $\mu(\bigcup_{i \in I} A_i) = \sum_{i \in I} \mu(A_i)$ for any countable index set I and measurable pairwise disjoint $A_i \subseteq S$. The *support* of probability measure μ is the smallest compact (measurable) set $spt(\mu) \subseteq S$ s.t. $\mu(spt(\mu)) = 1$. Given two probability measures μ_1 and μ_2 over S, the *product measure* is the unique probability measure such that $(\mu_1 \otimes \mu_2)(A_1 \times A_2) = \mu_1(A_1) \cdot \mu_2(A_2)$ for all measurable $A_1, A_2 \subseteq S$. A probability measure μ over $\mathbb{R}$ is absolutely continuous if $\mu(A) = 0$ for all A of Lebesgue measure 0; we call it a *continuous probability distribution* (CPD). $DPD(S)$ is the set of all discrete probability distributions, $PD(S)$ of all probability measures, and $CPD(S)$ of all CPDs over appropriate S.

Stochastic Automata. [12,15] are labelled transition systems equipped with *timers*: real-valued variables that can be reset to values sampled from CPDs, decrease synchronously with time, and guard the SA's edges. Formally:

Definition 1. *A* stochastic automaton *(SA) is a tuple $\langle \mathcal{L}, \ell_I, \mathcal{A}, \mathcal{V}, \mathcal{E} \rangle$ consisting of finite sets of locations $\mathcal{L}$ with initial location $\ell_I \in \mathcal{L}$, action labels $\mathcal{A}$ with $\mathcal{A} \cap \mathbb{R} = \varnothing$, timer variables $\mathcal{V}$, and edges $\mathcal{E} \subseteq \mathcal{L} \times \mathcal{P}(\mathcal{V}) \times \mathcal{A} \times \mathcal{P}(\mathcal{V}) \times \mathcal{L}$. Each timer x has an associated CPD μ_x with $\mu_x(\mathbb{R}_{\geqslant 0}) = 1$. We write $\ell \xrightarrow{G,a,R} \ell'$ for $\langle \ell, G, a, R, \ell' \rangle \in \mathcal{E}$; such an edge consists of a source location ℓ, a guard set $G \subseteq \mathcal{V}$, an action $a \in \mathcal{A}$, a reset set $R \subseteq \mathcal{V}$, and a target location ℓ'.*

Unless noted otherwise, we assume SA $\mathcal{M} = \langle \mathcal{L}, \ell_I, \mathcal{A}, \mathcal{V}, \mathcal{E} \rangle$ to be given in the remainder of this paper. Let $\boldsymbol{v} \colon \mathcal{V} \to \mathbb{R}$ be a timer valuation. Let valuation $\mathbf{0}$ be defined by, for all $t \in \mathcal{V}$, $\mathbf{0}(t) = 0$, and $\boldsymbol{v} - d$ for $d \in \mathbb{R}$ by $(\boldsymbol{v} - d)(t) = v(t) - d$.

In an SA's semantics, we start in ℓ_I with all timers having value 0. When in location ℓ, all timers synchronously decrease at rate 1 over time. If $\ell \xrightarrow{G,a,R} \ell'$ and all timers in g have *expired* (i.e. have value ≤ 0), then no further time can pass (maximal progress), and the edge must be taken: the timers in R are *reset* (i.e. assigned a new value randomly sampled from their associated CPD), and we continue in location ℓ'. The semantics also allows *discrete nondeterministic* choices between edges if there is another edge whose guard's timers have all expired. Formally, it is given in terms of an infinite-state but finitely branching transition system:

Definition 2. *A* timed probabilistic transition system *(TPTS) is defined as a tuple $\langle \mathcal{S}, s_I, \mathcal{A}', \mathcal{T} \rangle$ consisting of a measurable set of states $\mathcal{S}$ with initial state $s_I \in \mathcal{S}$, an alphabet $\mathcal{A}' = \mathbb{R}_{>0} \uplus \mathcal{A}$ partitioned into* delays *in $\mathbb{R}_{>0}$ and actions in $\mathcal{A}$, and transitions $\mathcal{T} \subseteq \mathcal{S} \times \mathcal{A}' \times PD(\mathcal{S})$. We write $s \xrightarrow{\alpha} \mu$ for $\langle s, \alpha, \mu \rangle \in \mathcal{T}$; such a transition consists of a source state s, label $\alpha \in \mathcal{A}'$, and measure over target states μ. For all $s \in \mathcal{S}$, we require $\mathcal{T}_s \stackrel{\text{def}}{=} \{ \langle s, \alpha, \mu \rangle \in \mathcal{T} \}$ to be finite and $|\mathcal{T}_s| = 1$ if $\exists \langle s, d, \mu \rangle \in \mathcal{T} \colon d \in \mathbb{R}_{>0}$ (i.e. states admitting delays are deterministic).*

Definition 3. *The TPTS* $[\![M]\!] \stackrel{\text{def}}{=} \langle \mathcal{L} \times (\mathcal{V} \to \mathbb{R}_\infty), \langle \ell_I, \mathbf{0} \rangle, \mathcal{A} \cup \mathbb{R}_{>0}, T_{[\![M]\!]} \rangle$ *is the* semantics *of an SA* $\mathcal{M}$ *with* $T_{[\![M]\!]}$ *being the smallest relation satisfying*

$$\frac{\ell \xrightarrow{G,a,R} \ell' \quad \bigwedge_{x_i \in G} \boldsymbol{v}(x_i) \leq 0}{\langle \ell, \boldsymbol{v} \rangle \xrightarrow{a} \{\, s' \mapsto 1 \,\} \otimes (\bigotimes_{x_i \in R} \mu_{x_i}) \otimes (\bigotimes_{x_i \notin R} \{\, \boldsymbol{v}(x_i) \mapsto 1 \,\})} \quad \text{(S1)}$$

and

$$\frac{0 < d = \min\{\, \max\{\, \boldsymbol{v}(x_i) \mid x_i \in G \,\} \mid \ell \xrightarrow{G,a,R} \ell' \,\}}{\langle \ell, \boldsymbol{v} \rangle \xrightarrow{d} \{\, \langle \ell, \boldsymbol{v} - d \rangle \mapsto 1 \,\}} \quad \text{(S2)}$$

with all resulting deadlock states receiving an ∞*-labelled self-loop.*

This definition is the "big time steps" semantics of SA [13]: Rule S2 admits only the delay that directly enables an edge, and no intermediate delays as in other "dense time steps" semantics. For the *closed* SA and properties we use, the two semantics are equivalent (in essence, weakly bisimilar in a time-abstract way).

Example 1. Figure 1 shows our running example SA $\mathcal{M}_{re}$. It has two timers x and y whose expiration times are sampled from the given uniform distributions. The two edges labelled a and τ out of locations ℓ_0 and ℓ_2 have empty guard sets: they must be taken immediately upon entering those locations. In ℓ_1, the outgoing edges can only and must be taken when the respective timer has expired. If y expires before x, we go to ℓ_2, and to ℓ_3 otherwise. Timers are only reset on the edge from ℓ_0 to ℓ_1 as denoted by the reset set $\text{R}(\{x, y\})$. We omit empty reset sets.

In general, SA are *compositional*: They can be equipped with a notion of *parallel composition* and an *open* semantics with dense time steps and without maximal progress. This facilitates elegant and compact models of systems consisting of concurrent components. For analysis, usually the composition's closed semantics which excludes parallel composition and synchronization on shared actions between SA is considered, and thus we use single closed SA in this paper.

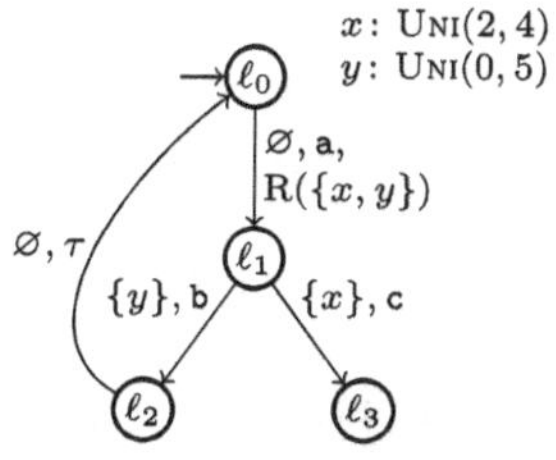

Fig. 1. Example SA $\mathcal{M}_{re}$.

Properties. Given an SA, we are interested in the probability of reaching any of a set of goal locations $\mathcal{G} \subseteq \mathcal{L}$, defined via the probability of reaching any state in $[\![\mathcal{G}]\!] \stackrel{\text{def}}{=} \{\, \langle \ell, \boldsymbol{v} \rangle \mid \ell \in \mathcal{G} \,\}$ in the TPTS semantics. In a TPTS, a *path* is an infinite sequence $s_0 \langle a_0, \mu_0 \rangle s_1 \langle a_1, \mu_1 \rangle \ldots \in (\mathcal{S} \times (\mathcal{A}' \times PD(\mathcal{S})))^\omega$ with $s_0 = s_I$ and $\forall i \colon s_i \xrightarrow{a_i} \mu_i \wedge s_{i+1} \in spt(\mu_i)$. Paths resolve all nondeterministic and probabilistic choices. A *strategy* resolves only the nondeterminism:

Definition 4. $\mathfrak{s} \colon \mathcal{S} \to \mathcal{A}' \times PD(\mathcal{S})$ *is a* strategy *if* $\mathfrak{s}(s) = \langle \alpha, \mu \rangle \Rightarrow s \xrightarrow{\alpha} \mu$.

We work with memoryless deterministic strategies. Randomised ones would be needed for e.g. multi-objective properties [20], and for reachability properties on SA, memoryless and history-dependent strategies are equally expressive [13].

Every strategy $\mathfrak{s}$ induces a probability measure $\mathbb{P}_{\mathfrak{s}}$ on the space of all paths. For a formal definition, see [41]. As is usual, we restrict to *non-Zeno* strategies: we require $\mathbb{P}_{\mathfrak{s}}(\Pi_\infty) = 1$, where Π_∞ is the set of paths whose sum of delays is ∞.

Definition 5. *Consider a TPTS that is the semantics of an SA with goal set* $\mathcal{G} \subseteq \mathcal{L}$. *Then* $\mathrm{P}_{\min}(\diamond \mathcal{G})$ *and* $\mathrm{P}_{\max}(\diamond \mathcal{G})$ *are the minimum and maximum* reachability probabilities *for* $\mathcal{G}$, *defined as* $\mathrm{P}_{\min}(\diamond \mathcal{G}) = \inf_{\mathfrak{s}} \mathbb{P}_{\mathfrak{s}}(\Diamond_{\mathcal{G}})$ *and* $\mathrm{P}_{\max}(\diamond \mathcal{G}) = \sup_{\mathfrak{s}} \mathbb{P}_{\mathfrak{s}}(\Diamond_{\mathcal{G}})$, *respectively, with* $\Diamond_{\mathcal{G}}$ *the set of paths that contain a state in* $[\![\mathcal{G}]\!]$.

Markov decision processes are finite TPTSs whose transitions lead into *discrete* probability distributions. They are the result of interval abstraction of SA.

Definition 6. *A* Markov decision process *(MDP) is a tuple* $\langle \mathcal{S}, s_I, \mathcal{A}, \mathcal{T} \rangle$ *with finite sets of states* $\mathcal{S}$ *with initial state* $s_I \in \mathcal{S}$, *action labels* $\mathcal{A}$, *and transitions* $\mathcal{T} \subseteq \mathcal{S} \times \mathcal{A} \times \mathit{DPD}(\mathcal{S})$. *For all* $s \in \mathcal{S}$, *we require non-empty* $\mathcal{T}_s \stackrel{\text{def}}{=} \{ \langle s, \alpha, \mu \rangle \in \mathcal{T} \}$.

We can define reachability probabilities for MDPs in a similar way as for TPTSs. On MDPs, they can be computed or approximated efficiently via linear programming, policy iteration, or various variants of value iteration (VI) [30].

3 Big-Steps Interval Abstraction for SA

Interval abstraction of models with continuous probability distributions was introduced for the analysis of SHA [21], where it turned the SHA into a *probabilistic* hybrid automaton (PHA) [40,42], which could in turn be analysed by combining a hybrid reachability checker with MDP model checking. Interval abstraction replaces sampling from a CPD μ by a combination of (1) discrete sampling from a finite set of intervals that partition (or, in general, cover) $\mathit{spt}(\mu)$, with interval I_i chosen with probability $\mu(I_i)$, followed by (2) a continuous nondeterministic choice to choose a concrete value from the sampled interval. Interval abstraction does not discretise the final outcome: Every value in $\mathit{spt}(\mu)$ can still be chosen, but instead of being fully random, the strategy now makes the choice within the sampled interval. Since we use absolutely continuous distributions, we can neglect singleton overlaps between the intervals (as they have probability 0), and will thus use closed intervals throughout. The intervals to use for the abstraction can be specified in different ways. In our implementation and examples, we specify the desired probability mass of each interval, but we could equally specify the interval width or just the intervals themselves.

Example 2. When abstracting uniform distribution $\textsc{Uni}(0,5)$ from Example 1 with per-interval probability of 0.5, we obtain intervals $[0, 2.5]$ and $[2.5, 5]$. For $\textsc{Exp}(1)$, the exponential distribution with rate 1, we obtain $[0, -\ln 0.5]$ and $[-\ln 0.5, \infty)$.

We apply interval abstraction to the semantics of Definition 3, but then propagate the intervals symbolically. We thus avoid (in effect: postpone) the explicit continuous nondeterministic choices and obtain a finite MDP. In this section, we specify our transformation formally and establish its relation to the TPTS semantics.

3.1 SA Interval Abstraction

Let $\mathfrak{I}$ denote the set of all closed intervals in $\mathbb{R}^{\infty}_{\geqslant 0}$ together with all intervals of the form $[l, \infty)$. Given interval I, let $lI = \inf(I)$ and $uI = \sup(I)$. A set of intervals $B \subseteq \mathfrak{I}$ is *non-overlapping* if for all $I_1, I_2 \in B$, $I_1 \cap I_2 \neq \varnothing$ implies either $uI_1 = lI_2$ or $uI_2 = lI_1$. Let $\mathscr{P}_{\mathrm{nol}}(\mathfrak{I})$ be the set of all non-overlapping subsets of $\mathfrak{I}$. A *timer-domain partition* is a function $\mathfrak{p}\colon \mathcal{V} \to \mathscr{P}_{\mathrm{nol}}(\mathfrak{I})$ s.t. for all $x \in \mathcal{V}$, $\bigcup \mathfrak{p}(x) = spt(\mu_x)$.

Given an SA $\mathcal{M}$ as usual and a timer-domain partition $\mathfrak{p}$, we define its *interval abstraction* as the MDP $\mathsf{ia}(\mathcal{M}) = (S_{\mathsf{ia}}, \langle \ell_I, \mathbf{0}_{\mathsf{ia}} \rangle, \mathcal{A} \cup \{\sigma\}, T_{\mathsf{ia}})$ where

- $\sigma \notin \mathcal{A}$ is a fresh new label indicating some unspecified sojourn time;
- $S_{\mathsf{ia}} = \mathcal{L} \times (\mathcal{V} \to \mathfrak{I} \cup \{ \lightning \})$ where, for each state $\langle \ell, b \rangle \in S_{\mathsf{ia}}$ and $x \in \mathcal{V}$, if $b(x) = \lightning$, timer x has already expired in this state and, otherwise, $b(x)$ is the interval containing all possible values timer x can take in this state with $lb(x)$ and $ub(x)$ denoting the interval's lower and upper bound, respectively, and for a set $C \subseteq \mathcal{V}$ of timers, we let $lb(C) = \max\{ lb(x) \mid x \in C \wedge b(x) \neq \lightning \}$ (the delay after which *all* timers *can* have expired) and $ub(C) = \max\{ ub(x) \mid x \in C \wedge b(x) \neq \lightning \}$ (the delay after which *all* timers *must* have expired);
- $\mathbf{0}_{\mathsf{ia}}$ maps every timer to $\lightning$ in the initial state; and
- T_{ia} is the smallest relation satisfying the following two inference rules:

$$\frac{\ell \xrightarrow{G,a,R} \ell' \quad \forall x \in G\colon b(x) = \lightning}{\langle \ell, b \rangle \xrightarrow{a} \pi^b_{R,\ell'}} \qquad \text{(R1)}$$

with the transitions' target distributions defined by

$$\pi^b_{R,\ell'}(\langle \ell'', b' \rangle) = \begin{cases} \prod_{x \in R} \mu_x(b'(x)) & \text{if } \ell'' = \ell', \\ & \forall x \notin R\colon b'(x) = b(x), \text{ and} \\ & \forall x \in R\colon b'(x) \in \mathfrak{p}(x) \\ 0 & \text{otherwise} \end{cases}$$

and

$$\frac{\ell \xrightarrow{G,a,R} \ell' \quad \forall \ell \xrightarrow{G',a',R'} \ell''\colon (\exists x \in G'\colon b(x) \neq \lightning) \wedge lb(G) \leq ub(G')}{\langle \ell, b \rangle \xrightarrow{\sigma} \{ \langle \ell, d^b_{\ell,G} \rangle \mapsto 1 \}} \qquad \text{(R2)}$$

with

$$d^b_{\ell,G}(x) = \begin{cases} \lightning & \text{if } x \in G \vee b(x) = \lightning \vee ub(x) \leq lb(G) \\ \left[\max\{0, lb(x) - \min\{ub(G') \mid \ell \xrightarrow{G',a',R'} \ell''\}\}, ub(x) - lb(G) \right] & \text{else} \end{cases}$$

with all resulting deadlock states receiving a σ-labelled self-loop.

Rule R1 creates an MDP transition to $\pi^b_{R,\ell'}$ whenever an edge in the SA would be enabled in the current state; the discrete distribution $\pi^b_{R,\ell'}$ samples an interval for every timer that is reset. Rule R2 creates σ-labelled transitions, which represent the symbolic passage of some time, as follows:

- The rule picks an edge $e = \ell \xrightarrow{G,a,R} \ell'$ out of the current location ℓ in the SA that should become enabled. If ℓ has no edges, then no symbolic time passes.
- First, *no* edge must be enabled in the current state—otherwise, by maximal progress, it would have to be taken immediately. This is condition "$\exists x \in G' \colon b(x) \neq \lightning$": In every guard set, at least one timer is not expired.
- Additionally, no edge may be *forced* to be taken earlier. This is condition "$lb(G) \leq ub(G')$": else, edge e' with guard G' *must* be taken after waiting for $ub(G')$ time units, while e needs *at least* delay $lb(G) > ub(G')$ to be enabled.
- If these conditions hold, then all timers in G *can* expire, and we let them do so. Function $d^b_{\ell,G}$ determines the new intervals of values that all timers not in G can then be in: The first case states that all timers are expired ($\lightning$) that (i) are in G, (ii) have previously expired, or (iii) whose upper bound is at most $lb(G)$ (which applies to timers not in G that have "low" values). In the second case, we have an unexpired timer $x \notin G$ that is not *forced* to expire now. Its lower bound is reduced by the maximum amount of time that could have passed, which is the lowest upper bound of the guard set of any edge out of ℓ, but not below 0. If the new lower bound is 0, x *could* have expired, but since $x \notin G$, it by choice *did not* expire.[1] The upper bound, in turn, is decreased by the lowest possible delay needed for all timers in G to expire.

Example 3. Applying interval abstraction to $\mathcal{M}_{re}$ of Example 1 with a per-interval probability of 0.5 results in the MDP of which an excerpt is shown in Fig. 2. In ℓ_0, initially, all timers are expired, so the a-labelled edge in the SA can be taken. It resets both timers, thus the corresponding transition in the MDP as per R1 leads to a distribution over the four combinations of intervals for x and y.

Location ℓ_1 has two edges, with two distinct non-empty guards, $\{y\}$ and $\{x\}$. By R2, we get up to two σ-labelled transitions out of each state in the second row of Example 1. For clarity, we instead label the ones where $\{y\}$ ($\{x\}$) was chosen to expire with δ_y (δ_x). Note that in ${\ell_1}^{[3,4]\times[0,2.5]}_{R2}$, only the expiration of $\{y\}$ is considered because the expiration of $\{x\}$ would violate the "$lb(G) \leq ub(G')$" requirement: $ub(\{y\}) = 2.5 \leq lb(\{x\}) = 3$, i.e. y *must* expire before x.

In the third row, the timers of the chosen guard set have expired, and the corresponding SA edge labelled b or c can be taken as per R1. Observe that timers not in the chosen guard set may have lower bound 0 now but are not

[1] Two timers expiring simultaneously has probability 0 because of our restriction to absolutely continuous distributions, so we can disregard this case here.

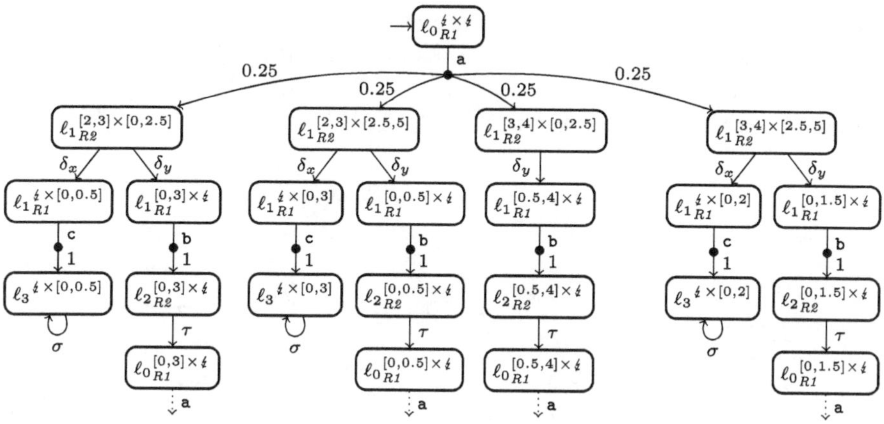

Fig. 2. Excerpt of $\mathsf{ia}(\mathcal{M}_{re})$ created with per-interval probability 0.5. The superscript intervals denote the possible valuations for timers x (on the left) and y (on the right).

considered expired, as explained earlier; thus e.g. in state $\ell_1{}_{R1}^{[0,0.5]\times\lightning}$ only the b-labelled transition is possible despite $lb(x) = 0$.

Finally, in the fourth row, we can either loop back to ℓ_0 from ℓ_2, or else we are stuck in ℓ_3. In the former case, the a-labelled transitions reset both timers, and thus lead into the same distribution (not shown in Example 1) as the a-labelled transition out of the initial state, looping back into the second row. Example 1 thus includes all 23 states of the interval abstraction and all transitions.

3.2 Correctness

In order to check the optimal probability of reaching a location ℓ (or a set of locations) in the SA, we compute the optimal probability of reaching a state $\langle \ell, g\rangle$ for any g in the interval abstraction MDP (see Sect. 2). Since the SA is finite, and the abstraction intervals are fixed for every timer, this MDP must also be finite—but it can be large when many different combinations of (post-delay) intervals can be reached. MDP model checking, however, has good practical scalability, solving models of hundreds of millions of states on standard hardware today.

The interval abstraction does not preserve reachability probabilities exactly, however; it rather introduces an overapproximation from two sources: First, MDP strategies can choose the precise value to use within the intervals whereas this choice is fully random in the SA semantics. Second, this choice is actually postponed from the moment the edge that does the sampling is taken to the point that one of the transitions generated by R2 must be chosen. This moves nondeterministic choices over (other) probabilistic choices, which also gives more information to the strategies.[2] Thus maximum (minimum) reachability probabilities in the MDP overapproximate (underapproximate) the corresponding probabilities in the SA's semantics:

[2] Consider tossing a coin to *then* nondeterminstically guess its value vs. the inverse.

Theorem 1. *Given an SA $\mathcal{M}$ as usual and a set of goal locations $\mathcal{G} \subseteq \mathcal{L}$, let $p_{min}^{\mathcal{M}} = \mathrm{P}_{\min}(\diamond \mathcal{G})$ and $p_{max}^{\mathcal{M}} = \mathrm{P}_{\max}(\diamond \mathcal{G})$ be the optimal reachability probabilities obtained from $[\![\mathcal{M}]\!]$ as per Definition 5, and let $p_{min}^{\mathsf{ia}(\mathcal{M})}$ and $p_{max}^{\mathsf{ia}(\mathcal{M})}$ be the analogous probabilities obtained from the MDP $\mathsf{ia}(\mathcal{M})$. Then*

$$p_{min}^{\mathsf{ia}(\mathcal{M})} \leq p_{min}^{\mathcal{M}} \quad \text{and} \quad p_{max}^{\mathcal{M}} \leq p_{max}^{\mathsf{ia}(\mathcal{M})}.$$

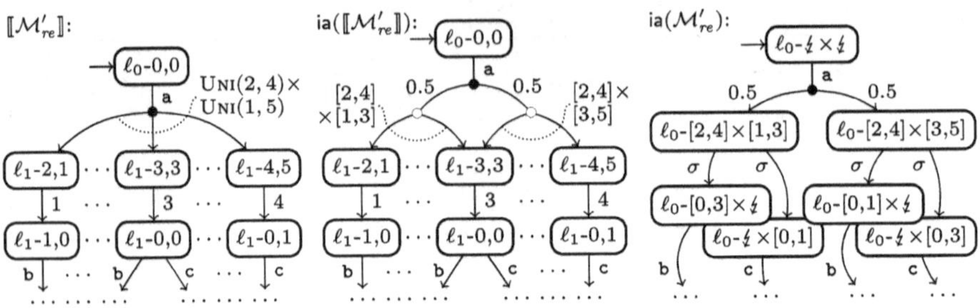

Fig. 3. Transformations of the semantics of the SA $\mathcal{M}'_{re}$.

Proof. (sketch). Our proof is inspired by that of [24, Thm. 4.22]. We illustrate the idea with a variant $\mathcal{M}'_{re}$ of $\mathcal{M}_{re}$ from Fig. 1 with CPDs slightly simplified to x: $\mathrm{U{\small NI}}(2,4)$ and y: $\mathrm{U{\small NI}}(1,3)$, and abstraction intervals $\mathfrak{p}'_{re}(x) = \{\, [2,4] \,\}$ and $\mathfrak{p}'_{re}(y) = \{\, [1,3], [3,5] \,\}$. We show an excerpt of $[\![\mathcal{M}'_{re}]\!]$ on the left of Fig. 3.

The interval abstraction directly maps an SA $\mathcal{M}$ to the MDP $\mathsf{ia}(\mathcal{M})$. We introduce an intermediate step: an interval abstraction $\mathsf{ia}([\![\mathcal{M}]\!])$ of $[\![\mathcal{M}]\!]$ over the same $\mathfrak{p}$. It is obtained from $[\![\mathcal{M}]\!]$ by replacing each continuous distribution μ_{x_i} in rule S1 of Definition 3 by a discrete distribution $\mu_{x_i}^{\mathsf{ia}}$ over the intervals in $\mathfrak{p}(x_i)$ followed by a continuous nondeterministic selection of a concrete value from the chosen interval such that $\mu_{x_i}^{\mathsf{ia}}(I) = \mu_{x_i}(I)$. We show $\mathsf{ia}([\![\mathcal{M}]\!]'_{re})$ in the middle of Fig. 3. Where a transition in $[\![\mathcal{M}]\!]$ leads into a continuous distribution over states, in $\mathsf{ia}([\![\mathcal{M}]\!])$, it now leads into a discrete distribution (solid dot •) over sets of states out of which one is chosen nondeterministically (outlined dots ∘).

The (reachable) states of $[\![\mathcal{M}]\!]$ and $\mathsf{ia}([\![\mathcal{M}]\!])$ are the same, as is the transition graph structure. Given a strategy $\mathfrak{s}$ for $[\![\mathcal{M}]\!]$, we can construct a randomised strategy $\mathfrak{s}_{[\![\mathsf{ia}]\!]}$ for $\mathsf{ia}([\![\mathcal{M}]\!])$ that induces the exact same probability measure: First, fix $\mathfrak{s}(s) = \mathfrak{s}_{[\![\mathsf{ia}]\!]}(s)$ for all states s. Then, $\mathfrak{s}_{[\![\mathsf{ia}]\!]}$ resolves all the new continuous nondeterministic choices of values from intervals randomly so that $\mathfrak{s}_{[\![\mathsf{ia}]\!]} \circ \mu_{x_i}^{\mathsf{ia}} = \mu_{x_i}$ everywhere. This is possible because $\mu_{x_i}^{\mathsf{ia}}$ and μ_{x_i} assign the same probabilities to all these intervals. $\mathsf{ia}([\![\mathcal{M}]\!])$ may have additional strategies not corresponding to any in $[\![\mathcal{M}]\!]$, and we thus have $p_{min}^{\mathsf{ia}([\![\mathcal{M}]\!])} \leq p_{min}^{\mathcal{M}}$ and $p_{max}^{\mathcal{M}} \leq p_{max}^{\mathsf{ia}([\![\mathcal{M}]\!])}$.

Observe that, by rule S2 of Definition 3, $\mathsf{ia}([\![\mathcal{M}]\!])$ contains no nondeterminism over delays, delay transitions are deterministic, and only connect states that are in the same SA location. We can thus collapse all delay transitions, i.e. delete them and copy the outgoing transitions of their successor states onto their source states, without affecting the trace distributions modulo delay steps of $\mathsf{ia}([\![\mathcal{M}]\!])$. Denote the result as $\mathsf{ia}([\![\mathcal{M}]\!])'$. Now, $\mathsf{ia}(\mathcal{M})$ weakly probabilistically simulates $\mathsf{ia}([\![\mathcal{M}]\!])'$ with the δ transitions being unobservable steps (that subsume the deterministic delays of $\mathsf{ia}([\![\mathcal{M}]\!])$ but also pull forward the subsequent nondeterministic choice of action-labelled transition). This gives us

$$p_{min}^{\mathsf{ia}(\mathcal{M})} \leq p_{min}^{\mathsf{ia}([\![\mathcal{M}]\!])'} = p_{min}^{\mathsf{ia}([\![\mathcal{M}]\!])} \quad \text{and} \quad p_{max}^{\mathsf{ia}([\![\mathcal{M}]\!])} = p_{max}^{\mathsf{ia}([\![\mathcal{M}]\!])'} \leq p_{max}^{\mathsf{ia}(\mathcal{M})}. \qquad \square$$

3.3 Refinement

As we make the abstraction intervals smaller, the strategy in $\mathsf{ia}([\![\mathcal{M}]\!])$ gets fewer new choices, and the overlap between intervals in $\mathsf{ia}(M)$ reduces. We thus expect that, the smaller the intervals are, the more closely we approximate true optimal probabilities. This would give rise to a refinement-based approach where we start with coarse intervals, and repeatedly split them until e.g. the max. probability to reach an unsafe state can be shown to be below a specified safety threshold.

However, although we conjecture that the abstraction error goes to zero as the width of the intervals goes to zero, it is not yet clear how to prove this. In particular, the effect of reducing interval width on the error is not monotonic: Consider another variant of $\mathcal{M}_{re}$ where x: $\textsc{Uni}(3,4)$ and y: $\textsc{Uni}(0,4)$, the τ-labelled edge is removed, and we are interested in $\mathrm{P}_{\max}(\diamond\{\ell_3\})$ (i.e. x wins the race). We have $\mathrm{P}_{\max}(\diamond\{\ell_3\}) = \frac{1}{8}$. With timer-domain partition $\mathfrak{p}_1$ given by $\mathfrak{p}_1(x) = \{[3,4]\}$ and $\mathfrak{p}_1(y) = \{[0,2],[2,4]\}$, we get max. probability $\frac{1}{2}$: Iff $[2,4]$ is sampled for y, which happens with probability $\frac{1}{2}$, we have a choice to let x expire first, because it is the only interval for y that overlaps with the one interval $[3,4]$ for x. With $\mathfrak{p}_2$ given by $\mathfrak{p}_2(x) = \{[3,4]\}$ and $\mathfrak{p}_2(y) = \{[0,1],[1,3],[3,4]\}$, which has smaller-or-equal intervals than $\mathfrak{p}_1$, we however get *higher* (i.e. worse) max. probability $\frac{3}{4}$ as x can be chosen to win when either $[1,3]$ or $[3,4]$ is sampled.

We suspect that monotonicity can be achieved with the right notion of "refinement", such as never removing interval bounds as in our example, which could open the way for a convergence proof. Generating intervals based on per-interval probabilities, however, does not guarantee such a kind of refinement.

4 Modelling and Model Exchange

Specifying SA mathematically as in Definition 1 or via drawings as in Fig. 1 is not convenient to build large SA models of real-life systems: We would want to use parallel composition, and have the ability to use discrete (e.g. Boolean and integer) variables to express data-dependent control flow. The only modelling language for SA providing these facilities so far was IOSA as used by FIG, which is an extension of the PRISM language [34]. It is specific to the input-output and weakly deterministic nature of the SA that FIG requires.

As part of our work on implementing model checking for general SA via interval abstraction, we also extended the MODEST language to support SA directly. MODEST is a high-level modelling language inspired by process algebra but with a syntax akin to common programming languages; it provides standard control flow constructs as well as exception handling, process declarations and calls to reuse code, and user-defined datatypes and functions. We mainly added a new `timer` type, and an "sa" option to request the maximal progress semantics of SA. Guard sets and resets can then be expressed using the standard MODEST guard expressions and assignment constructs. This makes modelling SA in MODEST slightly more flexible than our Definition 1 as different distributions can be used for the same timer in different assignments; this can be mapped back to Definition 1 by introducing additional timers if necessary. In Fig. 4, we show one way to express SA $\mathcal{M}_{re}$ of Example 1 in MODEST.

```
option "sa";
action a, b, c;
timer x, y;
do {
   a {= x = Uni(2, 4),
        y = Uni(0, 5) =};
   alt {
   :: when(y <= 0) b; tau
   :: when(x <= 0) c; stop
   }
}
```

Fig. 4. MODEST model for $\mathcal{M}_{re}$.

Implementing a new tool for the analysis of SA would now require implementing a parser for either IOSA (if its restrictions are acceptable) or MODEST, which is a large and tedious task in either case. To reduce such burden on tool developers, the JSON-based JANI format was designed as a model interchange format between tools. We also extended JANI with support for timers and a new `sa` model type, allowing MODEST SA models to be translated to JANI, which could then be read by other future SA analysis tools.

5 Experiments

To evaluate the scalability and performance of our new SA model checking approach, we created a prototype implementation and applied it to compute maximal and minimal reachability probabilities for three sets of example SA: Our running example and the small SA that distinguish strategy classes in [13], a set of (Markovian and non-Markovian) queueing models, and variants of the file server model from [25]. These examples cover all common features of SA that we support: discrete nondeterministic choices, discrete probability distributions over target locations, immediate and timer-guarded edges, including guards with multiple timers and races between edges, and timers following different CPDs.

5.1 Implementation

We implemented a prototype tool for our SA analysis approach in Rust. We chose Rust as a future-proof language for a new tool that delivers both high performance and a reasonable safety level, but note that our prototype is not yet highly optimised. It takes as input a stochastic automaton specified in a simple textual file format and for each timer x the per-interval probability p_I^x that determines how its distribution's support is abstracted into intervals. We extended the MODEST TOOLSET with a conversion from MODEST SA models into this format.

Using the given p_I^x for timer x with cumulative distribution function (CDF) F_x, the tool computes the intervals for each distribution by partitioning the support into $1/p_I^x$ intervals of equal probability mass p_I^x, with boundaries at the quantiles $F_x^{-1}(k \cdot p_I^x)$ for $k = 1, 2, \ldots$. If $1/p_I^x$ is not an integer, the last interval covers the remainder of the distribution's support extending to the upper end of the support, which may be ∞ for distributions such as exponential or Weibull, and might have a probability smaller than p_I^x. The tool currently supports timers adhering to uniform, exponential, Erlang, and Weibull distributions. Adding support for a new distribution is straightforward if there exists a way to compute its inverse cumulative distribution function F_x^{-1}, which is necessary to find intervals of probability mass p_I^x that partition the distribution's support. For the Erlang distribution, we currently use the `ruststat` library's implementation of the inverse CDF for the gamma distribution.

In addition to reachability probabilities, the tool also supports computing the probabilities of paths satisfying a $\neg avoid \mathbin{\mathsf{U}} \mathcal{G}$ until formula. After performing interval abstraction as described in Sect. 3.1, the tool uses standard VI on the resulting MDP to compute the minimum and maximum reachability probabilities. It uses an absolute difference of $\epsilon = 10^{-6}$ stopping criterion for VI. We plan to replace VI by the sound and floating point-safe version of interval iteration implemented in MCSTA [28] as the tool matures. In addition to returning the computed probabilities, the tool can also export the interval abstraction MDP in Graphviz `.dot` format.

5.2 Experimental Setup

Our experiments were performed on an AMD Ryzen 7 PRO 7730U (2.0–4.5 GHz) laptop with 32 GB of RAM running openSuse Leap 16 Linux. The tool was built with Rust's default release build profile. We measured the total tool runtime with the Linux `time` command; the tool additionally outputs the time it took for parsing, computing the intervals via CDF quantiles, constructing the interval abstraction MDP from the SA, and running VI for both minimum and maximum probability. The first two steps always took negligible time in our experiments. In our tables, we round all probabilities to three decimal places.

5.3 Experimental Results

Small SA. We first ran the tool on $\mathcal{M}_{re}$ from Example 1 and SA M_1, M_2, and M_5 from [13] (shown in [14, App. A]). Of the SA from [13], we chose the three where the optimal reachability probabilities of interest are nontrivial, i.e. not 0 and 1 for minimum and maximum, respectively, for the strategy class that our approach effectively uses (memoryless prophetic with full information).

Table 1. Results for the running example and the SA from [13].

Model	$\mathcal{M}_{re}$		M_1		M_2		M_5	
Property	p_{min}	p_{max}	p_{min}	p_{max}	p_{min}	p_{max}	p_{min}	p_{max}
True probability	1	1	0.25	0.75	0.03125	0.96875	0.125	0.875
Result	1.000	1.000	0.245	0.755	0.004	0.996	0.117	0.883
p_I	0.05		0.01		0.05		0.0125	
MDP states	2378		82657		467159		260395	
Total tool runtime	0.2 s		1.5 s		4.2 s		20.6 s	

The results are shown in Table 1. For $\mathcal{M}_{re}$, we use $\mathcal{G} = \{\ell_3\}$. It is easy to see that the minimum and maximum probabilities to reach ℓ_3 are both 1 (as there is a positive probability to go to ℓ_3 in each loop iteration, and we consider unbounded reachability), and our tool correctly approximates them (up to a small VI error).

Table 2. Results for M_1 from [13] with different per-interval probabilities.

p_I	0.5		0.1		0.05		0.01		0.005	
Property	p_{min}	p_{max}	p_{min}	p_{max}	p_{min}	p_{max}	p_{min}	p_{max}	p_{min}	p_{max}
Result	0.000	1.000	0.200	0.800	0.225	0.775	0.245	0.755	0.247	0.752
MDP states	45		957		3589		82657		325565	
Total tool runtime	0.08 s		0.07 s		0.09 s		1.6 s		13.1 s	

We also correctly under- and overapproximate the true probabilities for M_1, M_2, and M_5 (see [14, App. A] for the analytical calculations). These three automata are constructed in such a way that, in states of the abstraction MDP where the intervals for the model's timers overlap (e.g. $b(x) = [0, 0.5]$ and $b(y) = [0, 1]$), a prophetic strategy can force the goal to be reached (or not). Thus, the smaller we make per-interval probability, the more intervals we have at our disposal, and the lower is the probability (in the MDP) that the timers are reset to overlapping intervals, so we can more precisely approximate the actual probabilities. We investigate this effect for M_1 by varying p_I, with the results shown in Table 2. As expected, we see that, as p_I decreases, p_{max} decreases and p_{min} increases. Overall, on the models from [13], we obtain good approximations of the actual results in short runtimes despite the overapproximation from the abstraction.

Queues. Queuing systems, in particular non-Markovian ones, can be modelled by SA and are interesting because every location features a race condition between two timers for arrival and service. We consider two systems with bounded buffers so we get finite SA: An $M/M/1/100$ queue with

Exp(2)-distributed interarrival and *Exp*(0.5)-distributed service times and a *Weibull*/E_6/1/100 queue with *Weibull*(2, 1.5) interarrival and *Erlang*(6, 2.5) service times. The two SA's locations encode the number of jobs in the queue and the number of jobs completed so far. We compute the probability of reaching a queue overflow before a given number of jobs is completed. As there is no nondeterminism in these models, our bounds on p_{min} and p_{max} under- and overapproximate the one true probability, respectively, and we thus report our result as intervals.

Table 3. Results for *M*/*M*/1/100 and *Weibull*/E_6/1/100 queues with $p_I = 0.1$.

	M/M/1/100		Weibull/E_6/1/100	
Job bound	150	200	150	200
Result	[0.000, 0.979]	[0.000, 0.263]	[0.000, 0.924]	[0.000, 0.084]
SA locations	10202	15252	10202	15252
MDP states	2229270	3334270	2229270	3334270
Interval abstraction	46.3 s	67.3 s	47.2 s	64.9 s
Value iteration	44.7 s	70.0 s	46.0 s	66.7 s
Total tool runtime	95.0 s	143.8 s	96.8 s	137.4 s

The results of applying our tool to these SA are shown in Table 3. Both models have a buffer size of 100 and were analysed with a bound of 150 and 200 arrivals each. While for 150 arrivals the maximal probability for finishing all jobs without an overflow is relatively high for both models, it decreases rapidly for 200 arrivals. The table also shows the number of locations of the SA and the number of states of the interval abstraction MDP. What makes the MDPs rather large despite using only 10 intervals per timer is the fact that almost every location has two outgoing edges guarded by one timer—one for the next arrival and one for the service of the current job—and timers are reset on every edge. Despite having to generate and analyse MDPs with millions of states, the runtime from the prototype tool on a standard laptop remains below three minutes for the transformation and VI combined.

File Server. Finally, we check the model of a single-threaded file server with slow archival storage and queue length C introduced in [25]. The service times (timer x_{out}) depend on the file sizes and are assumed to be uniformly distributed over $[1, 3]$. A requested file is either on disk or in an archive with probability 0.8 and 0.2, respectively. Retrieving a file from the archive takes between 30 and 40 time units, which is continuously nondeterministic in the original model; we use a uniform distribution over $[30, 40]$ (timer x_a) instead. As a result, there is no more nondeterminism in the model. We additionally replace the discretely uniformly distributed initial queue length of the original model by starting with

Table 4. Results for the file server models compared to MCSTA's STA-based approach.

var.	C-n	SA interval abstraction (this paper)				MCSTA [25]			MODES
		intervals	result	states	time	result	states	time	result
unif.	3–5	$\frac{1}{8}/\frac{1}{10}/\frac{1}{2}$	[0.316, 0.359]	32690	0.2 s				
		$\frac{1}{16}/\frac{1}{10}/\frac{1}{2}$	[0.327, 0.350]	45460	0.4 s	[0.320, 0.362]	362389	3.0 s	0.340
		$\frac{1}{32}/\frac{1}{10}/\frac{1}{2}$	[0.331, 0.348]	228696	2.3 s				
	5–8	$\frac{1}{8}/\frac{1}{10}/\frac{1}{2}$	[0.177, 0.338]	158942	1.3 s				
		$\frac{1}{16}/\frac{1}{10}/\frac{1}{2}$	[0.210, 0.301]	229211	2.4 s	[0.197, 0.315]	1209231	8.9 s	0.255
		$\frac{1}{32}/\frac{1}{10}/\frac{1}{2}$	[0.224, 0.286]	1252056	17.4 s				
	5–10	$\frac{1}{8}/\frac{1}{10}/\frac{1}{2}$	[0.307, 0.512]	255844	2.3 s				
		$\frac{1}{16}/\frac{1}{10}/\frac{1}{2}$	[0.351, 0.469]	396175	4.2 s	[0.336, 0.485]	1633339	13.6 s	0.411
		$\frac{1}{32}/\frac{1}{10}/\frac{1}{2}$	[0.370, 0.451]	2314835	38.0 s				
exp.	3–5	$\frac{1}{8}/\frac{1}{2}/\frac{1}{2}$	[0.230, 0.373]	724557	5.3 s				
		$\frac{1}{10}/\frac{1}{4}/\frac{1}{2}$	[0.244, 0.369]	2130622	18.1 s	[0.261, 0.371]	1016802	2.8 s	0.311
		$\frac{1}{16}/\frac{1}{4}/\frac{1}{2}$	[0.267, 0.354]	19602981	175.5 s				
	5–8	$\frac{1}{4}/\frac{1}{4}/\frac{1}{2}$	[0.088, 0.472]	1624440	9.4 s	[0.206, 0.339]	3064195	8.2 s	0.284
		$\frac{1}{5}/\frac{1}{2}/\frac{1}{2}$	[0.108, 0.459]	2640688	16.3 s				
	5–10	$\frac{1}{4}/\frac{1}{4}/\frac{1}{2}$	[0.139, 0.659]	7612384	57.0 s	[0.304, 0.508]	4139557	12.0 s	0.437
		$\frac{1}{5}/\frac{1}{2}/\frac{1}{2}$	[0.171, 0.644]	14334125	110.6 s				

an empty queue. We ask for the probability of processing a given number n of arrivals without having a queue overflow; due to the absence of nondeterminism, our bounds on min. and max. probability again bound this one value.

Based on the original MODEST STA model file from [25], we created four MODEST models representing two variants, one where the interarrival times (timer x_{in}) are $Exp(\frac{1}{8})$-distributed and one where they are $Uni(0, 16)$-distributed, each as an STA and as an SA model. We first compute reference results using the MODES statistical model checker [10] with 10 million simulation runs for high accuracy. We can then compare the results and performance of our approach for SA with those of MCSTA's digital clocks-based approach for STA. Our tool is parametrised by a per-interval probability p_I^x for each timer x, which results in $\lceil 1/p_I^x \rceil$ intervals of possibly different widths but equal probability mass (except for the final one). MCSTA on the other hand takes a residual probability p_r and generates as many intervals of width 1 (as it uses digital clocks) as needed for the last (possibly open) interval to have probability mass $\leq p_r$. For the $Uni(0, 16)$ variant, using $p_I^{x_{in}} = \frac{1}{16}$, $p_I^{x_a} = \frac{1}{10}$, and $p_I^{x_{out}} = \frac{1}{2}$ with our tool (abbreviated $\frac{1}{16}/\frac{1}{10}/\frac{1}{2}$) results in the same intervals as MCSTA. For the $Exp(\frac{1}{8})$ variant, we cannot make the intervals match given the current tool parametrisations; here, MCSTA generates 25 intervals for x_{in} with its default $p_r = 0.05$, so we would need $p_I^{x_{in}} = \frac{1}{25}$ to generate the same *number* of intervals, which would however have very different bounds and cover much more of the support of $Exp(\frac{1}{8})$.

Our results are shown in Table 4, for different values of C and n. All results of our tool and MCSTA correctly bound the reference values. In our tool, runtime was again about equally divided between interval abstraction and VI. For the $Uni(0, 16)$ variant, we consistently obtain more accurate results in less than 30 % of the runtime compared to MCSTA. The MDPs we check are much smaller for the same intervals because we use big time steps while MCSTA's digital clocks approach has time steps of duration 1. In contrast to MCSTA, we can refine the intervals per timer. We found that refining intervals pays off most for x_{in}, significantly improving accuracy over MCSTA.

Results for the $Exp(\frac{1}{8})$ variant show that the choice of intervals matters, and in this case, MCSTA's width-1 intervals concentrated on lower delay values for x_{in} work better: To achieve similar accuracy, we need over 6× the runtime. Our interval abstraction MDPs are much larger despite using fewer intervals (as we do not get close to $p_I^{x_{in}} = \frac{1}{25}$ without running out memory). We suspect that generating less regular equal-probability intervals for $Exp(\frac{1}{8})$ leads to some very wide intervals that cause others to significantly expand in delay steps (reducing accuracy), as well as causing more combinations of differently-sized intervals in the state space instead of going back to the same bounds often (blowing up the MDP). We thus plan to implement MCSTA's interval generation method in our tool as well, which will allow it to perform better as for the $Uni(0, 16)$ variant.

6 Conclusion and Future Work

We have introduced the first model checking technique specifically designed for stochastic automata (SA), based on interval abstraction. It puts few restrictions on the features that can be used in SA models, and our experiments—though using a prototype implementation—indicate that it can offer much higher scalability than previous techniques for superclasses of SA.

Future Work. We plan to turn our prototype into a proper performance-optimised open-source tool connected to the MODEST TOOLSET or JANI. We can then use the MODEST TOOLSET's Kepler frontend [18] to analyse dynamic fault tree models, whose semantics is given in terms of SA (that are usually rather complex, with multiple timers, frequent resets, and hundreds of locations). To obtain not only upper/lower bounds on max./min. probabilities, we can use a game-based approach [27] to also obtain lower/upper bounds, which could be extended to automatic interval refinement akin to [33]. Finally, the presented analysis approach can be extended to deterministic delays and time-bounded properties.

Data Availability. An artifact that contains our tool and allows reproducing Sect. 5 is publicly available at DOI 10.5281/zenodo.19829349 [39].

References

1. Agha, G., Palmskog, K.: A survey of statistical model checking. ACM Trans. Model. Comput. Simul. **28**(1), 6:1–6:39 (2018). https://doi.org/10.1145/3158668
2. Baier, C., de Alfaro, L., Forejt, V., Kwiatkowska, M.: Model checking probabilistic systems. In: Handbook of Model Checking, pp. 963–999. Springer, Cham (2018). https://doi.org/10.1007/978-3-319-10575-8_28
3. Baier, C., Haverkort, B.R., Hermanns, H., Katoen, J.: Model-checking algorithms for continuous-time Markov chains. IEEE Trans. Software Eng. **29**(6), 524–541 (2003). https://doi.org/10.1109/TSE.2003.1205180
4. Baier, C., Haverkort, B.R., Hermanns, H., Katoen, J.P.: Performance evaluation and model checking join forces. Commun. ACM **53**(9), 76–85 (2010). https://doi.org/10.1145/1810891.1810912
5. Bellman, R.: A Markovian decision process. J. Math. Mech. **6**(5), 679–684 (1957)
6. Bohnenkamp, H.C., D'Argenio, P.R., Hermanns, H., Katoen, J.P.: MoDeST: a compositional modeling formalism for hard and softly timed systems. IEEE Trans. Software Eng. **32**(10), 812–830 (2006). https://doi.org/10.1109/TSE.2006.104
7. Bryans, J.W., Bowman, H., Derrick, J.: Model checking stochastic automata. ACM Trans. Comput. Log. **4**(4), 452–492 (2003). https://doi.org/10.1145/937555.937558
8. Buchholz, P., Kriege, J., Scheftelowitsch, D.: Model checking stochastic automata for dependability and performance measures. In: 44th Annual IEEE/IFIP International Conference on Dependable Systems and Networks (DSN 2014), pp. 503–514. IEEE Computer Society (2014). https://doi.org/10.1109/DSN.2014.53
9. Budde, C.E.: FIG: the finite improbability generator v1.3. SIGMETRICS Perform. Evaluation Rev. **49**(4), 59–64 (2022). https://doi.org/10.1145/3543146.3543160
10. Budde, C.E., D'Argenio, P.R., Hartmanns, A., Sedwards, S.: An efficient statistical model checker for nondeterminism and rare events. Int. J. Softw. Tools Technol. Transf. **22**(6), 759–780 (2020). https://doi.org/10.1007/S10009-020-00563-2
11. Budde, C.E., Dehnert, C., Hahn, E.M., Hartmanns, A., Junges, S., Turrini, A.: JANI: quantitative model and tool interaction. In: Legay, A., Margaria, T. (eds.) TACAS 2017. LNCS, vol. 10206, pp. 151–168. Springer, Heidelberg (2017). https://doi.org/10.1007/978-3-662-54580-5_9
12. D'Argenio, P.R.: Algebras and automata for timed and stochastic systems. Ph.D. thesis, University of Twente, Netherlands (1999). https://research.utwente.nl/en/publications/algebras-and-automata-for-timed-and-stochastic-systems
13. D'Argenio, P.R., Gerhold, M., Hartmanns, A., Sedwards, S.: A hierarchy of scheduler classes for stochastic automata. In: Baier, C., Dal Lago, U. (eds.) FoSSaCS 2018. LNCS, vol. 10803, pp. 384–402. Springer, Cham (2018). https://doi.org/10.1007/978-3-319-89366-2_21
14. D'Argenio, P.R., Hartmanns, A., Petri, A.: Effective stochastic automata model checking by interval abstraction (extended version). CoRR (2026, to appear)
15. D'Argenio, P.R., Katoen, J.P.: A theory of stochastic systems part I: stochastic automata. Inf. Comput. **203**(1), 1–38 (2005). https://doi.org/10.1016/J.IC.2005.07.001
16. D'Argenio, P.R., Monti, R.E.: Input/Output stochastic automata with urgency: confluence and weak determinism. In: Fischer, B., Uustalu, T. (eds.) ICTAC 2018. LNCS, vol. 11187, pp. 132–152. Springer, Cham (2018). https://doi.org/10.1007/978-3-030-02508-3_8
17. Delicaris, J., Stübbe, J., Schupp, S., Remke, A.: RealySt: A C++ tool for optimizing reachability probabilities in stochastic hybrid systems. In: Kalyvianaki, E.,

Paolieri, M. (eds.) 16th EAI International Conference on Performance Evaluation Methodologies and Tools (VALUETOOLS 2023). LNICSSITE, vol. 539, pp. 170–182. Springer (2023). https://doi.org/10.1007/978-3-031-48885-6_11

18. Dengler, G., Budde, C.E., Carnevali, L., Hartmanns, A.: Time-sensitive importance splitting. In: Prabhakar, P., Vandin, A. (eds.) 2nd International Joint Conference on Quantitative Evaluation of Systems and Formal Modeling and Analysis of Timed Systems (QEST+FORMATS 2025). LNCS, vol. 16143, pp. 21–41. Springer (2025). https://doi.org/10.1007/978-3-032-05792-1_2
19. Eisentraut, C., Hermanns, H., Katoen, J.-P., Zhang, L.: A semantics for every GSPN. In: Colom, J.-M., Desel, J. (eds.) PETRI NETS 2013. LNCS, vol. 7927, pp. 90–109. Springer, Heidelberg (2013). https://doi.org/10.1007/978-3-642-38697-8_6
20. Forejt, V., Kwiatkowska, M., Norman, G., Parker, D., Qu, H.: Quantitative multi-objective verification for probabilistic systems. In: Abdulla, P.A., Leino, K.R.M. (eds.) TACAS 2011. LNCS, vol. 6605, pp. 112–127. Springer, Heidelberg (2011). https://doi.org/10.1007/978-3-642-19835-9_11
21. Fränzle, M., Hahn, E.M., Hermanns, H., Wolovick, N., Zhang, L.: Measurability and safety verification for stochastic hybrid systems. In: Caccamo, M., Frazzoli, E., Grosu, R. (eds.) 14th ACM International Conference on Hybrid Systems: Computation and Control (HSCC 2011), pp. 43–52. ACM (2011). https://doi.org/10.1145/1967701.1967710
22. Frehse, G.: PHAVer: algorithmic verification of hybrid systems past HyTech. Int. J. Softw. Tools Technol. Transf. **10**(3), 263–279 (2008). https://doi.org/10.1007/S10009-007-0062-X
23. Ghasemieh, H., Remke, A., Haverkort, B.R.: Hybrid Petri nets with multiple stochastic transition firings. In: Haviv, M., Knottenbelt, W.J., Maggi, L., Miorandi, D. (eds.) 8th International Conference on Performance Evaluation Methodologies and Tools (VALUETOOLS 2014), pp. 217–224. ICST (2014). https://doi.org/10.4108/icst.valuetools.2014.258204
24. Hahn, E.M.: Model checking stochastic hybrid systems. Ph.D. thesis, Saarland University (2013). http://scidok.sulb.uni-saarland.de/volltexte/2013/5259/
25. Hahn, E.M., Hartmanns, A., Hermanns, H.: Reachability and reward checking for stochastic timed automata. Electron. Commun. Eur. Assoc. Softw. Sci. Technol. **70** (2014). https://doi.org/10.14279/TUJ.ECEASST.70.968
26. Hahn, E.M., Hartmanns, A., Hermanns, H., Katoen, J.P.: A compositional modelling and analysis framework for stochastic hybrid systems. Formal Methods Syst. Des. **43**(2), 191–232 (2013). https://doi.org/10.1007/S10703-012-0167-Z
27. Hahn, E.M., Norman, G., Parker, D., Wachter, B., Zhang, L.: Game-based abstraction and controller synthesis for probabilistic hybrid systems. In: 8th International Conference on Quantitative Evaluation of Systems (QEST 2011), pp. 69–78. IEEE Computer Society (2011). https://doi.org/10.1109/QEST.2011.17
28. Hartmanns, A.: Correct probabilistic model checking with floating-point arithmetic. In: Fisman, D., Rosu, G. (eds.) 28th International Conference on Tools and Algorithms for the Construction and Analysis of Systems (TACAS 2022). LNCS, vol. 13244, pp. 41–59. Springer (2022). https://doi.org/10.1007/978-3-030-99527-0_3
29. Hartmanns, A., Hermanns, H.: The Modest Toolset: an integrated environment for quantitative modelling and verification. In: Ábrahám, E., Havelund, K. (eds.) TACAS 2014. LNCS, vol. 8413, pp. 593–598. Springer, Heidelberg (2014). https://doi.org/10.1007/978-3-642-54862-8_51

30. Hartmanns, A., Junges, S., Quatmann, T., Weininger, M.: A practitioner's guide to MDP model checking algorithms. In: Sankaranarayanan, S., Sharygina, N. (eds.) 29th International Conference on Tools and Algorithms for the Construction and Analysis of Systems (TACAS 2023). LNCS, vol. 13993, pp. 469–488. Springer (2023). https://doi.org/10.1007/978-3-031-30823-9_24
31. Henzinger, T.A., Manna, Z., Pnueli, A.: What good are digital clocks? In: Kuich, W. (ed.) ICALP 1992. LNCS, vol. 623, pp. 545–558. Springer, Heidelberg (1992). https://doi.org/10.1007/3-540-55719-9_103
32. Howard, R.A.: Dynamic Programming and Markov Processes. MIT Press (1960)
33. Kwiatkowska, M., Norman, G., Parker, D.: Stochastic games for verification of probabilistic timed automata. In: Ouaknine, J., Vaandrager, F.W. (eds.) FORMATS 2009. LNCS, vol. 5813, pp. 212–227. Springer, Heidelberg (2009). https://doi.org/10.1007/978-3-642-04368-0_17
34. Kwiatkowska, M., Norman, G., Parker, D.: PRISM 4.0: verification of probabilistic real-time systems. In: Gopalakrishnan, G., Qadeer, S. (eds.) CAV 2011. LNCS, vol. 6806, pp. 585–591. Springer, Heidelberg (2011). https://doi.org/10.1007/978-3-642-22110-1_47
35. Kwiatkowska, M.Z., Norman, G., Parker, D., Sproston, J.: Performance analysis of probabilistic timed automata using digital clocks. Formal Methods Syst. Des. **29**(1), 33–78 (2006). https://doi.org/10.1007/S10703-006-0005-2
36. Kwiatkowska, M.Z., Norman, G., Segala, R., Sproston, J.: Automatic verification of real-time systems with discrete probability distributions. Theor. Comput. Sci. **282**(1), 101–150 (2002). https://doi.org/10.1016/S0304-3975(01)00046-9
37. Legay, A., Lukina, A., Traonouez, L.M., Yang, J., Smolka, S.A., Grosu, R.: Statistical model checking. In: Steffen, B., Woeginger, G. (eds.) Computing and Software Science. LNCS, vol. 10000, pp. 478–504. Springer, Cham (2019). https://doi.org/10.1007/978-3-319-91908-9_23
38. Marsan, M.A., Conte, G., Balbo, G.: A class of generalized stochastic Petri nets for the performance evaluation of multiprocessor systems. ACM Trans. Comput. Syst. **2**(2), 93–122 (1984). https://doi.org/10.1145/190.191
39. Petri, A.: Effective stochastic automata model checking by interval abstraction (artifact). Zenodo (2026). https://doi.org/10.5281/zenodo.19829349
40. Sproston, J.: Decidable model checking of probabilistic hybrid automata. In: Joseph, M. (ed.) FTRTFT 2000. LNCS, vol. 1926, pp. 31–45. Springer, Heidelberg (2000). https://doi.org/10.1007/3-540-45352-0_5
41. Wolovick, N.: Continuous probability and nondeterminism in labeled transition systems. Ph.D. thesis, Córdoba, Argentina (2012)
42. Zhang, L., She, Z., Ratschan, S., Hermanns, H., Hahn, E.M.: Safety verification for probabilistic hybrid systems. In: Touili, T., Cook, B., Jackson, P. (eds.) CAV 2010. LNCS, vol. 6174, pp. 196–211. Springer, Heidelberg (2010). https://doi.org/10.1007/978-3-642-14295-6_21

Verification of Parametric Markov Automata Under Time-Bounded Reachability

Kevin van de Glind(✉), Matthias Volk, and Tim A. C. Willemse

Eindhoven University of Technology, Eindhoven, The Netherlands
k.v.d.glind@tue.nl

Abstract. Analysis of Markov models is of high importance for formal verification. Until now, analysis of Markov Automata required them to be fully specified, which is a considerable restriction as rates may be unknown or influenced by uncertainty of the environment. We introduce parametric Markov Automata (pMA) to capture this uncertainty with parametric transition functions. On these parametrized models, two different synthesis problems for time-bounded reachability properties are considered: I) Does there exist a valuation in the parameter space such that the instantiated model satisfies/violates the property, and II) given the parameter space, how can it be partitioned into satisfying and violating regions?

Our approach comprises two steps: I) The pMA is discretized to a parametric Markov decision process (pMDP), and II) through analysis of the pMDP, bounds are obtained for the reachability probability in the parametric MA. This approach solves the above problems up to a specified precision, as the accumulated error terms can be made arbitrarily small.

We implemented the approach using the Storm model checker. Our experimental evaluation shows that the main performance bottlenecks originate from the discretization of the pMA.

1 Introduction

Markov Automata (MA) [13] are a compositional model for nondeterministic, continuous stochastically timed systems. They are able, as motivated in [4], to serve as semantics for other reliability models such as Generalized Stochastic Petri Nets [33] and Dynamic Fault Trees [12]. Analysis of MA is supported by various tools, such as Storm [28], Modest [24] and IMCA [17].

Until now, all algorithms for model checking MA required the model to be completely specified. In particular, each probabilistic and rate transition had to be precisely known prior to analysis. This can be an unrealistic assumption for real-world systems, as rates or probabilities may be subject to environmental changes or not precisely known. In order to reason about such uncertainty, we introduce *parametric Markov Automata (pMA)*, where transitions can

This work has been partially supported by NWO VENI Grant UNICORN (232.211).

M. Chen and M. Tschaikowski (Eds.): QEST+FORMATS 2026, LNCS 16913, pp. 389–409, 2027.
https://doi.org/10.1007/978-3-032-35298-9_22

be parametrized. This parametric extension of MA follows similar extensions of both discrete-time models [30] and continuous-time Markov chains [8,23].

Given a pMA and a region of parameter valuations, we solve two problems for time-bounded reachability properties. Firstly, *satisfaction verification* checks whether the valuations in the region satisfy the property. Secondly, *synthesis* [30] partitions the region into satisfying and violating subregions. We solve both problems by a two-step approach. First, we discretize the pMA and obtain a *parametric Markov decision process (pMDP)*. Second, after some transformations on the pMDP to make it suitable for further analysis, we perform *parameter lifting (PL)* [37] on this discretized model. PL encodes the choice of parameter valuation as non-deterministic choices, yielding an MDP that allows analysis via off-the-shelf methods. The proposed approach yields upper and lower bounds for the true reachability probability on the pMA, governed by three different error terms. These bounds are then used for a classical threshold synthesis scheme [8,30].

Related Work. Similar models to MA include *Continuous-Time MDPs (CT-MDPs)* [34] and *Interactive Markov Chains (IMC)* [29]. The former allows for states with a combination of nondeterminism and timing, the latter already assumes a clear separation between these two types of states. In the literature, these became known as probabilistic and Markovian states. Markov Automata generalize both and introduced the possibility of immediate choice with a successor distribution in probabilistic states [27]. Several approaches have been proposed to analyze time-bounded reachability properties on MA: some operate directly on the continuous model [1,4], while others first discretize the model and then carry out the analysis using step-bounded queries [18,27,35].

Parametric extensions are natural for Markov models. Model checking algorithms exist for discrete-time models [11,21,30] supported through e.g., Storm [28] and iscasMC [22], probabilistic timed automata [25] and continuous-time Markov chains [8,23] through PRISM-PSY [10]. A subset of parametric models are interval models, where each transition is assigned an interval of possible values instead. These are supported by e.g., PRISM [32] or Storm [28].

Contributions. This paper introduces parametric Markov automata. Our main contribution is an algorithm for time-bounded reachability on pMA. To perform analysis, we generalize the discretization result of [18] to pMA. Furthermore, under common mild assumptions—non-zenoness of the pMA, multi-affine parametric functions and a weak global condition—we transform the discretized pMA to an MDP, and solve both satisfaction verification and synthesis up to arbitrary precision. We implemented our approach using Storm and demonstrate it on a variety of pMA models.

Outline. Section 2 provides the necessary background on Markov Automata and discretization results on MA. We introduce pMA in Sect. 3 and formally define the satisfaction verification and synthesis problems. Section 4 covers the necessary transformations from pMA to a discretized model over which parameter lifting can be performed. Using PL, we solve the satisfaction verification and synthesis problems in Sect. 5. We experimentally evaluate our approach on a variety of benchmark models in Sect. 6 and conclude in Sect. 7.

2 Preliminaries

We briefly introduce Markov Automata and the discretization analysis technique [18] on which our approach is based.

Definition 1 (Markov Automaton). *A* Markov Automaton (MA) *is a tuple* $\mathcal{M} = (S, A, s_0, \rightarrow, \Rightarrow)$*, where:*

- S *is a finite set of* states.
- A *is a finite set of* actions.
- $s_0 \in S$ *is the* initial state.
- $\rightarrow \subseteq S \times A \times D(S)$ *is the* probabilistic *transition relation, where* $D(S)$ *is the set of discrete distributions over* S.
- $\Rightarrow \subseteq S \times \mathbb{R}_{>0} \times S$ *is the* rate *transition relation. The* rate *of a transition is given by* $R(s, s') = \sum_{(s,r,s') \in \Rightarrow} r$.

Under the *maximal progress assumption*—a common assumption for verification of MA [5]—we assume that probabilistic transitions happen without delay and therefore take precedence over rate transitions which happen instantaneously with probability zero. We can therefore partition the state space into *probabilistic* and *Markovian* states $S = PS \sqcup MS$. The two transition relations can then be restricted to probabilistic and Markovian states, respectively: $\rightarrow \subseteq PS \times A \times D(S)$ and $\Rightarrow \subseteq MS \times \mathbb{R}_{>0} \times S$. Figure 1b depicts an example MA.

For a Markovian state $s \in MS$ the *sojourn time* of the state is governed by an *exponential distribution* with rate $\lambda(s) = \sum_{s' \in S} R(s, s')$. Thus, the probability of leaving a state s within t units is given by $1 - e^{-\lambda(s)t}$. Afterwards, we transition to successor state $s' \in S$ with probability $p(s, s') = \frac{R(s,s')}{\lambda(s)}$. A probabilistic state $s \in PS$ is left immediately by choosing an available action $\alpha \in A$ with $(s, \alpha, \mu) \in \rightarrow$, and transitions to state s' with probability $p(s, s') = \mu(s')$.

An MA can therefore alternatively be defined as a tuple $\mathcal{M} = (S, A, s_0, \lambda, p)$, where we set $\lambda(s) = +\infty$ for all probabilistic states $s \in PS$, similar to [18].

A *scheduler* σ resolves non-determinism on an MA $\mathcal{M}$ by choosing available actions in probabilistic states, resulting in a deterministic model $\mathcal{M}_\sigma$. Schedulers $\sigma : PS \times \mathbb{R}_{\geq 0} \rightarrow A$ base this choice on the current state and the total time spent in the system. These belong to the class of total-time positional deterministic schedulers, denoted by $TP(\mathcal{M})$. Under a scheduler, one can define a probability measure on the MA; we refer to [36] for the details.

MA are a superclass of *Markov decision processes (MDPs)* and *continuous-time Markov chains (CTMCs)*. An MDP is a tuple (S, A, s_0, p) and can be seen as an MA where $MS = \emptyset$. Similarly, a CTMC can be viewed as an MA where $PS = \emptyset$. For a more detailed description of both models, we refer to [31].

Time-Bounded Reachability. In this paper, we focus the analysis on *time-bounded reachability properties* of the form $\mathbb{P}[\diamond^{\leq t}G] \sim \theta$, computing: "Is the maximal/minimal probability to reach a set of goal states $G \subseteq S$ within t time units greater/less than a given threshold θ?". For a given MA $\mathcal{M}$ and scheduler σ, the *time-bounded reachability probability* is denoted by $Pr^{\mathcal{M}_\sigma}[\diamond^{\leq t}G]$. Comparing this value with a user-specified threshold $\theta \in [0,1]$ yields a time-bounded reachability property. This can be formulated as a CSL model checking problem [2]:

$$\mathcal{M}_\sigma \models \mathbb{P}[\diamond^{\leq t}G] \sim \theta \quad \textit{iff} \quad Pr^{\mathcal{M}_\sigma}[\diamond^{\leq t}G] \sim \theta, \text{ for } {\sim} \in \{<, \leq, \geq, >\}, \theta \in [0,1].$$

For time-bounded reachability probabilities, total-time positional deterministic schedulers $TP(\mathcal{M})$ suffice [35] and we define the maximal/minimal probability:

$$Pmax[\mathcal{M}, \diamond^{\leq t}G] = \sup_{\sigma \in TP(\mathcal{M})} Pr^{\mathcal{M}_\sigma}[\diamond^{\leq t}G]$$
$$Pmin[\mathcal{M}, \diamond^{\leq t}G] = \inf_{\sigma \in TP(\mathcal{M})} Pr^{\mathcal{M}_\sigma}[\diamond^{\leq t}G]$$

Similarly to before, one can use $Pm[\mathcal{M}, \diamond^{\leq t}G] \leq \theta$ for $m \in \{min, max\}$ to define time-bounded reachability properties.

Discretization. Time-bounded reachability properties can be analysed using the *discretization method* [6,18,35], also called digitization [27]. This transformation allows one to reason about the approximate behaviour of the MA by analyzing the time-discretized model, yielding a (discrete-time) MDP.

Definition 2 (Discretized MA). *Let MA* $\mathcal{M} = (S, A, s_0, \rightarrow, \Rightarrow)$ *and discretization step* $\delta > 0$. *The* discretized MA *is an MDP* $\mathcal{M}^\delta = (S, A, s_0, P')$, *where:*

$$P'(s, s') = \begin{cases} (1 - e^{-\lambda(s)\delta})p(s, s'), & \textit{if } s \neq s' \\ (1 - e^{-\lambda(s)\delta})p(s, s) + e^{-\lambda(s)\delta}, & \textit{if } s = s' \end{cases}$$

An example of this transformation is shown in Fig. 1b-1c. The central idea relies on the fact that, for small δ, the probability of taking more than one Markovian step within time δ is small. The error made by this discretization can be bounded [18].

Theorem 1. *[18, Theorem 5.3] Let* $\mathcal{M}$ *be an MA,* $G \subseteq S, \lambda = \max_{s \in MS} \lambda(s)$, $t > 0$ *and fix* $\delta > 0$ *such that* $t/\delta \in \mathbb{N}$. *Then the error can be bounded as follows:*

$$Pmax[\mathcal{M}^\delta, \diamond^{\leq t/\delta}G] \leq Pmax[\mathcal{M}, \diamond^{\leq t}G] \leq Pmax[\mathcal{M}^\delta, \diamond^{\leq t/\delta}G] + \epsilon_\delta,$$

where $\epsilon_\delta = 1 - e^{-\lambda t}(1 + \lambda\delta)^{t/\delta}$ *denotes the* discretization error. *The same bounds hold for* $Pmin[\cdot]$.

Step-bounded reachability analysis on the discretized MA requires an adapted value iteration scheme, as probabilistic states are left immediately [18,27].

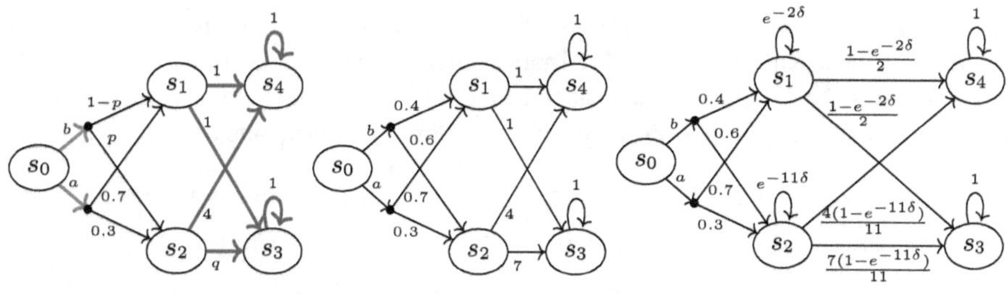

(a) Parametric MA (b) Instantiated MA (c) Discretized MA

Fig. 1. (a) Parametric MA with nondeterministic transitions in olive and rate transitions in teal, (b) instantiated MA for instantiation $p \mapsto 0.6, q \mapsto 7$, and (c) its discretization using discretization step δ. (Color figure online)

3 Parametric Markov Automata

In this section, we define *parametric MA* which extend MA with parametric rate- and probabilistic functions. We lift the discretization result of Theorem 1 to this parametric model in Lemma 1. Last, we provide the formal problem definition, defining the satisfaction verification and synthesis problems.
Parametric Markov Automata. We first introduce the necessary notation for parameters. Let $\mathcal{X} = \{x_1, ..., x_N\}$ be a finite set of *parameters*. A *(parameter) instantiation* is a mapping $\mathcal{X} \to \mathbb{R}$. A parametrized successor distribution is denoted by $D^{\mathcal{X}} : S \to [0, 1]$. A *parameter region* $\mathcal{R}$ is a closed and bounded subset, in the topological sense as described in for example [14], $\mathcal{R} \subset \mathbb{R}^{|\mathcal{X}|}_{\geq 0}$ with any $v \in \mathcal{R}$ called a *valuation*. Moreover, if $\mathcal{R}$ is a polytope, we write $V_{\mathcal{R}}$ to denote its vertices. A polynomial over $\mathcal{X}$ is a weighted sum over monomials $x_1^{a_1} ... x_n^{a_n}$ for $a_1, ..., a_n \in \mathbb{N}$ and it is called *multi-affine* if for each monomial $a_1, ..., a_n \leq 1$. The set of polynomials over $\mathcal{X}$ with real coefficients is denoted by $\mathbb{R}[\mathcal{X}]$. A polynomial $f \in \mathbb{R}[\mathcal{X}]$ is *monotone increasing/decreasing* on $\mathcal{R}$ with respect to $x \in \mathcal{X}$ if, respectively, $\frac{d}{dx} f|_{\mathcal{R}} \geq 0$ or $\frac{d}{dx} f|_{\mathcal{R}} \leq 0$. A polynomial f is called *monotone* on $\mathcal{R}$ if it is monotone with respect to each parameter. Multi-affine functions are monotone on any region $\mathcal{R}$.

Definition 3 (Parametric Markov Automaton). *A* parametric Markov Automaton (pMA) *over $\mathcal{X}$ is a tuple $\mathcal{M}^{\mathcal{X}} = (S, A, s_0, \to, \Rightarrow)$, where states S, actions A and initial state s_0 are as in Definition 1, and the two transition relations $\to \subseteq PS \times A \times D^{\mathcal{X}}(S)$ and $\Rightarrow \subseteq MS \times \mathbb{R}[\mathcal{X}] \times S$ are parametrized - over $\mathcal{X}$.*

Similarly to pMDPs and pCTMCs [23,30], pMAs can be instantiated. An example of a pMA and an instantiation can be seen in Fig. 1a-1b.

Definition 4 (Instantiated MA). *For any pMA $\mathcal{M}^{\mathcal{X}}$ and valuation $v \in \mathbb{R}^{|\mathcal{X}|}$, we obtain the* instantiated MA *$\mathcal{M}^{\mathcal{X}}[v]$ by applying the instantiation $x_i \mapsto v_i$.*

We use the same notation for an instantiation of a single rate or successor distribution with valuation v as for instantiating a pMA. A valuation v is called *well-defined* if the resulting instantiated MA $\mathcal{M}^{\mathcal{X}}[v]$ satisfies Definition 1. A well-defined valuation is *graph-preserving* if the following conditions are met:

$$\forall s \in MS, s' \in S, v \in \mathcal{R}: \qquad R(s,s') \not\equiv 0 \rightarrow R[v](s,s') \neq 0$$
$$\forall s \in PS, s' \in S, \alpha \in A, v \in \mathcal{R}: \qquad P(s,\alpha,s') \not\equiv 0 \rightarrow P[v](s,\alpha,s') \neq 0$$

In the remainder we assume each valuation is well-defined and graph-preserving.

A pMA is called *non-zeno* if, for every valuation, the instantiated MA does not have a *zeno cycle*, which is a cycle of probabilistic states one does not exit almost surely. For further details on zenoness of MA, we refer to [18,27]. We assume each pMA to be non-zeno, a common assumption on MA in the literature [7,27]. Moreover, in practice, non-zenoness, and the stronger occurrence of having any cycles of probabilistic states at all, is extremely rare [7] and existing MA models, found for example in the QCOMP benchmark suite [26], do not have, to our knowledge, cycles of probabilistic states.

In order to determine the exact extremal values of parametric functions, these are restricted to multi-affine polynomials. This is in line with the restriction on parametric rate functions in [8]. For the sake of presentation, we restrict attention to hyper-rectangular parameter spaces, a standard assumption in literature [8,30]. However, our methods extend to finite unions of hyper-rectangles.

Lifting Discretization. In order to solve time-bounded reachability queries on pMA, we generalize the discretization result of Theorem 1. The discretization of pMA is analogous to Definition 2, but instead yields a parametric MDP. The generalization is shown in Lemma 1, where, for any two parametric models $\mathcal{M}, \mathcal{M}'$ over the same parameters, we write $Pmax[\mathcal{M},\phi] \leq Pmax[\mathcal{M}',\phi]$ if for each valuation $v \in \mathcal{R}$ we have $Pmax[\mathcal{M}[v],\phi] \leq Pmax[\mathcal{M}'[v],\phi]$.

Lemma 1. *Let $\mathcal{M}^{\mathcal{X}}$ be a pMA over a region $\mathcal{R}$ and $t > 0$. Fix $\delta > 0$ as discretization step such that $t/\delta \in \mathbb{N}$. We get the following:*

$$Pmax[\mathcal{M}^{\mathcal{X}^{\delta}}, \diamond^{\leq t/\delta} G] \leq Pmax[\mathcal{M}^{\mathcal{X}}, \diamond^{\leq t} G] \leq Pmax[\mathcal{M}^{\mathcal{X}^{\delta}}, \diamond^{\leq t/\delta} G] + 1 - \inf_{v \in \mathcal{R}} [e^{-\lambda[v]t}(1+\lambda[v]\delta)^{t/\delta}]$$

The same results hold for $Pmin[\cdot]$.

Proof. Consider any valuation $v \in \mathcal{R}$. By Theorem 1 the following bound is obtained: $Pmax[\mathcal{M}^{\mathcal{X}^{\delta}}[v], \diamond^{\leq t/\delta} G] \leq Pmax[\mathcal{M}^{\mathcal{X}}[v], \diamond^{\leq t} G]$. Therefore, we immediately prove our lower bound. Similarly, we get:

$$\begin{aligned} Pmax[\mathcal{M}^{\mathcal{X}}[v], \diamond^{\leq t} G] &\leq Pmax[\mathcal{M}^{\mathcal{X}^{\delta}}[v], \diamond^{\leq t/\delta} G] + 1 - e^{-\lambda[v]t}(1+\lambda[v]\delta)^{t/\delta} \\ &\leq Pmax[\mathcal{M}^{\mathcal{X}^{\delta}}[v], \diamond^{\leq t/\delta} G] + 1 - \inf_{w \in R} [e^{-\lambda[w]t}(1+\lambda[w]\delta)^{t/\delta}] \end{aligned}$$

The proof for $Pmin[\cdot]$ is analogous. □

		Player	
		Cooperative	Adversarial
Nature	Cooperative	*Feasibility*	*Angelic-Aid*
	Adversarial	*Robust*	*Demonic*

(a) Different satisfaction relations.

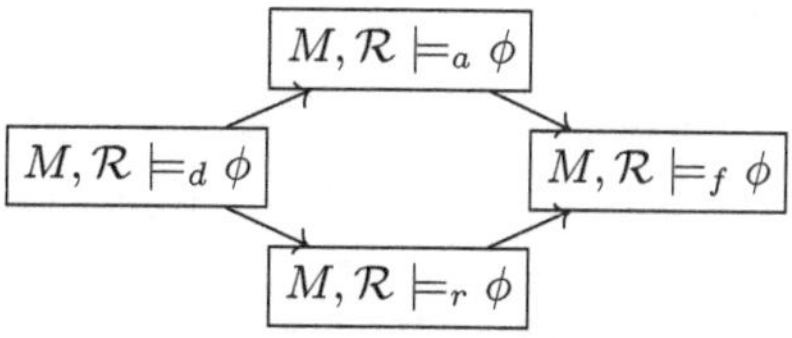

(b) Ordering of satisfaction relations; arrows correspond to implications.

Fig. 2. Satisfaction relations.

The infimum in the error bound requires a maximization of the maximum sojourn rate. From this point onward, for readability, we will denote pMA and its corresponding discretization by $\mathcal{M}$ and $\mathcal{M}^\delta$, respectively, whenever it is clear from the context that we are referring to the parametric model. In such cases, the discretization error will be written as $\epsilon_\delta = 1 - \inf_{v \in \mathcal{R}} [e^{-\lambda[v]t}(1 + \lambda[v]\delta)^{t/\delta}]$.

Problem Statement. Now that pMA are formally defined, we provide a formal problem definition. In contrast to MA, the analysis of pMA also needs to reason about the parameter valuations. We solve two main problems for time-bounded reachability properties: (1) *satisfaction verification* checks whether there exists a valuation in the region which satisfies/violates the given property, and (2) *synthesis* partitions the parameter space into satisfying and violating subregions.

Four different satisfaction relations are considered, distinguishing cases where the *player*—making the non-deterministic choices—and *nature*—assigning the parameter values—can either be cooperative or adversarial, similar to Definition 14 in [30] for discrete-time models.

Definition 5 (Satisfaction relations). *For a pMA $\mathcal{M}$, region $\mathcal{R}$ and time-bounded reachability property $\phi = \mathbb{P}[\diamond^{\leq t} G] \sim \theta$, we define the following satisfaction relations:*

$$\begin{aligned}
\textit{(Feasibility)} \quad & \mathcal{M}, \mathcal{R} \models_f \phi \quad \textit{iff} \quad \exists \sigma \in TP(\mathcal{M})\, \exists v \in \mathcal{R} : \mathcal{M}_\sigma[v] \models \phi \\
\textit{(Angelic-Aid)} \quad & \mathcal{M}, \mathcal{R} \models_a \phi \quad \textit{iff} \quad \forall \sigma \in TP(\mathcal{M})\, \exists v \in \mathcal{R} : \mathcal{M}_\sigma[v] \models \phi \\
\textit{(Robust)} \quad & \mathcal{M}, \mathcal{R} \models_r \phi \quad \textit{iff} \quad \exists \sigma \in TP(\mathcal{M})\, \forall v \in \mathcal{R} : \mathcal{M}_\sigma[v] \models \phi \\
\textit{(Demonic)} \quad & \mathcal{M}, \mathcal{R} \models_d \phi \quad \textit{iff} \quad \forall \sigma \in TP(\mathcal{M})\, \forall v \in \mathcal{R} : \mathcal{M}_\sigma[v] \models \phi
\end{aligned}$$

The bivariate table in Table 2a provides an intuition of the satisfaction relations and whether the player/nature has a cooperative or adversarial attitude towards satisfying the property. Moreover, there is a clear ordering on these satisfaction relations, as shown in Fig. 2b.

We now define the two main problems. Given a pMA $\mathcal{M}$, region $\mathcal{R}$, time-bounded reachability property ϕ and satisfaction relation $\clubsuit \in \{f, a, r, d\}$, the *satisfaction verification* problem is to decide whether $\mathcal{M}, \mathcal{R} \models_\clubsuit \phi$. The verification problems for discrete-time models considered in [30] are the robust and demonic

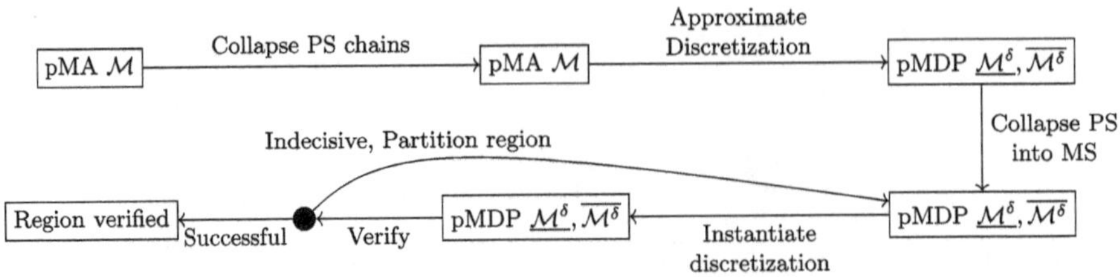

Fig. 3. Outline of the process for different synthesis problems.

cases of this problem. The *synthesis* problem is to find a partitioning $\mathcal{R} = T \sqcup F$ such that $\forall T' \subseteq T : \mathcal{M}, T' \models_\clubsuit \phi$ and $\forall F' \subseteq F : \mathcal{M}, F' \not\models_\clubsuit \phi$. This definition is equivalent to the approximate synthesis problem in [30]. As for each valuation in T, F the property must be satisfied/violated, feasibility and robust satisfaction relations yield the same result, and so do angelic-aid and demonic.

4 Model Transformations

Using Lemma 1, one could try to naively perform parameter synthesis on the discretized pMA. However, the discretized pMA requires an adapted value-iteration scheme [27], which combines unbounded reachability between PS states with time-bounded reachability in MS. Instead we transform the discretized pMA as outlined in Fig. 3, which allows us to apply existing analysis techniques for step-bounded reachability on pMDP. In the remainder of this section, we discuss these transformations in more detail.

Section 4.1 focuses on transforming and discretizing the pMA into a collapsed discretized pMA, where the instantaneous transitions from PS states are collapsed into Markovian states, using state-elimination approaches such as [20]. The adapted reachability scheme of the original model then yields the same result as step-bounded reachability on the transformed model. We could then perform parameter synthesis on the resulting pMDP. However, discretizing the pMA is expensive and would need to be performed for each subregion. Instead, Sect. 4.2 introduces a method to obtain approximate discretizations $\underline{\mathcal{M}^\delta}, \overline{\mathcal{M}^\delta}$ which include the discretization as a parameter in the discretized pMA, allowing the transformations covered in this section to be a preprocessing step.

After the preprocessing, we instantiate the discretization for each parameter region and perform the verification. The verification either terminates with a conclusive result, or we need to further partition the region and check each subregion individually. This process, including how to take approximation errors into account, is described in Sect. 5.

4.1 Collapse of Probabilistic States

The original adapted VI scheme [27] only counts steps in the MDP if they are taken in states corresponding to Markovian states in the MA. Therefore, we propose a method to eliminate *PS* using state elimination [20], such that all steps in the discretized model will be taken in *MS* and the alternative value-iteration scheme reduces to a step-bounded reachability query.

As time does not pass in a chain of probabilistic states and we assume non-zenoness, one can fix the policy in the first state of the chain and immediately get the distribution over Markovian states by eliminating the intermediate probabilistic states. Thus, a chain of *PS* can be reduced to a single state. In essence, this approach follows a similar idea to the MA to SAMA transformation in [16]. This transformation naturally comes at the cost of an exponential blow-up in the number of available actions for the given chain. However, in practice, chains of PS tend to be short, which controls this growth of available actions.

To eliminate the remaining *PS*, the pMA needs to be discretized, as pMA have a clear distinction between states with stochastic timing and non-deterministic choice. After discretization, we can collapse the PS into MS using well known elimination procedures [21]. Details of this algorithm can be found in [15]. Correctness of this approach immediately follows from the adapted value-iteration scheme in [27]. The cost of this transformation is again an exponential increase in the total number of actions with respect to the outdegree of Markovian states, which tends to be small in practice.

4.2 Approximate Discretization

Throughout the discretization of the pMA, in accordance with Definition 2, the parametric functions on the resulting pMDP will contain an exponential term. However, current methods [19,30] only allow for rational functions. We therefore need to represent the exponential via rational functions. We briefly describe two methods: I) *Minimizing/Maximizing the sojourn rates* to get the respective minimum/maximum reachability results, or II) *approximating the exponential* by a partial Taylor expansion [38]. Both of these result in over- and underapproximations of the final reachability probabilities.

Minimizing/Maximizing Sojourn Rates. The first method to obtain an over- and underapproximation of the reachability probability is to maximize/minimize the sojourn rates for each state independently from the probabilistic transitions of the same state. Clearly, maximizing these rates results in a higher probability of reaching the goal state within the set time bound, as states are left faster and the goal state can be reached earlier. In practice, this means that we set the sojourn rates of each state to $\max_{v \in V_{\mathcal{R}}} \lambda[v](s)$ or $\min_{v \in V_{\mathcal{R}}} \lambda[v](s)$. We use the assumption of multi-affine parametric rate functions here to guarantee that the extrema are attained at the vertices of the region. The main drawback of this approach is that intra-state dependencies between sojourn rate and successor distribution are lost, potentially leading to severe approximation errors.

Approximating the Exponential Function. To maintain intra-state dependencies, we approximate the exponential function by a partial Taylor expansion:

$$\forall N \in \mathbb{N}: \quad \sum_{k=0}^{2N+1} \frac{(-x)^k}{k!} \leq e^{-x} \leq \sum_{k=0}^{2N} \frac{(-x)^k}{k!} \tag{1}$$

The bounds in Eq. 1 are directly implied by the alternating series remainder theorem [38]. Based on this approximation, we introduce the *approximate discretized Markov Automaton.*

Definition 6 (Approximate discretized pMA). *Given discretization step $\delta > 0$, $N \in \mathbb{N}_{>0}$, and pMA $\mathcal{M} = (S, A, s_0, \rightarrow, \Rightarrow)$. The* approximate discretized parametric Markov Automaton *is given by $\mathcal{M}_N^\delta = (S, A, s_0, P')$, where:*

$$P'(s,s') = \begin{cases} (1 - \sum_{k=0}^{N} \frac{(-\lambda(s))^k}{k!})p(s,s'), & \text{if } s \neq s' \\ (1 - \sum_{k=0}^{N} \frac{(-\lambda(s))^k}{k!})p(s,s') + \sum_{k=0}^{N} \frac{(-\lambda(s))^k}{k!}, & \text{otherwise.} \end{cases}$$

Using Theorem 2, given below, the approximate discretization is guaranteed to bound the reachability probability of the pMA, and the error of this additional approximation is also bounded in terms of the discretized pMA.

Theorem 2. *Given pMA $\mathcal{M} = (S, A, s_0, \rightarrow, \Rightarrow), G \subseteq S, t > 0$. Fix discretization step $\delta > 0$ so $t/\delta \in \mathbb{N}$ and pick $K \in \mathbb{N}_{>0}$. Let $\underline{\mathcal{M}^\delta} = \mathcal{M}_{2K}^\delta$, $\overline{\mathcal{M}^\delta} = \mathcal{M}_{2K+1}^\delta$ and $\lambda = \sup_{v \in \mathcal{R}} \lambda[v]$. Then we have that:*

$$\begin{aligned} &Pmax[\mathcal{M}^\delta, \diamond^{\leq t/\delta} G] - \epsilon_{t/\delta}^{2K} \leq Pmax[\underline{\mathcal{M}^\delta}, \diamond^{\leq t/\delta} G] \\ \leq &Pmax[\mathcal{M}, \diamond^{\leq t} G] \\ \leq &Pmax[\overline{\mathcal{M}^\delta}, \diamond^{\leq t/\delta} G] + \epsilon_\delta \leq Pmax[\mathcal{M}^\delta, \diamond^{\leq t/\delta} G] + \epsilon_{t/\delta}^{2K+1} + \epsilon_\delta, \end{aligned}$$

where $\epsilon_{t/\delta}^M = (1 + \frac{(\lambda\delta)^M}{M!})^{t/\delta} - 1$ and ϵ_δ is the discretization error as per Lemma 1. Moreover, the same result holds for $Pmin[\cdot]$.

The proof of Theorem 2 can be found in the extended version of this paper [15].

The discretization error ϵ_δ dominates the error made by the approximate discretization. This can be seen in Fig. 4, where the resulting probabilities for pMA instantiations are plotted against the bounds obtained using the approximate discretization with N equal to $1, 2$ for $\overline{\mathcal{M}^\delta}, \underline{\mathcal{M}^\delta}$ respectively.

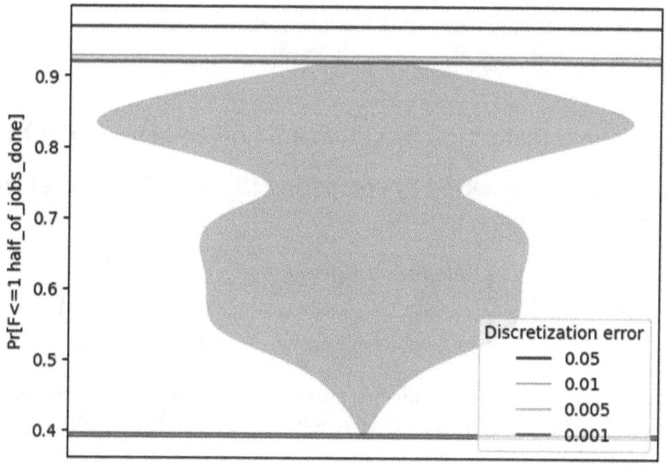

Fig. 4. Violinplot of probabilities for 10^4 pMA instantiations on Jobs benchmark with $\mathcal{R} = [1,3]^2$; horizontal lines are the approximate discretization bounds.

5 Solving Satisfaction Verification and Synthesis Problems

Section 4 focused on the necessary transformations to make step-bounded reachability on the discretized MA possible. This section focuses on the verification of regions on the resulting MDP using parameter lifting [37].

5.1 Parameter Lifting

Parameter Lifting (PL) is a technique which abstracts inter-state dependencies away with a transformation from pMDP to parameter-free *stochastic games (SG)* [37]. The extremal values over the parametric transition functions are precomputed, and nature, as a second player, chooses an extremal parameter value which, depending on the satisfaction relation, minimizes/maximizes the time-bounded reachability probability [30]. This relaxation introduces another error which becomes smaller when the absolute difference between extrema decreases.

As PL assumes that extrema are attained at the vertices of the region [37], we must ensure that this holds throughout the transformations from Sect. 4. Therefore, we assume that no two states in a chain of probabilistic states share a parametric function defined over the same variable. Moreover, to cover the collapse of PS in MS, we assume that the sets of parameters over rate- and probabilistic transition functions are disjoint. These assumptions can be enforced, if needed, by introducing auxiliary parameters. We show in [15] that I) by adding auxiliary parameters, one does not lose the guarantee of reasoning up to arbitrary precision over the original parameter space and II) these assumptions suffice to guarantee correctness of PL. None of the models we have come across from the literature are negatively impacted by these assumptions. Moreover, the same assumptions are used for pCTMC [8], motivating that these conditions are weak enough to model realistic scenarios.

5.2 Solving the Safisfaction Verification and Synthesis Problems

The problems can now be solved by comparing the bounds obtained through PL with the threshold θ, using a well-known refinement scheme [3,8,9,30]:

$$M, \mathcal{R} \models_{\clubsuit} \mathbb{P}[\diamond^{\leq t} G] \leq \theta = \begin{cases} \text{Satisfied, if } upper_{\clubsuit} \leq \theta, \\ \text{Violated, if } lower_{\clubsuit} > \theta, \\ \text{Inconclusive and needs further splitting, } Otherwise. \end{cases}$$

The bounds $upper_{\clubsuit}/lower_{\clubsuit}$ differ per satisfaction relation and are affected by the accumulated errors. Recall the three error terms: i) discretization error ϵ_δ, ii) the error of the approximate discretization, and iii) the PL error. We denote $Pr_{p,n}[\underline{\mathcal{M}^\delta}, \diamond^{\leq t/\delta} G]$ with $p, n \in \{\max, \min\}$ to indicate whether player and nature are trying to maximize/minimize the reachability probability in the SG, which results from applying PL on either pMDP $\underline{\mathcal{M}^\delta}$ or $\overline{\mathcal{M}^\delta}$.

The satisfaction verification for different satisfaction relations can now be derived using the bounds below. For feasibility and angelic-aid, due to the existential quantification over the valuations, one witness valuation $v \in \mathcal{R}$ as upper bound suffices to conclude that the entire region is satisfying. Similarly, for demonic and robustness, one counterexample suffices to conclude that the region is violating. Thus, the scheme can be optimized by terminating in such cases.

$$[lower_{\clubsuit}, upper_{\clubsuit}] \in \begin{cases} [Pr_{\min,\min}[\underline{\mathcal{M}^\delta}, \diamond^{\leq t/\delta} G], Pmin[\overline{\mathcal{M}^\delta}[v], \diamond^{\leq t/\delta} G] + \epsilon_\delta], \text{ if } \clubsuit = f, \\ [Pr_{\max,\min}[\underline{\mathcal{M}^\delta}, \diamond^{\leq t/\delta} G], Pmax[\overline{\mathcal{M}^\delta}[v], \diamond^{\leq t/\delta} G] + \epsilon_\delta], \text{ if } \clubsuit = a, \\ [Pmin[\underline{\mathcal{M}^\delta}[v], \diamond^{\leq t/\delta} G], Pr_{\min,\max}[\overline{\mathcal{M}^\delta}, \diamond^{\leq t/\delta} G] + \epsilon_\delta], \text{ if } \clubsuit = r, \\ [Pmax[\underline{\mathcal{M}^\delta}[v], \diamond^{\leq t/\delta} G], Pr_{\max,\max}[\overline{\mathcal{M}^\delta}, \diamond^{\leq t/\delta} G] + \epsilon_\delta], \text{ if } \clubsuit = d. \end{cases}$$

Bounds for the synthesis problem are derived similarly to the satisfaction verification. However, the early termination criterion does not apply anymore, as we aim to classify each valuation as violating or satisfying.

$$[lower_{\clubsuit}, upper_{\clubsuit}] \in \begin{cases} [Pr_{\min,\min}[\underline{\mathcal{M}^\delta}, \diamond^{\leq t/\delta} G], Pr_{\min,\max}[\overline{\mathcal{M}^\delta}, \diamond^{\leq t/\delta} G] + \epsilon_\delta], \text{ if } \clubsuit \in \{f, r\}, \\ [Pr_{\max,\min}[\underline{\mathcal{M}^\delta}, \diamond^{\leq t/\delta} G], Pr_{\max,\max}[\overline{\mathcal{M}^\delta}, \diamond^{\leq t/\delta} G] + \epsilon_\delta], \text{ if } \clubsuit \in \{a, d\}. \end{cases}$$

In practice, the synthesis is sample-guided, similar to synthesis methods for other models [8,30]. First, a relatively fast check is performed on the instantiated MA $\mathcal{M}[v]$, for an $v \in \mathcal{R}$, whose value then determines which inequality to check.

The bounds for both problems are affected by the discretization error ϵ_δ. While the parameter lifting error approaches zero as the region becomes smaller, the discretization error remains constant. We may therefore encounter cases where the outcome remains inconclusive for every subregion. To solve problems up to arbitrary precision, one must detect this situation and decrease the discretization step δ accordingly. A conservative method is presented in Sect. 6.

In summary, using the methodologies described in Sects. 4-5, we cover the entire process as depicted in Fig. 3.

Table 1. Benchmark statistics

Name	Type	$\lvert S\rvert$	$\lvert\mathcal{X}\rvert$	Property
erlang	pMA	187	1	$\mathbb{P}[\diamond^{\leq 5}\text{Goal}] \leq 0.2$
flexible-manufacturing3	pMA	1675	2	$\mathbb{P}[\diamond^{\leq 1}\text{M3\_fail}] \leq 0.05$
jobs5-2	pMA	117	2	$\mathbb{P}[\diamond^{\leq 3}\text{'half\_of\_jobs\_finished'}] \leq 0.65$
polling	pMA	990	2	$\mathbb{P}[\diamond^{\leq 3}\text{q1\_full}] \leq 0.55$
readers-writers5	pMA	842	2	$\mathbb{P}[\diamond^{\leq 2}\text{many\_requests}] \geq 0.01$
stream2-4	pMA	176	2	$\mathbb{P}[\diamond^{\leq 3}\text{'done'}] \leq 0.1$
kanban	pCTMC	4600	2	$\mathbb{P}[\diamond^{\leq 4}\text{done}] \leq 0.02$
polling9	pCTMC	6912	2	$\mathbb{P}[\diamond^{\leq 1}\text{first\_queue\_served}] \leq 0.03$
tandem	pCTMC	861	2	$\mathbb{P}[\diamond^{\leq 0.3}\text{first\_queue}] \leq 0.8$

6 Experimental Evaluation

Implementation. We implemented our approach, supporting both pCTMCs and pMAs, using the STORMPY Python interface of the STORM probabilistic modelchecker [28]. The implementation is publicly available on GitLab.[1]

The current implementation relies on explicit labels on goal-states and is able to check time-bounded reachability properties. Moreover, it can be easily extended to time-bounded until queries $\mathit{safe}\ U^{\leq t}\ G$, where $\mathit{safe}, G \subseteq S$ [3].

One can configure in which order to verify regions and when to decrease the discretization error, as it does not affect the theoretical guarantees. However, as it impacts performance, we briefly cover our choices. As the discretization error is the main component in performance degradation, we implemented a conservative method to decrease the discretization error. Intuitively, our approach uses that the difference between $Pr_{p,\max}[\cdot]$ and $Pr_{p,\min}[\cdot]$ tends to zero as the volume of the region decreases. Thus, if $Pr_{p,\min}[\cdot] + \epsilon_\delta > \theta$ whilst $Pr_{p,\min}[\cdot] \leq \theta$, no subregion will satisfy $Pr_{p,\max}[\cdot] + \epsilon_\delta \leq \theta$. Therefore, if $Pr_{p,\max}[\cdot] \leq \theta$ as well, splitting the region will not help and the discretization error should be reduced instead. The subregions that need to be checked after splitting are handled as a FIFO queue, ensuring that larger regions are handled first. We implemented an additional heuristic which tracks a frontier of the already decided parameter space, allowing to prioritize regions bordering decided regions.

Set-Up. All experiments were run on an HP Zbook G10 with a 13th Gen Intel(R) Core(TM) i7-13700H and 16 GB of RAM. The implementation ran on a single core. Unless specified otherwise, we terminate the synthesis algorithm when at least 99.99% of the parameter space is decided or the time limit of 15 min has been reached. The initial discretization error was set at $\theta/2$ where θ is the threshold probability as specified in Table 1. To obtain the approximate discretized MAs $\overline{\mathcal{M}^\delta}, \underline{\mathcal{M}^\delta}$, N has been set to 1 and 2 respectively. This suffices since, as motivated in Sect. 4.2, even for small N the introduced error is negligible compared to the discretization error. Lastly, in order to obtain an appropriate performance study of our underlying theoretical approach, all experiments were run without the frontier heuristic enabled.

Benchmarks. We use a number of Markov Automata and CTMCs included in the QCOMP benchmark suite [26]. Table 1 provides an overview of the models. We focus

[1] https://gitlab.tue.nl/k.v.d.glind/discretize-ma.

Table 2. PRISM-PSY compared to our method.

Model	Region	PRISM-PSY	Ours
kanban	$redo3 \in [0.1, 0.7], ok3 \in [0.2, 0.9]$	99.9%, 67 s	73.4%, TO
polling9	$gamma \in [10, 15], mu \in [0.1, 0.5]$	99.9%, 20 s	89.1%, TO
tandem	$p \in [0.01, 0.99], kappa \in [0.5, 1.5]$	not supported	84.1%, TO

on benchmarks where either a time-bounded property was specified or a natural extension from an unbounded reachability probability was possible. We adapted the models to include parameters and ensured that, for every region, the original model specification is included in this region.

Comparison to PRISM-PSY. We compare our method with PRISM-PSY [8] on pCTMCs. We solve the parameter synthesis problem and compare the decided fraction of the two tools within a time window of 15 min in Table 2. We see that PRISM-PSY outperforms our approach, mainly because it does not require discretization. However, our method focuses on the novel ability to analyse pMA, which is not possible in PRISM-PSY. Moreover, the implementation of PRISM-PSY is restricted to parametric rates of the form $R(s, s') = \text{c} \cdot x_i$ for parameter x_i and constant $c > 0$. We do not impose such conditions and are able to analyze the *tandem* model which involves parametric transition probabilities with fixed sojourn rates. This model cannot be analyzed with PRISM-PSY.

Satisfaction Verification of pMA. We check for a given benchmark model and parameter region whether the property holds under the specified satisfaction relation. The results are displayed in Table 3. All benchmarks produced a conclusive answer (column S/V), and most finished within a second of total runtime, which includes both model transformation and verification. For more complex models, model preprocessing becomes the main factor, displaying the limited scalability of the transformations presented in Sect. 4.1. Table 4 provides a breakdown of building and model checking times for varying region sizes and discretization errors. Verification times correlate with the maximum sojourn rate and discretization error, following our theoretical findings. The effect of the number of parameters on verification time stems from the additional nondeterministic choices added to the parameter-free SG. For an in-depth time breakdown, we refer the reader to the extended version [15].

Table 3. Satisfaction verification results. **S**atisfied/**V**iolated property. Extract of the full Table in [15].

Name	Region	Sat. relation	Time(s)	S/V
erlang	$R \in [7, 10]$	Feasibility	0.13	1/0
erlang	$R \in [7, 10]$	Angelic-aid	0.04	0/1
flexible-manufacturing3	$rate1 \in [0.5, 2.5], rate2 \in [0.5, 2.5]$	Feasibility	0.97	1/0
flexible-manufacturing3	$rate1 \in [0.5, 2.5], rate2 \in [0.5, 2.5]$	Angelic-aid	0.97	1/0
jobs5-2	$x_j1 \in [1, 3], x_j2 \in [2, 4]$	Feasibility	0.46	1/0
jobs5-2	$x_j1 \in [1, 3], x_j2 \in [2, 4]$	Angelic-aid	0.29	0/1
kanban	$redo3 \in [0.1, 0.7], ok3 \in [0.2, 0.9]$	Angelic-aid	11.85	1/0
polling9	$gamma \in [10, 15], mu \in [0.1, 0.5]$	Angelic-aid	5.37	1/0
polling	$inRate1 \in [2, 4], inRate2 \in [3, 6]$	Feasibility	5.23	1/0
polling	$inRate1 \in [2, 4], inRate2 \in [3, 6]$	Angelic-aid	6.88	1/0
readers-writers5	$readingRate \in [3, 7], rate2 \in [3, 10]$	Feasibility	0.15	1/0
readers-writers5	$readingRate \in [3, 7], rate2 \in [3, 10]$	Angelic-aid	0.15	1/0
stream2-4	$inRate \in [1, 7], processingRate \in [1, 7]$	Feasibility	0.10	1/0
stream2-4	$inRate \in [1, 7], processingRate \in [1, 7]$	Angelic-aid	0.25	1/0
tandem	$p \in [0.01, 0.99], kappa \in [0.5, 1.5]$	Angelic-aid	5.55	1/0

Table 4. Time-breakdown of checking a single region. MC time includes all 4 possible calls $Pr_{p,n}[\underline{\mathcal{M}^\delta}, \diamond^{\leq t/\delta} G]$. Extract of the full Table in [15].

Benchmark	Region	ϵ_δ	Build Time (s)	MC Time (s)
erlang	$R \in [1, 2]$	0.001	0.023	0.121
erlang	$R \in [1, 2]$	0.0001	0.023	1.359
erlang	$R \in [19, 20]$	0.001	0.024	12.517
erlang	$R \in [19, 20]$	0.0001	0.023	124.381
polling	$inRate1 \in [2, 4]$, $inRate2 \in [3, 6]$	0.001	5.259	8.109
polling	$inRate1 \in [2, 4]$, $inRate2 \in [3, 6]$	0.0001	5.309	78.210
jobs5-2-2params	$x_j1 \in [1, 3]$, $x_j2 \in [2, 4]$	0.001	0.245	0.562
jobs5-2-3params	$x_j1 \in [1, 3]$, $x_j2 \in [2, 4]$, $x_j3 \in [1, 3]$	0.001	0.254	0.654
jobs5-2-4params	$x_j1 \in [1, 3]$, $x_j2 \in [2, 4]$, $x_j3 \in [1, 3]$, $x_j4 \in [2, 4]$	0.001	0.246	1.166

Table 5. Parameter synthesis results. Fraction of parameter space **S**atisfying/**V**iolating/**U**nknown on property. Fractions are rounded to 3 decimals.

Name	Region	Sat. relation	Time(s)	S/V/U
erlang	$R \in [7, 10]$	Demonic	0.03	0.000/1.000/0.000
erlang	$R \in [7, 10]$	Robust	336.38	0.311/0.689/$< 10^{-4}$
flexible-manufacturing3	$rate1 \in [0.5, 2.5], rate2 \in [0.5, 2.5]$	Demonic	TO	0.687/0.291/0.022
flexible-manufacturing3	$rate1 \in [0.5, 2.5], rate2 \in [0.5, 2.5]$	Robust	TO	0.687/0.291/0.022
jobs5-2	$x_j1 \in [1, 3], x_j2 \in [2, 4]$	Demonic	0.31	0.000/1.000/0.000
jobs5-2	$x_j1 \in [1, 3], x_j2 \in [2, 4]$	Robust	TO	0.416/0.577/0.007
kanban	$redo3 \in [0.1, 0.7], ok3 \in [0.2, 0.9]$	Robust	TO	0.570/0.164/0.266
polling9	$gamma \in [10, 15], mu \in [0.1, 0.5]$	Robust	TO	0.884/0.007/0.109
polling	$inRate1 \in [2, 4], inRate2 \in [3, 6]$	Demonic	TO	0.199/0.784/0.017
polling	$inRate1 \in [2, 4], inRate2 \in [3, 6]$	Robust	TO	0.900/0.083/0.017
readers-writers5	$readingRate \in [3, 7], rate2 \in [3, 10]$	Demonic	TO	0.617/0.031/0.352
readers-writers5	$readingRate \in [3, 7], rate2 \in [3, 10]$	Robust	TO	0.617/0.031/0.352
stream2-4	$inRate \in [1, 7], processingRate \in [1, 7]$	Demonic	TO	0.494/0.496/0.010
stream2-4	$inRate \in [1, 7], processingRate \in [1, 7]$	Robust	TO	0.830/0.162/0.007
tandem	$p \in [0.01, 0.99], kappa \in [0.5, 1.5]$	Robust	TO	0.500/0.314/0.186

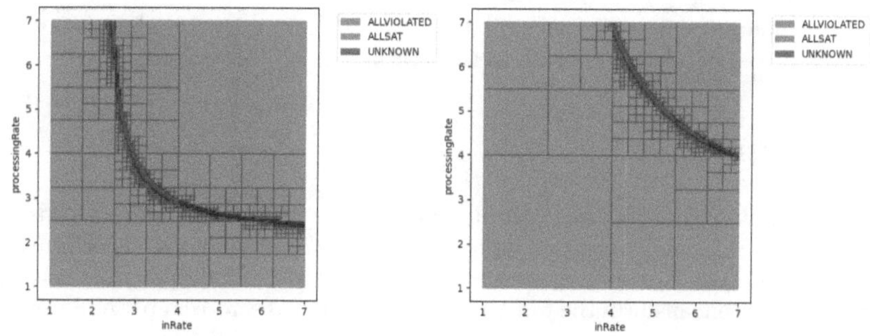

(a) Demonic (b) Robust

Fig. 5. Parameter synthesis on *stream2-4* under different satisfaction relations.

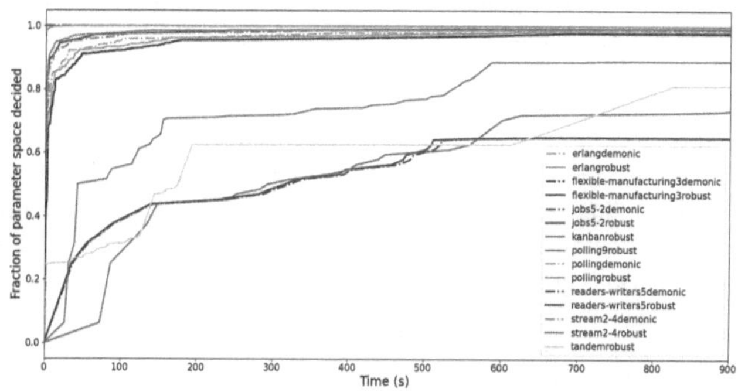

Fig. 6. Ratio of decided area over runtime for pMA benchmarks.

Parameter Synthesis for pMA. The synthesis results are displayed in Table 5. Note that the checked regions are exactly the same as in Table 3. For each benchmark, we checked robust and demonic satisfaction relations. In accordance with Fig. 2b, the fraction of satisfying valuations for the robust satisfaction relation is always at least as large as that for the demonic relation.

Figure 5 shows the space partitioning obtained for the *stream* benchmark. Each rectangle in the space gives the region on which this decision was based. We note that extensive portions of the parameter space can be resolved quickly, while more detailed analysis is necessary near the boundary between satisfaction and violation. Synthesis figures for all other benchmarks are provided in [15].

In Table 5, we observe that almost all benchmarks reach the timeout. This occurs for multiple reasons. First, determining whether a region satisfies a property requires a fixed number of model checking queries. As each subregion is progressively refined, accounting for an increasingly smaller portion of the overall parameter space, an equivalent computational effort is required for a diminished gain. This phenomenon is also observed in other methods, including work on discrete-time models [8,23,30]. Figure 6 illustrates this perceived stagnation, where the decided fraction over time is plotted. In theory one converges to the total area decided, in practice we observe an exponential deceleration of gain in decided parameter space. Second, as it is necessary to decrease the discretization error, the computational cost of each model-checking query is increased as well. Therefore, the perceived deceleration in Fig. 6 is exacerbated.

7 Conclusion

We introduced the notion of parametric Markov Automata, extending Markov Automata with parametric rate- and transition functions. We presented methods to perform satisfaction verification and parameter synthesis on these models. Our approach consists of two steps. First, we transform the pMA to an approximate discretized pMA. Second, we analyze the approximate behaviour of the pMA by applying existing parameter lifting techniques. The experimental evaluation indicated that both the predominant source of error and computational demand arise from the discretization method.

Future Work. Given our findings above, future research could investigate approaches which do not use the discretization method. This could for instance require the analysis of continuous-time stochastic games without discretization in order to perform parameter lifting. Alternatively, one could opt for an approach using uniformization, such as in [6], instead. However, this would most likely require many costly calls on the SG due to the prophetic scheduler.

A better choice of heuristics may optimize the total decided area for the parameter synthesis problem. That is, one may investigate how to use previous information to decide when to reduce the discretization error. Less conservative conditions on when to decrease the discretization error could potentially save expensive model checking queries on the discretized MA. However, it may lead to, potentially, checking regions with a lower discretization error than necessary.

We believe our method can also be extended to time-interval bounded queries. Here, the main difficulty lies in finding an appropriate approximate discretization such that lower and upper bounds are guaranteed.

Data Availability. The implementation and benchmark files are publicly available in the artifact at doi.org/10.5281/zenodo.19661781.

References

1. Ashok, P., Butkova, Y., Hermanns, H., Křetínský, J.: Continuous-time Markov decisions based on partial exploration. In: Lahiri, S.K., Wang, C. (eds.) ATVA 2018. LNCS, vol. 11138, pp. 317–334. Springer, Cham (2018). https://doi.org/10.1007/978-3-030-01090-4_19
2. Aziz, A., Sanwal, K., Singhal, V., Brayton, R.: Verifying continuous time Markov chains. In: Alur, R., Henzinger, T.A. (eds.) CAV 1996. LNCS, vol. 1102, pp. 269–276. Springer, Heidelberg (1996). https://doi.org/10.1007/3-540-61474-5_75
3. Braitling, B., Fioriti, L.M.F., Hatefi, H., Wimmer, R., Becker, B., Hermanns, H.: MeGARA: Menu-based game abstraction and abstraction refinement of Markov automata. In: QAPL. EPTCS, vol. 154, pp. 48–63 (2014). https://doi.org/10.4204/EPTCS.154.4
4. Butkova, Y., Fox, G.: Optimal time-bounded reachability analysis for concurrent systems. In: Vojnar, T., Zhang, L. (eds.) TACAS 2019. LNCS, vol. 11428, pp. 191–208. Springer, Cham (2019). https://doi.org/10.1007/978-3-030-17465-1_11

5. Butkova, Y., Hartmanns, A., Hermanns, H.: A modest approach to modelling and checking Markov automata. In: Parker, D., Wolf, V. (eds.) QEST 2019. LNCS, vol. 11785, pp. 52–69. Springer, Cham (2019). https://doi.org/10.1007/978-3-030-30281-8_4
6. Butkova, Y., Hatefi, H., Hermanns, H., Krčál, J.: Optimal continuous time Markov decisions. In: Finkbeiner, B., Pu, G., Zhang, L. (eds.) ATVA 2015. LNCS, vol. 9364, pp. 166–182. Springer, Cham (2015). https://doi.org/10.1007/978-3-319-24953-7_12
7. Butkova, Y., Wimmer, R., Hermanns, H.: Long-run rewards for Markov automata. In: Legay, A., Margaria, T. (eds.) TACAS 2017. LNCS, vol. 10206, pp. 188–203. Springer, Heidelberg (2017). https://doi.org/10.1007/978-3-662-54580-5_11
8. Ceska, M., Dannenberg, F., Paoletti, N., Kwiatkowska, M., Brim, L.: Precise parameter synthesis for stochastic biochemical systems. Acta Informatica **54**(6), 589–623 (2017). https://doi.org/10.1007/S00236-016-0265-2
9. Češka, M., Jansen, N., Junges, S., Katoen, J.-P.: Shepherding Hordes of Markov chains. In: Vojnar, T., Zhang, L. (eds.) TACAS 2019. LNCS, vol. 11428, pp. 172–190. Springer, Cham (2019). https://doi.org/10.1007/978-3-030-17465-1_10
10. Češka, M., Pilař, P., Paoletti, N., Brim, L., Kwiatkowska, M.: PRISM-PSY: precise GPU-accelerated parameter synthesis for stochastic systems. In: Chechik, M., Raskin, J.-F. (eds.) TACAS 2016. LNCS, vol. 9636, pp. 367–384. Springer, Heidelberg (2016). https://doi.org/10.1007/978-3-662-49674-9_21
11. Daws, C.: Symbolic and parametric model checking of discrete-time Markov chains. In: Liu, Z., Araki, K. (eds.) ICTAC 2004. LNCS, vol. 3407, pp. 280–294. Springer, Heidelberg (2005). https://doi.org/10.1007/978-3-540-31862-0_21
12. Dugan, J.B., Bavuso, S.J., Boyd, M.A.: Dynamic fault-tree models for fault-tolerant computer systems. IEEE Trans. Reliab. **41**(3), 363–377 (2002)
13. Eisentraut, C., Hermanns, H., Zhang, L.: On probabilistic automata in continuous time. In: LICS, pp. 342–351. IEEE Computer Society (2010). https://doi.org/10.1109/LICS.2010.41
14. Garling, D.J.: A Course in Mathematical Analysis: Volume 2, Metric and Topological Spaces, Functions of a Vector Variable, chap. 12. Cambridge University Press (2014)
15. van de Glind, K., Volk, M., Willemse, T.: Verification of parametric Markov automata under time-bounded reachability. CoRR **abs/2606.21444** (2026). https://doi.org/10.48550/arXiv.2606.21444
16. Gros, T.P.: Markov Automata taken by Storm. Master's thesis, Universität des Saarlandes (2018)
17. Guck, D., Han, T., Katoen, J.-P., Neuhäußer, M.R.: Quantitative timed analysis of interactive Markov chains. In: Goodloe, A.E., Person, S. (eds.) NFM 2012. LNCS, vol. 7226, pp. 8–23. Springer, Heidelberg (2012). https://doi.org/10.1007/978-3-642-28891-3_4
18. Guck, D., Hatefi, H., Hermanns, H., Katoen, J.P., Timmer, M.: Analysis of timed and long-run objectives for Markov automata. Log. Methods Comput. Sci. **10**(3) (2014). https://doi.org/10.2168/LMCS-10(3:17)2014
19. Hahn, E.M., Han, T., Zhang, L.: Synthesis for PCTL in parametric Markov decision processes. In: Bobaru, M., Havelund, K., Holzmann, G.J., Joshi, R. (eds.) NFM 2011. LNCS, vol. 6617, pp. 146–161. Springer, Heidelberg (2011). https://doi.org/10.1007/978-3-642-20398-5_12
20. Hahn, E.M., Hartmanns, A.: A comparison of time- and reward-bounded probabilistic model checking techniques. In: Fränzle, M., Kapur, D., Zhan, N. (eds.) SETTA 2016. LNCS, vol. 9984, pp. 85–100. Springer, Cham (2016). https://doi.org/10.1007/978-3-319-47677-3_6

21. Hahn, E.M., Hermanns, H., Zhang, L.: Probabilistic reachability for parametric Markov models. Int. J. Softw. Tools Technol. Transf. **13**(1), 3–19 (2011). https://doi.org/10.1007/S10009-010-0146-X
22. Hahn, E.M., Li, Y., Schewe, S., Turrini, A., Zhang, L.: iscasMc: a web-based probabilistic model checker. In: Jones, C., Pihlajasaari, P., Sun, J. (eds.) FM 2014. LNCS, vol. 8442, pp. 312–317. Springer, Cham (2014). https://doi.org/10.1007/978-3-319-06410-9_22
23. Han, T., Katoen, J.P., Mereacre, A.: Approximate parameter synthesis for probabilistic time-bounded reachability. In: RTSS, pp. 173–182. IEEE Computer Society (2008). https://doi.org/10.1109/RTSS.2008.19
24. Hartmanns, A., Hermanns, H.: The modest toolset: an integrated environment for quantitative modelling and verification. In: Ábrahám, E., Havelund, K. (eds.) TACAS 2014. LNCS, vol. 8413, pp. 593–598. Springer, Heidelberg (2014). https://doi.org/10.1007/978-3-642-54862-8_51
25. Hartmanns, A., Katoen, J.-P., Kohlen, B., Spel, J.: Tweaking the odds in probabilistic timed automata. In: Abate, A., Marin, A. (eds.) QEST 2021. LNCS, vol. 12846, pp. 39–58. Springer, Cham (2021). https://doi.org/10.1007/978-3-030-85172-9_3
26. Hartmanns, A., Klauck, M., Parker, D., Quatmann, T., Ruijters, E.: The quantitative verification benchmark set. In: Vojnar, T., Zhang, L. (eds.) TACAS 2019. LNCS, vol. 11427, pp. 344–350. Springer, Cham (2019). https://doi.org/10.1007/978-3-030-17462-0_20
27. Hatefi, H., Hermanns, H.: Model checking algorithms for Markov automata. Electron. Commun. Eur. Assoc. Softw. Sci. Technol. **53** (2012). https://doi.org/10.14279/TUJ.ECEASST.53.783
28. Hensel, C., Junges, S., Katoen, J.P., Quatmann, T., Volk, M.: The probabilistic model checker Storm. Int. J. Softw. Tools Technol. Transf. **24**(4), 589–610 (2022). https://doi.org/10.1007/S10009-021-00633-Z
29. Hermanns, H.: Interactive Markov Chains. LNCS, vol. 2428. Springer, Heidelberg (2002). https://doi.org/10.1007/3-540-45804-2
30. Junges, S., et al.: Parameter synthesis for Markov models: covering the parameter space. Formal Methods Syst. Des. **62**(1), 181–259 (2024). https://doi.org/10.1007/S10703-023-00442-X
31. Katoen, J.P.: The probabilistic model checking landscape. In: LICS, pp. 31–45. ACM (2016). https://doi.org/10.1145/2933575.2934574
32. Kwiatkowska, M., Norman, G., Parker, D.: PRISM 4.0: verification of probabilistic real-time systems. In: Gopalakrishnan, G., Qadeer, S. (eds.) CAV 2011. LNCS, vol. 6806, pp. 585–591. Springer, Heidelberg (2011). https://doi.org/10.1007/978-3-642-22110-1_47
33. Marsan, M.A., Conte, G., Balbo, G.: A class of generalized stochastic Petri nets for the performance evaluation of multiprocessor systems. ACM Trans. Comput. Syst. **2**(2), 93–122 (1984). https://doi.org/10.1145/190.191
34. Miller, B.L.: Finite state continuous time Markov decision processes with a finite planning horizon. SIAM J. Control **6**(2), 266–280 (1968). https://doi.org/10.1137/0306020
35. Neuhäußer, M.R.: Model checking nondeterministic and randomly timed systems. Ph.D. thesis, RWTH Aachen University (2010)
36. Neuhäußer, M.R., Stoelinga, M., Katoen, J.-P.: Delayed nondeterminism in continuous-time Markov decision processes. In: de Alfaro, L. (ed.) FoSSaCS 2009. LNCS, vol. 5504, pp. 364–379. Springer, Heidelberg (2009). https://doi.org/10.1007/978-3-642-00596-1_26

37. Quatmann, T., Dehnert, C., Jansen, N., Junges, S., Katoen, J.-P.: Parameter synthesis for Markov models: faster than ever. In: Artho, C., Legay, A., Peled, D. (eds.) ATVA 2016. LNCS, vol. 9938, pp. 50–67. Springer, Cham (2016). https://doi.org/10.1007/978-3-319-46520-3_4
38. Stewart, J., Clegg, D., Watson, S.: Calculus: Early Transcendentals. Cengage (2021)

Faster Algorithm for Achieving Minimal-Size Quantum Decision Diagrams

Juul Sanders[1(✉)], Sebastiaan Brand[1], Arend-Jan Quist[2], and Tim Coopmans[1]

[1] Delft University of Technology, Delft, The Netherlands
{juul.sanders,s.o.brand,t.j.coopmans}@tudelft.nl
[2] Leiden University, Leiden, The Netherlands
a.quist@liacs.leidenuniv.nl

Abstract. The decision diagram (DD) data structure enables fast linear-algebra calculations by bringing vectors into a normal form and subsequently merging equivalent ones, yielding a minimally-sized DD modulo the equivalence relation. A fruitful application area is quantum-circuit simulation, where the vectors represent quantum states. The Local Invertible Map Decision Diagram (LIMDD) type merges LIM-equivalent (typically Pauli-gate equivalent) vectors, can efficiently simulate Clifford circuits as well as some high-T-count circuits, and has theoretically been proven exponentially faster for simulation than other well-developed data structures, including other common DD variants. However, these exponential advantages have not fully materialized yet in existing implementations, for which the normal-form procedure, which is a highly complex algorithm, is either absent or only partially implemented. We here present a novel normal-form algorithm for Pauli-LIMDDs, achieving a worst-case speedup from $O(n^3)$ to $O(n^2)$ for an n-qubit DD node with a single child node while keeping the $O(n^3)$ run time in case of two distinct children nodes. We implement the algorithm as part of QolDDer, our Pauli-LIMDD simulator for quantum circuits, written from scratch in C/C++. The implementation realizes the theoretically-proven advantages of Pauli-LIMDDs on Clifford circuits, is significantly faster than the existing LIMDD simulators on such circuits, and on a public quantum-circuit data set often outperforms them by an order of magnitude. In the future, we envision that our work will enable further application and development of LIMDD variants, not only for quantum design tasks, but also for analysis of linear-algebra-based systems in general.

Keywords: Quantum computing · Decision diagrams

1 Introduction

Quantum computing promises computational advantages in domains such as cryptography [30], finance [23], optimization [15], and the simulation of quantum

M. Chen and M. Tschaikowski (Eds.): QEST+FORMATS 2026, LNCS 16913, pp. 410–428, 2027.
https://doi.org/10.1007/978-3-032-35298-9_23

physics [14] and chemistry [11]. For quantum computers to scale to practically useful tasks, various design challenges need to be solved, such as quantum circuit compilation, performance comparisons of quantum algorithms, and evaluating and mitigating the effects of noise. An important tool to tackle these challenges is the simulation of quantum circuits on classical computers [4,17,35].

Among the algorithms and data structures for quantum circuit simulation are decision diagrams (DD) [9], which have been proven to be successful in the simulation of quantum circuits and related tasks [7,16,21,26,31,36–38,41,43]. In quantum computing, states and transformations of states can be represented as pseudo-Boolean functions $f : \{0,1\}^n \to \mathbb{C}$, which are exponentially large in the number of qubits. Decision diagrams are a compact alternative that represents pseudo-Boolean functions as directed graphs. The nodes of these graphs represent sub-functions $f_a : \boldsymbol{x} \mapsto f(\boldsymbol{a}, \boldsymbol{x})$, for $\boldsymbol{a} \in \{0,1\}^m, m \leq n$. DDs achieve succinctness due to *canonicity*: by converting each node to a normal form, two nodes representing the same function are *merged*, avoiding storing multiple sub-functions if they are identical. DDs that offer stronger node merging also exist. For example, in quantum multiple-value decision diagrams (QMDDs [21], i.e., factored edge-valued DDs [34] with complex edge values), two nodes can be merged if they represent the same sub-function up to some scalar multiple.

Local-Invertible Map Decision Diagrams (LIMDDs) enable even stronger node merging: two nodes are merged if they represent the same sub-function up to some local invertible map (LIM), Kronecker products of invertible 2×2 matrices. Pauli-LIMDDs, where the LIMs consist of matrices from the Pauli group (a set of matrices well known in quantum physics), were proven to be exponentially faster for quantum-circuit simulation than QMDDs and Algebraic Decision Diagrams [6,13], and can, for specific circuits, also outperform common techniques such as matrix product states and Clifford+T simulation [38,39].

Practical implementations of LIMDDs have so far failed to fully realize their theoretical advantages, showcasing the difficulty of their (correct) implementation. The first cause is the complicated $\mathcal{O}(n^3)$ run time merge rule algorithm, solving a group-theoretic task: in practice, a fast implementation is needed to outperform the $\mathcal{O}(1)$ run time merging for QMDDs. In addition to this, canonicity, the crucial property for succinct and efficiently updatable decision diagrams, requires careful handling of many (edge) cases. To make matters worse, bugs in the canonicity implementation are hard to catch, as a non-canonical LIMDD can represent the correct quantum state (and thus pass all typical software tests) while still not being optimal in node number. Indeed, neither of the two existing proof-of-principle LIMDD implementations is fully canonical. First, MQT-LIMDD [40], built as an extension of the QMDD-simulator MQT-DDSIM [43], does not achieve minimal size in all cases. Second, LimTDD [16], which implements a generalization of Pauli-LIMDDs, deliberately does not include the canonicity property, voiding any guarantees against (exponential) blow-ups.

In this work, we present a novel algorithm for the Pauli-LIMDD merge rule, implemented as an open-source stand-alone library PauliLim [1]. The algorithm has less overhead and achieves a run time improvement from $\mathcal{O}(n^3)$ to $\mathcal{O}(n^2)$

for n-qubit nodes with a single child (one of the cases in which DDs achieve a succinctness advantage). We integrate this new algorithm into QolDDer, an open-source [2] from-scratch implementation of Pauli-LIMDDs. In an empirical evaluation, we compare QolDDer against the two previous LIMDD implementations [16,40], as well as against the well-established QMDD implementation MQT-DDSIM [43]. These empirical results corroborate that LIMDDs are able to achieve exponential advantages over QMDDs and underline that LIMDDs require canonicity to realize these advantages. Our work will enable further application and development of LIMDD variants, not only for quantum design tasks but also for analysis of linear-algebra-based systems in general.

2 Preliminaries

2.1 Quantum Computing

We briefly introduce quantum computing; see [22] for a full overview. The state of a single quantum bit (qubit) is a complex-valued 2-vector of unit norm: either $|0\rangle = \begin{bmatrix}1\ 0\end{bmatrix}^\mathsf{T}$ or $|1\rangle = \begin{bmatrix}0\ 1\end{bmatrix}^\mathsf{T}$, or a so-called *superposition* of both, $\begin{bmatrix}\alpha_0\ \alpha_1\end{bmatrix}^\mathsf{T} = \alpha_0 |0\rangle + \alpha_1 |1\rangle$ where α_0, α_1 are complex numbers satisfying $|\alpha_0|^2 + |\alpha_1|^2 = 1$ and where $|.|$ denotes the modulus of a complex number. The joint state of two qubits, $|\psi_A\rangle$ and $|\psi_B\rangle$, is obtained from the Kronecker product of their individual states; $\begin{bmatrix}\alpha_0\ \alpha_1\end{bmatrix}^\mathsf{T} \otimes \begin{bmatrix}\beta_0\ \beta_1\end{bmatrix}^\mathsf{T} = \begin{bmatrix}\alpha_0\beta_0 & \alpha_0\beta_1 & \alpha_1\beta_0 & \alpha_1\beta_1\end{bmatrix}^\mathsf{T}$. In general, the state of n qubits can be represented by a vector of unit norm, $|\psi\rangle \in \mathbb{C}^{2^n}$, and can be written in the *Shannon decomposition* $|0\rangle \otimes |\psi_0\rangle + |1\rangle \otimes |\psi_1\rangle$, where $|\psi_0\rangle, |\psi_1\rangle$ are (unnormalized) $(n-1)$-qubit states.

Quantum gates are (reversible) operations that map n-qubit states to n-qubit states, represented as $2^n \times 2^n$ unitary matrices. Several common gates are listed below. Two sets of gates that are particularly relevant to this work are the Pauli gates $\{I, X, Y, Z{=}P(\pi)\}$, and the Clifford gates, which are generated from $\{H, P(\frac{\pi}{2}), CX\}$ under matrix multiplication and Kronecker products.

$$I{=}\begin{bmatrix}1&0\\0&1\end{bmatrix}\ X{=}\begin{bmatrix}0&1\\1&0\end{bmatrix}\ Y{=}\begin{bmatrix}0&\text{-}i\\i&0\end{bmatrix}\ H{=}\frac{1}{\sqrt{2}}\begin{bmatrix}1&1\\1&\text{-}1\end{bmatrix}\ P(\theta){=}\begin{bmatrix}1&0\\0&e^{i\theta}\end{bmatrix}\ CX{=}\begin{bmatrix}1&0&0&0\\0&1&0&0\\0&0&0&1\\0&0&1&0\end{bmatrix}$$

The quantum state after the application of a gate is obtained through matrix-vector multiplication. When a gate is applied to a state with a greater number of qubits, the gate matrix is extended using Kronecker products with I, so on a 3-qubit state $|\psi\rangle$, applying X on the second qubit can be computed as $I \otimes X \otimes I\, |\psi\rangle$.

A quantum circuit is a sequence of gates (and measurements) performed on a register of qubits. A straightforward way to simulate quantum circuits is to compute the final state vector by performing matrix-vector multiplication for all gates in the circuit, starting from the initial state. Desired information, such as measurement outcomes, can then be extracted from this final state. The limiting factor of this approach is the exponential (2^n) size of n-qubit state vectors.

We finish with notions from Pauli stabilizer algebra [3,22]. By an n-qubit Pauli word $P \in \text{PAULI}_n$, we mean a length-n Kronecker product of the Pauli matrices, e.g. $X \otimes Y \otimes Z$ or $Z \otimes I$, abbreviated by omitting '$\otimes$' as XYZ and ZI. We will denote the $2^n \times 2^n$ identity matrix $I_2^{\otimes n}$ as I_n and write $\mathbb{I}$ when n is clear from the context. For this work, a useful representation of a Pauli gate is a binary tuple $(x, z) \in \{0, 1\}^2$, defined as $I \simeq (0, 0), Z \simeq (0, 1), X \simeq (1, 0), Y \simeq (1, 1)$. We extend this to Pauli words using a *reverse* order than is common [3], defining the binary vector of P as $\mathsf{revbin}(P_1 \otimes P_2 \otimes \dots P_n) = (x_n, x_{n-1}, \dots, x_1, z_n, z_{n-1}, \dots, z_1)$. For example, $\mathsf{revbin}(YZ) = (0, 1, 1, 1)$.

The n-qubit Pauli group is the closure of n-qubit Pauli words under matrix multiplication. A stabilizer group of a state $|\phi\rangle$ is the commutative subgroup $\text{Stab}(|\phi\rangle) = \{\lambda P \mid \lambda P |\phi\rangle = |\phi\rangle, P \in \text{PAULI}_n, \lambda \in \{\pm 1\}\}$ of the Pauli group. A generating set for a stabilizer group is a set S of signed Pauli words whose closure under multiplication, $\langle S \rangle$, is that stabilizer group. An example is $\{I, -Z\} = \langle -Z \rangle$, with generating set $\{-Z\}$ as $(-Z)^2 = I$. For a signed Pauli word $(-1)^s \cdot P$ with $s \in \{0, 1\}$, we define $\mathsf{revbin}(sP)$ as the concatenation of $\mathsf{revbin}(P)$ and the bit s, e.g., $\mathsf{revbin}(-YZ) = (0, 1, 1, 1, 1)$. Stabilizer generating sets $\{g_1, g_2, \dots, g_k\}$ can be written as binary matrix where the r-th row is the binary vector $\mathsf{revbin}(g_r)$. Such matrices can be brought in row-echelon form (REF) using Gaussian elimination, an elementary linear algebra algorithm, by a slight adjustment of vector addition (see the 'rowsum' algorithm in [3]). We will say that a stabilizer generating set is in REF if its binary representation is.

2.2 Decision Diagrams for Quantum Computing

Decision diagrams (DDs) [9] are a data structure that heuristically compresses the representation of (pseudo-)Boolean functions, i.e., functions of the form $\{0, 1\}^n \to \mathbb{D}$ for some domain $\mathbb{D}$. Quantum states $|\psi\rangle$ and gates U on n qubits can be viewed as pseudo-Boolean functions $\psi : \{0, 1\}^n \to \mathbb{C}$ and $U : \{0, 1\}^{2n} \to \mathbb{C}$, where $\psi(\boldsymbol{x})$ returns the element of the vector $|\psi\rangle$ at index $\boldsymbol{x}$, which can be encoded in decision diagrams. A crucial aspect of DDs is that the functions they represent can be manipulated *without uncompressing the data structure* [9]. It is also well known how to do matrix-vector multiplication (i.e., quantum circuit simulation in our context) in the DD representation [20].

In the context of quantum computing, various types of decision diagrams have been used, including binary decision diagrams (BDDs) [12,36], multi-terminal BDDs (MTBDDs) [26,37], quantum multiple-valued DDs (QMDDs) [7,21,43], (weighted) context-free-language ordered BDDs (CFLOBDDs) [31,32], and local invertible map DDs (LIMDDs) [16,38,40]. Because we focus on the effect of canonicity of LIMDDs (Sect. 2.3), we limit our scope to (canonical and non-canonical) LIMDDs as well as QMDDs, of which LIMDDs are a generalization.

Structurally, a DD that encodes a vector $|\psi\rangle$ is a directed acyclic graph whose internal nodes have two outgoing edges. An example of a QMDD or LIMDD node is shown in Fig. 1. Such

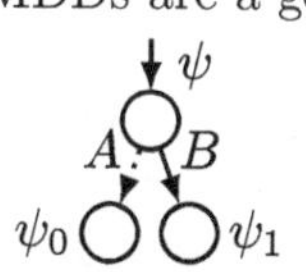

Fig. 1. DD node example.

a node represents the Shannon decomposition, $|\psi\rangle = |0\rangle \otimes A \cdot |\psi_0\rangle + |1\rangle \otimes B \cdot |\psi_1\rangle$, where the edge labels A, B are complex numbers for QMDDs, or local invertible maps (LIMs), which are Kronecker products of 2×2 invertible matrices, for LIMDDs. For a node v, we use $\mathsf{low}(v)$ and $\mathsf{high}(v)$ to refer to nodes corresponding to subvectors $|\psi_0\rangle$ and $|\psi_1\rangle$ respectively. A DD achieves succinctness by recognizing identical or equivalent subvectors and storing only one node for each unique subvector. This recognition of equivalent subvectors is done efficiently by storing DD nodes in a canonical form (details in Sect. 2.3). While in general every node has one or two child nodes, there is a single unique terminal node that represents the scalar 1.

QMDDs merge nodes which represent subvectors $|\phi\rangle, |\psi\rangle$ when these satisfy $|\phi\rangle = M \cdot |\psi\rangle$, where M is a complex number. For LIMDDs, this works likewise, but M is a LIM. This broader merging, based on Kronecker products of local (2×2) matrices, allows for even stronger compression of the representation. The original work on LIMDDs [38] mainly focuses on Pauli-LIMDDs, where LIMs are restricted to Pauli-LIMs: a complex number times a Pauli word.

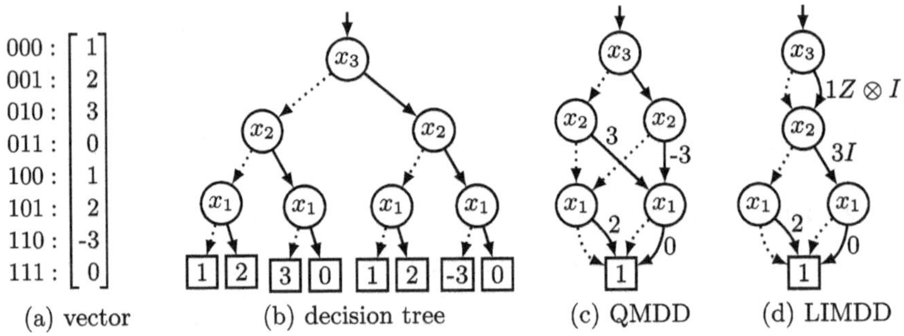

Fig. 2. In this figure, it is shown how a vector of size 2^n is represented as a decision tree of depth n. A QMDD (c) represents this vector more succinctly by factoring out scalars and storing these on edges. The LIMDD representation (d) compresses the data even further by putting Kronecker products of local (2×2) matrices on the edges (d).

Examples of both data types are found in Fig. 2. Here, to simplify the notation, edges with value 1 or $\mathbb{I}$ are left unlabeled. Note that in QMDDs, state vector values can be retrieved by following dashed/solid edges according to the vector index and multiplying all values on the path. Similarly, retrieving values from a LIMDD is done by multiplying the subvectors with the LIMs on the encountered edges. Observe that in this example, the LIMDD can merge the two x_2 nodes because $[1\ 2\ 3\ 0]^\mathsf{T} = (Z \otimes I)\,[1\ 2\ \text{-}3\ 0]^\mathsf{T}$.

We note that QMDDs can be seen as LIMDDs with LIMs restricted to λI_{2^n}, with $\lambda \in \mathbb{C}$. Hence, LIMDDs are strictly more succinct than QMDDs. A more succinct diagram directly benefits quantum-circuit simulation run time as well, as the run time of many DD operations scales with the number of nodes.

Pauli-LIMDDs, similar to the stabilizer formalism [3] and unlike QMDDs, can efficiently simulate Clifford circuits (i.e., circuits composed of Clifford gates). Moreover, it has been shown that there exist classes of circuits where LIMDDs are exponentially faster than both QMDDs and the extended stabilizer formalism for the universal Clifford+T gate set, such as W-state preparation [38]. LIMDDs have also been shown to be exponentially faster on certain classes of circuits than other simulation methods, such as matrix product states (MPS) [38,39].

2.3 Canonicity for Pauli-LIMDDs

While simulating quantum circuits, the DD algorithms create new diagram nodes. As mentioned in Section 2.2, DDs obtain their succinctness and resulting run time improvements for simulation by recognizing whether an equivalent node is already present in the diagram, so that the new node need not be added. Although one could implement this by checking the equivalence of the newly created node with each existing node in the diagram, the common (and faster) approach is to bring each node into canonical form. This reduces the equivalence check with existing nodes in the diagram to an identity check, which is typically implemented using a hash table for $\mathcal{O}(1)$ lookup in practice [9].

To define a canonical form for LIMDDs, the authors of [38] realized that two LIMDDs in general could represent Pauli-equivalent states, have all identical nodes merged, and still be different, due to the degrees of freedom listed below and visualized in Fig. 3.

(a) Because $0 \cdot \psi = 0 \cdot \varphi$, edges with a 0 label can in principle point to any node. This degree of freedom is made canonical by requiring $\mathsf{low}(v) = \mathsf{high}(v)$ whenever one or both has edge label 0.
(b) Because $|0\rangle \otimes |\psi_0\rangle + |1\rangle \otimes |\psi_1\rangle = (X \otimes I^k)(|1\rangle \otimes |\psi_0\rangle + |0\rangle \otimes |\psi_1\rangle)$, there is a degree of freedom regarding the order of $\mathsf{low}(v)$ and $\mathsf{high}(v)$ whenever X is in the LIM group (which it is in Pauli-LIMs). This order is made canonical by requiring $\mathsf{low}(v) \preccurlyeq \mathsf{high}(v)$ according to some total order over all nodes.
(c) Because $(I \otimes L)(|0\rangle \otimes |\psi_0\rangle + |1\rangle \otimes |\psi_1\rangle) = |0\rangle \otimes L\,|\psi_0\rangle + |1\rangle \otimes L\,|\psi_1\rangle$, any LIM L can be factored out of the children's edge labels. This is brought into a canonical form by requiring that the label of a low edge pointing to a node representing an n-qubit state is I_n (whenever it is not zero).
(d) (The *high determinism* rule) Two nodes can have different high labels and still be Pauli-equivalent (i.e., equivalent up to a Pauli word). For example, given a node v that represents $|0\rangle$, 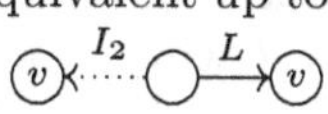

represents the (unnormalized) state $|0\rangle \otimes |0\rangle + |1\rangle \otimes |0\rangle$ for both $L = I_2$ and $L = Z$. The high label is brought into a canonical form by choosing the minimum high label according to a defined total order on Pauli-LIMs.

A *reduced* LIMDD is one that does not represent the vector $(0, 0, \ldots, 0)^\mathsf{T}$ and adheres to the above rules as well as the *merge rule*: it has no duplicate nodes. It is semi-reduced if it satisfies all except high determinism (d). It was proven in

[38] that the above rules eliminate all degrees of freedom: two reduced LIMDD have the same root node if and only if they represent equivalent states.

These rules thus establish a canonical form for nodes: given a newly created node v and edge with label L pointing to it, we want to find a *reduced* node v' and edge label L' such that v' satisfies the rules (a)-(d) above, and $L|v\rangle = L'|v'\rangle$, i.e., they represent the same quantum state. (If we were performing quantum circuit simulation, we would next check if v' is already present in the LIMDD.)

Given any node (v, L), finding a semi-reduced (v', L'), i.e., that satisfies the first three rules (a)-(c), is straightforward and can be done in time $\mathcal{O}(n)$; see Fig. 3 for how to update (v, L). For Pauli-LIMDDs, for the final rule (d), high determinism, an $\mathcal{O}(n^3)$ algorithm exists [38], which we improve upon in Sect. 3.

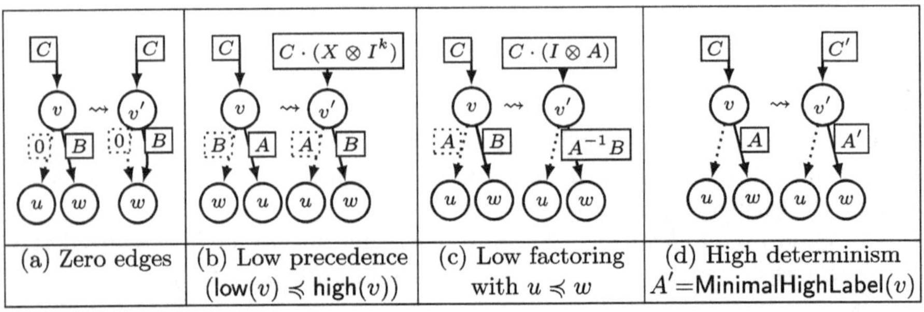

Fig. 3. Normalization rules that bring LIMDD nodes to their reduced form [38], assuming reduced children u, w. Reproduced and adjusted from [38, Fig. 11] (CC-BY 4.0).

2.4 Related Work: Implementations

One implementation of these Pauli-LIMDDs is MQT-LIMDD [40], which is an extension of the QMDD implementation MQT-DDSIM [43]. However, MQT-LIMDD implements canonicity partially, and does not achive minimal size LIMDDs in all cases. Aside from this implementation, LimTDDs [16] have extended LIMDDs to a more expressive group of LIMs, specifically all matrices generated under Kronecker and matrix products from $\{X, P(\frac{2\pi}{N})\}$ (and a complex scalar), where $N \in \mathbb{N}$ can be chosen freely. This XP group is a generalization of the one generated by the Pauli matrices, as setting $N = 2$ gives $P(\pi) = Z$, $P(\pi)X = iY$, and $XX = I$ which yields Pauli-LIMs. However, the implementation accompanying [16] does, by design, not guarantee canonicity and thus possibly yields LIMDDs that are not minimal in size.

3 Faster Algorithm for High Determinism

Here, we derive a new algorithm for transforming a semi-reduced Pauli-LIMDD node into one that satisfies the high determinism rule (Sect. 2.3) with the following advantages over the original algorithm of [38]: (a) For the case that the node has a single child (n qubits, m stabilizer generators), a run time of $\mathcal{O}(nm)$ instead of $\mathcal{O}(nm^2)$; (b) For the case of two distinct children, the same worst-case $\mathcal{O}(n^3)$ run time but with less overhead. Our advantage is gained by adjusting a well-known algorithm for computing intersections and unions of vector spaces.

3.1 High Determinism Algorithm: Problem Statement

Given n-qubit semi-reduced node $u :=$ 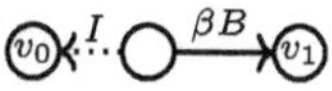

with $\beta \in \mathbb{C}, B \in \text{PAULI}_n$, it was shown in [38, Corollary 15] that the node $u_{\text{can}} :=$ 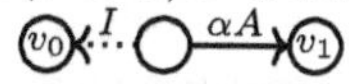

($\alpha \in \mathbb{C}, A \in \text{PAULI}_n$) is a valid choice for the reduced version of u, where

$$\alpha A = \min_{\substack{s,x\in\{0,1\},\\ g\in\text{Stab}(v_0),\, h\in\text{Stab}(v_1)}} \{(-1)^s \cdot \beta^{(-1)^x} gBh \mid x \neq 1 \text{ if } v_0 \neq v_1\}. \tag{1}$$

Indeed, u and u_{can} are Pauli-equivalent as [38, Theorem 14]

$$|u_{\text{can}}\rangle = (X \otimes \beta B)^x \cdot (Z^s \otimes g_{\min}) |u\rangle \tag{2}$$

where $g_{\min} \in \text{Stab}(v_0)$ achieves the minimum in Eq. (1). Canonicity is enforced by the minimization, defined through the following total order [38]. In the definition below, for a complex number $r \cdot e^{i\theta}$ in polar form, i.e. $r \in \mathbb{R}_{\geq 0}$ and $\theta \in [0, 2\pi)$ and P a Pauli word, we denote the vector $\boldsymbol{re^{i\theta}P}$ as the concatenation of revbin (P) and the tuple (r, θ), for example $\boldsymbol{3 \cdot e^{\pi/2} \cdot YZ} = (0, 1, 1, 1, 3, \pi/2)$.

Definition 1 (Pauli-LIM total order). *For $r, r' \in \mathbb{R}_{\geq 0}, \theta, \theta' \in [0, 2\pi)$ and P, P' Pauli words on the same number of qubits, we say that $re^{i\theta}P < r'e^{i\theta'}P'$ if $\boldsymbol{re^{i\theta}P} < \boldsymbol{r'e^{i\theta'}P'}$ in lexicographic ordering from left to right.*

For example, $\boldsymbol{-\frac{1}{2}X} = \boldsymbol{\frac{1}{2}e^{i\pi}X} = (1, 0, \frac{1}{2}, \pi) > (0, 1, 2, 0) = \boldsymbol{2e^0Z} = \boldsymbol{2Z}$.

Example 1. Consider the semi-reduced node v_{sem} in Fig. 4 and suppose that $\text{Stab}(v_0) = \langle XX, ZZ\rangle = \{II, XX, ZZ, -YY\}$ and $\text{Stab}(v_1) = \langle XX, YY\rangle = \{II, XX, -ZZ, YY\}$. By trying out all $4 \times 4 \times 2 \times 2 = 64$ combinations of g, h, s, x, we find that the minimum Eq. (1) is achieved by $(g_{\min}, h_{\min}, s, x) = (-YY, II, 1, 0)$ at $\alpha A = (-1)^1 \cdot i^{(-1)^0} (-YI) = iYI$, yielding reduced node v_{red}. We update the incoming edge label from XYZ to $XYZ \cdot (X \otimes iYI)^0 \cdot (Z^1 \otimes -YY) = XYZ \cdot -ZYY = YIX$ to keep the same represented quantum state.

The goal of the high determinism algorithm is to evaluate Eq. (1), i.e. to find αA given βB and $\text{Stab}(v_0)$, $\text{Stab}(v_1)$. In general, there are multiple tuples (g, h, s, x) that achieve the minimum of Eq. (1). However we show in [28, App.A] that any two such tuples (g, h, s, x), (g', h', s', x')

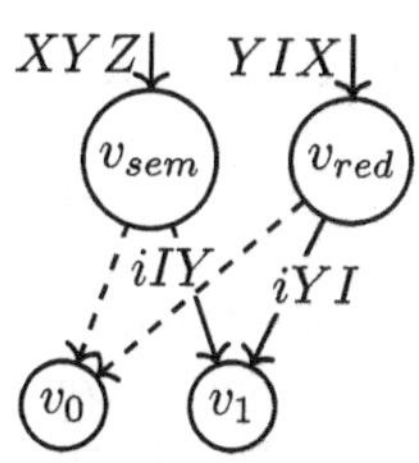

Fig. 4. High-label reduction example.

satisfy $g \cdot B \cdot h = \pm g' \cdot B \cdot h'$. Because the Pauli-LIM ordering is lexicographic, i.e., the scalars only matter if the Pauli words are equal, if we found g, h that is off by a factor of -1 of the true minimum, we correct for this by choosing s accordingly to get the right factor $(-1)^s$. This is why we may *greedily* perform the minimization in Eq. (1): We first find the Pauli word A as $\min_{g \in \mathrm{Stab}(v_0), h \in \mathrm{Stab}(v_0)} \mathsf{word}\,(gBh)$, where $\mathsf{word}\,(L)$ is the Pauli word of Pauli LIM L, e.g. $\mathsf{word}\,(+XY) = XY$ and $\mathsf{word}\,(-iZX) = ZX$. We give our algorithm for doing so in Sect. 3.2. Next, we find the minimal scalar α by trying the four combinations of $x, s \in \{0, 1\}$.

3.2 Finding the Minimal Pauli Word A

We here assume that we are given n-qubit stabilizer generating sets S_0, S_1 and an n-qubit Pauli word B. We will describe an algorithm to compute

$$(g_{\min}, h_{\min}) = \underset{g \in \langle S_0 \rangle, h \in \langle S_1 \rangle}{\operatorname{argmin}} \; \mathsf{word}\,(gBh)\,. \tag{3}$$

In the context of the high determinism rule, $\langle S_0 \rangle$ ($\langle S_1 \rangle$) is the stabilizer group of child node v_0 (v_1), but our algorithm is agnostic to the interpretation of S_0 and S_1, except for whether v_0 and v_1 are identical reduced nodes.

The algorithm from [38] to solve Eq. (3) consists of many individual steps, which include the calculation of intersections of stabilizer groups and cosets of stabilizer groups. We here sidestep having to calculate these by adjusting one part of the algorithm, *the Zassenhaus procedure*, to directly solve Eq. (3). Below, in Sect. 3.3, we will also re-use the adjusted algorithm's output to derive a faster and simpler Stabilizer Finding algorithm.

Given two sets of basis vectors $V, W \subset \mathbb{R}^m$ for some positive integer m, the Zassenhaus algorithm [19] finds bases for the vector spaces $V \cap W$ and $\langle V \cup W \rangle$ (where $\langle . \rangle$ denotes closure under vector addition). To do so, it defines a matrix of $2m$ columns, which we call the Zassenhaus matrix, with rows $\{v \circ v \mid v \in V\} \cup \{w \circ 000\ldots0 \mid w \in W\}$, where $\circ$ denotes vector concatenation and '$000\ldots0$' is the zero vector. The algorithm then brings the *left register* of the matrix (the leftmost m columns) into row-echelon form. To read off a basis for the union $\langle V \cup W \rangle$ after this, one takes all vectors in the left register that are not equal to the zero vector, while a basis for the intersection $V \cap W$ is the set of all (nonzero) vectors in the right register where the left vectors are zero.

We here expand the algorithm to *signed* Pauli words (recall that stabilizers can only have scalar ± 1) to create the Zassenhaus matrix M consisting of *three* registers: $M = \{\mathsf{word}\,(g) \circ g \circ (+I_{2^n}) \mid g \in S_0\} \cup \{\mathsf{word}\,(h) \circ (+I_{2^n}) \circ h \mid h \in S_1\}$. Thus, the left register contains length-$2n$ binary vectors, the middle only elements of $\langle S_0 \rangle$, and the right register only elements of $\langle S_1 \rangle$. We next bring the left register in REF by performing Gaussian elimination on M, where addition $(+)$ of rows is defined using Pauli-LIM multiplication $(\cdot)$ as $(\mathsf{word}\,(k), g, h) + (\mathsf{word}\,(k'), g', h') = (\mathsf{word}\,(k \cdot k'), g' \cdot g, h \cdot h')$. Because $\langle S_0 \rangle$ and $\langle S_1 \rangle$ are stabilizer groups, whose elements commute, this addition is commutative, as Gaussian elimination requires. As a consequence, the resulting matrix M' has the following properties:

(i) A basis of minimal size for the binary vector space $\langle \mathsf{word}\,(g \cdot h) \mid g \in \langle S_0 \rangle, h \in \langle S_1 \rangle \rangle$ is found as the set of vectors in the left register that are nonzero. This follows directly from the fact that the left register contains (the Pauli words of) all elements of both S_0 and S_1

(ii) Each left vector is the product of the middle and right Pauli stabilizers: each row (k, g, h) satisfies $k = \mathsf{word}\,(g \cdot h)$. This is an invariant of the Gaussian elimination algorithm and is hence straightforwardly proven by induction on the algorithm's steps.

To find $g_{\min}$ and $h_{\min}$ from Eq. (2), we now perform the standard algorithm for finding the minimal vector between a vector space V and a hyperplane $\{v_0 + v \mid v \in V\}$ for a given vector v_0 [38]. First, append a row $\mathsf{word}\,(B) \circ (+I_{2^n}) \circ (+I_{2^n})$ at the bottom of M. Next, add rows downward to minimize $\mathsf{word}\,(B)$ in the left register: iterating a column index c from 1 to $2n$, if the value of the bottom row b at c contains the value 1 and there is a row r which contains a leading entry 1 at column c, then replace b by $b + r$. Finally, the bottom row's left register will contain $\mathsf{word}\,(g_{\min} \cdot B \cdot h_{\min})$ by property (i) above, while by property (ii), $g_{\min}$ and $h_{\min}$ are found in the appended row's middle and right registers, respectively.

Example 2. Let $\mathrm{Stab}(v_0)$, $\mathrm{Stab}(v_1)$, $B = IY$ be as in Ex. 1. We first calculate $\mathsf{revbin}\,(B) = \mathsf{revbin}\,(IY) = 1010, \mathsf{revbin}\,(XX) = 1100, \mathsf{revbin}\,(ZZ) = 0011$ and $\mathsf{revbin}\,(YY) = 1111$.

Now, (a) bringing the left register of the Zassenhaus matrix in row-echelon form, after which we (b) minimize the appended row by adding rows from above depending on the column position of their leading entries, yields

$$\begin{pmatrix} 1100 & +XX & +II \\ 0011 & +ZZ & +II \\ 1100 & +II & +XX \\ 1111 & +II & +YY \\ \hdashline 1010 & +II & +II \end{pmatrix} \overset{(a)}{\rightarrow} \begin{pmatrix} 1100 & +XX & +II \\ 0011 & +ZZ & +II \\ 0000 & +XX & +XX \\ 0000 & -YY & +YY \\ \hdashline 1010 & +II & +II \end{pmatrix} \overset{(b)}{\rightarrow} \begin{pmatrix} 1100 & +XX & +II \\ 0011 & +ZZ & +II \\ 0000 & +XX & +XX \\ 0000 & -YY & +YY \\ \hdashline 0101 & -YY & +II \end{pmatrix}$$

The appended row shows $g_{\min}{=}{-YY}$ and $h_{\min}{=}{+II}$, recovering Ex. 1's result. Note that one could add the third and fourth row too while arriving at the same minimal string '0101' in the bottom row, resulting in the different values $g_{\min} = +XX$ and $h_{\min} = -ZZ$. However, such different values will lead to the same minimal high label $g_{\min} \cdot B \cdot h_{\min}$ modulo potentially a minus sign, which is accounted for separately (see Sect. 3.1 and [28, App.A]).

The run time of the algorithm above is dominated by the Zassenhaus procedure, taking time $\mathcal{O}(n \cdot (|S_0| + |S_1|)^2)$, i.e., at most $\mathcal{O}(n^3)$, identically to the original algorithm in [38]. However, in the case that u only has a single child node, i.e. $v_0 = v_1$ and hence $\langle S_0 \rangle = \langle S_1 \rangle$, the algorithm for finding A is simpler: note that $\min_{g,h \in \langle S_0 \rangle} \mathsf{word}\,(gBh) = \min_{g,h \in \langle S_0 \rangle} \mathsf{word}\,(\pm Bgh) = \min_{h \in \langle S_0 \rangle} \mathsf{word}\,(Bh)$ hence we only need to minimize B by multiplying with elements of $\langle S_0 \rangle$, i.e. performing the minimal-vector algorithm only, an $\mathcal{O}(n \cdot |S_0|)$ procedure.

As $|S_0| \leq n$, this improves the original algorithm's $\mathcal{O}(n^3)$ run time to $\mathcal{O}(n^2)$.

3.3 Finding the Stabilizer Group of a Reduced Node

So far, we have shown how to reduce a node by minimizing Eq. (1), provided that we have generator sets for the stabilizer groups of the node's children, in row-echelon form (REF). Here, we give a recursive algorithm to find the latter: find an REF generating set for the stabilizer group of a reduced node. Compared to the original algorithm [38], our algorithm has less overhead as it avoids computing group intersections multiple times by re-using the Zassenhaus result from the high determinism algorithm in Sect. 3.2, and has a speedup from $\mathcal{O}(n^3)$ to $\mathcal{O}(n^2)$ for reduced-nodes with a single child.

Example 3. Consider the node $v :=$ 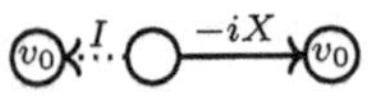 with single child node v_0 whose state is $|v_0\rangle = |0\rangle$. The stabilizer group $\text{Stab}(v_0) = \{I_2, Z\}$ which is generated as $\langle Z \rangle$. The goal is to find a generating set for $\text{Stab}(v)$ given v_0's generating set $\{Z\}$. Using the fact that $|v\rangle = |0\rangle \otimes |0\rangle - i\,|1\rangle \otimes |1\rangle$, a brute-force search over all 2×4^2 signed 2-qubit Pauli words reveals $\text{Stab}(v) = \{II, -YX, ZZ, -XY\}$.

The base case for our recursive algorithm is the leaf node, which has a stabilizer group consisting only of the number 1. For the inductive case, suppose that we have an n-qubit reduced node 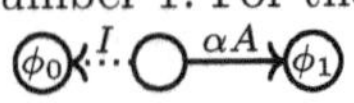 where $\alpha \in \mathbb{C}, A \in \text{PAULI}_{n-1}$ and ϕ_0, ϕ_1 are reduced nodes. Denote S_0 (S_1) for an REF (row-echelon form) generating set of node ϕ_0 (ϕ_1). Our goal is to find all λP such that $\lambda P\,|\phi\rangle = |\phi\rangle$, where $\lambda \in \{\pm 1\}$ and $P \in \text{PAULI}_n$.

We first expand $\lambda P = P_1 \otimes P'$ where $P_1 \in \{I, X, Y, Z\}$ (absorbing λ into the Pauli-LIM P') and distinguish cases $P_1 \in \{I, Z\}$ and $P_1 \in \{X, Y\}$. For the case $P_1 \in \{I, Z\}$ we write $P_1 = \left(\begin{smallmatrix} 1 & 0 \\ 0 & x \end{smallmatrix}\right)$ for $x \in \{1, -1\}$. Because $|\phi\rangle = |0\rangle \otimes |\phi_0\rangle + |1\rangle \otimes \alpha A\,|\phi_1\rangle$, in this case $P_1 \otimes P'\,|\phi\rangle = |\phi\rangle$ is equivalent to

$$P'\,|\phi_0\rangle = |\phi_0\rangle \quad \text{and} \quad P' \cdot \alpha A\,|\phi_1\rangle = x\alpha A\,|\phi_1\rangle \tag{4}$$

If $\alpha = 0$, then the solutions P' are precisely all stabilisers of $|\phi_0\rangle$. That is, the stabilizers of ϕ are $\{P_1 \otimes s \mid P_1 \in \{I, Z\}, s \in S_0\}$; we return $\{Z \otimes I_{n-1}\} \cup \{I \otimes s \mid s \in S_0\}$ as generating set. If $\alpha \neq 0$, we rewrite the latter equation of Eq. (4) as $x\sigma(A, P')P'\,|\phi_1\rangle = |\phi_1\rangle$ using the fact that $x = 1/x$ and where $\sigma(\cdot,\cdot) = 1$ if the arguments commute and -1 otherwise. So, a generating set for ϕ's stabilizers is

$$\left\{ \begin{pmatrix} 1 & 0 \\ 0 & \sigma(g, A) \cdot \gamma_g \end{pmatrix} \otimes g \mid g \in S' \right\} \tag{5}$$

where S' is an REF generating set for the set of stabilisers g of $|\phi_0\rangle$ such that $g\,|\phi_1\rangle = \gamma_g \cdot |\phi_1\rangle$ for some $\gamma_g \in \{\pm 1\}$.

If $v_0 = v_1$, then $S' = S_0$, otherwise S is found as $S' = \{g \mid (000\ldots 0, g, \gamma \cdot g) \in M'\}$, where M' is the Zassenhaus matrix of S_0 and S_1 (see Sect. 3.2).

Example 4. Let $\text{Stab}(v_0)$, $\text{Stab}(v_1)$ be as in Ex. 1. We already found their Zassenhaus matrix in Ex. 2. The relevant rows are $\begin{pmatrix} 0000 & +XX & +XX \\ 0000 & -YY & +YY \end{pmatrix} = \begin{pmatrix} 0000 & +XX & \gamma_{+XX} \cdot (+XX) \\ 0000 & -YY & \gamma_{-YY} \cdot (-YY) \end{pmatrix}$ from which we read off: $S' = \{+XX, -YY\}$ and $\gamma_{+XX} = +1$ and $\gamma_{-YY} = -1$.

Second, for the case $P_1 \in \{X, Y\}$ we write $P_1 = \begin{pmatrix} 0 & y^* \\ y & 0 \end{pmatrix}$ with $y \in \{1, i\}$. Then $P_1 \otimes P' |\phi\rangle = |\phi\rangle$ is expanded as

$$y^* \alpha P' A |\phi_1\rangle = |\phi_0\rangle \quad \text{and} \quad y P' |\phi_0\rangle = \alpha A |\phi_1\rangle \tag{6}$$

If $\alpha = 0$, then there are no solutions because that would imply that $|\phi_0\rangle = |\phi_1\rangle = 0$, which is not a reduced node by definition [38].

If $\alpha \neq 0$, then $|\phi_0\rangle$ and $|\phi_1\rangle$ are Pauli-isomorphic. But by assumption, ϕ_0 and ϕ_1 are reduced, hence $\phi_0 = \phi_1$.

Using the fact and multiplying both sides of the latter equation with $y^* P'$, Eq. (6) yields the same equation twice

$$y^* \alpha P' A |\phi_0\rangle = |\phi_0\rangle \tag{7}$$

where we used that $y^* = 1/y$ and that P' and A are their own inverse. Since all Pauli-LIMs which are stabilizers have scalar either 1 or -1 [22], while $y \in \{1, i\}$ and the multiplication of P' and A yields a scalar $\in \{\pm 1\}$, it follows that a solution in this case is only possible if $\alpha = \pm y$, which in particular implies that $\alpha \in \{\pm 1, \pm i\}$. Provided that that is the case, a solution to Eq. (7) is $P' = y/\alpha A$. That is, the second case yields

$$\begin{pmatrix} 0 & \alpha^* \\ \alpha & 0 \end{pmatrix} \otimes A \tag{8}$$

if $\alpha \in \{\pm 1, \pm i\}$, and no stabilizer generators otherwise. Since any generating set of a stabilizer group can be brought back to a form where only a single generator has $P_1 \in \{X, Y\}$ as the first Pauli entry [5], it suffices to return this one only.

Example 5. Let v, v_0 and $\alpha = -i, A = X$ be as in Ex. 3. We calculate a generating set for $\mathrm{Stab}(v)$ using the algorithm described above. As $\mathrm{Stab}(v_0)$ is generated by Z and $\sigma(Z, A) = \sigma(Z, X) = -1$, the first case yields $Z \otimes Z$. For the second case, as $\alpha = -i$, we find $y = i$, so a stabilizer of v is $y/\alpha \cdot Y \otimes A = -YX$. Thus $\mathrm{Stab}(v) = \langle ZZ, -YX \rangle$, which expands to $\mathrm{Stab}(v) = \{II, ZZ, -YX, -XY\}$, recovering the full stabilizer group given in Ex. 3.

The above algorithm returns a generating set for a node's stabilizer group. To return an REF generating set, we call Gaussian elimination on the output in case $\alpha \neq 0$. In the case of distinct children with $\alpha \neq 0$, the algorithm's run time is dominated by the Zassenhaus procedure, taking $\mathcal{O}(n \cdot (|S_0| + |S_1|)^2)$ time. In the case of a single child with $\alpha \neq 0$, the procedure takes $\mathcal{O}(n \cdot |S_0|)$ time to calculate $\sigma(s, A)$ for all $s \in S_0$. As S_0 is already in REF by assumption and our REF definition has the most significant qubit at the right (see $\mathsf{revbin}()$ definition in Sect. 2), the set in Eq. (5) is in REF already. Hence, Gaussian elimination only updates the single row appended in Eq. (8), taking $\mathcal{O}(n \cdot |S_0|)$ time. If $\alpha = 0$, then the output is already in REF, as the last row is $\mathsf{revbin}(Z \otimes I_{n-1}) = (0000\ldots001)$, i.e., it is majorized by $\mathsf{revbin}(I \otimes s)$ for any stabilizer $s \neq I_{2^n}$. As $|S_0|, |S_1| \leq n$, the run times are at most $\mathcal{O}(n^3), \mathcal{O}(n^2)$ and $\mathcal{O}(n^2)$.

4 Empirical Evaluation

We implemented our high determinism algorithm (Sect. 3) as an open-source C library called PauliLim [1], and we have created a from-scratch LIMDD implementation in C/C++ named QolDDer [2]. Together, these allow us to investigate the effects of canonicity on the simulation of quantum circuits. The implementations are separated to improve modularity and extensibility for future work. Below, we highlight two important design decisions of QolDDer.

As is typical for a DD implementation, to efficiently recognize equivalent nodes, QolDDer's nodes are stored in a hash table after bringing them into a canonical form [9]. We resolve hash collisions through chaining. Each node is assigned a unique ID based on its index in the table. These IDs are then used as canonical representations of the children when hashing a parent node.

While most of the information that uniquely defines a node (qubit number, children nodes, Pauli words of edge labels) is discrete and can be hashed directly; scalars need to be treated separately. Floating-point computations are subject to small rounding errors that can prevent nodes from merging (e.g., `1/sqrt(2)` yields 0.707...5, while `cos(pi/4)` yields 0.707...6). Most DD implementations that include floating points handle this by setting a threshold δ and consider two scalars α and α' equivalent when $|\alpha-\alpha'| \leq \delta$ [29,33,42]. In our case, with $\alpha, \alpha' \in \mathbb{C}$, we consider them equivalent when $|\alpha_{\text{real}} - \alpha'_{\text{real}}| \leq \delta$ and $|\alpha_{\text{imag}} - \alpha'_{\text{imag}}| \leq \delta$. We leave δ configurable, and set the default $\delta = 10^{-13}$; for 64-bit floats, δ should generally be greater than 10^{-15} [8]. To find δ-close scalars, we store them in a search tree. Additionally, in our implementation, we make a minor change to the LIMDD normalization rules that has been shown to mitigate numerical errors in QMDDs [7, Fig. 2c] while preserving LIMDD canonicity. Details are in [28, App.B].

The correctness of a LIMDD implementation has two components: representing the correct state vector at all times and representing it using reduced nodes only. We rigorously tested QolDDer on the first aspect by verifying that QolDDer's output, converted to a state vector, is identical to the output of MQT-DDSIM and LimTDD on the entire MQT-Bench benchmark set [24] (see Sect. 4.1) up to 10 qubits, and for the first 10 vector entries up to 20 qubits. We tested the second aspect by verifying PauliLim's correctness against a simple brute-force implementation that solves Eq. (1) on 10^5 randomly-generated 4 to 8-qubit instances of high edge labels and stabilizer groups.

4.1 Empirical Results

We investigated the effect of canonicity on simulation run time using two benchmark sets. First, we generate a set of random Clifford gates, RandCliff, as Pauli-LIMDDs are known to outperform QMDDs on such circuits [38,39]. It consists of randomly generated n-qubit Clifford circuits with $100n$ gates, ranging from $n = 10$ to 30 qubits, each with 10 instances. Gates are drawn uniformly from the Clifford generating set $\{H, P(\frac{\pi}{2}), CX\}$. Next, MQT-Bench [24] is an existing benchmark set comprised of a broad range of circuit types.

All benchmarks were run on an AMD Ryzen 7 9800X3D CPU, with 64 GB of available memory. Each run uses a 10-minute timeout and a 16 GB memory limit. Reproducible benchmarks are available online [27].

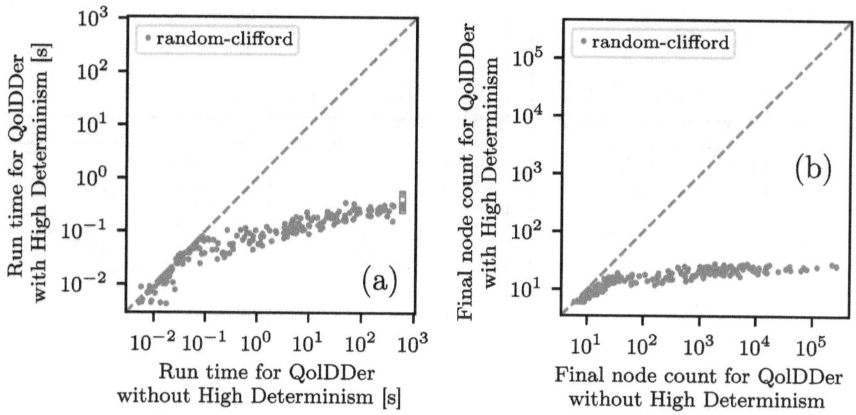

Fig. 5. Run times and node counts of canonical and semi-canonical versions of our Pauli-LIMDD simulator QolDDer on the RANDCLIFF benchmark set. Note the log-log scales, on which a monomial function ($x \mapsto x^k$) shows as a straight line. The dotted grey line visualizes equal run times or node counts. Open markers indicate timeouts.

By turning the use of high determinism on or off, we investigate the effect of simulation on RANDCLIFF with reduced versus semi-reduced Pauli-LIMDDs (see Sect. 2.3). Figure 5 shows a run time comparison. We observe that full canonicity results in significantly faster run times. On this log-log scale, the data points follow a trend that appears linear, or even sub-linear, indicating a polynomial run time improvement or better. To verify that the effect of canonical simulation is due to a reduction in node count, we show the number of nodes in the DD of the output state of the random Clifford circuits in Fig. 5b. Here we see that, indeed, the node count is significantly larger for semi-reduced Pauli-LIMDDs.

Next, in Fig. 6, we compare QolDDer's run time against the QMDD simulator MQT-DDSIM and two LIMDD simulators MQT-LIMDD and LimTDD (see also Sect. 2.4). Based on the observed sublinear scaling in the log-log plot of Fig. 6, we see that QolDDer is exponentially faster than QMDDs on RANDCLIFF. This is as expected, as LIMDDs can represent the output of Clifford circuits in polynomial space, but QMDDs cannot succinctly represent highly entangled states [39], which random Clifford circuits produce [18]. We also observe exponential speedup compared to MQT-LIMDD. We expect that this is due to the fact that their canonicity is only partially implemented; in theory, our algorithm should only show polynomial improvement (see Sect. 3). In Fig. 6c, we observe that QolDDer is a constant 25 times faster than LimTDD on some instances, while on larger instances LimTDD times out while QolDDer does not. This dif-

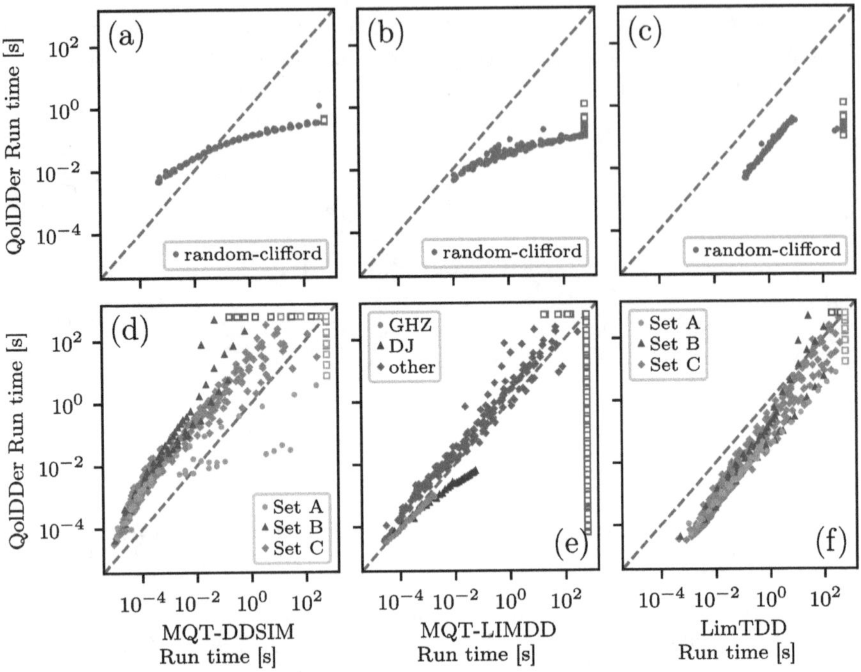

Fig. 6. Run time comparisons of our implementation (QolDDer) to the QMDD implementation (MQT-DDSIM) and the two LIMDD implementations (MQT-LIMDD and LimTDD) on log-log scales. As is shown by the legends, sub-figures (a-c) show the run time comparison for the RANDCLIFF benchmarks, while sub-figures (d-f) show these for the MQT-BENCH datasets. In the latter, the 27 circuit types have been highlighted according to different categories based on the observed trends. Open markers indicate timeouts or errors.

ference is likely due to QolDDer's canonicity, as we observed (not depicted) that LimTDD has much larger node numbers on this benchmark set.

In Fig. 6d-f, we observe that the advantages of canonicity do not hold on all circuits in MQT-BENCH. In Fig. 6d and f, we grouped the benchmarks into sets A = {Quantum Phase Estimation (QPE) exact, QPE inexact}, B = {Quantum Neural Networks (QNN), Quantum random walks, Grover}, and C = {all other} to highlight the observed trends. In Fig. 6e, only the Greenberger-Horne-Zeilinger state preparation (GHZ) and Deutsch-Josza (DJ) circuits have been highlighted.

In Fig. 6d, we observe that QolDDer significantly outperforms MQT-DDSIM on QPE circuits. However, we also observe that the latter is faster on QNN, Quantum random walks, and Grover algorithms ("Set B"). We speculate this is because the MQT-Bench benchmark set contains many circuits for which LIMs do not achieve more merging than QMDDs. In that case, the LIMs in the DD only add overhead. Nonetheless, we observe roughly a linear scaling on Sets B and C, indicating that QolDDer is only slower by a constant factor of 20.

In Fig. 6e, we see that QolDDer's performance is similar to MQT-LIMDD on MQT-Bench, with two types of outliers. First, on GHZ and DJ circuits, QolDDer shows improved scaling over MQT-LIMDD, although due to their sub-1-second run times, we draw no conclusions from these. Second, on various circuits, MQT-LIMDD times out or terminates with an error, indicating possible issues in the implementation. In Fig. 6f, we observe a constant-factor speedup of QolDDer over LimTDD on QPE (Set A; 30× faster). For others, this factor decreases as circuit size increases. We speculate this is due to the larger LIM group that LimTDD has (XP, a supergroup of the Pauli group), enabling more merging, despite not being canonical.

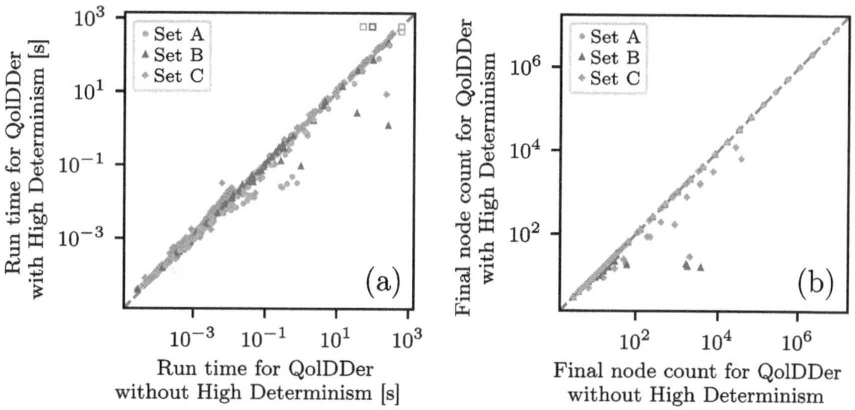

Fig. 7. Run times and node counts of reduced and semi-reduced versions of our Pauli-LIMDD simulator on the MQT-Bench benchmark set. Note the log-log axes scales. Open markers indicate timeouts.

To investigate the hypothesis that LIMDD canonicity does not speed up run times for MQT-Bench, we compared semi-reduced and reduced Pauli-LIMDDs in Fig. 7. We see little performance difference (Fig. 7a), which is explained by a limited gain in succinctness (Fig. 7b).

5 Discussion

We presented a novel, more efficient algorithm to bring LIMDD nodes into a canonical form, and implemented this algorithm as part of our LIMDD implementation QolDDer (both open source [1,2]). In our empirical evaluation, we have compared both run times and node counts for quantum circuit simulation with canonical and non-canonical LIMDDs. Our results corroborate that LIMDDs can achieve exponential advantages over QMDDs. Also, they show that, without canonicity, LIMDDs cannot always realize these advantages. On some benchmarks, however, ensuring canonicity is unnecessary overhead.

In the future, we will extend our canonicity algorithm beyond the Pauli group. Specifically, we expect that our algorithm is extendable to the XP group, which LimTDD uses [16]. We will also apply LIMDDs to a wider range of circuits, such as the universal Clifford+T circuits, for which LIMDDs have scaling guarantees [25]. Moreover, we will look into applications and tasks beyond quantum circuit simulation, such as quantum circuit equivalence checking, a task on which QMDDs have shown success [7,10]. In the classical domain, we will investigate the use of LIMDDs for linear-algebra tasks such as stochastic reasoning and Markov-chain analysis, investigating whether relevant tasks possess structure that the strong merge rule of LIMDDs could explicitly make use of.

Acknowledgments. SB is supported by the NWO Quantum Technology program (project number NGF.1582.22.035). AQ acknowledges the support received though the NWO NWA-ORC program, project Divide & Quantum (NWA.1389.20.241). TC is supported by an NWO Veni grant (VI.Veni.232.381).

References

1. PauliLim (2026). https://gitlab.tudelft.nl/ardeq-group/paulilim
2. QolDDer (2026). https://gitlab.tudelft.nl/ardeq-group/qoldder
3. Aaronson, S., Gottesman, D.: Improved simulation of stabilizer circuits. Phys. Rev. A-Atomic Mol. Opt. Phys. **70**(5), 052328 (2004)
4. Ardeshir-Larijani, E., Gay, S.J., Nagarajan, R.: Verification of concurrent quantum protocols by equivalence checking. In: Ábrahám, E., Havelund, K. (eds.) TACAS 2014. LNCS, vol. 8413, pp. 500–514. Springer, Heidelberg (2014). https://doi.org/10.1007/978-3-642-54862-8_42
5. Audenaert, K.M.R., Plenio, M.B.: Entanglement on mixed stabilizer states: normal forms and reduction procedures. New J. Phys. **7**(1), 170 (2005). http://stacks.iop.org/1367-2630/7/i=1/a=170
6. Bahar, R.I., et al.: Algebric decision diagrams and their applications. Formal Meth. Syst. Des. **10**(2), 171–206 (1997)
7. Brand, S., Laarman, A.: Q-Sylvan: a parallel decision diagram package for quantum computing. In: D'Souza, M., Komondoor, R., Srivathsan, B. (eds.) ATVA 2025. LNCS, vol. 16145, pp. 260–273. Springer, Cham (2025). https://doi.org/10.1007/978-3-032-08707-2_12
8. Brand, S., Quist, A.J., van Dijk, R.M., Laarman, A.: Numerical errors in quantitative system analysis with decision diagrams. In: Prabhakar, P., Vandin, A. (eds.) QEST+FORMATS 2025. LNCS, vol. 16143, pp. 371–388. Springer, Cham (2025). https://doi.org/10.1007/978-3-032-05792-1_20
9. Bryant, R.E.: Symbolic Boolean manipulation with ordered binary-decision diagrams. ACM Comput. Surv. (CSUR) **24**(3), 293–318 (1992)
10. Burgholzer, L., Wille, R.: Advanced equivalence checking for quantum circuits. IEEE Trans. Comput. Aided Des. Integr. Circuits Syst. **40**(9), 1810–1824 (2020)
11. Cao, Y., et al.: Quantum chemistry in the age of quantum computing. Chem. Rev. **119**(19), 10856–10915 (2019)
12. Chen, T.F., Jiang, J.H.R.: SliQSim: a quantum circuit simulator and solver for probability and statistics queries. In: In: Gurfinkel, A., Heule, M. (eds.) TACAS 2025. LNCS, vol. 15698, pp. 129–138. Springer, Cham (2025). https://doi.org/10.1007/978-3-031-90660-2_7

13. Fujita, M., McGeer, P.C., Yang, J.Y.: Multi-terminal binary decision diagrams: an efficient data structure for matrix representation. Formal Meth. Syst. Des. **10**(2), 149–169 (1997)
14. Georgescu, I.M., Ashhab, S., Nori, F.: Quantum simulation. Rev. Mod. Phys. **86**(1), 153–185 (2014)
15. Harwood, S., Gambella, C., Trenev, D., Simonetto, A., Bernal, D., Greenberg, D.: Formulating and solving routing problems on quantum computers. IEEE Trans. Quant. Eng. **2**, 1–17 (2021)
16. Hong, X., Dai, A., Gao, D., Li, S., Ji, Z., Ying, M.: LimTDD: a compact decision diagram integrating tensor and local invertible map representations. arXiv preprint arXiv:2504.01168 (2025)
17. Hong, X., Feng, Y., Li, S., Ying, M.: Equivalence checking of dynamic quantum circuits. In: Proceedings of the 41st IEEE/ACM International Conference on Computer-Aided Design, pp. 1–8 (2022)
18. Li, Y., Chen, X., Fisher, M.P.: Measurement-driven entanglement transition in hybrid quantum circuits. Phys. Rev. B **100**(13), 134306 (2019)
19. Luks, E.M., Rákóczi, F., Wright, C.R.: Some algorithms for nilpotent permutation groups. J. Symb. Comput. **23**(4), 335–354 (1997)
20. Clarke, E., McMillan, K., Campos, S., Hartonas-Garmhausen, V.: Symbolic model checking. In: Alur, R., Henzinger, T.A. (eds.) CAV 1996. LNCS, vol. 1102, pp. 419–422. Springer, Heidelberg (1996). https://doi.org/10.1007/3-540-61474-5_93
21. Miller, D.M., Thornton, M.A.: QMDD: a decision diagram structure for reversible and quantum circuits. In: 36th International Symposium on Multiple-Valued Logic (ISMVL'06), pp. 30–30. IEEE (2006)
22. Nielsen, M.A., Chuang, I.L.: Quantum Computation and Quantum Information. Cambridge University Press, Cambridge (2000)
23. Orús, R., Mugel, S., Lizaso, E.: Quantum computing for finance: overview and prospects. Rev. Phys. **4**, 100028 (2019)
24. Quetschlich, N., Burgholzer, L., Wille, R.: MQT bench: benchmarking software and design automation tools for quantum computing. Quantum **7**, 1062 (2023). https://doi.org/10.22331/q-2023-07-20-1062, MQT Bench is available at https://mqt-bench.app/
25. Quist, A.J., Coopmans, T., Laarman, A.: Exact quantum decision diagrams with scaling guarantees for Clifford+T circuits and beyond. arXiv:2602.17775 (2026)
26. Samoladas, V.: Improved BDD algorithms for the simulation of quantum circuits. In: Halperin, D., Mehlhorn, K. (eds.) ESA 2008. LNCS, vol. 5193, pp. 720–731. Springer, Heidelberg (2008). https://doi.org/10.1007/978-3-540-87744-8_60
27. Sanders, J., Brand, S., Coopmans, T.: QEST+FORMATS 2026 artifact for "faster algorithm for achieving minimal-size quantum decision diagrams" (2026). https://doi.org/10.5281/zenodo.19763549
28. Sanders, J., Brand, S., Quist, A.J., Coopmans, T.: Faster algorithm for achieving minimal-size quantum decision diagrams. arXiv preprint arXiv:2606.24789 (2026)
29. Sanner, S., McAllester, D.: Affine algebraic decision diagrams (AADDs) and their application to structured probabilistic inference. In: IJCAI. vol. 2005, pp. 1384–1390 (2005)
30. Shor, P.W.: Algorithms for quantum computation: discrete logarithms and factoring. In: FOCS, pp. 124–134. IEEE (1994)
31. Sistla, M., Chaudhuri, S., Reps, T.: Symbolic quantum simulation with Quasimodo. In: Enea, C., Lal, A. (eds.) CAV 2023. LNCS, vol 13966, pp. 213–225. Springer, Cham (2023). https://doi.org/10.1007/978-3-031-37709-9_11

32. Sistla, M., Chaudhuri, S., Reps, T.: Weighted context-free-language ordered binary decision diagrams. Proc. ACM Program. Lang. **8**(OOPSLA2), 1390–1419 (2024)
33. Somenzi, F.: CUDD: CU decision diagram package release 3.0.0 (2015). https://web.archive.org/web/20171208230728/, http://vlsi.colorado.edu/~fabio/CUDD/cudd.pdf
34. Tafertshofer, P., Pedram, M.: Factored edge-valued binary decision diagrams. Formal Meth. Syst. Des. **10**(2), 243–270 (1997)
35. Thanos, D., Coopmans, T., Laarman, A.: Fast equivalence checking of quantum circuits of Clifford gates. In: André, É., Sun, J. (eds.) ATVA 2023. LNCS, vol. 14216, pp. 199–216. Springer, Cham (2023). https://doi.org/10.1007/978-3-031-45332-8_10
36. Tsai, Y.H., Jiang, J.H.R., Jhang, C.S.: Bit-slicing the HilBert space: scaling up accurate quantum circuit simulation. In: 2021 58th ACM/IEEE Design Automation Conference (DAC), pp. 439–444. IEEE (2021)
37. Viamontes, G.F., Markov, I.L., Hayes, J.P.: High-performance QuIDD-based simulation of quantum circuits. In: Proceedings Design, Automation and Test in Europe Conference and Exhibition, vol. 2, pp. 1354–1355. IEEE (2004)
38. Vinkhuijzen, L., Coopmans, T., Elkouss, D., Dunjko, V., Laarman, A.: LIMDD: a decision diagram for simulation of quantum computing including stabilizer states. Quantum **7**, 1108 (2023)
39. Vinkhuijzen, L., Coopmans, T., Laarman, A.: A knowledge compilation map for quantum information. In: Proceedings of the AAAI Conference on Artificial Intelligence, vol. 40, pp. 19406–19414 (2026)
40. Vinkhuijzen, L., Grurl, T., Hillmich, S., Brand, S., Wille, R., Laarman, A.: Efficient implementation of LIMDDs for quantum circuit simulation. In: Caltais, G., Schilling, C. (eds.) SPIN 2023. LNCS, vol. 13872, pp. 3–21. Springer, Cham (2023). https://doi.org/10.1007/978-3-031-32157-3_1
41. Wei, C.Y., Tsai, Y.H., Jhang, C.S., Jiang, J.H.R.: Accurate BDD-based unitary operator manipulation for scalable and robust quantum circuit verification. In: Proceedings of the 59th Design Automation Conference, pp. 523–528 (2022)
42. Zulehner, A., Hillmich, S., Wille, R.: How to efficiently handle complex values? Implementing decision diagrams for quantum computing. In: 2019 IEEE/ACM International Conference on Computer-Aided Design (ICCAD), pp. 1–7. IEEE (2019)
43. Zulehner, A., Wille, R.: Advanced simulation of quantum computations. IEEE TCAD **38**(5), 848–859 (2018)

StochasticBarrier.jl: A Toolbox for Stochastic Barrier Function Synthesis

Rayan Mazouz[1(✉)], Frederik Baymler Mathiesen[2], Luca Laurenti[2], and Morteza Lahijanian[1]

[1] University of Colorado Boulder, Boulder, CO, USA
rayan.mazouz@colorado.edu
[2] Delft University of Technology, Delft, The Netherlands

QEST + FORMATS Artifacts Evaluation Available

QEST + FORMATS Artifacts Evaluation Functional

Abstract. We present `StochasticBarrier.jl`, an open-source Julia-based toolbox for generating Stochastic Barrier Functions (SBFs) for safety verification of discrete-time stochastic systems with additive Gaussian noise. `StochasticBarrier.jl` certifies linear, polynomial, and piecewise affine (PWA) systems. The latter enables verification for a wide range of system dynamics, including general nonlinear types. The toolbox implements a Sum-of-Squares (SOS) optimization approach, as well as methods based on piecewise constant (PWC) functions. For SOS-based SBFs, `StochasticBarrier.jl` leverages semi-definite programming solvers, while for PWC SBFs, it offers three engines: two using linear programming (LP) and one based on gradient descent (GD). Benchmarking `StochasticBarrier.jl` against the state-of-the-art shows that the tool outperforms existing tools in computation time, safety probability bounds, and scalability across over 30 case studies. Compared to its closest competitor, `StochasticBarrier.jl` is up to four orders of magnitude faster, achieves significant safety probability improvements, and supports higher-dimensional systems.

Keywords: Barrier Certificates · Stochastic Systems · Probabilistic Safety · Convex Optimization · Formal Verification

1 Introduction

Stochastic barrier functions (SBFs) are a class of tools used for the formal verification of stochastic systems [20,24]. These functions establish a lower bound on the probability of a system remaining within a safe set, using martingale theory [15,24]. A key advantage of SBFs over other verification methods is their capacity to analyze the time evolution of a system through a set of static constraints, eliminating the need to propagate system uncertainty. Instead, SBF synthesis methods solve a functional optimization problem, for which various approaches have been developed based on, e.g., Sum-of-Squares (SOS) polynomials [15,20,24] and piecewise (PW) functions [12,13,16]. Despite their potential, however, there has been limited effort to develop robust and efficient tools

M. Chen and M. Tschaikowski (Eds.): QEST+FORMATS 2026, LNCS 16913, pp. 429–443, 2027.
https://doi.org/10.1007/978-3-032-35298-9_24

for SBF synthesis, restricting their broader application in practical fields such as robotics and autonomous vehicles.

In this paper, we address this gap by introducing an open-source Julia-based toolbox that generates SBFs for *discrete-time stochastic systems* with *additive Gaussian noise* - `StochasticBarrier.jl`[1]. The toolbox supports a wide range of dynamics, namely, *linear*, *polynomial*, and *PW affine (PWA) inclusions*, the latter enabling the verification of non-linear, non-polynomial systems. The established SOS optimization method is implemented, as well as more recent approaches based on PW Constant (PWC) functions. For SOS optimization, existing semi-definite programming (SDP) solvers are leveraged. For PWC-based SBFs, three engines are included: two that utilize Linear Programming (LP) and one based on Gradient Descent (GD), as introduced in [13].

The toolbox is developed in Julia, a modern programming language designed for the scientific community. Julia combines the interactivity and dynamic typing of scripting languages, facilitating fast prototyping, with the high performance of compiled languages [18]. Also, Julia's support for parametric typing allows for customizable precision in numerical computations [11]. We evaluate `StochasticBarrier.jl` on various benchmarks and compare its performance against `PRoTECT` [29] and `StochasticBarrierFunctions` [24], two state-of-the-art tools, which, to the best of our knowledge, are the only available SBF tools for discrete-time stochastic systems. These existing tools exclusively support SOS-based approaches. In addition, the three PWC-SBF methods are compared against the SOS-method. Benchmarks include over 30 case studies across eight different systems with linear, polynomial, and PWA inclusion dynamics, ranging from *one* to *six* dimensions. The results demonstrate that `StochasticBarrier.jl` outperforms existing tools in three metrics: computation time, safety probability bounds, and scalability. Compared to its closest competitor, `StochasticBarrier.jl` is up to 1000× faster for successful synthesis, improves safety probability bounds from 0 to 1.0 in some cases, and handles systems with twice the dimensionality.

1.1 Related Work

Barrier certificates are synthesized to prove safety for dynamical systems in both deterministic [25,27] and stochastic settings [8,21]. Stochastic barrier functions (SBFs) are used to obtain a formal lower bound on the probability of remaining safe, starting from an initial set, under a well-defined probability measure. SBFs have been applied to verify safety for linear and polynomial systems [24], nonlinear systems [15], and data-driven systems [16].

Sum-of-Squares (SOS) optimization is a popular method for SBF synthesis [8,15,20,21,24]. Some works utilize SOSTOOLS [21] to obtain these SBFs. The main downside of SOSTOOLS-based SBFs is that the algorithms are embedded in MATLAB, which scales poorly when bridging to SDP. PRoTect [29], a Python based implementation, scales better, but is limited to low-degree SOS

[1] https://github.com/aria-systems-group/StochasticBarrier.jl.

barriers and systems of polynomial class. Our toolbox, which is implemented in Julia, provides significant speedups for intensive numerical optimization tasks and memory management [5,18].

Related tools for the verification of stochastic systems in Julia include *IntervalMDP.jl* [11], which performs CUDA-accelerated value iteration for Interval Markov Decision Processes (IMDPs). JuliaReach [4] performs set-based reachability for deterministic systems. A tool for Satisfiability modulo theories, Satisfiability.jl [26], provides high-level representations for SMT formulae. Finally, ModelVerification.jl [28] verifies DNNs and safety specifications. None of these verify safety for stochastic systems using SBFs. A platform for this purpose is provided in `StochasticBarrier.jl`.

The remainder of the paper is organized as follows. In Sect. 3, we provide a detailed overview of the toolbox, with instructions and examples on setting up system models and invoking SBF synthesis engines. In Sect. 4, we demonstrate the performance of `StochasticBarrier.jl` through a series of case studies and benchmark comparisons. Finally, we conclude the paper with a summary and closing remarks in Sect. 5.

2 Theoretical Framework

We provide an overview of the SBF theoretical framework. An outline of the problem formulation is provided, followed by a review of the implemented methods. For more detail, we refer the reader to [13,15].

2.1 Problem

Consider a stochastic process defined by difference equation

$$\mathbf{x}_{k+1} = f(\mathbf{x}_k) + \mathbf{w}_k, \tag{1}$$

where $\mathbf{x}_k \in \mathbb{R}^n$ is the state, $f : \mathbb{R}^n \to \mathbb{R}^n$ is the vector field representing the one-step dynamics of System (1), and $\mathbf{w}_k \in \mathbb{R}^n$ is an independent and identically distributed random variable, representing noise. The probability distribution of $\mathbf{w}_k$ is assumed to be zero-mean Gaussian, i.e., $\mathbf{w}_k \sim p_\mathbf{w} = \mathcal{N}(0, \Sigma)$, where $\Sigma = \text{diag}(\sigma_1^2, \ldots, \sigma_n^2)$. We define the stochastic kernel $T(X \mid x) := \int_{\mathbb{R}^\mathrm{w}} \mathbf{1}_X(f(x) + w)p_\mathbf{w}(dw)$. It follows from this definition of T that, given an initial condition $\mathbf{x}_0 = x_0 \in \mathbb{R}^n$, $\mathbf{x}_k$ is a Markov process [3]. Let $X_\mathrm{s} \subset \mathbb{R}^n$ be a bounded set representing the safe set, $X_0 \subseteq X_\mathrm{s}$ be the initial set, and $N \in \mathbb{N}$ be the time horizon. Then, *probabilistic safety* is defined as $P_\mathrm{s}(X_\mathrm{s}, X_0, N) = \inf_{x_0 \in X_0} \Pr^{x_0}[\forall k \in \{0, ..., N\}, \mathbf{x}_k \in X_\mathrm{s}]$.

To reason about $P_\mathrm{s}(X_\mathrm{s}, X_0, N)$, *Stochastic Barrier Function* (SBFs) can be used [15,24]. Function $B : \mathbb{R}^n \to \mathbb{R}_{\geq 0}$ is a SBF for System (1) if there exist scalars $\eta, \beta \geq 0$, such that

$$B(x) \geq 1 \qquad \forall x \in \mathbb{R}^n \setminus X_s, \tag{2a}$$

$$B(x) \leq \eta \qquad \forall x \in X_0, \tag{2b}$$

$$\mathbb{E}[B(f(x) + \mathbf{w}))] \leq B(x) + \beta \qquad \forall x \in X_s, \tag{2c}$$

which guarantees a lower-bound for the *probabilistic safety*

$$P_{\text{s}}(X_{\text{s}}, X_0, N) \geq 1 - (\eta + N\beta). \tag{3}$$

The goal of the SBF synthesis problem is to find a $B(x)$ that maximizes P_s, subject to Conditions (2a)-(2c). `StochasticBarrier.jl` solves this problem by computing a B along with a P_s lower bound.

From a computational perspective, solving this functional optimization problem is non-trivial. Several methods have been developed to ensure the convexity of the problem. One well-established approach assumes B is a polynomial of a given degree, which allows the problem to be formulated as a SOS optimization, ultimately leading to a semi-definite program [15,19,24]. A more recent approach, introduced in [13], assumes B is a piecewise constant (PWC) function, reducing the optimization problem to a linear program. Both methods are implemented in `StochasticBarrier.jl`.

2.2 Implemented SBF Methods

`StochasticBarrier.jl` supports polynomial (SOS) and PWC class-es of SBFs for System (1), where f is affine, polynomial, or bounded by piecewise affine (PWA) functions.

Sum-of-Squares (SOS). A multivariate polynomial $\lambda(x)$ with $x \in \mathbb{R}^n$ is an SOS polynomial if there exist polynomials λ_i, $i = 1, \ldots, r$, for $r \in \mathbb{N}$, such that $\lambda(x) = \sum_{i=1}^{r} \lambda_i^2(x)$. If $\lambda(x)$ is SOS, then $\lambda(x) \geq 0$ for all $x \in \mathbb{R}^n$. The set of SOS polynomials is denoted by $\Lambda(x)$, and a vector of SOS polynomials by $\mathcal{L}(x)$. Under the assumption that f in System (1) is a polynomial of degree one or higher, and let X_0, X_s be compact semi-algebraic sets, then, the following theorem [24] is implemented in `StochasticBarrier.jl` to reduce the SOS-SBF synthesis problem to a convex program.

Theorem 1. (SOS-SBF for Polynomial Systems [24]**).** *Let the safe set $X_{\text{s}} = \{x \in \mathbb{R}^n \mid h_{\text{s}}(x) \geq 0\}$, initial set $X_0 = \{x \in \mathbb{R}^n \mid h_0(x) \geq 0\}$, and unsafe set $X_{\text{u}} = \mathbb{R}^n \setminus X_{\text{s}} = \{x \in \mathbb{R}^n \mid h_{\text{u}}(x) \geq 0\}$, where h_s, h_0, h_u are vectors of polynomials, be given. Then, a SBF $B(x) \in \Lambda(x)$ for System (1) is obtained by solving the following SOS optimization problem*

$$\min_{\beta,\eta \geq 0} \quad \eta + N\beta \qquad \textit{subject to:} \tag{4a}$$

$$B(x) \in \Lambda(x), \tag{4b}$$

$$- B(x) - \mathcal{L}_0^T(x) h_0(x) + \eta \in \Lambda(x), \tag{4c}$$

$$B(x) - \mathcal{L}_{\text{u}}^T(x) h_{\text{u}}(x) - 1 \in \Lambda(x), \tag{4d}$$

$$- \mathbb{E}[B(f(x) + \mathbf{w})] + B(x) + \beta - \mathcal{L}_s^T(x) h_s(x) \in \Lambda(x), \tag{4e}$$

where $\mathcal{L}_0$, $\mathcal{L}_h$, and $\mathcal{L}_s$ are vectors of SOS polynomials of appropriate dimension. A valid solution to this optimization guarantees safety probability in Eq. (3).

PWA Inclusion Dynamics. Additionally to polynomial f, non-polynomial f are also supported in `StochasticBarrier.jl` by supplying piecewise affine (PWA) bounds of f. In particular, given a partition $Q = \{q_1, \dots, q_{|Q|}\}$ of the safe set X_s for every $q \in Q$, `StochasticBarrier.jl` supports f expressed via bounds

$$\underline{f}_q(x) = \underline{A}_q x + \underline{b}_q \quad \text{and} \quad \overline{f}_q(x) = \overline{A}_q x + \overline{b}_q \tag{5}$$

such that $\underline{f}_q(x) \leq f(x) \leq \overline{f}_q(x)$ for every $x \in q$. The following theorem shows a convex optimization framework for synthesis of SOS-SBF for such dynamics.

Theorem 2. (SOS-SBF for PWA Inclusion Systems [15]**).** *Consider function $B(x)$ and sets X_0, X_s, X_u as in Theorem 1 and X_s-partition $Q = \{q_1, \dots, q_{|Q|}\}$, where each $q \in Q$ is a semi-algebraic set $q = \{x \in \mathbb{R}^n \mid h_q(x) \geq 0\}$. Given PWA functions $\underline{f}_q$ and $\overline{f}_q$ over Q as defined in Eq.* (5)*, an SOS SBF $B(x)$ for System* (1)*, with general $f(x) \in [\underline{f}_q(x), \overline{f}_q(x)]$, can be obtained by solving the SOS optimization problem in* (4) *and replacing Constraint* (4e) *with $-\mathbb{E}[B(y+w) \mid x] + B(x) + \beta - \mathcal{L}_{q,x}^T(x) h_q(x) - \mathcal{L}_{q,y}^T(x)((\overline{f}_q(x) - y) \odot (y - \underline{f}_q(x))) \in \Lambda(x, y)$, for all $q \in Q$, where $\odot$ is the Schur product.*

In `StochasticBarrier.jl`, the implementation of partition Q is based on a grid, where each $q \in Q$ is defined as a hyper-rectangle. Then, provided partition Q and PWA functions $\underline{f}_q$ and $\overline{f}_q$, the tool performs the above optimization.

PWC Method. `StochasticBarrier.jl` also supports the class of PWC-SBFs. Consider a partition $Q = \{q_1, \dots, q_{|Q|}\}$ of X_s such that vector field $f(x)$ is continuous in each region $q \in Q$. Define PWC function $B : \mathbb{R}^n \to \mathbb{R}$ as

$$B(x) = b_i \ \text{ if } \ x \in q_i, \ \text{otherwise } B(x) = 1. \tag{6}$$

where $b_i \in \mathbb{R}$. Further, let $q_u = X_u$ and denote the bounds on the transition kernel T from partition $q_i \in Q$ by, for all $j \in \{1, \dots, |Q|, u\}$,

$$\underline{p}_{ij} = \inf_{x \in q_i} T(q_j \mid x), \quad \text{and} \quad \overline{p}_{ij} = \sup_{x \in q_i} T(q_j \mid x). \tag{7}$$

Then, the set of feasible values for T from q_i is given by $\mathcal{P}_i = \{p_i = (p_{i1}, \dots, p_{i|Q|}, p_{iu}) \in [0,1]^{|Q|+1}$ s.t. $\sum_{j=1}^{|Q|} p_{ij} + p_{iu} = 1, \ \underline{p}_{ij} \leq p_{ij} \leq \overline{p}_{ij} \quad \forall j \in \{1, \dots, |Q|, \mathrm{u}\}\}$. The following theorem formalizes the optimization problem for PWC class of SBFs for System (1).

Theorem 3. (PWC-SBF [13]**).** *Consider System* (1) *with safe set X_s, initial set X_0, and partition $Q = \{q_1, \dots, q_{|Q|}\}$ of X_s. Let $\mathcal{B}_Q$ be the class of PWC functions in the form of Eq.* (6) *and $\mathcal{P} = \mathcal{P}_1 \times \ldots \times \mathcal{P}_{|Q|}$, where each $\mathcal{P}_i$ is the set of feasible transitions. Then, $B^* \in \mathcal{B}_Q$ is a PWC-SBF if B^* is a solution to $B^* = \arg\min_{B \in \mathcal{B}_Q} \max_{(p_i)_{i=1}^{|Q|} \in \mathcal{P}} \ \eta + N\beta$ subject to,*

$$\forall i \in \{1, \dots, Q\}, \ \ b_i \geq 0, \ \ \sum_{j=1}^{Q} b_j \cdot p_{ij} + p_{iu} \leq b_i + \beta_i, \ \ 0 \leq \beta_i \leq \beta,$$

and for all $i : X_i \cap X_0 \neq \emptyset$, $b_i \leq \eta$. *Then,* B^* *guarantees the safety probability in Eq.* (3).

Table 1. Overview of SBF synthesis methods implemented in `StochasticBarrier.jl` for different classes of dynamics.

Dynamics $f(x)$	SBF $B(x)$	Solver Method
Linear	SOS	SDP
	PWC	Dual LP, CEGIS LP, GD
Polynomial	SOS	SDP
PWA Inclusion	SOS	SDP
(Eq. (5))	PWC	Dual LP, CEGIS LP, GD

Work [13] introduces three methods for synthesizing PWC SBFs: (i) an exact duality-based approach formulated as a zero-duality-gap Linear Program (LP), (ii) an exact *Counter-Example Guided Inductive Synthesis* (CEGIS) method that alternates between generating candidate barriers and counter-examples via two LPs, and (iii) a *gradient descent* (GD) method offering improved scalability and memory efficiency. `StochasticBarrier.jl` implements all three methods for System (1) with linear and PWA inclusion dynamics. The toolbox internally computes the kernel bounds in Eq. (7).

3 Overview of *StochasticBarrier.jl*

`StochasticBarrier.jl` is a Julia toolbox for the synthesis of SBFs for System (1). The tool includes multiple independent synthesis methods with the following features:

- Synthesis of SOS-SBFs for systems in the form of (1) with linear, polynomial and PWA inclusion dynamics,
- Computation of transition kernel bounds for System (1) with linear and PWA inclusion dynamics,
- Synthesis of PWC-SBFs through LP (dual and CEGIS) and GD optimization methods given the kernel bounds.

In Table 1, the SBF methods are outlined for different classes of dynamics f. The *solver method* refers to the convex optimization class. To instantiate an instance of SBF synthesis problem, the system model needs to be defined first, which consists of the dynamics and the relevant sets (safe, unsafe and initial). In Sect. 3.1, we show how to set up these system related parameters. Then, the SBF synthesis method needs to be specified, as described in Sect. 3.2.

3.1 System Setup

`StochasticBarrier.jl` dispatches and checks compatibility with the chosen synthesis algorithm based on the system type. We give examples about how to construct a linear, polynomial, and PWA inclusion system, and specify the safe, initial, and unsafe sets. The examples are for 2D-systems, but the user can easily set up an n-dimensional problem.

Linear Dynamics Example. With linear dynamics, both SOS and PWC classes of SBFs can be used.

SOS-SBF. To synthesize a SOS-SBF for System (1) with linear vector field $f(x) = Ax + b$, we first need to specify matrix A and bias vector b. The noise distribution is assumed to be zero-mean Gaussian $p_{\mathbf{w}} = \mathcal{N}(0, \sigma^2 I)$ characterized by standard deviation σ. An example setup is shown below for a 2D-linear system.

```
# System Linear: SOS
A = [0.95 0.0; 0.0 0.95], b = [0.0, 0.0], σ = [0.1, 0.1]
space = Hyperrectangle(low=[-1.0,-2.0], high=[1.0,2.0])
system = AdditiveGaussianLinearSystem(A, b, σ, space)
```

The `Hyperrectangle` struct of `LazySets.jl`, a Julia package for calculus with convex sets [6], is used to define the state space. A struct is passed as a system object (`AdditiveGaussianLinearSystem`) into `StochasticBarrier.jl`.

PWC-SBF. Using PWC-SBF requires an additional parameter to define a grid partitioning of the state space. The tool uses this grid to define the domain of PWC function in Eq. (6) and compute the transition kernel bounds in Eq. (7). The grid is parameterized by $\epsilon \in \mathbb{R}^n$, where the i-th entry is the grid cell's half-width in the i-th dimension. The `generate_partitions()` function takes the state space and ϵ as inputs, and returns the partitions. The system object `AdditiveGaussianLinearSystem`, along with the state partitions, are parsed into `transition_probabilities()`, which is a function in `StochasticBarrier.jl` that computes the transition kernel bounds. An example 2D-linear system setup is provided.

```
# System Linear: PWC
A = [0.50 0.0; 0.0 0.50], b = [0.0, 0.0], σ = [0.1, 0.1]
space = Hyperrectangle(low=[-1.0,-1.0], high=[0.5,0.5])
# Half-width vector
ϵ = [0.2, 0.2]
# Compute probability data
system = AdditiveGaussianLinearSystem(A, b, σ)
partitions = generate_partitions(space, ϵ)
probabilities = transition_probabilities(system, partitions)
```

In addition, the toolbox supports saving and loading transition probability bounds to the Network Common Data Form (NetCDF) [23]. See [14] [App. A] for more details on this feature.

Polynomial Dynamics Example. For polynomial systems, the tool only supports the SOS class of SBFs. To set up, first define a vector of symbol variables x, using the `@polyvar` macro of `DynamicPolynomials.jl` [9]. Then, the polynomial function f is defined symbolically. The programmatic construct of `AdditiveGaussianPolySystem` is passed as an object to the barrier synthesis method. An example is provided below for a 2D system.

```
# System Polynomial: SOS
@polyvar x[1:2]  # dim = 2
τ = 0.1
f = [x[1] + τ * x[2], x[2] + τ * (-x[1] + (1 - x[1])^2)]
σ = [0.02, 0.02]
space = Hyperrectangle(low=[-6.0, -6.0], high=[6.0, 6.0])
system = AdditiveGaussianPolySystem(f, σ, space)
```

PWA Inclusion Dynamics Example. Setting up the system model for PWA inclusion dynamics requires PWA bounds on dynamics f of System (1), as defined in Eq. (5). These bounds can be generated using Linear Bound Propagation [30], and stored in a NetCDF file. These files can be loaded into the tool, which offers the options of using SOS- or PWC-SBF methods.

SOS-SBF. Given a NetCDF file that contains the PWA bounds, the setup for the SOS-SBF option is as follows. First, the system flag and filename need to be defined, which together specify the path to the PWA bounds. Then, the `load_dynamics()` function reads this data. The bounds dataset, along with the standard deviation σ, are passed into the `AdditiveGaussianUncertainPWASystem` struct. See the example below for a 2D pendulum model.

```
# System
system_flag = "pendulum"
filename    = "path/to/$system_flag/data.nc"
dataset     = open_dataset(filename)
data        = load_dynamics(dataset)
σ = [0.01, 0.01]
system = AdditiveGaussianUncertainPWASystem(data, σ)
```

PWC-SBF. For the PWC-SBF option, the file and system specification are similar. In addition, the transition probability data must be computed using `transition_probabilities()`.

```
# Compute probability data
probabilities = transition_probabilities(system)
```

Specifying Initial and Unsafe Sets. To complete the system model setup, the initial set and unsafe region(s) need to be defined. This is done by using the `Hyperrectangle` struct. In the toolbox, the region $\mathbb{R}^n \setminus X$ is considered unsafe by default. The user can further declare unsafe regions in the domain. If more than one unsafe region is defined, the **UnionSet** function of **LazySets.jl** can be used to take their union, as shown in the example below.

```
initial = Hyperrectangle(low=[-0.2, -0.3], high=[0.0, 0.1])
unsafe1 = Hyperrectangle(low=[-0.55, 0.4], high=[-0.45, 0.6])
unsafe2 = Hyperrectangle(low=[0.45, 0.4], high=[0.55, 0.6])
unsafe_regions = UnionSet(unsafe1, unsafe2)
```

If domain X does not contain any unsafe regions, the variable **unsafe_regions** must be declared empty: **EmptySet(dim)**, where *dim* is the dimension of the system. With that, the system model setup is concluded, and one can call on various SBF synthesis methods, as explained below.

3.2 SBF Synthesis

All methods for synthesizing SBFs are called under **synthesize_barrier(alg, system, initial_region, unsafe_region; time_horizon=N)**, which is designed with a unified interface of input parameters. Julia's multiple dispatch will select the appropriate method based on the **alg** parameter, and each algorithm returns a result object, which contains, in addition to the barrier, the resulting safety certificate and computation time. The barrier type is determined by the algorithm. Specifying the time horizon is optional, with the default value set to $N = 1$.

SOS-SBF. To synthesize an SOS-SBF, **synthesize_barrier** is called with input **SumOfSquaresAlgorithm()** as the first parameter. The second parameter, the system model object defined in Sect. 3.1, can be any of the three system types.

```
N = 10
# Optimize: baseline 1 (sos approach)
res = synthesize_barrier(SumOfSquaresAlgorithm(), system,
    initial_region, unsafe_region; time_horizon = N)

# Result
barr = barrier(res)
η, β, ps = eta(res), beta(res), psafe(res, N)
computation_time = total_time(res)
```

The optimization returns a struct of values, which include the SOS-SBF B, η, and β values, and the computation time. The algorithm **SumOfSquaresAlgorithm()** accepts keyword arguments to configure, e.g. the underlying SDP solver and the degree of the SOS barrier. Below we list the default values and how to construct the object for customization.

```
alg = SumOfSquaresAlgorithm(;
    barrier_degree  = 4
    lagrange_degree = 2
    sdp_solver = Mosek.Optimizer)
```

Setting `SumOfSquaresAlgorithm(barrier_degree = 30)`, for example, overrides the barrier degree. The default SDP solver for SOS optimization is Mosek [2].

PWC-SBF. For PWC-SBFs, `synthesize_barrier` is called with first parameter `DualAlgorithm()` for the dual LP approach, `CEGISAlgorithm()` for the CEGIS approach, or `GradientDescentAlgorithm()` for the GD approach, respectively. The second parameter should be the result from `transition_probabilities`.

```
N = 10
# Optimize: method 1 (dual approach)
dual = synthesize_barrier(DualAlgorithm(), probabilities,
    initial_region, obstacle_region, time_horizon = N)

# Optimize: method 2 (CEGIS approach)
cegis = synthesize_barrier(CEGISAlgorithm(), probabilities,
    initial_region, obstacle_region; time_horizon = N)

# Optimize: method 3 (GD approach)
gd = synthesize_barrier(GradientDescentAlgorithm(), probabilities,
    initial_region, obstacle_region; time_horizon = N)
```

Similar to SOS, each algorithm returns a result struct, which, in addition to the PWC-SBF B, values for η, β and the computation time, also includes β_i, which is an upper bound on Eq. (2c) in each region i. Next, we highlight the core properties of each algorithm.

Linear Programming. As outlined in Sect. 2.2, amongst the LP classes, there is the dual LP method, and the CEGIS method which uses inner and outer LP optimizations. The dual LP method only has the solver as a configurable parameter.

```
alg = DualAlgorithm(;
    linear_solver = HiGHS.Optimizer)
```

The default LP solver across the toolbox is HiGHS solver [7], but is customizable to any LP solver supported by JuMP, Julia Mathematical Programming [10].

The `CEGISAlgorithm` can additionally be configured with the number of iterations.

```
alg = CEGISAlgorithm(;
    linear_solver = HiGHS.Optimizer
    num_iterations = 10
    distribution_guided = true)
```

The default is the distribution guided approach, where in each iteration one LP generates a candidate PWC-SBF that optimizes for the probability of safety, given a set of finite feasible distributions. Then, the second LP generates distribution witnesses that violate the safety probability guarantee of the candidate PWC-SBF. By default, it runs for a fixed number of iterations given by *num_iterations* (default = 10), or it can continue adaptively until no further counterexamples exist, guaranteeing finite-time convergence [13].

Projected Gradient Descent. The GD method solves the same problem as the LP approaches, utilizing the minimax formulation to define a convex loss function. The projected gradient descent method is subject to hyperparameters.

```
alg = GradientDescentAlgorithm(;
    num_iterations = 10000
    initial_lr = 1e-2
    decay = 0.9999
    momentum = 0.9)
```

The struct **GradientDescentAlgorithm** defines key parameters for running GD. The *num_iterations* parameter controls the maximum number of iterations (default: 10,000). The *initial_lr* parameter sets the initial learning rate for the algorithm (default: 0.01), determining the size of each update step. The *decay* parameter (default: 0.9999) reduces the learning rate over time to fine-tune convergence. Finally, *momentum* (default: 0.9) helps to accelerate convergence by incorporating a portion of the previous gradient step into the current step. Importantly, at termination, the values for η, β are guaranteed to be sound with respect to the returned barrier, although not absolutely optimal in the sense of Theorem 3.

`StochasticBarrier.jl` is also equipped with an independent interface to easily run benchmarks. See [14] [App. A] for details.

4 Benchmark Studies

In this section, we present benchmark results for various stochastic systems with linear, polynomial and PWA inclusion dynamics. We performed two main comparison studies. First, the SOS-SBF algorithm in

Table 2. Linear, polynomial and PWA inclusion stochastic dynamic systems for benchmarking in **StochasticBarrier.jl**.

Dynamics	Name	Dimension
Linear	Contraction Map 1 & 2	2
	Two Tank [22]	2
	Room Temperature [29]	3
	Quadrotor [13]	6
Polynomial	Thermostat [17]	1
	Van der Pol Oscillator [1]	2
PWA Inclusion	Pendulum NNDM [15]	2
	Unicycle [13]	4

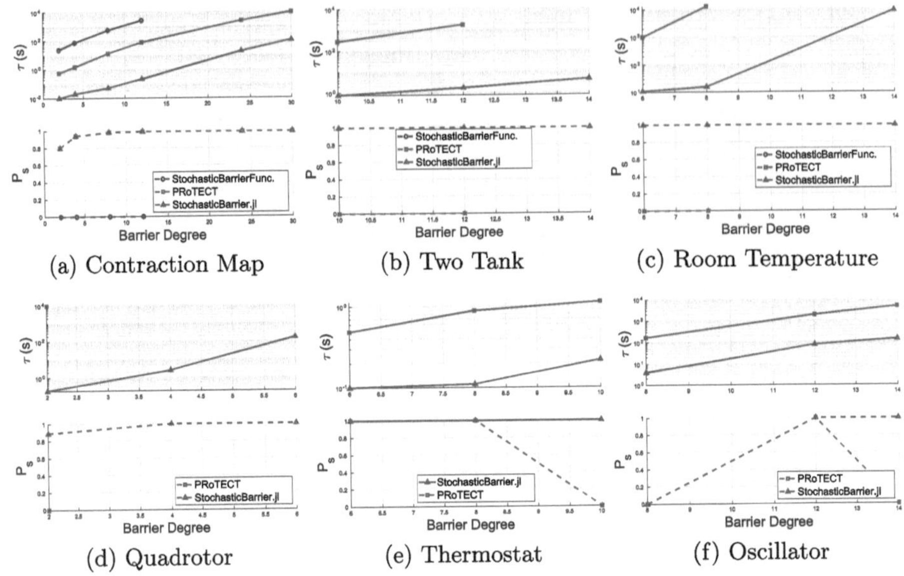

Fig. 1. Polynomial systems benchmarking of `StochasticBarrier.jl` *vs* StochasticBarrierFunctions [24] and PRoTECT [29]. Plots depict optimization time $\tau(s)$ and probability of safety P_s. If data points are not plotted, the algorithm ran out of memory.

`StochasticBarrier.jl` is compared against the state-of-the-art SOS toolboxes: `StochasticBarrierFunctions` [24] (MATLAB) and `PRoTECT` [29] (Python). The second study evaluates all the methods (SOS, DUAL-LP, CEGIS-LP, and GD) in `StochasticBarrier.jl` and compares their performance. All experiments are performed on a computer with a 3.9 GHz 8-Core CPU and 128 GB of memory. All methods use Mosek as the underlying SDP solver. The systems considered are summarized in Table 2 and details of the systems and their setups are provided in [14] [App. B.] For all benchmarks, time horizon $N = 10$ is used. The tool is available at https://github.com/aria-systems-group/StochasticBarrier.jl.

Comparison Against Other SOS-SBF Tools. Figure 1 shows performance of the SOS-SBF algorithm in `StochasticBarrier.jl` against state-of-the-art `StochasticBarrierFunc.` [24] and `PRoTECT` [29]. A detailed table of results can be found in [14] [App. D]. Note that as `StochasticBarrierFunc.` only supports linear systems, and `PRoTECT` only linear and polynomial systems, we could only compare the tools with systems of this dynamics class. We also note that the results from `PRoTECT` are random, and we present average results over 10 runs. We also note that P_s defined in Eq. (3) is computed differently in `PRoTECT`; we adjust it here to match our setting. This paper solely focuses on optimization-based SOS methods, i.e., disabling the objective is fundamentally different and not considered.

Across all benchmarks, `StochasticBarrier.jl` outperforms both `PRoTECT` and `StochasticBarrierFunc.` in terms of computation time and scalability, and tightness of safety probability. `StochasticBarrierFunc.` fails to produce non-trivial results and cannot handle higher-degree SOS-SBFs (beyond degree 12), often yielding a trivial $P_s \geq 0$ or exceeding available memory. `PRoTECT` performs better but remains significantly slower (by factors ranging from $5\times$ to $1000\times$) and often returns trivial safety bounds for complex systems. In contrast, `StochasticBarrier.jl`, scales efficiently to high-degree SOS-SBFs (up to 30), achieves non-trivial or high safety probabilities, and maintains orders-of-magnitude faster computation across all tested models (contraction map 1, two tank, room temperature, quadrotor, thermostat, and oscillator).

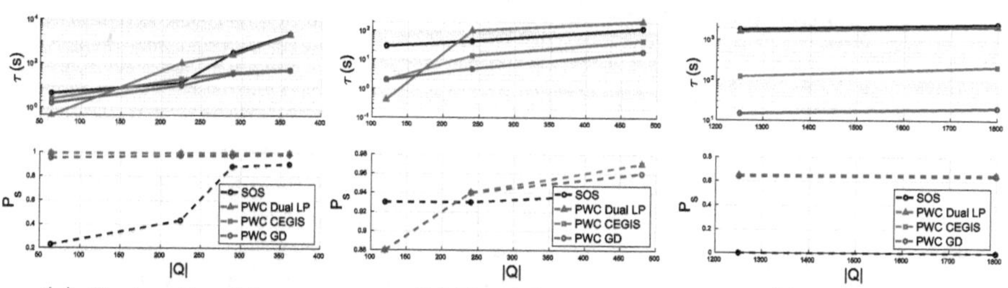

(a) Contraction Map (b) Pendulum (c) Unicycle

Fig. 2. Benchmark plots for the three PWC-SBF methods. For the pendulum and unicycle, the SOS results are also depicted for comparison. Plots depict total optimization time $\tau(s)$ and probability of safety P_s.

Comparisons of `StochasticBarrier.jl` Engines. Benchmark results for the three PWC-SBF methods implemented in `StochasticBarrier.jl` are summarized in Fig. 2, with detailed results presented in [14] [App. D]. The results show that PWC-SBFs consistently outperform SOS-SBFs, particularly for the contraction map and hybrid nonlinear systems. For the linear contraction map, the GD method achieves the fastest computation, especially as the number of partitions increases, while Dual LP and CEGIS yield tighter safety probability bounds that converge with higher partition counts. For PWA inclusion systems, PWC-SBFs are up to two orders of magnitude faster, and for the unicycle model, they provide non-trivial safety bounds where SOS fails. As expected, the Dual LP and CEGIS methods deliver the tightest bounds at the cost of higher computation time.

5 Conclusion

We introduced `StochasticBarrier.jl`, a Julia toolbox for generating barrier certificates and estimating safety probabilities of discrete-time stochastic dynamical systems. Benchmark results show that `StochasticBarrier.jl` is orders of magnitude faster and provides tighter safety bounds than existing MATLAB and Python implementations, with superior scalability to higher-dimensional systems and higher-degree SOS barriers. The toolbox also supports piecewise constant barriers, which consistently outperform SOS formulations. Current limitations include its focus on additive Gaussian noise. Future extensions will address non-Gaussian and non-additive noise.

References

1. Abate, A., et al.: ARCH-COMP20 category report: stochastic models. In: ARCH (2020)
2. ApS, M.: MOSEK optimization toolbox for MATLAB. User's Guide and Reference Manual, Version **4**(1) (2019)
3. Bertsekas, D.P., Shreve, S.E.: Stochastic Optimal Control: The Discrete-Time Case, vol. 5. Athena Scientific (2004)
4. Bogomolov, S., Forets, M., Frehse, G., Potomkin, K., Schilling, C.: JuliaReach: a toolbox for set-based reachability. In: Proceedings of the 22nd ACM International Conference on Hybrid Systems: Computation and Control, pp. 39–44 (2019)
5. Coleman, C., Lyon, S., Maliar, L., Maliar, S.: Matlab, Python, Julia: What to choose in economics? Comput. Econ. **58**, 1263–1288 (2021)
6. Forets, M., Schilling, C.: LazySets.jl: scalable symbolic-numeric set computations. In: Proceedings of the JuliaCon Conferences, vol. 1, no. 1, p. 11 (2021). https://doi.org/10.21105/jcon.00097
7. Huangfu, Q., Hall, J.J.: Parallelizing the dual revised simplex method. Math. Program. Comput. **10**(1), 119–142 (2018)
8. Jagtap, P., Soudjani, S., Zamani, M.: Formal synthesis of stochastic systems via control barrier certificates. IEEE Trans. Autom. Control **66**(7), 3097–3110 (2020)
9. Legat, B.: Multivariate polynomials in Julia. In: JuliaCon (2022). https://pretalx.com/juliacon-2022/talk/TRFSJY/
10. Lubin, M., Dowson, O., Dias Garcia, J., Huchette, J., Legat, B., Vielma, J.P.: JuMP 1.0: recent improvements to a modeling language for mathematical optimization. Math. Program. Comput. (2023). https://doi.org/10.1007/s12532-023-00239-3
11. Mathiesen, F.B., Lahijanian, M., Laurenti, L.: IntervalMDP. jl: accelerated value iteration for interval Markov decision processes, arXiv preprint arXiv:2401.04068 (2024)
12. Mathiesen, F.B., Romao, L., Calvert, S.C., Laurenti, L., Abate, A.: A data-driven approach for safety quantification of non-linear stochastic systems with unknown additive noise distribution, arXiv preprint arXiv:2410.06662 (2024)
13. Mazouz, R., Mathiesen, F.B., Laurenti, L., Lahijanian, M.: Piecewise stochastic barrier functions. Automatica **189**, 113020 (2026). https://doi.org/10.1016/j.automatica.2026.113020. https://www.sciencedirect.com/science/article/pii/S0005109826002049
14. Mazouz, R., Mathiesen, F.B., Laurenti, L., Lahijanian, M.: Stochasticbarrier. jl: a toolbox for stochastic barrier function synthesis, arXiv preprint arXiv:2602.20359 (2026)
15. Mazouz, R., Muvvala, K., Ratheesh, A., Laurenti, L., Lahijanian, M.: Safety guarantees for neural network dynamic systems via stochastic barrier functions. In: Advances in Neural Information Processing Systems (2022)

16. Mazouz, R., Skovbekk, J., Mathiesen, F.B., Frew, E., Laurenti, L., Lahijanian, M.: Data-driven permissible safe control with barrier certificates. In: 2024 IEEE 63rd Conference on Decision and Control (CDC), pp. 6844–6849. IEEE (2024)
17. Nejati, A., Soudjani, S., Zamani, M.: Compositional abstraction-based synthesis for continuous-time stochastic hybrid systems. Eur. J. Control. **57**, 82–94 (2021)
18. Perkel, J.M.: Julia: come for the syntax, stay for the speed. Nature **572**(7767), 141–142 (2019)
19. Prajna, S.: Barrier certificates for nonlinear model validation. Automatica **42**(1), 117–126 (2006)
20. Prajna, S., Jadbabaie, A., Pappas, G.J.: A framework for worst-case and stochastic safety verification using barrier certificates. IEEE Trans. Autom. Control **52**(8) (2007)
21. Prajna, S., Papachristodoulou, A., Parrilo, P.A.: Introducing SOSTOOLS: a general purpose sum of squares programming solver. In: Conference on Decision and Control, vol. 1. IEEE (2002)
22. Ramos, J.A., Dos Santos, P.L.: Mathematical modeling, system identification, and controller design of a two tank system. In: 2007 46th IEEE Conference on Decision and Control. IEEE (2007)
23. Rew, R., Davis, G.: NetCDF: an interface for scientific data access. IEEE Comput. Graphics Appl. **10**(4) (1990)
24. Santoyo, C., Dutreix, M., Coogan, S.: A barrier function approach to finite-time stochastic system verification and control. Automatica **125** (2021). https://doi.org/10.1016/j.automatica.2020.109439
25. Schneeberger, M., Dörfler, F., Mastellone, S.: SoS construction of compatible control Lyapunov and barrier functions. IFAC-PapersOnLine **56**(2), 10428–10434 (2023)
26. Soroka, E., Kochenderfer, M.J., Lall, S.: Satisfiability.jl: satisfiability modulo theories in Julia. J. Open Source Softw. **9**(100), 6757 (2024). https://doi.org/10.21105/joss.06757
27. Wang, L., Han, D., Egerstedt, M.: Permissive barrier certificates for safe stabilization using sum-of-squares. In: Annual American Control Conference (ACC), pp. 585–590. IEEE (2018)
28. Wei, T., et al.: Modelverification. jl: a comprehensive toolbox for formally verifying deep neural networks, arXiv preprint arXiv:2407.01639 (2024)
29. Wooding, B., Horbanov, V., Lavaei, A.: Protect: parallelized construction of safety barrier certificates for nonlinear polynomial systems, arXiv preprint arXiv:2404.14804 (2024)
30. Zhang, H., Weng, T.W., Chen, P.Y., Hsieh, C.J., Daniel, L.: Efficient neural network robustness certification with general activation functions. Adv. Neural Inf. Process. Syst. **31** (2018)

Author Index

M. Chen and M. Tschaikowski (Eds.): QEST+FORMATS 2026, LNCS 16913, pp. 445–446, 2027.
https://doi.org/10.1007/978-3-032-35298-9

Kolibri 360 GmbH
Ferdinand-Jühlke-Straße 7
99095 Erfurt, Deutschland
produktsicherheit@kolibri360.de